Handbook of
Graph Drawing
and Visualization

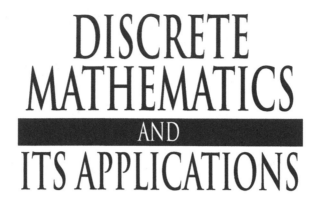

DISCRETE MATHEMATICS
AND
ITS APPLICATIONS

Series Editor
Kenneth H. Rosen, Ph.D.

DISCRETE MATHEMATICS AND ITS APPLICATIONS

Series Editor KENNETH H. ROSEN

Handbook of
Graph Drawing
and Visualization

Edited by

Roberto Tamassia

Brown University
Providence, Rhode Island, USA

CRC Press
Taylor & Francis Group
Boca Raton   London   New York

CRC Press is an imprint of the
Taylor & Francis Group, an **informa** business

A CHAPMAN & HALL BOOK

CRC Press
Taylor & Francis Group
6000 Broken Sound Parkway NW, Suite 300
Boca Raton, FL 33487-2742

First issued in paperback 2016

© 2014 by Taylor & Francis Group, LLC
CRC Press is an imprint of Taylor & Francis Group, an Informa business

No claim to original U.S. Government works

Version Date: 20130719

ISBN 13: 978-1-138-03424-2 (pbk)
ISBN 13: 978-1-58488-412-5 (hbk)

Library of Congress Cataloging-in-Publication Data

Handbook of graph drawing and visualization / editor, Roberto Tamassia.
    pages cm -- (Discrete mathematics and its applications)
  Includes bibliographical references and index.
  ISBN 978-1-58488-412-5 (hardcover : alk. paper)
  1. Graph theory--Handbooks, manuals, etc. 2. Graphic methods--Handbooks, manuals, etc. 3. Information visualization--Handbooks, manuals, etc. I. Tamassia, Roberto, 1960-

QA166.H358 2014
511'.5--dc23
                                        2013025835

Visit the Taylor & Francis Web site at
http://www.taylorandfrancis.com

and the CRC Press Web site at
http://www.crcpress.com

Contents

Preface

Objective

This handbook aims at providing a broad survey of the field of graph drawing. It covers topological and geometric foundations, algorithms, software systems, and visualization applications in business, education, science, and engineering.

The intended readership of this handbook includes:

- Practitioners and researchers in traditional and emerging disciplines of the physical, life, and social sciences interested in understanding and using graph drawing methods and graph visualization systems in their field.
- Information technology practitioners and software developers aiming to incorporate graph drawing solutions into their products.
- Researchers and students in graph drawing and information visualization seeking an up-to-date survey of the field.
- Researchers and students in related fields of mathematics and computer science (including graph theory, computational geometry, information visualization, software engineering, user interfaces, social networks, and data management) interested in using graph-drawing techniques in support of their research.

Organization

The chapters of this handbook are organized into four parts, as follows.

Topological and Geometric Foundations of Graph Drawing The first part (Chapters 1–4) deals with fundamental topological and geometric concepts and techniques used in graph drawing: planarity testing and embedding, crossings and planarization, symmetric drawings, and proximity drawings.

Graph Drawing Algorithms The second part (Chapters 5–14) presents an extensive collection of algorithms for constructing drawings of graphs. Some methods are designed to draw special classes of graphs (e.g., trees, planar graphs, or directed acyclic graphs) while other methods work for general graphs. Topics covered in this part include tree drawing algorithms, planar straight-line drawing algorithms, planar orthogonal and polyline drawing algorithms, spine and radial drawings, circular drawing algorithms, rectangular drawing algorithms, simultaneous embeddings, force-directed methods, hierarchical drawing algorithms, three-dimensional drawing algoritms, and labeling algorithms.

Graph Drawing Systems The third part begins by introducing the GraphML language for representing graphs and their drawings (Chapter 16). Next, it overviews three software systems for constructing drawings of graphs: OGDF, GDToolkit, and PIGALE (Chapters 17–19).

Applications of Graph Drawing The fourth part (Chapters 20–26) gives examples of the use of graph drawing methods for the visualization of networks in various important application domains: biological networks, computer security, data analytics, education, computer networks, and social networks.

Each chapter is intended to be self-contained and has its own bibliography.

Acknowledgments

I would like to thank all the authors of the chapters in this handbook and all the reviewers who have provided expert feedback on the initial drafts and revised versions of the chapters.

This handbook is a collective effort of the graph drawing research community, which has developed around the annual Symposium on Graph Drawing. I am grateful to the people who have founded with me this conference and have continued providing leadership for it: Franz Brandenburg, Giuseppe Di Battista, Peter Eades, Hubert de Fraysseix, Takao Nishizeki, Pierre Rosenstiehl, and Ioannis Tollis.

A huge thanks goes to Sunil Nair for proposing this handbook project and supporting its development. I truly appreciate his constant encouragement and patience. Help received from the entire CRC Press production team, and especially from Andre Barnett, Kari Budyk, Rachel Holt, Jim McGovern, and Shashi Kumar, is gratefully acknowledged. I would like to thank also my executive assistant Angel Murakami for her expert proofreading of the chapters.

I am indebted to Carlo Batini for introducing me to the problem of drawing graphs and inspiring me to pursue an academic path. A special thanks also goes to Franco Preparata, whose guidance and support, first as PhD advisor and then as colleague, have shaped by career.

Finally, warm thanks go to Isabel Cruz, Giuseppe Di Battista, Michael Goodrich, and Ioannis Tollis for their encouragement and support throughout this project.

Roberto Tamassia

About the Editor

Roberto Tamassia is the Plastech Professor of Computer Science and the Chair of the Department of Computer Science at Brown University. He is also the Director of Brown's Center for Geometric Computing. His research interests include analysis, design, and implementation of algorithms, applied cryptography, cloud computing, computational geometry data security, and graph drawing. He has published six textbooks and more than 250 research articles and books in the above areas and has given more than 70 invited lectures worldwide. He is a Fellow of the American Association for the Advancement of Science (AAAS), the Association for Computing Machinery (ACM), and the Institute of Electrical and Electronics Engineers (IEEE). He is the recipient of a Technical Achievement Award from the IEEE Computer Society for pioneering the field of graph drawing. He is listed among the 360 most cited computer science authors by Thomson Scientific, Institute for Scientific Information (ISI). He serves regularly on program committees of international conferences. His research has been funded by ARO, DARPA, NATO, NSF, and several industrial sponsors. He co-founded the Journal of Graph Algorithms and Applications (JGAA) and the Symposium on Graph Drawing. He serves as Co-Editor-in-Chief of JGAA. He received the PhD degree in electrical and computer engineering from the University of Illinois at Urbana-Champaign and the "Laurea" in Electrical Engineering from the "Sapienza" University of Rome.

1

Planarity Testing and Embedding

Maurizio Patrignani
Roma Tre University

1.1 Introduction

Testing the planarity of a graph and possibly drawing it without intersections is one of the most fascinating and intriguing algorithmic problems of the graph drawing and graph theory areas. Although the problem *per se* can be easily stated, and a complete characterization of planar graphs has been known since 1930, the first linear-time solution to this problem was found only in the 1970s.

Planar graphs play an important role both in the graph theory and in the graph drawing areas. In fact, planar graphs have several interesting properties: for example, they are sparse and 4-colorable, they allow a number of operations to be performed more efficiently than for general graphs, and their inner structure can be described more succinctly and elegantly (see Section 1.2.2). From the information visualization perspective, instead, as edge crossings turn out to be the main reason for reducing readability, planar drawings of graphs are considered clear and comprehensible.

In this chapter, we review a number of different algorithms from the literature for efficiently testing planarity and computing planar embeddings. Our main thesis is that all known linear-time planarity algorithms fall into two categories: cycle based algorithms and vertex addition algorithms. The first family of algorithms is based on the simple observation that in a planar drawing of a graph any cycle necessarily partitions the graph into the inside and outside portion, and this partition can be suitably used to split the embedding problem. Vertex addition algorithms are based on the incremental construction of the final planar drawing starting from planar drawings of smaller graphs. The fact that some algorithms were based on the same paradigm was already envisaged by several researchers [Tho99, HT08]. However, the evidence that all known algorithms boil down to two simple approaches is a relatively new concept.

The chapter is organized as follows: Section 1.2 introduces basic definitions, properties, and characterizations for planar graphs; Section 1.3 formally defines the planarity testing and embedding problems; Section 1.4 follows a historic perspective to introduce the main algorithms and a conventional classification for them. Some algorithmic techniques are common to more than one algorithm and sometimes to all of them. These are collected in Section 1.5. Finally, Sections 1.6 and 1.7 are devoted to the two approaches to the planarity testing problem, namely, the "cycle based" and the "vertex addition" approaches, respectively.

Algorithms for constructing planar drawings of graphs are discussed in Chapters 6 (straight-line drawings), 7 (orthogonal and polyline drawings), and 10 (rectangular drawings). Methods for reducing crossings in nonplanar drawings of graphs are discussed in Chapter 2.

1.2   Properties and Characterizations of Planar Graphs

1.2.1   Basic Definitions

A *graph* $G(V, E)$ is an ordered pair consisting of a finite set V of *vertices* and a finite set E of *edges*, that is, pairs (u, v) of vertices. If each edge is an unordered (ordered) pair of vertices, then the graph is *undirected* (*directed*). An edge (u, v) is a *self-loop* if $u = v$. A graph $G(V, E)$ is *simple* if E is not a multiple set and it does not contain self-loops. For the purposes of this chapter, we can restrict us to simple graphs.

The sets of edges and vertices of G can be also denoted $E(G)$ and $V(G)$, respectively. If edge $(u, v) \in E$, vertices u and v are said to be *adjacent* and (u, v) is said to be *incident* to u and v. Two edges are *adjacent* if they have a vertex in common.

A (rooted) *tree* T is a connected acyclic graph with one distinguished vertex, called the *root* r. A *spanning tree* of a graph G is a tree T such that $V(T) = V(G)$ and $E(T) \subseteq E(G)$.

Given two graphs $G_1(V_1, E_1)$ and $G_2(V_2, E_2)$, their *union* $G_1 \cup G_2$ is the graph $G(V_1 \cup V_2, E_1 \cup E_2)$. Analogously, their *intersection* $G_1 \cap G_2$ is the graph $G(V_1 \cap V_2, E_1 \cap E_2)$. A graph G_2 is a *subgraph* of G_1 if $G_1 \cup G_2 = G_1$.

Given a graph $G(V, E)$ and a subset V' of V, the subgraph *induced by* V' is the graph $G'(V', E')$, where E' is the set of edges of E that have both endvertices in V'. Given a graph $G(V, E)$ and a subset E' of E, the subgraph *induced by* E' is the graph $G'(V', E')$, where V' is the set of vertices incident to E'. A *subdivision* of an edge (u, v) consists of the insertion of a new node w and the replacement of (u, v) with edges (u, w) and (w, v). A graph G_2 is a *subdivision* of G_1 if it can be obtained from G_1 through a sequence of edge subdivisions.

A *drawing* Γ of a graph G maps each vertex v to a distinct point $\Gamma(v)$ of the plane and each edge (u, v) to a simple open Jordan curve $\Gamma(u, v)$ with endpoints $\Gamma(u)$ and $\Gamma(v)$. A drawing is *planar* if no two distinct edges intersect except, possibly, at common endpoints. A graph is *planar* if it admits a planar drawing. A planar drawing partitions the plane into connected regions called *faces*. The unbounded face is usually called *external face* or *outer face*. If all the vertices are incident to the outer face, the planar drawing is called *outerplanar* and the graph admitting it is an *outerplanar graph*. Given a planar drawing, the (clockwise) circular order of the edges incident to each vertex is fixed. Two planar drawings are *equivalent* if they determine the same circular orderings of the edges incident to each vertex (sometimes called *rotation scheme*). A *(planar) embedding* is an equivalence class of planar drawings and is described by the clockwise circular order of the edges incident to each vertex. A graph together with one of its planar embedding is sometimes referred to as a *plane* graph.

A *path* is a sequence of distinct vertices $v_1, v_2, \ldots, v_k$, with $k \geq 2$, together with the edges $(v_1, v_2), \ldots, (v_{k-1}, v_k)$. The *length* of the path is the number of its edges.

A *cycle* is a sequence of distinct vertices $v_1, v_2, \ldots, v_k$, with $k \geq 2$, together with the edges $(v_1, v_2), \ldots, (v_{k-1}, v_k), (v_k, v_1)$. The *length* of a cycle is the number of its vertices or the number of its edges.

An undirected graph G is *connected* if, for each pair of nodes u and v, G contains a path from u to v. A graph G with at least $k + 1$ vertices is *k-connected* if removing any $k - 1$ vertices leaves G connected. Equivalently, by Menger's theorem, a graph is k-connected if there are k independent paths between each pair of vertices [Men27]. 3-connected, 2-connected, and 1-connected graphs are also called *triconnected*, *biconnected*, and *simply connected* graphs, respectively. It is usual in the planarity literature to relax the definition of biconnected graph so to include *bridges*, i.e., graphs composed by a single edge between two vertices. A *separating k-set* is a set of k vertices whose removal disconnects the graph. Separating 1- and 2-sets are called *cutvertices* and *separation pairs*, respectively. Hence, a connected graph is biconnected if it has no cutvertices and it is triconnected if it has no separation pairs.

If a graph G is not connected, its maximal connected subgraphs are called the *connected components* of G. If G is connected, its maximal biconnected subgraphs (including bridges) are called the *biconnected components*, or *blocks* of G. Note that a cutvertex belongs to several blocks and that a biconnected graph has only one block. The graph whose vertices are the blocks and the cutvertices of G and whose edges link cutvertices to the blocks they belong to is a tree and is called the *block-cutvertex tree* (or *BC-tree*) of G (see Figure 1.1 for an example).

Given a biconnected graph G, its *triconnected components* are obtained by a complex splitting and merging process. The first linear-time algorithm to compute them was introduced in [HT73], while an implementation of it is described in [GM01]. The computation has two phases: first, G is recursively split into its *split components*; second, some split components are merged together to obtain triconnected components. The *split* operation is performed with respect to a pair of vertices $\{v_1, v_2\}$ of the biconnected (multi)graph G. Suppose the edges of G are divided into the equivalence classes $E_1, E_2, \ldots, E_k$ such that two edges are in the same class if both lie in a common path not containing a vertex in $\{v_1, v_2\}$ except, possibly, as an end point. If there are at least two such classes, then $\{v_1, v_2\}$ is a *split pair*. Let G_1 be the graph induced by E_1 and G_2 be the graph induced by E/E_1. A *split* operation consists of replacing G with G'_1 and G'_2, where G'_1 and G'_2 are obtained from G_1 and G_2 by adding the same *virtual edge* (v_1, v_2). The two copies of the virtual edge added to G_1 and G_2 are called *twin virtual edges*. Figure 1.2(b) shows the result of a split operation performed on the graph of Figure 1.2(a) with respect to split pair $\{2, 4\}$. The *split components* of a graph G are obtained by recursively splitting G until no split pair can be found in the obtained graphs. Figure 1.2(c) shows the split components of the graph of Figure 1.2(a). Split components are not unique and, hence, are not suitable for describing the structure of G.

Two split components sharing the same twin virtual edges (v_1, v_2) can be merged by identifying the two copies of v_1 and v_2 and by removing the twin virtual edges. Split components consisting of cycles are called *series split components*, while split components

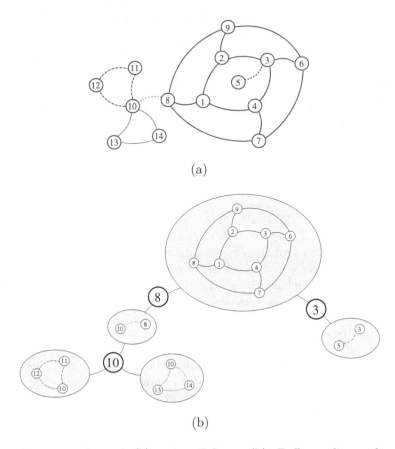

Figure 1.1 A connected graph (a) and its BC-tree (b). Different line styles are used for edges of different blocks.

that have only two vertices are called *parallel split components*. By recursively merging together series split components that share twin virtual edges we obtain *series triconnected components*, while by recursively merging together parallel split components that share twin virtual edges we obtain *parallel triconnected components*. Split components that are not affected by the merging operations described above are called *rigid triconnected components*. Figure 1.3(a) shows the triconnected components of the graph of Figure 1.2(a).

Triconnected components are unique and are used to describe the inner structure of a graph. In fact, a graph G can be succinctly described by its SPQR-tree $\mathcal{T}$, which provides a high-level view of the unique decomposition of the graph into its triconnected components [DT96a, DT96b, GM01]. Namely, each triconnected component corresponds to a node of $\mathcal{T}$. The triconnected component corresponding to a node μ of $\mathcal{T}$ is called the *skeleton of* μ. As there are parallel, series, and rigid triconnected components, their corresponding tree nodes are called P-, S-, and R-nodes, respectively. Triconnected components sharing a virtual edge are adjacent in $\mathcal{T}$. Usually, a fourth type of node, called Q-node, is used to represent an edge (u, v) of G. Q-nodes are the leaves of $\mathcal{T}$ and they don't have skeletons. Tree $\mathcal{T}$ is unrooted, but for some applications, it could be thought as rooted at an arbitrary Q-node. See Figure 1.3 for an example of SPQR-tree.

The connectivity properties of a graph have a strict relationship with its embedding properties. Triconnected planar graphs (and triconnected planar components) have a single

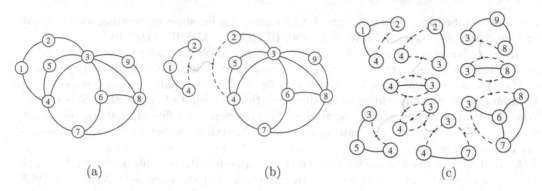

Figure 1.2 (a) A biconnected graph. (b) A split operation performed with respect to split pair $\{2, 4\}$. (c) The split components of the graph. Virtual edges are drawn dashed. Twin virtual edges are joined with dotted lines.

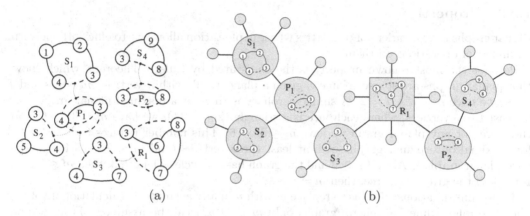

Figure 1.3 (a) The triconnected components of the same graph of Figure 1.2. (b) The corresponding SPQR-tree. Q-nodes are represented by empty circles.

embedding up to a flip (that is, up to a reversal of all their incidence lists) [Whi32]. The same holds for biconnected outerplanar graphs and their unique outerplanar embedding (adding a star on the outer face yields a triconnected plane graph).

A non-connected graph is planar if and only if all its connected components are planar. Thus, in the following, without loss of generality, we only consider the planarity of connected graphs. Also, a planar embedding of a graph implies a planar embedding for each one of its blocks, while, starting from a planar embedding of the blocks, a planar embedding for the whole graph can be found [Whi32]. Thus, since the blocks can be identified in linear time [Tar72], a common strategy, both to test planarity and to compute a planar embedding, is to divide the graph into its blocks and to tackle each block separately.

Finally, a graph is planar if and only if its triconnected components are planar [Mac37b]. More precisely, as parallel and series triconnected components are always planar, a graph is planar if and only if all its rigid triconnected components are planar. However, since dividing a graph into its triconnected components is a linear but rather laborious process [HT73, GM01], usually planarity algorithms do not assume that the input graph is triconnected.

Also, from a planar embedding of the triconnected components of a graph, a planar embedding of the whole graph can be obtained. This property can be exploited to explore the planar embeddings of a given graph when searching for some embedding with a specific property (see, for example, [MW99, MW00, BDBD00, GMW01, ADF⁺10]).

Given a plane (multi)graph G, its *plane dual* (or simply its *dual*) is the multigraph G^* such that G^* has one vertex for each face of G and two vertices of G^* are linked by one edge e^* if the corresponding faces in G share one edge e. Observe that the planar embedding of G induces a planar embedding of its dual and that the dual of the dual of G is G itself. Also, different embeddings of a planar graph G correspond to different dual graphs. Finally, a cycle in G corresponds to a minimal cut in G^* (anytime this property holds G and G^* are called *abstract dual*).

A graph $G(V, E)$ is k *colorable* if its vertices can be partitioned into k sets $V_1, V_2, \ldots, V_k$ in such a way that no edge is incident to two vertices of the same set. A graph $G(V, E)$ is *complete* if each vertex in V is adjacent to each other.

A graph $G(V, E)$ is *bipartite* if it is 2-colorable. A bipartite graph $G(V_1, V_2, E)$ is *complete* if each vertex in V_1 is adjacent to all vertices in V_2.

1.2.2 Properties

Planar graphs have a variety of properties whose exploitation allows us to efficiently perform a number of operations on them.

Perhaps the most renown property is the one stated by Euler's Theorem, which shows that planar graphs are sparse. Namely, given a plane graph with n vertices, m edges and f faces, we have $n - m + f = 2$. A simple corollary is that for a maximal planar graph with at least three vertices, where each face is a triangle $(2m = 3f)$, we have $m = 3n - 6$, and, therefore, for any planar graph we have $m \leq 3n - 6$. This number reduces to $m = 2n - 3$ for maximal outerplanar graphs with at least three vertices (and $m \leq 2n - 3$ for general outerplanar graphs). Also, if $n \geq 3$ and the graph has no cycle of length 3, then $m \leq 2n - 4$. Finally, if the graph is a tree, then $m = n - 1$.

These considerations allow us to replace m with n in any asymptotic calculation involving planar graphs, while for general graphs only $m \in O(n^2)$ can be assumed. From a more practical perspective, they allow us to decide the non-planarity of denser graphs without reading all the edges (which would yield a quadratic algorithm).

The Four Color Theorem [AH77, AHK77, RSST97] asserts that any planar graph is 4-colorable and settles a conjecture that was for more than a century the most famous unsolved problem in graph theory and perhaps in all of mathematics [Har69]. To stress how important this property is, it suffices to observe that, apart from being considered an important property of planar graphs, it has also been mentioned as the most notable property of the number 4.

While 3-colorability is NP-hard even on maximum degree four planar graphs [GJS76], every triangle-free planar graph is 3-colorable [Grö59] and such a 3-coloring can be found in linear-time [DKT09].

Determining whether the graph contains a k-clique, i.e., a set of k pairwise adjacent vertices, is polynomial for planar graphs, as no clique can have more than four vertices. This problem is polynomial even in the weighted case, where each vertex is associated with a weight and the sum of the weights of the pairwise adjacent vertices is requested to be at least k. Observe that both these problems are NP-complete on non-planar graphs.

Graph isomorphism is linear for planar graphs [HW74], while it is of unknown complexity for general graphs [GJ79].

The planar separator theorem [LT79] states that every planar graph $G = (V, E)$ admits a partition of its n vertices into three sets, $A, B,$ and C, such that the size of C is $O(\sqrt{n})$, the size of A and B is at most $\frac{2}{3}n$, and there is no edge with one endpoint in A and the other endpoint in B. Such a partition can be found in linear time and is the starting point of a hierarchical decomposition of the graph that may lead to efficient approaches to compute properties of the graph.

1.2.3   Characterizations

The first complete characterization of planar graphs is due to Kuratowski [Kur30] and states that a graph is planar if and only if it contains no subgraph that is a subdivision of K_5 or $K_{3,3}$, where K_5 is the complete graph of order 5 and $K_{3,3}$ is the complete bipartite graph with 3 vertices in each of the sets of the partition. An equivalent later result, recast in terms of graph minors, is Wagner's theorem that states that a graph G is planar if and only if it has no K_5 or $K_{3,3}$ as minor, that is, K_5 or $K_{3,3}$ cannot be obtained from G by contracting some edges, deleting some edges, and deleting some isolated vertices [Wag37a, HT65]. Observe that the two characterizations are different since a graph may admit K_5 as minor without having a subgraph that is a subdivision of K_5 (consider, for example, a graph of maximum degree 3).

Similarly, it can be proved that a graph is outerplanar if and only if it contains no subgraph that is a subdivision of K_4 or $K_{2,3}$. Trivially, a graph is a tree if it does not contain a subdivision (or a minor) of K_3.

If the graph is triconnected, a less renown but much simpler characterization can be formulated. Namely, a triconnected graph distinct from K_5 is planar if and only if it contains no subgraph that is a subdivision of $K_{3,3}$ [Wag37b, Hal43, Kel93, Lie01].

Given a graph G with no isolated vertices, the associated height-two *vertex-edge poset* $<_G$ has $V \cup E$ as elements, and $v <_G e$ if and only if $v \in V$, $e \in E$, and v is an endpoint of edge e. The smallest number of total orders the intersection of which yields the poset is called the *dimension* of the poset. Graph G is planar if and only if its corresponding vertex-edge poset has dimension at most three [Sch89]. Unfortunately, checking if a poset has dimension at most t is proved to be NP-complete for $t \geq 3$ and for $t \geq 4$ if the poset has height two [Yan82].

Edges traversing a bipartition of the vertices of G are called a *cocycle*. Observe that while a cycle is a collection of edges that covers each vertex an even (possibly zero) number of times, a cocycle is a collection of edges that intersects each cycle in an even number of edges. A *bicycle* is a collection of edges that is both a cycle and a cocycle. Planarity can be characterized in terms of the properties of the vector spaces of cycles [Mac37a], cocycles [APBL95, LS10], and bicycles [APBL95].

A further planarity characterization is expressed via Colin de Verdiére's graph invariant $\mu(G)$, which in turn is based on the maximum multiplicity of the second eigenvalue of certain Schrödinger operators defined by the graph [Col90, Col91], and states that a graph G is planar if and only if $\mu(G) \leq 3$.

Alternative characterizations can be found in the literature based on the existence of an abstract dual graph [Whi32], on the edge poset dimension [dO96], on the relationship among theta-graph minors [AŠ98], on the orientability of circuits [LH77, Che81], on the arrangements of pseudo-lines [TT97], or on DFS traversals of the graph [dR82, dR85, SH93, SH99, BM99, BM04].

1.3   Planarity Problems

The main planarity problem is the decision problem of recognizing planar graphs, that is, of deciding the planarity of the input graph. Both with the purpose of exhibiting a planarity certificate and of producing a planar embedding for information-visualization applications, planarity testing algorithms are usually coupled with planar embedding procedures, that sometimes, depending on the algorithmic approach, required a considerable research effort to be devised.

On the opposite, if the graph is not planar, the search for a non-planarity certificate is called Kuratowski subgraph isolation [CMS08], and the research concentrated on planarization algorithms that allow us to produce a planar graph where some degree-four vertices have been added to replace crossings [Lie01]. Since crossing number minimization is NP-complete [GJ79] planarization algorithms use heuristics to introduce a reduced number of dummy vertices.

Dynamic algorithms have also been devised for efficiently determining planarity and computing a planar embedding of graphs where edges and vertices are added or deleted one at a time [DT89, DT96b, GIS99, DBTV01].

Efficient algorithms for planarity testing in parallel have been investigated in [KR88, RR89, RR94].

1.3.1   Constrained Planarity

The problem of determining the planarity of a graph and of computing a possible embedding of it can be combined with additional constraints on the desired drawing that result in restrictions on the set of admissible planar embeddings [Tam98, GKM08]. Typical constraints ask for some vertices to be on the same face (usually the outer face), some vertex to have a specified circular ordering of its incident edges, some path to be drawn along a straight line, etc. In the easier cases, such constraints can be enforced by replacing sets of nodes and edges of the input graph with suitable gadgets, by launching an ordinary planarity algorithm, and by transferring the results back on the original graph. More complex cases require to efficiently explore the possible embeddings of the graph by considering their inner structures described by their BC-trees and SPQR-trees. In [GKM08], embedding constraints that restrict the admissible order of incident edges around a vertex are considered.

A very restrictive constraint is when the input graph G is partially embedded, i.e., when a subgraph H of G is provided with an embedding $\mathcal{H}$. In this case, the problem of determining a planar embedding of the whole graph that extends the embedding $\mathcal{H}$, if one exists, is linear [ADF+10]. Also, if the answer is negative, an obstruction taken from a collection of minimal non-planar instances can be produced in polynomial time [JKR11].

A constrained planarization is implied anytime an embedding that minimizes some quality measure is desired. As pointed out in [BM90, PT00, Piz05], the quality of a planar embedding can be measured in terms of the maximum distance of its vertices from the external face. Such a distance can be given in terms of different incidence relationships between vertices and faces. For example, if two faces are considered adjacent when they share a vertex, then the maximum distance to the external face is called *radius* [RS84]. If two vertices are adjacent when they are endpoints of an edge, then the maximum distance to the external face is called *width* [DLT84]. If two vertices are adjacent when they are on the same face and the external face is adjacent to all its vertices, then the maximum distance to the external face is called *outerplanarity* [Bak94]. If two faces are adjacent when they share an edge, then the maximum distance to the external face is called *depth* [BM88]. In [PT00, GM04], algorithms are proposed to minimize the maximum distance of the biconnected components of the graph from the external face, where two biconnected components are adjacent if they share a cutvertex. This measure, which is also called "depth," is a rougher indicator of the quality of the embedding but can be computed in linear time.

In [BM90], Bienstock and Monma present an algorithm to compute the planar embedding of an n-vertex planar graph with minimum maximum distance to the external face in $O(n^5 \log n)$ time, which is improved to $O(n^4)$ time in [ADP11]. The considered distance is the depth. However, it is possible to compute the radius, the width, and the outerplanarity of a graph by modifying and simplifying the algorithm for the minimum depth, since such distance measures are intrinsically simpler to compute than the depth [BM90]. The complexity bounds for computing such simpler distance measures is improved in [Kam07], where an algorithm that computes the outerplanarity of an n-vertex planar graph in $O(n^2)$ time is described. Simple variations of this algorithm can lead to compute the radius in $O(n^2)$ time and the width in $O(n^3)$ time [Kam07].

1.3.2 Deletion and Partition Problems

Deleting the minimum number of edges in order to obtain a planar graph is called *maximum planar subgraph* and proved to be NP-hard in [GJ79]. Analogously, deleting the minimum number of vertices in order to obtain a planar graph is called *maximum induced planar subgraph* and proved to be NP-hard in [Yan78].

The problem of partitioning the edges of a graph $G = (V, E)$ into k sets $E_1, \ldots, E_k$ in such a way that each graph $G_i = (V, E_i)$, with $i = 1, 2, \ldots, k$ is planar is called *graph thickness* and is shown to be NP-hard for $k = 2$ in [Man83].

1.3.3 Upward Planarity

If the input graph G is directed, adding the requirement that the drawing of G is upward, that is, that each edge is a curve of increasing y-coordinates, transforms the planarity problem into the upward planarity one, which was shown to be NP-complete in [GT01].

However, upward planarity testing turns out to be polynomial for several families of directed graphs:

1. If the digraph G is outerplanar. This problem was shown to be $O(n^2)$ in [Pap95].
2. If the digraph G is triconnected [BD91, BDLM94].
3. If the digraph G has a fixed embedding. An $O(n^2)$-time algorithm was introduced in [BDLM94], and the problem is linear in the case of embedded outerplanar graphs ([Pap95]).
4. If the digraph G is single-source. The $O(n^2)$-time algorithm described in [HL96] was improved to linear in [BDMT98].

1.3.4 Outerplanarity

Determining whether a graph is outerplanar and producing an outerplanar drawing of it is a problem that can be solved independently or by using a planarity algorithm as a subroutine. In fact, a graph $G = (V, E)$ is outerplanar if and only if the graph $G'(V', E')$ is planar, where $V' = V \cup \{v\}$ and E' is obtained from E by adding an edge (v_i, v) for each vertex $v_i \in V$.

Deleting the minimum number of vertices from a graph in order to make it outerplanar is NP-complete [Yan78].

1.4　History of Planarity Algorithms

Directly applying Kuratowski's characterization of planar graphs based on subdivisions would yield an exponential-time algorithm while Wagner's characterization based on minors would give a factorial-time algorithm. The first polynomial-time algorithms for planarity are due to Auslander and Parter [AP61], Goldstein [Gol63], and, independently, Bader [Bad64].

In 1974, Hopcroft and Tarjan [HT74] proposed the first linear-time planarity testing algorithm. This algorithm, also called "path-addition algorithm," starts from a cycle and adds to it one path at a time. However, the algorithm is so complex and difficult to implement that several other contributions followed their breakthrough. For example, about 20 years after [HT74], Mehlhorn and Mutzel [MM96] contributed a paper to clarify how to construct the embedding of a graph that is found to be planar by the original Hopcroft and Tarjan algorithm.

A different approach has its starting point in the algorithm presented by Lempel, Even, and Cederbaum [LEC67]. This algorithm, also called "vertex addition algorithm," is based on considering the vertices one-by-one, following an *st*-numbering; it has been shown to be implementable in linear time by Booth and Lueker [BL76], while a linear-time algorithm for computing the needed st-numbering was provided in [ET76]. Also in this case, a further contribution by Chiba, Nishizeki, Abe, and Ozawa [CNAO85] has been needed for showing how to construct an embedding of a graph that is found planar.

A further interesting algorithm [dOR06, dF08a, Bra09] is based on a characterization given by de Fraysseix and Rosenstiehl [dR82, dR85] in turn based on intuitions of Liu and Wu [Wu74, Ros80, Liu88, Liu89, Xu89]. For a long time, the algorithm has not been fully described in the literature but had a very efficient implementation in the Pigale software library [dO02].

However, although the planarity problem has been carefully studied in the above cited literature, the story of the planarity testing algorithms enumerates several more recent contributions. The motivations behind such relatively new papers are twofold. On one side, even if the known algorithms are combinatorially elegant, they are quite difficult to understand and to implement. On the other side, the researchers are interested in deepening the relationships between planarity and Depth First Search (DFS). Such relationships are clearly strong but, probably, up to now, not completely understood.

Two recent DFS-based planarity testing algorithms, whose similarities were stressed in [Tho99], are those presented by Shih and Hsu [SH93, SH99, Hsu03] and by Boyer and Myrvold [BM99, BM04].

The Shih-Hsu algorithm replaces biconnected portions of the graph with single nodes, called *C*-nodes, whose embedding is fixed.

The Boyer and Myrvold algorithm represents embedded biconnected portions of the graph with a data structure that allows the embeddings to be "flipped" in constant time.

1.5　Common Algorithmic Techniques and Tools

In this section, we introduce some definitions and common techniques used by the planarity testing algorithms. The most important technique, common to almost all the algorithms, is *Depth First Search*, or DFS. DFS is a method for visiting all the vertices of a graph G. It starts from an arbitrarily chosen vertex of G and continues moving from the current vertex to an adjacent one, as long as unexplored neighbors are found. When the current vertex has no unexplored neighbors, the traversal backtracks to the first vertex with an unexplored adjacent vertex.

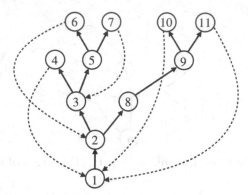

Figure 1.4 A DFS traversal of a graph. Thick lines represent the tree edges, while the back edges are drawn with dashed lines. Each vertex is identified with its DFS index.

The edges by which DFS discovers new vertices of G form a spanning tree T of G, called *Palm Tree*, or *DFS Tree*. The root of T is the vertex at which the traversal started. The edges of T are called *tree edges*, while the remaining edges of G are called *back edges* (or *co-tree edges*).

After performing a DFS traversal, each vertex v of G can be associated with a *DFS index*, $DFS(v)$, that is, the order in which v was reached during the DFS visit. The root of T has index one. For a tree edge (u, v), we have that $DFS(u) < DFS(v)$. On the contrary, a back edge is oriented from the end vertex with higher DFS index to the end vertex with lower DFS index. An example DFS is shown in Figure 1.4.

For each vertex v of G, we can also define two sets of edges, called $B_{in}(v)$ and $B_{out}(v)$. These sets contain, respectively, the back edges entering and exiting v. Note that each back edge in $B_{in}(v)$ connects v to a descendant in the DFS tree, while each back edge in $B_{out}(v)$ connects v to an ancestor. Given a tree edge $e = (u, v)$, its *returning edges* are those back edges that from a descendant of v (included v itself) go to an ancestor of u different from u itself. At last, the *lowpoint* of a vertex v, denoted by $lowpt(v)$, is the lowest DFS index of an ancestor of v reachable through a back edge from a descendant of v. Analogously, the *highpoint* of a vertex v, denoted by $highpt(v)$, is the highest DFS index of an ancestor of v reachable through a back edge from a descendant of v.

1.6   Cycle-Based Algorithms

The shared foundation of all algorithms in this section is an intuitive observation formalized in the Jordan curve theorem: every simple closed curve divides the plane into two connected regions, and hence there is no way to connect two points in both regions without crossing that curve.

Acyclic (undirected) graphs are forests and therefore planar. If a graph does contain a cycle, that cycle yields a simple closed curve in any planar drawing of it. Consequently, each of the remaining connected parts of the graph needs to be drawn entirely in one of the two connected regions bounded by the cycle. Deciding whether this is possible, and which region to choose, is the essence of planarity testing and embedding, respectively.

It will take three major steps to arrive at simple linear-time algorithms based on this observation. The first step consists in formalizing the approach in a recursive algorithm, the second step yields a linear-time realization of the algorithm, and the third step simplifies the second while adding a corresponding combinatorial characterization.

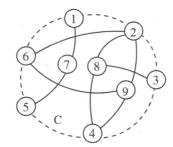

Figure 1.5 A biconnected graph G and a cycle C. The edges of C are drawn dashed.

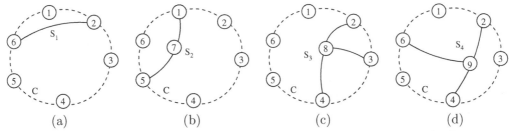

Figure 1.6 The four segments of graph G of Figure 1.5 separated by cycle C.

1.6.1   Adding Segments: The Auslander-Parter Algorithm

Algorithms based on the above cycle criterion were first proposed in [AP61] (see also [Gol63, Bad64, DETT99]).

To introduce the approach formally, consider a simple cycle C in a biconnected graph G. Recall that a graph is planar if and only if its biconnected components are, and that every edge of a biconnected graph is contained in at least one cycle. Each such cycle C yields a collection of connected, edge-induced subgraphs S_i, $i = 1, \ldots, k$ as follows. Either S_i is an edge that connects two vertices of C that are not consecutive (i.e., a *chord*), or S_i is induced by the edges of a connected component of $G \setminus C$ together with the edges connecting that component to C. Each S_i is called a *segment* and, because of biconnectivity, contains at least two vertices of C, referred to as the *attachments* of S_i. Note that vertices of C may be attachments of any number of segments. Figure 1.5 shows a biconnected graph G and a cycle C. The segments separated by C are depicted in Figure 1.6

A cycle C of G is said to be *separating* if it has at least two segments, while it is called *non-separating* otherwise. Of course, if G is a cycle, then C has no segments and is non-separating. In order to recur on subgraphs, the Auslander-Parter algorithm needs to pick a separating cycle.

LEMMA 1.1   [DETT99] Let G be a biconnected graph, let C be a non-separating cycle of G, and let S be its only segment. If S is not a path, then G has a separating cycle C' consisting of a subpath of C plus a path γ of S between two attachments.

Proof: Let u and v be two attachments of S that are consecutive in the circular ordering of C, let α be a subpath of C between u and v that does not contain any other attachment of S to C, and let β be the subpath of C between u and v different from α (see Figure 1.7 for an example). Since S is connected, there is a path γ in S between u and v. Let C' be the cycle obtained from C by replacing α with γ. We have that α is a segment of G with respect to C'. If S is not a path, let e be an edge of S not in γ. There is a segment

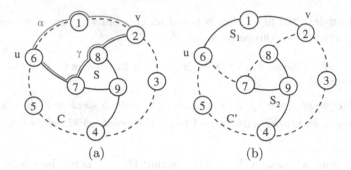

Figure 1.7 (a) A non-separating cycle C whose single segment S is not a path. Replacing subpath α with subpath γ as described in the proof of Lemma 1.1 yields a separating cycle C'.

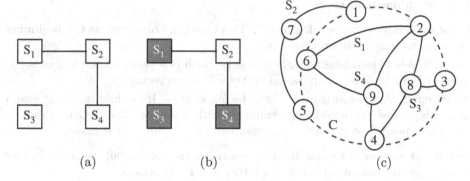

Figure 1.8 (a) The interlacement graph for the segments induced on graph G by the cycle C of Figure 1.5. (b) A possible bi-coloring of the interlacement graph. (c) The corresponding embedding choices for the segments of G, where segments colored black are placed inside C.

of C' distinct from α containing e. Therefore, if S is not a path, then C' has at least two segments and is thus a separating cycle of G. □

We have already argued that segments must be drawn entirely in one of the two regions created by the drawing of C. Two segments are said to be *compatible*, if they can be drawn in the same region of C, and *conflicting* otherwise. The following lemma shows that compatibility has a simple characterization.

LEMMA 1.2 Two segments are compatible, if and only if their attachments do not interleave.

The *interlacement graph* of the segments of G with respect to C is the graph whose vertices are the segments of G and whose edges are the pairs of interlacing segments. Figure 1.8(a) shows the interlacement graph for graph G and cycle C of Figure 1.5. If there are more than two pairwise incompatible segments, the graph is not planar, because there are only two regions in which they can be drawn. If G is planar, then the interlacement graph is bipartite and two-colorable, each color corresponding to one side of C (see Figures 1.8(b) and 1.8(c)). We can recursively check the planarity of all subgraphs obtained from the union of a segment S_i and C.

The AUSLANDER-PARTER algorithm is based on the following intuitive recursive characterization of planarity for biconnected graphs.

Theorem 1.1   *[DETT99] A biconnected graph G with a cycle C is planar if and only if the following two conditions hold:*

- *The interlacement graph of the segments of G with respect to C is bipartite.*
- *For each segment S of G with respect to C, the graph obtained by adding S to C is planar.*

Proof: If the graph is planar, it is easy to see that the two conditions hold by considering a planar drawing of it. If the two conditions hold, the proof is by construction and is based on the fact that compatible segments do not interleave (Lemma 1.2) and, hence, can be planarly arranged on the same side of C.   □

The algorithm has three cases:

Trivial case. Graph G is a single cycle C. This case can only occur at the beginning of the computation and terminates it.

Base case. Cycle C separates a single segment, which is a path. This terminates the current branch of the computation (there will be no recursion).

Recursive case. A separating cycle C can be found in G. If the interlacement graph is not bipartite, the algorithm terminates with a non-planarity. Otherwise, recursion is needed on the subgraphs composed by C and each segment.

Here it is not necessary to describe this algorithm in more detail, because, in fact, the subsequent ones are instantiations of this rather generic approach.

It can be shown that the number of recursions is $O(n)$ and that the interlacement graph has size $O(n^2)$, yielding an $O(n^3)$ time algorithm. Also, it is worth mentioning that for a graph that turns out to be planar, the embedding is constructed bottom-up, where planar embeddings may have to be flipped, depending on which region they are placed in. There is an interesting alternative approach presented by Demoucron, Malgrange, and Pertuiset [DMP64]. Instead of recursively testing segments for planarity, they start from a fixed embedding of one cycle, and incrementally add only a path connecting two attachments of a segment into a face of the current embedding. This approach requires a careful selection of (facial) cycles and paths and yields a quadratic-time algorithm but is the only algorithm known to us that does not require alterations of preliminary embeddings.

1.6.2   Adding Paths: The HOPCROFT-TARJAN Algorithm

The relative inefficiency of recursively testing augmented segments for planarity is caused by a lack of control over the instances obtained when selecting a cycle.

By exploiting the special structure of DFS trees, Hopcroft and Tarjan [HT74] (see also [Deo76, RND77, Eve79, Wil80]) were able to serialize the combination of trivially planar segments (namely, paths) in a bottom-up fashion.

Let us start from a *spine cycle*, i.e., a fundamental cycle consisting of a path of tree edges that start at the root of the DFS tree together with a single back edge returning to the root. Call the subgraph consisting of only the spine cycle G_0. Next, segments are added recursively, one path at a time, which is why the algorithm is often referred to as the *path-addition approach*.

To explain the order in which paths are selected, consider the subgraph G_i consisting of the spine cycle and the first i paths, and an edge e that is incident to but not contained in G_i.

Define the segment $S(e)$ of e to be the inclusion-maximal connected subgraph containing e, in which no vertex of G_i has degree larger than one. Moreover, define the vertex with the lowest DFS number in $S(e)$ to be the *lowpoint of the segment*. Since G is biconnected, $S(e)$ contains at least two vertices of G_i, which we call attachments as well. By the order in which paths are inserted, the lowpoint of $S(e)$ will always be an attachment.

Now assume that the DFS tree was re-built to determine lowpoints and biconnected components. When exploring the tree once again, but this time by traversing edges with lower lowpoints first, we are effectively performing a recursive traversal of segments in which segments with lower lowpoints are traversed first. This order is crucially important for our ability to test efficiently whether segments are conflicting, because it ensures that the attachments of a segment are visited in order of non-decreasing lowpoints. We can therefore place lowpoints on a stack and remove them from the top of the stack during backtracking, thus maintaining in the stack all attachments in the order in which they appear in the lower part of the segment-defining cycle not yet backtracked over. Recall that two segments are compatible if their attachments do not interleave.

Again, we do not go into further details, because the approach is further simplified below. We just note that the algorithm can actually be implemented to run in linear time, but that this is quite difficult and that it took many years until this test was complemented by an embedding phase [MM96] (which runs in linear time).

Part of the difficulty is in the absence of a characterization of planarity that is closely tied to the workings of the algorithm.

1.6.3 Adding Edges: The DE FRAYSSEIX-OSSONA DE MENDEZ-ROSENSTIEHL Algorithm

While we have argued that the test of Hopcroft and Tarjan implements that of Auslander and Parter by recursively building up segments one path at a time, it turns out that the original approach can be further simplified by interpreting it on an even more detailed level, adding one edge at a time.

This does not only simplify the algorithm, it also yields a characterization of planarity that provides a less procedural proof of correctness and a straightforward embedding. Therefore, following the approach of [Bra09], we first recall the characterization and then revisit the algorithm.

Consider a connected undirected graph which needs not to be biconnected, and let $G = (V, T \uplus B)$ be the directed graph obtained from a DFS, where T is the set of tree edges and B the set of back edges. We say that G is a *DFS-orientation* of the original graph. Note that this is not a procedural definition, since such an orientation is characterized by consisting of a rooted spanning tree such that each non-tree edge defines a directed cycle. As each back edge returns to an ancestor of its source, it implicitly defines a cycle, which is called *fundamental cycle*. A back edge (u, v) is a *return edge* for each tree edge of its fundamental cycle, with the exception of the first tree edge exiting v.

DEFINITION 1.1 [dOR06] Let $G = (V, T \uplus B)$ be a DFS-oriented graph. A partition $B = L \uplus R$ of its back edges into two classes, referred to as *left* and *right*, is called *left-right partition*, or *LR partition* for short, if for every vertex v with incoming tree edge e and outgoing edges e_1, e_2

- all return edges of e_1 ending strictly higher than $lowpt(e_2)$ belong to one class and

- all return edges of e_2 ending strictly higher than $lowpt(e_1)$ belong to the other class.

Intuitively, the partition of the back edges into classes L and R corresponds to orienting the fundamental cycles in such a way that those closed by back edges in L are counterclockwise while those closed by back edges in R are clockwise.

Theorem 1.2 *A graph is planar if and only if it admits an LR partition.*

Necessity of the condition of Theorem 1.2 is straightforward: given a DFS tree and a planar embedding of the graph it suffices to assign each back edge to the classes L or R, depending on whether the fundamental cycle it closes is counterclockwise or clockwise, respectively. Sufficiency is shown by constructing a planar embedding from a given LR partition. First, observe that in an LR partition it can be assumed that all return edges from a tree edge e that return to $lowpt(e)$ are on the same side. Such a LR partition is called *aligned*. If a partition is not aligned, an equivalent aligned partition can be found.

In order to obtain a planar embedding, the LR partition is extended to cover also outgoing tree edges and, for each vertex v, a linear nesting order is defined on its exiting tree edges. This order contains both right and left outgoing edges of v mixed together: restricted to the right outgoing tree edges it gives their clockwise order around v and restricted to the left outgoing tree edges it gives their counterclockwise order around v. The final embedding for each vertex v is obtained by suitably interleaving outgoing tree edges with back edges entering v.

The extension of the LR partition to tree edges is straightforward. If a tree edge has some return edges (i.e., its source is neither the root nor a cut vertex), it is assigned to the same side as one of its return edges ending at the highest return point. Otherwise, the side is arbitrary.

To determine the linear nesting order for tree edges outgoing v, suppose first that all back edges belong to R and consider a fork consisting of tree edge $e = (u, v)$ and outgoing tree edges e_1 and e_2 exiting v. If both e_1 and e_2 have some return edges, v is a branching point of at least two overlapping fundamental cycles sharing e. Since both cycles are clockwise (all edges belong to R), they must be properly nested in order to avoid edge crossings. As the root of the DFS tree is assumed to be on the outer face, we have to put e_2 clockwise after e_1 (i.e., inside the cycle defined by it) if and only if the lowpoint of e_1 is strictly lower than that of e_2. The same holds if both have the same lowpoint but only e_2 is *chordal*, i.e., has another return point above it. On the contrary, if both L and R are not empty, it can happen that both e_1 and e_2 are chordal. In this case the tie is broken arbitrarily, because in any planar embedding these two edges must be on different sides.

Let $e = (v, w)$ be a tree edge. We denote by $L(e)$ ($R(e)$, respectively) the sequence of incoming back edges entering v from descendants of w ordered in such a way that if $b_1 = (x_1, v)$ and $b_2 = (x_2, v)$ are two such back edges, and if (z, x), (x, y_1), and (x, y_2) is the fork of the two cycles closed by b_1 and b_2, then b_1 comes before b_2 in $L(e)$ ($R(e)$, respectively) if and only if (x, y_1) comes before (x, y_2) ((x, y_1) comes after (x, y_2), respectively) in the adjacency list of x.

DEFINITION 1.2 Given an LR partition and a vertex v, let $e_1^L, \ldots, e_l^L$ be the left outgoing tree edges of v, and $e_1^R, \ldots, e_r^R$ its right outgoing tree edges. If v is not the root, let u be its parent. The clockwise *left-right ordering*, or *LR ordering* for short, of the edges around v is defined as follows:

$$(u, v), L(e_l^L), e_l^L, R(e_l^L), \ldots, L(e_1^L), e_1^L, R(e_1^L), L(e_1^R), e_1^R, R(e_1^R), \ldots, L(e_r^R), e_r^R, R(e_r^R),$$

where (u, v) is absent if v is the root.

The following lemma shows the sufficiency of the left-right planarity criterion of Theorem 1.2 (the proof by contradiction can be found in [Bra09]).

LEMMA 1.3   Given an LR partition, its LR ordering yields a planar embedding.

Hence, the search for a planar embedding of the input graph boils down to the search for an LR partition of its back edges. Fortunately, from the definition of LR partition directly come two constraints that have to be satisfied by back edges in L and R classes. Let $b_1 = (u_1, v_1)$ and $b_2 = (u_2, v_2)$ be two back edges with overlapping fundamental cycles and let $(u, v), (v, w_1), (v, w_2)$ be their fork.

  1. b_1 and b_2 belong to different classes if $lowpt(w_2) < v_1$ and $lowpt(w_1) < v_2$.
  2. b_1 and b_2 belong to the same class if there is an edge $e' = (x, y)$, with $x \in C(b_1) \cap C(b_2)$ and $y \notin C(b_1) \cap C(b_2)$ such that $lowpt(y) < min\{v_1, v_2\}$.

Of course, if a pair of back edges is subject to both the constraints above, no LR partition can exist and hence the graph is non-planar. By exploiting the constraints a quadratic planarity test and embedding algorithm can be found immediately. Namely, build a *constraint graph*, analogous to the interlacement graph of the AUSLANDER-PARTER algorithm, where each back edge is a vertex and each constraint is an edge, labeled "-1" if the two back edges have to belong to different classes and labeled "$+1$" if they have to belong to the same class. After contracting "$+1$" edges, test if the constraint graph is bipartite.

In order to transform this quadratic-time algorithm into a linear one, the constraint graph cannot be explicitly built and the tentative assignment of back edges to the L and R classes may be changed several times during the computation, which is structured as a further traversal of the DFS tree. Details of the linear-time algorithm can be found in [dOR06, dF08a, Bra09].

1.7   Vertex Addition Algorithms

Given a planar drawing Γ of a graph $G(V, E)$, we could delete one vertex at a time from Γ to obtain a sequence of smaller planar drawings ending with a single isolated vertex. The intuition that this process could be suitably reversed yields the so-called "vertex addition" algorithms.

We classify in this family the LEMPEL-EVEN-CEDERBAUM, the SHIH-HSU, and the BOYER-MYRVOLD algorithms, although we know that some authors proposed a different classification for their approach. The similarities between the SHIH-HSU and the BOYER-MYRVOLD algorithms were already pointed out in [Tho99], while a common view encompassing all the three algorithms was envisaged by Haeupler and Tarjan in [HT08].

Vertex addition algorithms start from an initial graph G_1 composed by one isolated vertex v_1. At each step $i = 2, \ldots n$, a new vertex v_i is added to the graph and the subgraph $G_i(V_i, E_i)$, induced by the current vertices $V_i = \{v_1, \ldots, v_i\} \subseteq V$, is considered. Two kinds of operations are performed: first, G_i is checked for planarity; second, some data structures are updated in order to allow analogous checks to be efficiently performed at step $i + 1$.

A key feature, common to this family of algorithms, is that the order in which the vertices are added is not arbitrary. Let $\overline{G}_i(\overline{V}_i, \overline{E}_i)$ be the subgraph of G induced by the vertices $\overline{V}_i = V - V_i$ that have still to be added to the graph. All the algorithms based on vertex

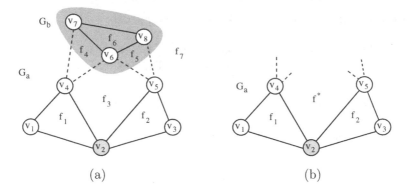

(a)   (b)

Figure 1.9 Properties of Lemma 1.4. (a) The embedding Γ where the connected subgraph G_b is highlighted. (b) The embedding Γ_a of G_a. By Property (α), v_6, v_7, and v_8 fall into f^*. By Property (β), f_1 and f_2 are also faces of Γ. By Property (γ), the cutvertex v_2 is on f^*.

addition require that $\overline{G}_i$ is connected for $i = 1, \ldots, n$, that is, the vertex addition order is a leaf-to-root order for some spanning tree of G. LEMPEL-EVEN-CEDERBAUM's algorithm, for example, requires that the vertices are added in the order given by an st-numbering; in the SHIH-HSU and in the BOYER-MYRVOLD algorithms, the order is that of a reverse DFS traversal of the graph. The importance of this requirement is stated by the following lemma.

LEMMA 1.4   Let $G(V, E)$ be a planar, connected graph and let $\{V_a, V_b\}$ be a bipartition of the vertices in V such that the graph $G_b(V_b, E_b)$ induced by V_b is connected. Consider any planar embedding Γ of G and denote by Γ_a the planar embedding Γ restricted to G_a. The following properties hold:

(α) Vertices of V_b are on the same face f^* of Γ_a.

(β) Each face f of Γ_a, with $f \neq f^*$, is also a face of Γ.

(γ) If G is biconnected, cutvertices of G_a are also incident to face f^* of Γ_a.

Proof: Property (α) trivially descends from the fact that G_b is connected and Γ is a planar embedding of G. Property (β) is also trivial. Suppose for a contradiction that $f \neq f^*$ is a face of Γ_a but not a face of Γ. Observe that f is a cycle of Γ and, since it is not a face of Γ, it contains at least one edge $e = (u, v)$ of Γ that is not an edge of Γ_a. If both u and v belong to V_a, we have a contradiction as e belongs to the graph induced by V_a but it is not in Γ_a. Otherwise, if one among u and v is not in V_a, we have again a contradiction since Property (α) ensures that $f = f^*$. This proves Property (β). Suppose that G is biconnected. If v is a cutvertex of G_a, then there is a face f of Γ_a that is incident at least two times on v. Since v is not a cutvertex of Γ, face f is a face of Γ_a but is not a face of Γ, and Property (β) ensures that f^* is the only face of Γ_a that has this property. $\square$

An example that shows the three properties of Lemma 1.4 is depicted in Figure 1.9. Property (α) was also proved in [Eve79, Lemma 8.10] for the special case of connected subgraphs induced by an st-numbering.

Let ψ be a function $\psi : V \to \{1, \ldots, n\}$ that assigns a different index to each vertex of G. We say that ψ is a *proper numbering* of G if for each i we have that the subgraph $\overline{G}_i(\overline{V}_i, \overline{E}_i)$ induced by $\overline{V}_i = \{v \mid \psi(v) > i\}$ is connected. In order to simplify the notation in the remaining part of this chapter, we denote by v_i the vertex for which $\psi(v_i) = i$. Vertex addition algorithms require that vertices are considered in the order imposed by a proper numbering, hence exploiting at each step the properties of Lemma 1.4. Namely, Property (α) guarantees that vertices and edges can be added to a single face f^* of Γ_i, which can be assumed to be the outer face. Property (β) implies that once a vertex or edge is closed inside an internal face of Γ_i it does not need to be considered again (this is a key point to ensure linearity). Finally, Property (γ) justifies the usual assumption, common to most vertex addition algorithms, that G is biconnected.

Properties (α) and (β) lead to the following lemma.

LEMMA 1.5 Let ψ be any proper numbering of a planar, connected graph G. Denote by G_i the subgraph of G induced by vertices in $V_i = \{v \mid \psi(v) \leq i\}$. There exists a sequence of planar embeddings Γ_i of G_i, with $i = 1, \ldots, n$, such that, for $i = 1, \ldots, n - 1$, all internal faces of Γ_i are also internal faces of Γ_{i+1}.

Proof: Let Γ_n be a planar drawing of G with v_n on the external face and let Γ_i, with $i = 1, \ldots, n - 1$, be the embedding of G_i obtained from Γ_n by removing the vertices v_j, with $j = i + 1, \ldots, n$. Vertex v_n is on the external face of Γ_n by definition. Since $\overline{G}_i$ is connected, vertex v_i is also on the external face of Γ_i for any $i = n - 1, n - 2, \ldots, 1$. Also, $V_a = \{v_1, \ldots, v_{i-1}\}$ and $V_b = \{v_i\}$ is a bipartition of the vertices of G_i of which Γ_i is a planar embedding and $G_b(V_b, \emptyset)$ is trivially connected. Lemma 1.4 applies and by Property (β) we have that all the faces of Γ_{i-1} with the exception of f^* are also faces of Γ_i. Since the external face of Γ_{i-1} is not a face of Γ_i, any other internal face f of Γ_{i-1} is also a face of Γ_i. Finally, as the external face of Γ_i contains v_i, which does not belong to G_{i-1}, face f is an internal face of Γ_i. $\qquad\square$

Provided that G is planar, Lemma 1.5 can be exploited for devising an incremental planarity algorithm that, starting from Γ_1, i.e., the trivial embedding of the isolated vertex v_1, computes Γ_i, with $i = 2, \ldots, n$, by adding at each step a vertex v_i on the outer face of Γ_{i-1}, until an embedding Γ_n of the whole graph is produced. Also, Lemma 1.4 provides an indication of what are the properties that these Γ_i should have. Namely, call *outer vertices of G_i* the cutvertices of G_i and the vertices of G_i adjacent to $v_{i+1}, v_{i+2}, \ldots, v_n$. Properties (α) and (γ) of Lemma 1.4 state that if G is biconnected, which can be assumed, each Γ_i necessarily has its outer vertices on the outer face.

Still, computing the sequence of Γ_i, with $i = 1, \ldots, n$, is not an easy task. First, G_i may be not connected. Second, it is easy to see that not any embedding of G_i with its outer vertices on the external face is equivalent to any other. In fact, given a planar graph G, there may exist a planar embedding Γ_i of G_i that has the outer vertices of G_i on the external face but is not obtainable from some planar embedding of G by vertex deletion (Figure 1.10 provides an example).

Hence, although we know that, starting from any proper numbering ψ of G, the planarity of G implies the existence of a sequence of planar embeddings Γ_i satisfying the conditions of Lemma 1.5, we do not know how to find such a sequence, and choosing a wrong embedding Γ_i along the way would lead to a failure of the whole process even if G is planar. The following lemma comes in help.

LEMMA 1.6 In any planar embedding of a biconnected graph G where vertices $v_1, v_2, \ldots, v_k$ share the same face, they appear in the same circular order up to a reversal.

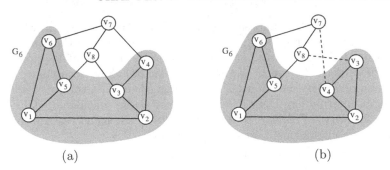

Figure 1.10 A planar graph G with subgraph G_6 highlighted. (a) and (b) show two planar embeddings of G_6, both with the outer vertices of G_6 on the external face. The embedding in (a) is compatible with a planar drawing of G while the embedding in (b) is not.

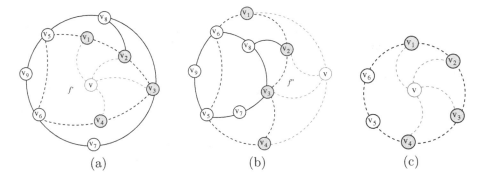

Figure 1.11 (a), (b) Two planar embeddings of a biconnected graph where vertices v_1, v_2, v_3, and v_4 (highlighted in the figure) share the same face. Vertex v is added as in the proof of Lemma 1.6.

Proof: The statement is trivial for $k = 2, 3$, since any circular sequence of 2 or 3 labels is equal to any other up to a reversal. Consider two planar embeddings Γ' and Γ'' of G such that vertices $v_1, v_2, \ldots, v_k$, with $k \geq 4$, share the face f' in Γ' and f'' in Γ'' (see Figure 1.11(a) and 1.11(b) for an example). The proof is based on the trivial observation that a dummy vertex v can be inserted into both f' and f'' and planarly connected to $v_1, v_2, \ldots, v_k$. Since G is biconnected, the cycle face f' is simple (see Figure 1.11(a)). Hence, the subgraph composed by the edges and vertices of f' and v is a wheel (dashed lines in Figs. 1.11(a), 1.11(b), and 1.11(c)) and admits a unique planar embedding up to a reversal. It follows that the circular order of the edges around v is the same in Γ' and in Γ'' up to a reversal.
□

Lemma 1.6 applied to each block of G_i is stated in [Eve79, Lemma 8.12] for the special case of subgraphs induced by st-numberings. When iteratively computing a planar embedding for G, the practical use of Lemma 1.6 is that, although in general no definitive choice can be made on the embedding of G_i, something can be said about the embedding of its blocks. Namely, apart from a possible flip, an embedding for them can be computed that is always compatible with a planar embedding of the whole graph, provided it exists. Surprisingly, this is the only thing that can be safely computed for the embedding of G_i. All the more so, this little amount of information suffices for computing analogous embeddings for the blocks of G_{i+1}, and, since $G_n = G$ is biconnected, at the last step a planar embedding Γ_n of the whole graph is obtained. Finally, the following lemma shows that if the process stops, the graph is not planar.

LEMMA 1.7 Let G be a graph and let ψ be any proper numbering of G. Denote by G_i, with $i = 1, \ldots, n$ the subgraph of G induced by vertices in $V_i = \{v \mid \psi(v) \leq i\}$ and by $B_i^1, B_i^2, \ldots, B_i^{b_i}$ the b_i blocks of G_i. For a given k, $1 \leq k \leq n-1$, let $\Gamma(B_k^j)$, $1 \leq j \leq b_k$, be arbitrary embeddings of B_k^j with the outer vertices of G_k on their outer faces. If the blocks of G_{k+1} cannot be embedded such that the outer vertices of G_{k+1} are on the outer face and $\Gamma(B_k^j)$, $1 \leq j \leq b_k$, are preserved up to a flip, then G is not planar.

Proof: Suppose for a contradiction that G is planar and that there is no planar embedding for all its blocks B_{k+1}^j, $1 \leq j \leq b_k$, such that the outer vertices of G_{k+1} are on the outer face and the blocks of G_k are embedded, up to a flip, as in $\Gamma(B_k^j)$, $1 \leq j \leq b_k$. Since G is planar, by Lemma 1.5, there is a pair of planar drawings Γ_k^* of G_k and Γ_{k+1}^* of G_{k+1}, both with their outer vertices on the outer face. By Lemma 1.6 the outer vertices of each block of G_k appear in the same order, up to a reversal, both in $\Gamma(B_k^j)$, $1 \leq j \leq b_k$, and in Γ_{k+1}^*. Hence, all embeddings $\Gamma(B_k^j)$ can be inserted into Γ_{k+1}^* yielding a planar embedding for the blocks B_{k+1}^j, $1 \leq j \leq b_k$, such that the outer vertices of G_{k+1} are on the outer face: a contradiction. $\square$

Lemma 1.7 proves the soundness of the vertex addition approach. In fact, it shows that iteratively building a planar embedding of the input graph G is not only a sufficient condition for the planarity of G, which is obvious, but also a necessary condition, as G is not planar if one step of the iterative process cannot be accomplished. Usually, in the vertex addition literature, the non-planarity of the input graph in case of failure of the proposed algorithms is proved by a complex case analysis, spread all over the description of the algorithm steps, aimed at identifying a subgraph isomorphic to K_5 or $K_{3,3}$ for each possible cause of failure. Instead, Lemma 1.7 provides a direct proof of the correctness of the approach that avoids the use of Kuratowski's theorem, as claimed in [HT08].

Observe that, since the internal faces of the blocks are preserved in the final embedding of G, at each iterative step of the vertex addition algorithms the embedded blocks may be flipped and composed together, but they are never inserted one into the other. Hence, all vertex addition algorithms make use of suitable data structures to describe the subgraph G_i that has been explored so far and in particular the embedding of its blocks. These data structures allow for permuting the blocks around the cutvertices and for flipping the blocks in constant time. In the LEMPEL-EVEN-CEDERBAUM algorithm, the data structure is Booth and Lueker's PQ-tree. The SHIH-HSU algorithm uses PC-trees. The BOYER-MYRVOLD algorithm uses the `bicomp` data structure. The purpose of these data structures is analogous: they allow us to flip a portion of the graph (a block) in constant time; they allow us to permute (or to leave undecided) the order of the blocks around a cutvertex until the blocks are merged together.

1.7.1 The LEMPEL-EVEN-CEDERBAUM Algorithm

The LEMPEL-EVEN-CEDERBAUM algorithm was the first one to exploit the vertex addition paradigm [LEC67] (see also [Eve79, BFNd04]). It is no surprise, therefore, that in order to ease the computation several simplifying assumptions are made. First, but this is usual, the input graph is assumed to be biconnected. Second, the description of the algorithm in [LEC67] only checks the planarity of the input graph, without actually computing a planar embedding if it exists. This gap was closed by Chiba, Nishizeki, Abe, and Ozawa [CNAO85] some decades later. Third, a proper numbering of the vertices of G is required that also ensures that G_i, the graph induced by V_i, is connected. Namely, given any edge (s, t) of a biconnected graph $G(V, E)$ with n vertices an *st-numbering* of G is a function $\psi : V \rightarrow \{1, \ldots, n\}$ that assigns to each vertex a different index, such that: (i) $\psi(s) = 1$; (ii) $\psi(t) = n$; and (iii) any vertex except s and t is adjacent both to a lower-numbered and to a higher-numbered vertex. This strong constraint, which implies that both the st-numbering and its reversal are proper numberings, fostered the search for a linear-time algorithm to actually compute an st-numbering of a biconnected graph. Such an algorithm was not known when the approach was introduced (the time complexity of the algorithm used in [LEC67] is $O(nm)$ [ET76]), and was finally found in [ET76].

Working of the algorithm

A *bush* is a single-source connected planar directed graph that admits a planar embedding, called a *bush form*, where all vertices of degree one are on the outer face.

Let G be a biconnected graph, let ψ be an st-numbering of G, and let G_i be the graph induced by vertices $\{v_1, \ldots, v_i\}$. Graph G can be assumed to be directed, where each edge is oriented from the vertex with the lower value to the vertex with higher value of ψ (see Figure 1.12(a) for an example). Denote by $\mathcal{B}_i$ the graph G_i augmented with the edges of G incident to the outer vertices of G_i. These edges are called *virtual edges*, while the leaves that they introduce in $\mathcal{B}_i$ are *virtual vertices*. Virtual vertices are labeled with the same indexes they have in G, and multiple instances of the same vertex are kept separate in $\mathcal{B}_i$. Since G_i is determined by an st-numbering, $\mathcal{B}_i$ is connected. Observe that a planar embedding of $\mathcal{B}_i$ with the virtual vertices on the outer face corresponds to a planar embedding of G_i with the outer vertices on the outer face. Hence, if G is planar, by Lemma 1.5 $\mathcal{B}_i$ is a bush. See Figure 1.12 for an example of a graph G_i and the corresponding bush $\mathcal{B}_i$. A bush form $\Gamma_{\mathcal{B}_i}$ is usually represented by drawing all the virtual vertices on the same horizontal line (dashed line of Figure 1.12(b)).

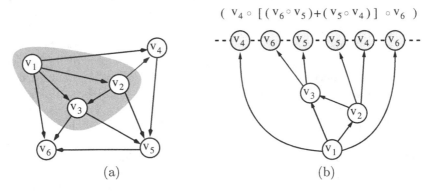

Figure 1.12 (a) A directed planar graph. Labels correspond to an st-numbering of the vertices. The highlighted area is the subgraph G_3 induced by $\{v_1, v_2, v_3\}$. Observe that, due to the st-numbering, both G_3 and $G - G_3$ are connected. (b) The bush form $\mathcal{B}_3$.

Bush form $\Gamma_{\mathcal{B}_i}$ contains a planar embedding of all biconnected components of G_i, and Lemma 1.7 ensures that such embeddings can be kept fixed up to a flip when searching for a planar drawing of G.

The strategy of the algorithm is that of focusing on the virtual vertices of $\mathcal{B}_i$ and of encoding the linear order that they have in $\Gamma_{\mathcal{B}_i}$ into a suitable algebraic expression $\varepsilon(\Gamma_{\mathcal{B}_i})$ that implicitly represents all their permutations compatible with a planar embedding of $\mathcal{B}_i$ with virtual vertices on the outer face.

The definition of $\varepsilon(\Gamma_{\mathcal{B}_i})$ can be inductively provided as follows. Let v be the source of $\Gamma_{\mathcal{B}}$. If $\Gamma_{\mathcal{B}}$ is a trivial bush form consisting of a single directed edge (v, u) then $\varepsilon(\Gamma_{\mathcal{B}}) = u$. Otherwise, if v is a cutvertex splitting $\Gamma_{\mathcal{B}}$ into bush forms $b_1, b_2, \ldots, b_k$, let $\varepsilon(b_1), \varepsilon(b_2), \ldots, \varepsilon(b_k)$ be the corresponding expressions for $b_1, b_2, \ldots, b_k$. The algebraic expression associated with $\Gamma_{\mathcal{B}}$ is $\varepsilon(\Gamma_{\mathcal{B}}) = (\varepsilon(b_1) \circ \varepsilon(b_2) \circ \ldots \circ \varepsilon(b_k))$. Observe that any permutation of $b_1, b_2, \ldots, b_k$ is compatible with a planar embedding of $\mathcal{B}$. Finally, if v is not a cut vertex of $\Gamma_{\mathcal{B}}$, consider the biconnected component b of $\Gamma_{\mathcal{B}}$ including v and let $u_1, u_2, \ldots, u_k$ be the cut vertices of $\mathcal{B}$ belonging to b. Observe that each subgraph of $\Gamma_{\mathcal{B}}$ routed at u_i, with $i = 1, \ldots, k$, is a bush form b_i. Let $\varepsilon(b_1), \varepsilon(b_2), \ldots, \varepsilon(b_k)$ be the corresponding expressions for $b_1, b_2, \ldots, b_k$. The algebraic expression associated with $\Gamma_{\mathcal{B}}$ is $\varepsilon(\Gamma_{\mathcal{B}}) = [\varepsilon(b_1) + \varepsilon(b_2) + \ldots + \varepsilon(b_k)]$. Observe that flipping the biconnected component b corresponds to flipping the expression $[\varepsilon(b_k) + \varepsilon(b_{k-1}) + \ldots + \varepsilon(b_1)]$.

Figure 1.13 illustrates an example of permutations and flipping in a bush form.

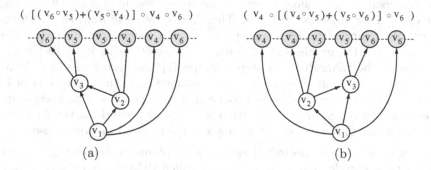

(a)　　　　　　　　　　　　　　　(b)

Figure 1.13 (a) A permutation of the bush form of Figure 1.12(b). (b) A flip of the bush form of Figure 1.12(b).

Given a bush form $\Gamma_{\mathcal{B}_i}$, the *reduction* operation changes the embedding of $\mathcal{B}_i$, by permuting bush forms attached to cut vertices and by flipping biconnected components, and produces a bush form $\Gamma'_{\mathcal{B}_i}$ where all virtual vertices labeled v_{i+1} are consecutively disposed. If this is not possible, then there is no way of adding vertex v_{i+1} to the embedding while keeping all outer vertices of G_{i-1} on the outer face, and by Lemma 1.7 the graph is not planar. If this is possible, then a *substitution* operation is performed on $\Gamma_{\mathcal{B}_i}$, obtaining a drawing $\Gamma_{\mathcal{B}_{i+1}}$. Namely, the virtual vertices labeled v_{i+1} are merged together, and for each edge (v_{i+1}, v_j) exiting v_{i+1} a new virtual vertex v_j is introduced and connected to v_{i+1}.

In the original description of the LEMPEL-EVEN-CEDERBAUM algorithm, these operations are not actually performed. Instead, it is shown that the reduction operation on $\Gamma_{\mathcal{B}_i}$ corresponds to an equivalent transformation on $\varepsilon(\Gamma_{\mathcal{B}_i})$ that produces an algebraic expression $\varepsilon(\Gamma'_{\mathcal{B}_i})$ where all the variables v_{i+1} are consecutive. Analogously, the substitution operation corresponds to the removal of the sequence of variables v_{i+1} which are replaced by $(v_{j_1} \circ v_{j_2} \circ \ldots \circ v_{j_k})$, where $v_{j_1}, \ldots, v_{j_k}$ are the vertices directly attached to v_{i+1}.

Data structures

The problem of efficiently identifying the flips and the permutations needed to reduce $\Gamma_{\mathcal{B}_i}$ (or, equivalently, needed to normalize $\varepsilon(\Gamma_{\mathcal{B}_i})$) is solved in [BL76], where the PQ-tree data structure is introduced. Intuitively, a PQ-tree is a data structure corresponding to the syntax tree of the expression $\varepsilon(\Gamma_{\mathcal{B}_i})$. Namely, a *PQ-tree* is a rooted, directed, ordered tree with three types of nodes: P-nodes, Q-nodes, and leaves. For each $\circ$ operation $(\epsilon_1 \circ \epsilon_2 \circ \ldots \circ \epsilon_k)$ in ε, the corresponding PQ-tree has a *P-node* with children $PQ(\epsilon_1), \ldots, PQ(\epsilon_k)$. Also, for each $+$ operation $(\epsilon_1 \circ \epsilon_2 \circ \ldots \circ \epsilon_k)$ in ε, the corresponding PQ-tree has a *Q-node* with children $PQ(\epsilon_1), \ldots, PQ(\epsilon_k)$. The children of a P-node can be arbitrarily permuted, while the order of the children of a Q-node can be reversed. In [BL76], it is shown how a bottom-up computation starting from all leaves labeled v_{i+1} is sufficient to compute a sequence of permutations and flips that consecutively disposes all v_{i+1} leaves. Only the smallest subtree that contains the v_{i+1} leaves is traversed.

1.7.2   The SHIH-HSU Algorithm

The SHIH-HSU algorithm either constructs a planar embedding of the input graph G or fails and outputs the information that G is not planar [SH93, SH99] (see also [Hsu01, Boy05]). The proper numbering ψ of the vertices of G used by the SHIH-HSU algorithm is obtained by a DFS traversal of G. Namely, vertices are considered in reverse DFS order, where the root r of the DFS tree has $\psi(r) = n$. Therefore, differently from the LEMPEL-EVEN-CEDERBAUM algorithm, although the graph $\overline{G}_i(\overline{V}_i, \overline{E}_i)$ induced by $\overline{V}_i = \{v \mid \psi(v) > i\}$ is always connected, the graph $G_i(V_i, E_i)$ induced by vertices in $V_i = \{v \mid \psi(v) \leq i\}$ is not guaranteed to be connected. At step 1, graph G_1 has vertex v_1 only. At a generic step i, with $i = 2, \ldots, n$, an embedding Γ_{G_i} is obtained from the embedding of $\Gamma_{G_{i-1}}$ by adding vertex v_i together with all edges connecting it to vertices with lower values of ψ. The strategy used by the SHIH-HSU algorithm is that of characterizing those configurations that determine a non-planarity, and by giving a recipe to build Γ_{G_i} otherwise.

As G_i is not necessarily connected, at each step i a planar embedding of each connected component of G_i is encoded into a data structure called PC-tree. A *PC-tree* $\mathcal{T}$ is a rooted, ordered tree with two types of nodes: P-nodes and C-nodes. While the neighbors of a P-node can be arbitrarily permuted, C-nodes come with a cyclic ordering of their adjacency list, which can only be reversed. Intuitively, P-nodes represent regular nodes of an embedded partial graph, while C-nodes represent biconnected components. Consider a planar embedding of a connected component C of graph G_i such that the outer vertices of C are on its outer face. The PC-tree $\mathcal{T}$ associated with C can be easily obtained from $\mathcal{C}$ by replacing each biconnected component of C with a C-node connected to the outer vertices of it in the same circular order as they appear on the border of the biconnected component. In order to simplify the tree, each C-node representing a trivial biconnected component composed of a single edge connecting two cutvertices v_j and v_k is replaced with a single edge attached to the two P-nodes corresponding to v_j and v_k. Let r be the *root* of the connected component C, i.e., the vertex of C with higher value of ψ. Observe that r is an outer vertex of C and always corresponds to a P-node of its PC-tree.

The PC-trees associated with G_i represent all the planar embeddings of G_i such that each connected component of G_i has its outer vertices on the outer face. In particular, if G_i is connected, the correspondence between its single PC-tree $\mathcal{T}$ and the PQ-tree of G_i used by the LEMPEL-EVEN-CEDERBAUM algorithm is apparent, since the former is obtained from the latter by removing leaves and replacing Q-nodes with C-nodes.

Working of the algorithm

At step 1, graph G_1 only has one isolated vertex labeled v_1 and its PQ-tree is a single P-node associated with vertex v_1.

At a generic step i, graph G_{i-1} has been already processed and its PC-trees have been computed. When the new vertex v_i is added, all tree edges and back edges connecting v_i to G_{i-1} are considered. Suppose that only a tree edge (v_i, u) exits from v_i. In this case, only the PC-tree corresponding to the connected component of G_{i-1} needs to be updated. Otherwise, if v_i has more than one child $u_1, u_2, \ldots, u_k$, then v_i is a cutvertex of G_i; a new P-node is introduced for it and suitably attached to the PC-trees of the connected components $C_1, C_2, \ldots, C_k$ of G_{i-1} containing $u_1, u_2, \ldots, u_k$, respectively, producing a single PC-tree for the new connected component of G_i. Consider a child u of v_i. The PC-tree $\mathcal{T}$ corresponding to the connected component containing u can be attached to v_i in a way that is independent of the PC-trees corresponding to the other children of v_i. Hence, for simplicity of description, we will assume that v_i has a single child u.

Let C be the connected component of G_{i-1} containing u. If no back edge from C attaches to v_i, then the P-node introduced for v_i is attached to the P-node representing r in $\mathcal{T}_i$, and step i concludes. Otherwise, suppose some vertex of C has some back edge to v_i. Since the input graph is biconnected, nodes of $\mathcal{T}_{i-1}$ either have *highpt* $= i$ or have *lowpt* $> i$, or both. Call *relevant node* each node w of $\mathcal{T}_{i-1}$ such that *highpt*$(w) = i$ and *lowpt*$(w) > i$. It is easy to see that the parent of w either is r or is a relevant node in its turn. Hence, relevant nodes form a subtree of the PC-tree rooted at r. By leveraging the relevant nodes subtree, it is possible to efficiently check the planarity of G_i and to compute the PC-tree updated with the P-node for u.

Namely, call *terminal nodes* the leaves of the subtree of the PC-tree composed by relevant nodes. We have the following lemma.

LEMMA 1.8 If $\mathcal{T}$ has more than two terminal nodes, then G_i (and hence G) is not planar.

Therefore, if G is planar, $\mathcal{T}_{i-1}$ has one or two terminal nodes and the relevant nodes subtree of $\mathcal{T}_{i-1}$ is either a path or a Y-shaped tree, respectively (see Figure 1.14).

Also, observe that an edge exiting a relevant node may be of five different types:

(i) a tree edge to another relevant node;

(ii) a back edge to v_i;

(iii) a tree edge to a subtree whose back edges are all type-(ii) edges; or

(iv) a back edge to a node v_j with $j > i$;

(v) a tree edge to a subtree whose back edges are all type-(iv) edges.

Subtrees attached to edges of type (iii) are called *i-subtrees*, while subtrees attached to edges of type (v) are called *i*-subtrees*. In Figure 1.14, i-subtrees are represented with black triangles and i*-subtrees with white triangles.

The SHIH-HSU algorithm either identifies a non-planarity or finds a planar arrangement of the back edges to v_i and the i-subtrees to produce a new C-node that represents the block determined by the additions of the back edges to v_i. The algorithm considers four main cases, depending on whether some relevant node is a C-node, and depending on whether $\mathcal{T}_{i-1}$ has one or two terminal nodes.

The easiest case is when $\mathcal{T}_{i-1}$ has exactly one terminal and all the relevant nodes are P-nodes. In fact, in this case all i-subtrees and back edges to i can always be embedded

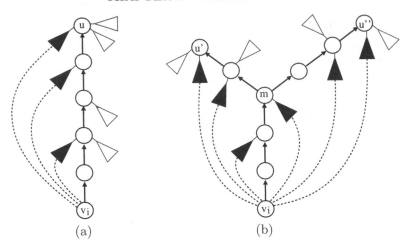

Figure 1.14   (a) Relevant nodes of $\mathcal{T}_{i-1}$ have one terminal. (b) Relevant nodes of $\mathcal{T}_{i-1}$ have two terminals.

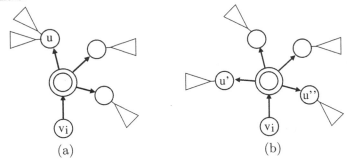

Figure 1.15   (a) The example of Figure 1.14(a) after contraction. The double border identifies C-nodes. (b) The example of Figure 1.14(b) after contraction.

on one side of $\mathcal{T}_{i-1}$, and the embedded part can be replaced with a C-node, as shown in Figure 1.15(a).

The case when $\mathcal{T}_{i-1}$ has exactly one terminal and some relevant node is a C-node, is analogous, with the difference that the constraints enforced by C-nodes (whose adjacency list can only be flipped) have to be taken into account and may cause a non-planarity whenever, no matter how they are flipped, they force one i-subtree (or back edge to v_i) to be outside the new block or one i*-subtree (or back edge to v_j, with $j > i$) to be inside it.

The most difficult case is when $\mathcal{T}_{i-1}$ has two terminal nodes u' and u''. Let m be their common ancestor, P be the unique path in $\mathcal{T}_{i-1}$ from u' to u'', and P' be the path from r to m. If all relevant nodes are P-nodes, then we have the following planarity criterion:

LEMMA 1.9   Graph G_i is planar if and only if any node internal to P' has edges of type (i), (ii), or (iii).

In fact, it is easy to see that if the conditions of Lemma 1.9 are satisfied a new block can be planarly embedded, its border being composed by path P and two paths from the two terminal nodes to v_i and containing all i-subtrees and back edges to v_i. Such a block is replaced by a C-node as shown in Figure 1.15(b). Again, if some relevant node is a C-node, its constraints on the embedding need to be taken into account and yield a more intricate, although not difficult, case.

Data Structures

A tricky point of the SHIH-HSU algorithm is when a newly identified block has to be replaced with a C-node. To understand why this operation is critical, consider that, in order to have a linear-time algorithm, each node of the PC-tree should have a pointer to the parent node. Such a pointer is used, for example, when, starting from the current vertex v_i, its incoming back edges are considered and i-subtrees are traversed moving from child to parent. This operation is needed to identify the relevant node subtree and its terminals. Observe that i*-subtrees cannot be traversed without losing linearity. Also, even identifying them by browsing the adjacency list of a relevant node would have the same result. If the block was naively replaced by a C-node structure as shown in Figure 1.15 the pointers to the parent of a possibly linear number of children would have to be updated.

Perhaps the easiest way to address this problem is that of encoding the neighborhood of C-nodes with a strategy analogous to that used in [BM99, BM04], which allows us to efficiently traverse the boundary of a block in parallel and flip it when needed [Hsu01, Hsu03, BFNd04]. A second approach, inspired by the analogous operation on Q-nodes of PQ-trees [LEC67], is that of borrowing the parent pointer from sibling to sibling. The two approaches turn out to be similar, since browsing siblings of a C-node in search for the parent pointer is equivalent to traversing the corresponding block border [Hsu01, Hsu03].

1.7.3 The BOYER-MYRVOLD Algorithm

The BOYER-MYRVOLD algorithm [BM99, BM04] (see also [Tho99, BCPD04, HT08]) has several features in common with the SHIH-HSU algorithm, so much so that the two have been sometimes identified [Tho99, HT08]. The proper numbering ψ of the vertices of G used by the BOYER-MYRVOLD algorithm is again a reverse DFS order. The general strategy is that of explicitly maintaining a "flexible" planar embedding of each connected component of G_i with the outer vertices on the outer face. This embedding is "flexible" in the sense that each block can be flipped in constant time, whatever its size, while the permutation of the blocks around cutvertices is left undecided. In order to achieve this, each block of G_i is maintained separately from the others in a special structure, and the cutvertex that has higher value of ψ in one block B, called the *root* of B, has a pointer to the corresponding cutvertex in the parent block.

Working of the algorithm

The algorithm described in [BM99] was simplified in [BM04]. First, we describe the primitive version in [BM99], which, in our opinion, is more intuitive. Second, we sketch the differences with [BM04].

The computation starts with an initial set of blocks corresponding to the tree edges of the DFS tree of G (see Figure 1.16). Hence, it could be argued that this is not a vertex addition algorithm, since all vertices are in place from the first iteration. Actually, a vertex v_j with index higher than the current iteration i is ignored until iteration j is reached. Vertices are considered in reverse DFS order, starting from v_1 and ending with the root v_n of the DFS tree (see Figure 1.16(a)). If vertex v_i has no incoming back edges, no operation is needed at iteration i.

So, for example, running the algorithm on the example of Figure 1.16(b) would not perform any operation at steps $1, 2, \ldots, 8$, as vertices $v_1, v_2, \ldots, v_8$ don't have incoming

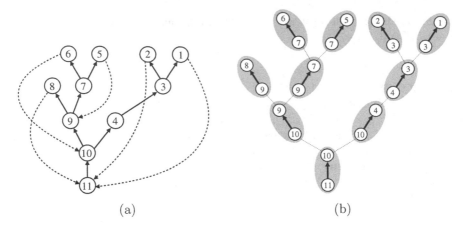

(a)   (b)

Figure 1.16 (a) The same DFS tree of Figure 1.4, where vertices are labeled with their reverse DFS index. (b) The BOYER-MYRVOLD algorithm starts by creating a block for each edge of the tree.

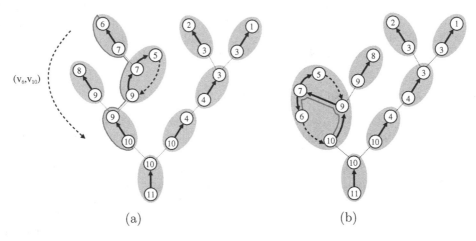

(a)   (b)

Figure 1.17 The BOYER-MYRVOLD algorithm on the DFS tree of Figure 1.16(a). (a) When vertex v_{10} is considered the back edge (v_6, v_{10}) needs to be embedded. (b) The embedding of back edge (v_6, v_{10}) corresponding to the choice of the red path from v_6 to v_{10} along the borders of blocks shown in (a).

back edges. Otherwise, if v_i has some incoming back edges, the strategy of the algorithm is that of deciding how to embed them by exploring the borders of the current blocks of G_{i-1}. To give an intuitive example Figure 1.17(a) represents G_9. At iteration 10 vertex v_{10} is considered by the algorithm and the back edge (v_6, v_{10}) needs to be embedded. In the embedding choice shown in Figure 1.17(b), the red path inside the closed face of the new block can be identified in Figure 1.17(a) as the red path going from v_6 to v_{10} along the borders of the blocks. The approach of the BOYER-MYRVOLD algorithm is that of first choosing suitable paths for the back edges returning to v_i and then using such paths to close a new block and update the data structures. Hence, each iteration has two phases: *Path searching* and *Block embedding* (in [BM99, BM04] these phases are called *Walkup* and *Walkdown*, respectively, but the tree is drawn upside down with respect to the convention used here).

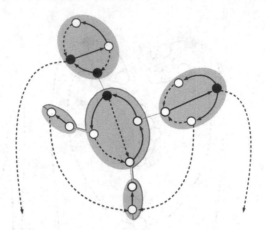

Figure 1.18 Properties of the admissible path to v_i. The lower vertex is the currently processed one while outer vertices of G_i are drawn black. The red path is not admissible, as it traverses an outer vertex of G_i (a cut vertex of G_i in this example).

Let's start from the Path searching phase. Suppose that some back edges enter v_i from vertices $u_1, u_2, \ldots, u_k$. For each j, with $j = 1, \ldots, k$, the algorithm searches for a path p_j from u_j to v_i with the following properties:

1. Vertices and edges of p_j are on the boundary of the blocks of G_{i-1}.
2. Each vertex of p_j that is the root of a block is followed by the corresponding cutvertex in the parent block, until v_i is reached.
3. Each vertex of p_j that is different from the entry point and the root of the current block is not an outer vertex of G_i (see Figure 1.18(a)).

Also, two paths (i.e., two embedding choices) may be incompatible with each other. Namely, let p_l and p_m be two paths to v_i and let b be a block b traversed by them. The following compatibility properties are enforced:

1. If p_l and p_m don't share edges of b, they do not share edges in any other block (see Figure 1.19(a)).
2. Paths p_l and p_m do not share edges of b if they traverse two other distinct outer blocks, where an *outer* block is one containing an outer vertex of G_i (see Figure 1.19(b)).
3. If p_l and p_m don't share edges of b and the root r_b of b is different from v_i, then r_b is not an outer vertex of G_i (see Figure 1.19(c)).

The above properties guarantee that when the new block is closed, no outer vertex of G_i falls inside a face of the block.

In order to be linear, the algorithm does not explicitly compute all the paths p_j, for $j = 1, \ldots, k$. In fact, if two paths share one edge, the second path can follow the same route toward v_1 used by the first one without the need of checking the above properties. Also, whenever a path enters a block b, it searches both sides of b in parallel, searching for the root r_b of b. In this way, the shorter admissible path to r_b is found by exploring at most twice the number of its edges. Since the edges used by the paths will be closed inside some face of the new block, they are never explored again in a subsequent iteration, and the total number of steps required by the algorithm for the computation of such paths is linear.

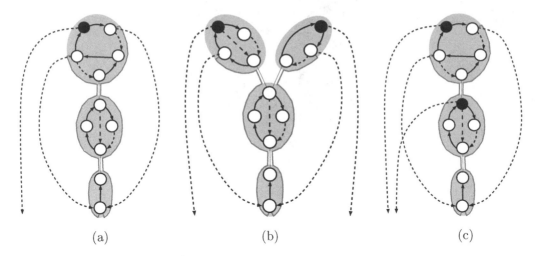

(a)   (b)   (c)

Figure 1.19 Compatibility properties of the paths to v_i. The lower vertex is the currently processed one while outer vertices of G_i are drawn black. (a) Two compatible paths not sharing edges in any block they traverse. (b) Two admissible paths coming from two distinct outer blocks. (c) Two non-admissible paths.

If the Path searching phase does not detect a non-planarity, the Block embedding phase starts. This is a simpler phase in which, starting from v_i and moving along the boundary of G_{i-1}, the blocks traversed by the paths are merged together and the back edges are embedded based on their corresponding paths to produce a planar embedding for G_i.

The simplified version of the algorithm described in [BM04] is based on the same two phases, Path searching and Block embedding. However, the check for the paths' compatibility, which in the primitive version were demanded to the first phase are moved to the second phase, which may, therefore, also detect a non-planarity.

Data Structures

The tricky point of the BOYER-MYRVOLD algorithm is when two blocks, traversed by a path, are merged together. It may happen that a path traverses the child block clockwise and the parent block counterclockwise, or viceversa. Fortunately, it can be shown that the properties of the paths guarantee that if one path does so all other paths comply with this embedding choice. However, in order to merge the two blocks, one of them needs to be flipped, and reversing the adjacency lists of all the vertices of the block may result in a linear-time operation that would yield a quadratic planarity algorithm. In order to solve this problem, the authors introduced a suitable data structure, called `bicomp`, that allows us to flip a block in $O(1)$-time, whatever its size. Such a data structure is based on circular lists that do not have a predefined orientation.

Namely, suppose that the list items of a circular list instead of having the usual `next` and `prev` pointers have two generic pointers `ref1` and `ref2` which could be used arbitrarily to store a reference to the next or previous list item. Suppose, also, that you maintain a reference to the last element encountered while traversing the list. If you want a reference to the next element you compare this reference with `ref1` and `ref2` and choose the one that is different from it. Hence, the circular list is traversed in the direction that is decided by the first step. If the circular order of the list has to be reversed, it suffices to begin the traversal in the opposite way.

Of course, if the clockwise direction of the adjacency list of each vertex of a block is independently chosen, this would not necessarily produce a planar embedding. However, it is not difficult to devise some convention to transfer the orientation of the adjacency list of one vertex to the adjacency lists of the neighboring vertices. For example, it may be prescribed that if in the adjacency list of vertex v_i the list item of vertex v_j uses `ref1` as `next` and `ref2` as `prev`, the same choice is made for the list item of v_i in the adjacency list of v_j (in [BM99, BM04] a less intuitive, but more practical, convention is adopted).

Hence, when two blocks are merged and their common cutvertex is identified, the two adjacency lists of the cutvertex can be suitably joined in such a way to implicitly reverse all the adjacency lists of one of the two blocks.

1.8 Frontiers in Planarity

1.8.1 Simultaneous Planarity

A recent variant of the planarity problem asks for the simultaneous embedding of two graphs on the same set of vertices V. Namely, a *simultaneous embedding* of $G_1 = (V, E_1)$ and $G_2 = (V, E_2)$ consists of two planar drawings Γ_1 and Γ_2 of G_1 and G_2, respectively, such that any vertex $v \in V$ is mapped to the same point in each of the two drawings. When Γ_1 and Γ_2 are required to be straight-line drawings, this problem is called *geometric simultaneous embedding*. When edges common to E_1 and E_2 are required to be represented by the same Jordan curve in Γ_1 and Γ_2 this problem is called *simultaneous embedding with fixed edges* (or SEFE, for short). The above definition can be easily generalized to k graphs $G_i = (V, E_i)$, with $i = 1, 2, \ldots, k$.

Geometric simultaneous embedding turns out to have limited usability, since testing whether two planar graphs admit such an embedding is NP-hard [EBGJ+07] and since a geometric simultaneous embedding does not always exist for two outerplanar graphs [BCD+07], for two trees [GKV09], and even for a tree and a path [AGKN12].

Conversely, for several classes of graphs the computation of a simultaneous embedding with fixed edges, if any, can be performed in polynomial time [EK05, DL07, Fra06, FGJ+08, JS09, ADF+10, HJL10, ADF+11], although the general problem is of unknown complexity. Refer to Chapter 11, "Simultaneous Embedding of Planar Graphs," for an in-depth exploration of this research area.

1.8.2 Clustered Planarity

The user's need of drawing some set of vertices near one to the other naturally leads to the requirement of drawing them inside the same simple closed region of the plane. This target is pursued by clustered planarity where the containment relationship among regions and vertices is described by an arbitrary hierarchy. More formally, a *clustered graph* $C(G, T)$ is a graph G and a rooted tree T whose leaves are the vertices of G. A *c-planar* drawing of $C(G, T)$ is such that G is planarly drawn and each internal node ν of T is drawn as a simple closed region $R(\nu)$ such that:

- $R(\nu)$ contains the drawing of the graph $G(\nu)$ induced by the vertices that are leaves of the subtree rooted at ν;
- $R(\nu)$ contains a region $R(\mu)$ if and only if μ is a descendant of ν in T;
- any two regions $R(\nu_1)$ and $R(\nu_2)$ do not intersect if ν_1 is not an ancestor or a descendant of ν_2; and
- an edge e does not cross the boundary of a region $R(\nu)$ more than once.

Restrictions on the c-planarity testing problem studied in the literature include: (i) assuming that each cluster induces a small number of connected components [FCE95b, FCE95a, Dah98, GJL$^+$02, GLS05, CW06, CDF$^+$08, JJKL08]; (ii) considering only *flat* hierarchies, where all clusters different from the root of T are children of the root [CDPP04, DF08b]; (iii) focusing on particular families of underlying graphs [CDPP04, CDPP05, JKK$^+$08]; and (iv) fixing the embedding of the underlying graph [DF08b, JKK$^+$08].

Although the general problem is of unknown complexity, it has been shown to be polynomial-time solvable in the following cases:

- If the subgraph $G(\nu)$ induced by each cluster ν is connected the clustered graph is called *c-connected*. The algorithm proposed in [FCE95b, FCE95a] is quadratic. Linear-time algorithms are described in [Dah98, CDF$^+$08]. The case when each cluster induces at most two connected components has been investigated in [JJKL08].

- The results [BKM98, Bie98] on "partitioned drawings" of graphs can be interpreted as linear-time c-planarity tests for non-connected flat clustered graphs with exactly two clusters. The same result (flat clustered planarity for non-connected graphs with exactly two clusters) is shown in [HN09] where the problem is modeled as a two-page book embedding.

- Gutwenger et al. presented a polynomial-time algorithm for c-planarity testing for *almost connected* clustered graphs [GJL$^+$02], i.e., graphs for which all nodes corresponding to the non-connected clusters lie on the same path in T starting at the root of T, or graphs in which for each non-connected cluster its parent cluster and all its siblings in T are connected.

- Cortese et al. studied the class of non-connected clustered graphs such that the underlying graph is a cycle and the clusters at the same level of T also form a cycle, where two clusters are considered adjacent if they are incident to the same edge [CDPP04, CDPP05]. The c-planarity testing and embedding problem is linear for this class of graphs.

- Goodrich et al. introduced a polynomial-time algorithm for producing planar drawings of *extrovert* clustered graphs [GLS05], i.e., graphs for which all clusters are connected or extrovert. A cluster μ with parent ν is extrovert if and only if ν is connected and each connected component of μ has a vertex with an edge that is incident to a cluster which is external to ν.

- Jelínková et al. presented a polynomial-time algorithm for testing the c-planarity of "k-rib-Eulerian" graphs [JKK$^+$08]. A graph is *k-rib-Eulerian* if it is Eulerian and it can be obtained from a 3-connected planar graph with k vertices, for some constant k, by replacing some edges with one or more paths in parallel.

1.8.3 Decomposition-Based Planarity

Since a graph is planar if and only if its triconnected components are planar, it is somehow surprising that all known linear-time planarity algorithms require at most the biconnectivity of the input graph. It could be asked whether the triconnectivity of the graph could be leveraged in order to obtain planarity algorithms that are easier to understand and to implement. A triconnected graph has several helpful properties with this respect: if it is planar, it admits a single planar embedding up to a flip (in contrast, a biconnected graph admits an exponential number of embeddings); if it is not planar and it is different from K_5, it contains a subdivision of $K_{3,3}$.

An intriguing research line in this direction is that of exploiting construction sequences: it is well known that a triconnected graph can be reduced by means of sequences of planarity-preserving transformations to graphs as simple as a wheel [Tut61] or as a K_4 [Tut66, BG69]. Such transformations, if reversed, yield construction sequences that could be possibly exploited to find a planar embedding of the input graph starting from a planar embedding of the reduced graph. The polynomial-time planarity algorithm described in [BSW70] uses the reduction sequences described in [Tut61]. The reduction sequences described in [Tut66, BG69] have been used to give a short proof of Kuratowski's theorem [Kel81, Tho81], while their application to planarity algorithms has been only recently investigated in [Sch12].

References

[ADF+10]   P. Angelini, G. Di Battista, F. Frati, V. Jelinek, J. Kratochvil, M. Patrignani, and I. Rutter. Testing planarity of partially embedded graphs. In M. Charikar, editor, *Symposium on Discrete Algorithms (SODA '10)*, pages 202–221, 2010.

[ADF+11]   P. Angelini, G. Di Battista, F. Frati, M. Patrignani, and I. Rutter. Testing the simultaneous embeddability of two graphs whose intersection is a biconnected graph or a tree. In *Workshop on Combinatorial Algorithms (IWOCA '10)*, volume 6460 of *LNCS*, pages 212–225, 2011.

[ADP11]   P. Angelini, G. Di Battista, and M. Patrignani. Finding a minimum-depth embedding of a planar graph in $o(n^4)$ time. *Algorithmica*, 60(4):890–937, 2011.

[AGKN12]   P. Angelini, M. Geyer, M. Kaufmann, and D. Neuwirth. On a tree and a path with no geometric simultaneous embedding. *Journal of Graph Algorithms and Applications*, 16(1):37–83, 2012. Special Issue on Selected Papers from GD '10.

[AH77]   K. Appel and W. Haken. Every planar map is four colourable, part I: discharging. *Illinois J. Math.*, 21:429–490, 1977.

[AHK77]   K. Appel, W. Haken, and J. Koch. Every planar map is four colourable, part II: Reducibility. *Illinois Journal of Mathematics*, 21:491–567, 1977.

[AP61]   L. Auslander and S. V. Parter. On imbedding graphs in the sphere. *Journal of Mathematics and Mechanics*, 10(3):517–523, 1961.

[APBL95]   D. Archdeacon, C. Paul Bonnington, and C. H. C. Little. An algebraic characterization of planar graphs. *Journal of Graph Theory*, 19(2):237–250, 1995.

[AŠ98]   D. Archdeacon and J. Šráň. Characterizing planarity using theta graphs. *Journal of Graph Theory*, 27(1):17–20, 1998.

[Bad64]   W. Bader. Das topologische Problem der gedruckten Schaltung und seine Lösung. *Electrical Engineering (Archiv für Elektrotechnik)*, 49(1):2–12, 1964.

[Bak94]   B. S. Baker. Approximation algorithms for NP-complete problems on planar graphs. *J. ACM*, 41:153–180, 1994.

[BCD+07]   P. Braß, E. Cenek, C. A. Duncan, A. Efrat, C. Erten, D. Ismailescu, S. G. Kobourov, A. Lubiw, and J. S. B. Mitchell. On simultaneous planar graph embeddings. *Computational Geometry*, 36(2):117–130, 2007.

[BCPD04]   J. M. Boyer, P. F. Cortese, M. Patrignani, and G. Di Battista. Stop minding your P's and Q's: Implementing a fast and simple DFS-based planarity testing and embedding algorithm. In Giuseppe Liotta, editor, *Graph Drawing (Proc. GD '03)*, volume 2912 of *LNCS*, pages 25–36, 2004.

[BD91]   P. Bertolazzi and G. Di Battista. On upward drawing testing of triconnected digraphs. In *Proc. 7th Annu. ACM Sympos. Comput. Geom.*, pages 272–280, 1991.

[BDBD00]   P. Bertolazzi, G. Di Battista, and W. Didimo. Computing orthogonal drawings with the minimum number of bends. *IEEE Transaction on Computers*, 49:826–840, August 2000.

[BDLM94] P. Bertolazzi, G. Di Battista, G. Liotta, and C. Mannino. Upward drawings
          of triconnected digraphs. *Algorithmica*, 6(12):476–497, 1994.

[BDMT98] P. Bertolazzi, G. Di Battista, C. Mannino, and R. Tamassia. Optimal
          upward planarity testing of single-source digraphs. *SIAM J. Comput.*,
          27(1):132–169, 1998.

[BFNd04] J. Boyer, C. Fernandes, A. Noma, and J. de Pina. Lempel, Even, and
          Cederbaum planarity method. In Celso Ribeiro and Simone Martins, edi-
          tors, *Experimental and Efficient Algorithms*, volume 3059 of *LNCS*, pages
          129–144. Springer, 2004.

[BG69]   D. W. Barnette and B. Grünbaum. On Steinitz's theorem concerning
          convex 3-polytopes and on some properties of 3-connected graphs. In
          Many Facets of Graph Theory, pages 27–40, 1969.

[Bie98]   T. C. Biedl. Drawing planar partitions III: Two constrained embedding
          problems. Technical Report RRR 13-98, RUTCOR Rutgen University,
          1998.

[BKM98]   T. C. Biedl, M. Kaufmann, and P. Mutzel. Drawing planar partitions
          II: HH-Drawings. In J. Hromkovic and O. Sýkora, editors, *Workshop on
          Graph-Theoretic Concepts in Computer Science (WG '98)*, volume 1517,
          pages 124–136. Springer, 1998.

[BL76]   K. Booth and G. Lueker. Testing for the consecutive ones property interval
          graphs and graph planarity using PQ-tree algorithms. *J. Comput. Syst.
          Sci.*, 13:335–379, 1976.

[BM88]   D. Bienstock and C. L. Monma. On the complexity of covering vertices
          by faces in a planar graph. *SIAM Journal on Computing*, 17:53–76, 1988.

[BM90]   D. Bienstock and C. L. Monma. On the complexity of embedding planar
          graphs to minimize certain distance measures. *Algorithmica*, 5(1):93–109,
          1990.

[BM99]   J. Boyer and W. Myrvold. Stop minding your P's and Q's: A simpli-
          fied O(n) planar embedding algorithm. In *10th Annual ACM-SIAM Sym-
          posium on Discrete Algorithms*, volume 1027 of *LNCS*, pages 140–146.
          Springer-Verlag, 1999.

[BM04]   J. Boyer and W. Myrvold. On the cutting edge: Simplified O(n) planarity
          by edge addition. *Journal of Graph Algorithms and Applications*, 8(3):241–
          273, 2004.

[Boy05]   J. Boyer. Additional PC-tree planarity conditions. In J. Pach, editor,
          Graph Drawing, volume 3383 of *LNCS*, pages 82–88. Springer, 2005.

[Bra09]   U. Brandes. The left-right planarity test. Manuscript submitted for pub-
          lication, 2009.

[BSW70]   J. Bruno, K. Steiglitz, and L. Weinberg. A new planarity test based on 3-
          connectivity. *IEEE Transactions on Circuit Theory*, 17(2):197–206, 1970.

[CDF+08] P. F. Cortese, G. Di Battista, F. Frati, M. Patrignani, and M. Pizzonia.
          C-planarity of c-connected clustered graphs. *Journal of Graph Algorithms
          and Applications*, 12(2):225–262, Nov 2008.

[CDPP04] P. F. Cortese, G. Di Battista, M. Patrignani, and M. Pizzonia. Clustering
          cycles into cycles of clusters. In János Pach, editor, *Proc. Graph Drawing
          2004 (GD '04)*, volume 3383 of *LNCS*, pages 100–110. Springer, 2004.

[CDPP05]   P. F. Cortese, G. Di Battista, M. Patrignani, and M. Pizzonia. Clustering cycles into cycles of clusters. *Journal of Graph Algorithms and Applications, Special Issue on the 2004 Symposium on Graph Drawing, GD '04*, 9(3):391–413, 2005.

[Che81]   C. C. Chen. On a characterization of planar graphs. *Bulletin of the Australian Mathematical Society*, 24:289–294, 1981.

[CMS08]   M. Chimani, P. Mutzel, and J. M. Schmidt. Efficient extraction of multiple Kuratowski subdivisions. In Seok-Hee Hong, Takao Nishizeki, and Wu Quan, editors, *Graph Drawing (GD 2007)*, volume 4875 of *LNCS*, pages 159–170. Springer, 2008.

[CNAO85]   N. Chiba, T. Nishizeki, S. Abe, and T. Ozawa. A linear algorithm for embedding planar graphs using PQ-trees. *J. Comput. Syst. Sci.*, 30(1):54–76, 1985.

[Col90]   Y. Colin de Verdière. Sur un nouvel invariant des graphes et un critère de planarité. *Journal of Combinatorial Theory, Series B*, 50(1):11–21, 1990.

[Col91]   Y. Colin de Verdière. On a new graph invariant and a criterion for planarity. In Neil Robertson and Paul D. Seymour, editors, *Graph Structure Theory*, volume 147 of *Contemporary Mathematics*, pages 137–148. American Mathematical Society, 1991.

[CW06]   S. Cornelsen and D. Wagner. Completely connected clustered graphs. *Journal of Discrete Algorithms*, 4(2):313–323, 2006.

[Dah98]   E. Dahlhaus. Linear time algorithm to recognize clustered planar graphs and its parallelization. In Claudio L. Lucchesi and Arnaldo V. Moura, editors, *Proc. Latin American Theoretical INformatics (LATIN '98)*, volume 1380 of *LNCS*, pages 239–248. Springer, 1998.

[DBTV01]   G. Di Battista, R. Tamassia, and L. Vismara. Incremental convex planarity testing. *Information Computation*, 169:94–126, August 2001.

[Deo76]   N. Deo. Note on Hopcroft and Tarjan planarity algorithm. *Journal of the Association for Computing Machinery*, 23:74–75, 1976.

[DETT99]   G. Di Battista, P. Eades, R. Tamassia, and I. G. Tollis. *Graph Drawing*. Prentice Hall, Upper Saddle River, NJ, 1999.

[dF08a]   H. de Fraysseix. Trémaux trees and planarity. *Electronic Notes in Discrete Mathematics*, 31:169–180, 2008.

[DF08b]   G. Di Battista and F. Frati. Efficient c-planarity testing for embedded flat clustered graphs with small faces. In Seok-Hee Hong, Takao Nishizeki, and Wu Quan, editors, *Proc. Graph Drawing 2007 (GD '07)*, volume 4875 of *LNCS*, pages 291–302. Springer, 2008.

[DKT09]   Z. Dvorak, K. Kawarabayashi, and R. Thomas. Three-coloring triangle-free planar graphs in linear time. In Claire Mathieu, editor, *SODA*, pages 1176–1182. SIAM, 2009.

[DL07]   E. Di Giacomo and G. Liotta. Simultaneous embedding of outerplanar graphs, paths, and cycles. *Int. J. Computational Geometry and Applications*, 17(2):139–160, 2007.

[DLT84]   D. Dolev, F. T. Leighton, and H. Trickey. Planar embedding of planar graphs. In Franco P. Preparata, editor, *VLSI Theory*, volume 2 of *Adv. Comput. Res.*, pages 147–161. JAI Press, Greenwich, Conn., 1984.

[DMP64]   G. Demoucron, Y. Malgrange, and R. Pertuiset. Graphes planaires: Re-
          connaissance et construction des représentations planaires topologiques.
          Revue Fran cais de Recherche Opérationelle, 8:33–47, 1964.

[dO96]   H. de Fraysseix and P. Ossona de Mendez. Planarity and edge poset
          dimension. *European Journal of Combinatorics*, 17(8):731–740, 1996.

[dO02]   H. de Fraysseix and P. Ossona de Mendez. P.I.G.A.L.E — Public Imple-
          mentation of a Graph Algorithm Library and Editor, 2002. SourceForge
          project page http://pigale.sourceforge.net/ (GPL License).

[dOR06]   H. de Fraysseix, P. Ossona de Mendez, and P. Rosenstiehl. Trémaux trees
          and planarity. *International Journal of Foundations of Computer Science*,
          17(5):1017–1029, 2006.

[dR82]   H. de Fraysseix and P. Rosenstiehl. A depth-first characterization of pla-
          narity. *Annals of Discrete Mathematics*, 13:75–80, 1982.

[dR85]   H. de Fraysseix and P. Rosenstiehl. A characterization of planar graphs
          by Trémaux orders. *Combinatorica*, 5(2):127–135, 1985.

[DT89]   G. Di Battista and R. Tamassia. Incremental planarity testing. In *Proc.
          30th Annu. IEEE Sympos. Found. Comput. Sci.*, pages 436–441, 1989.

[DT96a]   G. Di Battista and R. Tamassia. On-line maintenance of triconnected
          components with SPQR-trees. *Algorithmica*, 15:302–318, 1996.

[DT96b]   G. Di Battista and R. Tamassia. On-line planarity testing. *SIAM J.
          Comput.*, 25:956–997, 1996.

[EBGJ+07] A. Estrella-Balderrama, E. Gassner, M. Jünger, M. Percan, M. Schaefer,
          and M. Schulz. Simultaneous geometric graph embeddings. In S. H. Hong,
          T. Nishizeki, and W. Quan, editors, *Graph Drawing (GD '07)*, volume
          4875 of *LNCS*, pages 280–290, 2007.

[EK05]   C. Erten and S. G. Kobourov. Simultaneous embedding of planar graphs
          with few bends. *Journal of Graph Algorithms and Applications*, 9(3):347–
          364, 2005.

[ET76]   S. Even and R. E. Tarjan. Computing an st-numbering. *Theoret. Comput.
          Sci.*, 2:339–344, 1976.

[Eve79]   S. Even. *Graph Algorithms*. Computer Science Press, Potomac, Maryland,
          1979.

[FCE95a] Q. W. Feng, R. F. Cohen, and P. Eades. How to draw a planar clustered
          graph. In Ding-Zhu Du and Ming Li, editors, *Proc. Computing and Com-
          binatorics (COCOON '95)*, volume 959 of *LNCS*, pages 21–30. Springer,
          1995.

[FCE95b] Q. W. Feng, R. F. Cohen, and P. Eades. Planarity for clustered graphs.
          In *Proc. European Symposium on Algorithms (ESA '95)*, volume 979 of
          LNCS, pages 213–226. Springer, 1995.

[FGJ+08] J. J. Fowler, C. Gutwenger, M. Jünger, P. Mutzel, and M. Schulz. An
          SPQR-tree approach to decide special cases of simultaneous embedding
          with fixed edges. In I. G. Tollis and M. Patrignani, editors, *Graph Drawing
          (GD '08)*, volume 5417 of *LNCS*, pages 157–168, 2008.

[Fra06]   F. Frati. Embedding graphs simultaneously with fixed edges. In M. Kauf-
          mann and D. Wagner, editors, *Graph Drawing (GD '06)*, volume 4372 of
          LNCS, pages 108–113, 2006.

[GIS99]   Z. Galil, G. F. Italiano, and N. Sarnak. Fully dynamic planarity testing with applications. *Journal of the Association for Computing Machinery*, 46:28–91, January 1999.

[GJ79]   M. R. Garey and D. S. Johnson. *Computers and Intractability: A Guide to the Theory of NP-Completeness*. W. H. Freeman, New York, NY, 1979.

[GJL+02] C. Gutwenger, M. Jünger, S. Leipert, P. Mutzel, M. Percan, and R. Weiskircher. Advances in *C*-planarity testing of clustered graphs. In Stephen G. Kobourov and Michael T. Goodrich, editors, *Proc. Graph Drawing 2002 (GD '02)*, volume 2528 of *LNCS*, pages 220–235. Springer, 2002.

[GJS76]   M. R. Garey, D. S. Johnson, and L. Stockmeyer. Some simplified NP-complete graph problems. *Theoretical Computer Science*, 1(3):237–267, 1976.

[GKM08]   C. Gutwenger, K. Klein, and P. Mutzel. Planarity testing and optimal edge insertion with embedding constraints. *Journal of Graph Algorithms and Applications*, 12(1):73–95, 2008.

[GKV09]   M. Geyer, M. Kaufmann, and I. Vrt'o. Two trees which are self-intersecting when drawn simultaneously. *Discrete Mathematics*, 307(4):1909–1916, 2009.

[GLS05]   M. T. Goodrich, G. S. Lueker, and J. Z. Sun. C-planarity of extrovert clustered graphs. In P. Healy and N. S. Nikolov, editors, *Proc. Graph Drawing 2005 (GD '05)*, volume 3843 of *LNCS*, pages 211–222. Springer, 2005.

[GM01]   C. Gutwenger and P. Mutzel. A linear time implementation of SPQR-trees. In Joe Marks, editor, *Graph Drawing (GD 2000)*, volume 1984 of *LNCS*, pages 77–90. Springer, 2001.

[GM04]   C. Gutwenger and P. Mutzel. Graph embedding with minimum depth and maximum external face. In Giuseppe Liotta, editor, *Graph Drawing*, volume 2912 of *LNCS*, pages 259–272. Springer, 2004.

[GMW01]   C. Gutwenger, P. Mutzel, and R. Weiskircher. Inserting an edge into a planar graph. In *Proceedings of the Twelfth Annual ACM-SIAM Symposium on Discrete algorithms*, SODA '01, pages 246–255, Philadelphia, PA, USA, 2001. Society for Industrial and Applied Mathematics.

[Gol63]   A. J. Goldstein. An efficient and constructive algorithm for testing whether a graph can be embedded in the plane. In John R. Edmonds, Jr., editor, *Graphs and Combinatorics Conference*, Technical Report, page 2 unn. pp. Princeton University, 1963.

[Grö59]   H. Grötzsch. Ein dreifarbensatz fü dreikreisfreie netze auf der kugel. Wiss. Z. Martin-Luther-Univ. Halle-Wittenberg Math.-Natur. Reihe 8, 1959.

[GT01]   A. Garg and R. Tamassia. On the computational complexity of upward and rectilinear planarity testing. *SIAM J. Comput.*, 31(2):601–625, 2001.

[Hal43]   D. W. Hall. A note on primitive skew curves. *Bulletin of the American Mathematical Society*, 49(2):935–936, 1943.

[Har69]   F. Harary. *Graph Theory*. Addison-Wesley, Reading, Mass., 1969.

[HJL10]   B. Haeupler, K. R. Jampani, and A. Lubiw. Testing simultaneous planarity when the common graph is 2-connected. In *Proceedings of the 21st*

Symposium on Algorithms and Computation (ISAAC'10), volume 6507 of LNCS, pages 410–421. Springer Heidelberg/Berlin, 2010.

[HL96]   M. D. Hutton and A. Lubiw. Upward planar drawing of single-source acyclic digraphs. SIAM J. Comput., 25(2):291–311, 1996.

[HN09]   S. H. Hong and H. Nagamochi. Two-page book embedding and clustered graph planarity. Technical Report 2009-004, Department of Applied Mathematics & Physics, Kyoto University, 2009.

[Hsu01]   W. L. Hsu. PC-trees vs. PQ-trees. In Proceedings of the 7th Annual International Conference on Computing and Combinatorics, COCOON '01, pages 207–217, London, UK, 2001. Springer-Verlag.

[Hsu03]   W. L. Hsu. An efficient implementation fo the PC-Tree algorithm of Shih and Hsu's planarity test. Technical Report TR-IIS-03-015, Inst. of Inf. Science, Academia Sinica, 2003.

[HT65]   F. Haray and W. T. Tutte. A dual form of Kuratowski's theorem. Canad. Math. Bull., 8:17–20, 1965.

[HT73]   J. Hopcroft and R. E. Tarjan. Dividing a graph into triconnected components. SIAM J. Comput., 2(3):135–158, 1973.

[HT74]   J. Hopcroft and R. E. Tarjan. Efficient planarity testing. J. ACM, 21(4):549–568, 1974.

[HT08]   B. Haeupler and R. E. Tarjan. Planarity algorithms via PQ-trees (extended abstract). Electronic Notes in Discrete Mathematics, 31:143–149, 2008.

[HW74]   J. E. Hopcroft and J. K. Wong. Linear time algorithm for isomorphism of planar graphs (preliminary report). In Proceedings of the Sixth Annual ACM Symposium on Theory of Computing, STOC '74, pages 172–184, New York, NY, USA, 1974. ACM.

[JJKL08]   V. Jelinek, E. Jelinkova, J. Kratochvil, and B. Lidicky. Clustered planarity: Embedded clustered graphs with two-component clusters. In GD '08, volume 5417 of LNCS, pages 121–132, 2008.

[JKK+08]   E. Jelínková, J. Kára, J. Kratochvíl, M. Pergel, O. Suchý, and T. Vyskocil. Clustered planarity: Small clusters in Eulerian graphs. In Seok-Hee Hong, Takao Nishizeki, and Wu Quan, editors, Proc. Graph Drawing 2007 (GD '07), volume 4875 of LNCS, pages 303–314. Springer, 2008.

[JKR11]   V. Jelínek, J. Kratochvíl, and I. Rutter. A Kuratowski-type theorem for planarity of partially embedded graphs. In Proceedings of the 27th Annual ACM symposium on Computational Geometry, SoCG '11, pages 107–116, New York, NY, USA, 2011. ACM.

[JS09]   M. Jünger and M. Schulz. Intersection graphs in simultaneous embedding with fixed edges. Journal of Graph Algorithms and Applications, 13(2):205–218, 2009.

[Kam07]   F. Kammer. Determining the smallest k such that g is k-outerplanar. In L. Arge, M. Hoffmann, and E. Welzl, editors, ESA '07, volume 4698 of LNCS, pages 359–370, 2007.

[Kel81]   A. K. Kelmans. A new planarity criterion for 3-connected graphs. Journal of Graph Theory, 5:259–267, 1981.

[Kel93]   A. K. Kelmans. Graph planarity and related topics. In Neil Robertson and Paul Seymour, editors, Graph Structure Theory, Proceedings of the

AMS-IMS-SIAM Joint Summer Research Conference on Graph Minors, 1991, volume 147 of *Contemporary Mathematics*, pages 635–667, 1993.

[KR88]   P. N. Klein and J. H. Reif. An efficient parallel algorithm for planarity. *J. Comput. Syst. Sci.*, 37(2):190–246, 1988.

[Kur30]   K. Kuratowski. Sur le problème des courbes gauches en topologie. *Fund. Math.*, 15:271–283, 1930.

[LEC67]   A. Lempel, S. Even, and I. Cederbaum. An algorithm for planarity testing of graphs. In *Theory of Graphs: Internat. Symposium (Rome 1966)*, pages 215–232, New York, 1967. Gordon and Breach.

[LH77]   C. H. C. Little and D. A. Holton. A new characterization of planar graphs. *Bulletin of the American Mathematical Society*, 83(1):137–138, 1977.

[Lie01]   A. Liebers. Planarizing graphs – a survey and annotated bibliography. *Journal of Graph Algorithms and Applications*, 5(1):1–74, 2001.

[Liu88]   Y. Liu. A new approach to the linearity of testing planarity of graphs. *Acta Mathematicae Applicatae Sinica (English Series)*, 4(3):257–265, 1988.

[Liu89]   Y. Liu. Boolean approach to planar embeddings of a graph. *Acta Mathematica Sinica (New Series)*, 5(1):64–79, 1989.

[LS10]   C. H. C. Little and G. Sanjith. Another characterisation of planar graphs. *The Electronic Journal of Combinatorics*, 17(15), 2010.

[LT79]   R. J. Lipton and R. E. Tarjan. A separator theorem for planar graphs. *SIAM J. Appl. Math.*, 36:177–189, 1979.

[Mac37a]   S. MacLane. A combinatorial condition for planar graphs. *Fundamenta Mathematicae*, 28:22–32, 1937.

[Mac37b]   S. MacLane. A structural characterization of planar combinatorial graphs. *Duke Mathematical Journal*, 3:466–472, 1937.

[Man83]   A. Mansfield. Determining the thickness of graphs is NP-hard. *Proc. Math. Cambridge Philos. Soc.*, 93:9–23, 1983.

[Men27]   Karl Menger. Zur allgemeinen kurventheorie. *Fund. Math.*, 10:96–115, 1927.

[MM96]   K. Mehlhorn and P. Mutzel. On the embedding phase of the Hopcroft and Tarjan planarity testing algorithm. *Algorithmica*, 16:233–242, 1996.

[MW99]   P. Mutzel and R. Weiskircher. Optimizing over all combinatorial embeddings of a planar graph. In *Proceedings of the 7th International IPCO Conference on Integer Programming and Combinatorial Optimization*, pages 361–376, London, UK, 1999. Springer-Verlag.

[MW00]   P. Mutzel and R. Weiskircher. Computing optimal embeddings for planar graphs. In *Proceedings of the 6th Annual International Conference on Computing and Combinatorics*, COCOON '00, pages 95–104, London, UK, 2000. Springer-Verlag.

[Pap95]   A. Papakostas. Upward planarity testing of outerplanar dags. In R. Tamassia and I. G. Tollis, editors, *Graph Drawing (Proc. GD '94)*, volume 894 of *Lecture Notes Comput. Sci.*, pages 298–306. Springer-Verlag, 1995.

[Piz05]   M. Pizzonia. Minimum depth graph embeddings and quality of the drawings: An experimental analysis. In P. Healy and N. S. Nikolov, editors, *Graph Drawing '05*, volume 3843 of *LNCS*, pages 397–408, 2005.

[PT00]   M. Pizzonia and R. Tamassia. Minimum depth graph embedding. In M. Paterson, editor, *ESA '00*, volume 1879 of *LNCS*, pages 356–367, 2000.

[RND77]   E. M. Reingold, J. Nievergelt, and N. Deo. *Combinatorial Algorithms: Theory and Practice*. Prentice Hall, Englewood Cliffs, NJ, 1977.

[Ros80]   P. Rosenstiehl. Preuve algébrique du critère de planarité du Wu-Liu. *Annals of Discrete Mathematics*, 9:67–78, 1980.

[RR89]   V. Ramachandran and J. H. Reif. An optimal parallel algorithm for graph planarity. In *Proc. 30th Annu. IEEE Sympos. Found. Comput. Sci.*, pages 282–293, 1989.

[RR94]   V. Ramachandran and J. Reif. Planarity testing in parallel. *Journal of Computer and System Sciences*, 49:517–561, December 1994.

[RS84]   N. Robertson and P. D. Seymour. Graph minors. III. Planar tree-width. *Journal on Combinatorial Theory, Series B*, 36(1):49–64, 1984.

[RSST97]   N. Robertson, D. P. Sanders, P. D. Seymour, and R. Thomas. The four color theorem. *J. Combin. Theory Ser. B*, 70:2–4, 1997.

[Sch89]   W. Schnyder. Planar graphs and poset dimension. *Order*, 5:323–343, 1989.

[Sch12]   J. M. Schmidt. A planarity test via construction sequences. *CoRR*, abs/1202.5003, 2012.

[SH93]   W. K. Shih and W. L. Hsu. A simple test for planar graphs. In *Int. Workshop on Discrete Math. and Algorithms*, pages 110–122, 1993.

[SH99]   W. K. Shih and W. L. Hsu. A new planarity test. *Theor. Comp. Sci.*, 223, 1999.

[Tam98]   R. Tamassia. Constraints in graph drawing algorithms. *Constraints*, 3:87–120, April 1998.

[Tar72]   R. E. Tarjan. Depth-first search and linear graph algorithms. *SIAM J. Comput.*, 1(2):146–160, 1972.

[Tho81]   C. Thomassen. Kuratowski's theorem. *Journal of Graph Theory*, 5(3):225–241, 1981.

[Tho99]   R. Thomas. Graph planarity and related topics. In Jan Kratochvíl, editor, *Graph Drawing (Proc. GD '99)*, volume 1731 of *LNCS*, pages 137–144. Springer-Verlag, 1999.

[TT97]   Hisao Tamaki and Takeshi Tokuyama. A characterization of planar graphs by pseudo-line arrangements. In *Proc. 8th Annu. Internat. Sympos. Algorithms Comput.*, volume 1350 of *Lecture Notes Comput. Sci.*, pages 123–132. Springer-Verlag, 1997.

[Tut61]   W. T. Tutte. A theory of 3-connected graphs. *Indag. Math.*, 23:441–455, 1961.

[Tut66]   W. T. Tutte. *Connectivity in Graphs*. University of Toronto Press, 1966.

[Wag37a]   K. Wagner. Über eine Eigenschaft der ebenen Komplexe. *Mathematische Annalen*, 114:570–590, 1937.

[Wag37b]   K. Wagner. Über eine Erweiterung eines Satzes von Kuratowski. *Deutsche Mathematik*, 2:280–285, 1937.

[Whi32]   H. Whitney. Non-separable and planar graphs. *Transactions of the American Mathematical Society*, 34:339–362, 1932.

[Wil80]   S. G. Williamson. Embedding graphs in the plane – algorithmic aspects. *Annals of Discrete Mathematics*, 6:349–384, 1980.

[Wu74]   W. Wu. Planar embedding of linear graphs. *Kexue Tongbao*, 2:226–282, 1974. (In Chinese).

[Xu89]   W. Xu. Improved algorithm for planarity testing based on Wu-Liu's criterion. *Annals of the New York Academy of Science*, 576:641–652, 1989.

[Yan78]   M. Yannakakis. Node-and edge-deletion NP-complete problems. In *Proceedings of the Tenth Annual ACM Symposium on Theory of Computing*, STOC '78, pages 253–264, New York, NY, USA, 1978. ACM.

[Yan82]   M. Yannakakis. The complexity of the partial order dimension problem. *SIAM J. Algebraic Discrete Methods*, 3(3):351–358, 1982.

2

Crossings and Planarization

Christoph Buchheim
TU Dortmund

Markus Chimani
Friedrich-Schiller-Universität Jena

Carsten Gutwenger
TU Dortmund

Michael Jünger
University of Cologne

Petra Mutzel
TU Dortmund

2.1 Introduction

In many respects, crossing minimization is an exceptional problem in the wide range of optimization problems arising in automatic graph drawing. First of all, it is one of the most basic and natural problems among these, and, at the same time, very easy to formulate: given a graph, draw it in the plane with a minimum number of crossings between its edges. In fact, this problem is much older than automatic graph drawing. Crossing number problems were first examined by Turán when he worked in a brick factory during the Second World War. This work motivated him to search for crossing minimal drawings of the complete bipartite graph $K_{n,m}$, without success. Later, Zarankiewicz gave a rule for creating a drawing of $K_{n,m}$ with $\lfloor \frac{m}{2} \rfloor \lfloor \frac{m-1}{2} \rfloor \lfloor \frac{n}{2} \rfloor \lfloor \frac{n-1}{2} \rfloor$ crossings, but his proof of optimality was shown to be incorrect. Still today, this is an open question. The same is true for the crossing number of K_n.

Besides its theoretical relevance as a topological problem, crossing minimization has many practical applications. In automatic graph drawing, it is well known that the readability of a two-dimensional graph layout strongly depends on the number of edge crossings. This was verified by empirical studies of Purchase [Pur97]. In fact, the main information given by an abstract graph is whether two vertices are connected by an edge. This information should be easily recognizable. In particular, it should be easily possible to trace the edges in the drawing. This task is complicated by the presence of crossing edges, as they distract the concentration of the human viewer. See Figure 2.1 for a comparison.

Another important application for the crossing minimization problem is VLSI (very large scale integration) design. In this context, the problem was first discussed in depth. The

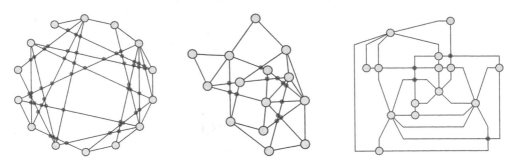

Figure 2.1 Different drawings of the same abstract graph with different numbers of edge crossings (51, 12, and 4, respectively). Most aesthetic criteria like few edge bends, uniform edge lengths, or a small drawing area favor the first two drawings. However, with respect to the number of edge crossings the last drawing is preferable.

aim of VLSI design is to arrange transistors on two-dimensional chips. Certain transistors need to be connected by wires, which have to be routed on the chip. Every crossing of two wires causes additional costs for realizing the chip, so that a small number of such crossings is desired to reduce these costs as far as possible.

An outstanding property of the crossing minimization problem is its hardness. It was shown by Garey and Johnson [GJ83] that this problem is NP-hard; see Section 2.3. However, several optimization problems arising in the area of automatic graph drawing are NP-hard and have nevertheless been solved in practice. In contrast, crossing minimization is extremely hard also practically. So far, exact approaches can only solve relatively sparse, medium sized instances within a reasonable running time; see Section 2.4. This is drastically shown by the fact that even the crossing numbers of the complete graphs K_n are unknown for $n \geq 13$.

Given the NP-hardness of the general problem, many restricted versions of crossing minimization have been considered, in the hope of finding polynomial time algorithms in these cases. However, in most cases, the problem remains NP-hard. Examples are bipartite drawings or linear embeddings; see Section 2.3. In practice, however, some of the resulting problems become easier as the degrees of freedom are reduced.

Besides considering special cases, it is natural to ask for approximation algorithms. However, up to now, only for graphs with bounded degree it was possible to find algorithms yielding provably near-optimal solutions; see Section 2.6. On the other hand, no negative results about approximability are known.

Currently used approaches to the general crossing minimization problem are of heuristic nature. The state-of-the-art approach for general crossing minimization is the planarization method, which is described in detail in Section 2.5. The main idea is to split up the problem into two steps: in the first step, a planar subgraph is computed. The aim in this step is to find a subgraph with as many edges as possible. In the second step, all edges not contained in this subgraph are reinserted into the drawing. Whenever an edge is inserted, the produced crossings are replaced by dummy vertices, so that the result is a planar graph again. Having added all edges in this way, a planar drawing algorithm can be used to compute a layout of the graph; see Chapters 6 and 7. After this, the dummy vertices are removed. For both steps of the planarization approach, a variety of possible algorithmic realizations has been discussed; see Section 2.5. This approach is also particularly interesting with respect to approximation algorithms, as it can be shown that certain insertion algorithms in fact approximate the crossing number in case of special graph classes; again see Section 2.6.

The second step of the planarization approach can also be realized in many different ways; see Section 2.5.3. Usually, it is again solved heuristically; edges are reinserted one after another, each with a minimal number of new edge crossings. It was shown recently that one can add a single edge optimally over all possible embeddings of the planar graph constructed so far.

After 60 years of research in different areas of mathematics and computer science, the crossing minimization problem is still far from being fully explored, both theoretically and practically. On the theoretical side, the most interesting open problems in our opinion are the crossing numbers of the complete graphs, including Turán's brick factory problem, as well as the approximability of crossing minimization. Practically, one can hope for new and better heuristic methods or faster exact approaches. At this point, we can only report on the status quo. We hope that parts of this chapter will become obsolete sooner or later.

2.2 Crossing Numbers

A drawing of a graph $G = (V, E)$ in the plane is a mapping of each vertex $v \in V$ to a distinct point and each edge $e = (v, w) \in E$ to a curve connecting the incident vertices v and w without passing through any other vertex. A common point of two edges in a drawing that is not an incident vertex is called a *crossing*. The *crossing number* $\mathrm{cr}(G)$ is defined to be the minimum number of crossings in any drawing of G.

In their paper "Which Crossing Number Is It, Anyway?", Pach and Tóth define two further possibilities on how to count the number of crossings in a graph (see [PT00]).

DEFINITION 2.1 Let $G = (V, E)$ be a simple graph.

1. The *pairwise crossing number* of G, denoted with $\mathrm{pcr}(G)$, is the minimum number of pairs of edges $(e_1, e_2) \in E \times E$, $e_1 \neq e_2$ such that e_1 and e_2 determine at least one crossing, over all drawings of G.

2. The *odd-crossing number* of G, denoted with $\mathrm{ocr}(G)$, is the minimum number of pairs of edges $(e_1, e_2) \in E \times E$, $e_1 \neq e_2$ such that e_1 and e_2 cross an odd number of times, over all drawings of G.

It is clear that $\mathrm{ocr}(G) \leq \mathrm{pcr}(G) \leq \mathrm{cr}(G)$, and we know that $\mathrm{cr}(G)$ cannot be arbitrarily large if $\mathrm{ocr}(G)$ is bounded. More precisely we have that $\mathrm{cr}(G) \leq 2(\mathrm{ocr}(G))^2$. Only after some years, an example was conceived by Pelsmajer et al. [PSŠ06], showing that there in fact exist graphs with $\mathrm{ocr}(G) \neq \mathrm{cr}(G)$. Yet, it is still unknown whether $\mathrm{pcr}(G) = \mathrm{cr}(G)$.

Another well-studied variant of the crossing minimization problem is the *rectilinear crossing number* $\mathrm{cr}_1(G)$, which is defined to be the minimum number of crossings in any drawing of a graph G where all edges are drawn as straight lines. Bienstock and Dean proved in [BD93] that for graphs with crossing number at most three, the rectilinear crossing number and the usual crossing number coincide. They could further show that there are graphs G_k such that $\mathrm{cr}_1(G_k)$ is arbitrarily large, even if $\mathrm{cr}(G_k)$ is only four.

As a generalization of the rectilinear crossing number, Bienstock introduced in [Bie91] the concept of the *t-polygonal crossing number*.

DEFINITION 2.2 Let $G = (V, E)$ be a graph. A *t-polygonal drawing* of G, for $t \geq 1$, is a good drawing where every edge is drawn as a *t-polygonal line*, i.e., a polygonal line with at most t segments. The *t-polygonal crossing number* $\mathrm{cr}_t(G)$ is defined as the minimum number of crossings in any t-polygonal drawing of G.

A *good drawing* is a drawing that satisfies the following conditions:

1. no edge crosses itself
2. adjacent edges do not cross one another
3. non-adjacent edges cross each other at most once

Bienstock also showed that there cannot be a polynomial time algorithm for producing optimal t-polygonal drawings of G unless $\mathrm{P} = \mathrm{NP}$ and that there is no fixed t such that $\mathrm{cr}(G) = \mathrm{cr}_t(G)$ for any graph G.

An even more restricted version of the crossing number problem is the *linear crossing number*: We call a drawing of a graph a *linear drawing* if all vertices lie on a straight line and edges are drawn as semicircles above and below this line. It is easy to see that the crossing number resulting from this drawing style is an upper bound for $\mathrm{cr}(G)$. Surprisingly, there is a further connection to the general crossing number problem, as was shown by Nicholson [Nic68]. He proved that any drawing in the plane with a minimum number of crossings can be converted into a linear drawing with an equivalent crossing structure such that all vertices are placed on a horizontal line and edges are drawn as a series of semicircles while successive semicircles lie on different sides of the horizontal line.

It is interesting to see that the complexity of the linear crossing number problem stays the same, even if we fix the ordering of the vertices of V (this is the so-called *fixed linear crossing minimization problem*). Masuda et al. proved in [MNKF90] that even this variant is NP-complete.

2.2.1   Known Bounds

No matter which definition or variant of the crossing number problem is used, its solution seems to be a difficult task. Even though crossing numbers have been investigated extensively in the past, useful theoretical results are rather limited. One of the first major results has been claimed in 1953 by Zarankiewicz (and, independently, by Urbaník) as a solution to Turán's brick factory problem, which in fact asks for the crossing number of the complete bipartite graph $K_{m,n}$.

$$\mathrm{cr}(K_{m,n}) = \lfloor \frac{m}{2} \rfloor \lfloor \frac{m-1}{2} \rfloor \lfloor \frac{n}{2} \rfloor \lfloor \frac{n-1}{2} \rfloor \qquad \text{(conjecture)} \qquad (2.1)$$

Over ten years later, an error in the induction argument of Zarankiewicz's proof was unveiled, which is still unremedied. Hence, the correctness of equation (2.1) is still unknown. The conjecture is derived from the following drawing rule for complete bipartite graphs $K_{m,n} = (A \cup B, E)$: place the vertices in vertex set A at coordinates $(i(-1)^i, 0)$ for all $i = 1, \ldots, m$ and the vertices of vertex set B at coordinates $(0, j(-1)^j)$ for all $j = 1, \ldots, n$. All edges are drawn as straight lines. Figure 2.2 shows a sample drawing of $K_{6,6}$ with 36 crossings. Even though the correctness of equation (2.1) could never be verified, the provided drawing rule gives us an upper bound $Z(m,n)$ for $\mathrm{cr}(K_{m,n})$. Recently, de Klerk et al. [KMP+06, KPS07] devised a method for computing asymptotic lower bounds for $\mathrm{cr}(K_{m,n})$ based on semidefinite programming. They show that

$$\lim_{n \to \infty} \frac{\mathrm{cr}(K_{m,n})}{Z(m,n)} \geq 0.8594 \frac{m}{m-1} \,.$$

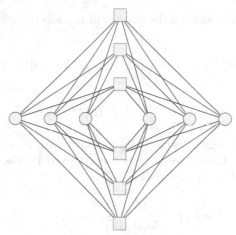

Figure 2.2 A drawing of $K_{6,6}$ with 36 crossings using Zarankiewicz's rule.

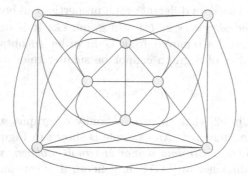

Figure 2.3 A drawing of K_8 with a minimum number of 18 crossings.

As for complete bipartite graphs, there is also a conjecture for the number of crossings of the complete graph K_n with n vertices.

$$\text{cr}(K_n) = \frac{1}{4} \lfloor \frac{n}{2} \rfloor \lfloor \frac{n-1}{2} \rfloor \lfloor \frac{n-2}{2} \rfloor \lfloor \frac{n-3}{2} \rfloor \qquad \text{(conjecture)} \qquad (2.2)$$

Constructions of corresponding drawings [GJJ68] show that also this conjecture yields an upper bound $Z(n)$ on $\text{cr}(K_n)$. For complete graphs on up to ten vertices, its correctness has been verified by Guy [Guy72]. Pan and Richter [PR07] have extended this verification to K_{11} and K_{12}. We show a sample drawing of K_8 with a minimum number of 18 crossings in Figure 2.3. For K_n, the best-known asymptotic lower bound is again due to de Klerk et al. [KMP+06]:

$$\lim_{n \to \infty} \frac{\text{cr}(K_n)}{Z(n)} \geq 0.83.$$

The crossing number of a graph G with n vertices cannot exceed the crossing number of the complete graph K_n, hence $Z(n)$ also marks an upper bound for general graphs. Unfortunately, it is the only known upper bound. A simple *lower bound* can be obtained from Euler's formula. Since any planar simple connected graph $G = (V, E)$ cannot have more than $3|V| - 6$ edges, clearly $\text{cr}(G) \geq |E| - 3|V| + 6$. If, in addition, G contains no triangle, then $\text{cr}(G) \geq |E| - 2|V| + 4$.

In 1983, Leighton used induction on the number of vertices to show the following theorem; see [Lei83].

Theorem 2.1 *Let $G = (V, E)$ be a simple graph. If $|E| \geq 4|V|$, we have*

$$cr(G) \geq \frac{1}{100} \frac{|E|^3}{|V|^2} . \tag{2.3}$$

Ajtai et al. obtained the same result independently with a smaller constant of $\frac{1}{375}$ in [ACNS82]. One of the best-known results has been derived by Pach and Tóth [PT97]. For any simple graph $G = (V, E)$, $cr(G)$ satisfies

$$cr(G) \geq \frac{1}{33.75} \frac{|E|^3}{|V|^2} - 0.9|V| . \tag{2.4}$$

Apart from bounds with respect to the number of vertices and edges, several approaches to obtain tight lower bounds based on different graph properties can be found in the literature.

A simple example is the *skewness* $sk(G)$ of a graph G. It is defined as the minimum number of edges that must be removed from G in order to obtain a planar subgraph. Clearly, the crossing number of a graph cannot be smaller than its skewness. Hence, we have that

$$cr(G) \geq sk(G) . \tag{2.5}$$

Cimikowski showed in [Cim92] that there is a family of graphs with skewness one, but an arbitrarily high crossing number. An example is shown in Figure 2.4. Computing the skewness is equivalent to the *maximum planar subgraph problem*, which was shown to be NP-hard by Liu and Geldmacher in [LG77] in general. For certain classes of graphs, i.e., complete and complete bipartite graphs, the skewness is known. We can derive it for K_n from Euler's formula and the observation that every maximal planar graph is also a maximum planar subgraph of K_n. Hence, the skewness for complete graphs K_n is given by

$$sk(K_n) = \frac{n(n-1)}{2} - 3n + 6, \tag{2.6}$$

and we can use similar arguments to derive the skewness for complete bipartite graphs as

$$sk(K_{m,n}) = mn - 2(m+n) + 4 . \tag{2.7}$$

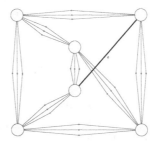

Figure 2.4 Construction of graphs with skewness one and arbitrarily high crossing number.

Another bound can be obtained from the *bisection width* bw(G). For any disjoint partition of the vertex set V into sets V_1 and V_2, we denote the edges (v_1, v_2) with $v_1 \in V_1$ and $v_2 \in V_2$ by $E(V_1, V_2)$. The bisection width bw(G) is defined as follows:

$$\text{bw}(G) = \min_{|V_1|, |V_2| \geq \frac{|V|}{3}} \{|E(V_1, V_2)|\}.$$

More intuitively, the bisection width is the minimum number of edges that must be removed from G in order to partition the graph into two separate components with nearly equal size. The first known bound for the crossing number based on the bisection width goes back to Leighton. He proved the following theorem [Lei84].

Theorem 2.2 *For any graph $G = (V, E)$ of bounded degree, we have*

$$\text{cr}(G) + |V| = \Omega(\text{bw}(G)^2) \,.$$

Pach, Shahrokhi, and Szegedy [PSS96] use the bisection width to show the following, more general, result, which can be used to derive a lower bound for cr(G).

Theorem 2.3 *Let $G = (V, E)$ be a simple graph with $|V| \geq 2$ vertices, and let $k \geq 1$ be an integer. If G has a drawing with at most k crossings, then*

$$|E| \leq 3|V|(10 \log_2 |V|)^{2k-2} \,.$$

A very similar parameter is the *cutwidth* cw(G). Let $\phi : V \to \{1, 2, \ldots, |V|\}$ be an injection. We define cw(G) as follows:

$$\text{cw}(G) = \min_{\phi} \max_i |\{(u, v) \in E : \phi(u) \leq i \leq \phi(v)\}|.$$

As a graphical interpretation, consider an injection of the vertices to the horizontal line and draw edges on one side of this line using semicircles. For each injection, we can "cut" the horizontal line between a pair of consecutive vertices such that the number of edges between each of the segments is maximized. The minimum value over all possible injections is the cutwidth. So far, the following relations are known (see [DV02], [PSS96], [SV94]; here $\delta(v)$ denotes the set of neighbors of vertex v):

$$\text{cr}(G) + \frac{1}{16} \sum_{v \in V} |\delta(v)|^2 \geq \frac{1}{40} \text{bw}^2(G), \qquad (2.8)$$

$$\text{cr}(G) + \frac{1}{16} \sum_{v \in V} |\delta(v)|^2 \geq \frac{1}{1176} \text{cw}^2(G). \qquad (2.9)$$

Unfortunately, the computation of both parameters bw(G) and cw(G) is NP-hard.

Both the bisection width and the cutwidth can be seen as a measure for the "non-planarity" of a graph. This applies also to the *thickness* $\Theta(G)$, which is defined as the minimum number of planar graphs whose union forms G. The only families of graphs whose thickness is known are complete graphs, complete bipartite graphs, and hypercubes. Mansfield proved in [Man83] that the determination of $\Theta(G)$ is NP-hard in general. There is a simple connection between thickness and crossing number:

$$\Theta(G) \leq \text{cr}(G) + 1$$

So far, all those bounds are only of limited use. Either their quality is poor or their computation often exceeds practical limits. The investigation of tighter bounds could help to improve practical applications and lead to more insight into the crossing minimization problem.

2.3   Complexity of Crossing Minimization

Crossing minimization is not only one of the most important problems arising in automatic graph drawing, it is also one of the hardest. This is true both in practice and in theory: until recently, not a single exact algorithm being able to solve instances of nontrivial size had been devised. In fact, even for a graph as small and regular as K_{13}, the minimal number of crossings is still unknown. For a discussion of exact crossing minimization approaches, see Section 2.4.

2.3.1   NP-hardness

On the theoretical side, it is a well-known fact that the general crossing minimization problem is NP-hard. More precisely, consider the following *crossing number problem*:

> Given a graph G and a nonnegative integer K, decide whether there is a drawing of G with at most K edge crossings.

In 1983, Garey and Johnson proved that this problem is NP-complete [GJ83]. In the following, we reproduce their proof. It is based on a transformation of the NP-complete *optimal linear arrangement problem*:

> Given a graph $G = (V, E)$ and a nonnegative integer K, decide whether there is a one-to-one function $f \colon V \to \{1, \ldots, |V|\}$ with $\sum_{(v,w) \in E} |f(v) - f(w)| \leq K$.

The corresponding optimization problem is thus to order the vertices of G such that the total length of edges is minimal, where the length of an edge is defined as the distance of the two adjacent vertices in this ordering.

As an intermediate step in the proof, Garey and Johnson show the NP-completeness of the bipartite version of the crossing number problem for multigraphs, the *bipartite crossing number problem*:

> Given a bipartite multigraph $G = (V_1, V_2, E)$ and a nonnegative integer K, decide whether there is a drawing of G inside the unit square such that all vertices of V_1 lie on the northern boundary, all vertices of V_2 lie on the southern boundary, and the number of edge crossings is at most K.

In the following, we will call such drawings *bipartite drawings* for short. It is interesting that, contrary to widespread belief, the NP-completeness of the bipartite crossing number problem for *simple* graphs was long open—it was shown only very recently [Sch12].

Theorem 2.4   *The crossing number problem is NP-complete.*

It is easy to see that this problem is in NP: for every edge of G, one can guess all crossings involving this edge, and their order along the edge. To answer the question whether such a guessed crossing configuration is feasible, one can place dummy vertices on all chosen crossings and test the resulting graph for planarity. Clearly, the result is positive if and only if the given crossing configuration can be realized by some drawing of G.

The proof of completeness consists of several reduction steps and is split up into three separate lemmas in the following.

LEMMA 2.1   The optimal linear arrangement problem can be reduced to the bipartite crossing number problem in polynomial time.

Proof: The rough idea of the reduction is as follows: every vertex is doubled and the linear ordering is modeled on two parallel layers (the northern and southern boundary of the unit square) at the same time, with edges leading from one layer to the other. By a large number of artificial edges connecting corresponding pairs of vertices, the ordering is forced to be the same on both layers. The distance between two adjacent vertices in the linear ordering problem is then essentially proportional to the number of artificial edges crossed.

More formally, the transformation is defined as follows: Let an instance for the optimal linear arrangement problem be given, consisting of a graph G and an integer K, and assume $V = \{v_1, \ldots, v_n\}$. We then construct an instance $G' = (V_1, V_2, E_1 \cup E_2)$ and K' of the bipartite crossing number problem as follows:

$$
\begin{aligned}
V_1 &= \{u_i \mid i = 1, \ldots, n\} \\
V_2 &= \{w_i \mid i = 1, \ldots, n\} \\
E_1 &= \{|E|^2 \text{ copies of } (u_i, w_i) \mid i = 1, \ldots, n\} \\
E_2 &= \{(u_i, w_j) \mid (v_i, v_j) \in E \text{ with } i < j\} \\
K' &= |E|^2 (K - |E|) + |E|^2 - 1
\end{aligned}
$$

This construction is obviously polynomial. We have to show that the graph G admits a linear ordering with total edge length at most K if and only if the bipartite multigraph G' admits a bipartite drawing with at most K' crossings.

If a linear ordering f of G with total edge length at most K exists, we construct a bipartite drawing as follows. We place vertex u_i on position $(f(v_i)/(n+1), 1)$ and vertex w_i on position $(f(v_i)/(n+1), 0)$. Furthermore, we draw all edges as straight lines; bundles of parallel edges are drawn as nearly straight lines without mutual crossings; see Figure 2.5 for an example.

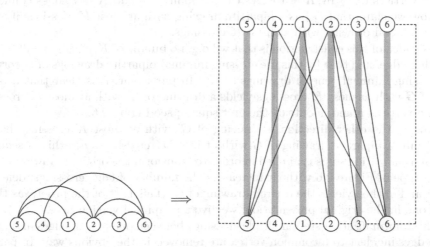

Figure 2.5   Reducing the optimal linear arrangement problem to the bipartite crossing number problem. Bold grey lines represent bundles of $|E|^2$ edges each.

In the constructed drawing, no artificial edge from E_1 will cross any other edge. Moreover, an edge $(u_i, w_j) \in E_2$ crosses exactly $|f(v_i) - f(v_j)| - 1$ bundles of $|E|^2$ edges each.

Consequently, the total number of such crossings is

$$\sum_{(u_i,w_j)\in E_2} |E|^2 \Big(|f(v_i) - f(v_j)| - 1\Big) = |E|^2 \sum_{(v,w)\in E} \Big(|f(v) - f(w)| - 1\Big) \le |E|^2(K - |E|) \,.$$

The remaining crossings in the constructed drawing can only occur between pairs of edges in E_2, so that their total number is at most $|E|^2 - 1$. Summing up, the total number of edge crossings in our drawing is at most $|E|^2(K - |E|) + |E|^2 - 1 = K'$.

For showing the other direction, assume that a bipartite drawing of G' with at most K' crossings is given. Then define $f(v_i)$ as the position of vertex u_i in the order of vertices on the northern boundary of the unit square. We claim that the linear ordering f leads to a total edge length of at most K. To see this, first observe that the order of vertices on both boundaries must be the same, as otherwise two bundles of $|E|^2$ edges each would cross each other, leading to $|E|^4 > K'$ crossings. Because of that, for each edge (v_i, v_j) with $i < j$, the distance $|f(v_i) - f(v_j)|$ is at most one more than the number of crossings of (u_i, w_j) with any edge bundle, so that

$$\sum_{(v,w)\in E} |f(v) - f(w)| \le |E| + K'/|E|^2 = |E| + (K - |E|) + 1 - 1/|E|^2 < K + 1 \,.$$

As the left-hand side of this inequality is integer, it is at most K.   □

LEMMA 2.2   The bipartite crossing number problem can be reduced to the general crossing number problem for multigraphs in polynomial time.

Proof: Let $G = (V_1, V_2, E)$ and K be an instance of the bipartite crossing number problem. We construct a multigraph G' as follows: we add two vertices u and w to G. Moreover, we connect u with all vertices of V_1 by $K+1$ edges each. Analogously, we connect w with all vertices of V_2 by $K + 1$ edges each. Finally, we add $K + 1$ edges connecting u and w. Now we claim that G has a bipartite drawing with at most K crossings if and only if G' has a general drawing with at most K crossings.

The basic idea of this construction is that w.l.o.g. no bundle of $K+1$ edges will be crossed by any other edge, and that by this the crossing minimal bipartite drawings of G correspond to the crossing minimal general drawings of G'. In particular, it is clear that a bipartite drawing of G with at most K crossings yields a drawing of G' with at most K crossings by placing the vertices u and w outside the unit square; see Figure 2.6.

For showing the other direction, a drawing of G' with at most K crossings has to be converted into a bipartite drawing of G with at most K crossings. For this, a sequence of so-called normalization steps is applied in order to transform the original drawing of G' into one of the type of Figure 2.6 without increasing the number of edge crossings; deleting the vertices u and w then yields the desired drawing of G. This part of the proof was the most technical one in the original presentation; we give a simplified version here.

In the first normalization step, multiple crossings between one pair of edges and crossings between edges incident to a common vertex are removed in the obvious way. In particular, every bundle of $K + 1$ edges connecting the same pair of vertices now defines a sequence of K regions in the drawing.

In a second step, one can obtain a drawing such that none of these bundle regions contains any vertex of G' or is crossed by any edge of G'. Indeed, for a fixed $v \in V_1$, consider the edge e connecting u and v that in the current drawing has the minimum number of crossings

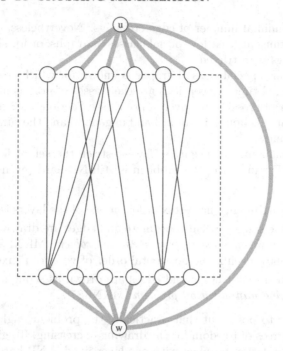

Figure 2.6 Reducing the bipartite crossing number problem to the general crossing number problem for multigraphs. Bold grey lines represent bundles of $K + 1$ edges each.

with other edges. Then one can reroute all edges (u, v), i.e., all edges parallel to e, along the same route as e. This yields a new drawing of G' with at most as many edge crossings as before. Repeating this for every $v \in V_1$ and analogously for w and every $v \in V_2$, we get a drawing without vertices in the bundle regions. Now it follows that no edge can cross any of these regions. The reason is that such an edge would have to cross all $K + 1$ edges of a bundle, as no vertices are contained in the bundle regions and multiple crossings were eliminated in the first normalization step.

Clearly, the drawing resulting from these two normalization steps is topologically equivalent to one of the type displayed in Figure 2.6.   □

LEMMA 2.3   The crossing number problem for multigraphs can be reduced to the crossing number problem for simple graphs.

Proof: For the given multigraph, place an artificial vertex in the middle of every edge. The result is a simple graph with the same crossing number as the original multigraph.   □

In the above scheme, we observe that the graph for which deciding on the crossing number is NP-hard requires two distinct vertices u and v of very high degree. One may think that they are central to the construction. Yet, using a different reduction strategy from optimal linear arrangement, Hliněný showed in [Hli06]:

Theorem 2.5   *The crossing number problem remains NP-complete even when restricted to cubic graphs, i.e., graphs where every vertex has degree 3.*

Theorem 2.4 shows that the crossing number problem is NP-complete. In particular, the crossing minimization problem is NP-hard, i.e., the problem of constructing a drawing of a given graph with a minimal number of edge crossings. Nevertheless, one might hope for polynomial time algorithms at least for special classes of graphs, or for situations where the class of allowed drawings is restricted.

However, no interesting special class of graphs is known for which crossing minimization can be done in polynomial time. Exceptions are the classes of graphs for which a constant bound c on the number of crossings is given a priori, see Section 2.3.2, but this is a purely theoretical result in that this bound is not at hand in general and the running time increases heavily with the constant c.

The results also remain mostly negative if we restrict the set of feasible drawings by additional conditions. For instance, the problem is still NP-hard (even for simple graphs) if we require that

- the drawing is bipartite and the vertex order on one of the layers is fixed [EW94].
- all vertices have the same vertical coordinate and edges are drawn as semicircles. This is the so-called *linear crossing minimization problem* [MKNF86]. This problem remains NP-hard even if the horizontal order of vertices is fixed [MNKF90].
- the vertices lie on the unit circle and edges are drawn as straight lines. This is the *circular crossing minimization problem* [MKNF87].

However, we would like to point out that practically the problem might become considerably easier with the degrees of freedom for the drawing decreasing. To give an example, the bipartite crossing minimization problem with one layer fixed is NP-hard but can be solved quickly in practice [JM97]. By now, also reasonably sized general multi-layer crossing minimization instances can be tackled effectively with integer linear and semidefinite programs; see [CHJM11] for an overview.

2.3.2   Fixed Parameter Tractability

In the last section, we reproduced a proof by Garey and Johnson showing that it is NP-hard to decide whether a given graph G can be drawn with at most K edge crossings. We can also consider the situation where K is not given as part of the input but as a fixed parameter. It is then easy to see that one can decide in polynomial time whether a drawing of G with at most K crossings exists: broadly speaking, one could check all possible configurations of the up to K crossings, replace the chosen crossings by dummy vertices, and check the resulting graph for planarity. We can answer the original question affirmatively if and only if we find any planar graph in this way.

Even if the above algorithm runs in polynomial time for fixed K, the obvious drawback is the strong increase in running time for increasing K: if implemented in the straightforward way, the runtime is $O(|V| \cdot |E|^{2K})$. For a long time, it was an open question whether the problem is *fixed-parameter tractable*, i.e., whether the problem can be solved in $O(f(K) \cdot |V|^c)$ running time for some function $f(K)$ that is independent from the instance and some constant c that is independent from K. This question was answered by Grohe in 2001 with $c = 2$ [Gro01]. However, the running time of Grohe's algorithm is $O(2^{2^{p(K)}}|V|^2)$, where p is a polynomial, and hence grows strongly with K. Thus, the relevance of this algorithm is rather theoretical than practical. Kawarabayashi and Reed [KR07] improved on this result by giving a linear algorithm, i.e., $c = 1$; yet $f(K)$ remains too large for any practical application.

2.4 Exact Crossing Minimization

Exact methods to solve the crossing minimization problem constitute the youngest research field we are discussing in this chapter. The development showcases various algorithm engineering aspects of algorithm development, as its iterative improvements were always based on the analysis of the bottlenecks of the earlier approaches. The first approach [BEJ$^+$05] already lay the setting used in the subsequent developments: it relies on mathematical programming in combination with branch-and-cut. Yet, its applicability was limited to very small graphs. By introducing column generation schemes into the branch-and-cut framework, its central ILP model, which we will describe in detail below, was later brought into the realm of applicability [CGM09, BCE$^+$08] to some real-world graphs. The currently best exact approach replaces a key concept (the so-called *simple* crossing number) of the first formulation by integrating multiple linear-ordering problems instead [CMB08]. This leads to a mathematically more complex model but offers the advantage of fewer variables on the one hand, and the possibility for even stronger column generation strategies, on the other hand. Together with other developments like strong upper bounds (cf. Section 2.5), preprocessing strategies like the *non-planar core reduction* [CG09], and efficient extraction of multiple Kuratowski subdivisions (see below) at once [CMS08], we are now in the position to compute the exact crossing number of sparse graphs with up to 100 vertices. Figure 2.4 gives an overview of the algorithmic progress over the last years, comparing the various algorithms on the way to the currently most successful one. For a more detailed description of all exact algorithms discussed in the following, see [Chi08].

A *linear program* (LP) is an optimization problem consisting of continuous variables, a linear objective function, and linear constraints. The "father" of linear programming, George B. Dantzig, proposed the following standard model:

$$\text{maximize } c^\top x$$
$$\text{subject to } Ax \leq b$$
$$x \geq 0$$

where $c \in \mathbf{R}^n$, $A \in \mathbf{R}^{m \times n}$, and $b \in \mathbf{R}^m$. The linear function $c^\top x : \mathbf{R}^n \to \mathbf{R}$ is called the *objective function* and the inequalities in the system $Ax \leq b$ are called *constraints*. A vector $\hat{x}$ that satisfies the system of inequalities $Ax \leq b$ is called a *feasible* solution of the LP. Moreover, $\hat{x}$ is called an *optimal solution* if $c^\top \hat{x} \geq c^\top x'$ for all feasible solutions x'.

Linear programs proved to provide a powerful tool for various optimization problems in the past and extensive research led to efficient algorithms able to solve them in polynomial time, *e.g.*, the simplex [Chv83], the ellipsoid [GLS88], and the interior point method [RT97]. However, additional constraints that require some or all of the variables to be integer, render the problem NP-complete in general [GJ79].

Anyway, *(mixed) integer linear programs* are widely used to solve NP-hard combinatorial optimization problems in conjunction with polyhedral combinatorics, which aims at describing combinatorial optimization problems as linear programs and solving these with special-purpose methods. A key feature therefore is the possibility to alternatively describe the convex hull of the feasible points and extreme rays of a problem by a system of linear inequalities. For an introduction into this field, the interested reader is referred to [Pul89].

Before introducing the ideas of the ILP formulation presented in this section, we have to mention *Kuratowski's theorem*, which is one of the most important results in the field of planarity testing providing a full characterization of planar graphs based on the complete graph K_5 and the complete bipartite graph $K_{3,3}$.

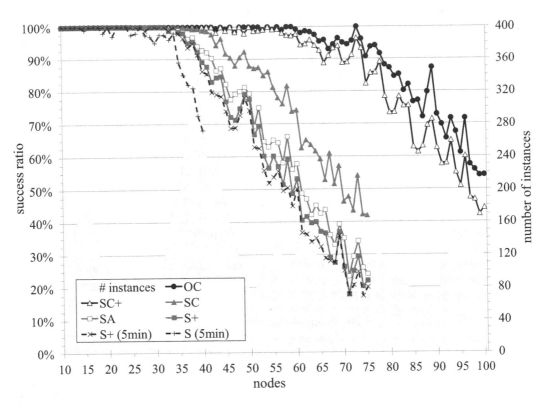

Figure 2.7 Success ratio (i.e., percentage of instances solved to provable optimality) in dependency to the graphs' size (number of vertices). Benchmark set: *Rome library* [DGL⁺97], see also Section 2.5. Different lines give different development steps of the algorithms; each instance was given 30 minutes of computation time unless specified otherwise. **S** is the first implementation of [BEJ⁺05], **S+** a more efficient reimplementation of the same algorithm. **SA** and **SC** denote the subdivision-based algorithms as considered in [CGM09], with algebraic pricing and the combinatorial column generation scheme, respectively. Finally, **SC+** and **OC** denote the latest implementations of the subdivision-based and the ordering-based ILPs, respectively, with combinatorial column generation and all further described improvements, as presented in [CMB08].

Theorem 2.6 *A finite graph is planar if and only if it contains no subgraph that is a subdivision of K_5 or $K_{3,3}$.*

We can obtain a subdivision S of a graph G by repeatedly replacing edges e by a path of length two. As a consequence of Theorem 2.6, at least two edges belonging to each Kuratowski subdivision have to cross. Based on this observation, we can try to address the crossing minimization problem using mathematical programming in the following way: we introduce a zero-one decision variable $x_{e,f}$ for each pair of edges $(e, f) \in E \times E$ that encode the crossings in an associated drawing: edges e and f cross each other if and only if $x_{e,f} = 1$. For each subdivision of K_5 or $K_{3,3}$, we can add constraints that force at least one of the involved variables to one.

Mutzel and Jünger [MJ01] pointed out the problems with this formulation. To our knowledge, there is no known polynomial time separation algorithm to identify the constraints of this type that are violated by a given fractional solution. Moreover, those constraints are not strong enough since it is not guaranteed that there is a realizing drawing if at least one of the involved crossing variables is one in every Kuratowski subdivision. Another severe problem of this formulation is the NP-hardness of the *realizability problem* [Kra91]:

Given a vector $x \in \{0, 1\}^{\binom{E}{2}}$, does there exist a drawing consistent with x?

In order to efficiently answer this question, we also need to know the *order* of the edge crossings for a particular edge e. This additional information can be exploited by the introduction of a dummy vertex for each crossing and the application of a linear-time planarity testing algorithm to test the existence of a realizing drawing in polynomial time. Despite all these drawbacks, it is interesting that under certain conditions the above-described constraints, as well as similar ones, in fact constitute *facets* of the polytope defined by the convex hull of the feasible solutions [Chi11].

2.4.1 Subdivision-Based Formulation

One way to work around the realizability problem is the reduction to *simple drawings*. We call a drawing simple if each edge crosses at most one other edge. As for planar graphs, we can find a bound for the maximum number of edges of graphs that admit a simple drawing. More precisely, Pach and Tóth show the following theorem [PT97]:

Theorem 2.7 *Let $G = (V, E)$ be a simple graph drawn in the plane so that every edge is crossed by at most k others. If $0 \le k \le 4$, then we have*

$$|E| \le (k + 3)(|V| - 2) .\tag{2.10}$$

They could further prove that this bound cannot be improved for $0 \le k \le 2$ and that for *any* $k \ge 1$ the following inequality holds:

$$|E| \le \sqrt{16.875k}|V| \approx 4.108\sqrt{k}|V| .\tag{2.11}$$

Furthermore, Bodlaender and Grigoriev prove in [BG04] that it is NP-complete to determine whether there is a simple drawing for a given graph G. If there is such a drawing, we denote the minimum number of crossings among all simple drawings of G by $\mathrm{crs}(G)$.

It is easy to see that $\mathrm{cr}(G) \le \mathrm{crs}(G)$. We cannot state equality because there are graphs G such that $\mathrm{crs}(G) > \mathrm{cr}(G)$. Consider the sample graph in Figure 2.8. The left drawing shows an optimal drawing with two crossings while the right drawing shows an optimal drawing among all simple drawings with three crossings.

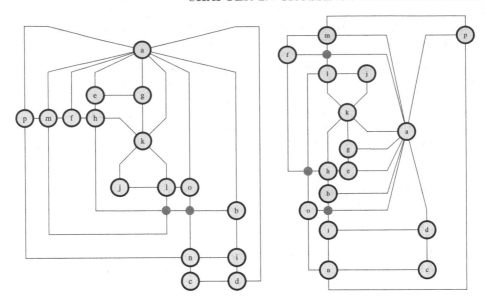

Figure 2.8   An optimal drawing of a graph with two crossings (left) and an optimal simple drawing of the same graph with three crossings (right). Both drawings were produced with the exact algorithm presented in this section.

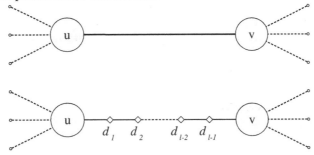

Figure 2.9   Edges are replaced with a path of length ℓ by inserting $\ell - 1$ dummy vertices.

Given an integer ℓ and a graph $G = (V, E)$ such that $\ell \geq |E|$, we can create a graph $G^* = (V^*, E^*)$ by replacing every edge $e \in E$ with a path of length ℓ. Figure 2.9 shows an example that illustrates this transformation. The graph G^* contains a total number of $|V| + (\ell - 1)|E|$ vertices and $\ell|E|$ edges.

It is easy to show that G can be drawn with n crossings if and only if there is a *simple drawing* of G^* with n crossings. Therefore, it is "sufficient" to solve the crossing minimization problem restricted to simple drawings in order to solve the "general" crossing minimization problem. Since the transformation obviously can be done in polynomial time, the NP-completeness of the corresponding decision problem for simple drawings follows immediately from the proof by Garey and Johnson; see Section 2.3. Since an edge $e = (u, v)$ never crosses itself or an adjacent edge in an optimal drawing it is sufficient to replace e with a path of length $|E| - |\delta(u)| - |\delta(v)| - 1$.

Let $G = (V, E)$ be a graph and let $D \subseteq E \times E$ be a set of unordered pairs of edges. We call D *realizable* if there is a drawing of G such that there is a crossing between edges e and f if and only if $(e, f) \in D$. Furthermore, D is called *simple* if for every $e \in E$ there is at most one $f \in E$ such that $(e, f) \in D$.

For every graph G and every simple D, G_D denotes the graph that is obtained by introducing a dummy vertex $d_{e,f}$ for each pair of edges $(e, f) \in D$: in other words, $d_{e,f}$ is the unique vertex when identifying the two vertices arising from subdividing both e and f. Note that G_D is only well defined if D is simple. For both edges e_1 and e_2 resulting from splitting e, we set $\hat{e}_1 = \hat{e}_2 = e$, analogously for f. Given a subgraph $H = (V', E') \subseteq G_D$, we denote with $\hat{H} \subseteq E$ the subset of original edges of G involved in the subgraph H of G_D, i.e., $\hat{H} = \{\hat{e} \mid e \in E'\} \subseteq E$. In the following, H will usually be a Kuratowski subdivision. We call the path corresponding to a single edge of the underlying K_5 or $K_{3,3}$ a *Kuratowski path*. By $\hat{H}^{[2]} \subset \hat{H}^2$ we will then denote the edge pairs (e, f) where e and f belong to different, nonadjacent Kuratowski paths. Hence, these are the only edge pairs that may actually form a crossing meaningful to the Kuratowski subdivision H.

COROLLARY 2.1   Let D be simple. Then D is realizable if and only if G_D is planar.

Using a linear-time planarity testing algorithm, we can test in time $O(|V| + |D|)$ whether D is realizable, and compute a realizing drawing if so.

DEFINITION 2.3   For a set of pairs of edges $D \subseteq E \times E$, we define

$$x_{e,f}^D = \begin{cases} 1 & \text{if } (e, f) \in D \\ 0 & \text{otherwise.} \end{cases}$$

PROPOSITION 2.1   Let D be simple and realizable. For an arbitrary set of pairs of edges $D' \subseteq E \times E$ of $G = (V, E)$ and any subdivision H of K_5 or $K_{3,3}$ in $G_{D'}$ the following inequality holds:

$$C_{D',H} : \sum_{(e,f) \in \hat{H}^{[2]} \setminus D'} x_{ef}^D \geq 1 - \sum_{(e,f) \in \hat{H}^{[2]} \cap D'} (1 - x_{ef}^D) . \tag{2.12}$$

Proof: Suppose inequality (2.12) is violated. Since every $x_{e,f}^D \in \{0, 1\}$, the left-hand side of the inequality must be zero and the right-hand side must be one, which means that

$$\begin{aligned} x_{e,f}^D &= 0 \quad \text{for all } (e, f) \in \hat{H}^{[2]} \setminus D', \text{ and} \\ x_{e,f}^D &= 1 \quad \text{for all } (e, f) \in \hat{H}^{[2]} \cap D'. \end{aligned}$$

It follows from the definition of x^D that $\hat{H}^{[2]} \cap D' = \hat{H}^{[2]} \cap D$, in other words, G_D agrees to $G_{D'}$ on the subgraph induced by $\hat{H}$, so that H is also a forbidden subgraph in G_D, i.e., a subdivision of K_5 or $K_{3,3}$. It follows from Kuratowski's Theorem that G_D is not planar. This contradicts the realizability of D by Corollary 2.1. $\square$

Theorem 2.8   *Let $G=(V,E)$ be a simple graph. A set of pairs of edges $D \subseteq E \times E$ is simple and realizable if and only if the following set of conditions holds:*

$$\begin{aligned} x_{e,f}^D &\in \{0, 1\} \quad && \forall \, e, f \in E, \ e \neq f \\ \sum_{f \in E} x_{e,f}^D &\leq 1 \quad && \forall \, e \in E \\ C_{D',H} \quad && && \textit{for every simple } D' \subseteq E \times E \\ && && \textit{and every forbidden subgraph } H \textit{ in } G_{D'} \end{aligned}$$

Proof: It is easy to see that the constraints from the second row are satisfied if and only if D is simple. It remains to show that a simple D is realizable if and only if the conditions $C_{D',H}$ from the last row hold. For a realizable D, every $C_{D',H}$ is satisfied according to Proposition 2.1.

We have to show that any set of pairs of edges D that is not realizable violates at least one of the constraints $C_{D',H}$. It follows from Corollary 2.1 that G_D is not planar if D is not realizable and we know from Theorem 2.6 that there exists a subdivision H of K_5 or $K_{3,3}$ in G_D. Let $D' = D$ and consider the constraint $C_{D,H}$:

$$C_{D,H} : \sum_{(e,f) \in \hat{H}^{[2]} \setminus D} x_{ef}^D \geq 1 - \sum_{(e,f) \in \hat{H}^{[2]} \cap D} (1 - x_{ef}^D) \qquad (2.13)$$

It follows from the definition of x^D that every $x_{e,f}^D \in \hat{H}^{[2]} \setminus D$ is zero, hence the left-hand side of inequality (2.13) is also zero. Since $\hat{H}^{[2]} \cap D \subseteq D$ we also know that $\sum_{(e,f) \in \hat{H}^{[2]} \cap D} (1 - x_{ef}^D)$ is zero and the right-hand side of $C_{D,H}$ is one. Thus, $C_{D,H}$ is violated. □

Since we can compute a corresponding drawing for a simple and realizable D in polynomial time, we can reformulate the crossing minimization problem for simple drawings as

> Given a graph $G = (V, E)$, find a simple realizable subset $D \subseteq E \times E$ of minimum cardinality.

This leads to the following ILP-Formulation. We use $x(F)$ as an abbreviation for the term $\sum_{(e,f) \in F} x_{e,f}$.

> minimize $x(E \times E)$

> subject to

> $\sum_{f \in E} x_{e,f} \leq 1$ $\forall\, e \in E$

> $x(\hat{H}^{[2]} \setminus D') - x(\hat{H}^{[2]} \cap D') \geq 1 - |\hat{H}^{[2]} \cap D'|$ for every simple D' and every forbidden subgraph H in $G_{D'}$

> $x_{e,f} \in \{0, 1\}$ $\forall\, e, f \in E$

Given a simple set of crossings D we can easily check if D is realizable by applying a planarity testing algorithm to G_D. If the answer is "no" we also get a forbidden subdivision H of G_D and we can separate an additional constraint $C_{D,H}$ according to the proof of Theorem 2.8 that excludes D.

2.4.2   Ordering-Based Formulation

The above subdivision-based formulation requires up to $\Omega(|E|^4)$ variables, as every edge may have to be subdivided into $\Omega(|E|)$ segments. The currently best-performing ILP model avoids this subdivision and instead considers *linear ordering problems* on each edge. Recall that the reason for the graph extension was to be able to model the order of the crossings, in order to obtain a tractable realizability problem. The *ordering-based* ILP formulation achieves this by explicitly computing an ordering of the edge crossings.

Consider an arbitrary, fixed *orientation* of the given graph G, i.e., for each undirected edge we decide on one of the two possible directions. As in the original (problematic) approach, we introduce $\Omega(|E|^2)$ many binary variables $x_{e,f}$, one for each edge pair e, f, which should be 1 if the two indexed edges cross. The objective function is simply the sum over all these variables. We then introduce binary variables $y_{e,f,g} \in \{0, 1\}$ for all edge triples e, f, g. This results in only $\Omega(|E|^3)$ additional variables. A variable $y_{e,f,g}$ should be 1 if and only if the edge e is crossed by both edges f and g, and the crossing with f occurs prior to the crossing with g, w.r.t. the fixed edge orientation. Conceptually, the variables $y_{e,\cdot,\cdot}$, when properly bound to their corresponding x variables, then form the variables of a linear-ordering problem with the additional property that some elements need not to be ordered at all. We can achieve this via

$$
\begin{aligned}
x_{e,f} &\geq y_{e,f,g} \\
x_{e,g} &\geq y_{e,f,g} \\
1 + y_{e,f,g} + y_{e,g,f} &\geq x_{e,f} + x_{e,g} \\
y_{e,f,g} + y_{e,g,f} &\leq 1 \\
y_{e,f,g} + y_{e,g,h} + y_{e,h,f} &\leq 2
\end{aligned}
$$

over the suitable edge indices. The first two constraints guarantee that the x variables (counting the crossing in the objective function) are set whenever a corresponding y variable is set. The third constraint ensures that whenever there are two crossings occurring on the same edge (here: on e), their relative order has to be specified. Then, we have to ensure in the fourth constraint that this order is unique. The last constraint, known as a *3-cycle constraint*, ensures that the order given by the y variables is in fact a linear, i.e., acyclic, order.

Using this setup it remains to introduce Kuratowski constraints much like the ones described for the subdivision-based formulation. Recall that D' in the subdivision-based formulation described a simple set of edge crossings. Similarly, we now consider a (technically more involved) *crossing shadow* $(\mathcal{X}, \mathcal{Y})$ instead. It can be thought of as a minimal description of a not-necessarily realizable crossing situation. I.e., $\mathcal{X}$ ($\mathcal{Y}$) lists x variables (y variables, respectively) that should be set. The "minimality" of this description is achieved by avoiding to list an x variable if a corresponding y variable in $\mathcal{Y}$ already induces that it has to be set. Similarly, we use the transitivity property of a linear ordering, and, e.g., do not include the variable $y_{e,f,h}$ if $y_{e,f,g}$ and $y_{e,g,h}$ are already in $\mathcal{Y}$. For a concise definition, we refer the reader to [CMB08]. Considering all possible crossing shadows $(\mathcal{X}, \mathcal{Y})$ and all thereby induced Kuratowski subdivisions H, we can require:

$$
x(\hat{H}^{[2]}) \geq 1 - \sum_{x' \in \mathcal{X}} (1 - x') - \sum_{y' \in \mathcal{Y}} (1 - y')
$$

2.4.3 Branch-and-Cut-and-Prize

For a practical implementation, we can omit variables in some cases. The graph G can be split up into its blocks first, which can be solved separately. The crossing number of G is equal to the sum of the crossing numbers of its blocks. Furthermore, it is easy to show that adjacent edges do not cross in an optimal drawing and no edge crosses itself, i.e., we can restrict ourselves to good drawings. Furthermore, we may apply more sophisticated preprocessing strategies like the *non-planar core reduction* [CG09], which further shrinks the graph based on its triconnectivity structure. Thereby, it may be necessary to introduce integer weights $w : E \to \mathbb{N}$ on the edges. A crossing between the edges e and f should then be counted as $w(e) \cdot w(f)$, which is easily achievable in both above ILP formulations by using these products as the coefficient for the respective x variables.

$L :=$ {initial problem}   {L denotes the list of unsolved problems}
repeat
   Choose a subproblem $\Pi \in L$ and set $L := L \setminus \{\Pi\}$
   repeat
      Let $\hat{x}$ be an optimal solution for the linear relaxation of Π
      if $\hat{x}$ is not feasible for Π **then**
         Separate violated inequalities and add them to the LP
      end if
   until no more violated inequalities can be found
   if no feasible solution for Π could be found **then**
      Split Π into subproblems and add them to L
   end if
until $L = \emptyset$
Print the best found feasible solution

Figure 2.10   An overview of the branch-and-cut approach.

Because of the exponential number of constraints, we cannot create them in advance and solve the ILP in a single step. A well-suited method for this class of ILPs is the *branch-and-cut approach*. The basic structure of a branch-and-cut based algorithm is outlined in Figure 2.10. The referred linear relaxation of Π can be easily obtained by dropping the integrality constraints, i.e., variables are allowed to be fractional.

In the case of zero-one integer linear programs, the set of unsolved subproblems L is organized as a binary tree, called the *branch-and-bound tree*. Each subproblem corresponds to a node in the tree and the list of unsolved problems L is represented by its leaves. If we need to split a problem Π into subproblems, we choose a fractional *branching variable* and create two new subproblems by setting the branching variable to zero and one, respectively.

Whenever we split a problem into two subproblems by setting the branching variable to zero and one, respectively, we can compute a local lower bound. This is the best value for the objective function that can be obtained subject to the assignments of values for the branching variables up to the root node. If this value is greater than the global upper bound, we can discard all descendants of the current subproblem since they can never improve the current feasible solution.

A severe problem of this approach is the *separation problem*: "Given a class of valid inequalities and a vector $z \in \mathbf{R}^n$, either prove that z satisfies all inequalities in the class, or find an inequality which is violated by z." Although we can easily separate violated inequalities for integral solution vectors according to the proof of Theorem 2.8, the problem becomes severe within the branch-and-cut framework since we have to deal with fractional values.

This problem can be solved heuristically by rounding variables to either zero or one, but one cannot guarantee that there is no violated inequality if the graph realizing the crossing D or $(\mathcal{X}, \mathcal{Y})$ is planar. In this case, we have to select a branching variable and split the current problem into two subproblems by setting the branching variable to 0 and 1, respectively.

The major bottleneck when following this approach then remains the large number of variables, rendering both approaches useless as such. Yet the concept of *column generation* turns out to allow drastic speed-ups of the algorithms. Conceptually, and somewhat similar to the separation approach, we start with a small subset of the variables. After solving the LP relaxation, we not only have to solve the separation problem ("is our solution too good because some constraints are missing?"), but also the pricing problem: "Is our solution too bad because some variables are missing?" Observe the difference in the obtained bounds when constraints or variables are missing, which, in general, leads to weaker bounding strategies than applicable to pure branch-and-cut approaches.

Following the traditional approach based on the Dantzig-Wolfe decomposition [DW60], we can solve the pricing problem in a purely algebraic way by computing the reduced costs of the variables not already in the model and adding them based on their sign. It turns out that this approach, denoted by *algebraic pricing*, already speeds up the computation, but we can do much better.

In a *combinatorial column generation scheme*, we refrain from computing reduced costs, but try to incorporate our problem-specific knowledge to obtain more efficient strategies. In particular, our special-purpose generation schemes allow us to overcome the aforementioned bounding problem, and retain the fact that the LP-relaxation always gives a lower bound to the problem, even when constraints and variables are missing.

We start with the observation that in most practical applications, most of the edges will not be crossed at all or are only involved in one crossing. On these edges, we do not have any ambiguity with the order of crossings, and the realizability problem is easy. The central idea for the combinatorial column generation scheme for both formulations can be roughly described as such: we start without any special constructions to avoid crossing-order ambiguities, i.e., we do not subdivide the edges for the subdivision-based formulation and do not introduce any y variables for the ordering-based formulation. Recall that the crossing order only becomes crucial when considering Kuratowski constraints; these are only generated via separation on a rounded solution. So, we use the branch-and-cut framework as outlined above. Whenever the separation routine considers a rounded solution where two or more edges cross the same edge, and their order is hence ambiguous, we introduce the necessary variables and constraints from the original model which are necessary to decide this order. Then, the LP relaxation is recomputed. We refrain from discussing the relatively technical details of which variables or subdivisions are necessary, and refer to [CGM09] and [CMB08] for the two formulations instead. The interesting part is that adding variables in such a way will never decrease the objective function.

Overall, the currently most efficient approach from the practical point of view is the ordering-based formulation, together with its combinatorial column generation scheme, the aforementioned preprocessing strategies and upper bounds obtained via the strong planarization heuristic that we will discuss in the following section. Furthermore, the separation routine is improved by not looking for single Kuratowski subdivisions in the rounded solution, but by applying an algorithm that obtains several such subdivisions in one pass requiring only linear time in input and output size [CMS08, CMS07]. This allows to solve sparse real-world graphs with up to 100 vertices to provable optimality within reasonable time bounds on average hardware; cf. Figure 2.4. Yet the subdivision-based formulation allows extensions to other crossing number concepts where a pair of edges crosses multiple times, e.g., the simultaneous crossing number [CJS08].

2.5 The Planarization Method

2.5.1 Overview

The most prominent and practically successful method for solving the crossing minimization problem heuristically is the *planarization approach*. This approach was introduced by Batini, Talamo, and Tamassia in [BTT84] and can be viewed as a general framework that addresses the problem with a two-step strategy. Each step aims at solving a particular optimization problem for which various solution methods are possible. Let $G = (V, E)$ be the graph for which we want to find a crossing minimal drawing. Then, the two steps to be executed are:

1. Compute a planar subgraph $P = (V, E_p)$ of G. The objective is to have as many edges in P as possible.

2. Reinsert the edges not contained in the planar subgraph, i.e., insert the edges in $E \setminus E_p$ into P. During this edge insertion process, edge crossings that occur when inserting an edge are replaced by dummy vertices with degree four, so that the graph remains planar. The objective is to keep the number of dummy vertices (and thus the number of crossings in the final drawing) as small as possible.

Figure 2.11 shows an example with the different stages of the approach. In this case, the planar subgraph contains all but one edge (edge (2,5) is missing) and the final drawing of G has only a single crossing.

The outcome of the planarization procedure is a planar graph $G_p = (V \cup V_d, E_p)$ such that every planar drawing of G_p implies a drawing of G with at most $|V_d|$ crossings. Hence, we also say that G_p is a *planarized representation* of G with (at most) $|V_d|$ crossings. We can obtain such a drawing of G as follows. First, we compute an embedding of G_p. Then, we have to distinguish two situations for each dummy vertex $v \in V_d$ (see Figure 2.12). If the corresponding edges of G, say, e and e', cross each other, then v in fact represents a crossing between e and e' in the drawing of G. Otherwise, e and e' just touch and we can save a crossing.

The two optimization problems we have to solve in the planarization approach are the *maximum planar subgraph problem* (MPSP) and the *edge insertion problem* (EIP). Both problems are NP-hard and are usually solved in practice by applying heuristic approaches. One reason for that is that even an optimal solution of MPSP in the first step and of EIP in the second step does not yield a crossing minimal solution in general. We show an example

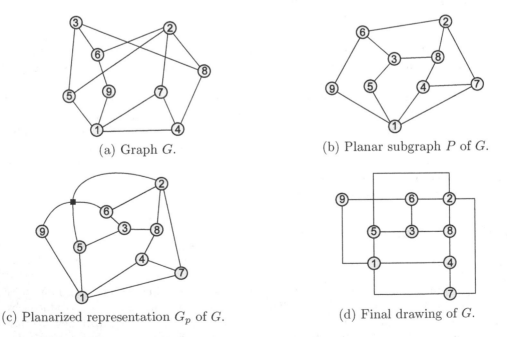

(a) Graph G.

(b) Planar subgraph P of G.

(c) Planarized representation G_p of G.

(d) Final drawing of G.

Figure 2.11 A sample application of the planarization method.

Figure 2.12 (a) The edges e and e' cross at dummy vertex v; (b) e and e' just touch at v.

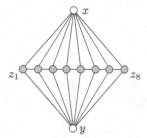

Figure 2.13 A wall with width 8.

(see [GMW05]) where the maximum planar subgraph contains all but one edge, but even an optimal solution of the edge insertion problem results in arbitrary many crossings, whereas a crossing minimal drawing has only two crossings.

We define a wall graph as follows. A *wall* with width k consists of the vertices $x, y, z_1, \ldots, z_k$, the edges (z_i, z_{i+1}) for $1 \leq i < k$, and the edges (x, z_i) and (y, z_i) for $1 \leq i \leq k$; see Figure 2.13 for an example of a wall with width 8. The vertices x and y are called the *poles* of the wall. A wall with width greater than 2 is a triconnected planar graph.

For an even number $m \geq 2$, the graph G_m is constructed in the following way; compare Figure 2.14(a). We start with a ring of walls $W_1, \ldots, W_6$ with width $m+1$, where the poles of adjacent walls in the ring are identified. We denote the pole vertices with $w_1, \ldots, w_6$ such that the poles of W_1 are w_1 and w_2, and so forth. For each wall W_j, the other two vertices on the boundary are denoted with u_j^i and u_j^e; see Figure 2.14(a). Moreover, the edges $e_1 = (u_1^e, w_3)$, $e_2 = (u_6^e, w_5)$, $e_3 = (u_2^i, u_3^i)$, and $e_4 = (u_5^i, u_4^i)$ are added, $m/2$ vertices are inserted by splitting edge (u_3^i, w_4) and $m/2$ vertices are inserted by splitting edge (w_4, u_4^i), and every created split vertex is connected with vertex w_1 by an edge h_j, $1 \leq j \leq m$. We want to insert edge (v_1, v_2) with $v_1 := u_1^i$ and $v_2 := u_6^i$, and we call the graph after addition of this edge G_m'.

By construction, G_m is triconnected and planar. In particular, G_m has only two embeddings which are mirror images of each other. It is easy to see that an optimal insertion of edge (v_1, v_2) crosses m edges, namely, $h_1, \ldots, h_m$, since passing through a wall would require at least $m+1$ crossings. On the other hand, there is a drawing of G_m' with only 2 crossings as shown in Figure 2.14(b). Here, only the two crossings e_1 with e_3 and e_2 with e_4 occur, independent of the choice of m.

In summary, this construction shows that the planarization approach may yield arbitrarily bad solutions even if both steps are solved optimally. On the other hand, practical experience has shown that it leads to excellent results in many applications even if each step is only solved heuristically. In the sequel, we address the two optimization problems—finding a planar subgraph and reinserting a set of edges—in detail, and discuss various solution methods.

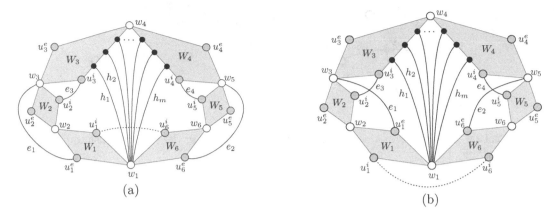

Figure 2.14   (a) The graph G_m; each shaded region represents a wall with width $m + 1$. The dashed edge (u_1^i, u_6^i) is the edge to be inserted. (b) A drawing of the graph G_m' with only two crossings.

2.5.2   Planar Subgraphs

In many practical applications, we expect that a graph $G = (V, E)$ can be made planar by removing only a few edges. Therefore, it is reasonable to use a planar subgraph with as many edges as possible as a starting point for crossing minimization. A *maximum planar subgraph* of G is a planar subgraph with the maximum number of edges among all planar subgraphs of G. If, in addition, a weight w_e is given for each edge of G, a *maximum weight planar subgraph* is a planar subgraph $P = (V, E')$ of G such that the sum of all edge weights $\sum_{e \in E'} w_e$ of P is maximum. Hence, a maximum planar subgraph is a special case of the weighted version with $w_e = 1$ for every edge $e \in G$. In both the weighted and the unweighted case, the problem of finding such a subgraph is NP-complete as shown in [LG77, GJ79].

Jünger and Mutzel [JM96] presented a branch-and-cut algorithm for finding a maximum weight planar subgraph. An overview of the branch-and-cut approach can be found in Section 2.4.

Let $\mathcal{P}_G$ be the set of all planar edge-induced subgraphs of G. For each planar subgraph $P = (V, F) \in \mathcal{P}_G$, we define its incidence vector $\chi^P \in \mathbf{R}^E$ by setting $\chi_e^P = 1$ if $e \in F$ and $\chi_e^P = 0$ if $e \notin F$. This yields a 1-1-correspondence of the planar subgraphs with certain $\{0, 1\}$-vectors in $\mathbf{R}^E$. The planar subgraph polytope $\mathcal{PLS}(G)$ of G is defined as the convex hull over all incidence vectors of planar subgraphs of G:

$$\mathcal{PLS}(G) := \mathrm{conv}\{\chi^P \in \mathbf{R}^E \mid P \in \mathcal{P}_G\} \, .$$

Let $w \in \mathbf{R}^E$ be a vector assigning a weight to each edge. The problem of finding a maximum weight planar subgraph can thus be written as the linear program

$$\max\{w^{\mathrm{T}} x \mid x \in \mathcal{PLS}(G)\},$$

since the vertices of the polytope $\mathcal{PLS}(G)$ are exactly the incidence vectors of the planar subgraphs of G. In order to apply linear programming techniques, $\mathcal{PLS}(G)$ has to be represented as the solution of an inequality system. Because of the NP-hardness of the problem, we cannot expect to find a full description of $\mathcal{PLS}(G)$. Jünger and Mutzel show several facet-defining inequalities of the polytope, including Kuratowski inequalities, which are based on the fact that a planar graph contains no subdivision of K_5 and $K_{3,3}$, and Euler inequalities, which are based on the maximal number of edges in a planar graph given by Euler's formula. Further facet-defining inequalities can be found in [JM96].

Require: graph $G = (V, E)$
Ensure: maximal planar subgraph P of G

$P :=$ a spanning tree of G
$F := E \setminus E(P)$
for all $e \in F$ **do**
  if $P \cup e$ is planar **then**
    $P := P \cup e$
  end if
end for

Figure 2.15   A simple algorithm for computing a maximal planar subgraph.

Using these inequalities, a branch-and-cut algorithm can be derived that adopts the planarity testing algorithm by Hopcroft and Tarjan [HT74] for cutting plane generation and as lower-bound heuristic. Computational results show that the algorithm is able to provide a provably optimal solution quite fast if the number of edges to be deleted is small. However, the method is quite complicated to understand and to implement. Moreover, if the number of deleted edges exceeds 10, the algorithm usually needs far too long to be acceptable for practical computation.

Since finding a maximum planar subgraph is hard, the problem of finding just a maximal planar subgraph has received much attention. A *maximal planar subgraph* of $G = (V, E)$ is a planar subgraph $P = (V, E \setminus F)$ of G such that adding any edge of F to P destroys the planarity, i.e., $P \cup e$ is not planar for every $e \in F$.

A widely used standard heuristic for finding a maximal planar subgraph is to start with a spanning tree of G, and to iteratively try to add the remaining edges one by one; see Figure 2.15. In every step, a planarity testing algorithm is called for the obtained graph. If the addition of an edge would lead to a nonplanar graph, then the edge is disregarded; otherwise, the edge is added permanently to the planar graph obtained so far. After $|F|$ planarity tests, we obtain a maximal planar subgraph P of G. Planarity can be tested in linear time; see Chapter 1, or [HT74, BL76, BM04]. Hence, the running time of the procedure is $\mathcal{O}((1 + |F|)(|V| + |E|))$.

This incremental approach can be made more efficient by using incremental planarity testing algorithms. Di Battista and Tamassia [DT96] presented an algorithm that tests in $\mathcal{O}(\log |V|)$ time if an edge can be added while preserving planarity, and that performs the required updates of the data structure when adding an edge in $\mathcal{O}(\log |V|)$ amortized time. The algorithm uses the data structures BC-tree and SPQR-tree equipped with efficient, dynamic update operations. A *BC-tree* represents the *block-cutvertex tree* of a connected graph G which consists of the interrelation of the blocks (B-nodes) and cutvertices (C-nodes) of G. It has an edge (c, B) if c is a cutvertex of G contained in block B. *SPQR-trees* have been introduced by Di Battista and Tamassia in [DT89]. They represent the decomposition of a biconnected graph into its triconnected components, which essentially consists of serial (expressed by S-nodes), parallel (P-nodes), and simple, triconnected structures (R-nodes). Additionally, Q-nodes represent the original edges of G. The specific structures of tree nodes are given by skeleton graphs that are associated with each node. Using these data structures, a maximal planar subgraph can be found in $\mathcal{O}(|E| \log |V|)$ time. SPQR-trees are also useful in a static environment for the representation of all planar embeddings of a graph. The static data structure can be built in linear time [HT73, GM01] using an algorithm for dividing a graph into its triconnected components.

The running time for incremental planarity testing has been improved by La Poutré [La 94] to $\mathcal{O}(\alpha(|E|, |V|))$ amortized time per query and update operation. This yields an almost linear time algorithm for the maximal planar subgraph problem that runs in $\mathcal{O}(|V| + |E| \cdot \alpha(|E|, |V|))$ time. Here, $\alpha(x, y)$ denotes the inverse Ackermann function, which means that $\alpha(x, y)$ is a function that grows extremely slowly. A linear time algorithm for finding a maximal planar subgraph is given by Djidjev [Dji95]. This algorithm uses BC- and SPQR-trees and applies a fast data structure for online planarity testing in triconnected graphs.

Jayakumar et al. [JTS89, JLM98] proposed a method for computing a planar subgraph that is based on PQ-trees. The PQ-tree data structure has been developed by Booth and Lueker [BL76] for solving the problem of finding permissible permutations of a set U. The permissible permutations are those in which certain subsets $S \subseteq U$ occur as consecutive subsequences. Drawbacks of this planar subgraph algorithm are that it cannot guarantee to find a maximal planar subgraph, and that the theoretical worst-case running time is $\mathcal{O}(|V|^2)$. However, in practice it is usually very fast and the quality of the results can be improved by introducing random events and calling the algorithm several times. The algorithm starts by computing an st-numbering of G, which determines the order in which the vertices are processed. A simple but useful randomization is to choose a random edge (s, t) for each run.

The trivial approach for finding a planar subgraph consists of computing a spanning tree. If G is a graph with n vertices and c components, then this approach has an approximation factor of $\frac{n-c}{3n-6c} > \frac{1}{3}$ for MPSP, since a spanning tree of G contains $n - c$ edges, and a planar graph with c components has at most $3n - 6c$ edges by Euler's formula. Surprisingly, we cannot guarantee a better approximation factor than that of the spanning tree approach if we demand that the computed subgraph must be maximal planar; see [DFF85]. Călinescu et al. [CFFK98] present an algorithm with approximation factor $4/9$ that runs in $\mathcal{O}(m^{3/2} n \log^6 n)$ time, where m is the number of edges of G. For the maximum weight planar subgraph problem, the simple approach is to compute a maximum weight spanning tree, which gives an approximation factor of $1/3$, and the best algorithm [CFKZ03] achieves an approximation factor of $1/3 + 1/72$.

2.5.3   Edge Insertion

The planar subgraph P computed in the first step of the planarization approach is a good starting point for finding a planarized representation G_p of G with few crossings. In practice, we expect that only a small number of edges has to be inserted into P in order to obtain G_p. However, the edge insertion step fixes the crossings in the final drawing, and the choice of the edge insertion technique may have a significant impact on the quality of the final solution. Ziegler and Mutzel [MZ99, Zie00] have shown that even a restricted variant of the edge insertion problem is NP-hard: The *constrained crossing minimization problem* (*CCMP*) asks for the minimum number of crossings required for inserting a set of edges into a fixed embedding. They also present a branch-and-cut algorithm to solve CCMP. However, experiments show that it can only solve instances to provable optimality if there are less than 10 edges to be inserted.

Gutwenger [GM04, Gut10] has conducted an extensive study on crossing minimization heuristics, including different methods for edge insertion. Figure 2.16 shows the general framework for edge insertion used in this study. It contains three essential parts leaving room for enhancement:

Single edge insertion: The edges are inserted into the planarized representation individually one after the other. The simple approach for inserting a single edge e

---

Require: planar subgraph $P = (V, E_P)$ of $G = (V, E)$
Ensure: planarized representation G_p^* of G

    Let $E \setminus E_P = \{e_1, \ldots, e_k\}$

    $best := \infty$
    for $i := 1$ **to** $nPermutations$ **do**
      Let σ be a randomly chosen permutation of $\{1, \ldots, k\}$
      $G_p := P$
      for $j := 1$ **to** k **do**
        Insert edge $e_{\sigma(j)}$ into G_p
      end for

      Determine a set $R \subseteq E$ of edges for which postprocessing shall be applied

      repeat
        for all $e \in R$ **do**
          Remove edge e from G_p
          Insert edge e into G_p
        end for
      until number of crossings in G_p has not decreased

      $current :=$ number of crossings in G_p
      if $current < best$ **then**
        $G_p^* := G_p$; $best := current$
      end if
    end for

Figure 2.16 Edge insertion with postprocessing and permutation.

is to fix an embedding Π of G_p and to insert e into Π. However, the choice of Π may have a considerable influence on the number of edges that e has to cross. A more sophisticated algorithm introduced by Gutwenger et al. [GMW05] is able to insert e with the minimum number of crossings among all embeddings of G_p.

Postprocessing: After all edges have been inserted, a simple postprocessing technique tries to improve the current solution. It determines a set of edges R which have one or more crossings and repeatedly tries to find a better insertion path for each of them by removing an edge from G_p and inserting it again. Variants for the choice of R include all edges, only the edges $e_1, \ldots, e_k$, or some portion of the edges with the most crossings (see [Gut10] for more details).

An alternative approach combines the edge insertion with the postprocessing. Instead of performing the remove-reinsert strategy after all edges have been inserted, we can perform this strategy after each edge insertion. The idea behind this variation is to keep the number of crossings low as early as possible. We call this strategy incremental postprocessing.

Permutation: The order in which the edges $e_1, \ldots, e_k$ are processed also affects the final number of crossings. Calling the complete edge insertion process several times with different, randomly chosen permutations of the edge list $e_1, \ldots, e_k$ may significantly improve the solution. The parameter $nPermutations$ in the algorithm determines the number of permutation rounds.

Apart from the choice of some parameters like the number of permutation rounds or the selection of the edges for postprocessing, the challenging part of the algorithm is the insertion of a single edge. We consider the two variants—insertion with *fixed* and with *variable* embedding—in more detail.

Fixed Embedding. Suppose, we want to insert edge $e = (v, w)$ into the planar graph G_p. Let Π be a fixed embedding of G_p. We construct the *extended dual graph* G^* of Π with respect to e as follows. The vertices of G^* are the faces of Π plus two new vertices v^* and w^* representing v and w. For each edge e' in G_p, we have an edge in G^* connecting the two faces separated by e' (if e' is a bridge, we have a self-loop in G_p). Additionally, we have an edge (v^*, f_v) for each face f_v adjacent to v, and (w^*, f_w) for each face f_w adjacent to w.

We observe that inserting e into Π corresponds to finding an (undirected) path from v^* to w^* in G^*. If such a path has length ℓ, then we can insert e with $\ell - 2$ crossings, since the first and the last edge on this path do not produce a crossing. Therefore, in order to insert e into Π with the minimum number of crossings, we have to find a shortest path from v^* to w^* in G^*. This is possible in linear time using a simple breadth-first search traversal starting at v^*. Figure 2.17 shows a nontrivial example. Here, we want to connect the vertices 1 and 2. The dashed vertices and edges belong to the extended dual graph. The optimal solution highlighted in bold crosses four edges.

Though we can easily find a crossing minimal solution if the embedding of G_p is fixed, the drawback of this method is that fixing an unfavorable embedding may result in an arbitrarily bad solution. Figure 2.18(a) gives an example of such a family of graphs G_k with embeddings Γ_k. The black fat lines in this figure denote bundles of $k + 1$ parallel

Figure 2.17 Edge insertion with fixed embedding by finding a shortest path in the extended dual graph.

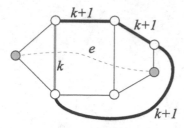

(a) Fixed embedding Γ_k.

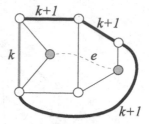

(b) Optimal Embedding.

Figure 2.18 A family of graphs G_k and embeddings Γ_k for which the insertion of an edge e requires k crossings more than the optimal solution.

edges, and the gray fat line a bundle of k parallel edges. Hence, inserting edge e into the given embedding requires at least $k+1$ crossings. On the other hand, it is possible to insert e with only one crossing by changing the embedding; see Figure 2.18(b). It is easy to see that this example can also be adapted to the case of simple graphs by splitting all the edges in each bundle.

Variable Embedding. Surprisingly, there exists also a linear time algorithm for finding an optimal embedding of G_p which allows to insert e with the minimum number of crossings. The algorithm by Gutwenger et al. [GMW05] uses the data structures BC-tree and SPQR-tree for the representation of all planar embeddings of a connected graph. These decomposition trees allow to enumerate all possible embeddings of a connected graph. Basically, we can

- put any subgraph which is only attached to the rest of the graph at a cutvertex into any face containing this cutvertex;
- arbitrarily permute parallel structures joined at a separation pair; and
- mirror any subgraph that is only attached to the rest of the graph at a separation pair.

Let G be a graph and $\mathcal{T}$ its SPQR-tree. We denote the skeleton of a node μ in $\mathcal{T}$ with $skeleton(\mu)$. Each edge e in a skeleton represents a subgraph of G called the *expansion graph* of e. Replacing each edge in a skeleton by its expansion graph yields G again.

In order to find the optimal insertion path for (v, w) in G, it is essentially sufficient to consider only the R-nodes in the SPQR-trees of the blocks of G. Assume first that G is already biconnected. Let $\mathcal{T}$ be its SPQR-tree and let $\mu_1, \ldots, \mu_k$ be the shortest path in $\mathcal{T}$ between a node μ_1 with $v \in skeleton(\mu_1)$ and a node μ_k with $w \in skeleton(\mu_k)$. For each R-node on this path, we expand its skeleton S in the following way. First, we make sure that we have a representative for both v and w. If one of these vertices, say, v, is not yet contained in S, then there is an edge whose expansion graph contains v and we split this edge introducing a representative for v. Then, we replace every edge that was not split with its expansion graph and compute an arbitrary embedding Π of the resulting graph. For this fixed embedding, we determine the ordered list of edges we have to cross when inserting an edge from the representative of v to the representative of w as described above for the fixed embedding scenario. If we do this for every R-node in $\mu_1, \ldots, \mu_k$, and if we join the resulting edge lists in the order they appear on the path from μ_1 to μ_k, then we obtain an optimal edge insertion path for inserting the edge (v, w).

If G is not biconnected, we determine the shortest path $B_1, c_1, \ldots, c_{k-1}, B_k$ in the BC-tree of G such that $v \in B_1$ and $w \in B_k$. Then, we find an optimal edge insertion path

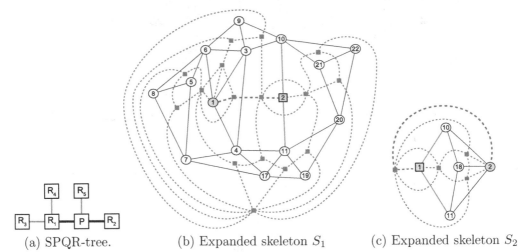

(a) SPQR-tree.   (b) Expanded skeleton S_1 (c) Expanded skeleton S_2

Figure 2.19   Edge insertion with variable embedding.

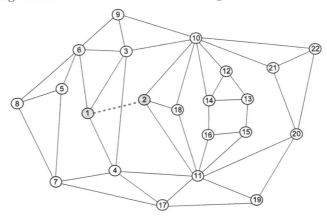

Figure 2.20   An embedding of the example graph that allows to embed edge $(1,2)$ with the minimum number of crossings.

for (c_{i-1}, c_i) in B_i for every $i = 1, \ldots, k$, where $c_0 := v$ and $c_k := w$. We can simply join the resulting insertion paths p_i to obtain an optimal insertion path $p_1, \ldots, p_k$ for inserting (v, w).

The corresponding embedding that allows to insert (v, w) with the minimum number of crossings is easy to find. We in fact insert the edge (v, w) into the planarized representation G_p by creating dummy vertices for edge crossings. The construction above guarantees that the resulting graph is planar. Then, we compute an embedding of this planar graph and remove the inserted edge(s) again.

Figure 2.19 continues our example for the insertion strategy with variable embedding. In this case, the graph is biconnected and the corresponding SPQR-tree has the structure depicted in Figure 2.19(a). The relevant path in the SPQR-tree is R_1, P, R_2, which contains two R-nodes. The expanded skeleton graphs S_1 for R_1 and S_2 for R_2 are shown in Figure 2.19(b) and (c). We need only a single crossing in S_1 and no crossing at all in S_2. Hence, an optimal solution will only cross a single edge, which is edge $(3,4)$ in our solution. Figure 2.20 shows an embedding that allows to insert $(1,2)$ with only one crossing.

2.5.4 Experimental Results

Recently, Gutwenger [Gut10] presented an extensive experimental study on the planarization approach for crossing minimization and analyzed the effect of pre- and postprocessing strategies, edge insertion, and permutations, including the nonplanar core reduction (NPC) as preprocessing, edge insertion with fixed (FIX) and variable (VAR) embedding, and various postprocessing strategies (all edges (ALL), only the inserted edges (INS), $x\%$ of the edges with the most crossings (MOSTx), and incremental postprocessing (INC)). The planar subgraph was computed using the PQ-tree-based algorithm with 100 random iterations. Two benchmark sets of graphs have been used in this study:

- The *Rome graphs* [DGL+97] are a collection of more than 11.000 graphs ranging from 10 to 100 vertices, which have been generated from a core set of 112 graphs used in real-life software engineering and database applications.

- The *Artificial graphs*[1] are a collection of nonplanar graphs with known crossing numbers. It contains 1946 graphs with up to 250 vertices and consists of cross products of cycles ($C_m \times C_n$), 5-vertex graphs with paths ($G_i \times P_n$), 5-vertex graphs with cycles ($G_i \times C_n$), and generalized Petersen graphs ($P(m, 2)$ and $P(m, 3)$).

Table 2.1 shows a ranking of some selected strategies, sorted by average number of crossings for graphs with 100 vertices.

rank	crossings	time [s]	EI	PRE	POST	PERM
1	26.71	9.387	VAR	NPC	INC	20
2	27.14	4.681	VAR	NPC	INC	10
3	28.49	1.857	VAR	NPC	ALL	20
4	28.69	0.727	FIX		INC	20
5	30.43	0.490	VAR	NPC	INC	1
6	30.52	0.221	FIX		ALL	20
7	32.66	0.105	VAR	NPC	ALL	1
8	33.33	0.098	FIX	NPC	ALL	1
9	33.96	0.067	FIX		INC	1
10	35.09	0.041	FIX		ALL	1
11	35.79	0.040	FIX		MOST25	1
12	38.38	0.037	FIX		MOST10	1
13	41.61	0.036	FIX		INS	1
14	45.47	0.034	FIX		NONE	1

Table 2.1 The ranking list of crossing minimization heuristics; the table shows average number of crossings and running times for graphs with 100 vertices.

Figure 2.21 compares the two edge insertion variants and some postprocessing strategies. It shows that VAR clearly dominates FIX and that postprocessing helps a lot. Although the INC strategy is rather time consuming, it justifies this by achieving excellent improvements. Using permutations also gives significant improvements, but not as much as postprocess-

---

[1]available at http://ls11-www.cs.uni-dortmund.de/people/gutweng/artificial-graphs.zip

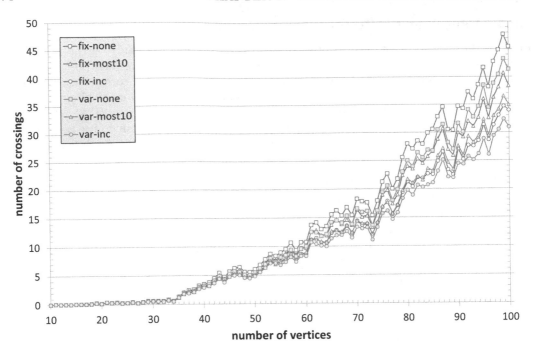

Figure 2.21   Average number of crossings for Rome graphs with various postprocessing strategies.

ing. Figure 2.22 demonstrates the effect of up to 500 permutations on graphs with 100 vertices. Performing only a few permutations achieves already good improvements; further permutations can still reduce the number of crossings, but the effect becomes smaller and smaller. The best result obtained for graphs with 100 vertices was 25.51 crossings with 500 permutations (NPC-VAR-INC).

We can judge the quality of the results better if we can compare them with the actual crossing numbers. Figure 2.23 shows the results for the artificial graphs, grouped by graph types. We observe that using edge insertion with variable embedding and incremental postprocessing already comes very close to the exact crossing numbers (OPT), whereas using fixed embedding without postprocessing achieves very bad results.

2.5.5   Beyond Edge Insertion

The central ingredient of the above discussed heuristic clearly is the efficient optimal edge insertion procedure that considers all possible embeddings. Starting from there, there are multiple extensions and generalizations.

Instead of considering only edges, we may also consider a vertex of the graph, together with all its incident edges. This problem is known as the *vertex-* or *star-insertion* problem. Reusing the ideas of the fixed-embedding edge insertion, we can easily find a BFS-based algorithm to insert a vertex into a fixed embedding of a planar graph. Yet, when considering the variable embedding setting, we cannot straightforwardly reuse many of the edge-insertion algorithm's methods, since vertex insertion does not offer the same degree of problem locality within each separate SPQR-skeleton. Chimani et al. [CGMW09] showed how to combine the SPQR-tree-based approach with a sophisticated dynamic programming scheme, to solve the vertex insertion problem in $O(\theta^2 \cdot |V|^3 \cdot |W|^2)$ time, where θ gives the

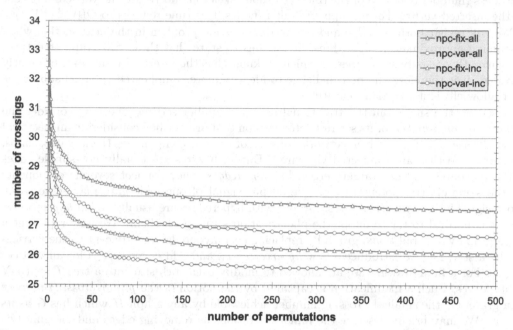

Figure 2.22 The effect of up to 500 permutations for Rome graphs with 100 vertices.

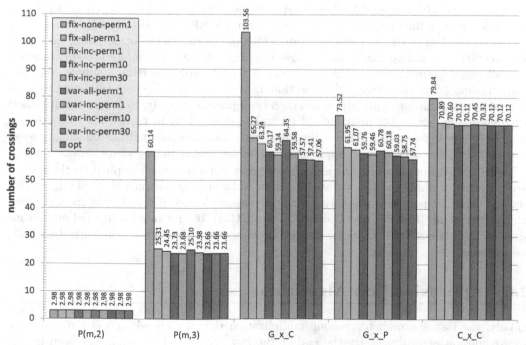

Figure 2.23 Average number of crossings for graphs with known crossing numbers.

thickness (number of edges) of the thickest P-node skeleton and W are the vertices adjacent to the inserted vertex. For a graph without P-nodes, this time reduces to $O(|V|^2 \cdot |W|^2)$.

We will revisit both the edge and the vertex insertion problem in the next section when discussing approximation algorithms. It remains to state that these two graph structures are currently the only structures for which we know that the insertion problem is efficiently solvable. It is therefore an open challenge to identify more complicated (planar) subgraphs that allow efficient insertion algorithms.

Based on the success in the traditional crossing number setting, one may consider the *minor crossing number*, or its set-restricted version that arises when considering an electrical network; see [CG07]. Such a network consists of several components (logic gates, chips, resistors,...) that are connected via wires. But such wires are usually not simple edges with one source and one target vertex, but *hyperedges*: they connect several components on the same electric potential, e.g., the output signal of one logic gate may serve as an input signal for several other gates. Graphs with hyperedges are usually called *hypergraphs*, and it is natural to try to adopt the planarization strategy to them. We may represent a hypergraph via a traditional graph by replacing each hyperedge ψ (adjacent to the vertices N) by a star, i.e., we introduce a new *hypervertex* v_ψ and add edges (v, v_ψ) for all vertices $v \in N$. For a final drawing, we are allowed to modify each such star into a tree T_ψ with N as its leaves. Such a modification is captured by the notion of the *minor crossing number* of a graph G: the smallest crossing number achievable by any graph H which has G as its minor. We may briefly describe the minor operations as removing edges and merging two adjacent vertices. It is easy to see that the expansion from a star to a tree can be obtained exactly by the inverse of the last operation. Therefore, the hypergraph crossing number is equivalent to the so-called W-*restricted minor crossing number*, where W is the set of hypervertices. By W-*restricted* we describe the constraint that the inverse minor operations may only be applied to the vertices W.

Chimani and Gutwenger [CG07] showed that inserting hyperedges (or, equivalently, inserting a vertex in the minor crossing number setting) is NP-hard, already when considering a fixed embedding of the (hyper)graph into which to insert. On the other hand, they show how to efficiently and optimally insert edges into a graph w.r.t. the minor crossing number, over all possible embeddings. This is equivalent to optimally inserting a simple edge into a hypergraph (note that during the insertion, other hyperedges become expanded to more general trees). This latter algorithm can then be applied iteratively, to (heuristically) insert a hyperedge by successive insertions of its star's edges. This in turn leads to a crossing number heuristic for electrical networks which generates drawings with astonishingly fewer crossings than the other known approaches, which are based on Sugiyama's framework.

The planarization strategy has also shown great potential when applied to the related issue of *upward drawings*, i.e., we want to draw a directed graph such that all edges point upward. Traditionally, this was solved via Sugiyama-style algorithms, but in the last years, Eiglsperger et al. [EKE03] and Chimani et al. [CGMW08] introduced algorithms reusing ideas of the planarization approach to find drawings of real-world graphs with drastically less crossings.

2.6   Approximation Algorithms

Finally, the last approach to crossing minimization that we will discuss is the search for approximation algorithms. By the end of the last century, this search has been mostly fruitless despite many attempts. It is still the case that on the one hand, no approximation algorithm for crossing minimization of general graphs with any type of guarantee could be found; on the other hand, the theoretical complexity of approximation is unknown. The only relevant case in which provably near-optimal solutions can be generated is the case of bounded degrees.

Recall the bisection width $\mathrm{bw}(G)$ of G as defined in Section 2.2.1. Bhatt and Leighton [BL84], later improved by Even et al. [EGS00], used a bisection approach for devising a polynomial-time algorithm with a quality guarantee for the number of crossings *plus the number of vertices*; the quality, however, depends on the quality of the incorporated approximation algorithm for the bisection width. As shown later [CY94], the latter problem can be approximated within a constant factor in polynomial time. Using this result, Bhatt and Leighton's algorithm yields a drawing with $O(\log^2 |V|(\mathrm{cr}(G) + |V|))$ edge crossings in polynomial time. In other words, the number of edge crossings *plus vertices* in the constructed drawing of G is at most a factor of $O(\log^2 |V|)$ away from the optimum. In fact, for bounded degree graphs satisfying $|E| \geq 4|V|$, this yields an $O(\log^2 |V|)$-approximation algorithm for the crossing minimization problem, as in this case the number of vertices is at most linear in the minimal number of crossings.

After laying semidormant for some time, the topic of approximation algorithms for crossing numbers received a lot of attention in recent years. The first decade of this millennium saw the first constant factor approximation algorithms in this area, although only for special graph classes, and only when assuming bounded degrees. Let Δ be the maximum degree in the following.

The first class of approximation algorithms are *insertion based*. They use the insertion algorithms presented in the previous section.

An *almost planar* or *near-planar* graph G is a graph that has an edge e such that $G' := G-e$, the graph obtained from removing e, is planar. In other words, G is a planar graph plus one additional edge. For such a graph, Hliněný and Salazar [HS06] showed that inserting e optimally into a planarly embedded G' [GMW05] (considering all possible embeddings, as described for the planarization heuristic) approximates the crossing number of G. The provably tight approximation factor of $\Delta/2$ was established by Cabello and Mohar [CM10] using a different proof strategy: the lower bound is obtained by analyzing its relation to the *facial distance*, i.e., a shortest insertion path with respect to the *minor-monotone crossing number* model, unknowingly using an algorithm first outlined in [CG07]. In this setting, the inserted edge not only crosses edges but may also cross through vertices, resembling a crossing solution in a graph that has G' as its minor.

Shortly after the aforementioned algorithm to optimally insert a vertex with its incident edges into a planar graph was presented in [CGMW09], Chimani et al. [CHM12] showed that this solution in fact approximates the crossing number of *apex graphs*. Similar to above, such a graph becomes planar when removing a specific vertex, together with its incident edges. The proof argues over different flip structures in the graph's SPQR-tree and thereby reuses some strategic elements of [HS06], as the stronger bounding techniques of [CM10] seem not applicable to the apex case. Consequently, the proven approximation guarantee of factor $|W| \cdot \Delta/2$ might not be tight (thereby, W are the vertices incident to the inserted vertex, and Δ is the maximum degree of the graph into which we insert). In the worst known example, the obtained crossing number is only $|W| \cdot \Delta/4$ times the optimal solution.

The second known class of approximation algorithms are *topology based*. Thereby, we assume that the given graph is embeddable—i.e., drawable without edge crossings—on some specific surface, more complex than the traditional plane. This class of graphs is then a superset of the class of planar graphs. By clever simplification strategies, usually based on cutting the surface with its embedded graph, the surface is simplified until a planar drawing is reached. Therein, the cut edges and vertices have to be reconnected cheaply to obtain a drawing of the original graph. Interestingly, the algorithms themselves, as well as estimating the number of the produced crossings, are relatively simply. The hard part is to show a matching lower bound in order to deduce the approximation factor.

In 2007, Gitler et al. showed in [GHLS08] that $cr(G)$ is approximable within a factor of $4.5\Delta^2$ when considering *projective graphs*, i.e., graphs that are embeddable in the *projective plane*. One may think of such a projective plane as a large circular area A, on which to draw the graph without any crossings. Any line leaving A re-enters A exactly at the opposite position. Consequently, any vertex drawn on the border of A is mirrored on the opposite side as well. The key idea of the approximation algorithm now is to take such a drawing, paste A on a regular plane, and connect the "jump points" cheaply outside of A. In order to prove that this strategy yields an approximation algorithm, one has to show a matching lower bound. This is established by proving that any nonplanar projective graph contains a diamond grid of certain size, which in turn induces a lower bound on the crossing number.

Hliněný and Salazar showed in [HS07] that $cr(G)$ is approximable within a factor of $12\Delta^2$ when considering (dense enough) *toroidal graphs*, i.e., graphs that are embeddable on the *torus*. Assume a graph is already drawn on a torus without any crossings (such an embedding can be found in linear time if it exists [Moh99]). We search for a shortest two-sided, non-separating loop around the torus (think, e.g., of a circular line "around" the thinner part of a torus) which only crosses through vertices of the graph, but not through edges. We then cut along this loop, effectively cutting the crossed vertices in two. The remaining surface can be thought of as a cylinder; when we cap its ends, it becomes topologically equivalent to a sphere and hence, for the purpose of drawings, to the plane. We denote this operation as *cut-and-cap*. For each pair of cut vertices, we remove the one with lower degree and route its incident edges to its twin, along the shortest path in the then-fixed embedding. In order to prove that this strategy yields an approximation algorithm, one has to show a matching lower bound for the number of crossings. This is established by proving that any toroidal graph of sufficient density contains a toroidal grid of certain minimal size as a minor. For toroidal grids of dimension $p \times q$ ($p \geq q \geq 3$), it is known that they require at least $(q-2)p/2$ crossings.

A torus is an (in fact, topologically, the unique) orientable surface of genus 1. Using the toroidal case as an inspiration, it is natural to try to generalize it to graphs embedded on any orientable surfaces of some fixed genus. Note that for every graph there is some g such that it is embeddable on a genus-g surface. The necessary basic tool of iteratively performing cut-and-cap operations on the surface's handles until we reach a sphere has already been investigated in [BPT06, DV06] in order to obtain upper bounds for the crossing number. Yet, there was no straightforward way to generalize the lower bound proof to higher genus. Only recently, Hliněný and Chimani [HC10] showed how to carefully choose the cycles to cut (both for the upper and the lower bound), such that the largest grid minor is retained within a factor depending only on the surface's genus and the graph's degree. They showed the following theorem:

Theorem 2.9   *Let G be a graph with maximum degree Δ and (densely enough) embeddable on an orientable surface of genus $g \geq 1$. There is an $O(n \log n)$ algorithm which generates a drawing of G in the plane with at most $3 \cdot 2^{3g+2} \cdot \Delta^2 \cdot cr(G)$ crossings. This is a constant factor approximation algorithm for bounded degree Δ and bounded genus g.*

These bounds are not known to be tight—in fact, they are likely not to be. Yet, some kind of density requirement (we refrain from defining the quite technical concise constraint here) will always be necessary in algorithms only performing surface cuts. Otherwise, the considered graph could even be a planar graph, awkwardly embedded on a higher genus surface.

Apart from considering restricted graph classes, one may also consider restricted crossing minimization problems, in order to obtain approximation results. For instance, for bipartite drawings with one layer fixed, Eades and Wormald [EW94] showed that there is a polynomial-time algorithm that produces drawings with at most three times as many edge crossings as necessary, for any graph G.

Acknowledgment

Markus Chimani was funded via a junior professorship by the Carl-Zeiss-Foundation.

References

[ACNS82]   M. Ajtai, V. Chvátal, M.M. Newborn, and E. Szemerédi. Crossing-free subgraphs. *Annals of Discrete Mathematics*, 12:9–12, 1982.

[BCE⁺08]   C. Buchheim, M. Chimani, D. Ebner, C. Gutwenger, M. Jünger, G. W. Klau, P. Mutzel, and R. Weiskircher. A branch-and-cut approach to the crossing number problem. *Discrete Optimization, Special Issue in Memory of George B. Dantzig*, 5(2):373–388, 2008.

[BD93]   D. Bienstock and N. Dean. Bounds for rectilinear crossing numbers. *J. Graph Theory*, 17(3):333–348, 1993.

[BEJ⁺05]   C. Buchheim, D. Ebner, M. Jünger, P. Mutzel, and R. Weiskircher. Exact crossing minimization. In P. Eades and P. Healy, editors, *Graph Drawing (Proc. GD '05)*, volume 3843 of *Lecture Notes in Computer Science*, pages 37–48. Springer-Verlag, 2005.

[BG04]   H. Bodlaender and A. Grigoriev. Algorithms for graphs embeddable with few crossings per edge. Research Memoranda 036, Maastricht : METEOR, Maastricht Research School of Economics of Technology and Organization, 2004. available at http://ideas.repec.org/p/dgr/umamet/2004036.html.

[Bie91]   D. Bienstock. Some provably hard crossing number problems. *Discrete Comput. Geom.*, 6(5):443–459, 1991.

[BL76]   K. Booth and G. Lueker. Testing for the consecutive ones property interval graphs and graph planarity using PQ-tree algorithms. *J. Comput. Syst. Sci.*, 13:335–379, 1976.

[BL84]   S. N. Bhatt and F. T. Leighton. A framework for solving VLSI graph layout problems. *J. Comput. Syst. Sci.*, 28:300–343, 1984.

[BM04]   J. M. Boyer and W. Myrvold. On the cutting edge: simplified $o(n)$ planarity by edge addition. *J. Graph Algorithms Appl.*, 8(3):241–273, 2004.

[BPT06]   K. Böröczky, J. Pach, and G. Tóth. Planar crossing numbers of graphs embeddable in another surface. *Internat. J. Found. Comput. Sci.*, 17:1005–1015, 2006.

[BTT84]   C. Batini, M. Talamo, and R. Tamassia. Computer aided layout of entity-relationship diagrams. *Journal of Systems and Software*, 4:163–173, 1984.

[CFFK98]   G. Călinescu, C. G. Fernandes, U. Finkler, and H. Karloff. A better approximation algorithm for finding planar subgraphs. *Journal of Algorithms*, 27(2):269–302, May 1998.

[CFKZ03]   G. Călinescu, C. G. Fernandes, H. Karloff, and A. Zelikovsky. A new approximation algorithm for finding heavy planar subgraphs. *Algorithmica*, 36(2):179–205, 2003.

[CG07]   M. Chimani and C. Gutwenger. Algorithms for the hypergraph and the minor crossing number problems. In *Proc. ISAAC '07*, volume 4835 of *LNCS*, pages 184–195. Springer, 2007.

[CG09]   M. Chimani and C. Gutwenger. Non-planar core reduction of graphs. *Discrete Mathematics*, 309(7):1838–1855, 2009.

[CGM09]   M. Chimani, C. Gutwenger, and P. Mutzel. Experiments on exact crossing minimization using column generation. *ACM Journal of Experimental Algorithmics*, 14(3):4.1–4.18, 2009.

[CGMW08] M. Chimani, C. Gutwenger, P. Mutzel, and H.-M. Wong. Layer-free upward crossing minimization. In *Proc. WEA '08*, volume 5038 of *LNCS*, pages 55–68. Springer, 2008.

[CGMW09] M. Chimani, C. Gutwenger, P. Mutzel, and C. Wolf. Inserting a vertex into a planar graph. In *Proc. SODA '09*, pages 375–383. ACM-SIAM, 2009.

[Chi08] M. Chimani. *Computing crossing numbers.* PhD thesis, TU Dortmund, 2008. http://hdl.handle.net/2003/25955.

[Chi11] M. Chimani. Facets in the crossing number polytope. *SIAM Journal on Discrete Mathematics*, 25(1):95–111, 2011.

[CHJM11] M. Chimani, P. Hungerländer, M. Jünger, and P. Mutzel. An SDP approach to multi-level crossing minimization. In *Proc. ALENEX'11*. SIAM, 2011.

[CHM12] M. Chimani, P. Hliněný, and P. Mutzel. Vertex insertion approximates the crossing number of apex graphs. *European Journal of Combinatorics*, 33(3):326–335, 2012.

[Chv83] V. Chvátal. *Linear Programming.* W. H. Freeman and Company, New York, 1983.

[Cim92] R. J. Cimikowski. Graph planarization and skewness. *Congressus Numerantium*, 88:21–32, 1992.

[CJS08] M. Chimani, M. Jünger, and M. Schulz. Crossing minimization meets simultaneous drawing. In *Proc. PacificVis '08*, pages 33–40, 2008.

[CM10] S. Cabello and B. Mohar. Crossing and weighted crossing number of near-planar graphs. *Algorithmica*, 2010. in print.

[CMB08] M. Chimani, P. Mutzel, and I. Bomze. A new approach to exact crossing minimization. In *Proc. ESA '08*, volume 5193 of *LNCS*, pages 284–296. Springer, 2008.

[CMS07] M. Chimani, P. Mutzel, and J. M. Schmidt. Efficient extraction of multiple Kuratowski subdivisions (TR). Technical Report TR07-1-002, June 2007, TU Dortmund, June 2007.

[CMS08] M. Chimani, P. Mutzel, and J. M. Schmidt. Efficient extraction of multiple Kuratowski subdivisions. In *Proc. GD '07*, volume 4875 of *LNCS*, pages 159–170. Springer, 2008.

[CY94] F. R. K. Chung and S.-T. Yau. A near optimal algorithm for edge separators. In *Proceedings of STOC'94*, pages 1–8, 1994.

[DFF85] M. E. Dyer, L. R. Foulds, and A. M. Frieze. Analysis of heuristics for finding a maximum weight planar subgraph. *European Journal of Operational Research*, 20(1):102–114, 1985.

[DGL+97] G. Di Battista, A. Garg, G. Liotta, R. Tamassia, E. Tassinari, and F. Vargiu. An experimental comparison of four graph drawing algorithms. *Comput. Geom. Theory Appl.*, 7:303–325, 1997.

[Dji95] H. N. Djidjev. A linear algorithm for the maximal planar subgraph problem. In *Proc. 4th Workshop Algorithms Data Struct.*, Lecture Notes Comput. Sci., pages 369–380. Springer-Verlag, 1995.

[DT89] G. Di Battista and R. Tamassia. Incremental planarity testing. In *Proc. 30th Annu. IEEE Sympos. Found. Comput. Sci.*, pages 436–441, 1989.

[DT96]   G. Di Battista and R. Tamassia. On-line planarity testing. *SIAM J. Comput.*, 25:956–997, 1996.

[DV02]   H. Djidjev and I. Vrt'o. An improved lower bound for crossing numbers. In *GD '01: Revised Papers from the 9th International Symposium on Graph Drawing*, pages 96–101, London, UK, 2002. Springer-Verlag.

[DV06]   H. Djidjev and I. Vrt'o. Planar crossing numbers of genus g graphs. In *Proc. ICALP '06*, volume 4051 of *LNCS*, pages 419–430. Springer, 2006.

[DW60]   G. B. Dantzig and P. Wolfe. Decomposition principle for linear programs. *Operations Research*, 8:101–111, 1960.

[EGS00]   G. Even, S. Guha, and B. Schieber. Improved approximations of crossings in graph drawing. In *Proc. STOC '00*, pages 296–305, 2000.

[EKE03]   M. Eiglsperger, M. Kaufmann, and F. Eppinger. An approach for mixed upward planarization. *J. Graph Algorithms Appl.*, 7(2):203–220, 2003.

[EW94]   P. Eades and N. C. Wormald. Edge crossings in drawings of bipartite graphs. *Algorithmica*, 11(4):379–403, 1994.

[GHLS08] I. Gitler, P. Hliněný, J. Leanos, and G. Salazar. The crossing number of a projective graph is quadratic in the face-width. *Electr. Journal of Combinatorics*, 15, 2008.

[GJ79]   M. R. Garey and D. S. Johnson. *Computers and Intractability: A Guide to the Theory of NP-Completeness*. W. H. Freeman, New York, NY, 1979.

[GJ83]   M. R. Garey and D. S. Johnson. Crossing number is NP-complete. *SIAM J. Algebraic Discrete Methods*, 4(3):312–316, 1983.

[GJJ68]   R. K. Guy, T. A. Jenkyns, and J.Schaer. The toroidal crossing number of the complete graph. *Journal of Combinatorial Theory*, 4:376–390, 1968.

[GLS88]   M. Grötschel, L. Lovász, and A. Schrijver. *Geometric Algorithms and Combinatorial Optimization*. Springer-Verlag, 1988.

[GM01]   C. Gutwenger and P. Mutzel. A linear time implementation of SPQR trees. In J. Marks, editor, *Graph Drawing (Proc. GD 2000)*, volume 1984 of *LNCS*, pages 77–90. Springer-Verlag, 2001.

[GM04]   C. Gutwenger and P. Mutzel. An experimental study of crossing minimization heuristics. In G. Liotta, editor, *11th Symposium on Graph Drawing 2003*, volume 2912 of *LNCS*, pages 13–24. Springer-Verlag, 2004.

[GMW05]   C. Gutwenger, P. Mutzel, and R. Weiskircher. Inserting an edge into a planar graph. *Algorithmica*, 41(4):289–308, 2005.

[Gro01]   M. Grohe. Computing crossing numbers in quadratic time. In *Proceedings of STOC'01*, 2001.

[Gut10]   C. Gutwenger. *Application of SPQR-Trees in the Planarization Approach for Drawing Graphs*. PhD thesis, TU Dortmund, 2010. http://hdl.handle.net/2003/27430.

[Guy72]   R. K. Guy. Crossing numbers of graphs. In *Graph Theory and Applications (Proceedings*, Lecture Notes in Mathematics, pages 111–124. Springer-Verlag, 1972.

[HC10]   P. Hliněný and M. Chimani. Approximating the crossing number of graphs embeddable in any orientable surface. In *Proc. SODA'10*, pages 918–927. SIAM, 2010. Proc. SODA '10.

[Hli06]   P. Hliněný. Crossing number is hard for cubic graphs. *Journal of Combinatorial Theory, Series B*, 96:455–471, 2006.

[HS06]   P. Hliněný and G. Salazar. On the crossing number of almost planar graphs. In *Proc. GD '05*, volume 4372 of *LNCS*, pages 162–173. Springer, 2006.

[HS07]   P. Hliněný and G. Salazar. Approximating the crossing number of toroidal graphs. In *Proc. ISAAC '07*, volume 4835 of *LNCS*, pages 148–159. Springer, 2007.

[HT73]   J. Hopcroft and R. E. Tarjan. Dividing a graph into triconnected components. *SIAM J. Comput.*, 2(3):135–158, 1973.

[HT74]   J. Hopcroft and R. E. Tarjan. Efficient planarity testing. *J. ACM*, 21(4):549–568, 1974.

[JLM98]   M. Jünger, S. Leipert, and P. Mutzel. A note on computing a maximal planar subgraph using PQ-trees. *IEEE Transactions on Computer-Aided Design of Integrated Circuits and Systems*, 17(7):609–612, 1998.

[JM96]   M. Jünger and P. Mutzel. Maximum planar subgraphs and nice embeddings: Practical layout tools. *Algorithmica*, 16(1):33–59, 1996. (special issue on Graph Drawing, edited by G. Di Battista and R. Tamassia).

[JM97]   Michael Jünger and Petra Mutzel. 2-layer straightline crossing minimization: Performance of exact and heuristic algorithms. *J. Graph Algorithms Appl.*, 1(1):1–25, 1997.

[JTS89]   R. Jayakumar, K. Thulasiraman, and M. N. S. Swamy. $O(n^2)$ algorithms for graph planarization. *IEEE Trans. Comp.-Aided Design*, 8:257–267, 1989.

[KMP+06]   E. de Klerk, J. Maharry, D. V. Pasechnik, R. B. Richter, and G. Salazar. Improved bounds for the crossing numbers of $K_{m,n}$ and K_n. *SIAM Journal on Discrete Mathematics*, 20(1):189–202, 2006.

[KPS07]   E. de Klerk, D. V. Pasechnik, and A. Schrijver. Reduction of symmetric semidefinite programs using the regular ∗-representation. *Mathematical Programming*, 109(2):613–624, 2007.

[KR07]   K. Kawarabayashi and B. Reed. Computing crossing number in linear time. In *Proc. STOC '07*, pages 382–380, 2007.

[Kra91]   J. Kratochvíl. String graphs. II: Recognizing string graphs is NP-hard. *J. Comb. Theory Ser. B*, 52(1):67–78, 1991.

[La 94]   J. A. La Poutré. Alpha-algorithms for incremental planarity testing. In *Proc. 26th Annu. ACM Sympos. Theory Comput.*, pages 706–715, 1994.

[Lei83]   F. T. Leighton. *Complexity issues in VLSI: optimal layouts for the shuffle-exchange graph and other networks*. MIT Press, 1983.

[Lei84]   F. T. Leighton. New lower bound techniques for VLSI. *Mathematical Systems Theory*, 17:47–70, 1984.

[LG77]   P. Liu and R. C. Geldmacher. On the deletion of nonplanar edges of a graph. In *Proc. 10th S.-E. Conf. on Combinatorics, Graph Theory and Computing*, pages 727–738, Boca Raton, FL, 1977.

[Man83]   A. Mansfield. Determining the thickness of a graph is np-hard. In *Mathematical Proceedings of the Cambridge Philosophical Society*, pages 9–23, 1983.

[MJ01]   Petra Mutzel and Michael Jünger. Graph drawing: Exact optimization
           helps! In M. Grötschel, editor, *The Sharpest Cut*, Series on Optimization.
           MPS - SIAM, 2001. Festschrift zum 60. Geburtstag von Manfred Padberg.

[MKNF86]   S. Masuda, T. Kashiwabara, K. Nakajima, and T. Fujisawa. An NP-hard
           crossing minimization problem for computer network layout. Technical
           Report SRC TR 86-80, Electrical Engineering Department and Systems
           Research Center, University of Maryland, 1986.

[MKNF87]   S. Masuda, T. Kashiwabara, K. Nakajima, and T. Fujisawa. On the NP-
           completeness of a computer network layout problem. In *Proceedings of the
           IEEE International Symposium on Circuits and Systems*, pages 292–295,
           1987.

[MNKF90]   S. Masuda, K. Nakajima, T. Kashiwabara, and T. Fujisawa. Crossing
           minimization in linear embeddings of graphs. *IEEE Trans. Comput.*,
           39(1):124–127, 1990.

[Moh99]   B. Mohar. A linear time algorithm for embedding graphs in an arbitrary
           surface. *SIAM J. Discrete Math.*, 12:6–26, 1999.

[MZ99]   P. Mutzel and T. Ziegler. The constrained crossing minimization problem.
           In J. Kratochvil, editor, *Graph Drawing (Proc. GD '99)*, volume 1731 of
           Lecture Notes in Computer Science, pages 175–185. Springer-Verlag, 1999.

[Nic68]   T. A. J. Nicholson. Permutation procedure for minimising the number of
           crossings in a network. *IEE Proceedings*, 115:21–26, 1968.

[PR07]   S. Pan and R. B. Richter. The crossing number of K_{11} is 100. *Journal of
           Graph Theory*, 56:128–134, 2007.

[PSS96]   J. Pach, F. Shahrokhi, and M. Szegedy. Applications of the crossing num-
           ber. *Algorithmica*, 16:111–117, 1996.

[PSŠ06]   M. Pelsmajer, M. Schaefer, and D. Štefankovič. Odd crossing number
           is not crossing number. In *Proc. GD '05*, volume 3843 of *LNCS*, pages
           386–396. Springer, 2006.

[PT97]   J. Pach and G. Tóth. Graphs drawn with few crossings per edge. *Combi-
           natorica*, 17(3):427–439, 1997.

[PT00]   J. Pach and G. Tóth. Which crossing number is it, anyway? *J. Comb.
           Theory Ser. B*, 80(2):225–246, 2000.

[Pul89]   W. R. Pulleyblank. Polyhedral combinatorics. In G. L. Nemhauser,
           A. H. G. Rinnooy Kan, and M. J. Todd, editors, *Optimization*, volume 1 of
           Handbooks in Operations Research and Management Science, pages 371–
           446. North-Holland, 1989.

[Pur97]   Helen Purchase. Which aesthetic has the greatest effect on human under-
           standing? In G. Di Battista, editor, *Graph Drawing (Proc. GD '97)*, vol-
           ume 1353 of *Lecture Notes Comput. Sci.*, pages 248–261. Springer-Verlag,
           1997.

[RT97]   C. Roos and T. Terlaky. Advances in linear optimization, 1997.

[Sch12]   M. Schaefer, 2012. personal communication, joint work with D.
           Štefankovič. Also mentioned in the unpublished manuscript *The Graph
           Crossing Number and Its Variants: A Survey*.

[SV94]   O. Sýkora and I. Vrt'o. On VLSI layout of the star graph and related
           networks. *The VLSI Journal*, 17:83–93, 1994.

[Zie00]   T. Ziegler. *Crossing Minimization in Automatic Graph Drawing*. PhD thesis, Max-Planck-Institut für Informatik, Saarbrücken, 2000.

Symmetric Graph Drawing

Peter Eades
University of Sydney

Seok-Hee Hong
University of Sydney

3.1 Introduction

Symmetry is one of the most important aesthetic criteria that clearly reveals the structure and properties of a graph. Graphs in textbooks on graph theory are normally drawn symmetrically. In some cases, a symmetric drawing may be preferred over a planar drawing. As an example, consider the two drawings of the same graph in Figure 3.1 (from [KK89]). The left drawing has five edge crossings, but eight symmetries (four rotations and four reflections). On the right is a planar drawing; it only has axial symmetry. Most people prefer the drawing on the left. As another example, the Petersen graph is normally drawn as in Figure 3.2. This drawing shows ten symmetries (five rotations and five reflections). In fact, it can be shown that a drawing of the Petersen graph can have at most ten symmetries, and Figure 3.2 is maximally symmetric.

Of course, every drawing has the trivial symmetry, the identity mapping on the plane. The aim of *symmetric graph drawing* is to draw a graph with nontrivial symmetry. More ambitiously, we aim to draw a graph with as much symmetry as possible.

Symmetries of a drawing of a graph G are clearly related to the automorphisms of G; intuitively, a symmetry of a drawing of G induces an automorphism of the graph. For example, in Figure 3.2, a rotational of the plane by $2\pi/5$ is a symmetry of the drawing and induces the automorphism $(0, 1, 2, 3, 4)(5, 6, 7, 8, 9)$. A reflection of the plane in a vertical axis induces the automorphism $(1, 4)(6, 9)(7, 8)(2, 3)$. The automorphism group of a graph G defines its "combinatorial symmetries." However, not every automorphism can be represented as a symmetry of a *drawing* of G. For example, the automorphism group of the Petersen graph has 120 elements, but, as mentioned above, a drawing can display only ten

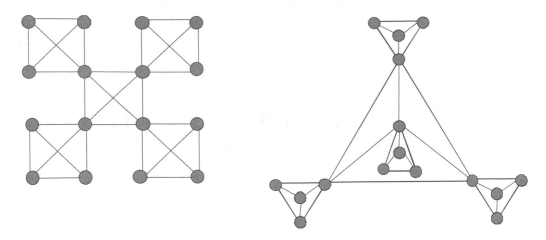

Figure 3.1 Two drawings of the same graph: a planar drawing with eight symmetries and five edge crossings, and a planar drawing with an axial symmetry [KK89].

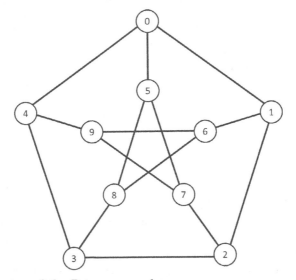

Figure 3.2 A drawing of the Petersen graph.

of these. Symmetric graph drawing involves determining those automorphisms of a graph G that can be represented as symmetries of a drawing of G.

This chapter describes a formal model for symmetric graph drawing in Section 3.2, and gives a characterization of subgroups of the automorphism group of a graph that can be displayed as symmetries of a drawing in Section 3.3. Most precise formulations of the symmetric graph drawing problem are NP-complete. Section 3.4 describes a proof of the NP-completeness of one such formulation and briefly reviews some heuristics for the general symmetric graph drawing problem. Of course, we want a drawing of a graph to satisfy other aesthetics as well as symmetry. In particular, it is useful to examine the problem of constructing a planar straight-line drawing of a planar graph, such that symmetry is maximized. Surprisingly, there is a linear-time algorithm for this problem; it is sketched in Section 3.5. The chapter concludes with a brief survey of some other approaches to symmetric graph drawing and some open problems.

3.2 Basic Concepts for Symmetric Graph Drawing

3.2.1 Drawing of a graph

A *graph* $G = (V, E)$ consists of a set V of vertices and a set E edges, that is, unordered pairs of vertices. Unless explicitly stated otherwise, we assume that the graph is *simple*, that is, it has no multiple edges and no self-loops.

A *drawing* D of a graph G consists of a point $D_V(u)$ in R^2 for every vertex $u \in V$, and a closed curve segment $D_E(u, v)$ in R^2 for every edge $(u, v) \in E$. The curve $D_E(u, v)$ has its endpoints at $D_V(u)$ and $D_V(v)$. Through most of this chapter, the curve $D_E(u, v)$ is a straight-line segment.

For an investigation of symmetric graph drawing, we must take a little care about the definition of the drawing of a graph. We allow two curves $D_E(u, v)$ and $D_E(u', v')$ to cross (share a point), but we have some non-degeneracy conditions as follows:[1]

ND1 The mapping D_V is injective. This excludes, for example, the ultra-symmetric case where all vertices are drawn at the origin.

ND2 A curve $D_E(u, v)$ must not contain a point $D_V(w)$ where $u \neq w \neq v$; in other words, an edge must not intersect with a vertex to which it is not incident.

ND3 Two curves must not overlap; that is, they must not share a curve of nonzero length. This excludes, for example, the axially symmetric case where all the vertices of the graph are drawn on the x axis.

ND4 If two curves share a point, then they must cross at this point; that is, they alternate in cyclic order around the crossing point.

Note that for straight-line drawings, ND2 implies both ND3 and ND4. Most of this chapter is concerned with straight-line drawings, and so discussions of degeneracy concentrate on ND1 and ND2.

3.2.2 Automorphisms of a graph

Basic concepts and terminology for permutation groups can be found in [Wie64].

An *isomorphism* from a graph $G_1 = (V_1, E_1)$ to a graph $G_2 = (V_2, E_2)$ is a one-one mapping β of V_1 onto V_2 that preserves adjacency, that is, $(u, v) \in E_1$ if and only $(\beta(u), \beta(v)) \in E_2$. An *automorphism* of a graph $G = (V, E)$ is an isomorphism of G onto itself, that is, a permutation of the vertex set that preserves adjacency. The *order* of an automorphism β is the smallest positive integer k such that β^k is the identity.

Any set of automorphisms of G that forms a group is called an *automorphism group* of G; the set of all automorphisms of G is denoted by $aut(G)$. The *size* of an automorphism group is the number of elements of the group.

We have defined an automorphism group A of a graph as a permutation group on the vertex set V of a graph $G = (V, E)$. It is easy to see that this defines a permutation group A' acting on the edge set E, and it is often convenient to regard A as acting on E. For

---

[1]The graph drawing literature is somewhat inconsistent about the precise details of the definition of a graph drawing. In some places, a drawing with these non-degeneracy conditions is called a *strict*, *clear*, and/or *proper*. In this chapter, however, we use the term "graph drawing" to includes these non-degeneracy conditions.

example, if $\beta \in A$ and $(u,v) \in E$, then we write "the edge $\beta(u,v)$" to denote the edge $(\beta(u), \beta(v))$.

A subset $B = \{\beta_1, \beta_2, \ldots, \beta_k\}$ of an automorphism group A *generates* A if every element of A can be written as a product of elements of B. We denote the group A generated by B by $\langle \beta_1, \beta_2, \ldots, \beta_k \rangle$. From the computational point of view, generators are important because they give a succinct way to represent an automorphism group. If we were to represent a permutation explicitly, then it may require $\Omega(n)$ space, where $n = |V|$. Thus, an explicit representation of an automorphism group of size k may take space $\Omega(kn)$. In many cases this is too large; for example, the space requirement may preclude a linear-time algorithm, merely because the representation of the output is super-linear. To avoid this problem, we usually represent a group by a set of generators; in general the set of generators is smaller than the group. Most of the groups discussed in this chapter are generated by one or two elements.

Many of the difficulties of symmetric graph drawing arise when vertices and/or edges are fixed by an automorphism. For this reason, we need a careful notion of "fix." Suppose that A is an automorphism group of $G = (V, E)$. The *stabilizer* of $u \in V$, denoted by $stab_A(u)$, is the set of automorphisms in A that fix u, that is,

$$stab_A(u) = \{\beta \in A \mid \beta(u) = u\}. \tag{3.1}$$

The definition can be extended to subsets of V: if $Y \subseteq V$, then

$$stab_A(Y) = \{\beta \in A \mid \forall y \in Y, \ \beta(y) \in Y\}. \tag{3.2}$$

Note that the stabilizer of a set fixes the set setwise.

For each automorphism β we denote $\{u \in V \mid \beta(u) = u\}$ by fix_β. The set of vertices that are fixed *elementwise* by every element of A is denoted by fix_A, that is,

$$fix_A = \{v \in V \mid \forall \ \beta \in A, \ \beta(v) = v \}. \tag{3.3}$$

Note that while $stab_A(Y)$ is a set of group elements, fix_A is a set of vertices. Further, the expression "fix the edge (u, v)" does not necessarily entail "fixing u and fixing v"; it could mean that u and v are swapped.

If $\beta \in A$ and $u \in V$, then the *orbit* of u under β, denoted by $orbit_\beta(u)$, is the set of images of u under $\langle \beta \rangle$, that is,

$$orbit_\beta(u) = \{\beta^i(u) \mid 0 \le i < k\}, \tag{3.4}$$

where β has order k. We can extend this definition to groups: the *orbit* of u under A is

$$orbit_A(u) = \{\beta(u) \mid \beta \in A\}. \tag{3.5}$$

Note that the orbits partition V. The following theorem is fundamental in finite group theory.

Theorem 3.1 (Orbit-stabilizer theorem [Arm88]) *Suppose that A is a group acting on a set X and let $x \in X$. Then $|A| = |orbit_A(x)| \times |stab_A(x)|$.*

The following corollary is helpful in the following sections.

COROLLARY 3.1   Suppose that A is a group acting on a set X.

- If A has no fixed points, then $|orbit_A(x)| = |A|$ for every $x \in X$.
- If A has one fixed point $w \in X$, then $|orbit_A(x)| = |A|$ for every $x \neq w \in X$.

3.2.3   Symmetries of a graph drawing

Symmetry is an intuitive notion that can be formally defined in many different ways. In this chapter we will concentrate on a standard mathematical notion of symmetry; other notions are discussed in Section 3.6.

An *isometry* is a mapping of the plane onto itself that preserves distances. A *symmetry* of a drawing $D = (D_V, D_E)$ of a graph $G = (V, E)$ is an isometry σ of the plane that maps the drawing onto itself, that is:

- for every vertex $u \in V$, there is a vertex $v \in V$ such that $\sigma(D_V(u)) = D_V(v)$, and

- for every edge $(u, v) \in E$, there is an edge $(a, b) \in E$ such that $\sigma(D_E(u, v)) = D_E(a, b)$.

Note that if σ is a symmetry of a drawing $D = (D_V, D_E)$ of a graph $G = (V, E)$, then $\beta = D_V^{-1} \sigma D_V$ is an automorphism of G. We say that D *displays* β. Given an automorphism β, if there is a drawing which displays β, then we say that β is *geometric*.

An automorphism group A is *geometric* if every element of A is displayed in a single drawing; in this case the drawing *displays* A.

To define the intuitive notion of "maximally symmetric drawing" of a graph, we need to decide what it means for one drawing to display more symmetry than another. Here we take a simple view: that if D displays A and D' displays A', then D is more symmetric than D' if A has a larger size than A'. This means that searching for a maximally symmetric drawing entails searching for a maximum size geometric automorphism group.

3.3   Characterization of Geometric Automorphism Groups

Suppose that a drawing D of a graph $G = (V, E)$ displays the automorphism group A. Let A' denote the group of symmetries of D. It is useful to note the group-theoretic relationship between A and A'. If D contains three non-collinear points, then A is isomorphic to A' because a motion of three non-collinear points in the plane uniquely determines an isometry. If all the vertices of the drawing lie on a single line, it may be the case that $|A'| = 4$ while $|A| = 2$, because the rotation by π gives the same automorphism as a reflection in the line. However, this is a pathological case, because the only graphs that have drawings on a single line are sets of paths; in general, we assume that A is isomorphic to A'.

Next, we consider the simple question: Given an automorphism group A of a graph G, is there a drawing of G that displays A? The answer is straightforward, since a symmetry of a finite set of points in the plane is relatively straightforward. The following theorem is an extension of results of Lipton et al. [LNS85], Manning et al. [MA86, MA88, AM88], and Lin [Lin92] to handle degeneracies.

Theorem 3.2   *Suppose that A is an automorphism group of a graph G. Then:*

  (a) A can be displayed as a reflection if and only if $|A| = 2$ and fix_A induces a set of disjoint paths.

  (b) A can be displayed as a rotation if and only if all the following conditions hold

      i. A has one generator ρ, and

      ii. $|fix_A| \leq 1$, and

      iii. if A fixes an edge, then $|fix_A| = 0$.

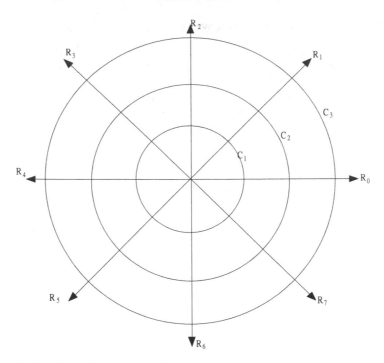

Figure 3.3   A circular grid.

(c) *A can be displayed as a dihedral group if and only if all the following conditions hold.*

    i. *A is dihedral; that is, it has two generators α and ρ such that $\alpha^2 = 1$, $\rho^k = 1$ for some $k > 1$, and $\alpha\rho = \rho^{-1}\alpha$.*

    ii. $|fix_A| \leq 1$.

    iii. *fix_α induces a set of disjoint paths.*

    iv. *If ρ fixes an edge, then $|fix_A| = 0$.*

The proof of Theorem 3.2 is an algorithm, stated below, that takes a graph G and an automorphism group A satisfying the conditions of the theorem, and draws G to display A. The drawing is on a *circular grid* as illustrated in Figure 3.3. An $m \times n$ circular grid has $n \geq 2$ equally spaced rays $R_0, R_1, \ldots, R_{n-1}$ from the origin, that is, the ray R_i makes an angle of $2\pi i/k$ to the x axis. There are $m \geq 1$ circles $C_1, C_2, \ldots, C_m$ centered at the origin, in increasing order of radius. However, the circles may not be equally spaced. The drawing algorithm below chooses a radius for each circle, and places vertices at the grid points, that is, at the intersection points between the circles and the rays.

To prove part (c) of Theorem 3.2, we need the following technical lemma.

LEMMA 3.1   Suppose that the radius of the circle C_i in the $m \times n$ circular grid is n^i, and there are three circular grid points that lie on a straight line ℓ. Then ℓ passes through the origin.

Proof: Suppose that u, v and w are circular grid points on C_i, C_j, and C_k, respectively, and that $i \geq j \geq k$.

Note the case $i = j = k$ is not possible.

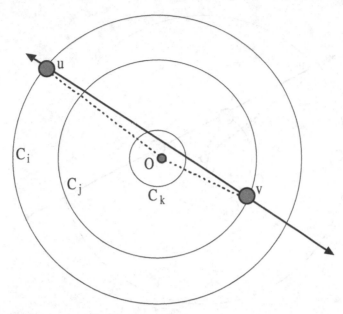

Figure 3.4 Three-in-a-line for the circular grid: first case.

First, consider the case that $i \geq j > k$, as in Figure 3.4. We show that the line segment between u and v cannot intersect C_k unless it passes through the origin.

Assume that such an intersection occurs. Then let $\theta = \angle Ouv$ and $\phi = \angle Ovu$. Since the line segment between u and v intersects C_k with $k < j$ and the radius of c_k is at least n times smaller than the radii of C_i and C_j, it follows that $\sin(\theta) \leq n^{-1}$ and $\sin(\phi) \leq n^{-1}$. One can deduce that for $n > 2$:

$$\sin(\theta + \phi) < 2n^{-1}. \tag{3.6}$$

However, considering the triangle uOv and noting that u and v are at circular grid points, we can see that $\theta + \phi$ is an integer multiple of $2\pi n^{-1}$. If both are nonzero, then $\theta + \phi \geq 2\pi n^{-1}$; this implies that $\sin(\theta + \phi) \geq 2n^{-1}$, contradicting the inequality above. It follows that both θ and ϕ are zero, and so the line segment between u and v passes through the origin.

For the case that $i > j = k$, as in Figure 3.5, a variation of the argument above can be used to show that ℓ passes through the origin.

$\square$

Next, we prove Theorem 3.2. We prove each of the parts (a), (b), and (c) in turn.

Part (a) Suppose that A is displayed as a reflection. Then every vertex u on the line of reflection is fixed by the reflection and thus $u \in fix_A$. Any cycle or vertex of degree more than two on this line violates the non-degeneracy conditions, thus fix_A induces a set of disjoint paths.

Conversely, suppose that $A = \{1, \alpha\}$ is an automorphism group such that fix_A induces a set of disjoint paths. We use the following algorithm.

First, draw $V - fix_A$ on a circle about the origin, so that the x coordinates of u and $\alpha(u)$ are the same, then draw the edges induced by $V - fix_A$ as straight lines. Note that, so far, the drawing is axially symmetric about the y axis and it is non-degenerate.

Next, note that the edges induced by $V - fix_A$ cross the y axis at a finite number of places. Draw fix_A on the y axis, one path at a time, in such a way that the vertices of fix_A avoid the edges induced by $V - fix_A$.

Finally, draw the edges between $V - fix_A$ and fix_A.

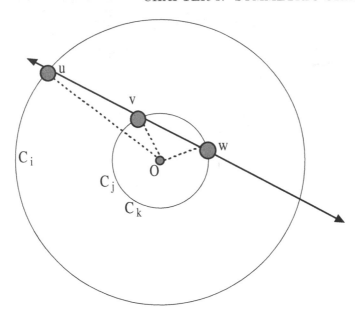

Figure 3.5   Three-in-a-line for the circular grid: second case.

This drawing displays α as a reflection in the y axis; note that it satisfies the non-degeneracy conditions.

 Part (b)   Suppose that A is displayed by a rotation. It is clear that A has one generator. A rotation fixes only one point of the plane, and thus A can fix at most one vertex. If A fixes an edge as well as a vertex w, then w must lie at the midpoint of the edge, and thus the drawing is degenerate.

Conversely, suppose that $A = \langle \rho \rangle$ is an automorphism group satisfying the conditions of the lemma. Assume, for the moment, that fix_A is empty. Our algorithm places every vertex u on a circle of radius one about the origin; it must choose the angle θ_u that the line between u and the origin makes with the x axis.

From Corollary 3.1, each orbit has the same size. Thus, there are n/k orbits $O_1, O_2, \ldots, O_{n/k}$, where $n = |V|$. We choose an element u_i from O_i and, for $j = 0, 2, \ldots, k-1$, place $\rho^j(u_i)$ so that $\theta_{\rho^j(u_i)} = 2\pi(i + jn/k)/n$. Effectively, this spaces the vertices equally around the circle so that the angle between consecutive elements of the same orbit is $2\pi/k$. Thus the drawing displays A with a rotation by $2\pi/k$. It is clear that this drawing is non-degenerate.

If fix_A is nonempty, then it has one element c, which we place at the origin. This preserves symmetry but introduces a possible degeneracy: the central fixed vertex may lie on an edge that forms a diameter of the circle. However, such an edge is fixed by a rotation by π and, from the conditions of part (b) of the lemma, cannot occur when fix_A is nonempty.

 Part (c)   Finally, suppose that A is dihedral. Suppose that $A = \langle \alpha, \rho \rangle$, where $\alpha^2 = 1$ and $\rho^k = 1$, with $k \geq 2$. If fix_ρ is nonempty, then it forms a trivial orbit of $\langle \rho \rangle$. Denote the nontrivial orbits of $\langle \rho \rangle$ by $O_1, O_2, \ldots, O_{n/k}$, where $n = |V - fix_A|$.

For $0 \leq i \leq k-1$, the automorphism ρ^i will be displayed as a rotation by $2\pi i/k$ about the origin, and the automorphism $\rho^{-i}\alpha\rho^i$ will be displayed as a reflection in the line through the origin at an angle of $2\pi i/k$ to the x axis.

We will use a circular grid with n rays $R_0, R_1, \ldots, R_{n-1}$ and n/k circles $C_1, C_2, \ldots, C_{n/k}$. First, we draw $fix_{\rho^{-i}\alpha\rho^i}$ for each i, starting with $i = 0$. We assume first that $fix_\rho = \emptyset$.

Since fix_α is a set of disjoint paths, we can draw it on the x axis so that each vertex is at a grid point of the circular grid and no vertex lies on any edge with which it is not adjacent. If k is even, then we must ensure that $\alpha(fix_{\rho^{-k/2}\alpha\rho^{k/2}}) = fix_{\rho^{-k/2}\alpha\rho^{k/2}}$; this is easily achieved.

Now consider $fix_{\rho^{-i}\alpha\rho^i}$.

Note that if $u \in fix_{\langle\alpha\rangle}$, then $\rho^i(u) \in fix_{\rho^{-i}\alpha\rho^i}$; in other words, if u is fixed by α then every vertex in $orbit_\rho(u)$ is fixed by a conjugate of α. We can draw $fix_{\rho^{-i}\alpha\rho^i}$ on $R_{ni/k}$ and $R_{n(k-i)/k}$ by rotating the drawing of fix_α by $2\pi i/k$. In this way, we draw $orbit_\rho(u)$ for every vertex $u \in fix_\alpha$.

Every other orbit is drawn on the innermost circle C_1. We use a similar method to that for cyclic groups, except that we display α. To do this, we choose a vertex u_1 from an orbit, and draw u_1 on ray R_1. Then draw $\alpha(u_1)$ on ray R_{n-1}. Next, we choose a vertex u_2 from another orbit and draw u_2 on R_2 and $\alpha(u_2)$ on R_{n-2}. This continues until we have placed one vertex from each orbit. To place the remaining vertices from these orbits, we just rotate by $2\pi/k$.

The resulting drawing displays A; we must show that it is not degenerate.

From Lemma 3.1, we can assume that if there is a degeneracy, then there is a vertex w lying on an edge (u, v) with $w \neq u, v$, and the line through u, w, and v passes through the origin. If $u \in fix_\alpha$, then since v and w are on the line through u and the origin, we must have $v, w \in fix_\alpha$. This is impossible since the layout method for fix_α precludes degeneracies. We can deduce that neither u nor v is in $fix_{\rho^{-i}\alpha\rho^i}$ for any i. Hence, we conclude that u and v are on C_1. The only possible degeneracy is if w is the central vertex, fixed by all automorphisms; thus, $fix_A \neq \emptyset$. However, $\langle\rho\rangle$ fixes the edge (u, v), contradicting the conditions of the theorem. This completes the proof of Theorem 3.2.

The proof of Theorem 3.2 essentially consists of an algorithm for the following problem.

Geometric Automorphism Drawing Problem (GADP)

Instance: A graph G, and an automorphism group A of G given as a set of at most 2 generators.

Output: If possible, a straight-line drawing of G that displays A.

COROLLARY 3.2   There is a linear-time algorithm that solves the Geometric Automorphism Drawing Problem.

Note that the resolution of the drawing obtained by the proof of Theorem 3.2 is poor in the dihedral case, because the radii of the circles in the circular grid used increase exponentially.

Theorem 3.2 does not solve the main problem in symmetric graph drawing: given a graph, find its largest geometric automorphism group. The next section shows that it is NP-complete to find such a group.

3.4   Finding Geometric Automorphisms

In this section, we discuss the complexity of computing geometric automorphisms, and, since the problem is NP-complete, we briefly mention heuristics.

The relationship between automorphisms of a graph and symmetries of drawings of the graph suggests that the problem of drawing a graph symmetrically is at least as hard as graph isomorphism. Manning [Man90] has shown a surprisingly stronger result: the problem is NP-hard. The intuition behind Manning's result comes from two directions. First, as noted in Section 3.3, the major difficulties in drawing graphs symmetrically arise from the

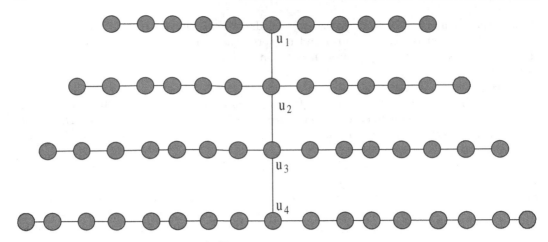

Figure 3.6   The auxiliary graph H.

fixed points of the automorphisms. Secondly, a result of Lubiw [Lub81] states that finding a fixed-point free automorphism of a graph is NP-complete.

In fact, Manning shows that a number of problems related to symmetric graph drawing are NP-hard. Here, we study just one of these problems: detecting whether a graph has an automorphism that can be displayed as a reflection.

Axial Geometric Automorphism Problem (AGAP)
Instance: A graph G.
Question: Is there an automorphism of G that can be displayed as a reflection?

Theorem 3.3   *The axial geometric automorphism problem is NP-complete.*

Proof: Lubiw [Lub81] showed that the following problem is NP-hard.

Fixed Point Free Automorphism Problem (FPFAP)
Instance: A graph G.
Question: Is there an automorphism of G with no fixed points?

We show that FPFAP reduces to AGAP.

Suppose that G is an instance of FPFAP with n vertices. We assume without loss of generality that G is connected and every vertex has degree at least 2. Define a graph H as follows: H has a path $P = (u_1, u_2, \ldots, u_{n+1})$. For $1 \leq i \leq n+1$, u_i is joined to two paths, each of length $n + i$. This is illustrated in Figure 3.6.

Now consider an automorphism β of H. It is clear that for $1 \leq i \leq n+1$, $\beta(u_i) = u_i$, and β either fixes or swaps the two paths joined to u_i. If a drawing D of H displays β, then it is displayed as a reflection, and P lies on the axis of reflection with the paths attached to each u_i on each side of the axis.

Now form a graph G' from H and G. The vertex set is the union of the vertex sets of H and G, plus extra vertices $w_0^v, w_1^v, \ldots, w_n^v$ for each vertex of v of G. For $2 \leq i \leq n$, join u_i to every vertex of G. For each vertex v of G, join w_0^v to v, and join all vertices $w_0^v, w_1^v, \ldots, w_n^v$ together to make a clique of size $n + 1$.

Note that G' can be formed in polynomial time.

We claim that G' has an axial geometric automorphism if and only if G has a fixed point free automorphism.

First, suppose that G has a fixed point free automorphism β. It is clear that one can extend β to G' to give an automorphism that satisfies part (a) of Theorem 3.2, and so G' has an axial geometric automorphism group.

Now suppose that G' has an axial geometric automorphism γ.

We claim that γ cannot map a vertex w of H to a vertex v of G, or to one of the new vertices w_i^v. This is because every vertex of G is adjacent to a clique of size $n + 1$, while no vertex of H has this property. Further, γ cannot map a vertex of G to one of the new vertices w_i^v, because each w_i^v is in a clique of size $n + 1$, and none of the original vertices have this property. Thus, γ restricted to H is an automorphism β of H; as mentioned above, the only drawing that displays β has P lying on the axis of reflection.

Also, γ restricted to G is an automorphism δ of G. Suppose that δ has a fixed point v. Recall that v is joined by an edge to a vertex u_i in P; this means that the induced subgraph fix_γ has a vertex of degree at least three. From Theorem 3.2, this is impossible. Thus δ is fixed-point-free.

Finally, note that AGAP is in NP, because one can guess an automorphism group, and, using Theorem 3.2, check whether it is geometric. $\qquad\square$

The NP-completeness results have led to a number of heuristic approaches; see [dF99, Kam89, Lin92, LNS85].

The most common are the generic *multidimensional scaling*, or *force directed* methods [dF99, Ead84, Kam88, Lin92]. Roughly speaking, this method projects a high-dimensional drawing of the graph into low dimensions. The first step is to define a distance function d between vertices, and then the graph is drawn in a high-dimensional space in such a way that the Euclidean distance in the high-dimensional space is equal or close to the distances defined by d. In some cases, this (high-dimensional) drawing is unique up to isometry; this implies that every automorphism of the graph is a symmetry of the drawing. In other words, it achieves maximum symmetry in the high dimension. The next step is to project the high-dimensional drawing into a low-dimensional space (either 2 or 3 dimensions) in such a way that the distances are preserved as much as possible.

As an example of such a method, de Fraysseix [dF99] uses the *Czekanovski-Dice* semi-distance for a graph $G = (V, E)$:

$$d(u, v) = \sqrt{1 - 2\frac{|N_u \cap N_v|}{|N_u| + |N_v|}}. \tag{3.7}$$

(A *semi-distance* $d : X \to R$ is a function that is almost a distance function: it satisfies two of the axioms of a distance function: $d(u, v) = d(v, u)$ and $d(u, v) + d(v, w) \geq d(u, w)$. However, it is possible that there are distinct elements $u, v \in X$ with $d(u, v) = 0$.) De Fraysseix uses projections defined by the principal components of the corresponding inner product matrix whose entries are defined by a pair s, t of vertices as follows:

$$W_{st} = \frac{1}{2}(d_s^2 + d_t^2 - d_V^2), \tag{3.8}$$

where for $w = s, t$,

$$d_w^2 = \frac{1}{n}\sum_{u \in V} d^2(w, u), \tag{3.9}$$

and

$$d_V^2 = \frac{1}{n}\sum_{w \in V} d_w^2. \tag{3.10}$$

These projections are remarkably successful in displaying two dimensional symmetry; see [dF99] for details.

It is common to look at such methods as a system of forces: for example, one can simulate a system of forces between vertices where the force exerted on u by v is proportional to the distance $d(u, v)$. A minimum energy configuration defines a drawing, and in many cases this drawing displays symmetries. For example, one can view the Tutte method [Tut63, DETT99] in this way. In fact, one of the reasons for the popularity of force directed methods is the fact that the drawings often display some symmetry. One can give some explanation (see [EL00, Lin92]) of why the approach works.

3.5   Symmetric Drawings of Planar Graphs

In this section, we describe a linear-time algorithm to draw planar graphs with no edge crossings and as much symmetry as possible.

The concept of geometric automorphism in Section 3.2.3 can be extended to planar drawings: an automorphism β of a graph G is *planar* if there is a planar drawing of G that displays β, and an automorphism group A is a *planar automorphism group* if there is a planar drawing which displays every element of A.

The problem of finding automorphisms of a planar graph can be solved in linear time (see [HW74, Won75]); however, it is clear that not all automorphisms are geometric. Further, not every geometric automorphism is planar. For example, the complete graph K_4 with four vertices has a dihedral geometric automorphism group of size eight, but this group is not planar. The largest planar automorphism group of K_4 has size six.

The following theorem summarizes the result.

Theorem 3.4   *There is a linear-time algorithm that constructs maximum planar automorphism group of a planar graph.*

The remainder of this section is a sketch of a proof of Theorem 3.4. The algorithm to prove the theorem uses a connectivity decomposition. We decompose the graph into connected components, then decompose each connected component into biconnected components, and finally decompose each biconnected component into triconnected components. Different algorithms are needed for triconnected, biconnected, one-connected, and disconnected graphs. Each uses the algorithms for higher connectivity as subroutines. Details of the proof can be found in [HE05, HE06, HE03, HME06].

In Section 3.5.5, we briefly describe the drawing algorithms.

3.5.1   Triconnected planar graphs

This section describes an algorithm for finding planar automorphism groups of maximum size for triconnected planar graphs.

The uniqueness of the faces of a triconnected planar graph $G = (V, E)$ means that an automorphism group A defines a permutation group acting on the set F of faces of G. Effectively this means that A defines an automorphism group of the dual G^* of G. We can regard A as acting on G^*, and write, for example, "the face $\beta(f)$" for some $\beta \in A$, $f \in F$.

It is well known that a triconnected planar graph can be represented as the skeleton of a polyhedron in three dimensions [SR34]. A more surprising and less well known result, due to Mani [Bab95, Man71], states that the automorphism group of a triconnected planar graph can be completely encapsulated in the symmetries of a polyhedron. The symmetry finding algorithm relies on this fundamental result.

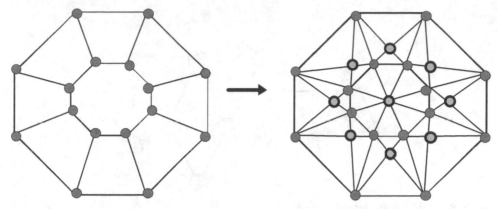

Figure 3.7   Example of star triangulation.

Theorem 3.5   *(Mani [Bab95, Man71]) Suppose that G is a triconnected planar graph. Then there is a convex polytope P in R^3 such that G is the skeleton of P and the full automorphism group of G is displayed by P.*

Mani's theorem leads to an elegant characterization of planar automorphisms of triconnected planar graphs.

Theorem 3.6   *Let G be a triconnected planar graph. An automorphism group of G is planar if and only if it is the stabilizer of a face of G.*

Proof: Every planar automorphism fixes the outside face. Further, if A stabilizes a face f then, using Theorem 3.5, a projection about f from the polyhedron to the plane gives a symmetric drawing.   □

An outline of the algorithm for the triconnected case of Theorem 3.2 is as follows.

> Algorithm `Max_PAG_tricon`
> 1. Find a face f of G such that the stabilizer $stab_A(f)$ of f in A is maximized.
> 2. Find the orbits of $stab_A(f)$.
> 3. Find generators of $stab_A(f)$.

For the first step, note that from Theorem 3.1, we must find an orbit (in the dual G^*) of minimum size. A linear-time algorithm of Fontet [Fon76] takes a triconnected planar graph G as input and outputs the orbits on vertices of $aut(G)$. Using Fontet's algorithm, we can compute the orbits of G^*, then choose an orbit of minimum size. Choose a face f in this minimum orbit; then f can be used as the outside face of an embedding that displays the maximum number of symmetries.

The next step is to find the orbits of $stab_A(f)$. This can be done by transformations of the graph, and then using Fontet's algorithm again. The first part of the transformation is *star triangulation*: we triangulate each internal face by inserting a new vertex in the face and joining it to each vertex of the face. This process is illustrated in Figure 3.7.

It is not difficult to show that the star triangulation takes linear time, and the new graph has exactly the same planar automorphism group as the original graph (see [HME06]).

Next, we transform the graph to ensure that the outside face f has more than three vertices. If f has three vertices v_0, v_1, v_3, we draw a hexagon surrounding G in the plane, with vertices $w_0, w_1, \ldots, w_5$ in clockwise order. Insert the edges $v_0w_0, v_0w_1, v_0w_2, v_1w_2, v_1w_3,$

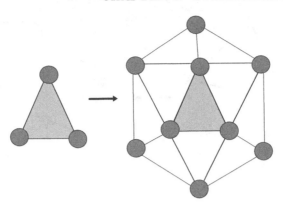

Figure 3.8   Adding an outside face.

$v_1 w_4$, $v_2 w_4$, $v_2 w_5$, and $v_2 w_0$. The transformation is shown in Figure 3.8. The transformation preserves automorphisms that fix f.

The transformed graph has a new outside face with more than three vertices, and all other faces are triangles. Now apply Fontet's algorithm to the transformed graph. The outside face must be fixed by all automorphisms, since all other faces have size three. Thus, Fontet's algorithm gives the orbits of $stab_A(f)$ in the transformed graph, and we can extract the orbits of A on the vertices of G.

The third and final step is to find generators of the planar automorphism group. Suppose that the vertices on the outside face f are $v_0, v_1, \ldots, v_{m-1}$, in clockwise order. If $v_0, v_1, \ldots, v_{m-1}$ are all fixed by A, then A is trivial. Otherwise, let v_i, v_j, v_k be three consecutive vertices in the same nontrivial orbit of A, where $j - i$ is as small as possible and v_k is the same as v_i if the orbit has size 2.

We need to introduce further terminology: a *flag* of an embedded graph is a triple (v, w, f), where v and w are adjacent vertices and f is a face that has the edge (v, w) on its boundary. The action of automorphisms on flags uniquely identifies them, as stated in the following lemma.

LEMMA 3.2   Let G be a triconnected planar graph. Let $F = (v, w, f)$ and $F' = (v', w', f')$ be flags of G. Then there is at most one automorphism of G that maps F onto F'. Moreover, there is a linear-time algorithm that finds that automorphism or determines that it does not exist.

Lemma 3.2 is folklore in graph automorphism theory; a proof is in [HME06].

We can apply Lemma 3.2 to find three possible automorphisms or prove that they do not exist. First, we compute three possible automorphisms α, ρ_1, ρ_2, as follows:

- α is the automorphism mapping the flag (v_i, v_{i+1}, f) onto the flag (v_j, v_{j-1}, f), if that automorphism exists. (That is, a reflection that exchanges v_i and v_j.)
- ρ_1 is the automorphism mapping the flag (v_i, v_{i+1}, f) onto the flag (v_j, v_{j+1}, f), if that automorphism exists. (That is, a rotation by $j - i$ positions.)
- ρ_2 does not exist in the case that $v_k = v_i$. Otherwise, ρ_2 is the automorphism mapping the flag (v_i, v_{i+1}, f) onto the flag (v_k, v_{k+1}, f), if that automorphism exists. (That is, a rotation by $k - i$ positions.)

This allows us to compute generators for A, as follows.

- If α does not exist, then ρ_1 exists and A is a cyclic group of size $m/(j-i)$ generated by the rotation ρ_1.

- If α exists but neither ρ_1 nor ρ_2 exists, then A is the group of size 2 generated by the reflection α.

- If α and ρ_1 exist, then A is the dihedral group of size $2m/(j-i)$ generated by the reflection α and the rotation ρ_1.

- Otherwise, α and ρ_2 exist, and A is the dihedral group of size $2m/(k-i)$ generated by the reflection α and the rotation ρ_2.

We summarize this section with the following lemma.

LEMMA 3.3   Algorithm `Max_PAG_tricon` computes generators for the largest planar automorphism group of a triconnected planar graph in linear time.

3.5.2   Biconnected planar graphs

If the input graph G is biconnected, then we break it into *triconnected components* and apply the algorithm for triconnected graphs in Section 3.5.1. However, this process is not as simple as it sounds.

We use a version of the "SPQR-tree" to represent the decomposition of a biconnected planar graph into triconnected components. Various versions of the SPQR tree appear in the literature; the version that we use is closely related to the original version of Tutte [Tut66].

It is useful to review the definition of triconnected components [HT73]. If G is triconnected, then G itself is the unique triconnected component of G. Otherwise, let u, v be a separation pair of G. We split the edges of G into two disjoint subsets E_1 and E_2, such that $|E_1| > 1$, $|E_2| > 1$, and the subgraphs G_1 and G_2 induced by E_1 and E_2 only have vertices u and v in common. Form the graph G_1' by adding an edge (called a *virtual edge*) between u and v; similarly, form G_2'. We continue the splitting process recursively on G_1' and G_2'. The process stops when each resulting graph reaches one of three forms: a triconnected simple graph, a set of three multiple edges (a triple bond), or a cycle of length three (a triangle). The triconnected components of G are obtained from these resulting graphs. They may be of three types:

1. a triconnected simple graph;

2. a *bond*, formed by merging the triple bonds into a maximal set of multiple edges;

3. a *polygon*, formed by merging the triangles into a maximal simple cycle.

The triconnected components of G are unique. See [HT73] for further details.

Now we can describe the SPQR tree. Each node v in the SPQR tree is associated with a graph *skeleton(v)*, corresponding to a triconnected component. There are several types of nodes in the SPQR tree, corresponding to the type of triconnected components described above. The edges of the SPQR tree are defined by the virtual edges, that is, if u and v are two nodes whose skeletons share a virtual edge, then u and v are connected in the SPQR tree.

The SPQR tree can be rooted at its center (if the tree has two centers, it can be rooted at either one). The motivation for using the rooted version is that the SPQR tree is unique for each biconnected planar graph [Bab95, DT92]. This means that the triconnected component corresponding to the root of the SPQR-tree is fixed by a planar automorphism group of a biconnected planar graph. Further, each leaf is mapped to a leaf. These two properties of the rooted SPQR tree are essential for our algorithm outlined below.

To state the algorithm, we need some more terminology. We say that a virtual edge e of $skeleton(v)$ is a *parent (child) virtual edge* if e corresponds to a virtual edge of u which is a parent (resp. child) node of v. We define a *parent separation pair* $s = (s_1, s_2)$ of v as the two endpoints of a parent virtual edge e.

The overall algorithm is composed of three steps.

Algorithm `MAX_PAG_bicon`

Step 1. Construct the SPQR-tree T of G.

Step 2. *Reduction*: For each level i of T (from the lowest level to the root level)

    (a) For each leaf node on level i, compute labels on the parent virtual edge in the leaf node.

    (b) For each leaf node on level i, label the corresponding virtual edge in the parent node with the labels.

    (c) Remove the leaf nodes on level i.

Step 3. Compute a maximum size planar automorphism group at the labeled center.

We briefly describe each step of the algorithm. The first step is to construct the SPQR-tree for the input biconnected planar graph. This can be done in linear time using the classical Hopcroft-Tarjan algorithm [HT73].

The second step, reduction, is the most important. This takes the rooted SPQR-tree of a biconnected graph, and proceeds up the SPQR-tree from the leaf nodes to the center level by level, computing labels. The labels consist of integer and boolean values that capture some information of the planar automorphisms of the leaf nodes. First, it computes the labels for the leaf nodes. Then, it labels the corresponding virtual edge in the parent node and delete each leaf node. The reduction process stops when it reaches the root.

The reduction process clearly does not decrease the planar automorphism group of the original graph. This is not enough; we need to also ensure that the planar automorphism group is not increased by reduction. This is the role of the labels. As a leaf v is deleted, the algorithm labels the virtual edge e of v in $skeleton(u)$ where u is a parent of v. Roughly speaking, the labels encode enough information about the deleted leaf to ensure that planar automorphisms of the labeled reduced graph can be extended to a planar automorphisms of the original graph.

We illustrate the basic idea of the algorithm with an example.

Consider the biconnected graph represented in Figure 3.9. Here the graph G has an SPQR tree with three leaves; these are triconnected components, G_1, G_2, and G_3, illustrated by shaded blobs. The remainder of the graph, G^*, is illustrated by a shaded oval. This is connected to the leaves by separation pairs $\{u_i, v_i\}$, for $i = 1, 2, 3..$

Intuitively, G can be drawn with an axial symmetry (a reflection in a horizontal line) as long as:

    1. L_1 is isomorphic to L_2 with an isomorphism that maps u_1 to u_2 and v_1 to v_2.

    2. L_3 has an axial planar automorphism that swaps u_3 with v_3.

    3. G^* has an axial planar automorphism that swaps u_3 with v_3, and maps u_1 to u_2 and v_1 to v_2.

To decide whether G can be drawn with an axial symmetry, we maintain a number of labels, including:

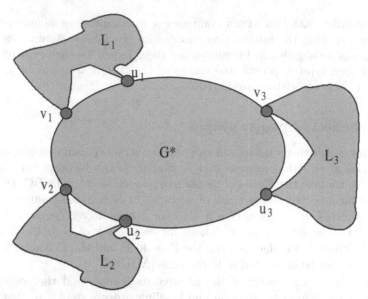

Figure 3.9 A biconnected graph.

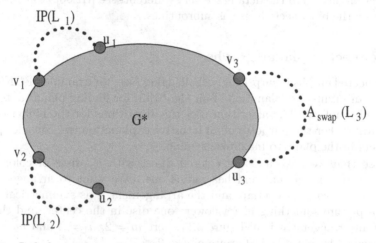

Figure 3.10 Labels on the reduced biconnected graph.

1. An *isomorphism code IP* that has the property that $IP(L_1) = IP(L_2)$ if and only if L_1 is isomorphic to L_2 with an isomorphism that maps u_1 to u_2 and v_1 to v_2.

2. A boolean *axial swap label A_{swap}* that has the property that $A_{swap}(L_3) = true$ if and only if L_3 has an axial planar automorphism that swaps u_3 with v_3.

These labels can be computed at Step 2(a) of Algorithm MAX_PAG_bicon, then transferred to the parent virtual edges in G^* at Step 2(b). Then Step 2(c) gives the labeled reduced graph illustrated in Figure 3.10.

The reduction then continues to the next iteration of Step 2, operating on the labeled reduced graph in Figure 3.10. This continues to the root of the SPQR tree.

In fact, the reduction step is much more complex than this example suggests. There are seven different kinds of labels and separate algorithms for computing these labels for each type of triconnected component. Details of these algorithms are in [HE05].

Step 3 of Algorithm MAX_PAG_bicon computes a maximum size planar automorphism group at the center, using the information encoded on the labels. Again this step is quite complex, with separate algorithms for computing these labels for each type of triconnected component and each type of center (the center of the SPQR tree can be a node or and edge). Details of these algorithms are in [HE05].

3.5.3   One-connected planar graphs

The algorithm for computing a maximum size planar automorphism group of one-connected planar graph uses a reduction process that is similar to the biconnected case. For one-connected graphs, we take the *block-cut vertex tree* (the *BC-tree*). The BC- tree defines the structure of the biconnected components of a graph. If G is a one-connected graph, then a maximal biconnected subgraph of G is a *block*, or a *biconnected component*. Two blocks share a *cut vertex*. The *BC-tree* has a *B-node* for each block of G and a *C-node* for each cut vertex of G. There is an edge between the B-node B and the C-node c if c is a vertex of B. The BC-tree can be computed in linear time [AHU83].

Again we can choose the center of the BC-tree as a root, and the rooted BC-tree is unique. This property allows a reduction and labeling process similar to that described in the previous section, although the details are very different; see [HE06]. The algorithm uses the algorithms for the biconnected case as subroutines.

3.5.4   Disconnected planar graphs

Drawing disconnected graphs is surprisingly challenging (see, for example, [FDK01]). In this section, we give an intuitive explanation of an algorithm for finding planar automorphisms of a disconnected graph G. The algorithm uses the algorithms for the higher-connectivity cases as subroutines. For the purposes of an intuitive explanation, we consider problems of arranging objects in the plane to maximize symmetry.

First, suppose that we have a set of colored discs, with n_j discs of color j, for $j = 1, 2, \ldots, m$. Each disc is circular and has radius one. We want to arrange the discs in the plane so that no two discs overlap, and the arrangement is as symmetrical as possible. We can make a picture something like a flower: one disc in the center, and the others as "petals." Such an arrangement is in Figure 3.11; here $m = 2$, $n_1 = 4$ and $n_2 = 6$, and the discs are arranged to have a dihedral group of size 6.

The center of the flower may be empty. In this case, all discs must be arranged as petals; if there are k petals, then n_j must divide k for $j = 1, 2, \ldots, m$. If the center of the flower has a disc of color i, $n_i - 1$ divides k, and for $j \neq i$, n_j divides k. We can deduce that the

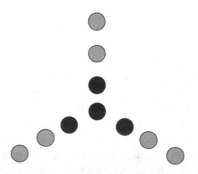

Figure 3.11   A symmetrical arrangement of circular discs.

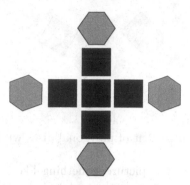

Figure 3.12 A symmetrical arrangement of polygonal discs.

Figure 3.13 Nesting of discs with holes.

maximum symmetry group is dihedral of size $2k$ as long as the following equation holds:

$$k = \max\{\gcd(n_1, n_2, \ldots, n_m), \max_{i=1}^{m} \gcd(n_1, n_2, \ldots, n_{i-1}, n_i - 1, n_{i+1}, \ldots, n_m)\}. \quad (3.11)$$

With some clever computation of the gcds, we can compute equation (3.11) and a maximally symmetric layout of the discs in time $O(n_1 + n_2 + \cdots + n_m)$.

Now consider a problem with a little more complexity. Suppose that we have colored polygonal discs, with n_j discs of color j, for $j = 1, 2, \ldots, m$. Each disc is a regular polygon; all discs of color i have s_i sides, and have radius one. Again, we can make a symmetric picture something like a flower, as in Figure 3.12; here $m = 2$, $n_1 = 5$, $s_1 = 4$, $n_2 = 4$ and $s_2 = 6$.

In this case, we can obtain a dihedral symmetry group of size $2k$ if k satisfies either:

$$k = gcd(n_1, n_2, \ldots, n_m), \quad (3.12)$$

(for the case where the center is empty), or for some i,

$$k = \gcd(s_i, n_1, n_2, \ldots, n_{i-1}, n_i - 1, n_{i+1}, \ldots, n_m)\} \quad (3.13)$$

(for the case where a disc with s_i sides is in the center).

Again, using some clever computation of the gcds and maximizing over i, we can compute a maximally symmetric layout of the discs in time $O(s_1 n_1 + s_2 n_2 + \cdots + s_m n_m)$.

Now consider a more complex problem: suppose that some of the discs have holes. We have n_j discs of color j, for $j = 1, 2, \ldots, m$. The outside of each disc is a regular polygon; all discs of color i have s_i sides. For some values of i, the all discs of color i have a circular hole in the middle. Further, each disc is shrinkable or expandable; this means that we can fit one disc inside another to make a kind of "nest," as in Figure 3.13.

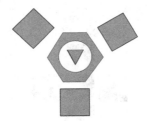

Figure 3.14 Symmetric arrangement of polygonal discs with holes.

Again, we can make a symmetric picture something like a flower, as in Figure 3.14; in this case, we can place a "nest" of discs in the center of the flower, as long as all but one of them have a hole.

Let H denote the set of colors of discs with holes. We can obtain a dihedral symmetry group of size $2k$ if there is a subset H' of H such that k satisfies one of the following:

$$k = \gcd\left(\gcd\{s_j : j \in H'\}, \gcd\{n_\ell : \ell \in H - H'\}\right) \tag{3.14}$$

(for the case where every discs in the center has a hole), or for some i,

$$k = \gcd\left(\gcd\{s_j : j \in H'\}, \gcd\{n_\ell : \ell \in H - H', \ell \neq i\}, s_i, n_i - 1\right), \tag{3.15}$$

for the case where there is a disc of color i, without a hole, in the center.

One can maximize over i and H' to compute a maximally symmetric layout of the colored polygonal discs, with and without holes, in time $O(s_1 n_1 + s_2 n_2 + \cdots + s_m n_m)$.

One can use such disc arrangement algorithms to construct maximally symmetric drawings of disconnected graphs. We can compute the connected components G_ℓ of a disconnected graph G and, using planar graph isomorphism algorithms, divide the components into isomorphism classes $N_1, N_2, \ldots, N_m$, where $|N_j| = n_j$. We compute maximal planar automorphism groups for G_j using the algorithm for connected graphs; assume for the moment that these groups are dihedral and the group for isomorphism class N_j has size $2s_j$. For the purposes of symmetric layout, the isomorphism class N_j is akin to a color class of polygonal disc with s_j sides. For some j, it is possible that the components in N_j has two faces fixed by their planar automorphism group. This is akin to a disc with a hole, because one fixed face can be the outside face and the other can be a central inside face.

There are some further complexities. First, some of the components may have no dihedral planar automorphism group: the group may be purely cyclic, or purely axial, or even trivial. This requires algorithms that are substantially more complex, but follow the same general pattern as above.

Secondly, the connected components may have several maximal planar automorphism groups, and the largest of these may not lead to the maximum planar automorphism group of the whole graph. An example is in Figure 3.15: the two pictures here show a graph with two drawings, one displaying 6 symmetries and one displaying 8 symmetries.

We say that a planar automorphism group A of G is *maximal* if A is not contained in another planar automorphism group of G. One must take all maximal groups into account when this graph is a connected component of a larger disconnected graph. Fortunately, this pathological case is relatively contained; the next Lemma explains why.

LEMMA 3.4   [HE03] A planar graph has at most 3 non-conjugate maximal planar automorphism groups.

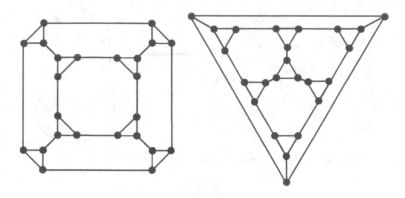

Figure 3.15 Display of two maximal planar automorphism groups.

This means that additional maximal planar automorphism groups only add a constant to the time complexity of the algorithms.

3.5.5 Drawing algorithms

The algorithms presented in the preceding sections take a planar graph as input and produce two outputs: a planar automorphism group of maximum size, and an embedding of the graph. In this section, we show how to use this information to construct a straight-line symmetric drawing of the graph. The drawing algorithms follow the same connectivity hierarchy.

For triconnected graphs, one could use the well-known barycenter algorithm of Tutte [Tut63, DETT99]. This algorithm draws symmetrically but unfortunately takes super-linear time. A much more complex algorithm, described in [HME06], runs in linear time. Note that the drawing can be "squashed" at a specified vertex on the outer face; that is, given an angle a and a vertex u on the outer face, we can adjust the drawing so that the angle at u on the outer face is at most a. The squashing can be done so that any axial symmetry that fixes a is preserved. This process, illustrated in Figure 3.16, is helpful for lower connectivity drawings.

For a biconnected planar graph, we use "augmentation": we increase the connectivity by adding new edges and new vertices to make it triconnected, while preserving the planar automorphism group. The easiest way to do this is to use the star triangulation method described in Section 3.5.1. Then we can apply the algorithm for constructing symmetric drawings of triconnected planar graphs with straight-line edges to construct a symmetric drawing.

Given an embedding of a one-connected planar graph, we use "attachment," as follows. First, we augment the biconnected component to make them triconnected, as above, and draw the triconnected components. Then we draw the root of the BC-tree; then we traverse the BC-tree "attaching" blocks as we go. We can scale blocks to fit inside faces of previously drawn blocks, using the "squash" operation described above.

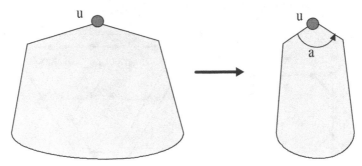

Figure 3.16   Squashing a triconnected component at u.

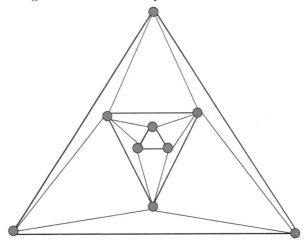

Figure 3.17   The graph G_3.

The drawing process takes linear time, and we can state the following result.

Theorem 3.7   *Given a planar graph G and a planar automorphism group A of G, we can construct a straight-line drawing of G that displays A in linear time.*

The drawings obtained in this way have poor resolution. Unfortunately, in the worst case, this is unavoidable, as the following example shows. Suppose that G_0 is a single triangle with vertices a_0, b_0, c_0. For $i > 0$, G_i is a planar graph with a triangular out-side face $\{a_i, b_i, c_i\}$. We form G_i from G_{i-1} by adding the face $\{a_i, b_i, c_i\}$ and the edges $(a_i, a_{i-1}), (a_i, b_{i-1}), (b_i, b_{i-1}), (b_i, c_{i-1}), (c_i, c_{i-1}), (c_i, a_{i-1})$. The graph G_3 is shown in Figure 3.17.

The graph G_k has $3k$ vertices and has a dihedral planar automorphism group of size 6. However, one can show that every straight-line drawing of G_k that displays this dihedral group requires exponential area; that is, if it has a minimum distance of one between vertices, then the area of the drawing is $\Omega(2^k)$.

3.6   Conclusion

This chapter describes the symmetric graph drawing problem, and discusses some of its qualitative and algorithmic aspects. In particular, we characterize those automorphism groups that can be displayed as symmetries of a graph drawing, we show that the gen-

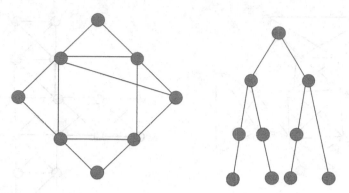

Figure 3.18 Almost symmetric drawings.

eral problem of finding such automorphisms is NP-complete, and we describe linear-time algorithms for finding and displaying such symmetries in the case where the input graph is planar.

In this section, we briefly mention some important aspects of symmetric graph drawing that have not been covered in this chapter and conclude with some open problems.

3.6.1 Further topics

Directed graphs. The model of symmetry needs some modification for directed graphs; for example, perhaps a *directed* geometric symmetry should either preserve the direction of every directed edge or reverse the direction of every directed edge. With a variety of modifications of the model, a number of algorithms have been developed for symmetrically directed graphs. Examples include algorithms for rooted trees [RT81, SR83], series-parallel digraphs [DETT99, HEL00], upward planar graphs [DTT92], and hierarchical graphs [ELT96].

Three-dimensional graph drawing is now well established and some attempts have been made to draw graphs symmetrically in 3D; see [HEQL98, HE00, Hon01].

Exact but exponential time algorithms often work well for small graphs. These include methods based on integer linear programming [BJ01, BJ03] and group theory [AHT07].

Approximation algorithms. The formal definition of the intuitive notion of symmetry display given in Section 3.2.3 is fairly strong. For example, it does not consider the drawings in Figure 3.18 to be symmetric at all. There have been several attempts to formalize the intuitive "approximate" symmetry such as shown in Figure 3.18. For example, Bachl [Bac99] gives a simple approach to approximate axial symmetry: if a graph has two large disjoint isomorphic subgraphs, then one can draw it so that a large part of the drawing displays axial symmetry. Finding such subgraphs is, of course, NP-complete; Bachl gives algorithms for some restricted cases. Other examples include [BJ03, CY02, CLY00].

3.6.2 Open problems

Here we list a couple of open problems in symmetric graph drawing.

Very very symmetric graph drawing. Consider the two drawings in Figure 3.19. The two drawings, according to the model in Section 3.2.3, have the same degree

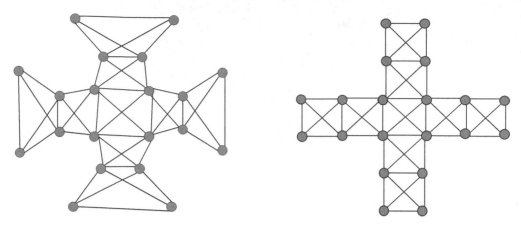

Figure 3.19   A symmetric drawing and a very very symmetric drawing.

of symmetry. However, intuitively the one on the right is more symmetric than the one on the left. The extra symmetry does not come from isometry of the plane; it arises in a more subtle way. Modeling this kind of "very very symmetric" drawing has not been done at this point. Further, algorithms to draw graphs very very symmetrically have not been designed.

An algorithmic version of Mani's Theorem.   Theorem 3.5 is one of the most beautiful results in graph drawing. It is not clear how to make Mani's proof [Bab95, Man71] into an algorithm. It would be very interesting to find a linear-time algorithm that takes a triconnected planar graph as input and draws it as the skeleton of a convex polyhedron so that every automorphism of the graph is a symmetry of the polyhedron.

Acknowledgments

This work has been supported by the Australian Research Council. Parts of this chapter were written when the authors were visiting the University of Kyoto under Grant-in-Aid 16092101 for Scientific Research on Priority Areas from the Ministry of Education, Culture, Sports, Science and Technology of Japan.

References

[AHT07]   David Abelson, Seok-Hee Hong, and Donald E. Taylor. Geometric automorphism groups of graphs. *Discrete Applied Mathematics*, 155(17):2211–2226, 2007.

[AHU83]   A. V. Aho, J. E. Hopcroft, and J. D. Ullman. *Data Structures and Algorithms*. Addison-Wesley, Reading, MA, 1983.

[AM88]   M. J. Atallah and J. Manning. Fast detection and display of symmetry in embedded planar graphs, 1988.

[Arm88]   M. A. Armstrong. *Groups and Symmetry*. Springer-Verlag, 1988.

[Bab95]   L. Babai. Automorphism groups, isomorphism, and reconstruction. In Groetschel Graham and Lovasz, editors, *Handbook of Combinatorics*, volume 2, chapter 27. Elsevier Science, 1995.

[Bac99]   Sabine Bachl. Isomorphic subgraphs. In Kratochvíl [Kra99], pages 286–296.

[BJ01]   Christoph Buchheim and Michael Jünger. Detecting symmetries by branch & cut. In Mutzel et al. [MJL02], pages 178–188.

[BJ03]   Christoph Buchheim and Michael Jünger. An integer programming approach to fuzzy symmetry detection. In Giuseppe Liotta, editor, *Graph Drawing*, volume 2912 of *Lecture Notes in Computer Science*, pages 166–177. Springer, 2003.

[CLY00]   Ho-Lin Chen, Hsueh-I Lu, and Hsu-Chun Yen. On maximum symmetric subgraphs. In Marks [Mar01], pages 372–383.

[CY02]   Ming-Che Chuang and Hsu-Chun Yen. On nearly symmetric drawings of graphs. In *IV*, pages 489–, 2002.

[DETT99] G. Di Battista, P. Eades, R. Tamassia, and I. G. Tollis. *Graph Drawing*. Prentice Hall, Upper Saddle River, NJ, 1999.

[dF99]   Hubert de Fraysseix. An heuristic for graph symmetry detection. In Kratochvíl [Kra99], pages 276–285.

[DT92]   G. Di Battista and R. Tamassia. On-line planarity testing. Report CS-92-39, Comput. Sci. Dept., Brown Univ., Providence, RI, 1992.

[DTT92]   G. Di Battista, R. Tamassia, and I. G. Tollis. Area requirement and symmetry display of planar upward drawings. *Discrete Comput. Geom.*, 7(4):381–401, 1992.

[Ead84]   P. Eades. A heuristic for graph drawing. *Congr. Numer.*, 42:149–160, 1984.

[EL00]   Peter Eades and Xuemin Lin. Spring algorithms and symmetry. *Theor. Comput. Sci.*, 240(2):379–405, 2000.

[ELT96]   P. Eades, X. Lin, and R. Tamassia. An algorithm for drawing a hierarchical graph. *Internat. J. Comput. Geom. Appl.*, 6:145–156, 1996.

[FDK01]   Karlis Freivalds, Ugur Dogrusöz, and Paulis Kikusts. Disconnected graph layout and the polyomino packing approach. In Mutzel et al. [MJL02], pages 378–391.

[Fon76]   M. Fontet. Linear algorithms for testing isomorphism of planar graphs. In *Proceedings Third Colloquium on Automata, Languages, and Programming*, pages 411–423, 1976.

[HE00]   Seok-Hee Hong and Peter Eades. An algorithm for finding three dimensional symmetry in trees. In Marks [Mar01], pages 360–371.

[HE03]   Seok-Hee Hong and Peter Eades. Symmetric layout of disconnected graphs.
          In Toshihide Ibaraki, Naoki Katoh, and Hirotaka Ono, editors, *ISAAC*, vol-
          ume 2906 of *Lecture Notes in Computer Science*, pages 405–414. Springer,
          2003.

[HE05]   Seok-Hee Hong and Peter Eades. Drawing planar graphs symmetrically, ii:
          Biconnected planar graphs. *Algorithmica*, 42(2):159–197, 2005.

[HE06]   Seok-Hee Hong and Peter Eades. Drawing planar graphs symmetrically, iii:
          One-connected planar graphs. *Algorithmica*, 44(1):67–100, 2006.

[HEL00]   Seok-Hee Hong, Peter Eades, and Sang Ho Lee. Drawing series parallel
          digraphs symmetrically. *Comput. Geom.*, 17(3-4):165–188, 2000.

[HEQL98] Seok-Hee Hong, Peter Eades, Aaron J. Quigley, and Sang Ho Lee. Drawing
          algorithms for series-parallel digraphs in two and three dimensions. In
          Graph Drawing, pages 198–209, 1998.

[HME06]   Seok-Hee Hong, Brendan D. McKay, and Peter Eades. A linear time algo-
          rithm for constructing maximally symmetric straight line drawings of tri-
          connected planar graphs. *Discrete & Computational Geometry*, 36(2):283–
          311, 2006.

[Hon01]   Seok-Hee Hong. Drawing graphs symmetrically in three dimensions. In
          Mutzel et al. [MJL02], pages 189–204.

[HT73]   J. Hopcroft and R. E. Tarjan. Dividing a graph into triconnected compo-
          nents. *SIAM J. Comput.*, 2(3):135–158, 1973.

[HW74]   J. E. Hopcroft and J. K. Wong. Linear time algorithm for isomorphism of
          planar graphs. In *Proc. of the Sixth Annual ACM Symposium on Theory
          of Computing*, pages 172–184, 1974.

[Kam88]   T. Kamada. *On Visualization of Abstract Objects and Relations*. PhD
          thesis, Department of Information Science, University of Tokyo, 1988.

[Kam89]   T. Kamada. Symmetric graph drawing by a spring algorithm and its ap-
          plications to radial drawing. Technical report, Department of Information
          Science, University of Tokyo, 1989.

[KK89]   T. Kamada and S. Kawai. An algorithm for drawing general undirected
          graphs. *Inform. Process. Lett.*, 31:7–15, 1989.

[Kra99]   Jan Kratochvíl, editor. *Graph Drawing, 7th International Symposium,
          GD'99, Stirín Castle, Czech Republic, September 1999, Proceedings*, vol-
          ume 1731 of *Lecture Notes in Computer Science*. Springer, 1999.

[Lin92]   X. Lin. *Analysis of Algorithms for Drawing Graphs*. PhD thesis, Depart-
          ment of Computer Science, University of Queensland, 1992.

[LNS85]   R. J. Lipton, S. C. North, and J. S. Sandberg. A method for drawing
          graphs. In *Proc. 1st Annu. ACM Sympos. Comput. Geom.*, pages 153–160,
          1985.

[Lub81]   Anna Lubiw. Some np-complete problems similar to graph isomorphism.
          SIAM J. Comput., 10(1):11–21, 1981.

[MA86]   J. Manning and M. J. Atallah. Fast detection and display of symmetry in
          outerplanar graphs. Technical Report CSD-TR-606, Department of Com-
          puter Science, Purdue University, 1986.

[MA88]   J. Manning and M. J. Atallah. Fast detection and display of symmetry in
          trees. *Congr. Numer.*, 64:159–169, 1988.

[Man71]   P. Mani. Automorphismen von polyedrischen graphen. *Math. Annalen*, 192:279–303, 1971.

[Man90]   J. Manning. *Geometric Symmetry in Graphs*. PhD thesis, Purdue Univ., 1990.

[Mar01]   Joe Marks, editor. *Graph Drawing, 8th International Symposium, GD 2000, Colonial Williamsburg, VA, USA, September 20-23, 2000, Proceedings*, volume 1984 of *Lecture Notes in Computer Science*. Springer, 2001.

[MJL02]   Petra Mutzel, Michael Jünger, and Sebastian Leipert, editors. *Graph Drawing, 9th International Symposium, GD 2001 Vienna, Austria, September 23-26, 2001, Revised Papers*, volume 2265 of *Lecture Notes in Computer Science*. Springer, 2002.

[RT81]   E. Reingold and J. Tilford. Tidier drawing of trees. *IEEE Trans. Softw. Eng.*, SE-7(2):223–228, 1981.

[SR34]   E. Steinitz and H. Rademacher. *Vorlesungen über die Theorie der Polyeder*. Julius Springer, Berlin, Germany, 1934.

[SR83]   K. J. Supowit and E. M. Reingold. The complexity of drawing trees nicely. *Acta Inform.*, 18:377–392, 1983.

[Tut63]   W. T. Tutte. How to draw a graph. *Proceedings London Mathematical Society*, 13(52):743–768, 1963.

[Tut66]   W. T. Tutte. *Connectivity in Graphs*. University of Toronto Press, 1966.

[Wie64]   Wielandt. *Finite Permutation Groups*. Academic Press, 1964.

[Won75]   J. K. Wong. *Isomorphism Problems Involving Planar Graphs*. PhD thesis, Cornell Univ., 1975.

4

Proximity Drawings

Giuseppe Liotta
University of Perugia

4.1 Introduction

In 1969, Gabriel and Sokal [GS69] presented a method for associating a graph to a set of geographic data points P by connecting points $x, y \in P$ with an edge if and only if the closed disk having the segment $\overline{xy}$ as diameter contained no other point of P. This geometric graph, called the *Gabriel graph* of P, is just one example of what have come to be called *proximity graphs*. Loosely speaking, a proximity graph is a geometric graph (i.e., a straight-line drawing) constructed from a set P of points in some metric space by connecting pairs of points that are deemed to be "sufficiently" close together. A set P can give rise to a variety of different proximity graphs depending upon the definition of closeness used.

Proximity graphs have applications in numerous areas where they are commonly used to describe the underlying "shape" of a set of points, including computer graphics, computational geometry, pattern recognition, computational morphology, numerical analysis, computational biology, and GIS (see, e.g., [OBS92, GO04]). A paper by Toussaint [Tou05] describes applications of proximity graphs in the context of instance-based learning and data-mining and a paper by Carreira-Perpinan and Zemel [CPZ04] to the field of clustering and manifold learning. Motivated by these many applications, a rich body of computational geometry literature has been devoted to the question of efficiently computing different types of proximity graphs of a given set of points. For exhaustive lists of references on the subject, the interested reader is referred to the above-mentioned paper by Toussaint [Tou05] and to the survey by Jaromczyk and Toussaint [JT92].

In this chapter, we shall look at proximity from the different perspective of graph drawing: The goal is computing a straight-line drawing of a given graph with the additional constraint that the drawing be a proximity graph. There is a strong connection between the graph drawing and the computational geometry point of views about proximity. Indeed, the (computational geometry) problem of analyzing the combinatorial properties of a given type of geometric graph naturally raises the (graph drawing) question of characterizing those graphs that admit the given type of straight-line drawing. This in turn leads to the investigation of the design of efficient algorithms for computing such a drawing when one exists.

We therefore will talk about the *proximity drawability problem*: Given a graph G and a definition of proximity, determine whether a set P of points exists such that the proximity graph of P is isomorphic with the given graph, and if so, compute such a set. Clearly, the set P, if it exists, gives rise to a straight-line drawing of G, called a *proximity drawing* of G, where each vertex of G is mapped to a distinct point of P and each edge to a straight-line segment between pairs of points of P. Proximity drawings have several interesting features. They are usually unaffected by changes in scale, since the measures of proximity used are based on relative distances between points. Also, adjacent vertices are drawn (relatively) more closely together than non-adjacent vertices, and vertices not incident to a particular edge are not drawn too close to the edge. Furthermore, neighbors of a given vertex tend to cluster together.

This chapter surveys some of the central problems, results, and research trends on proximity drawings. Although many of the ideas described here can be developed in the more general setting of a metric space, we shall most often assume that the drawings are to be made in Euclidean d-space (the only exception will be for Voronoi and Delaunay drawings).

The remainder of this chapter is structured as follows. In Section 4.2, various definitions of proximity drawings are given; in Section 4.3, the basic graph drawing literature on the proximity drawability problem is reviewed. Section 4.4 introduces extensions and relaxations of the definition of proximity drawing that make it possible to significantly enlarge the families of representable graphs. Some challenging open problems on proximity drawings are listed in Section 4.5. Finally, Section 4.6 concludes the chapter by briefly pointing at two research directions in the areas of sensor networks and of robust geometric computing where proximity graphs and drawings have received some attention in the last few years.

4.2   Proximity Rules and Proximity Drawings

At a first, broad approximation, the definition of closeness in a proximity drawing can be either based on the concept of *proximity region* or based on a *global proximity* measure. In a proximity region based drawing two or more vertices are adjacent if and only if some suitably defined region that describes the neighborhood of these vertices contains at most k other vertices, for a given integer value $k \geq 0$. Global proximity, by contrast, gives rise to proximity drawings where the overall sum of the lengths of the edges in the drawing is minimized. In the remainder of this section, we recall some of the most common definitions of region-based and of global proximity rules and drawings; unless stated otherwise, P denotes the set of vertices of a straight-line drawing Γ of some graph G.

4.2.1   Proximity Region Based Drawings

Let R be a function that associates to every set S of $k \geq 2$ points in Euclidean d-space E^d a subset $R(S)$ of E^d; $R(S)$ is called the *proximity region* or *region of influence* of S.

Now consider a straight-line drawing Γ of G in which the vertices are drawn at a set of locations P. Drawing Γ is an (h,k)-*proximity drawing* of G if Γ results from the following procedure: For every set $S \subset P$ of h vertices, edges are drawn between all pairs of vertices in S if and only if the proximity region $R(S)$ contains at most k vertices from $P - S$. While the proximity region can be any subset of the space in question, usually the regions chosen are homeomorphic to an open or closed ball of dimension equal to that of the space. Such drawings are referred to as *open* or *closed* proximity drawings , respectively. Examples of open/closed (h,k)-proximity drawings follow.

The *Gabriel region* [GS69] of two vertices x and y is defined to be the closed sphere (in d dimensions) having the segment $\overline{xy}$ as diameter. A *Gabriel drawing* is a closed $(2,0)$-proximity drawing where the region of influence is the Gabriel region. Indeed, a Gabriel drawing of G is a straight-line drawing of G having the property that two vertices x and y of the drawing form an edge if and only if the Gabriel region of x and y does not contain any other vertex. Figure 4.1 (a) shows a Gabriel drawing of a planar triangulated graph; in the figure, vertices p and q are adjacent because their Gabriel region does not contain any other vertex, while vertices u and v are not adjacent because their Gabriel region contains vertex q. If one changes the definition of Gabriel region by saying that the sphere defined by the two vertices x and y is an open set, then the corresponding proximity region is termed a *modified Gabriel region* and the associated drawing is called a *modified Gabriel drawing*. Figure 4.1 (b) shows a modified Gabriel drawing of a wheel graph of five vertices; note that vertices u and v are adjacent because the modified Gabriel region is an open set and hence it does not contain vertex q. Note that the graph of Figure 4.1 (b) does not have a Gabriel drawing [Cim92].

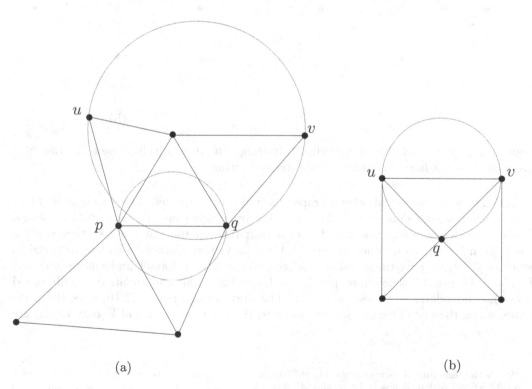

(a)   (b)

Figure 4.1   (a) A Gabriel drawing. (b) A modified Gabriel drawing of a graph that does not have a Gabriel drawing.

A *relative neighborhood drawing* of a graph G is an open $(2, 0)$-proximity drawing in which the region of influence of two points x and y is the intersection of the open disks of radius $d(x, y)$ centered at x and y. Thus, in a proximity drawing of G, x and y are adjacent if and only if there is no vertex whose distance to both x and y is less than the distance between x and y. The proximity region of x and y is called the *relative neighborhood region* or *lune* of x and y [Tou80].[1] A relative neighborhood drawing of a wheel graph consisting of a vertex of degree six adjacent to five vertices of degree three is depicted in Figure 4.2 (a): Since the relative neighborhood region is an open set, vertex w is not in the relative neighborhood region of vertices u and v and therefore the two vertices are adjacent.

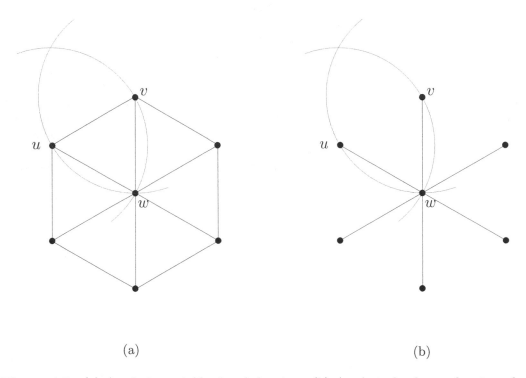

(a)   (b)

Figure 4.2   (a) A relative neighborhood drawing. (b) A relatively closest drawing of a tree that does not have a relative neighborhood drawing.

Variants of relative neighborhood graphs have also been studied. One example is the *k-relative neighborhood drawing* [CTL92] where the proximity constraint is relaxed by saying that two vertices are adjacent if and only if their lune contains at most k other vertices (for a given $k > 0$). As another example, if the relative neighborhood region is assumed to be a closed set, we have the so-called *relatively closest region* [Lan69] and *relatively closest drawing*. Figure 4.2 shows how proximity drawability differs for relative neighborhood drawings and relatively closest drawings. The drawing of Figure 4.2 (b) uses the same vertex set as the one of Figure 4.2 (a); however, the vertices u and v of Figure 4.2 (b) are

---

[1]While the term lune is commonly used in the computational geometry literature to denote the relative neighborhood region, it has to be recalled that in plane geometry the non-empty intersection of two disks of equal radius is called *symmetric lens*.

not adjacent because their relatively closest region, which is a closed set, contains vertex w. Note that a tree with one interior vertex and six leaves does not admit a relative neighborhood drawing [BLL96].

The Gabriel, modified Gabriel, relative neighborhood, and relatively closest drawings described above are all examples of members of a family of drawings called β-*drawings*. In 1985, Kirkpatrick and Radke [KR85, Rad88] introduced a family of closed $(2,0)$-proximity regions called β-*neighborhoods*, denoted by $R[x, y, \beta]$ and defined as follows (see also Figure 4.3):

1. For $\beta = 0$, $R[x, y, \beta]$ is the line segment $\overline{xy}$.
2. For $0 < \beta < 1$, $R[x, y, \beta]$ is the intersection of the two closed disks of radius $d(x, y)/(2\beta)$ passing through both x and y.
3. For $1 \leq \beta < \infty$, $R[x, y, \beta]$ is the intersection of the two closed disks of radius $\beta d(x, y)/2$ and centered on the line through x and y.
4. For $\beta = \infty$, $R[x, y, \beta]$ is the closed infinite strip perpendicular to the line segment $\overline{xy}$.

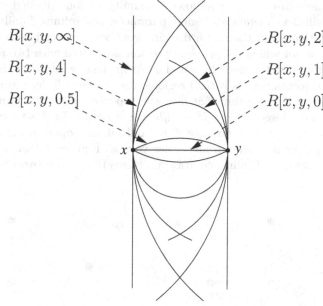

Figure 4.3 A set of $(2,0)$-proximity regions $R[x, y, \beta]$.

Obviously, one can also define the analogous regions $R(x, y, \beta)$ using open sets instead of closed sets ($R(x, y, 0)$ is defined to be the empty set). The Gabriel, modified Gabriel, relative neighborhood and relatively closest drawings mentioned above are obtained from the β-regions $R[x, y, 1]$, $R(x, y, 1)$, $R(x, y, 2)$ and $R[x, y, 2]$, respectively. The *closed strip drawings* are β-drawings that use the region $R[x, y, \infty]$. Similarly, the *open strip drawings* are β-drawings that use the region $R(x, y, \infty)$. The regions defined above are also referred to as *lune-based* β-regions. In the same papers, Kirkpatrick and Radke [KR85, Rad88] also describe *circle-based* β-regions: for each $\beta \geq 1$, the region associated with two vertices x and y is the union of the two disks of radius $\beta d(x, y)/2$ passing through both x and y and centered on the line through them.

In the $(2,0)$-proximity drawings described above, the proximity region chosen for a pair of vertices x, y is symmetric about the perpendicular bisector of the segment $\overline{xy}$. This guarantees a certain symmetry in the drawings produced. This symmetry, however, is not always desirable. Veltkamp [Vel92, Vel94, Vel95] introduced a family of proximity regions, called γ-*regions*, in which the proximity region may lack this symmetry and that generalize lune-based and circle-based β-regions. While Veltkamp takes advantage of this absence of symmetry in constructing object boundaries from a set of points in the context of visual processing and pattern recognition, the notion of γ-regions can be used from a graph drawing perspective to define γ-*drawings*. Another generalization of proximity drawings based on β-region are the so-called *empty region graphs*, recently introduced by Cardinal, Collette, and Langerman [CCL09]. An empty region graph is a proximity drawing where the proximity region of any pair of points u and v in the plane is a *template region*, that is a function mapping the pair u, v to a subset of the plane. In particular, the authors focus on the combinatorial properties of proximity graphs whose template regions are convex and symmetric.

Several $(2,0)$-proximity regions can be seen as special cases of either β-regions or γ-regions; however, there are some well-known proximity regions defined in the literature that cannot be classified as members of some parameterized infinite family. Among those that have been investigated in the graph drawing context, we recall here the *rectangle of influence* [IS85] region, for which the proximity region associated with two points x and y is the axis-parallel rectangle determined by x and y. As in the case of β-drawings, one can use either open or closed rectangles; as with β-regions, the choice will determine which graphs can be drawn. A proximity drawing that uses the (open or closed) rectangle of influence region is called (open or closed) *rectangle of influence drawing*. In this type of drawing two vertices x and y are connected by an edge if and only if the (open or closed) rectangle of influence of x and y does not contain any other vertex. Figures 4.4 (a) and (b) show an open and a closed rectangle of influence drawing, respectively; the two drawings have the same vertex set.

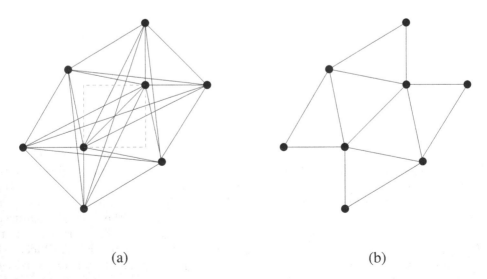

(a)   (b)

Figure 4.4   (a) An open rectangle of influence drawing; the dotted box represents the rectangle of influence of two vertices. (b) A closed rectangle of influence drawing.

The $(2,k)$-proximity drawing paradigm can also be used to produce drawings of directed graphs by associating with each *ordered* pair of points (x,y) a proximity region $R_{x,y}$. By allowing the region $R_{x,y}$ to be different from the region $R_{y,x}$, it is possible to produce drawings where the edge (x,y) is in the drawing, but not the edge (y,x). An early example of this is the *nearest neighbor drawing* (see, e.g, [PY92]), where each vertex $x \in P$ is connected to all vertices (or sometimes just one) of minimum distance from x. Although the nearest neighbor drawing is usually considered to be an undirected graph, the definition is inherently that of a directed graph. The proximity region $R_{x,y}$ in this case is the open disk of radius $d(x,y)$ centered at x.

Besides $(2,h)$-proximity drawings, there are many other meaningful and well-investigated families of proximity drawings. A *Delaunay drawing* [Del34] is an example of a closed $(3,0)$-proximity drawing: here triplets of points in P are connected into triangles if and only if the closed disk they determine contains no other points of P. Delaunay drawings make sense for planar triangulated graphs (a Delaunay drawing is commonly called a *Delaunay triangulation* in the computational geometry literature). A *Delaunay drawing of order h* (usually called a *higher order Delaunay triangulation* in the computational geometry literature [GHvK02]) is a $(3,h)$-proximity drawing of a planar triangulated graph, where for every triplet of vertices connected into a triangle the closed disk through the triplet contains at most h other points of P (for a given integer $h \geq 0$).

Related to Delaunay triangulations is another well-known proximity graph, namely, the *Voronoi diagram* (see, e.g., [PS90]). A Voronoi diagram of a set of points P is the geometric dual of the Delaunay triangulation of P, i.e., it is the straight-line drawing whose edges are the perpendicular bisectors of the edges of the Delaunay triangulation and whose vertices are the intersection points of these perpendicular bisectors. Equivalently, the Voronoi diagram of P, for a given metric, is a subdivision of the plane into regions such that each region is associated with a distinct point p of P and it contains all points of the plane that are closer to p than to any other elements of P. We can therefore define a new type of proximity drawing: A *Voronoi drawing* [LM03] of a graph G is a straight-line drawing of G that is also the Voronoi diagram of some set of points (also called *sites*).

Figures 4.5 (a) and (b) show examples of Voronoi drawings in the Euclidean and in the Manhattan metric, respectively. In the figure, the white points are the sites; for display purposes, the edges of infinite length of the Voronoi diagram have been replaced by edges of finite length and endvertices have been added.

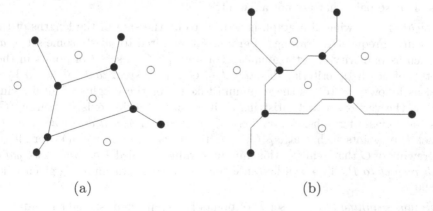

(a)   (b)

Figure 4.5 Voronoi drawings: (a) in the Euclidean metric and (b) in the Manhattan metric.

In our definition of (k, n)-proximity drawings, we have required that the sets S to which we associate proximity regions contain at least two points, since otherwise no edges can be formed. There is, however, a way in which proximity regions associated with single points can be used to create proximity drawings: pairs of points can be connected by an edge if the regions corresponding to the points intersect. We call such drawings *intersection drawings*. An example of such a proximity drawing would be a *sphere of influence drawing* of a graph. To produce this type of drawing, each point $x \in P$ has, as its proximity region, its *sphere of influence* [Tou88], namely, the disk centered at x of radius $r_x = \min\{d(x, y) : y \in P - \{x\}\}$. One can consider either open or closed sphere of influence drawings. An example of a sphere of influence graph is depicted in Figure 4.6; the drawing is valid for both the open and the closed model of proximity.

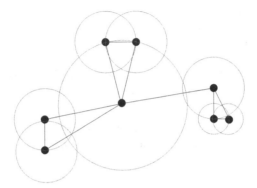

Figure 4.6   A sphere of influence drawing.

4.2.2   Global Proximity

Several graph drawing algorithms are designed to produce a representation of a graph that is as small as possible in some sense. For example, given a resolution rule (i.e., a minimum acceptable distance between any pair of graphic features in the drawing) one may want to optimize the area of the drawing or aim for minimum edge lengths. A proximity drawing that adopts a global measure of proximity is, in a sense, the smallest possible representation of a graph because it globally maximizes the closeness of adjacent vertices and the reciprocal distances of those pairs that are not adjacent.

The *weight* of a drawing of a graph is defined to be the sum of the lengths of the edges of the drawing. Frequently, drawings of graphs are required to satisfy some set of *aesthetic criteria* such as planarity or orthogonality. For a graph G, a set P of points in the plane, and a set $\mathcal{E}$ of aesthetic criteria, the *weight of G with respect to P*, denoted by $w_P(G)$, is defined as follows: $w_P(G)$ is the minimum taken over the weights of all drawings of G having P as the vertex set and satisfying $\mathcal{E}$; if no such drawing exists, then $w_P(G) = \infty$. Now let $\mathcal{C}$ be a class of graphs. A graph $G \in \mathcal{C}$ is *minimum weight drawable (for $\mathcal{C}$)* if there exists a set P of points such that $w_P(G)$ is finite and G minimizes $w_P()$ over all graphs in $\mathcal{C}$. Any drawing of G that achieves this minimum value is called a *minimum weight drawing of G with respect to P*. Two well-known examples of such drawings are given below (see, e.g., [PS90]).

A *minimum spanning tree* of a set P of points is a connected, straight-line drawing that has P as its vertex set and that minimizes the total edge length. So, letting $\mathcal{C}$ be the class of all trees, and letting $\mathcal{E}$ denote straight-line planar drawings, a tree G is minimum weight drawable for $\mathcal{C}$ if there exists a set P of points in the plane such that G minimizes $w_P()$ over all trees. This is equivalent to saying that G is isomorphic to a minimum spanning tree of P. Figure 4.7 shows a minimum weight drawing of a tree. A *minimum weight triangulation* of a set P is a triangulation of P having minimum total edge length. Letting $\mathcal{C}$ be the class of all planar triangulations, and letting $\mathcal{E}$ be as above, a planar triangulation G is minimum weight drawable for $\mathcal{C}$ if there exists a set P of points in the plane such that G is isomorphic to a minimum weight triangulation of P.

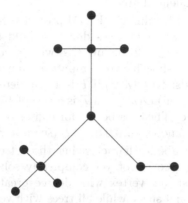

Figure 4.7   A minimum weight drawing of a tree.

We conclude this section by remarking that there are strong relations between global and region-based proximity rules. For example, it is well known that every minimum spanning tree on a set P of points is a subgraph of the Delaunay triangulation of P (see, e.g., [PS90]). Also, a significant research effort can be found in the computational geometry literature devoted to studying the relationships between $(2, k)$-proximity and minimum weight triangulations (see, e.g., [Kei94, WY01, CX01]).

4.3   Results

In this section, we survey some of the most relevant results on the proximity drawability problem for the types of proximity drawings described in the previous section.

4.3.1   Minimum Weight Drawings

If a graph admits a minimum weight drawing, it is called *minimum-weight drawable*; otherwise, it is called *minimum-weight forbidden*. As already mentioned above, most research on minimum weight drawings has focused on trees and on planar triangulated graphs.

The problem of testing whether a tree can be drawn as a Euclidean minimum spanning tree in the plane is essentially solved. Monma and Suri [MS92] proved that each tree with maximum vertex degree at most five can be drawn as a minimum spanning tree of some set of vertices by providing a linear-time (real RAM) algorithm. In the same paper, it is shown that every tree having at least one vertex with degree greater than six is minimum weight forbidden. As for trees having maximum degree equal to six, Eades and Whitesides [EW96b] showed that it is NP-hard to decide whether such trees can be drawn as minimum spanning trees.

One of the most challenging questions in the seminal paper by Monma and Suri [MS92] was about the area required by a minimum weight drawing of a tree. Namely, the construction by Monma and Suri used a grid of size $O(2^{n^2}) \times O(2^{n^2})$ and the authors conjectured an exponential lower bound for minimum weight drawings of trees with maximum vertex degree five (i.e., the existence of a tree T with n vertices such that any minimum weight drawing of T requires area at least $c^n \times c^n$ for some constant $c > 1$). This long-standing conjecture was only recently proved to be correct by Angelini et al. [ABC+11], who describe a tree T with n vertices having maximum degree five such that in any minimum weight drawing of T the ratio between the longest and the shortest edge is $2^{\Omega(n)}$, which implies that the drawing requires exponential area.

On the other hand, Frati and Kaufmann [FK11] proved that the exponential area lower bound of minimum weight drawings of trees does not hold for maximum vertex degree smaller than five. More precisely, let T be any tree with n vertices and maximum vertex degree four; Frati and Kaufmann show how to compute a minimum weight drawing of T with the following area upper bounds: (i) $O(n^{4.3})$ if T is a complete binary tree; (ii) $O(n^{11.3875})$ if T is an arbitrary binary tree; (iii) $O(n^{3.73})$ if T is a complete ternary tree; (iv) $O(n^{21.252})$ if T is an arbitrary ternary tree. The area bound for complete binary tree has been further reduced to $O(n^{3.8})$ by Di Giacomo et al. [DDLM12] (see also Section 4.4.3).

The 3-dimensional question about characterizing those trees that can be drawn as a Euclidean minimum spanning tree is not yet completely solved. In [LD95], it is shown that every tree having at least one vertex with degree greater than twelve is minimum weight forbidden in 3-dimensional space while all trees with vertex degree at most nine are drawable. King [Kin06] improved this last result by showing that all trees whose vertices have vertex degree at most ten can be realized as a Euclidean minimum spanning tree in 3-dimensional space. In general, the maximum vertex degree of a minimum weight drawable tree is bounded by the *kissing number*, i.e., by the maximum number of disjoint unit spheres that can be simultaneously tangent to a given unit sphere [RS95].

A significant research effort has also been devoted to drawing a planar triangulated graph G as a minimum weight triangulation of the points representing the vertices. However, the problem is still far from being solved. It may be worth recalling that, while computing a Euclidean minimum spanning tree of a set of points in the plane is solvable in polynomial time (see, e.g. [PS90]), the problem of computing a Euclidean minimum weight triangulation of a set of points in the plane is NP-hard [MR08].

In [LL96], it is shown that all maximal outerplanar triangulations are minimum-weight drawable, and a linear time (real RAM) drawing algorithm for constructing such a drawing is given. This naturally leads to investigation of the internal structure of minimum-weight drawable planar triangulated graphs. In [LL02] the authors examined the *endoskeleton*— or *skeleton*, for short—of planar triangulated graphs, that is, the subgraph induced by the internal vertices of the triangulation. They constructed skeletons that cannot appear in any minimum weight drawable triangulation and skeletons that guarantee minimum weight drawability. More precisely, the known results about of minimum weight drawable triangulations are as follows.

- In [LL02], the authors showed that any forest can be realized as the skeleton of some minimum weight triangulation. On the other end, Wang, Chin, and Yang [WCY00] gave examples of triangulations that do not admit a minimum weight drawing even if their skeleton is acyclic.
- In [LL02], it is also shown that any traingulation containing either the graph of Figure 4.8 (a) or the graph of Figure 4.8 (b) is not minimum weight drawable.

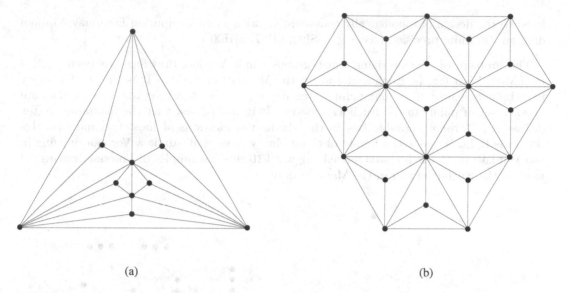

(a)   (b)

Figure 4.8 Two examples of triangulations that cannot be drawn as minimum-weight triangulations.

Another contribution of [LL02] is to study the relationship between Delaunay drawability and minimum weight drawability. The authors described graphs that do not admit a Delaunay drawing but do have a minimum weight drawing. One such example is the minimum weight drawing of Figure 4.9: As explained in the next section,the depicted graph violates a necessary condition for Delaunay drawability (see also Figure 4.11 (b)).

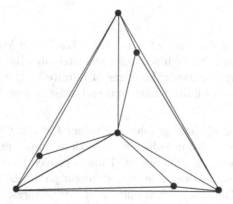

Figure 4.9 A minimum weight drawing of a Delaunay forbidden graph (see also Figure 4.11 (b).

4.3.2   Delaunay and Voronoi Drawings

The study of the combinatorial properties of Delaunay triangulations and of Voronoi diagrams (i.e., the Delaunay and the Voronoi drawability problems) has a long tradition in the computational geometry literature and is of particular interest because it is closely related to the design of topologically consistent algorithms for computing Delaunay/Voronoi diagrams in finite precision (see, e.g., [SI92, SH97, SIII00]).

The problem of characterizing which graphs admit Voronoi drawings has been studied in [LM03] both for the Euclidean and for the Manhattan metric. It is shown that every tree, independently of its maximum vertex degree, can be drawn as the Voronoi diagram of some set of points in the Euclidean metric. It is also proved that the maximum vertex degree of a Voronoi drawable tree in the Manhattan metric is at most five and that this bound is tight. Finally, the family of those binary trees that admit a Voronoi drawing in the Manhattan metric is characterized. Figure 4.10 shows examples of Voronoi drawings of trees in the Euclidean and in the Manhattan metric.

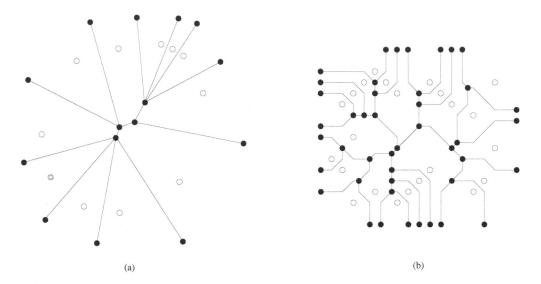

(a)   (b)

Figure 4.10 Two Voronoi drawings of trees: (a) a Euclidean Voronoi drawing and (b) a Manhattan Voronoi drawing. The white circles are sites and the black circles are vertices of the drawing. For display purposes, the edges of infinite length of the Voronoi diagrams have been replaced by edges of finite length and endvertices have been added.

An exact characterization of those graphs that admit Delaunay drawings remains a challenging open problem; however, some sufficient conditions and some necessary conditions for Delaunay drawability are known. Dillencourt [Dil90a] proved that if a graph G is maximal outerplanar, then G is Delaunay drawable. In a different paper, Dillencourt [Dil90b] studied the relationship between Delaunay drawability and 1-toughness. A graph G is *1-tough* if for any non-empty set S of vertices of G, the number of components obtained from G by removing the vertices of S and their incident edges is at most $|S|$. For example, the graph of Figure 4.11 (a) is not 1-tough because the removal of the four white vertices and of their incident edges results in a graph with five components. Dillencourt showed in [Dil90b] that every Delaunay drawable graph either (a) is 1-tough or (b) for any set S of vertices of G, the number of interior components obtained from G by removing the vertices of S and their incident edges is at most $|S| - 2$ (an interior component is a component that has no vertices in the outerface of G). This necessary condition is used by Dillencourt to construct examples of graphs that are nor Delaunay drawable. For example, neither graph of Figure 4.11 is Delaunay drawable: as already explained, the one of Figure 4.11 (a) violates the 1-toughness condition; the one of Figure 4.11 (b), violates the second necessary condition stated above because removing the four white vertices gives rise to three interior components. Another interesting property proved in [Dil90b] is that any Delaunay drawable graph has a perfect matching.

Based on the strict connection between the convex hull of a set of non-coplanar points on the surface of a sphere and a (2-dimensional) Delaunay triangulation [Bro79], the following equivalent definition of Delaunay drawable graphs was also given by Dillencourt [Dil96]: A planar triangulated graph G with triangular outerface is Delaunay drawable if and only if it is inscribable, i.e., it can be drawn in 3-dimensional space as the convex hull of a set of non-coplanar points on the surface of a sphere. If the outerface f of G is not triangulated, then G is Delaunay drawable if and only if the graph obtained from G by "stellating" f (i.e., by adding a vertex in f and connecting it to all vertices of f) is inscribable. Dillencourt and Smith [DS95] showed that every planar triangulated graph whose vertices all have degree three is inscribable (after having possibly stellated the outerface) and therefore Delaunay drawable. The same authors showed in [DS94] that any 4-connected planar graph is inscribable and that any triangulated graph with triangular outerface and without chords or non-facial triangles is Delaunay drawable.

The question whether Delaunay drawable graphs are Hamiltonian was posed by Mathieu [Mat87] and by O'Rourke [O'R87]. Examples of Delaunay drawable graphs that are not Hamiltonian can be found in papers by Dillencourt [Dil87, Dil89] and by Kantabutra [Kan83]. These examples suggested the question of the computational complexity of the Hamiltonicity of Delaunay drawable graphs. The question was answered by Dillencourt [Dil96], who proved that determining whether a Delaunay drawable graph is Hamiltonian is NP-complete. In the same paper it is also shown that there exist Delaunay drawable graphs that do not have a 2-factor (a 2-factor of a graph is a spanning collection of disjoint cycles). Finally, in the papers by Di Battista and Vismara [DV96], and by Sugihara and Hiroshima [SH97], the angles of Delaunay drawings were characterized.

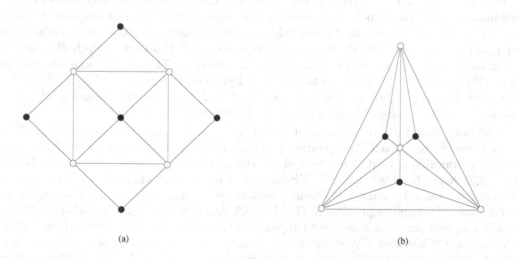

(a)

(b)

Figure 4.11 Two graphs that are not Delaunay drawable: (a) The graph is not 1-tough: removing the white vertices produces five components. (b) Removing the four white vertices produces three internal components.

4.3.3   Rectangle of Influence Drawings

The rectangle of influence drawability problem was first defined in [LLMW98], where both
the case that the rectangle of influence is an open set and the case that it is a closed set
are investigated. For both cases, characterization results are presented concerning cycles,
wheels, trees, outerplanar graphs, and cliques. As already observed, the set of representable
graphs can be quite different, depending on whether the open or the closed rectangle of
influence is used to define the proximity drawing. For example, Figure 4.12 (a) shows a
closed rectangle of influence drawing of a 4-cycle, which is not an open rectangle of influence
drawable graph. Figure 4.12 (b) gives an open rectangle of influence drawing of K_5 (i.e.,
the complete graph on five vertices), which is not a closed rectangle of influence drawable
graph.

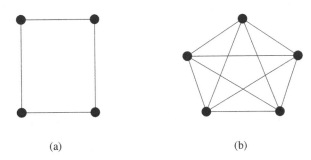

(a)   (b)

Figure 4.12   Examples of rectangle of influence drawings: (a) a closed rectangle of influence drawing of a 4-cycle; (b) an open rectangle of influence drawing of K_5. Note that a
4-cycle does not admit an open rectangle of influence drawings and that k_5 is not closed
rectangle of influence drawable.

4.3.4   Nearest Neighbor Drawings

Paterson and Yao [PY92] started the investigation of the combinatorial properties of nearest
neighbor graphs. Among other basic results, they proved that a nearest neighbor drawable
tree cannot branch too much: if the depth of the tree is high, then the tree contains some
long paths. More precisely, Paterson and Yao showed that if a tree of depth D is nearest
neighbor drawable, then it can have at most $O(D^9)$ vertices. The upper bound was reduced
to $O(D^6)$ by Eppstein [Epp92] and to $O(D^5)$ by Eppstein, Paterson, and Yao [EPY97]; this
last upper bound is tight since Paterson and Yao [PY92] had shown the existence of nearest
neighbor drawable graphs of depth D and $\Omega(D^5)$ vertices.

A precise characterization of nearest neighbor drawable graphs is still unknown. Eppstein, Paterson, and Yao [EPY97] conjectured that deciding whether a given graph is nearest
neighbor drawable is hard. The truth of the conjecture was proved by Eades and Whitesides [EW96a] who show that it is NP-hard to determine whether a graph G is nearest
neighbor drawable by using a mechanical device, called "logic engine," that simulates the
well-known NP-complete problem NOT-ALL-EQUAL-3SATISFIABILITY [GJ79] and that
provides a proof paradigm based on an approach first used by Bhatt and Cosmodakis [BC87].
Kitching and Whitesides [KW04] extend the technique to 3-dimensional space and, by using
a "3-dimensional" logic engine, prove that the mutual nearest neighbor drawability problem
is NP-hard in 3-dimensional space. It may be worth recalling that the logic engine paradigm
can be used to prove the hardness of other graph drawing problems such as, for example,

determining whether a graph is a subgraph of the hexagonal tiling or an induced subgraph of the square or hexagonal tilings [Epp09].

4.3.5 Sphere of Influence Drawings

Basic properties of sphere of influence drawable graphs are discussed by Harary et al. [HJLM93]. Harary et al. showed that if a graph G is open/closed sphere of influence drawable, an induced subgraph of G may not necessarily be open/closed sphere of influence drawable. This nonhereditary property greatly complicates the problem of characterizing sphere of influence drawable graphs. The conjecture of Harary et al. that K_9 does not admit an open sphere of influence drawing remains, to date, an open problem.

On the positive side, Jacobson, Lipman, and McMorris [JLM95] proved that if G is triangle-free and admits a sphere of influence drawing, then any subgraph of G is also drawable. Jacobson, Lipman, and McMorris exploited this result to characterize those trees that admit an open/closed sphere of influence drawing: A tree is open sphere of influence drawable if and only if it has a perfect matching; a tree is closed sphere of influence drawable if and only if it contains a $\{P_2, P_3\}$-factor (see, e.g., [Har69] for a definition of $\{P_2, P_3\}$-factor).

The number of edges of sphere of influence drawable graphs was independently studied by several researchers. Avis and Horton [AH82] proved that the number of edges of an open sphere of influence drawable graph cannot be larger than $29n$, where n is the number of vertices. An upper bound of $21n$ had also been already proven by Besicovitch [Bes45] in 1945, although he was not aware of the application of his result to the sphere of influence drawability problem. The bound of Besicovitch had later been improved by Reifenberg [Rei48] in 1948 and independently by Bateman and Erdős [BE51] in 1951, who showed an upper bound of $18n$ for the problem. Michael and Quint [MQ94b] had lowered the bound to $17.5n$. The best-known upper bound is due to Soss [Sos99a], who showed that any open/closed sphere of influence drawable graph can have at most $15n$ edges.

The study of sphere of influence drawings has also been extended to d-dimensional space and/or to different metrics (see, e.g., [GPS94, Sos99b, MQ99, MQ03]). The interested reader is also referred to the papers by Michael and Quint [MQ94a, MQ03] and to the work of Boyer, Lister, and Shader [BLS00] for more references and a list of open problems concerning the sphere of influence drawability problem.

4.3.6 β-Drawings

Kirkpatrick and Radke [KR85, Rad88] defined the open and closed β-regions ($R(x, y, \beta)$, $R[x, y, \beta]$) discussed in the previous section. From the graph drawing perspective, the central problem is that of determining, for a given graph G, the values of β such that G admits a β-drawing. For example, for $\beta < 2$, only connected graphs admit β-drawings; for $\beta > 1$, only planar graphs do. As mentioned previously, the open and closed β-drawings include several well-studied proximity drawings, including Gabriel, Modified Gabriel, relative neighborhood, and relatively closest drawings; indeed, the Gabriel region is the closed β-region for $\beta = 1$, the modified Gabriel region is the open β-region for $\beta = 1$, the relative neighborhood region is the open β-region for $\beta = 2$, and the relatively closest region is the closed β-region for $\beta = 2$. Some papers about these types of drawings are described below.

Toussaint [Tou80] studied the relationship between the graphs produced by relative neighborhood drawings and other proximity drawings. He showed that the relative neighborhood drawing on a set P of points is a supergraph of every minimum spanning tree of P and a subgraph of the Delaunay triangulation of P. Agarwal and Matoušek [AM92] showed that the number of edges of an n-vertex graph that has a relative neighborhood drawing in 3-dimensional space is $O(n^{4/3})$. Chazelle, Edelsbrunner, Guibas, Hershberger, Seidel, and Sharir [CEG$^+$94] showed that the maximum number of edges of an n-vertex graph that has a Gabriel drawing in d-dimensional space $(d \geq 3)$ is $\Omega(n^2)$. In [MS80], [Tou80], and [Lan69], the planarity of Gabriel drawable graphs, relative neighborhood graphs, and relatively closest drawable graphs were shown, respectively. Furthermore, in [Cim92] it was shown that a cycle with three vertices is not relatively closest drawable.

Particular attention has been devoted in the literature to β-drawings of trees. Matula and Sokal [MS80] gave a partial characterization of trees that admit Gabriel drawings. They proved that every tree with vertex degree at most three admits a Gabriel drawing, while no tree with vertex degree greater than six does. Urquhart [Urq83] gave the same two bounds on the vertex degree of relative neighborhood drawable trees. Cimikowski [Cim92] further extended the bounds to both modified Gabriel drawable and relatively closest drawable trees. Matula and Sokal [MS80] also conjectured that Gabriel drawable trees cannot have vertices of degree greater than four and cannot have two adjacent vertices of degree four.

The gaps left open in the above papers between the smallest and the largest vertex degree of a representable tree were the subject of a paper by Bose et al. [BLL96], who presented a complete characterization of those trees that admit Gabriel, Modified Gabriel, relative neighborhood, and relatively closest drawings. They showed that a tree admits a relative neighborhood and a relatively closest drawing if and only if its maximum vertex degree is at most five; also, a tree has a modified Gabriel drawing if and only if its maximum vertex degree is at most three. As for Gabriel drawability, they proved the truth of the conjecture by Matula and Sokal and characterized the family of representable trees by exhibiting families of forbidden subtrees and by showing that every tree that does not contain members of these families is Gabriel drawable. In the same paper, Bose et al. also presented linear-time algorithms to test whether a tree admits one of the above proximity drawings; it is shown that if such a drawing exists, one can be constructed in linear time in the real-RAM model.

As for other β-neighborhoods, Kirkpatrick and Radke [KR85] studied open strip drawable graphs (i.e., graphs that have β-drawings that use the open β-region $R(x, y, \infty)$) and showed that neither non-planar graphs nor triangulated planar graphs admit open strip drawings. A characterization of closed strip drawable graphs (i.e., proximity graphs that use the $R[x, y, \infty]$ region) can be found in the work by Bose et al. [BDLL95], where it is shown that a graph admits a closed strip drawing if and only if it is a binary forest other than one of the following: two non-adjacent vertices, a vertex and a non-adjacent edge, or two non-adjacent edges.

Bose et al. [BDLL95] also studied the proximity drawability of trees in the whole spectrum of β-proximity regions. Let $\mathcal{T}(\beta)$ $(\mathcal{T}[\beta])$ be the class of trees that have a proximity drawing where the proximity region is the open (closed) β-region and let $\mathcal{T}_k$ be the set of all finite trees of maximum vertex degree at most k. In [BDLL95], a complete characterization of $\mathcal{T}(\beta)$ for all β values such that $0 \leq \beta \leq \frac{1}{1-cos(\frac{2\pi}{5})} \simeq 1.45$ or such that $3.23 \simeq \frac{1}{cos(\frac{2\pi}{5})} < \beta < \infty$ is given. Also, a complete characterization of $\mathcal{T}[\beta]$ for all β values such that $0 \leq \beta < \frac{1}{1-cos(\frac{2\pi}{5})}$ or such that $\frac{1}{cos(\frac{2\pi}{5})} \leq \beta \leq \infty$ is presented. For all β values not in the above intervals, the authors give a partial characterization: They show that all trees in $\mathcal{T}_4$ and only trees in $\mathcal{T}_5$ belong to $\mathcal{T}(\beta)$ and $\mathcal{T}[\beta]$.

Table 4.1 summarizes the known results about families of trees that admit a β-drawing for different values of β in 2-dimensional space (for proofs and detailed description of recognition and drawing algorithms, see [BDLL95, BLL96]). In the table, $\overline{\mathcal{T}}$ denotes the family of trees that have at least two adjacent vertices of degree three and $\overline{\overline{\mathcal{T}}}$ denotes the family of "forbidden" graphs defined in [BLL96]. Figure 4.13 shows a β-drawing of a tree with all non-leaf vertices having degree four; the drawing is computed with the technique of [BDLL95] and assumes the value $\beta = 4$.

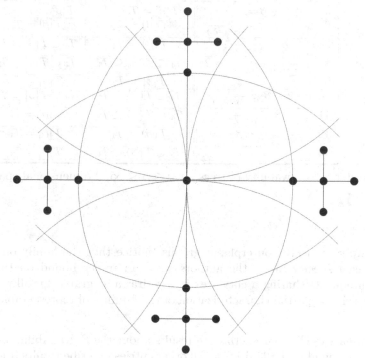

Figure 4.13   A β-drawing of a tree for $\beta = 4$ computed with the technique of [BDLL95].

The study of β-drawings of trees was also extended to 3-dimensional space. The definition of β-region recalled in the previous section can be straightforwardly extended to 3-dimensional space by considering open and closed 3-dimensional spheres instead of open and closed 2-dimensional spheres. In [LD95] it is shown that by using the third dimension the class of β-drawable trees becomes larger in many cases. For example, all trees having maximum vertex degree at most 4 are Gabriel drawable in 3-dimensional space, while this is not the case in the plane (see also Row 5 of Table 4.1); for $\beta = 2$ every tree having maximum vertex degree at most nine is drawable. The known results on β-drawability of trees in 3-dimensional space are summarized in Table 4.2, where the same notation of Table 4.1 is adopted; in the table, K_1 and K_2 denote the tree consisting of a single vertex and of a single edge, respectively.

Returning now to β-drawings in 2-dimensional space, the study of the β-drawability problem was extended from trees to outerplanar graphs by Lubiw and Sleumer [LS93], who showed that all maximal outerplanar graphs admit both a relative neighborhood drawing and a Gabriel drawing. They also proved that every biconnected outerplanar graph admits a relative neighborhood drawing. Lubiw and Sleumer also conjectured that any biconnected outerplanar graph admits a Gabriel drawing. This conjecture was settled in the affirmative in [LL97], where it is proved that indeed every biconnected outerplanar graph admits a β-drawing for all values of β such that $1 \leq \beta \leq 2$. In the same paper, the investigation was

	value of β	$\mathcal{T}(\beta)$	$\mathcal{T}[\beta]$
1	$\beta = 0$	$\mathcal{T}(\beta) = \{K_1, K_2\}$	$\mathcal{T}[\beta] = \mathcal{T}_2$
2	$0 < \beta < \frac{\sqrt{3}}{2}$	$\mathcal{T}(\beta) = \mathcal{T}_2$	$\mathcal{T}[\beta] = \mathcal{T}_2$
3	$\beta = \frac{\sqrt{3}}{2}$	$\mathcal{T}(\beta) = \mathcal{T}_2$	$\mathcal{T}[\beta] = \mathcal{T}_3 - \overline{\mathcal{T}}$
4	$\frac{\sqrt{3}}{2} < \beta < 1$	$\mathcal{T}(\beta) = \mathcal{T}_3$	$\mathcal{T}[\beta] = \mathcal{T}_3$
5	$\beta = 1$	$\mathcal{T}[\beta] = \mathcal{T}_4 - \overline{\overline{\mathcal{T}}}$	$\mathcal{T}[\beta] = \mathcal{T}_4 - \overline{\overline{\mathcal{T}}}$
6	$1 < \beta < \frac{1}{1-\cos(\frac{2\pi}{5})}$	$\mathcal{T}(\beta) = \mathcal{T}_4$	$\mathcal{T}[\beta] = \mathcal{T}_4$
7	$\beta = \frac{1}{1-\cos(\frac{2\pi}{5})}$	$\mathcal{T}(\beta) = \mathcal{T}_4$	$\mathcal{T}_4 \subset \mathcal{T}[\beta] \subset \mathcal{T}_5$
8	$\frac{1}{1-\cos(\frac{2\pi}{5})} < \beta < 2$	$\mathcal{T}_4 \subset \mathcal{T}(\beta) \subseteq \mathcal{T}_5$	$\mathcal{T}_4 \subset \mathcal{T}[\beta] \subseteq \mathcal{T}_5$
9	$\beta = 2$	$\mathcal{T}[\beta] = \mathcal{T}_5$	$\mathcal{T}[\beta] = \mathcal{T}_5$
10	$2 < \beta < \frac{1}{\cos(\frac{2\pi}{5})}$	$\mathcal{T}_4 \subset \mathcal{T}(\beta) \subseteq \mathcal{T}_5$	$\mathcal{T}_4 \subset \mathcal{T}[\beta] \subseteq \mathcal{T}_5$
11	$\beta = \frac{1}{\cos(\frac{2\pi}{5})}$	$\mathcal{T}_4 \subset \mathcal{T}(\beta) \subset \mathcal{T}_5$	$\mathcal{T}[\beta] = \mathcal{T}_4$
12	$\frac{1}{\cos(\frac{2\pi}{5})} < \beta < \infty$	$\mathcal{T}(\beta) = \mathcal{T}_4$	$\mathcal{T}[\beta] = \mathcal{T}_4$
13	$\beta = \infty$	$\mathcal{T}_3 \subset \mathcal{T}(\beta) \subset \mathcal{T}_4$	$\mathcal{T}[\beta] = \mathcal{T}_3$

Table 4.1 β-drawability of trees for $0 \leq \beta \leq \infty$, 2-dimensional space.

extended to simply connected outerplanar graphs (notice that the family of these graphs includes trees as a special case); the authors show an upper bound on the number of biconnected components sharing a cutvertex in a β-drawable graph, for all possible values of β, which gives rise to partial characterization of the families of representable outerplanar graphs.

Table 4.3 summarizes the characterization results about the β-drawability of outerplanar graphs that can be found in [LS93, LL97]. All other entries describe results from this paper. $\mathcal{CO}$, $\mathcal{BO}$, and $\mathcal{MO}$ are the set of all connected outerplanar, biconnected outerplanar, and maximal outerplanar graphs, respectively. $\mathcal{G}_{\mathcal{CO}}(\beta)$, $\mathcal{G}_{\mathcal{BO}}(\beta)$, and $\mathcal{G}_{\mathcal{MO}}(\beta)$ are the classes of connected outerplanar, biconnected outerplanar, and maximal outerplanar (β)-drawable graphs, respectively. Similarly, $\mathcal{G}_{\mathcal{CO}}[\beta]$, $\mathcal{G}_{\mathcal{BO}}[\beta]$, and $\mathcal{G}_{\mathcal{MO}}[\beta]$ are the classes of connected outerplanar, biconnected outerplanar, and maximal outerplanar [β]-drawable graphs, respectively. $\mathcal{G}_k$ denotes the class of graphs such that the number of biconnected components sharing a cut-vertex is at most k.

Little is known about the β-drawability of graphs that are not outerplanar. Irfan and Rahman [IR07] gave a sufficient condition for the β-drawability of 2-outerplanar graphs for values of β in the interval $1 < \beta < 2$; they also described examples of 2-outerplanar graphs that do not admit a β-drawing for $1 < \beta < 2$. In the same paper, Irfan and Rahman described and $O(n^2)$-time algorithm to test their sufficient condition on a given 2-outerplanar graph with n vertices. The time complexity of this test was later improved to $O(n)$ by Samee, Irfan, and Rahman [SIR08].

4.4   Variations of Proximity Drawings

Some generalizations and relaxations of proximity drawings have been described in the literature. This section recalls three of them.

	β	$\mathcal{T}(\beta)$ 3-D	$\mathcal{T}[\beta]$ 3-D
1	$\beta = 0$	$\mathcal{T}(\beta) = \{K_1, K_2\}$	$\mathcal{T}[\beta] = \mathcal{T}_2$
2	$0 < \beta < \frac{2}{3}$	$\mathcal{T}(\beta) = \mathcal{T}_2$	$\mathcal{T}[\beta] = \mathcal{T}_2$
3	$\beta = \frac{1}{2\sin^2(\frac{\pi}{3})} = \frac{2}{3}$	$\mathcal{T}(\beta) = \mathcal{T}_2$	$\mathcal{T}[\beta] = \mathcal{T}_3 - \mathcal{T}'$
4	$\frac{2}{3} < \beta < \frac{1}{2\sin^2(\arcsin\sqrt{\frac{2}{3}})}$	$\mathcal{T}(\beta) = \mathcal{T}_3$	$\mathcal{T}[\beta] = \mathcal{T}_3$
5	$\beta = 0.75$	$\mathcal{T}(\beta) = \mathcal{T}_3$	$\mathcal{T}_3 \subset \mathcal{T}[\beta] \subseteq \mathcal{T}_4$
6	$\frac{3}{4} < \beta < \frac{1}{2\sin^2(\frac{\pi}{4})} = 1$	$\mathcal{T}(\beta) = \mathcal{T}_4$	$\mathcal{T}[\beta] = \mathcal{T}_4$
7	$\beta = 1$	$\mathcal{T}(\beta) = \mathcal{T}_4$	$\mathcal{T}_4 \subset \mathcal{T}[\beta] \subset \mathcal{T}_6$
8	$1 < \beta < \frac{1}{2\sin^2(\frac{7\pi}{30})}$	$\mathcal{T}(\beta) = \mathcal{T}_6$	$\mathcal{T}[\beta] = \mathcal{T}_6$
9	$\frac{1}{2\sin^2(\frac{7\pi}{15})} \leq \beta < \frac{1}{2\sin^2(\frac{13\pi}{60})}$	$\mathcal{T}_6 \subseteq \mathcal{T}(\beta) \subseteq \mathcal{T}_7$	$\mathcal{T}_6 \subseteq \mathcal{T}[\beta] \subseteq \mathcal{T}_7$
10	$\frac{1}{2\sin^2(\frac{13\pi}{60})} \leq \beta < \frac{1}{2\sin^2(\frac{37\pi}{180})}$	$\mathcal{T}_6 \subseteq \mathcal{T}(\beta) \subseteq \mathcal{T}_8$	$\mathcal{T}_6 \subseteq \mathcal{T}[\beta] \subseteq \mathcal{T}_8$
11	$\frac{1}{2\sin^2(\frac{37\pi}{180})} \leq \beta < \frac{1}{2\sin^2(\frac{\pi}{5})}$	$\mathcal{T}_6 \subseteq \mathcal{T}(\beta) \subseteq \mathcal{T}_9$	$\mathcal{T}_6 \subseteq \mathcal{T}[\beta] \subseteq \mathcal{T}_9$
12	$\beta = \frac{1}{2\sin^2(\frac{\pi}{5})}$	$\mathcal{T}_6 \subseteq \mathcal{T}(\beta) \subseteq \mathcal{T}_9$	$\mathcal{T}_6 \subseteq \mathcal{T}[\beta] \subseteq \mathcal{T}_9$
13	$\frac{1}{2\sin^2(\frac{\pi}{5})} < \beta \leq \frac{1}{2\sin^2(\frac{7\pi}{36})}$	$\mathcal{T}_7 \subseteq \mathcal{T}(\beta) \subseteq \mathcal{T}_9$	$\mathcal{T}_7 \subseteq \mathcal{T}[\beta] \subseteq \mathcal{T}_9$
14	$\frac{1}{2\sin^2(\frac{7\pi}{36})} < \beta \leq \frac{1}{2\sin^2(\frac{67\pi}{360})}$	$\mathcal{T}_7 \subseteq \mathcal{T}(\beta) \subseteq \mathcal{T}_{10}$	$\mathcal{T}_7 \subseteq \mathcal{T}[\beta] \subseteq \mathcal{T}_{10}$
15	$\frac{1}{2\sin^2(\frac{67\pi}{360})} < \beta \leq \frac{1}{2\sin^2(\frac{16\pi}{90})}$	$\mathcal{T}_7 \subseteq \mathcal{T}(\beta) \subseteq \mathcal{T}_{11}$	$\mathcal{T}_7 \subseteq \mathcal{T}[\beta] \subseteq \mathcal{T}_{11}$
16	$\frac{1}{2\sin^2(\frac{16\pi}{90})} < \beta \leq \frac{1}{2\sin^2(\frac{61\pi}{360})}$	$\mathcal{T}_7 \subseteq \mathcal{T}(\beta) \subseteq \mathcal{T}_{12}$	$\mathcal{T}_7 \subseteq \mathcal{T}[\beta] \subseteq \mathcal{T}_{12}$
17	$\frac{1}{2\sin^2(\frac{61\pi}{360})} < \beta < \frac{1}{2\sin^2(\frac{\pi}{6})}$	$\mathcal{T}_7 \subseteq \mathcal{T}(\beta) \subseteq \mathcal{T}_{13}$	$\mathcal{T}_7 \subseteq \mathcal{T}[\beta] \subseteq \mathcal{T}_{13}$
18	$\beta = 2$	$\mathcal{T}_9 \subseteq \mathcal{T}(\beta) \subseteq \mathcal{T}_{13}$	$\mathcal{T}_9 \subseteq \mathcal{T}[\beta] \subseteq \mathcal{T}_{13}$
19	$2 < \beta < \frac{1}{\cos(\frac{61\pi}{180})}$	$\mathcal{T}_7 \subseteq \mathcal{T}(\beta) \subseteq \mathcal{T}_{13}$	$\mathcal{T}_7 \subseteq \mathcal{T}[\beta] \subseteq \mathcal{T}_{13}$
20	$\frac{1}{\cos(\frac{61\pi}{180})} \leq \beta < \frac{1}{\cos(\frac{16\pi}{45})}$	$\mathcal{T}_7 \subseteq \mathcal{T}(\beta) \subseteq \mathcal{T}_{12}$	$\mathcal{T}_7 \subseteq \mathcal{T}[\beta] \subseteq \mathcal{T}_{12}$
21	$\frac{1}{\cos(\frac{16\pi}{45})} \leq \beta < \frac{1}{\cos(\frac{67\pi}{180})}$	$\mathcal{T}_7 \subseteq \mathcal{T}(\beta) \subseteq \mathcal{T}_{11}$	$\mathcal{T}_7 \subseteq \mathcal{T}[\beta] \subseteq \mathcal{T}_{11}$
22	$\frac{1}{\cos(\frac{67\pi}{180})} \leq \beta < \frac{1}{\cos(\frac{7\pi}{18})}$	$\mathcal{T}_7 \subseteq \mathcal{T}(\beta) \subseteq \mathcal{T}_{10}$	$\mathcal{T}_7 \subseteq \mathcal{T}[\beta] \subseteq \mathcal{T}_{10}$
23	$\frac{1}{\cos(\frac{7\pi}{18})} \leq \beta < \frac{1}{\cos(\frac{2\pi}{5})}$	$\mathcal{T}_7 \subseteq \mathcal{T}(\beta) \subseteq \mathcal{T}_9$	$\mathcal{T}_7 \subseteq \mathcal{T}[\beta] \subseteq \mathcal{T}_9$
24	$\beta = \frac{1}{\cos(\frac{2\pi}{5})}$	$\mathcal{T}_6 \subseteq \mathcal{T}(\beta) \subseteq \mathcal{T}_9$	$\mathcal{T}_6 \subseteq \mathcal{T}[\beta] \subseteq \mathcal{T}_9$
25	$\frac{1}{\cos(\frac{2\pi}{5})} < \beta < \frac{1}{\cos(\frac{37\pi}{90})}$	$\mathcal{T}_6 \subseteq \mathcal{T}(\beta) \subseteq \mathcal{T}_9$	$\mathcal{T}_6 \subseteq \mathcal{T}[\beta] \subseteq \mathcal{T}_9$
26	$\frac{1}{\cos(\frac{37\pi}{90})} \leq \beta < \frac{1}{\cos(\frac{13\pi}{30})}$	$\mathcal{T}_6 \subseteq \mathcal{T}(\beta) \subseteq \mathcal{T}_8$	$\mathcal{T}_6 \subseteq \mathcal{T}[\beta] \subseteq \mathcal{T}_8$
27	$\frac{1}{\cos(\frac{13\pi}{30})} \leq \beta < \frac{1}{\cos(\frac{7\pi}{15})}$	$\mathcal{T}_6 \subseteq \mathcal{T}(\beta) \subseteq \mathcal{T}_7$	$\mathcal{T}_6 \subseteq \mathcal{T}[\beta] \subseteq \mathcal{T}_7$
28	$\frac{1}{\cos(\frac{7\pi}{15})} < \beta < \infty$	$\mathcal{T}(\beta) = \mathcal{T}_6$	$\mathcal{T}[\beta] = \mathcal{T}_6$
29	$\beta = \infty$	$\mathcal{T}_4 \subset \mathcal{T}(\beta) \subset \mathcal{T}_6$	$\mathcal{T}[\beta] = \mathcal{T}_4$

Table 4.2 β-drawability of trees for $0 \leq \beta \leq \infty$, 3-dimensional space.

4.4.1 Witness Proximity Drawings

Witness proximity has been introduced and studied in a series of papers by Aronov, Dulieu, and Hurtado [ADH, ADH11a, ADH11b]. These papers study both the computational geometry problem of computing witness proximity graphs on a given point set and the graph drawing question of defining a set of points whose witness proximity drawing represents a given combinatorial graph. We recall here only those results relative to the proximity drawability problem.

Witness proximity drawings are region of influence based proximity drawings where the adjacency between pairs of vertices depends on whether the proximity region of these vertices contains or does not contain a point form a second set, called the *witness points*.

	β	Connected	Biconnected	Maximal
1	$\beta = 1$	$\mathcal{G}_2 \not\subset \mathcal{G}_{CO}[1] \subset \mathcal{G}_4$	$\mathcal{G}_{BO}[1] = \{\mathcal{BO}\}$	$\mathcal{G}_{MO}[1] = \{\mathcal{MO}\}$
2	$1 < \beta < \frac{1}{1-\cos(\frac{2\pi}{5})}$	$\mathcal{G}_{CO}(\beta) \subset \mathcal{G}_4$	$\mathcal{G}_{BO}(\beta) = \{\mathcal{BO}\}$	$\mathcal{G}_{MO}(\beta) = \{\mathcal{MO}\}$
		$\mathcal{G}_{CO}[\beta] \subset \mathcal{G}_4$	$\mathcal{G}_{BO}[\beta] = \{\mathcal{BO}\}$	$\mathcal{G}_{MO}[\beta] = \{\mathcal{MO}\}$
3	$\beta = \frac{1}{1-\cos(\frac{2\pi}{5})}$	$\mathcal{G}_{CO}(\beta) \subset \mathcal{G}_4$	$\mathcal{G}_{BO}(\beta) = \{\mathcal{BO}\}$	$\mathcal{G}_{MO}(\beta) = \{\mathcal{MO}\}$
		$\mathcal{G}_4 \not\subset \mathcal{G}_{CO}[\beta] \subset \mathcal{G}_5$	$\mathcal{G}_{BO}[\beta] = \{\mathcal{BO}\}$	$\mathcal{G}_{MO}[\beta] = \{\mathcal{MO}\}$
4	$\frac{1}{1-\cos(\frac{2\pi}{5})} < \beta < 2$	$\mathcal{G}_4 \not\subset \mathcal{G}_{CO}(\beta) \subset \mathcal{G}_5$	$\mathcal{G}_{BO}(\beta) = \{\mathcal{BO}\}$	$\mathcal{G}_{MO}(\beta) = \{\mathcal{MO}\}$
		$\mathcal{G}_4 \not\subset \mathcal{G}_{CO}[\beta] \subset \mathcal{G}_5$	$\mathcal{G}_{BO}[\beta] = \{\mathcal{BO}\}$	$\mathcal{G}_{MO}[\beta] = \{\mathcal{MO}\}$
5	$\beta = 2$	$\mathcal{G}_4 \not\subset \mathcal{G}_{CO}(2) \subset \mathcal{G}_5$	$\mathcal{G}_{BO}(2) = \{\mathcal{BO}\}$	$\mathcal{G}_{MO}(2) = \{\mathcal{MO}\}$

Table 4.3 β-drawability of outerplanar graphs for $1 \leq \beta \leq 2$, 2-dimensional space.

Therefore, in a witness proximity drawing, we look at a set of points that represent the vertices and at a set of points that play the role of the witnesses. The existence/absence of an edge in the drawing depends on the location of the witness points (the set of witness points and the set of points representing the vertices of the graph in drawing may not coincide).

In a *positive witness proximity drawing* Γ, two vertices (x, y) are adjacent if and only if the proximity region of x and y contains at least one vertex that belongs to the set of witness points. In a *negative witness proximity drawing*, x and y are adjacent if and only if their region of influence does not contain any of the witness points (it may however contain other vertices of the graph that are not witnesses). It is worth noticing that the definition of witness proximity drawing includes the notion of $(h, 0)$-proximity drawing as a special case: A negative proximity drawing where the set of witness points coincides with the vertex set is in fact an $(h, 0)$-proximity drawing.

The computation of a witness proximity drawing requires to define the set of points representing the vertices and the set of witness points. For example, Figure 4.14 (a) shows a positive witness Gabriel drawing and Figure 4.14 (b) a negative witness Gabriel drawing; the two drawings have the same witness point q and the same set of vertices. In the figures, the Gabriel disk of u and v contains the witness point q, which makes u and v adjacent in the positive witness Gabriel drawing.

In [ADH11a], Aronov, Dulieu, and Hurtado studied *witness Delaunay drawings*. More specifically, they consider negative witness Delaunay drawings, which are proximity drawings where two vertices x and y are adjacent if and only if there exists an open disk whose boundary passes through x and y and does not contain any point of the witness set. It is proved that every tree admits a negative witness Delaunay drawing for suitable set of witness points and that the drawing can be computed in linear time, adopting the real RAM model of computation. As for forbidden graphs, it is proved that non-planar bipartite graphs never admit a negative witness Delaunay drawing. In the same paper, positive witness Delaunay drawings in the L_∞ metric are studied. These drawings, also called *square drawings*, are such that two vertices x and y are adjacent if and only if there exists an axis-aligned square whose boundary passes through x and y and that contains at least one witness point. It is proved in [ADH11a] that a graph admits a square drawing if and only if it is a permutation graph and that a square drawing of a permutation graph can be computed by using at most one witness point.

The witness generalization of Gabriel drawings is studied in [ADH]. The paper describes both graphs that do not admit a negative witness Gabriel drawing and graphs that are negative witness Gabriel drawable. It is proved that all graphs containing $K_{3,3,3,3}$ as an

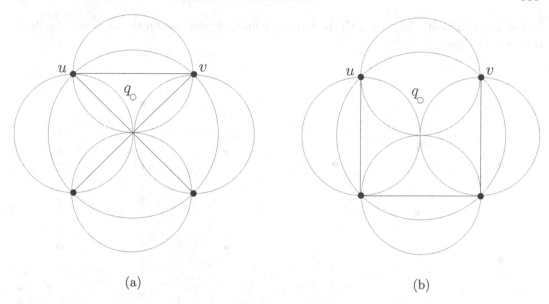

(a)   (b)

Figure 4.14 Two witness Gabriel drawings where q is the witness point: (a) Positive witness Gabriel drawing. (b) Negative witness Gabriel drawing.

induced subgraph do not have a negative witness Gabriel drawing. It is also proved that all trees are negative witness Gabriel drawable.

Positive witness rectangle of influence drawings are explored in [ADH11b]. In this paper, Aronov, Dulieu, and Hurtado show that a tree admits a positive witness rectangle of influence drawing if and only if it has no three independent edges. The paper also gives necessary conditions for positive witness rectangle of influence drawability of general graphs. Namely, a graph that has a positive witness rectangle of influence drawing has at most two non-trivial connected components (a connected component is non-trivial if its number of vertices is larger than one). If the graph has exactly two components, then each component has diameter three; if the graph has one component, it has diameter six. Finally, a characterization of the positive witness rectangle of influence drawable graphs having exactly two non-trivial components is given: A graph belongs to this family if and only if it is a disjoint union of zero or more isolated vertices and two co-interval graphs.

4.4.2 Weak Proximity Drawings

We recall that, a $(2,0)$-proximity drawing Γ is a straight-line drawing such that: (i) for each edge (x,y) of Γ, the proximity region of x and y does not contain any other vertex, and (ii) for each pair of non-adjacent vertices x,y of Γ, the proximity region of x and y contains at least one other vertex. In this section, we shall call such drawings *strong* proximity drawings.

A relaxation of strong proximity drawings, called *weak proximity drawings*, was first introduced and studied in [DLW06]. A weak proximity drawing of a graph G is one that ignores requirement (ii). In other words, if x,y is *not* an edge of the graph, then no requirement is placed on the proximity region of x and y in the weak drawing. For example, Figure 4.15 (a) shows a weak proximity drawing of a tree. Here, the proximity region of any two points x and y is the disk having x and y as antipodal points. Note that the drawing is not a strong drawing, as no edges between neighbors of the degree six vertex are included.

The strong proximity drawing with the same proximity region and on the same set of points is shown in Figure 4.15 (b).

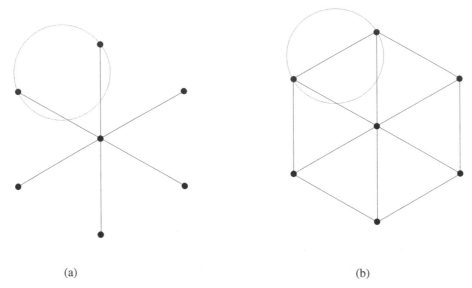

(a)   (b)

Figure 4.15   (a) A weak proximity drawing and (b) a strong proximity drawing.

For purposes of graph visualization, there are several reasons for studying weak proximity drawings. We summarize the ones that, in our opinion, are the most relevant.

- Strong proximity drawability may appear too restrictive for graph drawing purposes. By relaxing (ii), a graph G can no longer be reconstructed from the locations of its vertices in a weak drawing; however, many graphs that do not admit strong drawings can be drawn weakly. For example, a tree that has a vertex of degree greater than five has no strong 2-dimensional β-drawing for any β (see also Table 4.1). Thus, the drawing in Figure 4.15 (a) illustrates a graph that is *weak* but not *strong* drawable for the Gabriel region.

- A visibility drawing of a graph is a drawing such that (e.g., see [DETT99, KW01]) vertices are mapped to horizontal segments and edges are mapped to vertical segments that intersect only adjacent vertex segments. Of course, a necessary condition for drawing an edge is that the vertex segments corresponding to its end-vertices are visible in the vertical direction. If this condition is also sufficient, then we have a *strong visibility drawing*; otherwise, we have a *weak visibility drawing*. In the field of visibility drawing, the coordinated study of both strong and weak types of drawings led to deep and practical results.

- Weak proximity can be considered as an "edge-vertex resolution rule" in the sense that a vertex cannot enter the region of influence of an edge. Thus, the study of weak proximity can contribute to the body of drawing strategies that adopt a resolution rule (e.g., see [DETT99, KW01]).

- The weak proximity model may well be sufficient for many drawing applications, particularly ones that do not require recovery of the graph solely from the positions of its vertices. For example, weak proximity drawings have been receiving increasing interest for their applications to wireless network design, where distributed topology control can be based on proximity structures constructed from given geometric graphs by deleting those edges that do not satisfy a given proximity rule. The resulting graph is a weak proximity drawing because its edges satisfy the given proximity rule, whereas pairs of non-adjacent vertices may or may not contain other vertices in their proximity region. Papers devoted to the study of weak proximity graphs defined in the context of sensor networks include [CWL02, LCWW03, PS04, KL10]. See also Section 4.5 for more discussion and some other references about proximity and wireless ad-hoc networks.

In particular the research in [DLW06] focused on 2-dimensional weak β-drawing; the following results were proved.

General graphs: Any graph G is weak β-drawable for all β in the range 0 to some upper bound that is a function either of the number of vertices or of the maximum vertex degree of G.

Planar graphs: For any value of β such that $1 < \beta \leq \infty$, strong and weak β-drawings of triangulated planar graphs coincide. It was also shown how to interpret any straight-line drawing algorithm for planar triangulated graphs as an algorithm for constructing weak proximity drawings.

Trees: An algorithm was presented to draw any tree as a weak β-drawing for any value of β less than two. It was shown that for $2 \leq \beta < \infty$, the weak and the strong proximity models give rise approximately to the same class of 2-dimensional β-drawable trees. Finally, the NP-hardness of deciding whether a tree has a weak proximity drawing for $\beta = \infty$, where the region of influence is an open strip, was proved.

Table 4.4 schematically compares the known results on weak β-drawability against those on strong β-drawability for trees. Each row corresponds to a different interval of β and reports the maximum vertex degree k that a tree can have to admit a strong or weak β-drawing for some values of β in the interval. Of particular interest is the value $\beta = 2$, where remarkable differences in the drawable trees can be noticed, depending on whether the region of influence is an open set (in which case it coincides with the relative neighborhood region) or a closed set (in which case it coincides with the relatively closest region).

	value of β	strong β-drawability	weak β-drawability
1	$0 \leq \beta < 2$	$k \leq 5$	$k = \infty$
2	$\beta = 2$	$k = 5$	$k = \infty$ (w-(β)-draw.), $k = 5$ (w-$[\beta]$-draw.)
3	$2 < \beta \leq \infty$	$k \leq 5$	$k \leq 5$

Table 4.4 Comparing weak β-drawability of trees vs. strong β-drawability of trees. In the table, w-(β)-drawable means that the tree has a weak β-drawing where the β-region is an open set and w-$[\beta]$-drawable means that the tree has a weak β-drawing where the β-region is a closed set.

The advantage of using a weak model of proximity was also highlighted in [LL97], where it was proved that, in contrast with the results in Table 4.3, every connected outerplanar graph admits a weak Gabriel drawing, a weak relative neighborhood drawing, and a weak β-drawing for any given β such that $1 < \beta < 2$.

A comparison of strong and weak β-drawings in terms of area requirement can be found in [LTTV97] and in the work by Penna and Vocca [PV04]. Penna and Vocca [PV04] extended the study of weak proximity β-drawings to 3-dimensional space and proved several polynomial area/volume bounds for families of graphs for which a strong proximity drawing is either not admitted or requires exponential area. In general however, there exist families of graphs for which a Gabriel drawing in 2-dimensional space requires exponential area both for the strong and the weak model of proximity [LTTV97].

Weak nearest neighbor graphs were studied by Eades and Whitesides [EW96a], who showed that the problem of deciding whether a graph admits a weak nearest neighbor drawing is NP-hard. Thus, nearest neighbor drawability is NP-hard both in the weak and in the strong proximity model (see also Section 4.3.4).

Weak rectangle of influence drawings were first studied by Biedl, Bretscher, and Meijer [BBM99]. They showed that a planar graph admits a weak closed rectangle of influence drawing if and only if it admits a planar embedding where the outerface is not a 3-cycle and such that there is no separating 3-cycle; they call a separating 3-cycle a *filled triangle* and call the family of graphs with no filled triangles *NF3-graphs*. In the same paper, Biedl, Bretscher, and Meijer also showed that every $NF3$-graph admits an open weak rectangle of influence drawing but left as open the question of characterizing the open weak rectangle of influence drawable graphs. There are several subsequent papers that present partial answers to this question.

Miura, Matsuno, and Nishizeki [MMN09] characterize those *triangulated plane graphs* (i.e., maximal planar graph with a given planar embedding) that admit an open weak rectangle of influence drawing; the characterization gives rise to a linear time testing algorithm. In addition, the paper gives a sufficient condition for the weak open rectangle of influence drawability of *inner triangulated plane graphs* (i.e., planar graphs with a given planar embedding and all triangular faces, except the external face that has more than three vertices). This sufficient condition is expressed in terms of labeling of angles of a suitable subgraph, called *frame graph*. The frame graph of an inner triangulated plane graph G is obtained by removing all vertices and edges in the proper inside of every maximal filled triangle of G. Testing the sufficient condition, and eventually constructing an open rectangle of influence drawing of G, can be executed in $O(n^{1.5} \log n)$ time. The computed drawing has area $(n - 1) \times (n - 1)$ and it has the property that every edge of the frame graph is *oblique*, i.e., it is neither vertical nor horizontal. Alamdari and Biedl [AB12] further elaborate on the ideas by Miura, Matsuno, and Nishizeki and characterize the inner triangulated plane graphs that admit a weak open rectangle of influence drawing such that no two vertices of the frame graph have the same x-coordinate or the same y-coordinate. The characterization by Alamdari and Biedl yields an $O(n^{1.5} \log n)$-time testing and drawing algorithm. A recent paper by Alamdari and Biedl [AB] generalizes the characterization for non-aligned frames to all planar graphs with a fixed planar embedding. The paper also shows that if the planar embedding is not fixed, then deciding if a given planar graph has an open weak rectangle of influence drawing is NP-complete. NP-completeness holds even for open weak rectangle of influence drawings with non-aligned frames.

A significant research effort has also been devoted to the area required by weak open and closed rectangle of influence drawings. The construction by Biedl, Bretscher, and Meijer [BBM99] gives rise to weak closed and open rectangle of influence drawings with n vertices on an integer grid of size $(n - 1) \times (n - 1)$. Sadavisam and Zhang [SZ11] show that an integer grid of size at most $(n - 3) \times (n - 3)$ is always sufficient and sometimes necessary to compute a weak closed rectangle of influence drawing of an *irreducible triangulation*, i.e., a maximal $NF3$-graph. In the same paper, they also proved an expected area of $\left(\frac{22n}{27} + \sqrt{n}\right) \times \left(\frac{22n}{27} + \sqrt{n}\right)$ for a weak closed rectangle of influence drawing of a random irreducible triangulation. Miura and Nishizeki [MN05] prove that the convex grid drawing computed by the algorithm of Miura, Nakano, and Nishizeki [MNN00, MNN06] is in fact a weak open rectangle of influence drawing; this result implies that a four connected planar graph with n vertices has a weak open rectangle of influence drawing in area $\lceil \frac{n-1}{2} \rceil \times \lfloor \frac{n-1}{2} \rfloor$. Zhang and Vaidya [ZV09a, ZV09b] further improve this bound as follows: (i) An irreducible triangulation with n vertices taken uniformly at random has a weak open rectangle of influence drawing whose area is asymptotically $\frac{11n}{27} \times \frac{11n}{27}$ with high probability, up to an additive error of $O(\sqrt{n})$; (ii) A quadriangulation with n vertices taken uniformly at random has a weak open rectangle of influence drawing whose area is asymptotically $\frac{13n}{27} \times \frac{13n}{27}$ with high probability, up to an additive error of $O(\sqrt{n})$. Both results are proved as applications of previous techniques by Fusy [Fus06, Fus09].

4.4.3 Approximate Proximity Drawings

As discussed in Section 4.3, proximity drawability imposes severe restrictions on the families of the representable graphs; for example, the tables of Section 4.3.6 show families of β-drawable graphs whose maximum vertex degrees are all bounded by small constant values. In order to overcome these restrictions on the combinatorial structure of the drawable graphs, recent papers study straight-line drawings of graphs that are "good approximations" of proximity drawings.

Di Giacomo et al. [DDLM12] investigate drawings that approximate the global proximity rule; in particular, they study approximate minimum weight drawings of trees in the 2-dimensional space. A $(1+\varepsilon)$-*EMST drawing* is a planar straight-line drawing of a tree such that, for any fixed $\varepsilon > 0$, the distance between any two vertices is at least $\frac{1}{1+\varepsilon}$ the length of the longest edge in the path connecting them. Therefore, $(1+\varepsilon)$-EMST drawings are good approximations of Euclidean minimum spanning trees. Figure 4.16 shows a $(1+\varepsilon)$-EMST drawing of a tree for $\varepsilon = 0.5$. In the figure, the ratio between the distance $d(u, v)$ and the length of the longest edge along the path between u and v is $\frac{d(u,v)}{|e_T(u,v)|} = 0.714$, which is larger than $\frac{1}{1+\varepsilon} = 0.667$. Note that the tree of the figure does not admit a minimum weight drawing (a Euclidean minimum spanning tree cannot have two adjacent vertices both having degree six).

While it is known that all trees with maximum vertex degree five have a Euclidean minimum spanning tree realization [MS92] and it is NP-hard deciding whether trees of maximum vertex degree six admit one [EW96b], in [DDLM12] it is shown that every tree T has a $(1+\varepsilon)$-EMST drawing for any given $\varepsilon > 0$ and that this drawing can be computed in linear time in the real RAM model of computation.

Also, while Angelini et al. [ABC+11] have proved that EMST drawings of trees with vertex degree at most five may require exponential area, Di Giacomo et al. describe polynomial area approximation schemes for $(1+\varepsilon)$-EMST drawings: Any tree with n vertices and maximum vertex degree Δ admits a $(1+\varepsilon)$-EMST drawing whose area is $O(n^{c+f(\varepsilon,\Delta)})$, where c is a positive constant and $f(\varepsilon, \Delta)$ is a polylogarithmic function that tends to infinity as ε tends to zero. As already mentioned in Section 4.3.1, a byproduct of the techniques of [DDLM12] ia that the polynomial area upper bound for minimum weight drawings of complete binary trees by Frati and Kaufmann [FK11] is improved from $O(n^{4.3})$ to $O(n^{3.8})$.

Evans et al. [EGK+12], introduce and study approximations of $(h, 0)$-proximity drawings called $(\varepsilon_1, \varepsilon_2)$-*proximity drawings*. Intuitively, given a definition of proximity region and two real numbers $\varepsilon_1 \geq 0$ and $\varepsilon_2 \geq 0$, an $(\varepsilon_1, \varepsilon_2)$-proximity drawing of a graph is a planar straight-line drawing Γ such that: (i) For every pair of adjacent vertices u, v, their proximity region "shrunk" by the multiplicative factor $\frac{1}{1+\varepsilon_1}$ does not contain any vertices of Γ; and

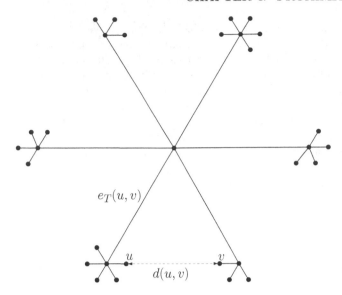

Figure 4.16   A $(1+\varepsilon)$-EMST drawing Γ of a tree with maximum vertex degree 6 for $\varepsilon = 0.5$. For the two highlighted vertices u and v, we have that $\frac{d(u,v)}{|e_T(u,v)|} = 0.714 \geq \frac{1}{1+\varepsilon} = 0.667$.

(ii) For every pair of non-adjacent vertices u, v, their proximity region "blown-up" by the factor $(1 + \varepsilon_2)$ contains some vertices of Γ other than u and v. More formally, let D be a disk with center c and radius r, and let ε_1 and ε_2 be two nonnegative real numbers. The ε_1-*shrunk disk of* D is the disk centered at c and having radius $\frac{r}{1+\varepsilon_1}$; the ε_2-*expanded disk of* D is the disk centered at c and having radius $(1+\varepsilon_2)r$. An $(\varepsilon_1, \varepsilon_2)$-proximity drawing is a planar straight-line drawing where the proximity region of two adjacent vertices is defined by using ε_1-shrunk disks, while the region of influence of two non-adjacent vertices uses ε_2-expanded disks.

Figure 4.17 is an example of an $(\varepsilon_1, \varepsilon_2)$-Gabriel drawing for $\varepsilon_1 = 0$ and $\varepsilon_2 = 0.7$. Note that the drawing is not a Gabriel drawing: For example, the dotted disk in the figure is a Gabriel disk (and its emptiness would imply an edge), while the solid one is its 0.7-expanded version. In fact, the tree of Figure 4.17 is not Gabriel drawable (see also Table 4.1).

In [EGK$^+$12], it is proved that one can arbitrarily approximate a proximity drawing of any planar graph for some of the most-studied definitions of proximity. Namely, it is shown that for any positive values of $\varepsilon_1, \varepsilon_2$ an embedded planar graph admits both an $(\varepsilon_1, \varepsilon_2)$-Gabriel drawing and an $(\varepsilon_1, \varepsilon_2)$-Delaunay drawing and an $(\varepsilon_1, \varepsilon_2)$-$\beta$-drawing ($1 \leq \beta \leq \infty$) that preserve the given embedding. These results are proved to be, in a sense, tight since it is shown that for each of the above types of proximity rules there are embedded planar graphs that do not have an embedding preserving $(\varepsilon_1, \varepsilon_2)$-proximity drawing with either $\varepsilon_1 = 0$ or $\varepsilon_2 = 0$.

Note that both the strong and the weak proximity drawings described in Sections 4.2.1 and 4.4.2 are special cases of $(\varepsilon_1, \varepsilon_2)$-proximity drawings. Namely, an $(\varepsilon_1, \varepsilon_2)$-proximity drawing is a strong proximity drawing if $\varepsilon_1 = \varepsilon_2 = 0$; also, an $(\varepsilon_1, \varepsilon_2)$-proximity drawing is a weak proximity drawing if $\varepsilon_1 = 0$ and $\varepsilon_2 = \infty$. Therefore, $(0, \varepsilon_2)$-proximity drawings make it possible to study weak and strong proximity drawability in a unified framework: As the value of ε_2 increases, $(0, \varepsilon_2)$-proximity drawings approach weak proximity drawings.

Several questions can be asked within this unifying framework. For example, not all trees have a Gabriel drawing [BLL96], while all trees have a weak Gabriel drawing [DLW06].

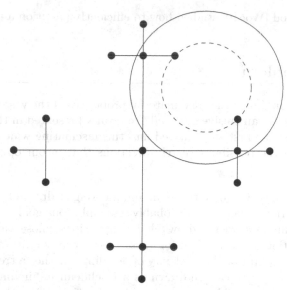

Figure 4.17 A $((0,0.7)$-Gabriel drawing of a tree that does not have a Gabriel drawing.

What is the minimum threshold value such that if ε_2 is larger than this threshold all trees are drawable? Evans et al. [EGK+12] answer this question by proving that every tree has a $(0, \varepsilon_2)$-Gabriel drawing for any given value of ε_2 such that $\varepsilon_2 \geq 2$. In the same paper, it is also proved that for each value of ε_2 such that $0 \leq \varepsilon_2 < 2$, there exists a tree T such that T does not have a $(0, \varepsilon_2)$-Gabriel drawing.

All biconnected outerplanar graphs have a Gabriel drawing [LL97], while a connected outerplanar graph where a cut vertex is shared by more than four biconnected components is not Gabriel drawable (see also Section 4.3). For a contrast, it is shown in [EGK+12] that every outerplanar graph without vertices of degree one admits a $(0, \varepsilon_2)$-Gabriel drawings for any arbitrarily chosen positive value of ε_2.

The study of approximate rectangle of influence drawings has also been recently initiated in [DLM], where it is proved that all planar graphs have an open/closed $(\varepsilon_1, \varepsilon_2)$-rectangle of influence drawing for $\varepsilon_1 > 0$ and $\varepsilon_2 > 0$, while there are planar graphs that do not admit an open/closed $(\varepsilon_1, 0)$-rectangle of influence drawing and planar graphs that do not admit a $(0, \varepsilon_2)$-rectangle of influence drawing. In the same paper, it is shown that all outerplanar graphs have an open/closed $(0, \varepsilon_2)$-rectangle of influence drawing for any $\varepsilon_2 \geq 0$. Concerning area bounds, it is shown that if $\varepsilon_2 > 2$ an open/closed $(0, \varepsilon_2)$-rectangle of influence drawing of an outerplanar can be computed in polynomial area. For values of ε_2 such that $\varepsilon_2 \leq 2$, a drawing algorithm is described that computes $(0, \varepsilon_2)$-rectangle of influence drawings of binary trees in area $O(n^{c+f(\varepsilon_2)})$, where c is a positive constant, $f(\varepsilon_2)$ is a polylogarithmic function that tends to infinity as ε_2 tends to zero, and n is the number of vertices of the input tree.

We conclude the section by recalling a different approach, studied by Hurtado et al. [HLW10], to approximate a proximity drawing. Given a graph G, the idea is to first partition G into subgraphs such that each subgraph is proximity drawable and then compute a drawing Γ of G such that each subdrawing of Γ representing a subgraph of the partition is a proximity drawing. In particular, Hurtado et al. showed different drawing techniques that receive as input a tree T with a partition into subtrees of bounded degree and produce as output a drawing of T such that the subdrawing of each subtree is a minimum spanning tree. In a companion paper, Wood [Woo10] studied how to efficiently partition a tree into subtrees of bounded degree.

4.5 Open Problems

To date, a full understanding of the combinatorial properties of the vast majority of proximity drawable graphs is still an elusive goal and the results presented in the previous sections can be regarded as just the first steps moved into this fascinating wide-open research area. We list below some of the possible research directions that in our opinion are among the most interesting.

Minimum Weight Drawings: Characterizing minimum weight drawable triangulations seems to be a serious challenge; a probably less ambitious goal could be to characterize those minimum weight drawable triangulations whose skeleton is a tree. Another interesting open problem for these types of proximity drawings is determining the computational complexity of deciding whether a tree with vertices of degree at most twelve can be drawn as a Euclidean minimum spanning tree in 3-dimensional space. Also, as described in Section 4.3.1, the algorithm by Monma and Suri [MS92] requires $O(2^{n^2}) \times O(2^{n^2})$ area for a 2-dimensional minimum weight drawing of a tree with n vertices and vertex degree at most five. Angelini et al. [ABC$^+$11] establish an $\Omega(2^n) \times \Omega(2^n)$ lower bound for these trees and conjecture that there is a tree requiring $\Omega(2^{n^2}) \times \Omega(2^{n^2})$. Proving/disproving this conjecture is a fascinating question.

Delaunay and Voronoi Drawings: Characterizing Delaunay drawable graphs is one of the oldest open problems in this area. It would also be interesting to better understand the combinatorial relationship between minimum weight and Delaunay drawable triangulations. Indeed, while Figure 4.9 shows a Delaunay forbidden graph that is minumum weight drawable, it is not known whether there exist Delaunay drawable graphs that are minimum weight forbidden. Another research direction is to study graphs that admit a Delaunay drawing of order h for some $h > 0$; good starting points for this problem are the papers by Abrego et al. [ÁMFM$^+$11] and by Bose et al. [BCH$^+$10], devoted to the combinatorial properties of higher-order proximity graphs. Finally, a complete characterization of (positive or negative) witness Delaunay drawable graphs is another fascinating question.

β-Drawings: The entries of Tables 4.1, 4.2, and 4.3 show gaps in the characterization of strong β-drawable trees and outerplanar graphs. Each of these gaps motivates further research. Also, little is known about the β-drawability properties of general graphs; for example, finding a complete characterization of β drawable k-outerplanar graphs for a given constant k such that $k \geq 2$ is an interesting problem. It would be also interesting to investigate area/volume bounds for strong and weak proximity drawings, also in the unifying framework of $(0, \varepsilon_2)$-proximity drawings. Finally, a natural question is to extend the study of (positive/negative) witness proximity drawability to the whole spectrum of possible β values.

Sphere of Influence Drawings: There are examples of non-planar graphs that admit a sphere of influence drawing. However, the result by Soss [Sos99a] proves that a sphere of influence drawable graph always has a number of edges that is linear in the number of the vertices. It is however not known whether the upper bound of $15n$ by Soss is tight; Toussaint [Tou05] reports on a conjecture of Avis, who claims that such a tight upper bound could be $9n$. What about approximate sphere of influence drawings? Or witness sphere of influence drawings?

Rectangle of Influence Drawings: Except for the classes of graphs described in [LLMW98], very little is known about recognizing which graphs have admit an (open or closed) strong rectangle of influence drawing. Also, as mentioned in the previous section, it would be interesting to characterize which planar graphs have a weak open rectangle of influence drawing. Similar characterizations can also be studied either in the witness proximity or in the approximate proximity models.

Other Proximity Rules: Several well-known proximity rules are still unexplored from a graph drawing point of view. For example, one could study the γ-drawability problem (see Section 4.2.1) or other proximity rules, not mentioned in the previous sections. A very limited list includes *α-complexes* (see, e.g., [Ede95] and also [SLL+08] for preliminary results on α-drawability), *sphere-of-attraction graphs* (see, e.g., [MW00]), *class-cover catch digraphs* (see, e.g., [PMDS03]), and *maximum weight triangulations* (see, e.g., [WCY99, QW04, QW06]).

4.6 Beyond this Chapter

We conclude this chapter by briefly pointing at two research directions in the areas of sensor networks and of robust geometric computing where proximity graphs and drawings have received some attention in the last few years.

Proximity Drawings and Ad-Hoc Networks: Different types of proximity graphs have attracted the interest of network engineers. Indeed, topology control and management, i.e., how to maintain network connectivity while consuming the minimum possible power, has emerged as one of the most important issues in wireless networks.

A wireless sensor network can be modeled as a set of points in the plane where each sensor s can communicate directly with each other sensor that is within its power range; this model gives rise to a proximity graph called a *unit distance graph*, where the proximity region for a sensor s is a circle of radius one centered at s, and there is an edge connecting s to another sensor t if and only if t is within the power range of S. However, the unit distance graph may be too dense for the limited memory of the sensors in the network; also, in order to reduce energy consumption, it is desirable that each sensor communicates directly with only a few of the sensors that are within its range.

An increasing number of topology control algorithms have thus been presented in the literature that are based on proximity graphs that are sparser than the unit distance graph, have small vertex degree, can be computed locally in a distributed manner, and are good spanners (a straight-line drawing Γ of a graph G is a *k-spanner* if for every pair of vertices u and v of G their geometric distance in Γ is at most k times the graph theoretic distance of u and v in G). A limited list of these structures includes *k-localized Delaunay triangulations* (see, e.g., [LCWW03]), *local minimum spanning trees* (see, e.g., [LHS03, CISRS05]), *partial Delaunay triangulations* (see, e.g., [LSW04]), *directed relative neighborhood graphs*, and *directed local minimum spanning trees* [LH04]. The interested reader is also referred to [BM04, BDEK06, BDL+11, CKLS10, CKX11, GLN02, Kan09, KPX10, NS07, Li04] for a limited list of references on geometric spanners and applications of proximity graphs to wireless networks. See also [CBF+06] for a paper that studies the drawability of a graph as a local minimum spanning tree.

We only remark here that all the proximity graphs mentioned above are constructed by pruning those edges of the unit distance graph which do not satisfy a given proximity rule; hence, the resulting proximity drawing guarantees closeness among adjacent vertices while there is no constraint on pairs of non-adjacent vertices. In other words, these structures inherently adopt a weak model of proximity.

Finally, there is general consensus that the knowledge of the combinatorial properties of the communication network is a basic requirement for the design of efficient localized routing algorithms (see, e.g., [BMSU01, KWZ03, LSW05]). Unlike traditional wired and cellular networks, the movement of wireless devices during the communication could change the network topology to some extent: Understanding what types of networks (proximity drawings) can result is therefore a natural question to ask. See, for example, [PS04], where the edge complexity of locally Delaunay triangulations is studied.

Proximity Drawings and Geometric Checkers: The intrinsic structural complexity of the implementation of geometric algorithms makes the problem of formally proving the correctness of the code unfeasible in most of the cases. This has been motivating research on *checkers*. A checker is an algorithm that receives as input a geometric structure and a predicate stating a property that should hold for the structure. The task of the checker is to verify whether the structure satisfies or not the given property. Here, the expectation is that it is often easier to evaluate the quality of the output than the correctness of the software that produces it. Different papers (see, e.g., [DLPT98, MNS+99]) have agreed on the basic features that a "good" checker should have:

Correctness: The checker should be correct beyond any reasonable doubt. Otherwise, one would incur into the problem of checking the checker.

Simplicity: The implementation should be straightforward.

Efficiency: The expectation is to have a checker that is not less efficient than the algorithm that produces the geometric structure.

Robustness: The checker should be able to handle degenerate configurations of the input and should not be affected by errors in the flow of control due to round-off approximations.

Geometric checkers can be quite naturally studied in the context of proximity drawings. Suppose one is given a straight-line drawing Γ of a graph together with some proximity rule $\mathcal{R}$. A *proximity drawing checker* for Γ is an algorithm that either certifies that Γ satisfies the proximity rule $\mathcal{R}$ or reports evidence that Γ does not satisfy $\mathcal{R}$.

One possible approach to solve this problem is to compute the proximity graph on the vertex set of Γ by applying the proximity rule $\mathcal{R}$ and then verify whether the computed drawing coincides with Γ. For example, suppose that Γ is a drawing of a binary tree and one wants to check whether Γ is a minimum weight drawing. One could compute the Euclidean minimum spanning tree of the vertices of Γ in $O(n \log n)$ time [PS90] and verify whether the computed graph coincides with Γ. However, can one perform the check in $o(n \log n)$ time? Also, what if the proximity graph on the vertex set of Γ is not unique? Linear-time checkers for Delaunay and Voronoi drawings can be found in [DLPT98, MNS+99]. Aronov,

Dulieu, and Hurtado [ADH] show an $O(n^2 \log m)$-time algorithm that receives as input a straight-line drawing Γ with n vertices and m edges and checks whether Γ is a negative witness Gabriel drawing for some set of witness points. If the answer is affirmative, the algorithm also returns the witness points.

Acknowledgments

This chapter extends and updates an early survey on proximity drawings co-authored by Giuseppe Di Battista, William Lenhart, and me [DLL95]. I thank Boris Aronov, Ferran Hurtado, and Sue H. Whitesides for their insights and constructive comments on earlier versions of this chapter. Work supported in part by MIUR of Italy under project AlgoDEEP prot. 2008TFBWL4.

References

[AB]   Soroush Alamdari and Therese Bied. Open rectangle-of-influence draw-
            ings of non-triangulated planar graphs. In W. Didimo and M. Patrignani,
            editors, *Graph Drawing (Proc. 20th International Symposium, GD 2012)*,
            Lecture Notes Comput. Sci. to appear.

[AB12]   Soroush Alamdari and Therese C. Biedl. Planar open rectangle-of-
            influence drawings with non-aligned frames. In Marc J. van Kreveld
            and Bettina Speckmann, editors, *Graph Drawing (Proc. 19th Interna-
            tional Symposium, GD 2011)*, volume 7034 of *Lecture Notes Comput.
            Sci.*, pages 14–25. Springer-Verlag, 2012.

[ABC+11]   Patrizio Angelini, Till Bruckdorfer, Marco Chiesa, Fabrizio Frati, Michael
            Kaufmann, and Claudio Squarcella. On the area requirements of Eu-
            clidean minimum spanning trees. In Frank Dehne, John Iacono, and
            Jörg-Rüdiger Sack, editors, *Algorithms and Data Structures (Proc. 12th
            International Symposium, WADS 2011)*, volume 6844 of *Lecture Notes
            Comput. Sci.*, pages 25–36. Springer-Verlag, 2011.

[ADH]   Boris Aronov, Muriel Dulieu, and Ferran Hurtado. Witness Gabriel
            graphs. *Computational Geometry.* to appear.

[ADH11a]   Boris Aronov, Muriel Dulieu, and Ferran Hurtado. Witness (Delaunay)
            graphs. *Comput. Geom.*, 44(6-7):329–344, 2011.

[ADH11b]   Boris Aronov, Muriel Dulieu, and Ferran Hurtado. Witness rectangle
            graphs. In Frank Dehne, John Iacono, and Jörg-Rüdiger Sack, editors,
            *Algorithms and Data Structures (Proc. 12th International Symposium,
            WADS 2011)*, volume 6844 of *Lecture Notes Comput. Sci.*, pages 73–85.
            Springer-Verlag, 2011.

[AH82]   D. Avis and J. Horton. Remarks on the sphere of influence graphs. *Ann.
            New York Acad. Sci.*, 440:323–327, 1982.

[AM92]   Pankaj K. Agarwal and J. Matoušek. Relative neighborhood graphs in
            three dimensions. *Comput. Geom. Theory Appl.*, 2(1):1–14, 1992.

[ÁMFM+11]   Bernardo M. Ábrego, Ruy Fabila Monroy, Silvia Fernández-Merchant,
            David Flores-Peñaloza, Ferran Hurtado, Vera Sacristan, and Maria
            Saumell. On crossing numbers of geometric proximity graphs. *Comput.
            Geom.*, 44(4):216–233, 2011.

[BBM99]   T. Biedl, A. Bretscher, and H. Meijer. Rectangle of influence drawings
            of graphs without filled 3-cycles. In *Graph Drawing (Proc. GD '99)*,
            volume 1731 of *Lecture Notes Comput. Sci.*, pages 359–368. Springer-
            Verlag, 1999.

[BC87]   S. Bhatt and S. Cosmadakis. The complexity of minimizing wire lengths
            in VLSI layouts. *Inform. Process. Lett.*, 25:263–267, 1987.

[BCH+10]   Prosenjit Bose, Sébastien Collette, Ferran Hurtado, Matias Korman, Ste-
            fan Langerman, Vera Sacristan, and Maria Saumell. Some properties of
            higher order Delaunay and Gabriel graphs. In *Proceedings of the 22nd
            Annual Canadian Conference on Computational Geometry, CCCG 2010*,
            pages 13–16, 2010.

[BDEK06]   Prosenjit Bose, Luc Devroye, William S. Evans, and David G. Kirk-
            patrick. On the spanning ratio of Gabriel graphs and beta-skeletons.
            SIAM J. Discrete Math., 20(2):412–427, 2006.

[BDL+11]   Prosenjit Bose, Luc Devroye, Maarten Löffler, Jack Snoeyink, and Vishal Verma. Almost all Delaunay triangulations have stretch factor greater than $pi/2$. *Comput. Geom.*, 44(2):121–127, 2011.

[BDLL95]   P. Bose, G. Di Battista, W. Lenhart, and G. Liotta. Proximity constraints and representable trees. In R. Tamassia and I. G. Tollis, editors, *Graph Drawing (Proc. GD '94)*, volume 894 of *Lecture Notes Comput. Sci.*, pages 340–351. Springer-Verlag, 1995.

[BE51]   P. Bateman and P. Erdös. Geometrical extrema suggested by a lemma of Besicovitch. *American Mathematical Monthly*, 58:306–314, 1951.

[Bes45]   A.S. Besicovitch. A general form of the covering principle and relative differentiation of additive functions. *Proceedings of the Cambridge Philosophical Society*, 41:103–110, 1945.

[BLL96]   P. Bose, W. Lenhart, and G. Liotta. Characterizing proximity trees. *Algorithmica*, 16:83–110, 1996. (special issue on Graph Drawing, edited by G. Di Battista and R. Tamassia).

[BLS00]   E. Boyer, L. Lister, and B. Shader. Sphere of influence graphs using the sup-norm. *Mathematical and Computer Modelling*, 32(10):1071–1082, 2000.

[BM04]   Prosenjit Bose and Pat Morin. Online routing in triangulations. *SIAM J. Comput.*, 33(4):937–951, 2004.

[BMSU01]   Prosenjit Bose, Pat Morin, Ivan Stojmenovic, and Jorge Urrutia. Routing with guaranteed delivery in ad hoc wireless networks. *Wireless Networks*, 7(6):609–616, 2001.

[Bro79]   K. Q. Brown. Voronoi diagrams from convex hulls. *Inform. Process. Lett.*, 9(5):223–228, 1979.

[CBF+06]   Pier Francesco Cortese, Giuseppe Di Battista, Fabrizio Frati, Luca Grilli, Katharina Anna Lehmann, Giuseppe Liotta, Maurizio Patrignani, Ioannis G. Tollis, and Francesco Trotta. On the topologies of local minimum spanning trees. In Thomas Erlebach, editor, *Combinatorial and Algorithmic Aspects of Networking, Third Workshop, CAAN 2006*, volume 4235 of *Lecture Notes in Computer Science*, pages 31–44. Springer, 2006.

[CCL09]   Jean Cardinal, Sébastien Collette, and Stefan Langerman. Empty region graphs. *Comput. Geom.*, 42(3):183–195, 2009.

[CEG+94]   B. Chazelle, H. Edelsbrunner, L. Guibas, J. Hershberger, R. Seidel, and M. Sharir. Selecting heavily covered points. *SIAM J. Comput.*, 23:1138–1151, 1994.

[Cim92]   R. J. Cimikowski. Properties of some Euclidian proximity graphs. *Pattern Recogn. Lett.*, 13(6):417–423, June 1992.

[CISRS05] Julien Cartigny, François Ingelrest, David Simplot-Ryl, and Ivan Stojmenovic. Localized LMST and RNG based minimum-energy broadcast protocols in ad hoc networks. *Ad Hoc Networks*, 3(1):1–16, 2005.

[CKLS10]   Siu-Wing Cheng, Christian Knauer, Stefan Langerman, and Michiel H. M. Smid. Approximating the average stretch factor of geometric graphs. In Otfried Cheong, Kyung-Yong Chwa, and Kunsoo Park, editors, *Algorithms and Computation - 21st International Symposium, ISAAC 2010*, volume 6506 of *Lecture Notes in Computer Science*, pages 37–48. Springer, 2010.

[CKX11]   Shiliang Cui, Iyad A. Kanj, and Ge Xia. On the stretch factor of Delaunay triangulations of points in convex position. *Comput. Geom.*, 44(2):104–109, 2011.

[CPZ04]   Miguel Á. Carreira-Perpiñán and Richard S. Zemel. Proximity graphs for clustering and manifold learning. In *Neural Information Processing Systems, NIPS 2004*, 2004.

[CTL92]   M. S. Chang, C. Y. Tang, and C. T. Lee. Solving the Euclidean bottleneck matching problem by k-relative neighborhood graphs. *Algorithmica*, 8:177–194, 1992.

[CWL02]   G. Calinescu, P. Wan, and X. Li. Distributed construction of planar spanners and routing for ad hoc wireless networks. In *Proc. 21st Annual Joint Conference of the IEEE Computer and Communication Societies (INFOCOM 02)*, 2002.

[CX01]   Siu-Wing Cheng and Yin-Feng Xu. On β-skeleton as a subgraph of the minimum weight triangulation. *Theoretical Computer Science*, 262:459–471, 2001.

[DDLM12]   Emilio Di Giacomo, Walter Didimo, Giuseppe Liotta, and Henk Meijer. Drawing a tree as a minimum spanning tree approximation. *J. Comput. Syst. Sci.*, 78(2):491–503, 2012.

[Del34]   B. Delaunay. Sur la sphère vide. A la memoire de Georges Voronoi. *Izv. Akad. Nauk SSSR, Otdelenie Matematicheskih i Estestvennyh Nauk*, 7:793–800, 1934.

[DETT99]   G. Di Battista, P. Eades, R. Tamassia, and I. G. Tollis. *Graph Drawing*. Prentice Hall, Upper Saddle River, NJ, 1999.

[Dil87]   M. B. Dillencourt. A non-Hamiltonian, nondegenerate Delaunay triangulation. *Inform. Process. Lett.*, 25:149–151, 1987.

[Dil89]   M. B. Dillencourt. An upper bound on the shortness exponent of inscribable polytopes. *J. Combin. Theory Ser. B*, 46(1):66–83, February 1989.

[Dil90a]   M. B. Dillencourt. Realizability of Delaunay triangulations. *Inform. Process. Lett.*, 33(6):283–287, February 1990.

[Dil90b]   M. B. Dillencourt. Toughness and Delaunay triangulations. *Discrete Comput. Geom.*, 5:575–601, 1990.

[Dil96]   M. B. Dillencourt. Finding Hamiltonian cycles in Delaunay triangulations is NP-complete. *Discrete Applied Mathematics*, 64(3):207–217, 1996.

[DLL95]   G. Di Battista, W. Lenhart, and G. Liotta. Proximity drawability: a survey. In R. Tamassia and I. G. Tollis, editors, *Graph Drawing (Proc. GD '94)*, volume 894 of *Lecture Notes Comput. Sci.*, pages 328–339. Springer-Verlag, 1995.

[DLM]   Emilio Di Giacomo, Giuseppe Liotta, and Henk Meijer. The approximate rectangle of influence drawability problem. In W. Didimo and M. Patrignani, editors, *Graph Drawing (Proc. 20th International Symposium, GD 2012)*, Lecture Notes Comput. Sci. to appear.

[DLPT98]   Olivier Devillers, Giuseppe Liotta, Franco P. Preparata, and Roberto Tamassia. Checking the convexity of polytopes and the planarity of subdivisions. *Comput. Geom. Theory Appl.*, 11:187–208, 1998.

[DLW06]   Giuseppe Di Battista, Giuseppe Liotta, and Sue Whitesides. The strength
            of weak proximity. *J. Discrete Algorithms*, 4(3):384–400, 2006.

[DS94]   M. B. Dillencourt and W. D. Smith. Graph-theoretical conditions for in-
            scribability and Delaunay realizability. In *Proc. 6th Canad. Conf. Com-
            put. Geom.*, pages 287–292, 1994.

[DS95]   M. B. Dillencourt and W. D. Smith. A linear-time algorithm for testing
            the inscribability of trivalent polyhedra. *Internat. J. Comput. Geom.
            Appl.*, 5:21–36, 1995.

[DV96]   G. Di Battista and L. Vismara. Angles of planar triangular graphs. *SIAM
            J. Discrete Math.*, 9(3):349–359, 1996.

[Ede95]   Herbert Edelsbrunner. The union of balls and its dual shape. *Discrete
            & Computational Geometry*, 13:415–440, 1995.

[EGK+12]   William Evans, Emden R. Gansner, Michael Kaufmann, Giuseppe Liotta,
            Henk Meijer, and Andreas Spillner. Approximate proximity drawings. In
            Marc J. van Kreveld and Bettina Speckmann, editors, *Graph Drawing
            (Proc. 19th International Symposium, GD 2011)*, volume 7034 of *Lecture
            Notes Comput. Sci.*, pages 166–78. Springer-Verlag, 2012.

[Epp92]   D. Eppstein. The diameter of nearest neighbor graphs. Tech. Report
            92-76, Dept. Inform. Comput. Sci., Univ. California, Irvine, CA, July
            1992.

[Epp09]   David Eppstein. Isometric diamond subgraphs. In Ioannis G. Tollis and
            Maurizio Patrignani, editors, *Graph Drawing (Proc. 16th International
            Symposium, GD 2008)*, volume 5417 of *Lecture Notes in Computer Sci-
            ence*, pages 384–389, 2009.

[EPY97]   David Eppstein, Mike Paterson, and F. Frances Yao. On nearest-neighbor
            graphs. *Discrete & Computational Geometry*, 17(3):263–282, 1997.

[EW96a]   P. Eades and S. Whitesides. The logic engine and the realization problem
            for nearest neighbor graphs. *Theoret. Comput. Sci.*, 169:23–37, 1996.

[EW96b]   P. Eades and S. Whitesides. The realization problem for Euclidean min-
            imum spanning trees is NP-hard. *Algorithmica*, 16:60–82, 1996. (special
            issue on Graph Drawing, edited by G. Di Battista and R. Tamassia).

[FK11]   Fabrizio Frati and Michael Kaufmann. Polynomial area bounds for mst
            embeddings of trees. *Comput. Geom.*, 44(9):529–543, 2011.

[Fus06]   Éric Fusy. Counting d-polytopes with d+3 vertices. *Electr. J. Comb.*,
            13(1), 2006.

[Fus09]   Éric Fusy. Transversal structures on triangulations: A combinatorial
            study and straight-line drawings. *Discrete Mathematics*, 309(7):1870–
            1894, 2009.

[GHvK02]   Joachim Gudmundsson, Mikael Hammar, and Marc J. van Kreveld.
            Higher order Delaunay triangulations. *Comput. Geom.*, 23(1):85–98,
            2002.

[GJ79]   M. R. Garey and D. S. Johnson. *Computers and Intractability: A Guide
            to the Theory of NP-Completeness*. W. H. Freeman, New York, NY,
            1979.

[GLN02]   Joachim Gudmundsson, Christos Levcopoulos, and Giri Narasimhan.
            Fast greedy algorithms for constructing sparse geometric spanners. *SIAM
            J. Comput.*, 31(5):1479–1500, 2002.

[GO04]   J. E. Goodman and J. O'Rourke, editors. *Handbook of Discrete and Computational Geometry, 2nd Edition*. CRC Press, 2004.

[GPS94]   L. Guibas, J. Pach, and M. Sharir. Sphere of influence graphs in higher dimensions. *Colloquia Mathematica Societatis János Bolyai*, 63:131–137, 1994.

[GS69]   K. R. Gabriel and R. R. Sokal. A new statistical approach to geographic variation analysis. *Systematic Zoology*, 18:259–278, 1969.

[Har69]   F. Harary. *Graph Theory*. Addison-Wesley, Reading, Mass., 1969.

[HJLM93]   F. Harary, M.S. Jacobson, M.J. Lipman, and F.R. Morris. On abstract sphere of influence graphs. *Mathematical and Computer Modelling*, 17(11):77–83, 1993.

[HLW10]   Ferran Hurtado, Giuseppe Liotta, and David R. Wood. Proximity drawings of high-degree trees. *CoRR*, abs/1008.3193, 2010.

[IR07]   Mohammad Tanvir Irfan and Md. Saidur Rahman. Computing beta-drawings of 2-outerplane graphs. In M. Kaykobad and Md. Saidur Rahman, editors, *Workshop on Algorithms and Computation 2007 - Proceedings of First WALCOM*, pages 46–61. Bangladesh Academy of Sciences (BAS), 2007.

[IS85]   M. Ichino and J. Sklansky. The relative neighborhood graph for mixed feature variables. *Pattern Recognition*, 18(2):161–167, 1985.

[JLM95]   M.S. Jacobson, M.J. Lipman, and F.R. Morris. Trees that are sphere of influence graphs. *Appl. Math. Letters*, 8(6):89–93, 1995.

[JT92]   J. W. Jaromczyk and G. T. Toussaint. Relative neighborhood graphs and their relatives. *Proc. IEEE*, 80(9):1502–1517, September 1992.

[Kan83]   V. Kantabutra. Traveling salesman cycles are not always subgraphs of Voronoi duals. *Inform. Process. Lett.*, 16:11–12, 1983.

[Kan09]   Iyad A. Kanj. On spanners of geometric graphs. In Jianer Chen and S. Barry Cooper, editors, *Theory and Applications of Models of Computation, 6th Annual Conference, TAMC 2009*, volume 5532 of *Lecture Notes in Computer Science*, pages 49–58. Springer, 2009.

[Kei94]   M. Keil. Computing a subgraph of the minimum weight triangulation. *Comput. Geom. Theory Appl.*, 4:13–26, 1994.

[Kin06]   James A. King. Realization of degree 10 minimum spanning trees in 3-space. In *Proceedings of the 18th Annual Canadian Conference on Computational Geometry, CCCG 2006*, 2006.

[KL10]   Sanjiv Kapoor and Xiang-Yang Li. Proximity structures for geometric graphs. *Int. J. Comput. Geometry Appl.*, 20(4):415–429, 2010.

[KPX10]   Iyad A. Kanj, Ljubomir Perkovic, and Ge Xia. On spanners and lightweight spanners of geometric graphs. *SIAM J. Comput.*, 39(6):2132–2161, 2010.

[KR85]   D. G. Kirkpatrick and J. D. Radke. A framework for computational morphology. In G. T. Toussaint, editor, *Computational Geometry*, pages 217–248. North-Holland, Amsterdam, Netherlands, 1985.

[KW01]   M. Kaufmann and D. Wagner, editors. *Drawing Graphs*, volume 2025 of *Lecture Notes in Computer Science*. Springer-Verlag, 2001.

[KW04]   Matthew Kitching and Sue Whitesides. The three dimensional logic engine. In János Pach, editor, *Graph Drawing (Proc. 12th International*

Symposium, GD 2004), volume 3383 of *Lecture Notes in Computer Science*, pages 329–339. Springer, 2004.

[KWZ03]   Fabian Kuhn, Roger Wattenhofer, and Aaron Zollinger. Worst-case optimal and average-case efficient geometric ad-hoc routing. In *Proceedings of the 4th ACM Interational Symposium on Mobile Ad Hoc Networking and Computing, MobiHoc 2003*, pages 267–278. ACM, 2003.

[Lan69]   P. M. Lankford. Regionalization: theory and alternative algorithms. *Geogr. Anal.*, 1:196–212, 1969.

[LCWW03]   X.Y. Li, G. Calinescu, P.J. Wan, and Y. Wang. Localized Delaunay triangulation with application in ad hoc wireless networks. *IEEE Transactions on Parallel and Distributed Systems*, 14:1035–1047, 2003.

[LD95]   Giuseppe Liotta and Giuseppe Di Battista. Computing proximity drawings of trees in the 3-dimensional space. In *Proc. 4th Workshop Algorithms Data Struct.*, volume 955 of *Lecture Notes Comput. Sci.*, pages 239–250. Springer-Verlag, 1995.

[LH04]   Ning Li and Jennifer C. Hou. Topology control in heterogeneous wireless networks: Problems and solutions. In *INFOCOM*, 2004.

[LHS03]   Ning Li, Jennifer C. Hou, and Lui Sha. Design and analysis of an mst-based topology control algorithm. In *INFOCOM*, 2003.

[Li04]   X.Y. Li. Applications of computational geometry in wireless networks. In X. Cheng, X. Huang, and D.-Z. Du, editors, *Ad Hoc Wireless Networking*, pages 197–264. Kluwer Academic Publisher, 2004.

[LL96]   W. Lenhart and G. Liotta. Drawing outerplanar minimum weight triangulations. *Inform. Process. Lett.*, 57(5):253–260, 1996.

[LL97]   W. Lenhart and G. Liotta. Proximity drawings of outerplanar graphs. In S. North, editor, *Graph Drawing (Proc. GD '96)*, volume 1190 of *Lecture Notes Comput. Sci.*, pages 286–302. Springer-Verlag, 1997.

[LL02]   W. Lenhart and G. Liotta. The drawability problem for minimum weight triangulations. *Theoretical Computer Science*, 270:261–286, 2002.

[LLMW98]   G. Liotta, A. Lubiw, H. Meijer, and S.H. Whitesides. The rectangle of influence drawability problem. *Comput. Geom. Theory and Applications*, 10(1):1–22, 1998.

[LM03]   G. Liotta and H. Meijer. Voronoi drawings of trees. *Comput. Geom. Theory and Applications*, 24(3):147–178, 2003.

[LS93]   A. Lubiw and N. Sleumer. Maximal outerplanar graphs are relative neighborhood graphs. In *Proc. 5th Canad. Conf. Comput. Geom.*, pages 198–203, 1993.

[LSW04]   Xiang-Yang Li, Ivan Stojmenovic, and Yu Wang. Partial Delaunay triangulation and degree limited localized bluetooth scatternet formation. *IEEE Transactions on Parallel and Distributed Systems*, 15(4):350–361, 2004.

[LSW05]   Xiang-Yang Li, Wen-Zhan Song, and Weizhao Wang. A unified energy efficient topology for unicast and broadcast. In *Proc. MobiCom'05*, 2005.

[LTTV97]   G. Liotta, R. Tamassia, I. G. Tollis, and P. Vocca. Area requirement of Gabriel drawings. In *Algorithms and Complexity (Proc. CIAC' 97)*, volume 1203 of *Lecture Notes Comput. Sci.*, pages 135–146. Springer-Verlag, 1997.

[Mat87]   C. Mathieu. Some problems in computational geometry. *Algorithmica*, 2:131–134, 1987.

[MMN09]   Kazuyuki Miura, Tetsuya Matsuno, and Takao Nishizeki. Open rectangle-of-influence drawings of inner triangulated plane graphs. *Discrete & Computational Geometry*, 41(4):643–670, 2009.

[MN05]   Kazuyuki Miura and Takao Nishizeki. Rectangle-of-influence drawings of four-connected plane graphs. In Seok-Hee Hong, editor, *Asia-Pacific Symposium on Information Visualisation, APVIS 2005*, volume 45 of *CRPIT*, pages 75–80, 2005.

[MNN00]   Kazuyuki Miura, Takao Nishizeki, and Shin-Ichi Nakano. Convex grid drwaings of four-connected plane graphs. In D. T. Lee and Shang-Hua Teng, editors, *Algorithms and Computation, 11th International Conference, ISAAC 2000*, volume 1969 of *Lecture Notes in Computer Science*, pages 254–265. Springer, 2000.

[MNN06]   Kazuyuki Miura, Shin-Ichi Nakano, and Takao Nishizeki. Convex grid drawings of four-connected plane graphs. *Int. J. Found. Comput. Sci.*, 17(5):1031–1060, 2006.

[MNS⁺99]   K. Mehlhorn, S. Näher, M. Seel, R. Seidel, T. Schilz, S. Schirra, and C. Uhrig. Checking geometric programs or verification of geometric structures. *Comput. Geom. Theory Appl.*, 12(1–2):85–103, 1999.

[MQ94a]   T.S. Michael and T. Quint. Sphere of influence graphs: a survey. *Congressus Numerantium*, 105:153–160, 1994.

[MQ94b]   T.S. Michael and T. Quint. Sphere of influence graphs: Edge density and clique size. *Mathematical and Computer Modelling*, 127(7):19–24, 1994.

[MQ99]   T.S. Michael and T. Quint. Sphere of influence graphs in general metric spaces. *Mathematical and Computer Modelling*, 29(7):45–53, 1999.

[MQ03]   T.S. Michael and T. Quint. Sphere of influence graphs and the l_∞ metric. *Discrete Applied Mathematics*, 127:447–460, 2003.

[MR08]   Wolfgang Mulzer and Günter Rote. Minimum-weight triangulation is NP-hard. *J. ACM*, 55(2), 2008.

[MS80]   D. W. Matula and R. R. Sokal. Properties of Gabriel graphs relevant to geographic variation research and clustering of points in the plane. *Geogr. Anal.*, 12(3):205–222, 1980.

[MS92]   C. Monma and Subhash Suri. Transitions in geometric minimum spanning trees. *Discrete Comput. Geom.*, 8:265–293, 1992.

[MW00]   F. R. McMorris and C. Wang. Sphere of attraction graphs. *Congressus Numerantium*, 142:149–160, 2000.

[NS07]   Giri Narasimhan and Michiel H. M. Smid. *Geometric spanner networks*. Cambridge University Press, 2007.

[OBS92]   Atsuyuki Okabe, Barry Boots, and Kokichi Sugihara. *Spatial Tessellations: Concepts and Applications of Voronoi Diagrams*. John Wiley & Sons, Chichester, UK, 1992.

[O'R87]   J. O'Rourke. Computational geometry column 2. *SIGACT News*, 18(2):10–12, 1987. Also in Computer Graphics 21(1987), 155–157.

[PMDS03]   C. E. Priebe, D. J. Marchette, J. DeVinney, and D.A. Socolinsky. Classification using class cover catch digraphs. *Journal of Classification*, 20(1):3–23, 2003.

[PS90]   F. P. Preparata and M. I. Shamos. *Computational Geometry: An Intro-duction.* Springer-Verlag, 3rd edition, October 1990.

[PS04]   R. Pinchasi and S. Smorodinsky. On locally Delaunay geometric graphs. In *Proc. 20th ACM Symposium on Computational Geometry (SoCG'04)*, pages 378–382, 2004.

[PV04]   Paolo Penna and Paola Vocca. Proximity drawings in polynomial area and volume. *Comput. Geom. Theory and Applications*, 29(2):91–116, 2004.

[PY92]   M. S. Paterson and F. F. Yao. On nearest-neighbor graphs. In *Proc. 19th Internat. Colloq. Automata Lang. Program.*, volume 623 of *Lecture Notes Comput. Sci.*, pages 416–426. Springer-Verlag, 1992.

[QW04]   Jianbo Qian and Cao An Wang. A linear-time approximation scheme for maximum weight triangulation of convex polygons. *Algorithmica*, 40(3):161–172, 2004.

[QW06]   Jianbo Qian and Cao An Wang. Progress on maximum weight triangulation. *Comput. Geom.*, 33(3):99–105, 2006.

[Rad88]   J. D. Radke. On the shape of a set of points. In G. T. Toussaint, editor, *Computational Morphology*, pages 105–136. North-Holland, Amsterdam, Netherlands, 1988.

[Rei48]   E.R. Reifenberg. A problem on circles. *Mathematical Gazette*, 32:290–292, 1948.

[RS95]   G. Robins and J. S. Salowe. Low-degree minimum spanning trees. *Discrete Comput. Geom.*, 14:151–165, 1995.

[SH97]   Kokichi Sugihara and Tetsuya Hiroshima. How to draw a Delaunay diagram with a given topology. In *Abstracts 13th European Workshop Comput. Geom.*, pages 13–15. Universität Würzburg, 1997.

[SI92]   K. Sugihara and M. Iri. Construction of the Voronoi diagram for 'one million' generators in single-precision arithmetic. *Proc. IEEE*, 80(9):1471–1484, September 1992.

[SIII00]   K. Sugihara, M. Iri, H. Inagaki, and T. Imai. Topology-oriented implementation - an approach to robust geometric algorithms. *Algorithmica*, 27(1):5–20, 2000.

[SIR08]   Md. Abul Hassan Samee, Mohammad Tanvir Irfan, and Md. Saidur Rahman. Computing *beta* -drawings of 2-outerplane graphs in linear time. In Shin-Ichi Nakano and Md. Saidur Rahman, editors, *Algorithms and Computation, Second International Workshop, WALCOM 2008*, volume 4921 of *Lecture Notes in Computer Science*, pages 81–87. Springer, 2008.

[SLL+08]   Svetlana Stolpner, Jonathan Lenchner, Giuseppe Liotta, David Bremner, Christophe Paul, Marc Pouget, and Stephen K. Wismath. A note on alpha-drawable k-trees. In *Proceedings of the 20th Annual Canadian Conference on Computational Geometry*, 2008.

[Sos99a]   M. Soss. On the size of the Euclidean sphere of influence graph. In *Proc. 11th Canad. Conf. Comput. Geom.*, 1999.

[Sos99b]   M. Soss. The size of the open sphere of influence graph in L_∞ metric spaces. In *Graph Drawing (Proc. GD '98)*, volume 1547 of *Lecture Notes Comput. Sci.*, pages 458–459. Springer-Verlag, 1999.

[SZ11]   Sadish Sadasivam and Huaming Zhang. Closed rectangle-of-influence
            drawings for irreducible triangulations. *Comput. Geom.*, 44(1):9–19,
            2011.

[Tou80]   G. T. Toussaint. The relative neighbourhood graph of a finite planar set.
            Pattern Recogn., 12:261–268, 1980.

[Tou88]   G. T. Toussaint. A graph-theoretical primal sketch. In G. T. Toussaint,
            editor, *Computational Morphology*, pages 229–260. North-Holland, Ams-
            terdam, Netherlands, 1988.

[Tou05]   G. Toussaint. Geometric proximity graphs for improving nearest neigh-
            bor methods in instance-based learning and data mining. *International
            Journal of Comput. Geom. and Applications*, 15(2):101–150, 2005.

[Urq83]   R. B. Urquhart. Some properties of the planar Euclidean relative neigh-
            bourhood graph. *Pattern Recogn. Lett.*, 1:317–332, 1983.

[Vel92]   R. C. Veltkamp. The γ-neighborhood graph. *Comput. Geom. Theory
            Appl.*, 1(4):227–246, 1992.

[Vel94]   R. C. Veltkamp. *Closed Object Boundaries from Scattered Points*, volume
            885 of *Lecture Notes Comput. Sci.* Springer-Verlag, 1994.

[Vel95]   R. C. Veltkamp. Boundaries through scattered points of unknown density.
            Graphics Models and Image Processing, 57(6):441–452, November 1995.

[WCY99]   Cao An Wang, Francis Y. L. Chin, and Bo-Ting Yang. Maximum weight
            triangulation and graph drawing. *Inf. Process. Lett.*, 70(1):17–22, 1999.

[WCY00]   C. An Wang, F. Y. Chin, and B. Yang. Triangulations without minimum
            weight drawing. *Information Processing Letters*, 74(5–6):183–189, 2000.

[Woo10]   David R. Wood. Partitions and coverings of trees by bounded-degree
            subtrees. *CoRR*, abs/1008.3190, 2010.

[WY01]   C. A. Wang and B. Yang. A lower bound for β-skeleton belonging to
            minimum weight triangulations. *Comput. Geom. Theory Appl.*, 19:35–
            46, 2001.

[ZV09a]   Huaming Zhang and Milind Vaidya. On open rectangle-of-influence and
            rectangular dual drawings of plane graphs. *Discrete Mathematics, Algo-
            rithms and Applications*, 1(3):319–333, 2009.

[ZV09b]   Huaming Zhang and Milind Vaidya. On open rectangle-of-influence draw-
            ings of planar graphs. In Ding-Zhu Du, Xiaodong Hu, and Panos Parda-
            los, editors, *Combinatorial Optimization and Applications*, volume 5573
            of *Lecture Notes in Computer Science*, pages 123–134. Springer Berlin /
            Heidelberg, 2009.

5

Tree Drawing Algorithms

Adrian Rusu
Rowan University

5.1 Introduction

Tree drawing is concerned with the automatic generation of geometric representations of relational information, often for visualization purposes. The typical data structure for modeling hierarchical information is a tree whose vertices represent entities and whose edges correspond to relationships between entities. Visualizations of hierarchical structures are only useful to the degree that the associated diagrams effectively convey information to the people that use them. A good diagram helps the reader understand the system, but a poor diagram can be confusing.

The automatic generation of drawings of trees finds many applications, such as software engineering (program nesting trees, object-oriented class hierarchies), business administration (organization charts), decision support systems (activity trees), artificial intelligence (knowledge-representation isa hierarchies), logic programming (SLD-trees), website design and browsing (structure of a website), biology (evolutionary trees), and chemistry (molecular drawings).

Algorithms for drawing trees are typically based on some graph-theoretic insight into the structure of the tree. The input to a tree drawing algorithm is a tree T that needs to be drawn. The output is a *drawing* Γ, which maps each node of T to a distinct point in the plane, and each edge (u, v) of T to a simple Jordan curve with endpoints u and v.

T is an *ordered tree* if the children of each node are assigned a fixed left-to-right order. For any node u in T, its *leftmost child* (*rightmost child*) is the one that comes first (last) in the left-to-right ordering of the children of u in T. The *leftmost path* p of T is the maximal path consisting of nodes that are leftmost children, except the first one, which is the root of T. The last node of p is called the *leftmost* node of T. Two nodes of T are *siblings* if they have the same parent. The *subtree* of T rooted at a node v consists of v and all the descendants of v. T is the *empty tree* if it has zero nodes in it.

Let v be a node of an ordered tree. Then $n(v)$, $p(v)$, $l(v)$, $r(v)$, and $s_1(v), \ldots, s_i(v)$, are the number of nodes in the subtree rooted at v, parent, leftmost child, rightmost child, and siblings of v, respectively.

The rest of the chapter is organized as follows. After motivating the need for tree drawing algorithms and providing drawing conventions and aesthetics in this section, we describe the main approaches for tree drawing algorithms in subsequent sections. We then present some of the most representative algorithms for drawing binary and general trees.

5.1.1   Drawing Conventions

A *drawing convention* is a basic rule that a drawing must satisfy to be admissible [DETT99]. A list of the most used drawing conventions for drawing trees and their significance is given below (see Figure 5.1):

Polyline Drawings

A *polyline drawing* is a drawing in which each edge is drawn as a connected sequence of one or more line segments, where the meeting point of consecutive line segments is called a *bend* (see Figure 5.1(a)).

Orthogonal Drawings

An *orthogonal drawing* is one in which each edge is drawn as a chain of alternating horizontal and vertical segments (see Figure 5.1(b)).

Upward and Non-Upward Drawings

An *upward drawing* is defined as a drawing where no child is placed higher in the y-direction than its parent (see Figure 5.1(a),(c)). A *non-upward drawing* is a drawing that is not upward (see Figure 5.1(b),(d)).

Grid Drawings

A *grid drawing* is one in which each vertex is placed at integer coordinates. Assuming that the plane is covered by *horizontal* and *vertical channels*, with unit distance between two consecutive channels, the meeting point of a horizontal and a vertical channel is called a *grid-point*. The computer screen can be viewed as a grid of pixels placed at integer coordinates. Grid drawings guarantee at least unit distance separation between the nodes of the tree, and the integer coordinates of the nodes and edge-bends allow the drawings to be rendered in a (large-enough) grid-based display surface, such as a computer screen, without any distortions due to truncation and round-off errors. The smallest rectangle with horizontal and vertical sides parallel to the axes that covers the entire grid drawing is called the *enclosing rectangle*.

Planar Drawings

A *planar drawing* is a drawing in which edges do not intersect each other in the drawing (for example, the drawings (a), (b), and (c) in Figure 5.1 are planar drawings, and the drawing (d) is a non-planar drawing). Planar drawings are normally easier to understand than non-planar drawings, i.e., drawings with edge-crossings. Since any tree

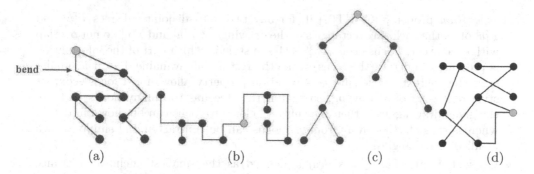

bend ⟶

(a)   (b)   (c)   (d)

Figure 5.1 Various kinds of drawings of the same tree: (a) polyline, (b) orthogonal, (c) straight-line, (d) non-planar. Also note that the drawings shown in Figures (a) and (c) are upward drawings, whereas the drawings shown in Figures (b) and (d) are not. The root of the tree is shown as a shaded circle, whereas other nodes are shown as black circles.

admits a planar drawing, it is desirable to obtain planar drawings for trees.

Straight-line Drawings

The so-called *straight-line* tree drawings have each edge drawn as a straight-line segment (see Figure 5.1(c)). It is natural to draw each edge of a tree as a straight-line between its end-nodes. Straight-line drawings are easier to understand than polyline drawings.

The experimental study of the human perception of graph drawings has concluded that minimizing the number of edge crossings and minimizing the number of bends increases the understandability of drawings of graphs [TDB88, Pur97, PCJ97, Pur00]. Ideally, the drawings should have no edge crossings, i.e., they should be planar drawings and should have no edge-bends, i.e., they should be straight-line drawings.

5.1.2 Aesthetics

Aesthetics specify graphic properties of the drawing that we would like to apply as much as possible. Most of the tree drawing algorithms have concentrated on drawing trees in as small as possible area with user-controlled aspect ratio. A list of the most important aesthetics of drawings of trees is given below:

- **Area**: The *area* of a grid drawing is defined as the number of grid points contained in its enclosing rectangle. Drawings with small area can be drawn with greater resolution on a fixed-size page. Note that we cannot discuss the area of non-grid drawings (i.e., drawings that have the nodes placed at real coordinates), since, by placing the nodes closer or farther, such a drawing can be scaled down or up by any value.

- **Aspect Ratio**: The *aspect ratio* of a grid drawing is defined as the ratio of the length of the shortest side to the length of the longest side of its enclosing rectangle. An aspect ratio is considered *optimal* if it is equal to 1. Giving the users control over the aspect ratio of a drawing allows them to display the drawing in different kinds of displays surfaces with different aspect ratios. The optimal use of the screen space is achieved by minimizing the area of the drawing and by providing user-controlled aspect ratio.

- **Subtree Separation**: Let $T[v]$ be the subtree rooted at node v of tree T. $T[v]$ consists of v and all the descendants of v. A drawing of T has the *subtree-*

separation property [CGKT97] if, for any two node-disjoint subtrees $T[u]$ and $T[v]$ of T, the enclosing rectangles of the drawings of $T[u]$ and $T[v]$ do not overlap with each other. Focus+context [SB94] is a style in which part of the information is presented in detail (the focus) while the rest is still available, but at a smaller size (the context). The subtree-separation property allows for a focus+context style rendering of a drawing, so that if the tree has too many nodes to fit in the given drawing area, then the subtrees closer to focus can be shown in detail, whereas those farther away from the focus can be contracted and simply shown as filled-in rectangles.

- **Closest Leaf**: The *closest leaf* is defined as the smallest euclidean distance between the root of the tree and a leaf in the drawing [RS08].

- **Farthest Leaf**: The *farthest leaf* is defined as the largest euclidean distance between the root of the tree and a leaf in the drawing [RS08].

The aesthetics closest leaf and farthest leaf help determine whether the algorithms place leaves close or far from the root. It is important to minimize the distance between the root and the leaves of the tree, especially in the case when the user needs to visually analyze the information contained in the levels close to the root and levels close to the leaves, without the information in between. Such a case appears in particular for algorithms where a change at the top level (root) of the tree generates modifications at the bottom levels (leaves) of the tree (for example, usual operations—find, insert, remove—on binary search trees, splay trees, or B^+ trees).

Other well-known aesthetics that have been used in various tree drawing studies are as follows [DETT99]:

- **Size**: the longest side of the smallest rectangle with horizontal and vertical sides covering the drawing.

- **Total Edge Length**: the sum of the lengths of the edges in the drawing.

- **Average Edge Length**: the average of the lengths of the edges in the drawing.

- **Maximum Edge Length**: the maximum among the lengths of the edges in the drawing.

- **Uniform Edge Length**: the variance of the edge lengths in the drawing.

- **Angular Resolution**: the smallest angle formed by two edges incident on the same node.

- **Symmetry**: visual identification of symmetries in the drawing.

It is widely accepted [DETT94, DETT99, Pur97, PCJ97] that small values of the size, total edge length, average edge length, maximum edge length, and uniform edge length are related to the perceived aesthetic appeal and visual effectiveness of the drawing. High angular resolution is desirable in visualization applications and in the design of optical communication networks. For binary trees, the degree of a node is at most three, hence a trivial upper bound on the angular resolution is 120°. Given a symmetric drawing, a conceptual understanding of the entire tree can be built up from that of a smaller subtree, replicated a number of times.

5.2   Level-Based Approach

The *level-based approach* can be used on both binary and general trees, and it is characterized by the fact that in the drawings produced, the nodes at the same distance from the root are horizontally aligned. Algorithms based on this approach are usually simple to understand and implement and produce intuitive drawings that exhibit clear display of symmetries. However, these algorithms have two disadvantages: the drawing has an area of $\Omega(n^2)$ and, for balanced trees with many nodes, the width is much larger than the height.

Level-based algorithms have been designed previously [Blo93, RT81, BJL02, Wal90]. The algorithms described in [BJL02, Wal90] achieve better area, but they do not exhibit the subtree separation property.

A recursive algorithm for binary trees [RT81], which exhibits the subtree separation property, uses the following steps: draw the subtree rooted at the left child, draw the subtree rooted at the right child, place the drawings of the subtrees at horizontal distance 2, and place the root one level above and halfway between the children. If there is only one child, place the root at horizontal distance 1 from the child. A drawing produced by this algorithm is provided in Figure 5.2.

Figure 5.2 Drawing of the Fibonacci tree with 88 nodes, generated by the level-based algorithm of [RT81].

By using a geometric transformation (cartesian → polar), level drawings yield *radial drawings*, where nodes are placed on concentric circles by level (see Figure 5.3).

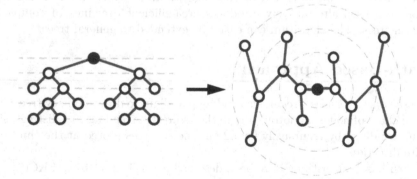

Figure 5.3 Example of a transformation from a level drawing to a radial drawing. Figure taken from [CT].

Radial drawings are often used in drawing graphs, even though they do not always guarantee planarity. Several algorithms for radial drawings of trees have been designed, and some of them have also been used in various applications [Ber81, Ead92, CPM+98, CPP00, BM03, Bac07].

5.3 H-V Approach

The *horizontal-vertical approach* can be used on both binary and general trees. In this approach, a divide-and-conquer strategy is used to recursively construct an upward, orthogonal, and straight-line drawing of a tree, by placing the root of the tree in the top-left corner, and the drawings of its left and right subtrees one next to the other (*horizontal composition*) or one below the other (*vertical composition*) (see Figure 5.4). The resulting drawing also exhibits the subtree separation property within an $O(n \log n)$ area.

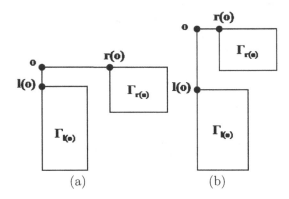

Figure 5.4 General H-V approach. (a) *Horizontal composition*: the drawings of the subtrees rooted at the children of o are placed one next to the other. (b) *Vertical composition*: the drawings of the subtrees rooted at the children of o are placed one below the other.

Various H-V algorithms can be obtained, depending on which layout is used and what other conditions are imposed on the drawing. An algorithm using this approach has been developed for binary trees [CDP92]. This algorithm places the drawing of the subtree with the greater width one unit below the drawing of the subtree with the smaller width (see Figure 5.4(b)). A modification of this algorithm, in which vertical and horizontal combinations are used alternatively, produces area-efficient drawings of complete, AVL, and Fibonacci trees. The algorithm can easily be extended to general trees.

5.4 Path-Based Approach

The *path-based approach* uses a recursive winding paradigm to draw a binary tree T by laying down a small chain of nodes monotonically in the x-direction leading to a distinguished node v, and then "winding" by recursively laying out the subtrees rooted at the children of v in the opposite direction.

Several path-based algorithms have been designed [CGKT02, GR03a, SKC00].

Recursively, for every subtree rooted at a node v, a parameter A is fixed, so that, if $n(v) \leq A$, then the drawings of the subtrees rooted at the children of v are placed one next to the other, as in Figure 5.5 (a). Otherwise, the subtree looks like Figure 5.5 (b), where v_1 is the root of the subtree, $v_{i+1} = r(v_i)$ for $i \geq 1$, $k \geq 1$ is the first index for which $n(v_k) > n - A$ and $n(v_{k+1}) \leq n - A$, T_i is the subtree rooted at $l(v_i)$, $T' = l(v_k)$, and $T'' = r(v_k)$. In the second case, depending on whether an upward or a non-upward drawing is to be obtained, the drawings are placed as in Figures 5.6(a) and 5.6(b), respectively.

The user controls the aspect ratio by modifying parameter A.

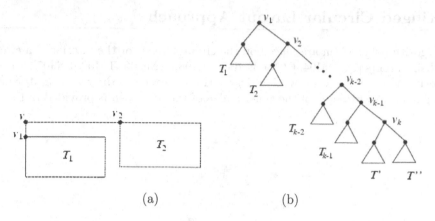

Figure 5.5 (a) When $n(v) \leq A$, the subtrees are placed one next to the other. (b) When $n(v) > A$, the tree is divided into subtrees $T_1, T_2, \ldots, T_{k-2}, T_{k-1}, T', T''$.

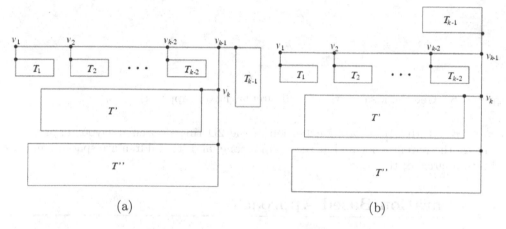

Figure 5.6 (a) Upward drawing of binary tree T. (b) Non-upward drawing of binary tree T.

A drawing of the Fibonacci tree with 88 nodes produced by the algorithm of Chan et al. [CGKT02], with the value for the parameter A at one of the extremes, is provided in Figure 5.7. This algorithm produces the best worst-case theoretical bound on area for path-based algorithms: $O(n \log \log n)$.

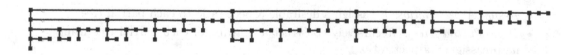

Figure 5.7 Drawing of Fibonacci tree with 88 nodes produced by the path-based algorithm [CGKT02], with parameter A at one of the extremes: $A = 88$.

5.5   Ringed Circular Layout Approach

In these algorithms, children are placed on the circumference or the interior of a circle centered at their parents [GADM04, CC99, Ead92, MH98, MMC99, TM02, RSJ07]. In general, these algorithms are used to draw high-degree trees. However, the resulting drawings are often not planar. An example of the general idea of the approach is provided in Figure 5.8.

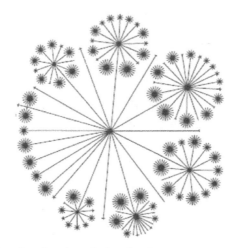

Figure 5.8   General idea for the ringed circular layout approach.

Cone trees [RMC91] are a 3D extension of the 2D ringed circular layout approach. In cone trees, the parent is located at the tip of a cone, and its children are spaced equally on the bottom circle of the cone.

5.6   Separation-Based Approach

The *separation-based approach* can be used on both binary and general trees. Separation-based algorithms have been designed [GR02, GR03b, GR03c, RS07]. In this approach, a divide-and-conquer strategy is used to recursively construct a drawing of a tree, by performing the following actions at each recursive step:

- *Find a Separator Edge or a Separator Node*: A *separator edge* (*node*) of a tree T with $degree(T) = d$ is an edge (node), which, if removed, divides T into at most d smaller, partial, trees. It has been shown that every tree contains a separator edge or a separator node [GR03c, Val81].
- *Split Tree*: Split T into at most d partial trees by removing a separator edge or a separator node.
- *Assign Aspect Ratios*: Preassign a desirable aspect ratio to each partial tree.
- *Draw Partial Trees*: Recursively construct a drawing of each partial tree using its preassigned aspect ratio.
- *Compose Drawings*: Arrange the drawings of the partial trees, and draw the nodes and edges that were removed from the tree to divide it, such that the drawing of the tree thus obtained meets certain aesthetics.

5.7 Algorithms for Drawing Binary Trees

A *binary tree* is one where each node has at most two children. Most of the research on drawing trees targets binary trees; hence, in this section, several algorithms for drawing binary trees are presented.

Binary trees have a strong connection to real-life applications. For instance, binary trees represent programs in combinatory logic, which is under investigation as an approach to nanostructure synthesis and control [Mac03]. The idea is to use molecular processes to implement the combinatory logic tree substitution operations, so that the molecular reorganization of the trees results in the desired structure or process. Visualization of these binary trees could improve the investigator's ability in interpreting the substitution operations involved in combinatory logic.

5.7.1 Theoretical Results

We summarize some known theoretical results on planar grid drawings of binary trees. (See Table 5.1.)

Drawing Type	Area	Aspect Ratio	Reference
upward orthogonal polyline	$O(n \log \log n)$	$\Theta(\log^2 n/(n \log \log n))$	[GGT96]
(non-upward) orthogonal polyline	$O(n)$	$\Theta(1)$	[Lei80, Val81]
upward orthogonal straight-line	$O(n \log n)$	$[1, n/\log n]$	[CDP92, CGKT02]
(non-upward) orthogonal straight-line	$O(n \log \log n)$	$\Theta(\log^2 n/(n \log \log n))$	[CGKT02, SKC00]
upward polyline	$O(n)$	$[n^{-\epsilon}, n^\epsilon]$	[GGT96]
upward straight-line	$O(n \log \log n)$	$\Theta(\log^2 n/(n \log \log n))$	[SKC00]
(non-upward) straight-line	$O(n)$	$[n^{-\epsilon}, n^\epsilon]$	[GR04]

Table 5.1 Bounds on the areas and aspect ratios of various kinds of planar grid drawings of an n-node unordered binary tree. Here, ϵ is an arbitrary constant, such that $0 < \epsilon < 1$.

Let T be an n-node binary tree. Garg et al. [GGT96] present an algorithm for constructing an upward polyline drawing of T with $O(n)$ area, and any user-specified aspect ratio in the range $[n^{-\epsilon}, n^\epsilon]$, where ϵ is any constant, such that $0 < \epsilon < 1$. It also shows that $n \log \log n$ is a tight bound for the area of upward orthogonal polyline drawings, i.e., any binary tree can be drawn in this fashion in $O(n \log \log n)$ area, and there exists a family of binary trees that requires $\Omega(n \log \log n)$ area in any such drawing. Leiserson [Lei80] and Valiant [Val81] present algorithms for constructing a (non-upward) orthogonal polyline drawing of T with $O(n)$ area. Chan et al. [CGKT02] give an algorithm for constructing an upward orthogonal straight-line drawing of T with $O(n \log n)$ area, and any user-specified aspect ratio in the range $[1, n/\log n]$. It also shows that $n \log n$ is a tight bound for such drawings. Shin et al. [SKC00] give an algorithm for constructing an upward straight-line drawing of T with $O(n \log \log n)$ area. Chan et al. [CGKT02] and Shin et al. [SKC00] show that T admits a non-upward planar straight-line orthogonal grid drawing with height $O(n/A) \log A$ and width $O(A + \log n)$, where $2 \le A \le n$ is any user-specified number. This result also implies that we can draw any binary tree in this fashion in area $O(n \log \log n)$ (by setting $A = \log n$). If T is a Fibonacci tree (AVL tree and complete binary tree), then Crescenzi et al. [CDP92] and Trevisan [Tre96] (Crescenzi et al. [CPP98, CDP92], respectively) give algorithms for constructing an upward straight-line drawing of T with $O(n)$ area. Garg and Rusu [GR04] present an algorithm for constructing a (non-upward) straight-line drawing of T with $O(n)$ area, and any user-specified aspect ratio in the range $[n^{-\epsilon}, n^{\epsilon}]$, where ϵ is any constant, such that $0 < \epsilon < 1$. This is trivially a tight bound, as any straight-line drawing of a binary tree with n nodes requires $\Omega(n)$ area.

Table 5.2 summarizes the results for order-preserving algorithms.

Drawing Type	Area	Aspect Ratio	Ref.
Complete tree			
upward straight-line order-preserving	$\Theta(n)$	$O(1)$	[CDP92]
Fibonacci tree			
upward straight-line order-preserving	$\Theta(n)$	$O(1)$	[Tre96]
Special balanced binary tree such as red-black			
upward straight-line order-preserving	$O(n(\log \log n)^2)$	$n/\log^2 n$	[SKC00]
Logarithmic tree			
upward straight-line order-preserving	$\Theta(n)$	$O(1)$	[CP98]
Binary tree			
upward orthogonal polyline order-preserving	$O(n \log n)$	$\Theta(\log^2 n/(n \log \log n))$	[Kim95, GGT96]
non-upward orthogonal polyline order-preserving	$O(n)$	$(9a + 8)/(9b + 8)$	[DT81]
upward orthogonal straight-line order-preserving	$\Theta(n^2)$	$O(1)$	[CDP92, Fra07]
non-upward orthogonal straight-line order-preserving	$O(n^{1.5})$	$O(\sqrt{(n)}/n)$	[Fra07]
upward polyline order-preserving	$O(n \log n)$	$\log n/n$	[Kim04]
	$O(n \log n)$	$\Theta(\log^2 n/(n \log \log n))$	[GGT96, CDP92]
non-upward polyline order-preserving	$O(n \log \log n)$	$(n \log \log n)/\log^2 n$	[GR03a]
upward straight-line order-preserving	$\Theta(n \log n)$	$n/\log n$	[GR03a]
non-upward straight-line order-preserving	$O(n \log n)$	$[1, n/\log n]$	[GR03a]
	$O(n \log \log n)$	$(n \log \log n)/\log^2 n$	[GR03a]

Table 5.2 Bounds on the areas and aspect ratios of various kinds of order-preserving planar grid drawings of an n-node ordered tree. Here, $ab \le kn$, where k is some constant.

Shin et al. [SKC00] have shown that a special class of balanced binary trees, which includes k-balanced, red-black, $BB[\alpha]$, and (a, b) trees, admits order-preserving planar upward straight-line grid drawings with area $O(n(\log\log n)^2)$. Crescenzi et al. [CDP92], Crescenzi and Penna [CP98], and Trevisan [Tre96] give order-preserving planar upward straight-line grid drawings of complete, logarithmic, and Fibonacci trees, respectively, with area $O(n)$. Dolev and Trickey [DT81] prove that binary trees admit $\Theta(n)$ area order-preserving orthogonal drawings. Kim [Kim95] shows an upper bound of $O(n\log n)$ area for upward order-preserving orthogonal drawings of ternary trees (trees whose nodes have at most three children), result that immediately extends to binary trees. This area bound is optimal, as Garg et al. [GGT96] demonstrate a lower bound of $O(n\log n)$ area for such drawings of binary trees. Crescenzi et al. [CDP92] give an algorithm that achieves $O(n^2)$ area for upward orthogonal straight-line order-preserving drawings of binary trees. Frati [Fra07] proves that this bound is optimal. Frati [Fra07] also gives the best known upper bound of $O(n^{1.5})$ area for non-upward orthogonal straight-line order-preserving drawings of binary trees. It is unknown whether this is an optimal bound, as the trivial $O(n)$ is the lower bound currently known. Garg et al. [GGT96] provides an algorithm that constructs an upward polyline order-preserving drawing of a binary tree with $O(n\log n)$ area, which is the optimal bound for such drawings [CDP92]. Kim [Kim04] improves the number of bends from $O(n)$ to $O(n/\log n)$, while matching the area bound. Garg and Rusu [GR03a] show that a binary tree admits an order-preserving planar straight-line grid drawing with $O(n\log\log n)$ area. In addition, they show that a binary tree admits an order-preserving upward planar straight-line drawing with optimal $O(n\log n)$ area.

A variety of results exist for other kinds of drawings. Di Battista et al. [DETT99] and Frati [Fra09] have given a survey of these results.

5.7.2 Experimental Analysis

Experimental studies provide insight into the behavior of tree drawing algorithms beyond their targetted aesthetic criteria. In a comprehensive experimental study [RS08], separation-based algorithm by Garg and Rusu [GR04], path-based algorithm by Chan et al. [CGKT02], level-based algorithm by Reingold and Tilford [RT81], and ringed circular layout algorithm by Teoh and Ma [TM02] were compared on a large suite of seven types of binary trees of various sizes, based on ten quality measures: area, aspect ratio, size, total edge length, average edge length, maximum edge length, uniform edge length, angular resolution, closest leaf, and farthest leaf. As the specific algorithms compared are intended to be representative of their respective approaches, it is expected that the results generally apply to other algorithms using the same approach and even extend to trivial extensions to general trees.

This experimental analysis includes some interesting findings:

- The performance of a drawing algorithm on a tree-type is not a good predictor of the performance of the same algorithm on other tree-types: some of the algorithms perform best on a tree-type, and worst on other tree-types.
- Reingold-Tilford algorithm [RT81] scores worse in comparison to the other chosen algorithms for almost all ten aesthetics considered.
- The intuition that low average edge length and area go together is contradicted in only one case.
- The intuitions that average edge length and maximum edge length, uniform edge length and total edge length, and short maximum edge length and close farthest leaf go together are contradicted for unbalanced binary trees.
- With regards to area, of the four algorithms studied, three perform best on different types of trees.

- With regards to aspect ratio, of the four algorithms studied, three perform well on trees of different types and sizes.

- Not all algorithms studied perform best on complete binary trees even though they have one of the simplest tree structures.

- The level-based algorithm of Reingold-Tilford [RT81] produces much worse aspect ratios than algorithms designed using other approaches.

- The path-based algorithm of Chan et al. [CGKT02] tends to construct drawings with better area at the expense of worse aspect ratio.

5.7.3   Unordered Trees

In this section, we present the algorithm of [GR04] in more detail. This algorithm uses a separation-based approach (therefore, we call it *Separation*), and achieves optimal linear area for planar straight-line grid drawings, while at the same time, giving the user control over the aspect ratio. In addition, the drawings produced by this algorithm exhibit the subtree separation property.

Let T be a tree with root o. Let n be the number of nodes in T. A *partial tree* of T is a connected subgraph of T.

For some trees, the algorithm designates a special *link* node u^* that has at most one child.

Let T be a tree with link node u^*. A planar straight-line grid drawing Γ of T is a *feasible* drawing of T, if it has the following three properties:

- **Property 1**: The root o is placed at the top-left corner of Γ.

- **Property 2**: If $u^* \neq o$, then u^* is placed at the bottom boundary of Γ. Moreover, u^* can move downward in its vertical channel by any distance without causing any edge-crossings in Γ.

- **Property 3**: If $u^* = o$, then no other node or edge of T is placed on or crosses the vertical and horizontal channels occupied by o. Moreover, u^* (i.e., o) can move upward in its vertical channel by any distance without causing any edge-crossings in Γ.

Let A and ϵ be two numbers, where ϵ is a constant, such that $0 < \epsilon < 1$, and $n^{-\epsilon} \leq A \leq n^{\epsilon}$. A is called the *desirable aspect ratio* for T.

Theorem 5.1 *[Separator Theorem [Val81]] Every binary tree T with n nodes, where $n \geq 2$, contains an edge e, called a* separator edge, *such that removing e from T splits it into two non-empty trees with n_1 and n_2 nodes, respectively, such that for some x, where $1/3 \leq x \leq 2/3$, $n_1 = xn$, and $n_2 = (1-x)n$. Moreover, e can be found in $O(n)$ time.*

The algorithm takes ϵ, A, and T as input and uses a divide-and-conquer strategy to recursively construct a feasible drawing Γ of T, by performing the following actions at each recursive step:

- *Split Tree*: Split T into at most five partial trees by removing at most two nodes and their incident edges from it. Each partial tree has at most $(2/3)n$ nodes. Based on whether the separator edge is on the leftmost path of T or not, there are two general cases, which are shown in Figure 5.9.

- *Assign Aspect Ratios*: Correspondingly, assign a desirable aspect ratio A_k to each partial tree T_k. The value of A_k is based on the value of A and the number of nodes in T_k.

- *Draw Partial Trees*: Recursively construct a feasible drawing of each partial tree T_k with A_k as its desirable aspect ratio.

- *Compose Drawings*: Arrange the drawings of the partial trees, and draw the nodes and edges that were removed from T to split it, such that the drawing Γ of T is a feasible drawing. Note that the arrangement of these drawings is done based on the cases shown in Figure 5.9. In each case, if $A < 1$, then the drawings of the partial trees are stacked one above the other, and if $A \geq 1$, then they are placed side-by-side.

Remark: The drawing Γ constructed by the algorithm may not have aspect ratio exactly equal to A, but it fits inside a rectangle with area $O(n)$ and aspect ratio A.

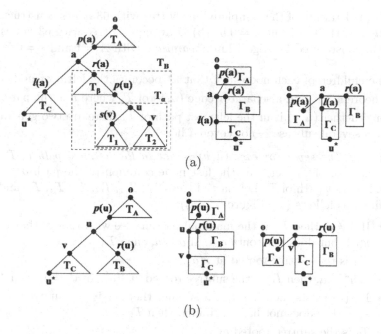

Figure 5.9 (a) Drawing T in Case 1 (when the separator (u, v) is not in the leftmost path of T). (b) Drawing T in Case 2 (when the separator (u, v) is in the leftmost path of T). For each case, first the structure of T for that case is shown, then its drawing when $A < 1$, and then its drawing when $A \geq 1$. For simplicity, $p(a)$ and $p(u)$ are shown to be in the interior of Γ_A, but actually, either they are the same as o, or if $A < 1$ ($A \geq 1$), then they are placed at the bottom (right) boundary of Γ_A. For simplicity, Γ_A, Γ_B, and Γ_C are shown as identically sized boxes, but in actuality, they may have different sizes.

Figure 5.10 (a) shows a drawing of a complete binary tree with 63 nodes constructed by algorithm *Separation*, with $A = 1$ and $\epsilon = 0.5$. Figure 5.10 (b) shows a drawing of a tree with 63 nodes, consisting of a single path, constructed by algorithm *Separation*, with $A = 1$ and $\epsilon = 0.5$.

Split Tree

The splitting of tree T into partial trees is done as follows:

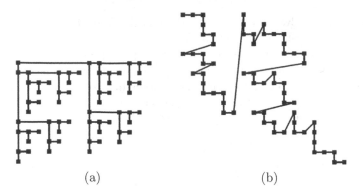

(a)   (b)

Figure 5.10 (a) Drawing of the complete binary tree with 63 nodes constructed by Algorithm *Separation*, with $A = 1$ and $\epsilon = 0.5$. (b) Drawing of a tree with 63 nodes, consisting of a single path, constructed by Algorithm *Separation*, with $A = 1$ and $\epsilon = 0.5$.

- Order the children of each node such that u^* becomes the leftmost node of T.
- Using Theorem 5.1, find a separator edge (u, v) of T, where u is the parent of v.
- Based on whether (u, v) is in the leftmost path of T, there are two general cases (each with several subcases—not covered here):

  - *Case 1: The separator edge (u, v) is not in the leftmost path of T.* Let o be the root of T. Let a be the last node common to the path $o \rightsquigarrow v$, and the leftmost path of T. Let partial trees T_A, T_B, T_C, T_α, T_β, T_1, and T_2 be defined as follows (see Figure 5.9 (a)):

    * If $o \neq a$, then T_A is the maximal partial tree with root o, that contains $p(a)$, but does not contain a. If $o = a$, then $T_A = \emptyset$.
    * T_B is the subtree rooted at $r(a)$.
    * If $u^* \neq a$, then T_C is the subtree rooted at $l(a)$. If $u^* = a$, then $T_C = \emptyset$.
    * If $s(v)$ exists, i.e., if v has a sibling, then T_1 is the subtree rooted at $s(v)$. If v does not have a sibling, then $T_1 = \emptyset$.
    * T_2 is the subtree rooted at v.
    * If $u \neq a$, then T_α is the subtree rooted at u. If $u = a$, then $T_\alpha = T_2$. Note that T_α is a subtree of T_B.
    * If $u \neq a$ and $u \neq r(a)$, then T_β is the maximal partial tree with root $r(a)$, that contains $p(u)$, but does not contain u. If $u = a$ or $u = r(a)$, then $T_\beta = \emptyset$. Again, note that T_β belongs to T_B.

    Nodes a and u and their incident edges are being removed to split T into at most five partial trees T_A, T_C, T_β, T_1, and T_2. $p(a)$ is designated as the link node of T_A, $p(u)$ as the link node of T_β, and u^* as the link node of T_C. Arbitrarily select a leaf of T_1, and a leaf of T_2, and designate them as the link nodes of T_1 and T_2, respectively.

  - *Case 2: The separator edge (u, v) is in the leftmost path of T.* Let o be the root of T. Let partial trees T_A, T_B, and T_C be defined as follows (see Figure 5.9 (b)):

    * If $o \neq u$, then T_A is the maximal partial tree with root o, that contains $p(u)$, but does not contain u. If $o = u$, then $T_A = \emptyset$.

* If $r(u)$ exits, i.e., u has a right child, then T_B is the subtree rooted at $r(u)$. If u does not have a right child, then $T_B = \emptyset$.
* T_C is the subtree rooted at v.

Node u and its incident edges are being removed to split T into at most three partial trees T_A, T_B, and T_C. $p(u)$ is designated as the link node of T_A, and u^* as the link node of T_C. Arbitrarily select a leaf of T_B and designate it as the link node of T_B.

Assign Aspect Ratios

Let T_k be a partial tree of T, where for Case 1, T_k is either T_A, T_C, T_β, T_1, or T_2, and for Case 2, T_k is either T_A, T_B, or T_C. Let n_k be the number of nodes in T_k.

Definition: T_k is a *large* partial tree of T if:

* $A \geq 1$ and $n_k \geq (n/A)^{1/(1+\epsilon)}$, or
* $A < 1$ and $n_k \geq (An)^{1/(1+\epsilon)}$,

and is a *small* partial tree of T otherwise.

In Step *Assign Aspect Ratios*, a desirable aspect ratio A_k is assigned to each non-empty T_k as follows: Let $x_k = n_k/n$.

* If $A \geq 1$: If T_k is a large partial tree of T, then $A_k = x_k A$, otherwise (i.e., if T_k is a small partial tree of T) $A_k = n_k^{-\epsilon}$.
* If $A < 1$: If T_k is a large partial tree of T, then $A_k = A/x_k$, otherwise (i.e., if T_k is a small partial tree of T) $A_k = n_k^{\epsilon}$.

Intuitively, the above assignment strategy ensures that each partial tree gets a good desirable aspect ratio.

Draw Partial Trees

If $A \geq 1$, then the values of A_A and A_β (A_A and A_β are the desirable aspect ratios for T_A and T_β, respectively) are being changed to $1/A_A$ and $1/A_\beta$, respectively. This is done so because later in Step *Compose Drawings*, when constructing Γ, if $A \geq 1$, then the drawings of T_A and T_β are rotated by $90°$. Drawing T_A and T_β with desirable aspect ratios $1/A_A$ and $1/A_\beta$, respectively, compensates for the rotation, and ensures that the drawings of T_A and T_β that eventually get placed within Γ are those with desirable aspect ratios A_A and A_β, respectively.

Next, each non-empty partial tree T_k, $k \in \{A, B, C, \alpha, \beta, 1, 2\}$, is drawn recursively with A_k as its desirable aspect ratio. The base case for the recursion happens when T_k contains exactly one node, in which case, the drawing of T_k is simply the one consisting of exactly one node.

Compose Drawings

Let Γ_k denote the drawing of a partial tree T_k constructed in Step *Draw Partial Trees*. We now describe the construction of a feasible drawing Γ of T from the drawings of its partial trees in Case 1.

In Case 1, first a drawing Γ_α of the partial tree T_α is constructed by composing Γ_1 and Γ_2 as shown in Figure 5.11, then a drawing Γ_B of T_B is constructed by composing Γ_α and Γ_β as shown in Figure 5.12, and finally Γ is constructed by composing Γ_A, Γ_B, and Γ_C as shown in Figure 5.9 (a).

In the general case ($u \neq a$ and $T_1 \neq \emptyset$), Γ_α is constructed as follows (see Figure 5.11):

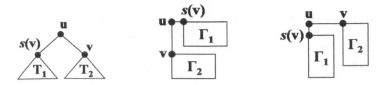

Figure 5.11 Drawing T_α in the general case ($u \neq a$ and $T_1 \neq \emptyset$). First, the structure of T_α is shown, then its drawing when $A < 1$, and then its drawing when $A \geq 1$. For simplicity, Γ_1 and Γ_2 are shown as identically sized boxes, but in actuality, their sizes may be different.

- If $A < 1$, then Γ_1 is placed above Γ_2 such that the left boundary of Γ_1 is one unit to the right of the left boundary of Γ_2; u is placed in the same vertical channel as v and in the same horizontal channel as $s(v)$.

- If $A \geq 1$, then Γ_1 is placed one unit to the left of Γ_2, such that the top boundary of Γ_1 is one unit below the top boundary of Γ_2; u is placed in the same vertical channel as $s(v)$ and in the same horizontal channel as v.

Draw edges $(u, s(v))$ and (u, v).

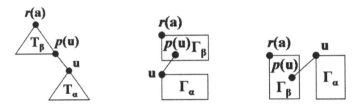

Figure 5.12 Drawing T_B in the general case ($T_\beta \neq \emptyset$). First, the structure of T_B is shown, then its drawing when $A < 1$, and then its drawing when $A \geq 1$. For simplicity, $p(u)$ is shown to be in the interior of Γ_β, but actually, it is either same as $r(a)$, or if $A < 1$ ($A \geq 1$), then is placed on the bottom (right) boundary of Γ_β. For simplicity, Γ_β and Γ_α are shown as identically sized boxes, but in actuality, their sizes may be different.

In the general case ($T_\beta \neq \emptyset$), Γ_B is constructed as follows (see Figure 5.12):

- if $A < 1$, then Γ_β is placed one unit above Γ_α such that the left boundaries of Γ_β and Γ_α are aligned.

- If $A \geq 1$, then first Γ_β is rotated clockwise by $90°$ and then flipped right-to-left, then Γ_β is placed one unit to the left of Γ_α such that the top boundaries of Γ_β and Γ_α are aligned.

Draw edge $(p(u), u)$.

In general Case 1, Γ is constructed from Γ_A, Γ_B, and Γ_C as follows (see Figure 5.9 (a)):

- If $A < 1$, then Γ_A, Γ_B, and Γ_C are stacked one above the other, such that they are separated by unit distance from each other, and the left boundaries of Γ_A and Γ_C are aligned with each other and are placed one unit to the left of the left boundary of Γ_B; a is placed in the same vertical channel as o and $l(a)$, and in the same horizontal channel as $r(a)$.

- If $A \geq 1$, then first Γ_A is rotated clockwise by $90°$ and flipped right-to-left. Then, Γ_A, Γ_C, and Γ_B are placed from left-to-right in that order, separated by unit distances, such that the top boundaries of Γ_A and Γ_B are aligned with each other, and are one unit above the top boundary of Γ_C. Then, Γ_C is moved down until u^* becomes the lowest node of Γ; a is placed in the same vertical channel as $l(a)$ and in the same horizontal channel as o and $r(a)$.

Draw edges $(p(a), a)$, $(a, r(a))$, and $(a, l(a))$.
In general Case 2, Γ is constructed by composing Γ_A, Γ_B, and Γ_C, using a procedure similar to the one of Case 1 (see Figure 5.9(b)).

Theorem 5.2 *Let T be a binary tree with n nodes. Given two numbers A and ϵ, where ϵ is a constant, such that $0 < \epsilon < 1$, and $n^{-\epsilon} \leq A \leq n^{\epsilon}$, a planar straight-line grid drawing of T with $O(n)$ area and aspect ratio A, can be constructed in $O(n \log n)$ time. Moreover, Γ has the subtree-separation property.*

Proof: Designate any leaf of T as its link node. Construct a drawing Γ of T by invoking Algorithm *Separation* with T, A, and ϵ as input. Γ will be a planar straight-line grid drawing contained entirely within a rectangle with $O(n)$ area and aspect ratio A, and which exhibits the subtree separation property. $\qquad\square$

5.7.4 Ordered Trees

In this Section, we present two algorithms of [GR03a] in detail. The first algorithm (we call it *Fixed Spine*) shows that a binary tree admits an order-preserving *upward* planar straight-line grid drawing with *optimal* $O(n \log n)$ area. The second algorithm (we call it *Arbitrary Spine*), shows that a binary tree admits an order-preserving planar straight-line grid drawing with width $O(A + \log n)$, height $O((n/A) \log A)$, and area $O(n \log n)$, for any given $2 \leq A \leq n$. Setting $A = \log n$, it results in an area of $O(n \log \log n)$. Both algorithms take $O(n)$ time to construct the drawings.

Let T be an ordered tree. Each node of T has at most two children, called its *left* and *right* children, respectively.

Let α be a positive integer. An order-preserving planar straight-line grid drawing of T is an α-*drawing* of T, if it has the following two properties:

- **Property 1**: No node is placed to the left of, or above the root of, T.
- **Property 2**: The vertical and horizontal separations between the root and its rightmost child are equal to α and one units, respectively.

A *left-corner* drawing of an ordered tree is an order-preserving planar straight-line grid drawing, where no node of the tree is placed to the left of, or above its root. Note that an α-drawing is also a left-corner drawing.

The *mirror-image* of T is the ordered tree obtained by reversing the counterclockwise order of edges incident on each node.

A *spine* of T is a path $v_0 v_1 v_2 \ldots v_m$, where $v_0, v_1, v_2, \ldots, v_m$ are nodes of T, that is defined recursively as follows (see Figure 5.13):

- v_0 is the same as the root of T;
- v_{i+1} is a child of v_i, such that the subtree rooted at v_{i+1} has the maximum number of nodes among all the subtrees that are rooted at the children of v_i.

A *non-spine* node of T is one that does not belong to its spine.

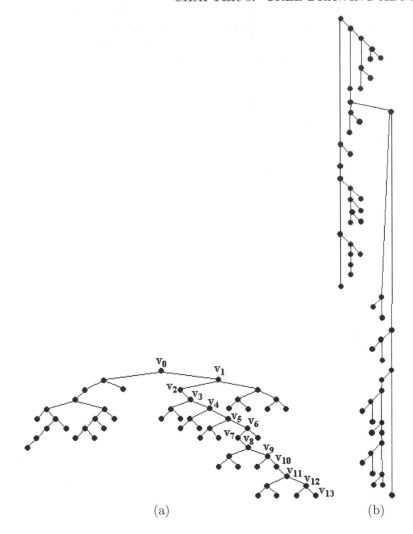

Figure 5.13   (a) A binary tree T with spine $v_0v_1\ldots v_{13}$. (b) The order-preserving planar upward straight-line grid drawing of T constructed by *fixed spine* algorithm.

Algorithm Fixed Spine

For simplicity, throughout this section, it is assumed that each non-leaf node has exactly two children. The algorithm can be simply extended to cover the case where a non-leaf node has only one child.

The *fixed spine* drawing algorithm uses a path-based approach to obtain an order-preserving upward planar straight-line grid drawing with optimal ($O(n\log n)$) area of an ordered binary tree T. In each recursive step, it breaks T into several subtrees, draws each subtree recursively, and then combines their drawings to obtain an upward α-drawing $D(T)$ of T, where α is a positive integer given as a parameter to the algorithm.

Let $P = v_0v_1v_2\ldots v_m$ be a spine of T.

There are two cases (see Figures 5.14 and 5.15):

- *Case 1: v_1 is the left child of v_0* (see Figure 5.14(a)).
  Let L be the subtree rooted at v_1, s be the non-spine child of v_0, and R be the

subtree rooted at s. v_0 is placed at the origin. 1-drawings $D(L)$ and $D(R)$ of L and R are recursively constructed. $D(R)$ is placed such that s is one unit to the right of, and α units below v_0. $D(L)$ is placed such that v_1 is in the same vertical channel as v_0, and is one unit below $D(R)$ (see Figure 5.14(b)).

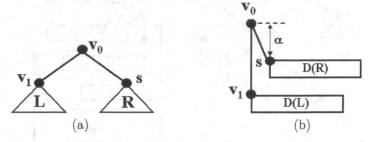

(a)   (b)

Figure 5.14 (a) The structure of a binary tree T in *Case 1*, where v_1 is the left child of v_0. (b) The drawing of T in *Case 1*. For simplicity, $D(L)$ and $D(R)$ are shown as identically sized boxes, but in actuality, they may have different sizes.

- *Case 2: v_1 is the right child of v_0* (see Figure 5.15(a)).
  Let $k \geq 1$ be the smallest integer, such that v_k is either a leaf, or has a non-spine node as its left child.
  There are two subcases:

  - *v_k has a non-spine node as its left child*: Let $s_0, s_1, \ldots, s_k$ be the non-spine children of $v_0, v_1, \ldots, v_k$, respectively. Let L, A, and B be the subtrees rooted at s_0, s_k, and v_{k+1}, respectively. Let $R_1, R_2, \ldots, R_{k-1}$ be the subtrees rooted at $s_1, s_2, \ldots, s_{k-1}$, respectively. T is drawn as shown in Figure 5.15(b). v_0 is placed at the origin. v_1 is placed one unit to the right of, and α units below, v_0. 1-drawings $D(L), D(A), D(R_1), D(R_2), \ldots, D(R_{k-1})$ of $L, A, R_1, R_2, \ldots, R_{k-1}$, respectively, are recursively constructed. $D(R_1)$ is placed one unit to the right of, and one unit below, v_1. For each i, where $2 \leq i \leq k-1$, v_i and $D(R_i)$ are placed such that v_i is in the same horizontal channel as the bottom of $D(R_{i-1})$ and is in the same vertical channel as v_{i-1}, and $D(R_i)$ is one unit to the right of, and one unit below, v_i. Node v_k is placed in the same vertical channel as v_{k-1}, and in the same horizontal channel as the bottom of $D(R_{k-1})$. $D(A)$ is placed one unit below v_k, such that s_k is in the same vertical channel as v_k. $D(L)$ is placed one unit below $D(A)$, such that s_0 is in the same vertical channel as v_0. Let $\beta = h(D(A)) + h(D(L)) + 2$, where $h(D(A))$ and $h(D(L))$ denote the heights of $D(A)$ and $D(L)$, respectively. Let G be the drawing with the maximum width among $D(L), D(A), D(R_1), D(R_2), \ldots, D(R_{k-1})$. Let W be the width of G. A β-drawing of the mirror image of B is recursively constructed, and then flipped right-to-left to obtain a drawing $D(B)$ of B. $D(B)$ is placed such that v_{k+1} is one unit below v_k, and $\max\{W + 3, \text{width of } D(B)\}$ units to the right of v_0.

  - *v_k is a leaf*: T is drawn in a similar fashion as in the previous subcase, except that $D(A)$ and $D(B)$ do not exist.

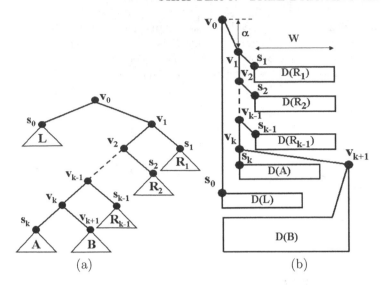

Figure 5.15 The structure of a binary tree T in *Case 2*, where v_1 is the right child of v_0: (a) v_k has a non-spine node as its left child; (b) the drawing of T, when v_k has a non-spine node as its left child. For simplicity, $D(A), D(L), D(R_1), \ldots, D(R_{k-1})$ are shown as identically sized boxes, but in actuality, they may have different sizes.

Theorem 5.3   *An ordered binary tree with n nodes admits an order-preserving upward planar straight-line grid drawing with height at most n, width $O(\log n)$, and optimal $O(n \log n)$ area, which can be constructed in $O(n)$ time.*

Proof: Let T be an n-node ordered binary tree. Using the above algorithm, construct a 1-drawing $D(T)$ of T in $O(n)$ time. As discussed above, $D(T)$ will be an order-preserving upward planar straight-line grid drawing of T with height at most n, width $O(\log n)$, and optimal $O(n \log n)$ area. □

LEMMA 5.1   A left-corner drawing of an n-node ordered binary tree with area $O(n \log n)$, height $O(\log n)$, and width at most n, can be constructed in $O(n)$ time.

Proof: First a 1-drawing of the mirror image of T is constructed using Theorem 5.3, then it is rotated clockwise by $90°$, and then it is flipped right-to-left. □

Algorithm Arbitrary Spine

For any user-defined number A, where $2 \le A \le n$, algorithm *Arbitrary Spine* uses a path-based approach to construct an order-preserving planar straight-line grid drawing of T with $O((n/A) \log A)$ height and $O(A + \log n)$ width. Thus, by setting the value of A, users can control the aspect ratio of the drawing. This implies that, by setting $A = \log n$, such a drawing can be constructed with area $O(n \log \log n)$.

An order-preserving planar straight-line grid drawing of a binary tree T is called a *feasible drawing*, if the root of T is placed on the left boundary and no node of T is placed between the root and the upper-left corner of the enclosing rectangle of the drawing. Note that a left-corner drawing is also a feasible drawing.

Let n be the number of nodes in T. Let $2 \leq A \leq n$ be any number given as a parameter to the algorithm.

Figure 5.16 shows the drawing of the tree of Figure 5.13(a) constructed by algorithm *Arbitrary Spine* with $A = \sqrt{n}$, using Lemma 5.1.

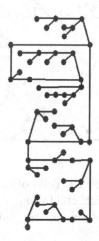

Figure 5.16 Drawing of the tree with $n = 57$ nodes of Figure 5.13(a) constructed by the Algorithm *Arbitrary Spine* with $A = \sqrt{n} = \sqrt{57} = 7.55$, using Lemma 5.1.

In each recursive step, the algorithm constructs a feasible drawing of a subtree T' of T. If T' has at most A nodes in it, then it constructs a left-corner drawing of T' using Lemma 5.1 such that the drawing has width at most m and height $O(\log m)$, where m is the number of nodes in T'. Otherwise, i.e., if T' has more than A nodes in it, then it constructs a feasible drawing of T' as follows:

1. Let $P = v_0 v_1 v_2 \ldots v_q$ be a spine of T'.

2. Let m_i denote the number of nodes in the subtree of T' rooted at v_i, where $0 \leq i \leq q$. Let v_k be the node of P with the value for k such that $m_k > m - A$ and $m_{k+1} \leq m - A$ (since T' has more than A nodes in it, and $m_0, m_1, \ldots, m_q$ is a strictly decreasing sequence of numbers, such a k exists).

3. See Figures 5.17 and 5.18. Let T_i denote the subtree rooted at the non-spine child of v_i, where $0 \leq i \leq k - 1$. Assume, for simplicity, that v_k and v_{k+1} are not leaves (the algorithm can be easily extended to handle the case, where v_k or v_{k+1} is a leaf). Let T^* and T^+ denote the subtrees rooted at the non-spine children of v_k and v_{k+1}, respectively. Let T'' denote the subtree rooted at v_{k+1}. Let T''' denote the subtree rooted at v_{k+2}.

4. Place v_0 at the origin.

5. There are two cases:

   - $k = 0$: Recursively construct a feasible drawing D^* of T^*. Recursively construct a feasible drawing D^+ of the mirror image of T^+. Recursively construct a feasible drawing D''' of the mirror image of T'''. Let s_0 be the root of T^* and s_1 be the root of T^+.

     T' is drawn as shown in Figure 5.17. If s_0 is the left child of v_0, then D^* is placed one unit below v_0, with its left boundary aligned with v_0 (see

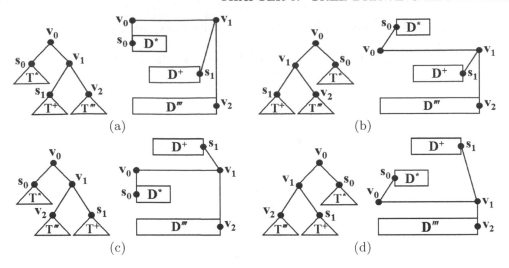

Figure 5.17   Case $k = 0$: (a) s_0 is the left child of v_0, and s_1 is the left child of v_1; (b) s_0 is the right child of v_0, and s_1 is the left child of v_1; (c) s_0 is the left child of v_0, and s_1 is the right child of v_1; (d) s_0 is the right child of v_0, and s_1 is the right child of v_1.

Figure 5.17(a,c)). If s_0 is the right child of v_0, then D^* is placed one unit above, and one unit to the right of v_0 (see Figure 5.17(b,d)). Let W^*, W^+, and W''' be the widths of D^*, D^+, and D''', respectively. Place v_1 in the same horizontal channel as v_0 to its right at the distance $\max\{W^* + 2, W^+ + 2, W'''\}$ from it. Let B_0 and C_0 be the lowest and highest horizontal channels, respectively, occupied by the subdrawing consisting of v_0 and D^*. If s_1 is the left child of v_1, then D^+ is flipped right-to-left, and placed one unit below B_0, and one unit to the left of v_1 (see Figure 5.17(a,b)). If s_1 is the right child of v_1, then D^+ is flipped right-to-left, and placed one unit above C_0, and one unit to the left of v_1 (see Figure 5.17(c,d)). Let B_1 be the lowest horizontal channel occupied by the subdrawing consisting of v_0, D^*, v_1 and D^+. Flip D''' right-to-left, and place it one unit below B_1, such that its right boundary is aligned with v_1 (see Figure 5.17).

- $k > 0$: For each T_i, where $0 \leq i \leq k - 1$, construct a left-corner drawing D_i of T_i using Lemma 5.1.

  Recursively construct feasible drawings D^* and D'' of the mirror images of T^* and T'', respectively.

  T' is drawn as shown in Figure 5.18. If T_0 is rooted at the left child of v_0, then D_0 is placed one unit below v_0, with its left boundary aligned with v_0. If T_0 is rooted at the right child of v_0, then D_0 is placed one unit above, and one unit to the right of v_0. Place each D_i and v_i, where $1 \leq i \leq k - 1$, such that:

  - v_i is in the same horizontal channel as v_{i-1} and is one unit to the right of D_{i-1}, and

  - if T_i is rooted at the left child of v_i, then D_i is placed one unit below v_i, with its left boundary aligned with v_i, otherwise (i.e., if T_i is rooted at the right child of v_i) D_i is placed one unit above, and one unit to the right of v_i.

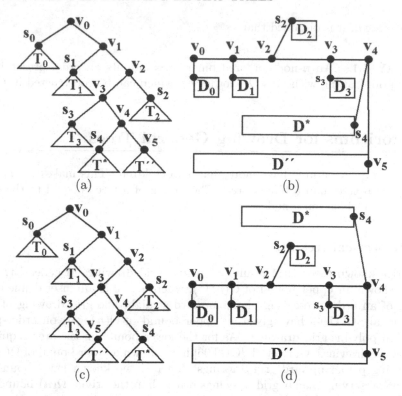

Figure 5.18 Case $k > 0$: Here $k = 4$, s_0, s_1, and s_3 are the left children of v_0, v_1, and v_3, respectively, s_2 is the right child of v_2, T_0, T_1, T_2, T_3, and T'' are the subtrees rooted at v_0, v_1, v_2, v_3, and v_5, respectively, s_4 is the non-spine child of v_4, and T^* is the subtree rooted at s_4; (a) s_4 is the left child of v_4; (c) s_4 is the right child of v_4. For simplicity, boxes D_0, D_1, D_2, D_3 are drawn with same size, but in actuality, they may have different sizes.

Let B_{k-1} and C_{k-1} be the lowest and highest horizontal channels, respectively, occupied by the subdrawing consisting of $v_0, v_1, v_2, \ldots, v_{k-1}$ and $D_0, D_1, D_2, \ldots, D_{k-1}$. Let d be the width of the subdrawing consisting of $v_0, v_1, v_2, \ldots, v_{k-1}$ and $D_0, D_1, D_2, \ldots, D_{k-1}$. Let W^* and W'' be the widths of D^* and D'', respectively.

Place v_k to the right of and in the same horizontal channel as v_{k-1}, such that the horizontal distance between v_k and v_0 is equal to $\max\{d+1, W^*+2, W''\}$. If T^* is rooted at the left-child of v_k, then D^* is flipped right-to-left, and placed one unit below B_{k-1}, and one unit left of v_k (see Figure 5.18(b)). If T^* is rooted at the right-child of v_k, then D^* is flipped right-to-left, and placed one unit above C_{k-1}, and one unit to the left of v_k (see Figure 5.18(d)). Let B_k be the lowest horizontal channel occupied by the subdrawing consisting of $v_1, v_2, \ldots, v_k$, and $D_0, D_1, \ldots, D_{k-1}, D^*$. Flip D'' right-to-left, and place it one unit below B_k, such that its right boundary is aligned with v_k (see Figure 5.18(b,d)).

Theorem 5.4 *Let T be an ordered binary tree with n nodes. Let $2 \leq A \leq n$ be any number. T admits an order-preserving planar straight-line grid drawing with width $O(A + \log n)$, height $O((n/A) \log A)$, and area $O((A + \log n)(n/A) \log A) = O(n \log n)$, which can be constructed in $O(n)$ time.*

Setting $A = \log n$, it is obtained that:

COROLLARY 5.1   An n-node ordered binary tree admits an order-preserving planar straight-line grid drawing with area $O(n \log \log n)$, which can be constructed in $O(n)$ time.

5.8   Algorithms for Drawing General Trees

In a general tree, a node may have more than two children. This makes it more difficult to draw a general tree than a binary tree. The *degree* of a tree is equal to the maximum number of edges incident on a node.

5.8.1   Theoretical Results

We summarize known theoretical results on planar grid drawings of general trees. Chan [Cha02] has shown an upper bound of $O(n^{1+\epsilon})$, where $\epsilon > 0$ is any user-defined constant, on the area of an order-preserving planar upward straight-line grid drawing of a general tree. Garg et al. [GGT96] have given an upper bound of $O(n \log n)$ on order-preserving planar upward polyline grid drawings. As for the lower bound on the area-requirement of order-preserving drawings, Garg et al. [GGT96] have shown a lower bound of $\Omega(n \log n)$ for order-preserving planar upward grid drawings. There is no known lower bound for non-upward order-preserving planar grid drawings other than the trivial $\Omega(n)$ bound. Garg et al. [GGT96] show that any tree with degree d admits a non-order-preserving planar upward polyline grid drawing with height $h = O(n^{1-\alpha})$ and area $O(n + dh \log n)$, where $0 < \alpha < 1$ is any user-specified constant. This result implies that any tree with degree $O(n^{\beta})$, where $0 \le \beta < 1$ is any constant, can be drawn in this fashion in $O(n)$ area with aspect ratio $O(n^{\gamma})$, where γ is any user-defined constant, such that $\max\{0, 2\beta - 1\} < \gamma < 1$. Garg and Rusu [GR03c] show that any tree with degree $O(n^{\delta})$, where $0 \le \delta < 1/2$ is any constant, admits a non-order-preserving planar non-upward straight-line drawing with area $O(n)$, and any user-specified aspect ratio in the range $[1, n^{\alpha}]$, where $0 \le \alpha < 1$ is any constant.

Table 5.3 summarizes these results.

A variety of results are available for other kinds of drawings. Di Battista et al. [DETT99] and Frati [Fra09] have given a survey of these results.

5.8.2   Unordered Trees

In this section, we briefly sketch a bottom-up algorithm developed using the ringed circular layout approach [TM02]. This algorithm (called *Rings*) is space-efficient for high-degree trees, however, the resulting drawing is straight-line but not planar.

The subtrees rooted at the children of the root of the tree are drawn recursively as circles placed in concentric rings around the center of the circle to ensure efficient use of space. The children of the root are divided into multiple categories according to their size. One ring is assigned to each category, so the outer rings consist of the largest trees, while the inner rings consist of the smallest ones (see Figure 5.19). In this way, a tree containing more information is allocated more space, thus showing more distinguishable edges and allowing more structural information to be shown in context.

The relationship below can be established between the number of children circles in the outermost ring and the percentage of area taken up by the ring.

Tree Type	Drawing Type	Area	Aspect Ratio	Ref.
Tree with degree $O(n^\delta)$, for any constant $0 \leq \delta < 1/2$	non-upward straight-line non-order-preserving	$\Theta(n)$	$[1, n^\alpha]$	[GR03c]
Tree with degree $O(n^\beta)$, for any constant $0 \leq \beta < 1$	upward polyline non-order-preserving	$\Theta(n)$	$[1, n^\gamma]$	[GGT96]
General	upward polyline order-preserving	$\Theta(n \log n)$	$n/\log n$	[GGT96]
	upward straight-line order-preserving	$O(n^{1+\epsilon})$	n	[Cha02]
	non-upward straight-line order-preserving	$O(n^{1+\epsilon})$	n	[Cha02]
		$O(n \log n)$	$n/\log n$	[GR03a]

Table 5.3 Bounds on the areas and aspect ratios of various kinds of planar straight-line grid drawings of an n-node tree. Here, α, γ, and ϵ are arbitrary user-defined constants, such that $0 \leq \alpha < 1$, $0 \leq \gamma < 1$, and $0 < \epsilon < 1$.

$$f(n) = \frac{(R_2)^2}{(R_1)^2} = \frac{(1 - \sin(\theta))^2}{(1 + \sin(\theta))^2} = \frac{(1 - \sin(\frac{\pi}{n}))^2}{(1 + \sin(\frac{\pi}{n}))^2} \tag{5.1}$$

here, $f(n)$ is the fraction of the area left after n circles have been placed in the ring. The basic steps of the algorithm are presented below:

Algorithm *Rings*

    Sort the children by their number of children;

    Find the smallest k for which the sum of the number of children of the first k children expressed as a fraction of the total number of grandchildren is greater or equal to $f(k)$;

    Place first k children in the outermost ring;

    Place the rest of the children in the same way in the inner rings;

end Algorithm.

Visual cues like color and transparency are also used to enhance structural information, as well as to highlight specific information (such as information importance or relevance). Adjacent concentric rings are rotated in opposite directions to decrease the occlusion of a particular branch (see Figure 5.20).

A binary tree adaptation of the *Rings* algorithm [RS08] places the children of a node in either the same vertical or horizontal channel, starting with the same horizontal channel at the root (depth 0), and alternates between vertical and horizontal channel placement for every following depth in the tree. In addition, the length of the edge connecting a subtree to its parent is set to $depth(subtree(v)) + 1$, where $depth(subtree(v))$ is the depth of the subtree rooted at node v. This ensures that enough space is made available to draw the rest of the subtree, which is consistent with other rings-based algorithms. A drawing produced by the binary tree adaptation of the *Rings* algorithm is provided in Figure 5.21.

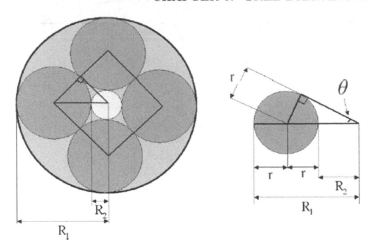

Figure 5.19 Layout of the ringed circular layout algorithm of [TM02]. The four larger rings represent the largest children of the parent node, and the inner ring represents the area left for the rest of the children.

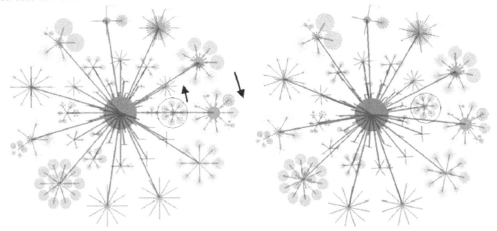

Figure 5.20 Rotation strategy to decrease occlusion. Figure taken from [TM02].

In order to allow for real-time interaction, a top-down variation of the *Rings* algorithm, called *FastRings* [RSJ07], trades space for time. In *FastRings*, all nodes of the tree are considered to be equivalent and assigned same size circles. This allows the algorithm to start drawing the tree much sooner, when only the first level of children is available. The drawing can be refined later by filling up the circles from the first level once new information becomes available. Experiments show that *FastRings* increases the speed of constructing entire drawings by 51%, and is twelve times faster in producing first drawings.

5.8.3   Ordered Trees

In this section, we briefly sketch an algorithm for constructing a (non-upward) order-preserving planar straight-line grid drawing of a general ordered tree with n nodes with $O(n \log n)$ area in $O(n)$ time [GR03a]. This algorithm uses a path-based approach.

Let T be an ordered tree with n nodes. In each recursive step, the algorithm breaks T into several subtrees, draws each subtree recursively, and then combines their drawings to

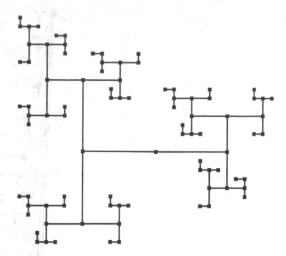

Figure 5.21 Drawing of the Fibonacci tree with 88 nodes, generated by the binary tree adaptation of the *Rings* algorithm.

obtain an α-drawing $D(T)$ of T, where α is a positive integer given as a parameter to the algorithm.

Let $P = v_0 v_1 v_2 \ldots v_m$ be a spine of T (see Section 5.7.4 for the definition of spine). The general structure of T is shown in Figure 5.22(a). Let $s_0, s_1, \ldots, s_i, v_1, s_{i+1}, s_{i+2}, \ldots, s_p$ be the left-to-right order of the children of v_0, where the list $s_0, s_1, \ldots, s_i$ is empty if v_1 is the leftmost child of v_0, and the list $s_{i+1}, s_{i+2}, \ldots, s_p$ is empty if v_1 is the rightmost child of v_0. Let A_k denote the subtree rooted at the node s_k, where $0 \le k \le p$. Let $t_0, t_1, \ldots, t_j, v_2, t_{j+1}, t_{j+2}, \ldots, t_r$ be the left-to-right order of the children of v_1, where the list $t_0, t_1, \ldots, t_j$ is empty if v_2 is the leftmost child of v_1, and the list $t_{j+1}, t_{j+2}, \ldots, t_r$ is empty if v_2 is the rightmost child of v_1. Let B_k denote the subtree rooted at the node t_k, where $0 \le k \le r$. Let C denote the subtree rooted at v_2.

T is drawn as follows (see Figure 5.22(b)):

1. Recursively construct 1-drawings $D(A_0), \ldots, D(A_p)$ of $A_0, \ldots, A_p$, respectively, and $D(B_0), \ldots, D(B_r)$ of $B_0, \ldots, B_r$, respectively.

2. Place v_0 at the origin.

3. Place $D(A_{i+1}), \ldots, D(A_p)$ one above the other at unit vertical separations from each other, such that $D(A_p)$ is at the top, $D(A_{i+1})$ is at the bottom, $s_{i+1}, \ldots, s_p$ are in the same vertical channel, and s_p is α units below, and one unit to the right of v_0.

4. Place $D(B_{j+1}), \ldots, D(B_r)$ one above the other at unit vertical separations from each other, such that $D(B_r)$ is at the top, $D(B_{j+1})$ is at the bottom, $t_{j+1}, \ldots, t_r$ are in the same vertical channel, and t_r is one unit below $D(A_{i+1})$, and one unit to the right of s_{i+1}.

5. Place v_1 in the same horizontal channel as the bottom of $D(B_{j+1})$, and one unit to the right of v_0.

6. Place $D(B_0), \ldots, D(B_j)$ one above the other at unit vertical separations from each other, such that $D(B_j)$ is at the top, $D(B_0)$ is at the bottom, $t_0, \ldots, t_j$ are in the same vertical channel, and t_j is one unit below, and one unit to the right of v_1.

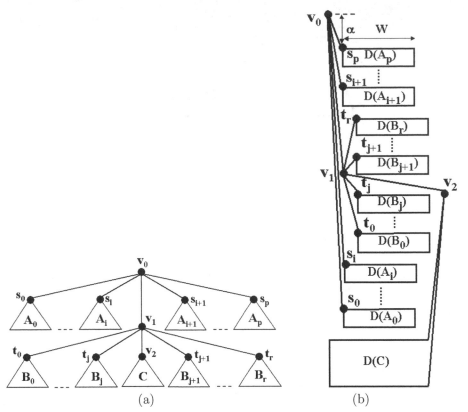

Figure 5.22 (a) The structure of a general tree T. (b) The drawing of T constructed by the algorithm of Section 5.8.3. For simplicity, $D(A_0), \ldots, D(A_p), D(B_0), \ldots, D(B_r)$ are shown as identically sized boxes, but in actuality they may have different sizes.

7. Place $D(A_0), \ldots, D(A_i)$ one above the other at unit vertical separations from each other, such that $D(A_i)$ is at the top, $D(A_0)$ is at the bottom, $s_0, \ldots, s_i$ are in the same vertical channel, and s_i is one unit below $D(B_0)$, and in the same vertical channel as v_1.

8. Let $\beta = h(D(B_0)) + \ldots + h(D(B_j)) + h(D(A_0)) + \ldots + h(D(A_i)) + i + j + 2$, where $h(D(B_0)), \ldots, h(D(B_j)), h(D(A_0)), \ldots, h(D(A_i))$ denote the heights of $D(B_0), \ldots, D(B_j), D(A_0), \ldots, D(A_i)$, respectively. Recursively construct a β-drawing of the mirror image of C, and flip it right-to-left to obtain a drawing $D(C)$ of C. Let W be the width of G, which is the drawing with the maximum width among $D(A_0), \ldots, D(A_p), D(B_0), \ldots, D(B_r)$. Place $D(C)$ such that v_2 is one unit below v_1, and $\max\{W + 3,\ \text{width of } D(C)\}$ units to the right of v_0.

Theorem 5.5   *An ordered tree with n nodes admits a (non-upward) order-preserving planar straight-line grid drawing with $O(n \log n)$ area, $O(\log n)$ width, and height at most n, which can be constructed in $O(n)$ time.*

Proof: Let T be an n-node ordered tree. Using the above algorithm, construct a 1-drawing $D(T)$ of T in $O(n)$ time. As discussed above, $D(T)$ is an order-preserving planar straight-line grid drawing of T with height at most n, width $O(\log n)$, and area $O(n \log n)$.

$\square$

LEMMA 5.2 A left-corner drawing (see Section 5.7.4 for the definition of a left-corner drawing) of an n-node ordered tree with area $O(n \log n)$, height $O(\log n)$, and width at most n, can be constructed in $O(n)$ time.

Proof: First a 1-drawing of the mirror-image of T is constructed using Theorem 5.5, then it is rotated clockwise by 90°, and then it is flipped right-to-left. □

5.9 Other Tree Drawing Methods

Drawing trees is one of the best studied areas in graph drawing, initiated more than forty years ago [Knu68]. Any tree accepts a planar drawing, hence most tree drawing algorithms achieve this aesthetic. Several tree drawing strategies exist that allow one to create drawings with small area, user-controlled aspect ratio, relatively high angular resolution, a small number of bends, and in efficient time.

We conclude the chapter by introducing several algorithms and techniques that do not fit the general approaches described in the previous sections.

Hyperbolic tree [LRP95] (see Figure 5.23) simulates the distortion effect of fisheye lens (enlarge the focus and shrink the rest).

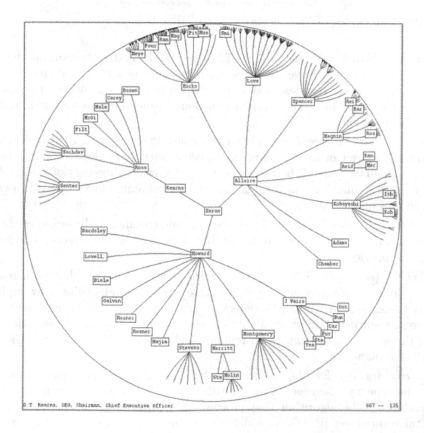

Figure 5.23 Screenshot of Hyperbolic tree, taken from [LRP95].

Pad++ [BHS+97] (See Figure 5.24) displays the nodes as thumbnails of pages of information. It institutes a focus+context style by enlarging the focus node and allowing other nodes to be in view.

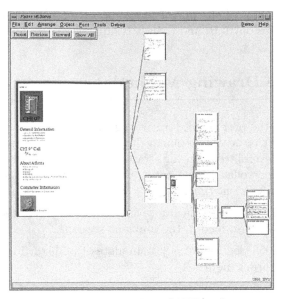

Figure 5.24   Screenshot of Pad++, taken from [BHS+97].

Botanical tree [KvdWW01] (see Figure 5.25) is based on the observation that people can easily see the branches, leaves, and their arrangement in a botanical tree, despite the large number of elements. Non-leaf nodes are mapped to branches and child nodes to sub-branches. Continuing branches are emphasized, long branches are contracted, and sets of leaves are shown as fruit.

A *layered drawing* of a tree T is a planar straight-line drawing of T such that the vertices are drawn on a set of layers. Some applications such as phylogenetic evolutions and programming language parsing benefit from layered upward drawings of trees. Alam et al. [ASRR08] (see Figure 5.26) provide algorithms for minimum-layer upward drawings of both ordered and unordered trees.

Space tree [PGB02] (see Figure 5.27) allows dynamic rescaling of branches of the tree to best fit the available screen space. Branches that do not fit on the screen are summarized by a triangular preview.

Quad [RYC08] (see Figure 5.28) allows the user to specify a preferred angular resolution, and then employs a best-effort delivery to generate a planar straight-line drawing in which all angles between edges are above the specified angular coefficient. When a node has too many children, resulting in an impossibility of achieving angles above the specified angular coefficient, the algorithm distributes all remaining children evenly among the three quads of the Cartesian plane.

Adaptive tree drawing [RCJ06] is a system that first analyzes the input tree to classify it as a specific type and then selects an algorithm to draw it with respect to user-specified quality measures. The algorithm that is selected to draw a given tree is based on an experimental comparison [RJSC06], which orders the performance of the algorithms for each quality measure.

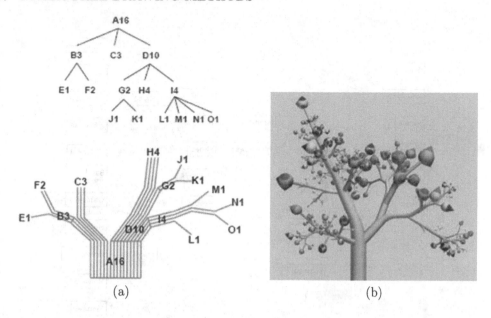

(a) (b)

Figure 5.25 (a) Node and link diagram (top) and corresponding strands model (bottom). (b) Screenshot of Botanical tree, taken from [KvdWW01].

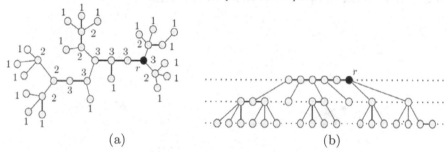

(a) (b)

Figure 5.26 (a) A tree with root r and layer-labelings. (b) A minimum-layer upward drawing of the tree in (a). Figure taken from [ASRR08].

Hexagonal tree drawing [BBB$^+$09] (see Figure 5.29) allows drawings of degree-6 trees on the hexagonal grid, which consists of equilateral triangles.

Most of the tree drawing algorithms draw trees on unbounded planes, and few of them draw trees on regions that are bounded by rectangles. However, certain applications, such as a graphics software by which one would like to draw a tree inside a star-shaped polygon, require trees to be drawn on regions which are bounded by general polygons [BR04] (see Figure 5.30).

A comparative experiment with five tree visualization systems, some which do not draw trees as node-link diagrams, was performed in [Kob04]. Subjects performed tasks relating to the structure of a directory hierarchy and to attributes of files and directories. Task completion times, correctness, and user satisfaction were measured, and video recordings of subjects interaction with the systems were made. The study showed the merits of distinguishing structure and attribute-related tasks, for which some systems behave differently.

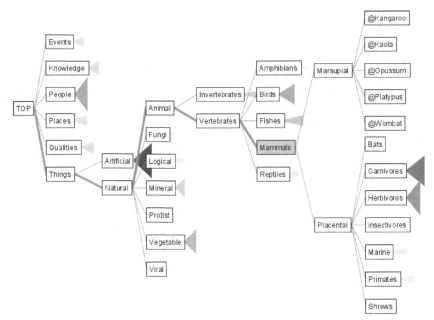

Figure 5.27   Screenshot of Space tree, taken from [PGB02].

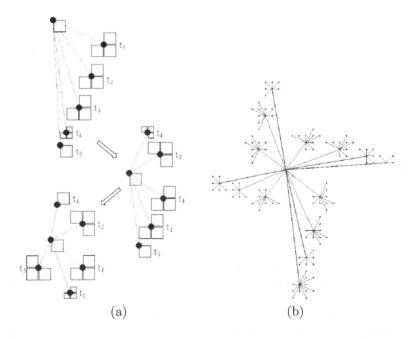

Figure 5.28   (a) Subtrees are distributed into three quads of the Cartesian plane when the angular coefficient cannot be met by using only one or two quads. (b) Screenshot of a drawing generated using Quad algorithm, with user-specified angular resolution of 45°. Figure taken from [RYC08].

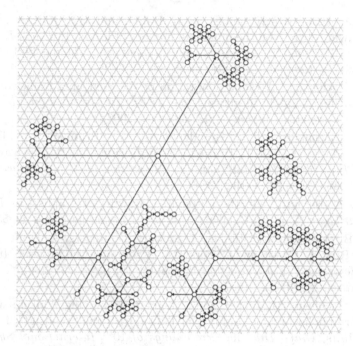

Figure 5.29 A planar straight-line drawing of a tree with outdegree five on the hexagonal grid. Figure taken from [BBB$^+$09].

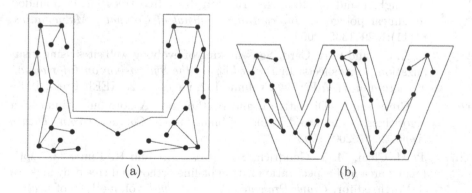

(a)   (b)

Figure 5.30 (a) Drawing of a 31-node complete binary tree inside a U-shaped rectiliniar polygon. (b) Drawing of a 31-node complete binary tree inside a W-shaped polygon. Figure taken from [BR04].

References

[ASRR08]   M.J. Alam, M.A.H. Samee, M.M. Rabbi, and M.S. Rahman. Upward drawing of trees on the minimum number of layers. In *Proceedings of the 2nd Workshop on Algortihms and Computation*, volume 4921 of *Lecture Notes in Computer Science*, pages 88–99, 2008.

[Bac07]   C. Bachmaier. A radial adaptation of the Sugiyama framework for visualizing hierarchical information. *IEEE Transactions on Visualization and Computer Graphics*, 13(3):583–594, 2007.

[BBB+09]   C. Bachmaier, F.J. Brandenburg, W. Brunner, A. Hofmeier, M. Matzeder, and T. Unfried. Tree drawings on the hexagonal grid. In *Proceedings 16th International Symposium on Graph Drawing*, pages 372–383. Springer-Verlag, 2009.

[Ber81]   M. A. Bernard. On the automated drawing of graphs. In *Proc. 3rd Caribbean Conf. on Combinatorics and Computing*, pages 43–55, 1981.

[BHS+97]   B.B. Bederson, J.D. Hollan, J. Stewart, D. Rogers, A. Druin, D. Vick, L. Ring, E. Grose, and C. Forsythe. A zooming web browser. In *Human Factors and Web Development*, chapter 19, pages 255–266. New Jersey: Lawrence Erlbaum, 1997.

[BJL02]   C. Buchheim, M. Jünger, and S. Leipert. Improving Walker's algorithm to run in linear time. In Michael T. Goodrich and Stephen G. Kobourov, editors, *Graph Drawing (Proceedings of 10th International Symposium on Graph Drawing, 2002)*, volume 2528 of *Lecture Notes in Computer Science*, pages 344–353. Springer, 2002.

[Blo93]   A. Bloesch. Aesthetic layout of generalized trees. *Software Practice and Experience*, 23(8):817–827, 1993.

[BM03]   M. Bernard and S. Mohammed. Labeled radial drawing of data structures. In *Proceedings 7th International Conference on Information Visualisation*, pages 479–555. IEEE Computers Society, 2003.

[BR04]   A. Bagheri and M. Razzazi. How to draw free trees inside bounded rectilinear polygons. *International Journal of Computer Mathematics*, 81(11):1329–1339, 2004.

[CC99]   E. A. Chi and S. K. Card. Sensemaking of evolving web sites using visualization spreadsheets. In *Proceedings of the Symposium on Information Visualization (InfoViz '99)*, volume 142, pages 18–25. IEEE Press, 1999.

[CDP92]   P. Crescenzi, G. Di Battista, and A. Piperno. A note on optimal area algorithms for upward drawings of binary trees. *Comput. Geom. Theory Appl.*, 2:187–200, 1992.

[CGKT97]   T. M. Chan, M. T. Goodrich, S. R. Kosaraju, and R. Tamassia. Optimizing area and aspect ratio in straight-line orthogonal tree drawings. In S. North, editor, *Graph Drawing (Proc. GD '96)*, volume 1190 of *Lecture Notes Comput. Sci.*, pages 63–75. Springer-Verlag, 1997.

[CGKT02]   T. Chan, M. Goodrich, S. Rao Kosaraju, and R. Tamassia. Optimizing area and aspect ratio in straight-line orthogonal tree drawings. *Computational Geometry: Theory and Applications*, 23:153–162, 2002.

[Cha02]   T. M. Chan. A near-linear area bound for drawing binary trees. *Algorithmica*, 34(1):1–13, 2002.

[CP98]   P. Crescenzi and P. Penna. Strictly-upward drawings of ordered search trees. *Theoretical Computer Science*, 203(1):51–67, 1998.

[CPM+98]   E. H. Chi, J. Pitkow, J. Mackinlay, P. Pirolli, and R. Gossweiler. Visualizing the evolution of Web ecologies. In *Proceedings of the Human Factors in Computing Systems*, pages 400–407, 1998.

[CPP98]   P. Crescenzi, P. Penna, and A. Piperno. Linear-area upward drawings of AVL trees. *Comput. Geom. Theory Appl.*, 9:25–42, 1998. (special issue on Graph Drawing, edited by G. Di Battista and R. Tamassia).

[CPP00]   E.H. Chi, P. Pirolli, and J. Pitkow. The Scent of a Site: A system for analyzing and predicting information scent, usage, and usability of a Web site. In *Proceedings of the Human Factors in Computing Systems*, pages 161–168, 2000.

[CT]   Isabel F. Cruz and Roberto Tamassia. Graph drawing tutorial.

[DETT94]   G. Di Battista, P. Eades, R. Tamassia, and I. G. Tollis. Algorithms for drawing graphs: an annotated bibliography. *Comput. Geom. Theory Appl.*, 4:235–282, 1994.

[DETT99]   G. Di Battista, P. Eades, R. Tamassia, and I. G. Tollis. *Graph Drawing*. Prentice Hall, Upper Saddle River, NJ, 1999.

[DT81]   D. Dolev and H.W. Trickey. On linear area embedding of planar graphs. Technical report, Stanford University, Stanford, USA, 1981.

[Ead92]   P. D. Eades. Drawing free trees. *Bulletin of the Institute for Combinatorics and its Applications*, 5:10–36, 1992.

[Fra07]   F. Frati. Straight-line orthogonal drawings of binary and ternary trees. In Seok-Hee Hong and Takao Nishizeki, editors, *15th International Symposium on Graph Drawing*, volume 4875 of *Lecture Notes in Computer Science*, pages 76–87, 2007.

[Fra09]   F. Frati. *Small Screens and Large Graphs: Area-Efficient Drawings of Planar Combinatorial Structures*. PhD thesis, Computer Science and Engineering, Roma Tre University, 2009.

[GADM04]   S. Grivet, D. Auber, J.-P. Domenger, and G. Melancon. Bubble tree drawing algorithm. In *International Conference on Computer Vision and Graphics*, pages 633–641. Springer Verlag, 2004.

[GGT96]   A. Garg, M. T. Goodrich, and R. Tamassia. Planar upward tree drawings with optimal area. *Internat. J. Comput. Geom. Appl.*, 6:333–356, 1996.

[GR02]   A. Garg and A. Rusu. Straight-line drawings of binary trees with linear area and arbitrary aspect ratio. In Michael T. Goodrich and Stephen G. Kobourov, editors, *Graph Drawing (Proceedings of 10th International Symposium on Graph Drawing, 2002)*, volume 2528 of *Lecture Notes in Computer Science*, pages 320–331. Springer, 2002.

[GR03a]   A. Garg and A. Rusu. Area-efficient order-preserving planar straight-line drawings of ordered trees. *International Journal of Computational Geometry and Applications*, 13(6):487–505, 2003.

[GR03b]   A. Garg and A. Rusu. A more practical algorithm for drawing binary trees in linear area with arbitrary aspect ratio. In Giuseppe Liotta, editor, *Graph Drawing (Proceedings of 11th International Symposium on Graph Drawing, 2003)*, volume 2912 of *Lecture Notes in Computer Science*, pages 159–165. Springer, 2003.

[GR03c]   A. Garg and A. Rusu. Straight-line drawings of general trees with linear
           area and arbitrary aspect ratio. In *Proceedings 2003 International Con-
           ference on Computational Science and Its Applications*, volume 2669 of
           Lecture Notes in Computer Science, pages 876–885. Springer, 2003.

[GR04]   A. Garg and A. Rusu. Straight-line drawings of binary trees with lin-
           ear area and arbitrary aspect ratio. *Journal of Graph Algorithms and
           Applications*, 8(2):135–160, 2004.

[Kim95]   Sung Kwon Kim. Simple algorithms for orthogonal upward drawings of
           binary and ternary trees. In *Proc. 7th Canad. Conf. Comput. Geom.*,
           pages 115–120, 1995.

[Kim04]   S.K. Kim. Order-preserving, upward drawing of binary trees using fewer
           bends. *Discrete Applied Mathematics Journal*, 143(1–3):318–323, 2004.

[Knu68]   D. E. Knuth. *Fundamental Algorithms*, volume 1 of *The Art of Computer
           Programming*. Addison-Wesley, Reading, MA, 1st edition, 1968.

[Kob04]   Alfred Kobsa. User experiments with tree visualization systems. In *Pro-
           ceedings of the IEEE Symposium on Information Visualization*, INFOVIS
           '04, pages 9–16. IEEE Computer Society, 2004.

[KvdWW01] E. Kleiberg, H. van de Wetering, and J.J. Van Wijk. Botanical visual-
           ization of huge hierarchies. In *Proceedings of the IEEE Symposium on
           Information Visualization*, 2001.

[Lei80]   C. E. Leiserson. Area-efficient graph layouts (for VLSI). In *Proc. 21st
           Annu. IEEE Sympos. Found. Comput. Sci.*, pages 270–281, 1980.

[LRP95]   J. Lamping, R. Rao, and P. Pirolli. A focus+context technique based
           on hyperbolic geometry for visualizing large hierarchies. In *Proc. ACM
           Conf. on Human Factors in Computing Systems (CHI)*, 1995.

[Mac03]   B. MacLennan. Molecular combinatory computing for nanostructure syn-
           thesis and control. In *Proceedings 3rd IEEE Conference on Nanotechnol-
           ogy*, volume 2 of *IEEE Press*, pages 179–182, 2003.

[MH98]   G. Melacon and I. Herman. Circular drawing of rooted trees. Technical
           Report INS-9817, Netherland National Research Institute for Mathemat-
           ics and Computer Sciences, 1998.

[MMC99]   G. Melacon, J.D. Mackinlay, and S. K. Card. Cone trees: animated 3D
           visualization of hierarchical information. In *Human Factors in Comput-
           ing Systems, CHI'99 Conference Proceedings*, pages 189–194. ACM Press,
           1999.

[PCJ97]   H. C. Purchase, R. F. Cohen, and M. I. James. An experimental study of
           the basis for graph drawing algorithms. *ACM J. Experim. Algorithmics*,
           2(4), 1997.

[PGB02]   C. Plaisant, J. Grosjean, and B.B. Bederson. Spacetree: supporting
           exploration in large node link tree, design evolution and empirical eval-
           uation. In *Proceedings of the IEEE Symposium on Information Visual-
           ization*, pages 57–64, 2002.

[Pur97]   Helen Purchase. Which aesthetic has the greatest effect on human un-
           derstanding? In G. Di Battista, editor, *Graph Drawing (Proc. GD '97)*,
           volume 1353 of *Lecture Notes Comput. Sci.*, pages 248–261. Springer-
           Verlag, 1997.

[Pur00]   Helen C. Purchase. Effective information visualisation: A study of graph drawing aesthetics and algorithms. *Interact. Comput.*, 13(2):147–162, 2000.

[RCJ06]   A. Rusu, C. Clement, and R. Jianu. Adaptive binary trees visualization with respect to user-specified quality measures. In *Proceedings 10th International Conference on Information Visualisation*, pages 469–474. IEEE Computers Society, 2006.

[RJSC06]   A. Rusu, R. Jianu, C. Santiago, and C. Clement. An experimental study on algorithms for drawing binary trees. In *Proceedings 5th Asia Pacific Symposium on Information Visualization*, volume 60 of *Conference in Research and Practice in Information Technology*, pages 85–88. Australian Computer Society Inc., 2006.

[RMC91]   G. G. Robertson, J. D. Mackinlay, and S. K. Card. Cone trees: Animated 3D visualizations of hierarchical information. In *Proc. ACM Conf. on Human Factors in Computing Systems*, pages 189–193, 1991.

[RS07]   A. Rusu and C. Santiago. A practical algorithm for planar straight-line grid drawings of general trees with linear area and arbitrary aspect ratio. In *Proceedings 11th International Conference on Information Visualisation*, pages 743–750. IEEE Computers Society, 2007.

[RS08]   A. Rusu and C. Santiago. Grid drawings of binary trees: An experimental study. *Journal of Graph Algorithms and Applications*, 12(2):131–195, 2008.

[RSJ07]   A. Rusu, C. Santiago, and R. Jianu. Real-time interactive visualization of information hierarchies. In *Proceedings 11th International Conference on Information Visualisation*, pages 117–123. IEEE Computers Society, 2007.

[RT81]   E. Reingold and J. Tilford. Tidier drawing of trees. *IEEE Trans. Softw. Eng.*, SE-7(2):223–228, 1981.

[RYC08]   A. Rusu, C. Yao, and A. Crowell. A planar straight-line grid drawing algorithm for high degree general trees with user-specified angular coefficient. In *Proceedings 12th International Conference on Information Visualisation*, pages 600–609. IEEE Computers Society, 2008.

[SB94]   M. Sarkar and M. H. Brown. Graphical fisheye views. *Commun. ACM*, 37(12):73–84, 1994.

[SKC00]   C.-S. Shin, S. K. Kim, and K.-Y. Chwa. Area-efficient algorithms for straight-line tree drawings. *Comput. Geom. Theory Appl.*, 15:175–202, 2000.

[TDB88]   R. Tamassia, G. Di Battista, and C. Batini. Automatic graph drawing and readability of diagrams. *IEEE Trans. Syst. Man Cybern.*, SMC-18(1):61–79, 1988.

[TM02]   S. T. Teoh and K. L. Ma. Rings: A technique for visualizing large hierarchies. In Michael T. Goodrich and Stephen G. Kobourov, editors, *Graph Drawing (Proceedings of 10th International Symposium on Graph Drawing, 2002)*, volume 2528 of *Lecture Notes in Computer Science*, pages 268–275. Springer, 2002.

[Tre96]   L. Trevisan. A note on minimum-area upward drawing of complete and Fibonacci trees. *Inform. Process. Lett.*, 57(5):231–236, 1996.

[Val81]   L. Valiant. Universality considerations in VLSI circuits. *IEEE Trans. Comput.*, C-30(2):135–140, 1981.

[Wal90]   J. Q. Walker II. A node-positioning algorithm for general trees. *Softw. – Pract. Exp.*, 20(7):685–705, 1990.

6

Planar Straight-Line Drawing Algorithms

Luca Vismara

6.1   Introduction

Planar straight-line drawings have been an early subject of investigation in combinatorial mathematics. A classic result states that every planar graph admits a planar straight-line drawing. Namely, if a graph can be drawn with no crossings using edges of arbitrary shape (e.g., polygonal lines or curves), then it can be drawn with no crossings using only straight-line edges (see Figure 6.1). The proof of this result was independently discovered by Steinitz and Rademacher [SR34], Wagner [Wag36], Fary [Fár48], and Stein [Ste51].

All the above classic constructions focus on establishing the existence of planar straight-line drawings but do not address the area of the drawing or the arithmetic precision required for representing the coordinates of the vertices. Indeed, following the constructions in these papers one obtains drawings of area exponential in the length of the shortest edge, which are unsuitable in practice.

Algorithms for constructing planar straight-line grid drawings, where the edges have integer coordinates, were developed by de Fraysseix, Pach, and Pollack [dFPP90] (shift method) and by Schnyder [Sch90] (realizer method). They independently showed that every n-vertex planar graph has a planar straight-line grid drawing with $O(n)$ height and $O(n)$ width, resulting in $O(n^2)$ area. These bounds are asymptotically tight in the worst case as can be shown with the example of Figure 6.2.

Convex drawings are planar straight-line drawings where all the faces are drawn as convex polygons (see Figure 6.1(c)). We say that a planar graph is convex planar if it admits a convex drawing. In another classic work, Tutte [Tut60, Tut63] showed how to construct a

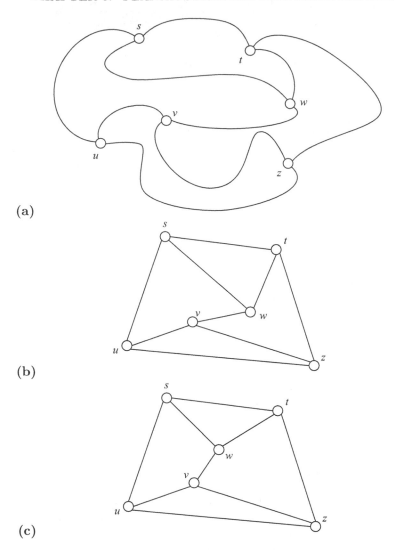

Figure 6.1 Examples of planar drawings of the same graph: (a) planar drawing with curved edges; (b) planar straight-line drawing; (c) planar convex drawing.

convex planar drawing of every triconnected planar graph. His method places the vertices of the external face on an arbitrary convex polygon and computes the coordinates of the remaining vertices by solving a system of linear equations.

The rest of this chapter is organized as follows. Basic definitions are introduced in Section 6.2. Tutte's classic algebraic method for convex drawings is presented in Section 6.3. Area bounds for planar straight-line grid drawings computed by the shift method and by the realizer method are summarized in Section 6.4. Canonical orderings of planar graphs are discussed in Section 6.5. Section 6.6 describes the shift method and Section 6.7 describes the realizer method.

For further details on the subject of planar drawings of graphs, we refer the reader to the book by Nishizeki and Rahman [NR04] and the survey by Di Battista and Frati [DF13]. See also the work by Cruz and Garg [CG95] for a declarative approach to the construction of planar drawings.

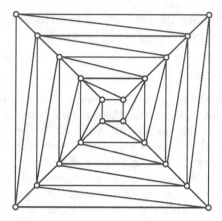

Figure 6.2 Planar straight-line grid drawing of graph S_5 consisting of five nested cycles of four vertices. This drawing has height 9 and width 9. In general, graph S_k has $4k$ vertices and requires height and width proportional to k in any planar-straight-line grid drawing.

6.2 Preliminaries

6.2.1 Planar Drawings

In the context of this chapter, a *drawing* of a graph G is a mapping of each vertex v of G to a distinct point $P(v) = (v_x, v_y)$ of the plane[1] and of each edge (u, v) of G to a simple Jordan curve with endpoints $P(u)$ and $P(v)$. A *straight-line drawing* is a drawing in which every edge is mapped to a straight-line segment; more formally, a straight-line drawing is an injective function $f : v \in V \to (v_x, v_y) \in \mathbb{R}^2$.

A drawing is *planar* if no two edges intersect, except, possibly, at common endpoints. A graph is planar if it has a planar drawing. Two planar drawings of a planar graph G are *equivalent* if, for each vertex v, they have the same circular clockwise sequence of edges incident with v. Hence, the planar drawings of G are partitioned into equivalence classes. Each of those classes is called an *embedding* of G. An *embedded* planar graph (also *plane* graph) is a planar graph with a prescribed embedding. A triconnected planar graph has a unique embedding, up to a reflection. A planar drawing divides the plane into topologically connected regions delimited by cycles; these cycles are called *faces*. The *external* face is the cycle delimiting the unbounded region; all the other faces are *internal*. Two equivalent planar drawings have the same faces. Hence, one can refer to the faces of an embedding. A vertex or edge of a plane graph is said to be *external* if it belongs to the external face, and *internal* otherwise.

A *maximal planar graph* is a planar graph with the maximal number of edges, i.e., adding an edge between any two vertices destroys its planarity. Note that in a maximal planar graph all faces consist of three edges. An *outerplanar graph* is a planar graph that admits a planar drawing with all its vertices on the same (say, the external) face; such a drawing is called an outerplanar drawing.

Let G be a plane graph; the *dual graph* G^* of G is defined as follows: (i) each face f of G has a *dual vertex* f^* in G^*; (ii) each vertex v of G has a *dual face* v^* in G^*; (iii) let e be

---

[1]We will use interchangeably (v_x, v_y) and $(x(v), y(v))$ to denote the coordinates of $P(v)$.

an edge of G and let f_1 and f_2 be the two faces of G incident with e (note that f_1 and f_2 may not be distinct); e has a *dual edge* $e^* = (f_1^*, f_2^*)$ in G^*.

6.2.2   Convex Drawings

A *polygon* is a finite set of segments such that every segment endpoint is shared by exactly two segments and no subset of segments has the same property. A polygon is *simple* if there is no pair of nonconsecutive segments sharing a point. A simple polygon is *convex* if its interior is a convex set. A simple polygon is *strictly convex* if its interior is a strictly convex set, i.e., no 180° angle is allowed. A *convex drawing* of a planar graph G is a planar straight-line drawing of G in which all faces are drawn as convex polygons (see Figure 6.3(a)). A *strictly convex drawing* of a planar graph G is a planar straight-line drawing of G in which all faces are drawn as strictly convex polygons (see Figure 6.3(b)). A planar graph is said to be *(strictly) convex planar* if it admits a (strictly) convex drawing.

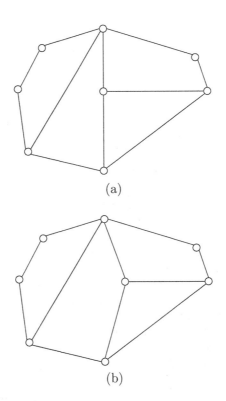

(a)

(b)

Figure 6.3   (a) A convex drawing of a biconnected planar graph G. (b) A strictly convex drawing of a biconnected planar graph G.

6.2.3   Connectivity

We recall some basic definitions on connectivity. A *separating k-set* of a graph is a set of k vertices whose removal disconnects the graph; separating 1-sets and 2-sets are called *cutvertices* and *separation pairs*, respectively. A graph is *k-connected* if it contains more than k vertices and no separating $(k-1)$-set; 1-connected, 2-connected, and 3-connected graphs are called *connected, biconnected,* and *triconnected,* respectively. A *separating edge* of a graph is an edge whose removal disconnects the graph.

The *biconnected components* of a connected graph (also called *blocks*) are its maximal biconnected subgraphs and its separating edges. The *triconnected components* of a biconnected graph G are defined as follows [HT73].

If G is triconnected, then G itself is the unique triconnected component of G. Otherwise, let $\{u, v\}$ be a separation pair of G. We partition the edges of G into two disjoint subsets E_1 and $E_2, |E_1| \geq 2, |E_2| \geq 2$, such that the subgraphs G_1 and G_2 induced by them have only vertices u and v in common. Graphs $G'_1 = G_1 + (u, v)$ and $G'_2 = G_2 + (u, v)$ are called the *split graphs* of G with respect to $\{u, v\}$ (multiple edges are allowed); edge (u, v) in G'_1 and G'_2 is called a *virtual edge*. Dividing G into split graphs G'_1 and G'_2 is called *splitting*. Reassembling split graphs G'_1 and G'_2 into G, is called *merging*. Note that only split graphs that resulted from the same splitting operation can be merged together. We continue the splitting process recursively on G'_1 and G'_2 until no further splitting is possible. Each resulting graph is either a triconnected simple graph, or a set of three multiple edges (called *"triple bond"* in [HT73]), or a cycle of length three (called *"triangle"* in [HT73]). The *triconnected components* of G are obtained from these graphs by merging the "triple bonds" into maximal sets of multiple edges (called *"bonds"* in [HT73]), and the "triangles" into maximal simple cycles (called *"polygons"* in [HT73]). When merging "triple bonds" into "bonds" and "triangles" into "polygons," virtual edges with both endvertices in common are removed; we refer to the remaining virtual edges at the end of the merging process as the *virtual edges of the triconnected components*. Note that, although the graphs obtained at the end of the splitting process depend on the order of the splittings, the triconnected components of G are unique. See [HT73] for further details.

In the rest of the chapter, we denote by n, m, and l the number of vertices, edges, and faces of a plane graph, respectively; we always assume $n \geq 3$. Unless otherwise specified, graphs are assumed to be simple, i.e., without self-loops and multiple edges. Often, we do not distinguish between a vertex (edge) of G and the point (segment) representing it.

We recall *Euler's formula*, which holds for every plane graph, and two bounds for the number of edges and faces of a plane graph (the equalities hold for maximal planar graphs), which easily follow from it:

$$n + l = m + 2 \tag{6.1}$$

$$m \leq 3n - 6 \tag{6.2}$$

$$l \leq 2n - 4 \tag{6.3}$$

Let $P_1 = (x_1, y_1)$ and $P_2 = (x_2, y_2)$ be two points on the plane; the *Manhattan distance* between P_1 and P_2 is defined as $|x_1 - x_2| + |y_1 - y_2|$.

A $w \times h$ *integer grid* is a grid of integer points of width w and height h; note that a $w \times h$ integer grid contains $(w + 1) \times (h + 1)$ integer points. A *grid drawing* is an injective function $f : v \in V \to (v_x, v_y) \in \mathbb{Z}^2$. The *area* of a grid drawing is the number of integer points contained in the smallest integer grid containing the drawing. In the rest of the chapter, we will often omit "integer" before "grid" for brevity.

6.3   Real-Coordinate Drawings

In a classic paper, Tutte [Tut60, Tut63] presented a method for constructing strictly convex drawings of triconnected plane graphs by solving a system of linear equations that place each internal vertex at the barycenter of its neighbors. Hence, this method is referred to as the *barycenter method*.

Initially, the vertices of the external face are placed at the vertices of a strictly convex polygon, P. We refer to the vertices not on the external face as internal vertices.

For a vertex v, let $N(v)$ be the set of neighbors of v and $d(v)$ the degree of v, i.e., $d(v) = |N(v)|$. The position of an internal vertex v is determined by the following linear equations:

$$x(v) = \frac{1}{d(v)} \sum_{w \in N(v)} x(w) \tag{6.4}$$

$$y(v) = \frac{1}{d(v)} \sum_{w \in N(v)} y(w) \tag{6.5}$$

Tutte showed that the above system of linear equations admits a unique solution that corresponds to a strict convex drawing of the graph. An example of a drawing constructed with the barycenter method is shown in Figure 6.4.

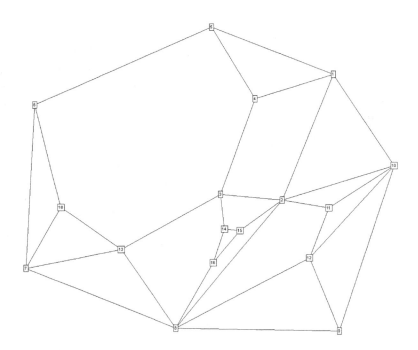

Figure 6.4   Planar convex drawing obtained with Tutte's barycenter method. Drawing created by the PIGALE tool (see Chapter 18).

Combinatorial characterizations of convex and strictly convex planar graphs and methods for constructing convex and strictly convex drawings appear in papers by Tutte [Tut60, Tut63], Thomassen [Tho80, Tho84], Chiba, Yamanouchi, and Nishizeki [CYN84], Chiba, Onoguchi, and Nishizeki [CON85], and Djidjev [Dji95]. Note that the above methods compute drawings with real coordinates for the vertices.

6.4   Grid Drawings

The drawings generated by Tutte's algorithm presented in Section 6.3 exhibit some drawbacks:

- they require high-precision real arithmetic relative to the size of the input graph, and therefore cannot be used even for graphs of moderate size; and
- in the produced drawings, the ratio of the largest distance to the smallest distance between vertices is very large (exponential in the size of the graph), i.e., vertices are represented by arbitrarily close points, or, equivalently, if the graph is drawn on an integer grid, then the grid has exponential size.

Motivated by these drawbacks, Rosenstiehl and Tarjan [RT86] posed the question whether every planar graph has a planar straight-line drawing on an $O(n^k) \times O(n^k)$ integer grid for some fixed constant k, where n is the number of vertices of the graph. As we will see, the question was answered in the positive and various algorithms were presented over the years. Selected algorithms are summarized in Table 6.1.

[CP95, dFPP90]	$(2n - 4) \times (n - 2)$	shift
[CN98]	$\lfloor \frac{2}{3}(n-1) \rfloor \times 4 \lfloor \frac{2}{3}(n-1) \rfloor - 1$	shift
[Bra08]	$\lceil \frac{4}{3}n \rceil \times \lceil \frac{2}{3}n \rceil$	shift
[Sch90]	$(2n - 5) \times (2n - 5)$	realizer
	$(n - 2) \times (n - 2)$	

Table 6.1 Width and height of the drawing achieved by selected planar straight-line grid drawing algorithms that use the shift method or the realizer method. We denote with n the number of vertices of the graph.

The algorithms listed in Table 6.1 are designed for drawing maximal plane graphs but can actually be used to draw general plane graphs: it is sufficient to transform the input plane graph into a maximal plane graph by adding a linear number of extra edges, draw the resulting graph, and then remove the segments corresponding to the extra edges from the obtained drawing. These algorithms are based on two different methods, called the *shift method* and the *realizer method*, and are described in Sections 6.6 and 6.7, respectively.

6.5 Canonical Orderings

In this section, we recall the definitions of canonical ordering of maximal plane graphs, as given by de Fraysseix, Pach, and Pollack [dFPP90], and of triconnected plane graphs, as given by Kant [Kan96].

DEFINITION 6.1 Let G be a maximal plane graph with n vertices, and let u_0, u_1, u_2 be the external vertices of G in counterclockwise order. A *canonical ordering* of G (see Figure 6.5) is an ordering $v_1, \ldots, v_n$ of the vertices of G such that the following conditions are verified:

1. $v_1 = u_1$, $v_2 = u_2$.
2. For $3 \leq k \leq n$, let G_k be the plane subgraph of G induced by vertices $v_1, \ldots, v_k$ and let C_k be the external face of G_k. Vertex v_k is on face C_k. Also, if $k < n$, vertex v_k has at least one neighbor in $G - G_k$.

3. For each $3 \leq k \leq n-1$, subgraph G_k is biconnected and internally maximal (i.e., all internal faces of G_k are triangles).

4. $v_n = u_0$.

LEMMA 6.1   **[dFPP90]** Each maximal plane graph has a canonical ordering, which can be computed in linear time and space.

A canonical ordering of G yields an incremental construction of graph G starting from edge (v_1, v_2). In step k $(3 \leq k \leq n)$, vertex v_k and the edges between v_k and its neighbors in C_{k-1} are added to the current graph G_{k-1}. For each $3 \leq k \leq n$, we denote by $v_1 = w_1, w_2, \ldots, w_t = v_2$ the sequence of vertices of C_{k-1}, when traversed in clockwise order. For the sake of enhancing intuition, we visualize $w_2, \ldots, w_{t-1}$ as arranged from left to right above (v_1, v_2) in the plane. For each $3 \leq k \leq n$, let $w_p, \ldots, w_q$ be the subsequence of vertices of C_{k-1} that are adjacent to v_k (note that $p+1$ may be equal to q). After v_k has been added to G_{k-1}, vertices $w_{p+1}, \ldots, w_{q-1}$ (if any) are no longer external; we say that vertex v_k *covers* these vertices.

A canonical ordering $v_1, \ldots, v_n$ of graph G defines a spanning tree of graph $G - \{v_1, v_2\}$, called *cover tree*, which consists of all edges (u, v) such that u covers v. We set v_n as the root of the cover tree. Thus, the children of a vertex u in the cover tree are the vertices covered by u. (See Figure 6.6.) We define the *cover forest* associate with a canonical ordering as its cover tree together with the single-vertex trees v_1 and v_2.

The definition of canonical ordering can be generalized to triconnected plane graphs as follows. A biconnected plane graph G is said to be *internally triconnected* if for any separation pair $\{u, v\}$ of G, u and v are external vertices and each connected component of $G \setminus \{u, v\}$ contains an external vertex; in other words, G is internally triconnected if and only if the graph obtained from G by adding a new vertex and connecting it to all the external vertices of G is triconnected.

DEFINITION 6.2   Let G be a triconnected plane graph with n vertices, (u_1, u_2) be an external edge of G, and $u_0 \neq u_1, u_2$ be an external vertex of G. A *canonical ordering* of G is an ordering $v_1, \ldots, v_n$ of the vertices of G that can be partitioned into subsequences $V_1, \ldots, V_h$, where $V_k = \{v_{s_k}, \ldots, v_{s_k + d_k}\}$, $1 \leq k \leq h$, $1 = s_1 < s_2 < \cdots < s_h < s_{h+1} = n+1$, $d_k = s_{k+1} - s_k - 1$, such that the following conditions are verified:

1. $v_1 = u_1$, $v_2 = u_2$, and $V_1 = \{v_1, v_2\}$.

2. Let G_k be the plane subgraph of G induced by $V_1 \cup \cdots \cup V_k$, $1 \leq k \leq h$, and C_k be the external face of G_k. For each $2 \leq k \leq h-1$, one of the following cases occurs:

   (a) $V_k = \{v_{s_k}\}$ is a vertex of C_k (and has at least one neighbor in $G - G_k$);

   (b) $V_k = \{v_{s_k}, \ldots, v_{s_k + d_k}\}$ is a subpath of C_k, and each vertex v_i, $s_k \leq i \leq s_k + d_k$, has degree two in G_k (and has at least one neighbor in $G - G_k$).

3. Each subgraph G_k, $2 \leq k \leq h-1$, is biconnected and internally triconnected.

4. $v_n = u_0$ and $V_h = \{v_n\}$.

LEMMA 6.2   **[Kan96]** Each triconnected plane graph has a canonical ordering, which can be computed in linear time and space.

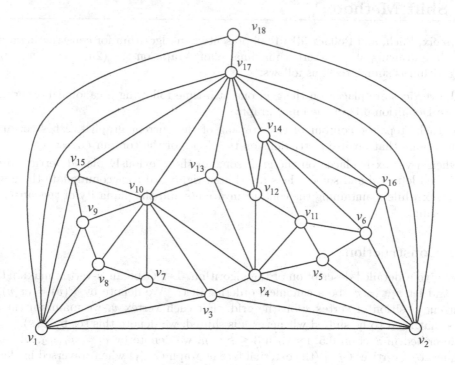

Figure 6.5   A canonical ordering of a maximal plane graph.

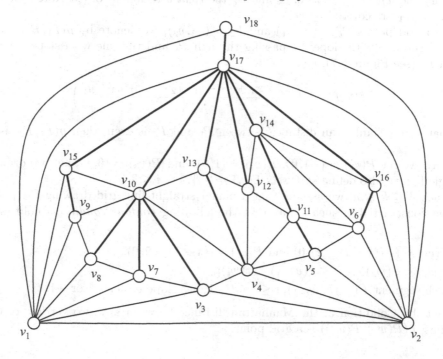

Figure 6.6   Cover tree induced by a canonical ordering of a maximal plane graph. The edges of the tree are drawn with thick lines.

6.6   Shift Method

de Fraysseix, Pach, and Pollack [dFPP90] presented an algorithm for constructing a planar straight-line drawing of an n-vertex maximal plane graph on the $(2n - 4) \times (n - 2)$ grid. The algorithm is summarized as follows:[2]

- the vertices are placed on the grid one at a time following a canonical ordering (see Definition 6.1) of the input graph;

- at each step, the contour of the drawing of the current graph satisfies certain invariants that involve restrictions on the slopes of the contour edges;

- when a vertex is placed on the grid, some of the previously placed vertices are shifted leftward and some others are shifted rightward to accommodate the new vertex while maintaining the contour invariants and the planarity of the current drawing.

6.6.1   Construction

We now give a detailed description of the algorithm. Let G be an n-vertex maximal plane graph, and let $v_1, \ldots, v_n$ be a canonical ordering of G. We denote by $P(v) = (x(v), y(v))$ the current position of vertex v on the grid. For each vertex v, we maintain the set of vertices that need to be shifted whenever v is shifted; we denote this set by $L(v)$.

As described in Section 6.5, for each $3 \le k \le n$, we denote by $v_1 = w_1, w_2, \ldots, w_t = v_2$ the sequence of vertices C_{k-1} (the external face of graph G_{k-1}) when traversed in clockwise order, and by $w_p, \ldots, w_q$ the subsequence of vertices of C_{k-1} that are adjacent to vertex v_k. We call w_p the *left attachment* of v_k and w_q the *right attachment* of v_k. Note that vertices $w_{p+1}, \ldots, w_{q-1}$ are covered by v_k.

For two grid points $P_1 = (x_1, y_1)$ and $P_2 = (x_2, y_2)$, we denote by $\mu(P_1, P_2)$ the intersection of the line with slope $+1$ passing through P_1 and the line with slope -1 passing through P_2 (see Figure 6.7), i.e.,

$$\mu(P_1, P_2) = \left(\frac{x_2 + x_1 + y_2 - y_1}{2}, \frac{x_2 - x_1 + y_2 + y_1}{2} \right) \tag{6.6}$$

Note that if the Manhattan distance between P_1 and P_2 is even, then $\mu(P_1, P_2)$ is a grid point.

Initially, we set $P(v_1) = (-1, 0)$, $P(v_2) = (1, 0)$, and $P(v_3) = (0, 1)$, i.e., we draw G_3 as a triangle Γ_3; we also define *shift sets* $L(v_i) = \{v_i\}, 1 \le i \le 3$.

For each $4 \le k \le n$, we assume that a planar straight-line grid drawing Γ_{k-1} of G_{k-1} has been constructed in such a way that the following *contour conditions* hold (see Figure 6.8):

1. $P(v_1) = (-((k-1) - 2), 0)$ and $P(v_2) = ((k-1) - 2, 0)$;
2. $x(w_1) < x(w_2) < \cdots < x(w_{t-1}) < x(w_t)$;
3. each segment $P(w_i)P(w_{i+1}), 1 \le i \le t - 1$, has slope either $+1$ or -1.

Note that, by Condition 3, the Manhattan distance between any two vertices of C_{k-1} is even; thus, $\mu(P(w_p), P(w_q))$ is a grid point.

---

[2]Our description of the algorithm, which uses left shifts and right shifts, is slightly different from the one given in [dFPP90], which uses only right shifts, but is conceptually equivalent.

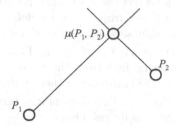

Figure 6.7 Definition of point $\mu(P_1, P_2)$ as the intersection of the line with slope $+1$ passing through P_1 and the line with slope -1 passing through P_2.

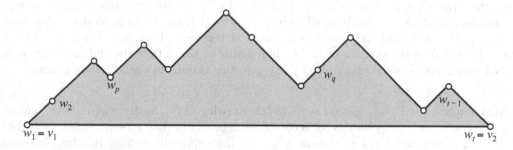

Figure 6.8 Schematic illustration of a drawing of Γ_{k-1} that satisfies the contour conditions, i.e., the external face is drawn as a polygon consisting of a horizontal edge and a chain of segments with slope $+1$ or -1 between endpoints $P(v_1) = (-((k-1)-2), 0)$ and $P(v_2) = ((k-1)-2, 0)$.

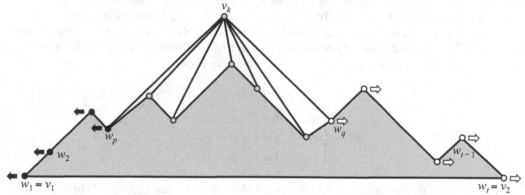

Figure 6.9 Schematic illustration of the addition of vertex v_k to drawing Γ_{k-1} to obtain drawing Γ_k. Contour vertices $w_1, \ldots, w+p$ (black-filled) are shifted by one unit to the left and contour vertices $w_q, \ldots, w_t$ (white-filled) are shifted by one unit to the right. When a contour vertex is shifted, we also shift all the vertices in its shift set (not shown). Finally, vertex v_k is placed at point $\mu(P(w_p), P(w_q))$. Drawing Γ_k satisfies the contour conditions, i.e., the external face is drawn as a polygon consisting of a horizontal edge and a chain of segments with slope $+1$ or -1 between endpoints $P(v_1) = (-k-2, 0)$ and $P(v_2) = (k-2, 0)$.

We now show how to add point $P(v_k)$ to Γ_{k-1} and obtain a planar straight-line drawing Γ_k of G_k (see Figure 6.9):

Step 1 For each $v \in \bigcup_{i=1}^{p} L(w_i)$, set $x(v) = x(v) - 1$. This step translates leftward by 1 the vertices of the external face from w_1 to the left attachment w_p of v_k plus all the other vertices in the shift sets of these vertices.

Step 2 For each $v \in \bigcup_{i=q}^{t} L(w_i)$, set $x(v) = x(v) + 1$. This step translates rightward by 1 the vertices of the external face from the right attachment v_q of v_k to w_t plus all the other vertices in the shift sets of these vertices.

Step 3 Set $P(v_k) = \mu(P(w_p), P(w_q))$. This step places vertex v_k so that it can be joined with straight-line edges to its neighbors.

Step 4 Set $L(v_k) = \{v_k\} \cup (\bigcup_{i=p+1}^{q-1} L(w_i))$. This step defines shift set $L(v_k)$ as the union of v_k and the shift sets of the vertices covered by v_k.

Steps 1, 2, and 3 ensure that points $P(w_p), \ldots, P(w_q)$ are all visible from $P(v_k)$, i.e., segments $P(v_k)P(w_i), p \leq i \leq q$, can be added to Γ_{k-1} without introducing crossings. Conditions 1–3 above are clearly satisfied in Γ_k. By Step 4, we obtain inductively that each set $L(u)$ is the subtree of the cover forest rooted at vertex u. Thus, sets $L(w_1), \ldots, L(w_t)$, form a partition of the vertices of G_{k-1}. It remains to prove that the shift operations in Steps 1 and 2 preserve the planarity of Γ_{k-1}, and this is done in the following lemma.

LEMMA 6.3   Let Γ_j be a planar straight-line drawing of G_j, as described above, and let $v_1 = w_1', w_2', \ldots, w_{t'}' = v_2$ be the sequence of vertices of C_j. Let s be an index such that $1 \leq s \leq t'$. If, for each $1 \leq i \leq s$ (resp., $s \leq i \leq t'$), we shift the vertices in $L(w_i')$ leftward (resp., rightward) by a positive integer number ρ, then the resulting straight-line drawing is still planar.

Proof: By induction on j. For Γ_3 the lemma is trivially true. We now suppose that the lemma is true for $\Gamma_{j-1}, j \geq 4$, and prove that it is true for Γ_j. We use the notation from the algorithm description above; namely, $v_1 = w_1, w_2, \ldots, w_t = v_2$ is the sequence of vertices of C_{j-1}, and w_p and w_q are the leftmost and rightmost neighbors of v_j in C_{j-1}, respectively. We denote by ζ the difference between the number of vertices of C_{j-1} and the number of vertices of C_j, i.e., $\zeta = (q - p - 1) - 1 \geq -1$. Thus, we have:

$$t' = t - \zeta$$

$$w_i' = \begin{cases} w_i & \text{for } i = 1, \ldots, p \\ v_j & \text{for } i = p+1 \\ w_{i+\zeta} & \text{for } i = p+2, \ldots, t' \end{cases}$$

Note, in particular, that $w_{p+2}' = w_q$. We prove the claim for the rightward shift; the proof for the leftward shift is symmetric.

If $s > p + 2$, then v_j and its neighbors $w_p, \ldots, w_q$ in C_{j-1} do not move. Thus, by the induction hypothesis, Γ_j is planar.

If $s \leq p$, then v_j and its neighbors $w_p, \ldots, w_q$ in C_{j-1} shift rigidly rightward by ρ. Thus, by the induction hypothesis, Γ_j is planar.

If $s = p+1$, we apply the induction hypothesis to Γ_{j-1} with $s = p+1$; thus, the planarity of Γ_{j-1} is preserved. Vertex v_j and its neighbors $w_{p+1}, \ldots, w_q$ in C_{j-1} shift rigidly rightward by ρ, while w_p does not move. Point $P(w_p)$ is clearly still visible from points $P(v_j)$ and $P(w_{p+1})$, and thus, Γ_j is planar.

If $s = p+2$, we apply the induction hypothesis to Γ_{j-1} with $s = q$; thus, the planarity of Γ_{j-1} is preserved. Vertex v_j and its neighbors $w_p, \ldots, w_{q-1}$ in C_{j-1} do not move, while w_q

shifts rightward by ρ. Point $P(w_q)$ is clearly still visible from points $P(v_j)$ and $P(w_{q-1})$, and thus Γ_j is planar. □

In the end, we obtain a planar straight-line drawing of G in which $P(v_1) = (-(n-2), 0)$ and $P(v_2) = (n-2, 0)$. By Condition 3 above, $P(v_n) = (0, n-2)$. Therefore, G is drawn on the $(2n-4) \times (n-2)$ grid.

Figures 6.10 through 6.19 show several steps of the execution of the algorithm on the graph and canonical ordering of Figure 6.5. The final drawing is shown in Figure 6.20.

6.6.2 Implementation

A straightforward implementation of the shift method results in an $O(n^2)$-time algorithm. In their paper, de Fraysseix, Pach, and Pollack [dFPP90] were able to reduce this time bound to $O(n \log n)$. An optimal $O(n)$-time implementation of the shift method was presented by Chrobak and Payne [CP95], and this is the implementation we describe below.

The crucial observation is that, when vertex v_k is placed on the grid, it is not necessary to know the exact positions of w_p and w_q. If their y-coordinates and their x-offset, i.e., $x(w_q) - x(w_p)$, are known, then $y(v_k)$ and the x-offset between v_k and w_p can be computed; namely, by Eq. 6.6, we have

$$y(v_k) = \frac{x(w_q) - x(w_p) + y(w_q) + y(w_p)}{2}, \tag{6.7}$$

$$x(v_k) - x(w_p) = \frac{x(w_q) - x(w_p) + y(w_q) - y(w_p)}{2}. \tag{6.8}$$

The algorithm consists of three phases. In the first phase, we compute a canonical ordering of the input graph. In the second phase, we add vertices one at a time, according to that canonical ordering: for each added vertex v_k, we compute its y-coordinate and x-offset $x(v_k) - x(w_p)$, update the x-offset of w_q (from its previous value $x(w_q) - x(w_{q-1})$) to $x(w_q) - x(v_k)$, and possibly update the x-offset of w_{p+1}. In the third phase, we suitably traverse the graph starting from v_1 and compute the final x-coordinates of the vertices by accumulating offsets.

We now describe the data structure used to implement the algorithm. For each $4 \le k \le n$, the family of sets $L(w_1), \ldots, L(w_t)$ for vertices $w_1, \ldots, w_t$ of C_{k-1} can be viewed as an ordered forest F of trees $L(w_i)$ rooted at vertex w_i, $1 \le i \le t$. When vertex v_k is added and set $L(v_k)$ is created (see Step 4 above), a new tree $L(v_k)$ of F is created out of trees $L(w_{p+1}), \ldots, L(w_{q-1})$ by making v_k the parent of $w_{p+1}, \ldots, w_{q-1}$ (in this order from left to right). A standard way to represent an ordered forest F is by means of a binary tree T: the roots of the trees of F are all considered siblings; the root of T corresponds to the root of the first tree of F; if n_T is a node of T corresponding to a node n_F of F, then the left child of n_T corresponds to the leftmost child of n_F (if any), and the right child of n_T corresponds to the next sibling of n_F (if any).

In our context, the root of T corresponds to $v_1 = w_1$, its right child corresponds to w_2, its right child's right child corresponds to w_3, and so on; thus, the rightmost leaf corresponds to $w_t = v_2$. Tree $L(w_i)$, $1 \le i \le t$, is represented by the node corresponding to w_i and its left subtree. The subtree of T rooted at the node corresponding to w_i represents $\bigcup_{j \ge i} L(w_j)$. For brevity, in the rest of the section, we refer with the same symbol to a vertex of G, the corresponding node of F, and the corresponding node of T.

If u is an ancestor of v in T, the x-offset between v and u is defined as $\Delta x(v, u) = x(v) - x(u)$. If u is the parent of v, we simply use the term x-offset of v and the symbol $\Delta x(v)$. With each vertex v of G, we store the following information:

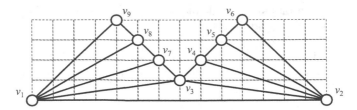

Figure 6.10   Drawing Γ_9 of graph G_9 which consists of vertices $v_1, v_2, \ldots, v_9$.

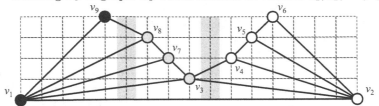

Figure 6.11   Preparing to add vertex v_{10} to drawing Γ_9. Vertex v_{10} has left attachment v_9 and right attachment v_4: the black-filled vertices are shifted to the left by one unit; the gray-filled vertices do not move; and the white-filled vertices are shifted to the right by one unit.

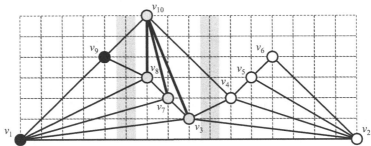

Figure 6.12   Addition of vertex v_{10} and its incident edges, which yields drawing Γ_{10}. Vertex v_{10} covers vertices v_8, v_7, and v_3.

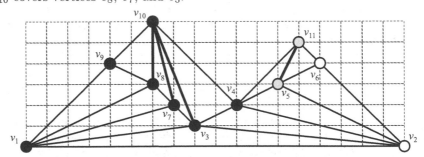

Figure 6.13   Drawing Γ_{11} obtained by adding vertex v_{11} and its incident edges after shifting the black-filled vertices to the left and the white-filled vertices to the right. Vertex v_{11} covers vertex v_5.

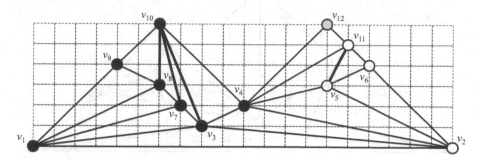

Figure 6.14 Drawing Γ_{12}.

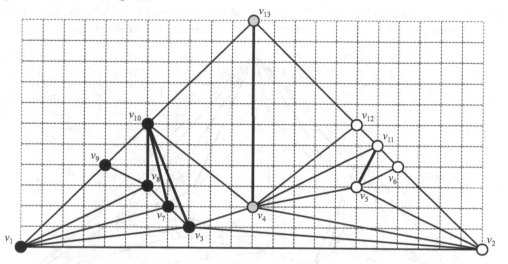

Figure 6.15 Drawing Γ_{13}.

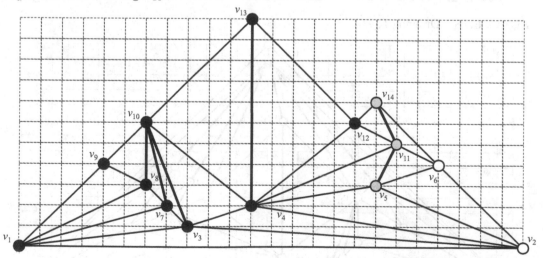

Figure 6.16 Drawing Γ_{14}.

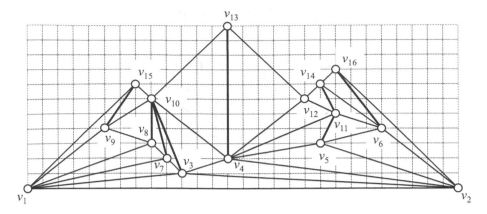

Figure 6.17   Drawing Γ_{16}. Note that we have skipped drawing Γ_{15}. Also, here and in the next two figures we do not fill the vertices to denote the amount of shifting.

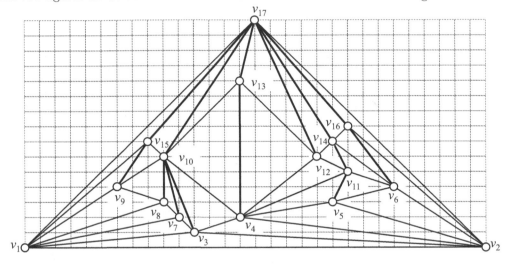

Figure 6.18   Drawing Γ_{17}.

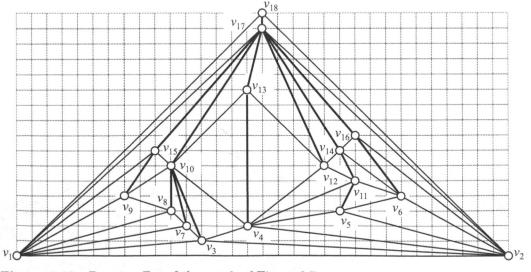

Figure 6.19   Drawing Γ_{18} of the graph of Figure 6.5.

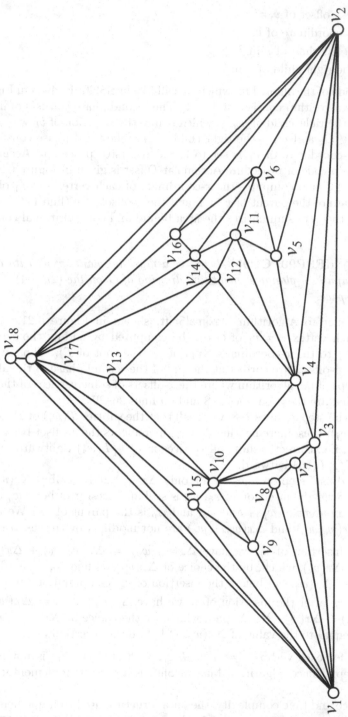

Figure 6.20 Planar straight-line grid drawing of the graph of Figure 6.5 constructed with the shift method by de Fraysseix, Pach, and Pollack (Algorithm MaximalShift shown in Figure 6.21). The graph has $n = 18$ vertices and the drawing has width $2n - 4 = 32$ and height $n - 2 = 16$. Note that the drawing is the same as that of Figure 6.19 except that it has been rotated counterclockwise by 90 degrees and the grid lines have been omitted for better readability.

- $\Delta x(v)$, the x-offset of v;
- $y(v)$, the y-coordinate of v;
- $left(v)$, the left child of v in T;
- $right(v)$, the right child of v in T.

The pseudo-code of the algorithm, which we call MaximalShift, is given in Figure 6.21. The first phase of the algorithm consists of line 1. The second phase consists of lines 2–30; note that, on line 23, the right child of w_{q-1}, which before the insertion of v_k was w_q, is set to *nil* since now $w_{q-1} \notin C_k$ and w_q is the right child of v_k. The third phase consists of lines 31–32, where the x-coordinate of v_1 is set to 0 and recursive procedure AccumulateOffset is called. The pseudo-code of procedure AccumulateOffset is given in Figure 6.22. It performs a preorder visit of T and computes the x-coordinate of each vertex $v \neq v_1$ of G as the sum of the x-coordinate of the parent of v in T and the x-offset of v (line 1).

The following theorem summarizes the area bound and computational complexity of the shift method.

Theorem 6.1   **[dFPP90, CP95]** *Let G be a maximal plane graph with n vertices. The shift method computes a planar straight-line drawing of G on the $(2n - 4) \times (n - 2)$ grid in $O(n)$ time and space.*

Proof: We refer to Algorithm MaximalShift, shown in Figure 6.21. Clearly, the x-coordinate of each vertex $v \neq v_1$ of G can be computed by adding the x-offset $\Delta x(v, v_1)$ between v and v_1 to the x-coordinate $x(v_1)$ of v_1 (the root of T). Thus, we only have to prove that those x-offsets are correct at the end of the second phase of the algorithm. Note that the only steps of the algorithm where the x-offsets of some vertices of the current graph G_{k-1} are modified are those on lines 7–8 and on lines 26–29.

For the "stretch" step on lines 7–8, we recall that the subtree $T(w_i)$ of T rooted at w_i represents $\bigcup_{j \geq i} L(w_j)$; thus, incrementing $\Delta x(w_i)$ increments the x-offset between each vertex of $T(w_i)$ and v_1, i.e., correctly shifts all vertices in $\bigcup_{j \geq i} L(w_j)$ rightward, or, equivalently, all vertices in $\bigcup_{j < i} L(w_j)$ leftward.

During the "adjust" step on lines 26–29, only $\Delta x(w_q)$ and possibly $\Delta x(w_{p+1})$ are modified. Note that, after the insertion of v_k, w_p is still an ancestor of both w_q and w_{p+1} in T: namely, v_k is the parent of w_{p+1} and w_q, and w_p is the parent of v_k. We now prove that the values of $\Delta x(w_q, w_p)$ and $\Delta x(w_{p+1}, w_p)$ are not modified by the insertion of v_k.

- After the insertion of v_k we have $\Delta x(w_q, w_p) = \Delta x(w_q, v_k) + \Delta x(v_k, w_p) = \Delta x(w_q) + \Delta x(v_k)$, which, by the choice of $\Delta x(w_q)$ on line 26, is clearly equal to the value of $\Delta x(w_q, w_p)$ before the insertion of v_k, computed on line 10.
- If $p + 1 \neq q$, after the insertion of v_k we have $\Delta x(w_{p+1}, w_p) = \Delta x(w_{p+1}, v_k) + \Delta x(v_k, w_p) = \Delta x(w_{p+1}) + \Delta x(v_k)$, which, by the choice of $\Delta x(w_{p+1})$ on line 28, is clearly equal to the value of $\Delta x(w_{p+1})$ before the insertion of v_k.

It follows that, for each vertex $v \neq v_1 \in G_{k-1}$, x-offset $\Delta x(v, v_1)$ is not modified during the "adjust" step. Hence, algorithm MaximalShift is a correct implementation of the shift method.

As for its space and time complexity, the data structure used to implement the algorithm clearly takes $O(n)$ space. By Lemma 6.1, the first phase takes $O(n)$ time and space. The time complexity of the body (lines 6–29) of the main for loop is dominated by the computation of $\Delta x(w_q, w_p)$ on line 10, which takes $O(\deg(v_k))$; thus, by Eq. 6.2, the second phase globally takes $O(n)$ time. The third phase clearly takes $O(n)$ time since at the end of the second phase T has n nodes. $\square$

Input: A maximal plane graph G with n vertices
Output: A planar straight-line drawing of G on the $(2n - 4) \times (n - 2)$ grid
1: compute a canonical ordering $v_1, \ldots, v_n$ of G
2: $(\Delta x(v_1), y(v_1), \text{left}(v_1), \text{right}(v_1)) \leftarrow (0, 0, \text{nil}, v_3)$
3: $(\Delta x(v_3), y(v_3), \text{left}(v_3), \text{right}(v_3)) \leftarrow (1, 1, \text{nil}, v_2)$
4: $(\Delta x(v_2), y(v_2), \text{left}(v_2), \text{right}(v_2)) \leftarrow (1, 0, \text{nil}, \text{nil})$
5: **for** $4 \le k \le n$ **do**
6:     /* stretch the $L(w_p)$-to-$L(w_{p+1})$ and $L(w_{q-1})$-to-$L(w_q)$ gaps */
7:     $\Delta x(w_{p+1}) \leftarrow \Delta x(w_{p+1}) + 1$
8:     $\Delta x(w_q) \leftarrow \Delta x(w_q) + 1$
9:     /* compute $\Delta x(w_q, w_p)$ */
10:     $\Delta x(w_q, w_p) \leftarrow \Delta x(w_{p+1}) + \cdots + \Delta x(w_q)$
11:     /* compute $\Delta x(v_k)$ and $y(v_k)$; see Eqs. 6.8 and 6.7 */
12:     $\Delta x(v_k) \leftarrow (\Delta x(w_q, w_p) + y(w_q) - y(w_p))/2$
13:     $y(v_k) \leftarrow (\Delta x(w_q, w_p) + y(w_q) + y(w_p))/2$
14:     /* add v_k to T */
15:     $\text{right}(w_p) \leftarrow v_k$
16:     **if** $p + 1 \neq q$ **then**
17:         $\text{left}(v_k) \leftarrow w_{p+1}$
18:     **else**
19:         $\text{left}(v_k) \leftarrow \text{nil}$
20:     **end if**
21:     $\text{right}(v_k) \leftarrow w_q$
22:     **if** $q - 1 \neq p$ **then**
23:         $\text{right}(w_{q-1}) \leftarrow \text{nil}$
24:     **end if**
25:     /* adjust $\Delta x(w_q)$ and $\Delta x(w_{p+1})$ */
26:     $\Delta x(w_q) \leftarrow \Delta x(w_q, w_p) - \Delta x(v_k)$
27:     **if** $p + 1 \neq q$ **then**
28:         $\Delta x(w_{p+1}) \leftarrow \Delta x(w_{p+1}) - \Delta x(v_k)$
29:     **end if**
30: **end for**
31: $x(v_1) \leftarrow 0$
32: AccumulateOffset$(v_1, x(v_1))$

Figure 6.21   Algorithm MaximalShift.

Input: A vertex v of T and an integer x
1: **if** $v \neq \text{nil}$ **then**
2:     $x(v) \leftarrow x + \Delta x(v)$
3:     AccumulateOffset$(\text{left}(v), x(v))$
4:     AccumulateOffset$(\text{right}(v), x(v))$
5: **end if**

Figure 6.22   Procedure AccumulateOffset.

6.6.3   Refinements and Variations

Chrobak and Nakano [CN98] and Brandenburg [Bra08] refined the shift method by de Frays-seix, Pach, and Pollack [dFPP90], thus reducing the area of the drawing.

In the original shift method [dFPP90], we have seen that at each step the drawing satisfies the contour conditions. In the refinement by Chrobak and Nakano [CN98], these conditions are relaxed: $x(w_i) \leq x(w_{i+1}), 1 \leq i \leq t - 1$ and the equality may hold only when $y(w_i) < y(w_{i+1})$. Thus, each contour segment $P(w_i)P(w_{i+1})$ belongs to one of the following four types:

vertical $x(w_i) = x(w_{i+l})$ and $y(w_i) < y(w_{i+l})$;

upward $x(w_i) < x(w_{i+l})$ and $y(w_i) < y(w_{i+l})$;

horizontal $x(w_i) < x(w_{i+l})$ and $y(w_i) = y(w_{i+l})$;

downward $x(w_i) < x(w_{i+l})$ and $y(w_i) > y(w_{i+l})$.

The presence of vertical contour segments allows to avoid some shifts, thus obtaining a more compact drawing. The authors present a new combinatorial structure, called a *domino chain*, which allows to partition the vertices into *stable* and *unstable*; a stable vertex v_k can be added to G_{k-1} with edge (w_p, v_k) drawn as a vertical segment and no shift is necessary. Namely, the method avoids making any shifts in approximately $\frac{n}{3}$ steps and results in a drawing of size $\lfloor \frac{2}{3}(n-1) \rfloor \times 4\lfloor \frac{2}{3}(n-1) \rfloor - 1$.

Brandenburg further improves the shifting strategy and also rotates the drawing to choose the best base edge. This refinement of the shift method results in a drawing of size $\lceil \frac{4}{3}n \rceil \times \lceil \frac{2}{3}n \rceil$. Also, this height and width are necessary if the drawing is constrained to be enclosed by an isosceles right-angled triangle.

Kant [Kan96] presents an algorithm based on the shift method for constructing convex drawings of triconnected plane graphs on the $(2n - 4) \times (n - 2)$ grid. The size of the grid is reduced to $(n - 2) \times (n - 2)$ in a successive algorithm by Chrobak and Kant [CK97].

6.7   Realizer Method

An alternative method for drawing maximal planar graphs on an integer grid was presented by Schnyder [Sch90]. The origins of the approach can be found in [Sch89], where it was used to characterize planar graphs as the graphs whose incidence relation is the intersection of at most three total orders[3] (see Theorems 4.1 and 6.2 of [Sch89]).

6.7.1   Realizers

DEFINITION 6.3   A *realizer* of a maximal plane graph G is a triplet of rooted directed spanning trees of G with the following properties[4] (see Figure 6.23):

---

[3] More formally, a graph $G = (V, E)$ is planar if and only if the order dimension of the poset $(V \cup E, \prec)$, where incidence relation $\prec$ is defined by $v \prec e \Leftrightarrow v \in V, e \in E, v \in e$, is at most 3. The *order dimension* of a poset is the minimum cardinality of its realizers. A *realizer* of a poset $(X, \prec)$ is a nonempty set of total orders on X whose intersection is $\prec$.

[4] This definition of a realizer of a maximal plane graph is slightly different from the one given in [Sch90], as we consider also the external edges; our definition allows to reduce the number of special cases and to generalize the concept of realizer to triconnected plane graphs.

1. In each spanning tree, the edges of G are directed from children to parent.

2. The sinks (roots) of the spanning trees are the three external vertices of G.

3. Each internal edge of G is contained in one spanning tree.

4. Each external edge of G is contained in two spanning trees and it has different directions in the two trees.

5. Consider the edges of G with the directions they have in the three spanning trees (the external edges are considered twice):

   (a) Each non-sink vertex v of G has exactly three *outgoing edges*; the circular order of the outgoing edges around v induces a circular order of the spanning trees around v; all the non-sink vertices of G have the same circular order of the spanning trees.

   (b) For each vertex of G, the *incoming edges* that belong to the same spanning tree appear consecutively between the outgoing edges of the other two spanning trees (for the sink of each spanning tree the first and last incoming edges are coincident with the two outgoing edges).

6. For the sink of each spanning tree, all the incoming edges belong to that spanning tree.

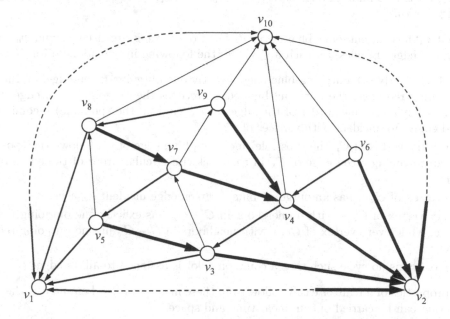

Figure 6.23   A realizer of a maximal plane graph whose vertices are numbered according to a canonical ordering. The edges are thick for the green spanning tree, medium for the blue spanning tree, and thin for the red spanning tree. Note the 2-colored edges on the external face.

Let T_b, T_g, and T_r be the spanning trees forming a realizer of a maximal plane graph G (see Figure 6.23). We assign a color to the edges of G contained in T_b, T_g, and T_r, say, *blue*, *green*, and *red*, respectively. In the figures, we use dark grey for blue, light grey for green, and medium grey for red. According to Property 3 of the realizers, each internal edge of

G is assigned one color and is said to be 1-*colored*, while the three external edges of G are assigned two colors and are said to be 2-*colored*.

In the proof of the following lemma, we present a mechanism for constructing a realizer of a maximal plane graph G based on a canonical ordering of G; this is different from the mechanism based on edge labelings presented in [Sch90].

LEMMA 6.4   Each maximal plane graph has a realizer, which can be computed in linear time and space.

Proof: Let G be a maximal plane graph. A realizer of G can be constructed by assigning colors and directions to the edges of G as follows:

1. a canonical ordering of G is computed;

2. v_1, v_2, and v_n are the sinks of the blue, green, and red tree, respectively;

3. (v_1, v_2) is an outgoing blue edge for v_2 and an outgoing green edge for v_1;

4. for each $3 \leq k \leq n$, let $c_l, \ldots, c_r$ be the consecutive neighbors of v_k on C_{k-1} from left to right; (v_k, c_l) is an outgoing blue edge for v_k; (v_k, c_r) is an outgoing green edge for v_k; each edge $(v_k, c_i), l < i < r$, is an outgoing red edge for c_i (see Figure 6.23);

5. (v_n, v_1) is also an outgoing red edge for v_1, and (v_n, v_2) is also an outgoing red edge for v_2.

Note that v_1 has no outgoing blue edge, v_2 has no outgoing green edge, and v_n has no outgoing red edge. Besides, for each $3 \leq k \leq n$, the following invariants hold for G_k:

- v_k has exactly one outgoing blue edge, exactly one outgoing green edge, and no outgoing red edge; the outgoing blue edge precedes the outgoing green edge in the clockwise circular order of the edges of C_k, and all the (possible) incoming red edges are incident with vertices of C_{k-1};

- for every vertex of C_k the (possible) incoming blue edge of C_k follows the (possible) incoming green edge of C_k in the clockwise circular order of the edges of C_k;

- no vertex of C_{k-1} has an outgoing blue or green edge incident with v_k;

- every vertex of C_{k-1} with no neighbor in $G - G_k$ has exactly one outgoing red edge, while every vertex of C_{k-1} with neighbors in $G - G_k$ has no outgoing red edge;

- G_k contains no cycle such that a common color is assigned to all its edges.

All the properties of a realizer easily follow from these invariants. By Lemma 6.1, the above construction can be carried out in linear time and space. □

From the construction in the proof of Lemma 6.4, it follows that, for every realizer of a maximal plane graph G, all internal edges of G are 1-colored, while the three external edges are 2-colored. Also, for each vertex of G, the colors of the three outgoing edges appear in the following counterclockwise circular order: blue, green, red. Set $\{b, g, r\}$ will be considered accordingly ordered in the rest of the chapter.

In the rest of the section, we consider a maximal plane graph G equipped with a realizer $\{T_b, T_g, T_r\}$. We denote v_1, v_2, and v_n by s_b, s_g, and s_r, respectively. For each vertex v of G, the *blue path* $p_b(v)$ is the path of G along T_b from v to s_b. In the same way, we define the *green path* $p_g(v)$ as the path of G along T_g from v to s_g and the *red path* $p_r(v)$ as the path of G along T_r from v to s_r. Note that $p_i(s_i), i \in \{b, g, r\}$, is a degenerate path consisting only of s_i. The subpath of $p_i(v), i \in \{b, g, r\}$, from v to the ancestor u of v in T_i is denoted by $p_i(v, u)$. The parent of vertex v in $T_i, i \in \{b, g, r\}$, is denoted by $par_i(v)$. The lowest common ancestor of vertices u and v in $T_i, i \in \{b, g, r\}$, is denoted by $lca_i(u, v)$.

LEMMA 6.5   For each vertex v of G, $p_b(v)$, $p_g(v)$, and $p_r(v)$ have only vertex v in common.

Proof: W.l.o.g., suppose, for a contradiction, that $p_b(v)$ and $p_g(v)$ have a vertex u in common, and that $p_b(v, u) - \{u, v\}$ and $p_g(v, u) - \{u, v\}$ have no vertex in common with each other and with $p_r(v)$. Vertex u has both a blue and a green incoming edge; thus, by Property 6 of the realizers, we have $u \neq s_b, s_g$. Let R be the subgraph of G bounded by $p_b(v, u)$ and $p_g(v, u)$; from the circular order of the outgoing edges at v, we have that $p_b(v, u)$ (resp., $p_g(v, u)$) follows the boundary of R counterclockwise (resp., clockwise). Thus, by Property 5 of the realizers at u and by the planarity of G, $par_b(u) \in R$ (the same is true for $par_g(u)$). Still by the planarity of G, $p_b(par_b(u))$ leaves R at a vertex w; two cases are possible: (i) $w \in p_g(v, u) - \{u, v\}$, but this contradicts Property 5 of the realizers at w, or (ii) $w \in p_b(v, u)$, but this contradicts the acyclicity of T_b.   □

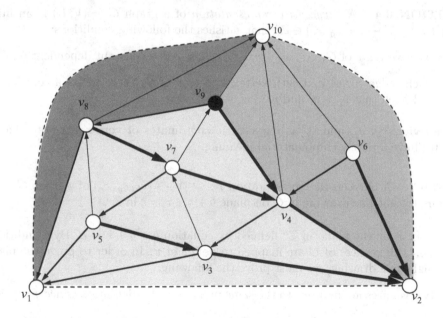

Figure 6.24   The blue (medium), green (thick), and red (thin) paths for vertex v_9 and corresponding blue (medium shaded), green (dark shaded) and red (light shaded) regions of vertex v_9. The coordinates of v_9 in the barycentric representation are the number of faces in the blue, green, and red region, respectively, i.e., $(4, 2, 9)$.

For each vertex v of G, the *blue region* $R_b(v)$ is the subgraph of G bounded by $p_g(v)$, $p_r(v)$ and (s_g, s_r). In the same way, the *green region* $R_g(v)$ is the subgraph of G bounded by $p_b(v)$, $p_r(v)$ and (s_r, s_b), and the *red region* $R_r(v)$ is the subgraph of G bounded by $p_b(v)$, $p_g(v)$ and (s_b, s_g) (see Figure 6.24). Note that $R_b(s_g) = R_b(s_r)$ is a degenerate region consisting only of (s_g, s_r). In the same way, $R_g(s_r) = R_g(s_b) = (s_r, s_b)$ and $R_r(s_b) = R_r(s_g) = (s_b, s_g)$.

LEMMA 6.6   Let u and v be two distinct vertices of G. If $u \in R_k(v), k \in \{b, g, r\}$, then $R_k(u) \subset R_k(v)$.

Proof: W.l.o.g, we assume $k = r$. Two cases are possible: (i) $u \notin p_b(v) \cup p_g(v)$, or (ii) $u \in p_b(v) \cup p_g(v)$. We consider only the first case; the second is similar. By the planarity of G and by Property 5 of the realizers, $p_b(u)$ has no vertex in common with $p_g(v)$, and $p_g(u)$ has no vertex in common with $p_b(v)$. Thus, the region $R_r(v) - R_r(u)$ bounded by $p_b(u, lca_b(u, v))$, $p_g(u, lca_g(u, v))$, $p_b(v, lca_b(u, v))$, and $p_g(v, lca_g(u, v))$ is nonempty; hence, $R_r(u) \subset R_r(v)$.   □

Note that, for each $k \in \{b, g, r\}$, the inclusion partial order of the k-regions induces a partial order on the vertices of G defined by $u \prec_k v \Leftrightarrow R_k(u) \subset R_k(v)$. Partial order $\prec_k$ is represented by tree $T_k, k \in \{b, g, r\}$ of the realizer of G. Also, for each edge (u, w) and each vertex $v \neq u, w$ of G, by the planarity of G, (u, w) is in some region $R_k(v), k \in \{b, g, r\}$; hence, $u \prec_k v$ and $w \prec_k v$. Any choice of three linear extensions of $\prec_b$, $\prec_g$, and $\prec_r$, produces a realizer of the poset defined in footnote (3) on page 212.

6.7.2   Barycentric Representation

DEFINITION 6.4   A *barycentric representation* of a graph $G = (V, E)$ is an injective function $f : v \in V \to (v_b, v_g, v_r) \in \mathbb{Z}^3$ that satisfies the following conditions:

1. For each vertex v of G, $v_b + v_g + v_r = c$, where c is a constant dependent on G.

2. For each edge (u, w) and each vertex $v \neq u, w$ of G, there exists a coordinate $i \in \{b, g, r\}$ such that $v_i > u_i$ and $v_i > w_i$.

One can view v_b, v_g, and v_r as barycentric coordinates of vertex v. Note that these coordinates have a purely combinatorial meaning.

LEMMA 6.7   A barycentric representation $f : v \in V \to (v_b, v_g, v_r)$ of a graph $G = (V, E)$ is a planar straight-line drawing of G on plane $b + g + r = c$ in $\mathbb{Z}^3$.

Proof: Let π be the plane in $\mathbb{Z}^3$ defined by equation $b + g + r = c$. By Condition 1 of Definition 6.4, all vertices of G are mapped to points of π. In order to prove the planarity of the straight-line drawing, we must prove the following:

- No two vertices are mapped to the same point of π. By definition, since f is injective.

- No vertex overlaps an edge. Let (u, w) be an edge of G and let $max_i = \max\{u_i, v_i\}, i \in \{b, g, r\}$. Let λ_b be the line of π defined by equation $b = max_b$, i.e., the line of π passing through the endpoint of segment $f(u)f(w)$ with maximum b-coordinate and perpendicular to the b axis. Lines λ_g and λ_r are defined in a similar way, and, together with λ_b, identify a (closed) triangle T containing $f(u)f(w)$. Suppose, for a contradiction, that there exists a vertex $v \neq u, w$ of G such that $f(v)$ overlaps $f(u)f(w)$. Clearly, $f(v)$ is contained by T, i.e., $v_i \leq max_i$, for each $i \in \{b, g, r\}$. But this contradicts Condition 2 of Definition 6.4.

- No two edges cross. Let $e_1 = (u, w)$ and $e_2 = (x, y)$ be two nonincident edges of G, and let T_1 and T_2 be the two (closed) triangles containing $f(u)f(w)$ and $f(x)f(y)$, respectively, identified as above. Suppose, for a contradiction, that $f(u)f(w)$ and $f(x)f(y)$ cross. Then, either T_1 contains $f(x)$ or $f(y)$, or T_2 contains $f(u)$ or $f(v)$. But this again contradicts Condition 2 of Definition 6.4.

$\square$

Lemma 6.7 implies that only a planar graph can have a barycentric representation. For each vertex v of G, we denote by $l_b(v)$, $l_g(v)$, and $l_r(v)$ the number of faces in $R_b(v)$, $R_g(v)$, and $R_r(v)$, respectively. Note that $0 \leq l_b(v), l_g(v), l_r(v) \leq 2n - 5$ and

$$l_b(v) + l_g(v) + l_r(v) = 2n - 5.$$

We have that these values yield barycentric coordinates (see Figures 6.24 and 6.25), as shown by the following lemma.

LEMMA 6.8 Let $G = (V, E)$ be a maximal plane graph equipped with a realizer. Function $f : v \in V \to (l_b(v), l_g(v), l_r(v))$ is a barycentric representation of G.

Proof: The injectivity of f follows from Lemma 6.6. Condition 1 of Definition 6.4 is trivially satisfied since for each vertex v, $v_b + v_g + v_r = 2n - 5$. As for Condition 2, let (u, w) and $v \neq u, w$ be an edge and a vertex of G, respectively. W.l.o.g., let $u \in R_r(v)$; by the planarity of G, $w \in R_r(v)$, as well. By Lemma 6.6, $R_r(u) \subset R_r(v)$ and $R_r(w) \subset R_r(v)$. Hence, $v_r > u_r$ and $v_r > w_r$. $\square$

Let Γ be the planar straight-line drawing resulting from the barycentric representation of Lemma 6.8. By that lemma and by Lemma 6.7, Γ is a planar straight-line drawing of G on plane $b + g + r = 2n - 5$ in $\mathbb{Z}^3$. In particular, vertices s_b, s_g, and s_r are mapped to points $(2n - 5, 0, 0)$, $(0, 2n - 5, 0)$, and $(0, 0, 2n - 5)$, respectively. A planar straight-line drawing of G on the $(2n - 5) \times (2n - 5)$ grid in $\mathbb{Z}^2$ can be obtained by projecting Γ, e.g., by dropping, for each vertex v, the red coordinate v_r, as illustrated in Figure 6.26.

As for the time and space complexity, by Lemma 6.4, a realizer of G can be constructed in linear time and space. The coordinates of the vertices of G can also be computed in linear time and space.

It is possible to obtain more compact drawings by relaxing the constraints imposed on the vertex coordinates by Definition 6.4. Given two ordered pairs (a, b) and (c, d), the $>_{lex}$ relation is defined by $(a, b) >_{lex} (c, d) \Leftrightarrow a > c \vee (a = c \wedge b > d)$.

DEFINITION 6.5 A *weak barycentric representation* of a graph $G = (V, E)$ is an injective function $f : v \in V \to (v_b, v_g, v_r) \in \mathbb{Z}^3$ that satisfies the following conditions:

1. For each vertex v of G, $v_b + v_g + v_r = c$, where c is a constant dependent on G.

2. For each edge (u, w) and each vertex $v \neq u, w$ of G, there exist two consecutive coordinates i and j in the circularly ordered set $\{b, g, r\}$ such that $(v_i, v_j) >_{lex} (u_i, u_j)$ and $(v_i, v_j) >_{lex} (w_i, w_j)$.

The following lemma can be proved similarly to Lemma 6.7 and implies that only a planar graph can have a weak barycentric representation.

LEMMA 6.9 [Sch90] A weak barycentric representation $f : v \in V \to (v_b, v_g, v_r)$ of a graph $G = (V, E)$ is a planar straight-line drawing of G on plane $b + g + r = c$ in $\mathbb{Z}^3$.

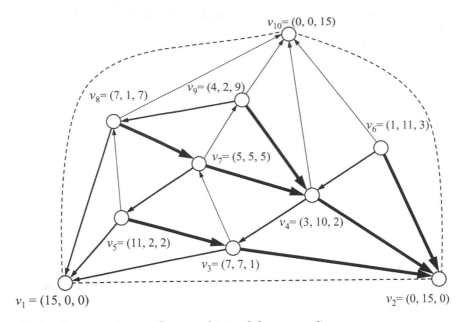

Figure 6.25 Barycentric coordinates obtained from a realizer.

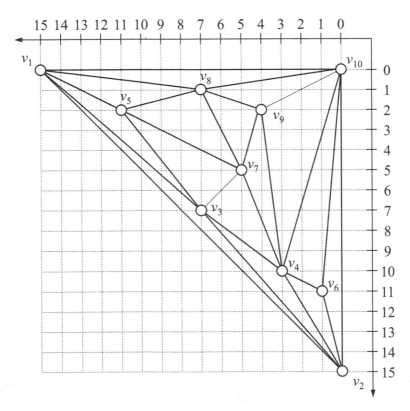

Figure 6.26   Planar straight-line grid drawing obtained from the barycentric coordinates of Figure 6.25 by dropping the third (red) coordinate. The horizontal and vertical axes are shown reversed to maintain the visual correspondence with the drawing of Figure 6.25.

For each vertex v of G, we denote by $n_b(v)$, $n_g(v)$, and $n_r(v)$ the number of vertices in $R_b(v) - p_r(v)$, $R_g(v) - p_b(v)$, and $R_r(v) - p_g(v)$, respectively. Note that $0 \le n_b(v), n_g(v), n_r(v) \le n - 2$ and $n_b(v) + n_g(v) + n_r(v) = n - 1$.

LEMMA 6.10   Let u and v be two distinct vertices of G, and let i and j be two consecutive coordinates in the circularly ordered set $\{b, g, r\}$. If $u \in R_i(v)$, then $(n_i(v), n_j(v)) >_{lex} (n_i(u), n_j(u))$.

Proof: W.l.o.g, we assume $i = r$ and thus $j = b$. Two cases are possible:

1. $u \notin p_g(v)$; by Lemma 6.6, $R_r(u) \subset R_r(v)$, and thus $p_g(u)$ is in $R_r(v)$; since $u \in p_g(u)$, we have $u \notin R_r(u) - p_g(u)$ while $u \in R_r(v) - p_g(v)$; thus, $R_r(u) - p_g(u) \subset R_r(v) - p_g(v)$; hence, $n_r(v) > n_r(u)$;

2. $u \in p_g(v)$; two subcases are possible:

   (a) $R_r(u) - p_g(u) \subset R_r(v) - p_g(v)$; hence, $n_r(v) > n_r(u)$;

   (b) $R_r(u) - p_g(u) = R_r(v) - p_g(v)$ (this subcase occurs if $par_b(u) = par_b(v)$); hence, $n_r(v) = n_r(u)$; however, $u \in R_b(v)$ and $u \notin p_r(v)$; by the same argument used for Case 1, $n_b(v) > n_b(u)$.

Thus, $(n_r(v), n_b(v)) >_{lex} (n_r(u), n_b(u))$.   □

LEMMA 6.11   Let $G = (V, E)$ be a maximal planar graph equipped with a realizer. Function $f : v \in V \rightarrow (n_b(v), n_g(v), n_r(v))$ is a weak barycentric representation of G.

Proof: Injectivity of f follows from Lemma 6.10. Condition 1 of Definition 6.5 is trivially satisfied, since, for each vertex v, $v_b + v_g + v_r = n - 1$. As for Condition 2, let (u, w) and $v \ne u, w$ be an edge and a vertex of G, respectively. W.l.o.g., let $u \in R_r(v)$; by the planarity of G, $w \in R_r(v)$, as well. Hence, by Lemma 6.10, $(v_r, v_b) >_{lex} (u_r, u_b)$ and $(v_r, v_b) >_{lex} (w_r, w_b)$.   □

6.7.3   Implementation

Let Γ be the straight-line drawing of G resulting from the weak barycentric representation of Lemma 6.11. By that lemma and by Lemma 6.9, Γ is a planar straight-line drawing of G on plane $b + g + r = n - 1$ in $\mathbb{Z}^3$. In particular, vertices s_b, s_g, and s_r are mapped to points $(n - 2, 1, 0)$, $(0, n - 2, 1)$, and $(1, 0, n - 2)$, respectively. A planar straight-line drawing of G on the $(n - 2) \times (n - 2)$ grid in $\mathbb{Z}^2$ can be obtained by projecting Γ, e.g., by dropping, for each vertex v, the red coordinate v_r.

We now consider the time and space complexity. By Lemma 6.4, a realizer of G can be constructed in linear time and space. Next, we show that the coordinates for the vertices of G can be computed in linear time and space. In particular, we show how to compute, for each vertex v of G, coordinate v_r; coordinates v_b and v_g can be computed similarly.

From the planarity of G and Property 5 of the realizers, it follows that, for each vertex $u \ne v \in R_r(v)$, (i) the subtree $T_r(u)$ of T_r rooted at u is contained by $R_r(v)$, and (ii) $p_r(u)$ has exactly one vertex w in common with $p_b(v) \cup p_g(v)$ (note that $u \in T_r(w)$).

First, we compute, for each vertex v of G, the number of its descendants in T_r, including v itself, and store it in variable $numdesc_r(v)$; this can be done by a postorder visit of T_r.

Second, we compute, for each vertex v of G, the number of its ancestors in T_g, including v itself, and store it in variable $numanc_g(v)$; this can be done by a preorder visit of T_g. Finally, we compute, for each vertex v of G, $\sum_{w \in p_b(v)} numdesc_r(w)$ and $\sum_{w \in p_g(v)} numdesc_r(w)$; this can be done by a preorder visit of T_b and T_g, respectively.

For each vertex v of G, the number $n_r(v)$ of vertices in $R_r(v) - p_g(v)$, i.e., coordinate v_r, is given by the expression

$$\sum_{w \in p_b(v)} numdesc_r(w) + \sum_{w \in p_g(v)} numdesc_r(w) - numdesc_r(v) - numanc_g(v)$$

It follows that the coordinates for the vertices of G can be computed by a constant number of traversals of T_b, T_g, and T_r, and thus globally in $O(n)$ time. Furthermore, the additional variables used in the tree traversals clearly take $O(n)$ space.

Thus, we obtain the following theorem that summarizes the area bound and computational complexity of the realizer method.

Theorem 6.2   [Sch90] *Let G be a maximal plane graph with n vertices. The realizer method computes a planar straight-line drawing of G on the $(n-2) \times (n-2)$ grid in $O(n)$ time and space.*

6.7.4   Refinements and Variations

Zhang and He [ZH03] discovered some new properties of Schnyder's realizers and were able to further reduce the grid size (in most cases).

Di Battista, Tamassia, and Vismara [DTV99] extend the realizer method to construct in linear time a convex grid drawing of a triconnected plane graph on the $(f-1) \times (f-1)$ grid, where f is the number of faces of the graph. The same result had been claimed by Schnyder and Trotter [ST92] without proof and is independently obtained by Felsner [Fel01] with different techniques. A method that further improves the grid size was developed by Bonichon, Felsner, and Mosbah [BFM07].

Acknowledgment

Roberto Tamassia contributed to the writing of this chapter.

References

[BFM07] Nicolas Bonichon, Stefan Felsner, and Mohamed Mosbah. Convex drawings of
3-connected plane graphs. *Algorithmica*, 47:399–420, 2007.

[Bra08] Franz J. Brandenburg. Drawing planar graphs on $\frac{8}{9}n^2$ area. *Electronic Notes in
Discrete Mathematics*, 31:37–40, 2008.

[CG95] I. F. Cruz and A. Garg. Drawing graphs by example efficiently: Trees and planar
acyclic digraphs. In R. Tamassia and I. G. Tollis, editors, *Graph Drawing (Proc.
GD '94)*, volume 894 of *Lecture Notes Comput. Sci.*, pages 404–415. Springer-
Verlag, 1995.

[CK97] M. Chrobak and G. Kant. Convex grid drawings of 3-connected planar graphs.
Internat. J. Comput. Geom. Appl., 7(3):211–223, 1997.

[CN98] Marek Chrobak and S. Nakano. Minimum-width grid drawings of plane graphs.
Comput. Geom. Theory Appl., 11:29–54, 1998.

[CON85] N. Chiba, K. Onoguchi, and T. Nishizeki. Drawing planar graphs nicely. *Acta
Inform.*, 22:187–201, 1985.

[CP95] M. Chrobak and T. Payne. A linear-time algorithm for drawing planar graphs.
Inform. Process. Lett., 54:241–246, 1995.

[CYN84] N. Chiba, T. Yamanouchi, and T. Nishizeki. Linear algorithms for convex drawings of planar graphs. In J. A. Bondy and U. S. R. Murty, editors, *Progress in
Graph Theory*, pages 153–173. Academic Press, New York, NY, 1984.

[DF13] Giuseppe Di Battista and Fabrizio Frati. Drawing trees, outerplanar graphs,
series-parallel graphs, and planar graphs in a small area. In J. Pach, editor,
Thirty Essays on Geometric Graph Theory, pages 121–166. 2013.

[dFPP90] H. de Fraysseix, J. Pach, and R. Pollack. How to draw a planar graph on a grid.
Combinatorica, 10(1):41–51, 1990.

[Dji95] H. N. Djidjev. On drawing a graph convexly in the plane. In R. Tamassia and
I. G. Tollis, editors, *Graph Drawing (Proc. GD '94)*, volume 894 of *Lecture Notes
Comput. Sci.*, pages 76–83. Springer-Verlag, 1995.

[DTV99] G. Di Battista, R. Tamassia, and L. Vismara. Output-sensitive reporting of
disjoint paths. *Algorithmica*, 23(4):302–340, 1999.

[Fár48] I. Fáry. On straight lines representation of planar graphs. *Acta Univ. Szeged.
Sect. Sci. Math.*, 11:229–233, 1948.

[Fel01] Stefan Felsner. Convex drawings of planar graphs and the order dimension of
3-polytopes. *Order*, 18:19–37, 2001.

[HT73] J. Hopcroft and R. E. Tarjan. Dividing a graph into triconnected components.
SIAM J. Comput., 2(3):135–158, 1973.

[Kan96] G. Kant. Drawing planar graphs using the canonical ordering. *Algorithmica*,
16:4–32, 1996. (special issue on Graph Drawing, edited by G. Di Battista and R.
Tamassia).

[NR04]   Takao Nishizeki and Md. Saidur Rahman. *Planar Graph Drawing*. World Scientific, 2004.

[RT86]   P. Rosenstiehl and R. E. Tarjan. Rectilinear planar layouts and bipolar orientations of planar graphs. *Discrete Comput. Geom.*, 1(4):343–353, 1986.

[Sch89]   W. Schnyder. Planar graphs and poset dimension. *Order*, 5:323–343, 1989.

[Sch90]   W. Schnyder. Embedding planar graphs on the grid. In *Proc. 1st ACM-SIAM Sympos. Discrete Algorithms*, pages 138–148, 1990.

[SR34]   E. Steinitz and H. Rademacher. *Vorlesungen über die Theorie der Polyeder*. Julius Springer, Berlin, Germany, 1934.

[ST92]   W. Schnyder and W. T. Trotter. Convex embeddings of 3-connected plane graphs. *Abstracts of the AMS*, 13(5):502, 1992.

[Ste51]   S. K. Stein. Convex maps. *Proc. Amer. Math. Soc.*, 2(3):464–466, 1951.

[Tho80]   C. Thomassen. Planarity and duality of finite and infinite planar graphs. *J. Combin. Theory Ser. B*, 29(2):244–271, 1980.

[Tho84]   C. Thomassen. Plane representations of graphs. In J. A. Bondy and U. S. R. Murty, editors, *Progress in Graph Theory*, pages 43–69. Academic Press, New York, NY, 1984.

[Tut60]   W. T. Tutte. Convex representations of graphs. *Proceedings London Mathematical Society*, 10(38):304–320, 1960.

[Tut63]   W. T. Tutte. How to draw a graph. *Proceedings London Mathematical Society*, 13(52):743–768, 1963.

[Wag36]   K. Wagner. Bemerkungen zum Vierfarbenproblem. *Jahresbericht der Deutschen Mathematiker-Vereinigung*, 46:26–32, 1936.

[ZH03]   Huaming Zhang and Xin He. Compact visibility representation and straight-line grid embedding of plane graphs. In *Algorithms and Data Structures*, volume 2748 of *Lecture Notes in Computer Science*, pages 493–504. Springer, 2003.

7

Planar Orthogonal and Polyline Drawing Algorithms

Christian A. Duncan
Quinnipiac University

Michael T. Goodrich
University of California, Irvine

7.1 Introduction

One can assess the quality of a drawing of a graph in many different ways. Many important criteria deal with the aesthetics, readability, of the drawing. For example, the size of the drawing, roughly measured as the ratio between the farthest two objects of the drawings and the closest two, is a measure of how much information can be displayed at one time. The aesthetic that is of biggest concern in this chapter is that of *angular resolution*. Essentially, we are concerned with how close together edges that stem from the same vertex are to each other. The smaller the angle the more likely are the chances that the distinct edges become one. Clearly, a high-degree vertex, one with many edges extending out of it, will inevitably have a small angle between at least one pair of edges. So, the goal is to make the resolution determined to some extent by the degree of the vertex.

Optimizing angular resolution in drawings has been addressed by countless researchers. The two approaches we focus on in this chapter are to draw the graph orthogonally, that is using only vertical and horizontal line segments for the edges. Orthogonal drawings have the benefit that the smallest angle is at most $\pi/2$ and that the resulting graphs are often quite pleasing to the eye because of the few edge directions employed, but they also have the disadvantage that no vertex can have degree more than four. The study of orthogonal graphs also has the advantage of being of interest to VLSI design, because many wires are routed similarly. There are many different approaches to drawing orthogonal graphs. Early results draw the graph using few bends but sacrificed size or running-time efficiency. Improved techniques, involving computing a visibility representation, yielded orthogonal drawings in linear time with few bends and small size. By using network flows, we can draw (embedded) graphs with the guaranteed minimum number of bends possible in the smallest area allowable, but the run-time performance goes up to near quadratic time.

When using graphs containing vertices with degree more than four, one can no longer apply standard orthogonal drawing techniques. More general polyline drawing techniques, however, do exist. The goal is usually to focus directly on the sizes of angles created rather than the types of edges allowed. Thus, during the drawing, we can route edges in any orientation so long as the angle does not go below some fixed threshold. The most successful approaches all seem to work by taking a vertex and assigning exit ports, which are adequately spaced, such that edges are routed from the start vertex through distinct ports to the destination vertex. These techniques typically produce the layout by creating a canonical ordering on the vertices and adding the vertices into the drawing based on this ordering, while constantly maintaining the routing requirements of the edges. Using this approach, one can guarantee, for example, that a drawing can be made in linear time with good angular resolution, good size bounds, and using at most one bend per edge.

Before going into the details of the different approaches, we first present some basic terminology and general techniques in Section 7.2. Section 7.3 describes some standard approaches to drawing orthogonal graphs. Section 7.4 describes work done on more general polyline drawings. We conclude our chapter in Section 7.5 with a brief summary of the main results presented.

7.2 Preliminaries

We begin with a few basic definitions of some general graph terminology along with some more detailed descriptions of techniques useful for constructing drawings of graphs.

7.2.1 Definitions

Although common in nearly any book on graph algorithms, we borrow notation predominantly from [DETT99]. A *(simple) graph* $G = (V, E)$ is a finite set V of *vertices* and a finite set E of *edges*, where each edge is an unordered pair $e = (u, v)$ of vertices. A *multigraph* is a graph where the edges are multisets, that is two edges may have the same pair of vertices. For each edge $e = (u, v)$, we say that e is *incident* to u and v. We also say that u and v are *neighbors*. The *degree of a vertex* is the number of edges incident to it. The *maximum degree of a graph* is the maximum degree among all vertices in V. A *(simple) path* p of G is a sequence of distinct vertices of G, $(v_1, v_2, \ldots, v_k)$ such that for $1 \leq i < k$, $(v_i, v_{i+1}) \in E$. A *(simple) cycle* c *of* G is a path such that $v_1 = v_k$ with $k > 1$. A graph is *acyclic* if it has no cycles. A graph is *connected* if for every pair of vertices $u, v \in V$, there is a path from u to v. For any $k > 0$, a graph is *k-connected* if the removal of any $k - 1$ vertices from the graph still leaves the graph connected. We often refer to 2-connected graphs as *biconnected* and 3-connected graphs as *triconnected*.

We may also define many of our terms based on giving each edge a specific direction. A *directed graph* (*digraph*) is a graph where each *directed edge* $e = (u, v)$ is an ordered pair, where we consider u to be the *origin* and v to be the *destination* of the edge. In addition, $e = (u, v)$ is an *incoming edge* of v and an *outgoing edge* of u. The *indegree* of a vertex v is the number of its incoming edges, and the *outdegree* of a vertex v is the number of its outgoing edges. A *source* is a vertex with no incoming edges, i.e., with indegree 0. A *sink* is a vertex with no outgoing edges, i.e., with outdegree 0. A *directed path* of G is a path of G, $(v_1, v_2, \ldots, v_k)$, such that for $1 \leq i < k$, (v_i, v_{i+1}) is a directed edge in E. A *directed acyclic graph* (*DAG*) is a directed graph that has no cycles.

A *drawing* Γ of a graph $G = (V, E)$ is essentially a mapping of each vertex $v \in V$ to a distinct point $\Gamma(v)$ and of each edge $e = (u, v) \in E$ to a simple open Jordan curve $\Gamma(e)$, which has $\Gamma(u)$ and $\Gamma(v)$ as its endpoints. If G is directed, it is common to draw the edge with an arrow toward the destination vertex. When the drawing is understood from the context, we often leave out the Γ notation. For example, we may say that an edge e is made of horizontal and vertical segments rather than the drawing $\Gamma(e)$.

A *planar graph* is a graph G that admits a *planar drawing* Γ, a drawing with no edges intersecting, except for edges that share a common vertex v and only at that vertex. A *planar embedding*, or, simply, *embedding*, of a graph is the collection of (counter-clockwise) circular orderings of incident edges around every vertex induced by a planar drawing. A *plane graph* is a graph that has been associated with a specific planar embedding. A *maximal planar graph* is a graph where the addition of any edge $e \notin E$ causes the graph to be non-planar. Maximally planar graphs have the property that every face is a triangle, a cycle of three edges. For notation, we often refer to planar graphs with maximum degree k as k-*planar graphs*, in particular, we deal with many cases of 4-planar graphs.

A *straight-line drawing* of a graph is a drawing where every edge is a straight-line segment. A *polyline drawing* of a graph is a drawing Γ such that every edge $e = (u, v) \in E$ is represented as a connected sequence of line segments $\overline{p_1 p_2}, \overline{p_2 p_3}, \ldots, \overline{p_{k-1} p_k}$, where $p_1 = \Gamma(u)$ and $p_k = \Gamma(v)$ are the endpoints of the edge. We refer to $p_2, \ldots, p_{k-1}$ as *bend points* of the drawing of the edge. An *orthogonal drawing* of a graph is a polyline drawing where every edge is an alternating sequence of horizontal and vertical line segments. A *grid drawing* is a drawing of the graph where each vertex and each bend point has integer coordinate values, effectively being placed on an integer grid. The *area of a grid drawing* is the area of the smallest enclosing axis-aligned rectangle containing the drawing. For a given drawing of G, the *angular resolution of a vertex* v is the smallest angle between two distinct edges incident to v and the *angular resolution* of G is the minimum angular resolution among all vertices.

An *st-graph* is a DAG with one source and one sink. A *planar st-graph* is an *st*-graph that has a planar embedding with the source s and sink t located on the external face.

DEFINITION 7.1 Given a planar *st*-graph G, the *dual planar st-graph* $G^* = (V^*, E^*)$ is a digraph with the following properties:

- V^* is the set of faces in G with the addition that the external face $(s, \ldots, t, \ldots, s)$ is broken into two parts s^* representing the portion of the face from s to t and t^* representing the portion from t to s.

- For every edge $e \in E$, we have an edge $e^* = (f, g) \in E^*$ where f is the face to the left of e and g is the face to the right of e.

In the construction of an orthogonal drawing of a graph G discussed in Sections 7.2.3 and 7.3.1, the dual graph coupled with the following special ordering of vertices play a critical role in the creation of an intermediate visibility representation of G.

DEFINITION 7.2 Let $G = (V, E)$ be a directed acyclic graph. A *topological ordering* $T(G)$ is an assignment of integer values $T(v)$ to each vertex $v \in V$ such that for every directed edge $(u, v) \in E$, we have that $T(u) < T(v)$. The *size of the topological ordering* $s(T)$ is $\max_{v \in V} T(v) - \min_{u \in V} T(u)$. An *optimal topological ordering* $T^*(G)$ is a topological ordering with the smallest size, $s(T^*) = \min_{T(G)} s(T)$.

Require: $G = (V, E)$ be a Directed Acyclic Graph
Ensure: $T(G)$ is an optimal topological ordering
    {Compute the indegree for every vertex}
    for all $v \in V$ **do**
      $in(v) \leftarrow 0$
    end for
5: **for all** $(u, v) \in E$ **do**
      increment $in(v)$
    end for
    {Identify all sinks}
    $S_0 \leftarrow \emptyset$
10: **for all** $v \in V$ **do**
      if $in(v) = 0$ **then**
        $S_0.\text{add}(v)$
      end if
    end for
15: $n \leftarrow 0$
    repeat
      {Mark all current sinks and remove them from G}
      $S_{n+1} \leftarrow \emptyset$
      for all $v \in S_n$ **do**
20:     $T(v) \leftarrow n$
        for all $(v, u) \in E$ **do** {remove v from the graph}
          decrement $in(u)$
          if $in(u) = 0$ **then** {u is a new sink}
            $S_{n+1}.\text{add}(u)$
25:       **end if**
        end for
        increment n
      end for
    until S_n is empty {No more sinks}

Figure 7.1 Algorithm for computing an optimal topological ordering of a DAG.

In our definition, it is possible for two vertices u and v to have the same value if there is no directed path between u and v. Note, this is basically a partial ordering where the optimal size is the length of the longest chain in the partial order. Topological orderings are discussed in most standard graph and algorithms textbooks. See, for example, [CLR90, GT02].

Computing an optimal topological ordering in linear time is fairly straightforward. We assign every sink vertex a number 0, remove these vertices and their edges from the graph, and repeat the process with a number one larger until there are no vertices left. Figure 7.1 describes the process in more detail.

This common algorithm proves useful for the construction of orthogonal graphs via a visibility representation. However, there are other more difficult, but equally useful, orderings. We next discuss one such ordering, the canonical ordering.

7.2.2 Canonical Ordering and Shifting Sets

In [dFPP90], de Fraysseix, Pach, and Pollack describe a technique for embedding a plane graph on a grid. Their technique uses an incremental approach that is built around a particular ordering of the vertices known as a canonical ordering. Initially defined for maximal plane graphs, Kant [Kan96] later extended it to triconnected plane graphs and Gutwenger and Mutzel [GM98] to biconnected plane graphs. In this section, we define and describe the canonical ordering of [dFPP90, CDGK01] as well as the shifting sets derived from this ordering, which are needed in the polyline drawing method described in Section 7.4.

DEFINITION 7.3 Let G be a maximal plane graph on m vertices. Let $\pi = (v_1, v_2, \ldots, v_n)$ be an ordering of the vertices of G. For $1 \le k \le n$, let G_k be the plane subgraph of G induced by the vertices of $v_1, \ldots, v_k$ and let $C_k = (v_1 = w_1, w_2, \ldots, w_m = v_2)$ be the cycle forming the external face of G_k. We call π a *canonical ordering* of G if

1. v_1, v_2, and v_n are the external vertices of G in counter-clockwise order,

2. for $2 < k < n$, G_k is 2-connected and internally maximal, i.e., every internal face is a triangle, and

3. for $2 < k < n$, v_k is a vertex of C_k and has at least one neighbor in $G - G_k$.

de Fraysseix, Pach, and Pollack [dFPP90] prove the following theorem, which was later extended to triconnected plane graphs by Kant [Kan96]:

Theorem 7.1 *Every maximal plane graph has a canonical ordering that can be found in linear time and space.*

The canonical ordering has the property that all of the neighbors of v_{k+1} in G_{k+1} lie on C_k. Intuitively, the ordering is constructed in reverse order by starting with the initial external triangular face and repeatedly removing a vertex $v_{k+1} \notin \{v_1, v_2\}$ that has at most two neighbors on C_{k+1} creating the new graph G_k and external face C_k. See Figure 7.2.

Once constructed, the canonical ordering π leads to an incremental approach for constructing a drawing of G. Here, we start with the triangle v_1, v_2, v_3 and repeatedly add the next vertex v_{k+1} to the graph of G_k by adding edges for v_{k+1} to its neighbors in C_k forming G_{k+1} and C_{k+1}. The vertices of C_k that are no longer on C_{k+1} are said to be *covered* by v_{k+1}. Since the neighbors of v_{k+1} are all continuous on the cycle C_k, we can label them as $w_l, w_{l+1}, \ldots, w_r$. We refer to the two vertices w_l and w_r as the leftmost and rightmost neighbors of v_{k+1} in C_k. Since all the neighbors of v_{k+1} except the leftmost and rightmost neighbors are covered by v_{k+1}, we know that the cycle $C_{k+1} = (v_1 = w_1, w_2, \ldots, w_l, v_{k+1}, w_r, \ldots w_m = v_2)$. See Figure 7.3.

Starting with de Fraysseix, Pach, and Pollack, several authors have used this canonical ordering (or a variant) to build a graph incrementally. However, to place the vertices effectively, onto a grid location for example, one must also repeatedly shift the vertices in G_k to create a proper location for v_{k+1}. Typically, the approach is to increase the space between the leftmost and rightmost neighbors of v_{k+1}. However, shifting these two vertices also forces other vertices to shift to avoid creating edge crossings.

To solve the problem of determining which vertices must shift together, we also define a shifting set associated with each vertex on the current external face. See Cheng et al. [CDGK01].

DEFINITION 7.4 For a given canonical ordering $\pi = (v_1, v_2, \ldots, v_n)$, we define for $3 \le k \le n$, the *shifting set* $M_k(w_i) \subseteq V$ for each vertex $w_i \in C_k$ on the external face of G_k as follows. $M_3(v_3) = \{v_3\}$, $M_3(v_2) = M_3(v_3) \cup \{v_2\}$, $M_3(v_1) = M_3(v_2) \cup \{v_1\}$. For $3 \le k < n$, let w_l and w_r be the leftmost and rightmost neighbors of v_{k+1} in C_k. Then,

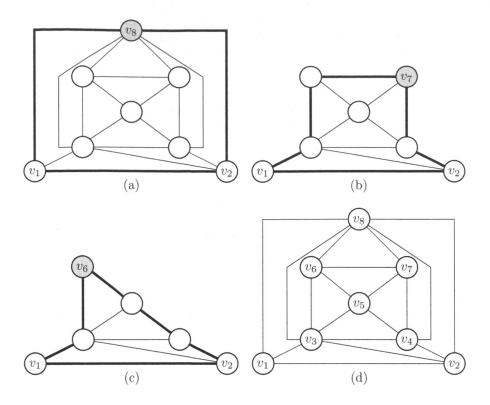

Figure 7.2 An illustration showing the creation of the canonical ordering in reverse order. (a) The first vertex v_8 is about to be removed with the external cycle highlighted. (b) Removal of the next vertex, v_7. (c) Removal of vertex v_6. (d) The final canonical ordering of the vertices.

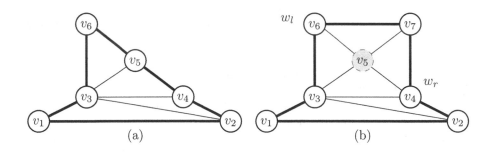

Figure 7.3 Inserting a vertex using the canonical ordering. This example does not follow the vertex placement techniques employed by the standard algorithms used to produce good area drawings. (a) The graph G_6, with its external cycle C_6 drawn in bold. (b) The graph G_7 after inserting vertex v_7. The covered vertex v_5 is lightened. The leftmost and rightmost neighbors are $w_l = v_6$ and $w_r = v_4$. The new external cycle C_7 is therefore $(v_1, v_3, v_6, v_7, v_4, v_2)$.

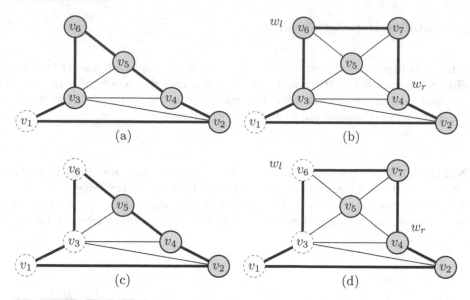

Figure 7.4 The incremental construction of a shifting set. The vertices for each set shown are highlighted. (a) The shifting set for $M_6(v_3)$. (b) After inserting v_7, the shifting set for $M_7(v_3)$. This simply merges in the new vertex. (c) The shifting set for $M_6(v_5)$. (d) After inserting v_7, the shifting set for $M_7(v_7)$. Since $w_{l+1} = v_5$, this set is the union of $M_6(v_5)$ and v_7.

- for $i \leq l$, $M_{k+1}(w_i) = M_k(w_i) \cup \{v_{k+1}\}$,
- for $j \geq r$, $M_{k+1}(w_j) = M_k(w_j)$, and
- $M_{k+1}(v_{k+1}) = M_k(w_{l+1}) \cup \{v_{k+1}\}$.

From this definition, one can show that the following properties of the shifting set hold for all $3 \leq k \leq n$ for the incremental drawing algorithms described in Section 7.4:

1. $w_j \in M_k(w_i)$ iff $j \geq i$,
2. $M_k(w_1) \supset M_k(w_2) \supset \cdots \supset M_k(w_m)$,
3. For $1 \leq i \leq m$ and a planar drawing of G_k, if we shift all vertices in $M_k(w_i)$ by distance $\delta_i \geq 0$ to the right, then the resulting drawing of G_k remains planar.

In other words, the shifting set for a vertex w_i on the external face is just the set of all vertices that need to be shifted to the right to maintain planarity if w_i is shifted to the right.

Note that $M_{k+1}(w_i)$ is undefined for $l < i < r$, since these covered vertices are no longer on the external face. See Figure 7.4.

A careful examination of the set reveals that a vertex w_i that is covered by v_{k+1} shifts by δ units if and only if v_{k+1} shifts by δ units. That is, for $k' > k$, $w_i \in M_{k'}(v)$ iff $v_{k+1} \in M_{k'}(v)$. This property of the shifting set is exploited during the incremental embedding algorithms that use a canonical ordering to ensure that shifts do not produce crossings.

7.2.3   Visibility Representations

Orthogonal drawings, and even general drawings, of planar graphs often start by computing a visibility representation of the graph. Before going into the details of using a visibility representation to compute an orthogonal drawing, presented in Section 7.3.1, we first explain the general approach of computing such a representation.

DEFINITION 7.5   Given a graph $G = (V, E)$, a *visibility representation* Γ, for G maps every vertex $v \in V$ to a horizontal *vertex segment* $\Gamma(v)$ and every edge $(u, v) \in E$ to a vertical *edge segment* $\Gamma(u, v)$ such that each vertical edge segment $\Gamma(u, v)$ has its endpoints lying on the horizontal vertex segments $\Gamma(u)$ and $\Gamma(v)$ and no other segment intersections or overlaps occur.

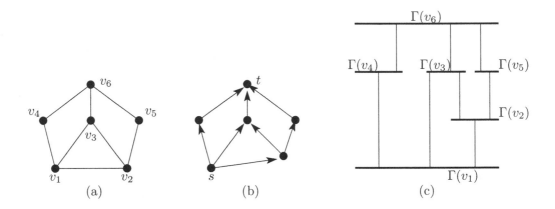

Figure 7.5   (a) A simple graph G (b) An st-ordering of G (c) A visibility representation of G.

See Figure 7.5 for one example of a visibility representation. Otten and van Wijk [OvW78] introduced the visibility representation. With varying improvements, several researchers have proved that every planar graph has such a representation, which can be found in linear time [OvW78, DHVM83, RT86, TT86]. In general, we have the following theorem about computing a visibility representation:

Theorem 7.2   *[TT86] A graph admits a visibility representation if and only if it is planar. Furthermore, a visibility representation for a planar graph can be constructed in linear time.*

Figure 7.6 describes an algorithm to compute the visibility representation of a given graph. After making the graph biconnected by adding dummy edges [FM98], we compute an st-ordering on the graph creating a planar st-graph and its dual graph G^*. The location of the vertex-segments and edge-segments are then determined by a topological ordering of the st-graph and its dual with the former serving to determine y-values and the latter to determining x-values. Figure 7.7 shows an example construction.

Require: $G = (V, E)$ be a plane graph
Ensure: Γ is a visibility representation of G on the integer grid of size $O(n^2)$
    Make G biconnected by adding "dummy" edges {See [FM98]}
    Select an edge (s, t) on the external face
    Compute a planar st-graph on G {For simplicity, we refer to it as G}
    Create the dual planar st-graph G^*
5: Compute the optimal topological ordering $T_x = T(G^*)$ {See Figure 7.1}
    Compute the optimal topological ordering $T_y = T(G)$
    for all $v \in V$ **do** {Assigning positions to the horizontal vertex segments}
        Let f_l be the face to the left of the leftmost outgoing edge of v
        Let f_r be the face to the right of the rightmost outgoing edge of v
10:       $\{f_l$ and f_r are vertices in the dual graph $G^*\}$
        $\Gamma(v).y \leftarrow T_y(v)$
        $\Gamma(v).\mathsf{xmin} \leftarrow T_x(f_l)$
        $\Gamma(v).\mathsf{xmax} \leftarrow T_x(f_r) - 1$
    end for
15: **for all** $e = (u, v) \in E$ **do** {Assigning positions to the vertical edge segments}
        Let f_l be the face to the left of e $\{f_l$ is a vertex in $G^*\}$
        $\Gamma(e).x \leftarrow T_x(f_l)$
        $\Gamma(e).\mathsf{ymin} \leftarrow T_y(u)$
        $\Gamma(e).\mathsf{ymax} \leftarrow T_y(v)$
20: **end for**
    Remove any added "dummy" edges

Figure 7.6 Algorithm for constructing a visibility representation of a plane graph.

7.2.4 Network Flows

Network flows, useful in many areas of graph theory and graph drawing, are particularly useful in finding drawings of orthogonal graphs with a minimum number of bends. We describe this use in Section 7.3.2. Beforehand, we discuss the general structure of a network flow, borrowing notation from Goodrich and Tamassia [GT02].

A *(single-source single-sink) flow network* N is a connected directed graph of *arcs* and *nodes*[1] with the following properties:

- Each arc e has a positive integer *capacity* $c(e)$ and a nonnegative integer *cost* $w(e)$;
- There exists a *source* node, s, such that s has no incoming arcs;
- There exists a *sink* node, t, such that t has no outgoing arcs;
- All other *non-terminal* nodes have at least one incoming and one outgoing arc.

Figure 7.8(a) shows one particular flow network. The network is viewed as transporting some *commodity* from the source to the sink by flowing along the arcs. A *flow f* for some network N is an assignment to each arc e of some (integer) *flow value* $f(e)$ such that the following two rules apply:

- *Capacity rule*: The (positive) flow for each arc does not exceed the capacity. For each arc $e \in N$, $0 \le f(e) \le c(e)$.

---

[1]We use the terms *arc* and *node* for a flow network instead of the analogous terms edge and vertex to help differentiate between a flow network and a graph, which is to be drawn using the flow network.

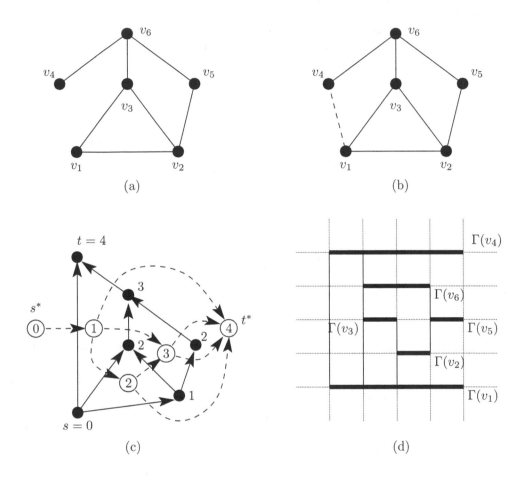

Figure 7.7 (a) A simple graph G. (b) G after augmenting to make it biconnected. (c) The st-planar graph of G (solid) and the dual graph G^* (dashed). The two topological orderings from these graphs are shown labeled by their nodes. (d) The visibility representation of G computed from these orderings.

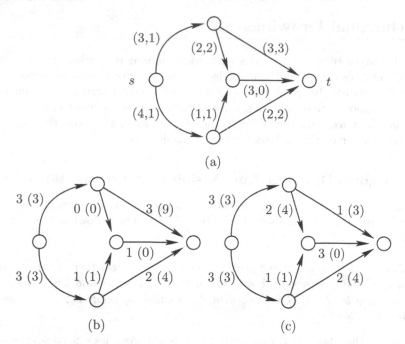

(a)

(b)   (c)

Figure 7.8 (a) A (single-source single-sink) flow network N with arcs labeled with the pair $(c(e), w(e))$ (capacity, cost). (b) A maximum flow of value 6 for flow network N. Each arc of N is labelled with its flow and, in parentheses, the cost of the flow on that arc. The total cost of this flow is 20. (c) A minimum-cost maximum flow for N. Note, the value of this flow is still 6 but the cost is now 18.

- *Conservation rule*: The flow coming *in* to a non-terminal node is the same as the flow going *out* of the node.

  For each non-terminal node $v \in N$, with $v \neq s, t$,

$$\Sigma_{e \in \mathsf{inarc}(v)} f(e) = \Sigma_{e \in \mathsf{outarc}(v)} f(e).$$

The *value* of the flow $v(f)$ is the total flow leaving the source node, which because of the conservation rule is the same as the flow entering the sink node. That is, $v(f) = \Sigma_{e \in \mathsf{outarc}(s)} f(e)$. For a given flow f, the cost of the flow on a given arc e is the cost of the arc $w(e)$ times the amount of flow on that arc $f(e)$. The *cost* of the flow $w(f)$ is the sum of the costs of each arc. That is, $w(f) = \Sigma_{e \in N} w(e) f(e)$.

The *maximum flow problem for N* is to find a flow f^* with maximum value among all possible flows of N. The *minimum-cost flow problem for N* is to find the minimum cost flow among all possible maximum flows in N. Figure 7.8 shows a maximum flow that does not have minimum cost as well as a minimum-cost maximum-flow solution.

There are several methods for solving flow networks, which are beyond the scope of this chapter. Their running times often depend on combinations of the number of nodes in the network, the capacity of the edges in the network, and the cost of the edges in the network. For details, see [CLR90, GT02].

Of particular relevance are minimum cost flow algorithms with running time that depends on the value of computed flow [CK12, GT97]. We use such an algorithm in Section 7.3.2 to compute a planar orthogonal drawing with the minimum number of bends.

7.3   Orthogonal Drawings

One highly effective way to draw graphs with good angular resolution is to use only edges that are rectilinear, or orthogonal. Such edges consist of alternating sequences of vertical and horizontal segments. In graph representations where each vertex is a point and where two edges are not allowed to overlap, a necessary condition for a graph to have an orthogonal drawing is that the maximum vertex degree be at most four. However, the introduction of rectangular regions for vertices allows for larger graph degrees.

7.3.1   Orthogonal Drawings from Visibility Representations

Given a 4-planar graph $G = (V, E)$, one can construct a good *orthogonal drawing* using the *visibility representation* discussed in Section 7.2.3. The following theorem is due to Tamassia and Tollis [TT89]:

Theorem 7.3   *Let G be a 4-plane graph. If G is biconnected, there exists an orthogonal grid drawing of G using $O(n^2)$ area with at most $2n + 4$ bends and where only two edges have more than two bends. If G is connected, the number of bends is $2.4n + 2$ and no edge has more than four bends.*

The version of the algorithm used to prove this theorem uses a *constrained visibility representation*. The additional constraint is that each (horizontal) vertex segment other than the source and sink have two (vertical) edge segments incident to its leftmost endpoint, with one being above and the other below the vertex segment. We describe the simpler, but slightly less effective, algorithm that uses a regular visibility representation. First, we compute a visibility representation $\Gamma(G)$. For each vertex $v \in V$, place the vertex at a single point on the horizontal vertex segment $\Gamma(v)$, determined below. The routing of the edges incident to v and the location of v on the vertex segment are based on various cases. Since each vertex has at most 4 incident edges and accounting for symmetry and subcases with smaller vertex degrees, Figure 7.9 shows the six possible cases along with the resulting edge routings and vertex placements. A careful study of the cases shows that no edge has more than two bends per endpoint, resulting in no more than four bends total. This creates an orthogonal shape, discussed in the next section, for G. To help improve the size and number of bends one can do a few heuristics to straighten out various edges. Finally, using the compaction technique described in the next section or similar more efficient techniques, one can convert the orthogonal shape into an orthogonal drawing using the smallest area. Figure 7.10 shows an example of an orthogonal drawing constructed from a visibility representation.

7.3.2   Network Flow Algorithms

Tamassia [Tam87] showed that by using a network flow algorithm one could construct orthogonal drawings of embedded 4-planar graphs with a minimum number of bends.

The fact that the graph is given with its embedding is significant. Formann et al. [FHH+93] and Garg and Tamassia [GT01] showed that the problem of determining whether a drawing with no bends exists is NP-hard for 4-planar graphs. The strategy in their proof deals with the difficulty of assigning an order of the edges around vertices of degree 4. It is interesting to note that the problem is polynomial when the maximum degree is 3 [DLV98].

Tamassia's algorithm originally ran in $O(n^2 \log n)$ time. However, an improvement for certain types of planar flow networks (see Section 7.2.4) presented by Garg and Tamas-

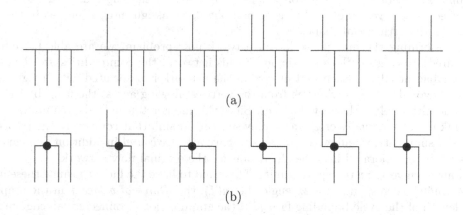

(a)

(b)

Figure 7.9 (a) The six possible cases for horizontal vertex segments intersecting with its 4 incident vertical edge segments in a visibility representation, accounting for symmetry. (b) The vertex placement and edge routings for each of the cases.

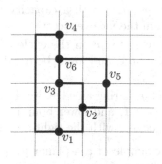

Figure 7.10 An orthogonal drawing from the visibility representation of Figure 7.7.

sia [GT97] reduced the running time to $O(n^{7/4}\sqrt{\log n})$. Recently, Cornelsen and Karren-bauer obtained a running time of $O(n^{3/2} \log n)$ [CK12].

Let $G = (V, E)$ be an embedded planar graph having maximum degree 4. We can compute a drawing of G with the minimum number of bends in two phases. First, we compute an *orthogonal shape* for G. Here we only define the bends of the edges and angles between adjacent edges at a vertex of G. In the second phase, we assign integer lengths to the edge segments of the orthogonal shape.

By transforming the first phase into a network flow problem, we are able to compute the required drawing's orthogonal shape. In this network, the commodities are the angles between adjacent edges. Each unit of flow in the network is associated with a right angle in the orthogonal shape, originating from the vertices, flowing across the faces by the edge bends, and ultimately sinking at the faces. Since this interpretation leads to a multi-source, multi-sink flow, we actually create a dummy source and sink that connect to the respective nodes. For simplicity, we allow certain arcs to have a *lower bound* in addition to a capacity. This is easily incorporated into the algorithms for the original flow network.

We want each vertex v to supply 4 units of flow and to have the faces consume these units. Here, 4 "units" correspond to a 2π angle. Let $d(f)$, the *degree of a face* f in the graph G, be the length of the cycle bounding face f. If the graph is not biconnected, an edge may be counted twice on the same face. The *consumption rate* of each face is designated by $\sigma(f)$ with

$$\sigma(f) = \begin{cases} 2d(f) - 4 & \text{if } f \text{ is an internal face} \\ 2d(f) + 4 & \text{if } f \text{ is the external face} \end{cases}$$

From Euler's formula, we know that $\Sigma_f \sigma(f) = 4n$, which is the total number of units supplied by the vertices. Our network N has three types of nodes and four types of arcs with the following described attributes:

- Non-terminal nodes correspond to the vertices and faces of G;
- A source node s and sink node t serve to supply and consume the commodity;
- For every vertex v, arcs of type (s, v) with a capacity of 4, cost 1, and lower bound 4 act to supply the vertex v with its commodity;
- For every face f, arcs of type (f, t) with a capacity of $\sigma(f)$ and cost 1 act to consume the commodity from the face vertices;
- From every face f and every vertex v on the cycle of f, we use an arc of type (v, f) with a capacity of 4, cost 1, and lower bound 1. This arc flow represents the angle at vertex v in face f;
- For every pair of faces f and g sharing an edge, we designate an arc of type (f, g) having a capacity of $+\infty$, cost 1, and lower bound 0. This arc flow represents the number of bends along edge e with the right angle *inside* of the face f.

Figure 7.11 shows a detailed example of a 4-planar graph, its network model, and the minimum cost solution. We now take a closer look at an interpretation of the network from the source side. At every vertex v the network supplies the vertex with 4 units, all of which must, by the conservation rule, flow across the (v, f) arcs. Since each unit corresponds to $\pi/2$ radians, this guarantees that the sum of the angles around a vertex, which is equivalent to the sum of the flow leaving v along these arcs, is 2π.

From the sink side, by the conservation rule, we know that the sum of the units at the vertices and the bends of a face is equal to $2d(f) - 4$ units for an internal face and $2d(f) + 4$

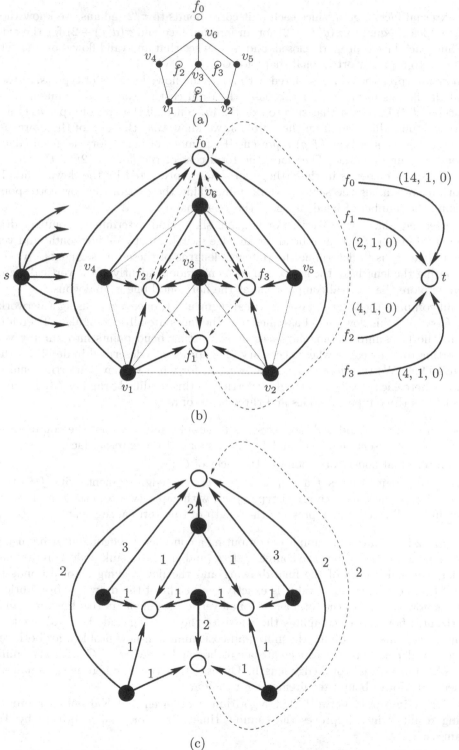

Figure 7.11 (a) A simple planar graph G with maximum degree 4. (b) The network N associated with G. The arcs from the source s to the vertex nodes have label $(4, 1, 4)$, i.e., capacity 4, cost 1, and a lower bound of 4. The vertex to face arcs are drawn as solid lines with label $(4, 1, 1)$. The face to face arcs are drawn bi-directional with both directions having label $(+\infty, 1, 0)$. (c) A minimum cost max flow with the arc labels reflecting the flow. Some vertices are omitted and some edges are partially drawn for better readability.

for the external face. Again, since each unit corresponds to $\pi/2$ radians, we know the sum of these angles is equal to $\pi(d(f) - 2)$ for an internal face and $\pi(d(f) + 2)$ for the external face. Thus, each face is properly closed, and we can see that any valid flow ϕ on the network corresponds to a proper orthogonal shape for G.

We now interpret the cost associated with a specific flow. For arcs of type (s, v) the cost is 1 and the flow is fixed. So, for this case, the total cost is exactly $4n$. Similarly, all the arcs of type (f, t) have cost that sum to exactly $4n$. Since all the arcs of type (v, f) have to release the commodity sent from the source s, we know that the sum of these arcs is also $4n$. Finally, the arcs of type (f, g) represent the number of bends for the given edge with each bend costing one unit. Therefore, the total cost of the flow is $12n + B$, where B is the total number of bends in the orthogonal shape represented by the flow. Since $12n$ is fixed for all flows along the same network, minimizing the cost of the flow corresponds to minimizing the number of bends in the orthogonal shape.

In the second phase, we take this orthogonal shape and determine a compact drawing for the actual graph. Since each bend for an edge switches between horizontal and vertical lines, our strategy is to determine the (integer) lengths of these line segments. We do this by computing the lengths of the horizontal segments independently of the vertical segments. We shall explore the vertical computation as the horizontal one is analogous.

We can compute the length of each vertical segment by, once again, using a network flow model. However, this flow model assumes that the faces are all rectangular. Therefore, we first split the faces into rectangular faces by converting bend points into dummy vertices and inserting dummy edges where necessary. This process is described in detail in Chapter 5 of [DETT99]. We therefore explain the solution for when we have an orthogonal representation where each face is a rectangle, referring to this modified graph as G'. In this case, our model has three types of nodes and three types of arcs.

- A source node s and sink node t serve to supply and consume the commodity and also represent the "left" and "right" regions of the external face;

- Non-terminal nodes correspond to the faces of G';

- For every pair of faces f and g sharing a *vertical* edge segment, with f to the left of g, we designate an arc of type (f, g), with capacity $+\infty$, cost 1, and lower bound 1. The arc flow represents the length of this vertical segment.

Figure 7.12 illustrates an example of computing a compact orthogonal drawing using this network flow approach. Since the source node s (and similarly sink node t) represents the entire left vertical border of the final drawing and the flow leaving s corresponds to the height of this border, the flow value is exactly the height of the drawing. In addition, the cost of the flow is equal to the total length of all vertical segments in the drawing. Similarly, the horizontal flow model computes the width of the drawing and the total length of all horizontal segments. By solving the minimum-cost minimum-flow problem for both vertical and horizontal networks, we can create an orthogonal drawing of G with the minimum height, width, area, and total edge length. Observe that the flow here is the smallest flow that meets the lower bound requirements for each arc.

Using their improved network flow algorithm, Cornelsen and Karrenbauer proved the following result, which improves the running time of the original algorithm by Tamassia [Tam87]:

Theorem 7.4 *[CK12] Let G be an embedded 4-planar graph with n vertices. A planar orthogonal drawing of G with the minimum number of bends can be computed in $O(n^{3/2} \log n)$ time.*

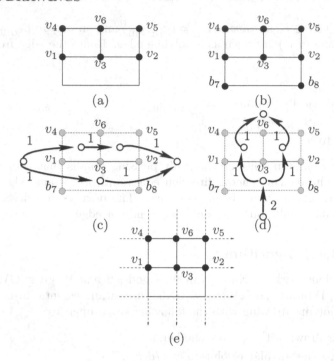

Figure 7.12 (a) An orthogonal drawing with the orthogonal representation described by Figure 7.11c. (b) The same drawing with the two bend points temporarily converted to vertices so that each face is rectangular. (c) The network flow for computing the vertical segments along with the solution. (d) The network flow for computing the horizontal segments along with the solution. (e) The final compact solution with the horizontal and vertical segments determined from the two flows and the inserted dummy vertices removed.

7.4 Polyline Drawings

When one wishes to draw planar graphs having maximum degree more than 4 with good angular resolution and with vertices as single points, clearly orthogonal drawings do not suffice. There have been various other approaches to creating planar polyline drawings with good angular resolution, many of these results extend the work of Kant [Kan96], including work by Goodrich and Wagner [GW00], Gutwenger and Mutzel [GM98], Cheng et al. [CDGK01], and Duncan and Kobourov [DK03]. The general approach is to use an incremental insertion method to add vertices one at a time using a canonical ordering and continually maintain the proper angular resolution qualities and other specific restrictions.

7.4.1 Mixed-Model Algorithm

The approach of Gutwenger and Mutzel [GM98] is similar to the approaches taken by [GW00, CDGK01, DK03], which are discussed in the next subsection. However, unlike those approaches which rely on the graph being either maximal, tri-connected, or having artificial edges added to make them maximal, the approach by Gutwenger and Mutzel uses an ordering that is defined for biconnected graphs. The benefits are significant in the sense that such artificial edges, once removed, often create unexpected artifacts. In their *mixed-model algorithm*, they take a given biconnected plane graph $G = (V, E)$, and using this new ordering, assign for each edge $e \in E$, an *inpoint* $e_{in} = (x_{in}, y_{in})$ and an *outpoint* $e_{out} = (x_{out}, y_{out})$. Then each edge $e = (v, w)$ is drawn as a polyline edge. Route the edge from v to w in the following manner:

- from v to e_{out},
- from e_{out} vertically to point $b = (x_{out}, y_{in})$,
- from b horizontally to e_{in},
- and finally to w.

This approach results in quite aesthetically pleasing graphs that combine a mixture of good angular resolution via general direction edges and orthogonal edges. However, the results require, in general, three bends per edge. The next section describes a technique that achieves similar results but with only one bend per edge.

7.4.2   One Bend Algorithm

Building off previous work by Kant [Kan96], Goodrich and Wagner [GW00], and Cheng et al. [CDGK01], Duncan and Kobourov [DK03] use an incremental insertion approach to create a planar polyline drawing with the following key properties:

- each edge is drawn with at most one bend;
- each vertex v has angular resolution $\Theta(1/d(v))$;
- all vertices and bend points lie on an $O(n) \times O(n)$ grid.

The incremental approach uses the canonical ordering and the shifting set described in Section 7.2.2.

7.4.3   Vertex Regions

In [dFPP90], de Fraysseix, Pach, and Pollack present an algorithm to draw an n-vertex plane graph with straight-line edges on an $O(n) \times O(n)$ integer grid. Chrobak and Payne [CP95] show how to implement the algorithm in linear time. In this algorithm, each new vertex v_{k+1} is inserted above its neighbors $w_l, \ldots, w_r$, and after proper shifting, edges are drawn as straight-line segments from the location of v_{k+1} to each neighbor of v_{k+1}. In the approach used in [GW00, CDGK01], each vertex is associated with a diamond-shaped region where edges are routed through ports along the boundary of the region before connecting to the vertices. This creates bends in the edges but allows better control over the angles that are formed by the edges around vertices. To reduce the overall grid size, Duncan and Kobourov [DK03] use slightly altered vertex regions. Each vertex is surrounded by six vertex regions of two types, free regions and port regions, which alternate around the vertex. The regions are bounded by rays extending from v in various directions, with $0°$ indicating a positive vertical direction. See Figure 7.13.

DEFINITION 7.6   Let $v \in V$ have degree $d = d(v)$. The *vertex regions* associated with v are of two types, *free regions* and *port regions*. Free regions have the property that only one edge extends from v to another vertex through that region. Port regions are bounded on one side by a horizontal or vertical line segment with a number of (integer coordinate) ports, and each edge going through a port region of v from v to any other vertex passes through a unique port. Moreover, every edge is drawn as two line segments. The first, starting at one endpoint v, connects to a port in the port region of v, and the second connects from

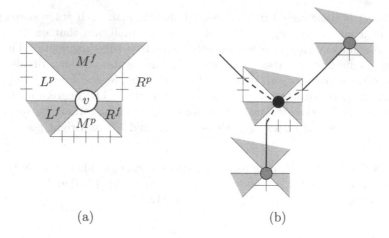

(a)   (b)

Figure 7.13   (a) The vertex regions around a particular vertex v. Notice that each port region can have a different number of ports. (b) Edges extending from a (darkened) vertex. The port edge segment is drawn dashed and the free edge segment is drawn solid.

that port to the other vertex w passing through one of w's free regions. The six regions associated with v are defined as follows:

- Free region M^f lies between $-45°$ and $45°$;
- Free region R^f lies between $90°$ and $135°$;
- Free region L^f lies between $-135°$ and $-90°$;
- Port region M^p lies between L^f and R^f;
- Port region L^p lies between L^f and M^f; and
- Port region R^p lies between R^f and M^f.

The algorithm proceeds similar to the standard embeddings that use the canonical ordering. In particular, one starts with an initial face v_1, v_2, v_3 and then repeatedly inserts the next vertex v_{k+1} by finding its leftmost and rightmost neighbors, w_l and w_r, on the current external face shifting the space between these vertices so that the lines connecting v_{k+1} to w_l and w_r intersect at a grid location. To ensure good angular resolution, one must introduce some bends, which requires a slight alteration in the approach.

Except for the initial horizontal edge (v_1, v_2), we route each edge (v_i, v_j) through a port of one of the two vertices. In the process, each edge consists of two edge segments. One segment, the *port segment*, extends from v_i to one of v_i's ports, lying entirely in one of v_i's port regions. The other, *free segment*, extends from this port to v_j passing through one of v_j's free regions. See Figure 7.13(b).

The ports are arranged in such a way that the angle between successive ports and v is $O(1/d(v))$. By Definition 7.6, since for every vertex v each free segment associated with v lies inside a free region boundary, each free region has exactly one free segment passing through it, each port segment associated with v lies inside a port region and passes through a unique port, the resulting angular resolution at v is $O(1/d(v))$. For compactness, port segments, which are essentially bend points, can also coincide with the destination vertex, effectively creating a free edge segment of zero length. That is, if we have an edge (u, v) that goes through u's port p, we may have a situation where p coincides with v. This is not necessary but allows for smaller grid size in the end.

The embedding is constructed in incremental stages, with each stage corresponding to the insertion of a new vertex v_{k+1}. At each stage, we maintain that each vertex except those on the current external face has exactly three free edge segments. The remaining edge segments connect to a vertex v through port segments. We can divide the current degree of v into three parts: $d_l(v)$, $d_r(v)$, and $d_m(v)$. The degree $d_l(v)$ corresponds to the current number of port edge segments using the L^p region. The degrees $d_r(v)$ and $d_m(v)$ are defined similarly for the R^p and M^p regions. At each insertion, we route port edge segments involving the new vertex v_{k+1} through maximal left and right ports.

DEFINITION 7.7   Let a vertex v have coordinates (v_x, v_y). Then, the *maximal left port of* v, $L_{\max}^p(v)$, has coordinates $(v_x - d_l(v) + 1, v_y + d_l(v))$ if $d_l(v) > 0$ and (v_x, v_y) otherwise. Define the *maximal right port of* v, $R_{\max}^p(v)$, similarly.

7.4.4   The Embedding

Initially, the first three vertices have integer coordinates $v_1 = (0, 0)$, $v_2 = (4, 0)$, and $v_3 = (2, 1)$. In subsequent stages, we insert the next vertex v_{k+1} maintaining the following invariants:

- All vertices and ports lie on the integer grid.
- Let $C_k = (w_1 = v_1, w_2, \ldots, w_m = v_2)$ be the exterior face of G_k with $w_i(x)$ corresponding to the x-coordinate of w_i. Then $w_1(x) < w_2(x) < \ldots < w_m(x)$. In other words, the vertices of the exterior face are strictly x-monotonic.
- Let $e = (w_i, w_{i+1})$ be an edge on the external face. The free edge segment of e has a slope of ± 1. The port edge segment of e passes through a maximal port.
- Every vertex v has at most one free edge segment crossing each free region, and each port segment goes to a unique port.

When we insert a new vertex v_{k+1}, we must create enough space so that the two neighbors w_l and w_r can "see" the new vertex through their maximal right and left ports, which are typically already used. Thus, we must shift these vertices over to create space and also to ensure that the intersection of these ports lies on a grid location, for the new vertex. Of course, we cannot simply shift these vertices, we must shift other vertices to be sure that we do not produce any crossings. Therefore, to shift a vertex w, we shift all vertices in its shifting set, defined in Section 7.2.2, and also most of the ports. See Figure 7.14.

DEFINITION 7.8   For $\delta \geq 0$ and a vertex $w_i \in C_k$, define a *regular-shift by δ units of* w_i as shifting all vertices in $M_k(w_i)$ by δ units to the right, including all associated ports. Define the *right-shift by δ units on* w_i as a regular-shift of w_i *except* that the ports in the L^p region of w_i are *not* shifted. Similarly, define the *left-shift by δ units on* w_i as a regular shift of w_{i+1} *and* additionally shifting the ports in the R^p region of w_i.

Notice that left-shifting a vertex w_i is nearly identical to right-shifting its neighbor w_{i+1} except for the ports that are moved.

Assume that G_k has been embedded and that the invariants hold. We now look at the specific insertion of a new vertex v_{k+1} to create G_{k+1} while maintaining the invariants. For a vertex $w \in C_k$, recall that the current number of port edge segments using R^p is $d_r(w)$ and for L^p is $d_l(w)$. If $d_r(w_l) = 0$, we perform a left-shift of 2 units on w_l; otherwise, we perform a left-shift of 1 unit on w_l. This frees a space for a new maximal port in the

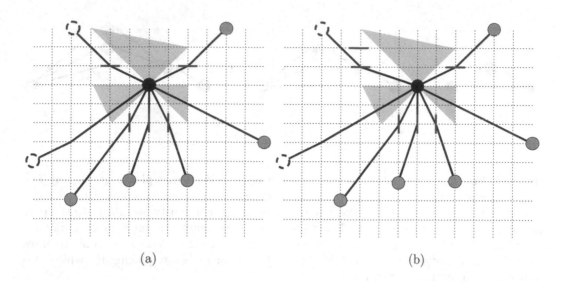

(a)

(b)

Figure 7.14 (a) A (darkened) vertex and its neighbors before a right shift of one unit. (b) And after a right shift of one unit. The other vertices that are part of the shifting set are highlighted, while those that are not are drawn dashed. Notice that the left port region remains in place creating a location for one more port.

R_p region of w_l. Similarly, if $d_l(w_r) = 0$, we perform a right-shift of 2 units on w_r, and otherwise, we perform a right-shift of 1 unit.

Let l be the line of slope $+1$ passing through w_l's newly created maximal right port. Let r be the line of slope -1 passing through w_r's newly created maximal left port. We place v_{k+1} at the intersection of lines l and r. If l and r intersect at a non-grid location, we simply perform a regular-shift of 1 unit on w_r. Observe that we therefore perform at most 5 shifts per insertion.

We now route the edges as follows. The edge from w_l to v_{k+1} goes from w_l to $R^p_{\max}(w_l)$ and then to v_{k+1} through its free region L^f. The edge from w_r to v_{k+1} goes from w_r to $L^p_{\max}(w_r)$ and then to v_{k+1} through its free region R^f. The remaining edges are from v_{k+1} to w_i for $l < i < r$. These edges are routed from v_{k+1} to nearly consecutive ports on the M^p region of v_{k+1} and then to w_i through its free region M^f. We locate the horizontal line segment containing the ports of M^p exactly $\lceil (r-l)/2 \rceil$ units below v_{k+1}. Duncan and Kobourov [DK03] prove that this guarantees that each port is above each neighbor vertex w_i. In the case that $r - l$ is even, there is exactly one port per edge routed, and the ports are mapped consecutively. In the case of an odd value, we must skip one port in the region, which is easy to identify [DK03]. Figure 7.15 shows the insertion of five vertices of a planar graph using this algorithm.

Duncan and Kobourov prove that this algorithm properly maintains the previous invariants leading to the following theorem:

Theorem 7.5 *[DK03] For a given plane graph $G = (V, E)$, there is a linear-time algorithm that constructs a planar polyline drawing of G with grid size $5n \times 5n/2$ using at most one bend per edge and with an angular resolution no less than $1/2d(v)$ for every vertex $v \in V$.*

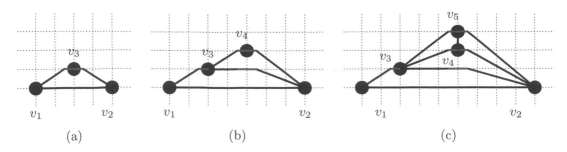

Figure 7.15 The insertion of the first five vertices of a particular planar graph. (a) The initial configuration with 3 vertices. Note that the port edge segment connecting v_1 to v_3 connects to v_1's port which is at the same location as v_3. For clarity, we illustrate the port slightly outside this location. (b) Insertion of v_4. This requires a left-shift of 1 unit for v_3 and a right-shift of 1 unit for v_2 before placing v_4. (c) Insertion of v_5. This requires a left-shift of 1 unit for v_3 and a right-shift of 1 unit for v_5 before placing v_5, which also connects to the *covered* vertex v_4.

7.5   Conclusion

When angular resolution is a desired criterion in drawing a graph, many techniques exist to accommodate it. If the graph is known to be 4-planar or if one is willing to use rectangular regions instead of points for vertices, one can efficiently construct aesthetically pleasing orthogonal drawings [Tam87, TT89, GT97, CK12]. This body of work uses network flows to compute an orthogonal shape with the minimum number of bends and to compact the representation into an orthgonal drawing with minimal height and width.

In addition, several polyline drawing strategies exist that allow one to create good drawings with relatively high angular resolution, a small number of bends, and good area bounds even when the maximum degree of the graph is greater than four [Kan96, GM98, GW00, CDGK01, DK03]. These all extend the incremental insertion algorithm using a canonical ordering initially employed by de Fraysseix, Pach, and Pollack [dFPP90]. The mixed-model approach, employed by Kant [Kan96] and Gutwenger and Mutzel [GM98], uses primarily orthogonal edges but must still connect vertices using some segments whose slopes depend on the degree of the vertex. The works of Cheng et al. [CDGK01] and Duncan and Kobourov [DK03] use an optimal one bend per edge but with one of the two segments of each edge having arbitrary slope. Unlike the purely orthogonal representations, the set of slopes determined by the edges in these polyline drawings is possibly large.

References

[CDGK01] C. C. Cheng, C. A. Duncan, M. T. Goodrich, and S. G. Kobourov. Drawing planar graphs with circular arcs. *Discrete and Computational Geometry*, 25(3):405–418, 2001.

[CK12] Sabine Cornelsen and Andreas Karrenbauer. Accelerated bend minimization. *Journal of Graph Algorithms and Applications*, 16(3):635–650, 2012.

[CLR90] T. H. Cormen, C. E. Leiserson, and R. L. Rivest. *Introduction to Algorithms*. MIT Press, Cambridge, MA, 1990.

[CP95] M. Chrobak and T. Payne. A linear-time algorithm for drawing planar graphs. *Inform. Process. Lett.*, 54:241–246, 1995.

[DETT99] G. Di Battista, P. Eades, R. Tamassia, and I. G. Tollis. *Graph Drawing*. Prentice Hall, Upper Saddle River, NJ, 1999.

[dFPP90] H. de Fraysseix, J. Pach, and R. Pollack. How to draw a planar graph on a grid. *Combinatorica*, 10(1):41–51, 1990.

[DHVM83] P. Duchet, Y. Hamidoune, M. Las Vergnas, and H. Meyniel. Representing a planar graph by vertical lines joining different levels. *Discrete Math.*, 46:319–321, 1983.

[DK03] C. A. Duncan and S. G. Kobourov. Polar coordinate drawing of planar graphs with good angular resolution. *Journal of Graph Algorithms and Applications*, 7(4):311–332, 2003.

[DLV98] G. Di Battista, G. Liotta, and F. Vargiu. Spirality and optimal orthogonal drawings. *SIAM J. Comput.*, 27(6):1764–1811, 1998.

[FHH⁺93] M. Formann, T. Hagerup, J. Haralambides, M. Kaufmann, F. T. Leighton, A. Simvonis, Emo Welzl, and G. Woeginger. Drawing graphs in the plane with high resolution. *SIAM J. Comput.*, 22:1035–1052, 1993.

[FM98] S. Fialko and P. Mutzel. A new approximation algorithm for the planar augmentation problem. In *Proceedings of the 9th Annual ACM-SIAM Symposium on Discrete Algorithms (SODA '98)*, pages 260–269. ACM Press, 1998.

[GM98] C. Gutwenger and P. Mutzel. Planar polyline drawings with good angular resolution. In S. Whitesides, editor, *Graph Drawing (Proc. GD '98)*, volume 1547 of *Lecture Notes Comput. Sci.*, pages 167–182. Springer-Verlag, 1998.

[GT97] A. Garg and R. Tamassia. A new minimum cost flow algorithm with applications to graph drawing. In S. C. North, editor, *Graph Drawing (Proc. GD '96)*, volume 1190 of *Lecture Notes Comput. Sci.*, pages 201–216. Springer-Verlag, 1997.

[GT01] A. Garg and R. Tamassia. On the computational complexity of upward and rectilinear planarity testing. *SIAM J. Computing*, 31(2):601–625, 2001.

[GT02] Michael T. Goodrich and Roberto Tamassia. *Algorithm design: foundations, analysis, and Internet examples*. John Wiley and Sons, Inc., New York, NY, 2002.

[GW00] M. T. Goodrich and C. G. Wagner. A framework for drawing planar graphs with curves and polylines. *Journal of Algorithms*, 37(2):399–421, 2000.

[Kan96]   G. Kant. Drawing planar graphs using the canonical ordering. *Algorithmica*, 16:4–32, 1996. Special issue on Graph Drawing, edited by G. Di Battista and R. Tamassia.

[OvW78]   R. H. J. M. Otten and J. G. van Wijk. Graph representations in interactive layout design. In *Proc. IEEE Internat. Sympos. on Circuits and Systems*, pages 914–918, 1978.

[RT86]   P. Rosenstiehl and R. E. Tarjan. Rectilinear planar layouts and bipolar orientations of planar graphs. *Discrete Comput. Geom.*, 1(4):343–353, 1986.

[Tam87]   R. Tamassia. On embedding a graph in the grid with the minimum number of bends. *SIAM J. Comput.*, 16(3):421–444, 1987.

[TT86]   R. Tamassia and I. G. Tollis. A unified approach to visibility representations of planar graphs. *Discrete Comput. Geom.*, 1(4):321–341, 1986.

[TT89]   R. Tamassia and I. G. Tollis. Planar grid embedding in linear time. *IEEE Trans. Circuits Syst.*, CAS-36(9):1230–1234, 1989.

Spine and Radial Drawings

Emilio Di Giacomo
University of Perugia

Walter Didimo
University of Perugia

Giuseppe Liotta
University of Perugia

8.1 Introduction

A *layered drawing* of a graph is a drawing such that the vertices are constrained to lie on geometric layers that can be lines, circles, or other kinds of curves. Partitioning the vertices into distinct layers can be an effective way to emphasize some structural properties of the graph; in many cases this is required in some real-world applications to convey the so-called *semantic constraints* [Sug02].

In this chapter, we concentrate on layered drawings of undirected graphs, where the edges are not constrained to be monotone in a given direction. Conversely, this is typically a basic requirement in the layered drawings of directed graphs or hierarchies, where all edges must flow in a common direction (usually the vertical one), according to their orientation. Layered drawings algorithms for directed graphs are extensively investigated in Chapter 13.

Although it is theoretically interesting to study layered drawings where the layers can be curves of any type, it is rather difficult to extract properties and design algorithms if the layers do not have a quite "regular" shape. Indeed, most of the literature assumes that the layers are either parallel straight lines or concentric circles, which is also the most common requirement in real-world application domains. Therefore, we only give an overview of the results on layered drawings where layers are straight lines or circles. We call the first family of drawings *spine drawings* and the second family *radial drawings*.

The remainder of this chapter is structured as follows. We first give formal definitions that are needed in the chapter and describe a unified investigation framework for spine and radial drawings (Section 8.2). Then, we investigate the results on spine and radial drawings in a general scenario (Section 8.3); this scenario has the only requirement that the vertices are placed on layers. Results on scenarios that consider additional constraints are investigated in Section 8.4. Finally, we mention some topological and geometric problems related to the spine and radial drawability of a graph (Section 8.5) and we give conclusions (Section 8.6).

8.2   A Unified Framework for Spine and Radial Drawings

8.2.1   Definitions

A *drawing* Γ of a graph G is a geometric representation of G such that each vertex u of G is mapped to a distinct point p_u of the plane and each edge (u, v) of G is drawn as a simple Jordan curve with end-points p_u and p_v. Drawing Γ is *planar* if two distinct edges never intersect except at common end-vertices. G is *planar* if it admits a planar drawing. A planar drawing Γ of G partitions the plane into topologically connected regions called the *faces*. The unbounded face is called the *external face* and the other faces are called *internal faces*. The *boundary* of a face is its delimiting circuit (not necessarily a simple cycle) described by the circular list of its edges and vertices. The *boundary* of the external face, also called the *external boundary*, is the circular list of edges and vertices delimiting the unbounded region. If the graph is biconnected, the boundary of each face is a simple cycle. An *embedding* of a planar graph G is an equivalence class of planar drawings that determine the same set of faces, i.e., the same set of face boundaries. A planar graph G with a given embedding is called an *embedded planar graph*. In this chapter, we only deal with planar graphs and planar drawings. From a practical point of view, if a graph is not planar, one can think of applying a planarization algorithm on it in order to find a planar embedding with dummy vertices that replace crossings [DETT99].

A drawing Γ of G such that the edges are represented as a polygonal chain is a *polyline drawing*. A *bend* along an edge e of Γ is a common point between two consecutive straight-line segments that form e. If every edge of Γ has at most b bends, Γ is a *b-bend drawing* of G. A 0-bend drawing is also called a *straight-line drawing*.

Let γ_1 and γ_2 be two curves. Curves γ_1 and γ_2 are *parallel* if every normal to one curve is a normal to the other curve and the distance between the points where the normals cut the two curves is a constant. Examples of parallel curves are parallel straight lines or concentric circles. A *set of layers* is a set of pairwise parallel curves; each curve in the set is called a *layer*. Given a set of layers it is possible to order the layers according to the order they are encountered while walking along a straight line normal to all of them. More precisely, let $\mathcal{C}$ be a set of layers, and let l_n be a normal to all the layers in $\mathcal{C}$. Let p_i be the intersection point between l_n and $\gamma_i \in \mathcal{C}$ and let p_j be the intersection point between l_n and $\gamma_j \in \mathcal{C}$. Given an orientation for l_n, we have that γ_i is before γ_j if p_i is encountered before p_j while walking along l_n according to the given orientation, γ_i is after γ_j otherwise. In the following, given a set of layers denoted as $\gamma_0, \ldots, \gamma_{k-1}$, we always assume that γ_i is before γ_{i+1} for each $0 \le i \le k - 1$.

DEFINITION 8.1   Let $G = (V, E)$ be a planar graph, and let $\mathcal{C} = \{\gamma_0, \ldots, \gamma_{k-1}\}$ be a set of layers, with $k \le n$. A *k-layered drawing* of G on $\mathcal{C}$ is a polyline planar drawing Γ of G such that each vertex $v \in V$ is represented in Γ as a point $p_v \in \gamma_i$ $(0 \le i \le k - 1)$.

An example of a 4-layered drawing is shown in Figure 8.1. A k-layered drawing will be simply called a *layered drawing* when we are not interested in the number of layers.

Let Γ be a k-layered drawing of a graph G, and let $e = (u, v)$ be an edge of G such that u is drawn in Γ on layer γ_i and v is drawn in Γ on layer γ_j $(0 \le i, j \le k - 1)$. The *span* of e

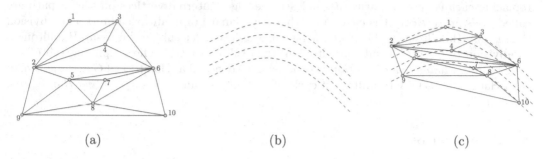

(a)   (b)   (c)

Figure 8.1 (a) A planar graph G. (b) A set $\mathcal{C}$ of four layers. (c) A 4-layer, 0-bend drawing of G on $\mathcal{C}$

in Γ is $|i - j|$. An *intra-layer edge* is an edge with span equal to 0, i.e., an edge connecting vertices that are on the same layer. A *long edge* is an edge with span greater than 1.

In the following, we shall consider two special cases of layered drawings: *spine drawings*[1] and *radial drawings*, which are defined as follows.

DEFINITION 8.2 A *k-spine drawing* of a planar graph is a planar k-layered drawing such that the layers are horizontal straight lines, called *spines*.

DEFINITION 8.3 A *k-radial drawing* of a planar graph is a planar k-layered drawing such that the layers are concentric circles.

For a k-spine drawing, we denote the set of layers as $\mathcal{C} = \{L_0, \ldots, L_{k-1}\}$ and we assume that they are ordered from the highest to the lowest, i.e., L_0 is the topmost line and L_{k-1} is the bottommost one. For a k-radial drawing, we denote the set of layers as $\mathcal{C} = \{C_0, \ldots, C_{k-1}\}$, and we assume that they are ordered from the more external to the innermost, i.e., C_0 is the circle with the largest radius and C_{k-1} is the one with the smallest radius. When we are not interested in distinguishing between spine and radial drawings, we will generically denote the layers as $\mathcal{C} = \{\gamma_0, \ldots, \gamma_{k-1}\}$. If a planar graph G admits a k-spine, b-bend drawing (k-radial, b-bend drawing), we say that G is *k-spine, b-bend drawable* (*k-radial, b-bend drawable*).

We conclude this section with some definitions about Hamiltonicity that will be used in the following. A *Hamiltonian cycle* of G is a simple cycle that contains all vertices of G. A graph G that admits a Hamiltonian cycle is said to be *Hamiltonian*. A planar graph G is *sub-Hamiltonian* if either G is Hamiltonian or G can be augmented with dummy edges (but not with dummy vertices) to a graph that is Hamiltonian and planar. We denote by $\mathsf{aug}(G)$ a planar Hamiltonian graph obtained by G by possibly adding edges (if G is Hamiltonian then $\mathsf{aug}(G) = G$). A *subdivision* of a graph G is a graph obtained from G by

---

[1]Drawings on a set of horizontal layers are often called *layered drawings* in the literature. Since in this chapter we use the term *layered drawing* to denote the more general case of a drawing on any set of parallel curves, we use the term *spine drawings* when the layers are straight lines. This term is taken from the theory of book embeddings, which can be regarded as drawings on a single horizontal line, usually called the *spine* of the book embedding.

replacing each edge by a path with at least one edge. Internal vertices on such a path are called *division vertices*. It is easy to see that any planar graph always admits a subdivision that is sub-Hamiltonian. Let G be a planar graph and let $\mathsf{sub}(G)$ be a sub-Hamiltonian subdivision of G (if G is sub-Hamiltonian, then $\mathsf{sub}(G) = G$). The graph $\mathsf{aug}(\mathsf{sub}(G))$ is called a *Hamiltonian augmentation of G* and will be denoted as $\mathsf{Ham}(G)$ (if G is Hamiltonian then $\mathsf{Ham}(G) = G$). A Hamiltonian cycle of $\mathsf{Ham}(G)$ is called an *augmenting Hamitonian cycle* of G.

8.2.2 Scenarios

In the following, we are interested in characterizing k-spine, b-bend drawable and k-radial, b-bend drawable graphs for different values of k and b. We are also interested in the drawability testing problems, i.e., in studying the complexity of deciding whether a given planar graph is k-spine, b-bend drawable (k-radial, b-bend drawable). More precisely, we consider the following two problems.

Characterization Problem. Let k and b be two given integers. What is the largest class of k-spine, b-bend drawable (k-radial, b-bend drawable) graphs?

Drawability Testing Problem. Let k and b be two given integers and let G be a planar graph. What is the complexity of deciding whether G is k-spine, b-bend drawable (k-radial, b-bend drawable)?

The study of these two problems is motivated by the fact that, for aesthetic reasons, one can be interested in keeping the number of layers and the number of edge bends in a layered drawing as small as possible. Observe that every planar graph G with n vertices is k-spine 0-bend drawable, for some value of $k \leq n$. Indeed, it is known that G admits a planar straight-line drawing Γ [Fár48], and at most n distinct horizontal parallel layers are sufficient to intersect all vertex-points in Γ. Furthermore, since G also admits a planar straight-line drawing on an integer grid of size $O(n) \times O(n)$ [dPP90], G is always k-spine 0-bend drawable within an $O(n^2)$ area. With analogous considerations, every planar graph with n vertices is k-radial 0-bend drawable for some value of $k \leq n$.

The Characterization Problem and the Drawability Testing Problem can be studied within different scenarios, depending on the additional constraints that one can define. We first consider the two problems without any additional constraint. We will refer to this scenario as the *general scenario*. We then consider the same problems with some of the following additional constraints:

Intra-layer edges not allowed. Many results in the literature assume that there is no intra-layer edge in a layered drawing. For example, avoiding intra-layer edges in a k-layered drawing could be important to put in evidence a k-partite structure of the graph. Indeed, a k-layered drawing of a graph $G = (V, E)$ implicitly defines a partition of the set V into k sets $V_0, V_1, \ldots, V_{k-1}$, where each set V_i is the set of vertices drawn on layer γ_i. Layered drawings with no intra-layer edges will be called *upright drawings*.

Assigned vertex partitioning. In some cases, the partitioning of the vertices can be given as a part of the input. In these cases, the vertex partition determined by the layered drawing has to preserve the one given in the input. Layered drawings where the partition of the vertices is given will be called *partitioned layered drawings*.

Long edges not allowed. Edges that span more than one level are more difficult to follow by the human eye than edges connecting vertices on consecutive layers.

Thus another common constraint in a layered drawing is to avoid long edges. Layered drawings with no long edges will be called *proper drawings*.

Assigned layers. In the general scenario, we are assuming that only the number and the type (spines or circles) of layers are given. However, one can consider the case when also the distance between every two consecutive layers is given as part of the input. Having the distances of the layers assigned as a part of the input may change the answer to both the Characterization and the Drawability Testing Problem.

8.3 Results in the General Scenario

8.3.1 Spine Drawings in the General Scenario

We start by considering the easiest case for k-spine drawings, i.e., the case when $k = 1$. A trivial result is that if only 0 bends per edge are allowed we can only draw forests of paths, and therefore, the drawability test can be executed in $O(n)$ time, where n is the number of vertices of the input graph.

PROPOSITION 8.1   A planar graph is 1-spine, 0-bend drawable if and only if it is a forest of paths.

If one bend per edge is allowed the problem of computing a 1-spine, 1-bend drawing of a planar graph G is equivalent to that of computing a book embedding of G on two pages. A *book embedding* of a graph $G = (V, E)$ consists of a total order $<_\sigma$ of V and a partition of E into p sets, called *pages*, such that there are no two edges (u, v) and (w, z) in the same page with $u <_\sigma w <_\sigma v <_\sigma z$. The *pagenumber* of a graph G is the minimum value p for which G admits a book embedding with p pages.

A book embedding can be seen as a drawing of G where: (i) all vertices are drawn along a straight line, called the *spine*, according to the total order $<_\sigma$, (ii) each edge is assigned to one among p half-planes having the spine as a common boundary, (iii) no two edges in the same page cross (see Figure 8.2). It is not difficult to prove that if two edges can be drawn without crossings on a half-plane with the endvertices on the boundary of the half-plane, then they can be drawn without crossings as two polylines with one bend on the same half-plane and with the end-vertices in the same position (see also Figure 8.2). Since a straight line on a plane define two half-planes we have the following lemma.

LEMMA 8.1   A planar graph is 1-spine, 1-bend drawable if and only if it has pagenumber two.

Bernhart and Kainen [BK79] prove that a graph has pagenumber at most two if and only if it is sub-Hamiltonian. If a graph G admits a book embedding on two pages, then let $v_0, v_1, \ldots, v_{n-1}$ be the vertices of G ordered according to the total ordering $<_\sigma$ of the book embedding. An augmenting Hamiltonian cycle of G is $(v_0, v_1), (v_1, v_2), \ldots, (v_{n-2}, v_{n-1})$, (v_{n-1}, v_0) where each edge (v_i, v_{i+1}) is either an edge of G or a dummy edge that can be added to G without violating planarity (see Figure 8.3 for an example). Conversely, if G is sub-Hamiltonian there exists an augmenting Hamiltonian cycle H (possibly obtained by adding some edges) in G. Choose an embedding Ψ of G with an edge e of H on the external face. By removing e we have a path P containing all the vertices of G. Define the total

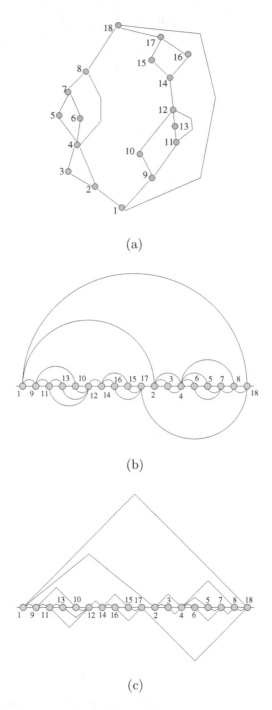

(a)

(b)

(c)

Figure 8.2 (a) A planar graph G. (b) A book embedding of G on two pages: the total order of the vertices is the left-to-right order of the vertices along the horizontal line, while the two pages are represented by the two half-planes defined by the same line. (c) A 1-spine, 1-bend drawing of G.

order $<_\sigma$ according to the order the vertices of G are encountered while walking along P. The edges in P can be assigned to one of the two pages. Edge e can also be assigned to the same page as the edges in P. All the remaining edges of G are either inside or outside H in the embedding Ψ. Those that are inside H are assigned to the same page as all the edges of H, those that are outside are assigned to the other page. There cannot be two edges (u, v) and (w, z) in the same page such that $u <_\sigma w <_\sigma v <_\sigma z$, because otherwise there would be a crossing in the embedding Ψ.

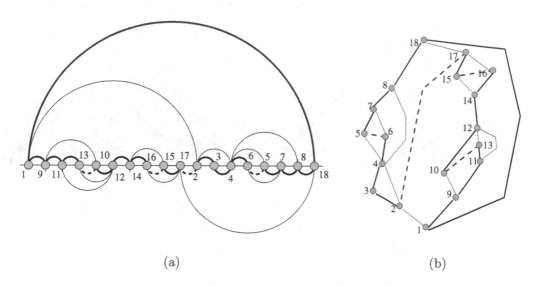

(a)   (b)

Figure 8.3 An augmentation of the planar graph of Figure 8.2 to a planar Hamiltonian graph.

Based on the result of Bernhart and Kainen and on Lemma 8.1, we have that the class of graphs that admit a planar 1-spine, 1-bend drawing is the class of sub-Hamiltonian graphs. Since testing sub-Hamiltonicity is $\mathcal{NP}$-complete [Wig82], we have that testing a graph for 1-spine, 1-bend drawability is $\mathcal{NP}$-complete, too.

Theorem 8.1 *A planar graph is 1-spine, 1-bend drawable if and only if it is sub-Hamiltonian.*

Although Theorem 8.1 gives a complete characterization of 1-spine, 1-bend drawable graphs, such graphs cannot be recognized efficiently; thus it is worth investigating some specific families of graphs that are sub-classes of the sub-Hamiltonian graphs and that can be recognized efficiently. Among them we recall here: outerplanar graphs [BK79] (that coincide with the graphs having pagenumber one), series-parallel graphs [DDLW06, RM95], planar bipartite graphs [ddMP95], square grids, and X-trees [CLR87].

If two bends per edge are allowed, then every planar graph is drawable on one spine. This result is a consequence of a result by Kaufmann and Wiese [KW02] about point-set embeddability. Given a planar graph $G = (V, E)$ and a set S of points in the plane such that $|S| = |V| = n$, a *point-set embedding* of G onto S is a planar drawing of G such that each vertex of G is represented as a point of S. Kaufmann and Wiese [KW02] prove that

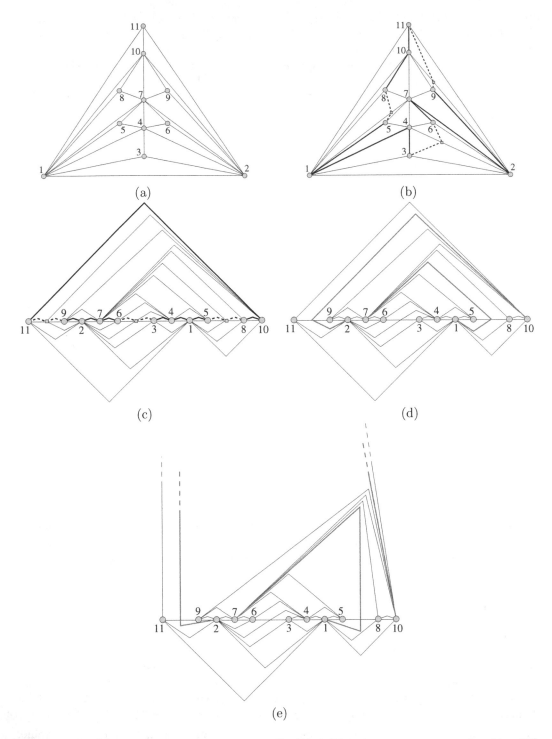

Figure 8.4 (a) A non-Hamiltonian graph G. (b) A Hamiltonian augmentation $\mathsf{Ham}(G)$ of G. (c) A 1-spine, 1-bend drawing of $\mathsf{Ham}(G)$. (d) A 1-spine, 3-bend drawing of G; the edges with 3 bends are highlighted. (e) A 1-spine, 2-bend drawing of G obtained by rotating the segments of the edges that had 3 bends in the previous picture.

every planar graph G admits a point-set embedding on any given set of points such that every edge of G is represented as a polyline with at most 2 bends. Such a drawing can be computed in $O(n \log n)$ time. In order to compute a planar 1-spine, 2-bend drawing of a planar graph G, it is sufficient to choose a set of n collinear points and then apply the Kaufmann and Wiese algorithm. As a consequence, the following theorem holds.

Theorem 8.2 *Every planar graph is 1-spine, 2-bend drawable.*

Although the paper by Kaufmann and Wiese is about point-set embeddings and does not mention book embeddings, their drawing technique can be regarded as an extension of the technique used to compute a 1-spine, 1-bend drawing of a Hamiltonian graph. Kaufmann and Wiese compute a Hamiltonian augmentation $\mathsf{Ham}(G)$ of the input graph G such that each edge of G is subdivided at most once (see Figure 8.4). Since $\mathsf{Ham}(G)$ is Hamiltonian it admits a 1-spine, 1-bend drawing by Theorem 8.1. An edge $e = (u, v)$ that has been subdivided by a division vertex w, is represented in the 1-spine, 1-bend drawing of $\mathsf{Ham}(G)$ by two edges (u, w) and (w, v) each one drawn with at most one bend. Removing the division vertex w we obtain another bend on edge e at the point p_w where w was drawn. This removal would give rise to at most three bends per edge (see Figure 8.4). However, it is possible to remove this third bend by suitably rotating the segments incident to p_w (see Figure 8.4). The drawing technique described above requires to compute a Hamiltonian augmentation $\mathsf{Ham}(G)$ of G. Kaufmann and Wiese describe a Hamiltonian augmentation technique that runs in $O(n)$ time and subdivides each edge at most once. Details about different Hamiltonian augmentation techniques are given in Section 8.5.1.

We conclude this discussion about 1-spine, 2-bend drawings by further remarking the connection between them and book embeddings. The 1-spine, 2-bend drawing of the input graph G is obtained from a 1-spine, 1-bend drawing of the Hamiltonian graph $\mathsf{Ham}(G)$. An edge $e = (u, v)$ that has been subdivided by a division vertex w, is represented in the 1-spine, 1-bend drawing of $\mathsf{Ham}(G)$ by two edges (u, w) and (w, v) that may be on the two different half-planes defined by the spine. This means that a 1-spine, 2-bend drawing of a planar graph G can be seen as a book embedding of G on two pages, where each edge is not required to be on one page only but is allowed to cross the spine at most once. A book embedding where edges are allowed to cross the spine is also called a *topological book embedding*. Therefore, Theorem 8.2 implies that every planar graph has a topological book embedding on two pages where each edge crosses the spine at most once. Since two bends are sufficient to draw all planar graphs on a single spine, it does not make sense to further investigate 1-spine, b-bend drawings for $b > 2$.

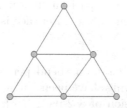

Figure 8.5 An outerplanar graph that does not admit a 2-spine, 0-bend drawing.

Consider now the case when two spines are given. It is immediate to see that if a graph admits a 2-spine, 0-bend drawing, then it is outerplanar (i.e., it admits a planar embedding such that all the vertices are on the external face). Indeed, in a 2-spine, 0-bend drawing each vertex is either a topmost vertex or a bottommost vertex, and therefore, since the edges are straight lines, it is on the external face. Observe however that not all outerplanar graphs admits a 2-spine, 0-bend drawing. The graph in Figure 8.5 is the smallest (in terms of number of vertices) outerplanar graph that is not 2-spine, 0-bend drawable.

Some preliminary results about 2-spine, 0-bend drawability were presented by Felsner et al. [FLW03], who characterize trees that are 2-spine, 0-bend drawable. They prove that a tree T admits a 2-spine, 0-bend drawing if and only if there exists a path P in T such that removing P from T we are left with a collection of vertex disjoint paths (see Figure 8.6). A characterization of (outer)planar graphs that admit a planar 2-spine, 0-bend drawing has been given by Cornelsen et al. [CSW04]. They first consider biconnected outerplanar graphs and prove that a biconnected outerplanar graph G admits a 2-spine, 0-bend drawing if and only if its internal faces induce a path in the dual graph of G (see Figure 8.6). The *dual graph* G^* of a planar graph G is a multigraph that has a vertex for each face of G and an edge between two vertices f and g if the two faces represented by f and g share an edge. For general simply connected outerplanar graphs Cornelsen et al. [CSW04] describe a decomposition of an outerplanar graph G into components like paths, trees and biconnected outerplanar components and describe necessary and sufficient conditions that these components must satisfy for the 2-spine, 0-bend drawability of G. Therefore, the outerplanar graphs whose components satisfy these conditions are exactly the planar graphs that are 2-spine, 0-bend drawable. The necessary and sufficient conditions described in [CSW04] cannot be shortly summarized. Intuitively, they guarantee that each single component is 2-spine, 0-bend drawable and that the vertices shared by different components are drawn so that the drawings of the different components can be merged together. Finally, Cornelsen et al. [CSW04] prove that the necessary and sufficient condition above can be tested in $O(n)$ time, thus proving that 2-spine, 0-bend drawability can be tested in linear time.

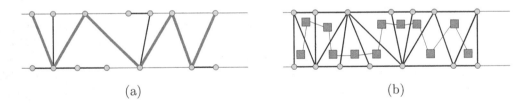

(a)   (b)

Figure 8.6   (a) A 2-spine, 0-bend drawable tree. The removal of the highlighted path leaves a set of paths. (b) A 2-spine, 0-bend drawing of a biconnected outerplanar graph G such that the inner faces of G induce a path in the dual graph of G. In the picture, the node of the dual graph corresponding to the outer face is not shown.

Drawings on two spines with at most one bend per edge are the subject of [DDLS06] where, in fact, k-spine, 1-bend drawings have been studied. In [DDLS06], a k-spine, 1-bend drawing is considered as an extension of a 2-page book embedding where the spines are more than one, and it is proved that, for any fixed $k \geq 2$, not all planar graphs are k-spine, 1-bend drawable. The proof is based on the observation that, if a graph admits a k-spine, 1-bend drawing, it must exist a special cycle, called *cutting cycle* (see Figure 8.7), removing which we are left with $(k-1)$-spine, 1-bend drawable subgraphs. The cutting cycle is actually a sequence of vertices that may or may not correspond to an actual cycle in the graph. Instead, the sequence of vertices is such that, if dummy edges are inserted between non-adjacent vertices that are consecutive in the sequence, then the resulting drawing is still a k-spine, 1-bend drawing. The following lemma holds.

LEMMA 8.2 If G is a maximal planar graph that is k-spine, 1-bend drawable for $k \geq 2$, then there exists a simple cycle C in G such that $G \setminus C$ is $(k-1)$-spine, 1-bend drawable.

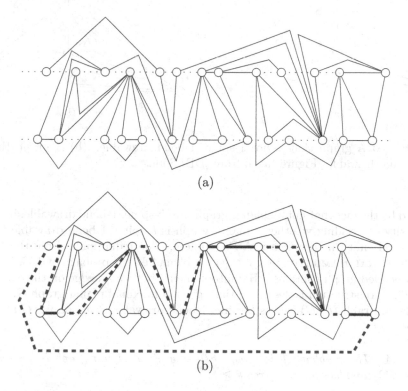

(a)

(b)

Figure 8.7 (a) A planar 2-spine, 1-bend drawing of a planar graph G. (b) A cutting cycle of G. Figure taken from [DDLS06].

We can use now the necessary condition expressed by Lemma 8.2 to construct, for any fixed $k \geq 1$, a maximal planar graph N^k that is not k-spine, 1-bend drawable. Graph N^1 is the graph shown in Figure 8.8 (it is the same graph as in Figure 8.4), which is a non-Hamiltonian graph. Graph N^k is obtained from N^1 by replacing each black vertex with a copy of N^{k-1} and triangulating the result (see Figure 8.8).

The proof that N^k is not k-spine, 1-bend drawable is by induction on k. N^1 is not 1-spine, 1-bend drawable by Theorem 8.1 because it is not Hamiltonian. Let N^{k-1} be not $(k-1)$-spine, 1-bend drawable and assume by contradiction that N^k is k-spine, 1-bend drawable. By Lemma 8.2 there exists a simple cycle C in N^k whose removal leaves us with $(k-1)$-spine, 1-bend drawable subgraphs. Since each copy of N^{k-1} is not $(k-1)$-spine, 1-bend drawable, then C contains at least one vertex for each copy of N^{k-1}. Also, since each copy of N^{k-1} is inside a triangle of white vertices we have that also all white vertices must be in C. However, this would imply that N^1 is Hamiltonian.

Theorem 8.3   *For each integer $k \geq 1$, there exists a planar graph that is not k-spine,*
1-bend drawable.

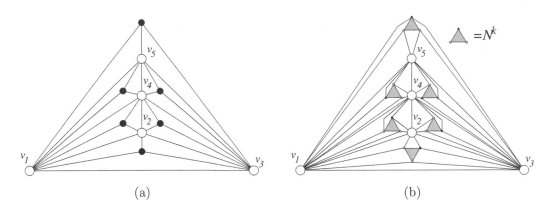

(a)   (b)

Figure 8.8   (a) A graph that is not 1-spine, 1-bend drawable. (b) A graph that is not
k-spine, 1-bend drawable. Figure taken from [DDLS06].

Motivated by the fact that not all planar graphs are k-spine, 1-bend drawable, in [DDLS06]
the complexity of deciding whether a planar graph is k-spine, 1-bend drawable is studied,
and it is proved that this problem is $\mathcal{NP}$-complete. The reduction is from the MAXIMAL
PLANAR EXTERNAL HAMILTONIAN CIRCUIT problem, i.e., the problem of deciding whether
a planar embedded graph contains a Hamiltonian circuit with an edge on the external face.
In [DDLS06], a construction is described that, given a maximal planar graph G, produces a
maximal planar graph $H^k(G)$ that is k-spine, 1-bend drawable if and only if G is externally
Hamiltonian.

Theorem 8.4   *The problem of deciding whether a given planar graph is k-spine, 1-bend*
drawable is $\mathcal{NP}$-complete for any fixed $k \geq 1$.

For the special case of $k = 2$, a complete characterization of 2-spine, 1-bend drawable
graphs is given in [DDLS06]. In this case, the necessary condition expressed by Lemma 8.2
can be better detailed. Namely, after removing the cutting cycle, we are left with a set
of disjoint paths whose endvertices are adjacent (or can be made adjacent) to the cutting
cycle and that satisfy some additional properties (see Figure 8.9). It can be proved that
this necessary condition is also sufficient. Graphs whose vertices can be covered by a
cycle and a set of vertex-disjoint paths whose end-vertices are connected to the cycle are
called *(sub-)Hamiltonian-with-handles graphs* in [DDLS06], which appears as an extension
of (sub-)Hamiltonian graphs.

Theorem 8.5   *A planar graph is 2-spine, 1-bend drawable if and only if it is sub-Hamiltonian-*
with-handles.

Theorem 8.4 says that it is $\mathcal{NP}$-complete to recognize sub-Hamiltonian-with-handles
graphs. However, there are subclasses of sub-Hamiltonian-with-handles graphs that can
be recognized in polynomial time. For example, in [DDLS06] it has been proved that 2-
outerplanar graphs are sub-Hamiltonian-with-handles and hence 2-spine, 1-bend drawable.

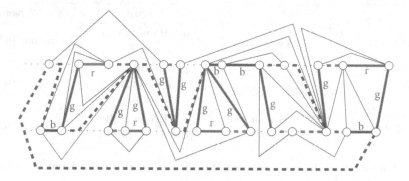

Figure 8.9 The planar graph of Figure 8.7 covered by a cycle (thick blue edges) and a set of vertex-disjoint paths (thick red edges) whose end-vertices are connected to the cycle (thick green edges). The dashed edges are dummy edges. Labels b, g, and r denote the color of the solid thick edges. The color of the dashed thick edges can be easily inferred.

A complete characterization of the family of k-spine, 1-bend drawable graphs is still missing, but Theorem 8.3 tells us that this family is a proper subclass of planar graphs.

A characterization is still missing also for k-spine, 0-bend drawable graphs, where some preliminary results have been obtained only for trees. Felsner et al. [FLW03] proved that, for any fixed k, it is possible to construct a tree that is not k-spine, 0-bend drawable. To produce such a tree, Felsner et al. [FLW03] introduce the notion of *strictness* of a tree T defined as follows. A tree T is 2-strict if it contains a vertex of degree greater than or equal to three. T is k-strict if it contains a vertex v adjacent to at least three vertices u_1, u_2, and u_3 such that the subtrees rooted at u_1, u_2, and u_3 are $(k-1)$-strict. In [FLW03] it is proved that a k-strict tree is not $(k-1)$-spine, 0-bend drawable. The proof is by induction. A 2-strict tree is not 1-spine, 0-bend drawable since it is not a path. If a tree is k-strict, then the three subtrees rooted at u_1, u_2 and u_3 are $(k-1)$-strict and require at least k spine to be drawn. In this case, there is no location for v on the k spines that allows it to connect to the three subtrees without creating a crossing. Based on this result about the strictness of a tree we have that the complete ternary tree of height[2] $k+1$ is not k-spine, 0-bend drawable because it is $(k+1)$-strict.

An interesting result shown in [FLW03] is that the strictness of a tree T is closely related to the pathwidth; more precisely, we have that the strictness s of T and the pathwidth p of T are such that $p \leq s \leq p+1$. This implies that if a tree has pathwidth p, then it is not k-spine, 0-bend drawable for $k < p$. The relationship between the pathwidth p of a tree T and the k-spine, 0-bend drawability of T has been further investigated by Suderman [Sud04] who proved that every tree with pathwidth h has a k-spine, 0-bend drawing with $p \leq k \leq \lceil 3p/2 \rceil$. Suderman [Sud04] also describes a linear-time drawing algorithm that computes a k-spine, 0-bend drawing for a tree with pathwidth p, where $k = \lceil 3p/2 \rceil$. In his paper, Suderman studies layered drawings of trees with pathwidth p not only within the general scenario, but also in several different constrained scenarios. We will describe these results in Section 8.4.

A summary of the results described in this section is presented in Table 8.1 (for the Characterization Problem) and in Table 8.2 (for the Drawability Testing Problem).

---

[2]The *height* is measured as the number of vertices on the path from the root to the deepest leaf.

	0 *bends*	1 *bend*	2 *bends*
1 *spine*	paths	sub-Hamiltonian [BK79]	planar [KW02]
2 *spines*	subclass of outerplanar [CSW04]	sub-Hamiltonian-with-handles [DDLS06]	planar [KW02]
$k > 2$ *spines*	OPEN (not all trees [FLW03])	OPEN (not all planar [DDLS06])	planar [KW02]

Table 8.1 Summary of the results about the Characterization Problem for different numbers of spines and bends.

	0 *bends*	1 *bend*	2 *bends*
1 *spine*	$O(n)$	$\mathcal{NP}$-complete [BK79]	always true
2 *spines*	$O(n)$ [CSW04]	$\mathcal{NP}$-complete [DDLS06]	always true
$k \geq 2$ *spines*	OPEN	$\mathcal{NP}$-complete [DDLS06]	always true

Table 8.2 Summary of the results about the Drawability Testing Problem for different numbers of spines and bends.

8.3.2 Radial Drawings in the General Scenario

In this section, we consider k-radial, b-bend drawings, and start with the case of a single circle. If no bend per edge is allowed, then the class of planar graphs that can be drawn on a circle trivially coincides with the class of outerplanar graphs, which can be recognized in linear time.

Theorem 8.6 *A planar graph is* 1-*radial,* 0-*bend drawable if and only if it is outerplanar.*

Drawings on one circle and at most one bend per edge have been studied in [DDLW05] where it is proved that every planar graph admits a planar 1-bend drawing on a semi-circle and therefore it is 1-radial, 1-bend drawable. More generally, in [DDLW05] it has been shown that, for every planar graph G, it is possible to define a linear ordering L of the vertices of G, called *curve embedding*, such that G admits a planar 1-bend drawing on any concave curve Λ where the vertices appear along Λ in the same order as in L. This rather surprising result says that, although not all planar graph can be drawn with one bend on a single spine, it is sufficient to "curve" this spine in order to support all of them. Thus, in one sense, a circle is "more powerful" than any number of spines, because, for any $k > 0$, we know that there are planar graphs that are not k-spine, 1-bend drawable.

The algorithm described in [DDLW05] to compute a 1-bend drawing on a semi-circle Λ of a maximal planar graph G uses the canonical ordering defined by de Fraysseix, Pach, and Pollack [dPP90]. Let G be a maximal embedded planar graph with external boundary u, v, w; a *canonical ordering* of G is an ordering $v_1 = u, v_2 = v, v_3, \ldots, v_{n-1}, v_n = w$ of the vertices of G such that for every $4 \leq k \leq n$:

- the subgraph G_{k-1} of G induced by $v_1, v_2, \ldots, v_{k-1}$ is biconnected and the external boundary C_{k-1} of G_{k-1} contains edge (u, v).

- v_k is on the external face of G_{k-1}, and its neighbors in G_{k-1} form a subpath of the path $C_{k-1} - (u, v)$ (see Figure 8.10).

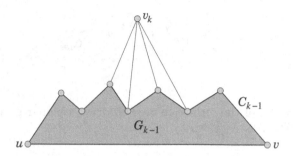

Figure 8.10 Illustration of the properties of the canonical ordering. Figure taken from [DDLW05].

Once a canonical ordering has been computed, the drawing algorithm in [DDLW05] first draws G_3 by placing vertices v_1, v_2, and v_3 at three arbitrary points of Λ in the order v_1, v_3, and v_2; the edges between them are drawn as straight-line segments. Vertices $v_4, v_5, \ldots, v_n$ are added one per step. At step k vertex v_k is placed and a planar 1-radial, 1-bend drawing Γ_k of G_k is computed. The algorithm guarantees that the following invariants hold for Γ_k:

- the clockwise order of the vertices along Λ is equal to the clockwise order they have on the external boundary C_k of G_k;
- each vertex c on the external boundary of C_k is drawn on Λ so that there exists two points α_c and β_c, such that no point of an edge (i.e., no vertex and no internal point of an edge) is encountered going clockwise along Λ between α_c and c and between c and β_c. The arc of Λ between α_c and c is called the *left safe region* of c while the arc of Λ between c and β_c is called the *right safe region* of c.

After the drawing of G_k has been computed, vertex v_{k+1} has to be added to the drawing. By the properties of the canonical ordering, v_{k+1} is adjacent to a set of vertices $w_1, w_2, \ldots, w_h$ that are consecutive on the external boundary of G_k and, by the first invariant, are consecutive along Λ. Vertex v_{k+1} is placed in the right safe region of w_1 (i.e., between w_1 and β_{w_1}). By the second invariant, this arc is "free," i.e., it does not contain any vertex or any crossing between an edge and Λ. Edge (w_1, v_{k+1}) is drawn as a straight-line segment, while each edge $e_i = (v_{k+1}, w_i)$ $(i = 2, \ldots, h)$ is drawn as a polyline with one bend by suitably choosing two intersection points between e_i and Λ. The first intersection point is a point of the arc of Λ between v_{k+1} and β_{w_1}, while the second is a point of the left safe region of w_i (i.e., it is a point between α_{w_i} and w_i). This choice of the two intersection points guarantees that edges $e_2, e_3, \ldots, e_h$ can be drawn without crossings. For an illustration of the incremental technique described above, see Figure 8.11. For an example of a 1-radial, 1-bend drawing of a planar graph, see Figure 8.12.

Theorem 8.7 *Every planar graph is 1-radial, 1-bend drawable.*

A 1-bend drawing on a semi-circle can be seen as an extension of a book embedding on two pages and, indeed, in [DDLW05] planar 1-bend drawings on a semi-circle are used to give an alternative proof of Theorem 8.2. Informally speaking, a planar 1-bend drawing on a semi-

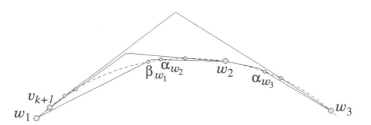

Figure 8.11   Illustration of the technique used to draw a planar graph on a semi-circle. Figure taken from [DDLW05].

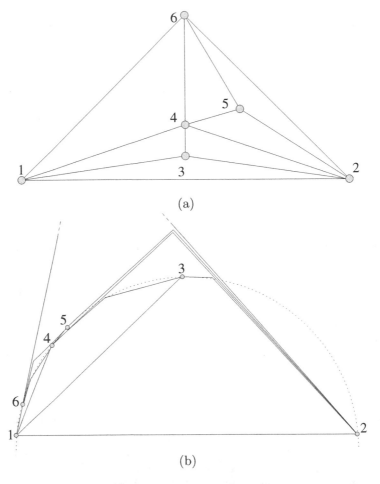

Figure 8.12   (a) A planar graph G (where the vertices are numbered according to a canonical ordering of G). (b) A 1-radial, 1-bend drawing. The linear ordering of the vertices along the (semi)-circle is different from the canonical ordering. Figure taken from [DDLW05].

circle is a topological book embedding where the spine is "bent." By "straightening" this "bent" spine, one can obtain a topological book embedding on two pages. More precisely, according to the algorithm presented in [DDLW05], each edge is either straight-line or it crosses the circle in two points (other than its endvertices). If we consider these two intersection points as two division vertices, then each edge (real or obtained by subdividing a real edge with two division vertices) is either straight-line and completely inside the (semi)-circle or it is bent and completely outside the (semi)-circle. A topological book embedding on two pages can now be computed by assigning edges inside the circle to one page (for example to the one corresponding to the half-plane below the spine), and edges outside the circle to the other page (for example, to the one corresponding to the half-plane above the spine). In the obtained topological book embedding each edge crosses the spine at most twice. However, the 1-bend drawing on a semi-circle is such that one of this spine crossing can be avoided. For an illustration, see Figure 8.13, for more details see [DDLW05].

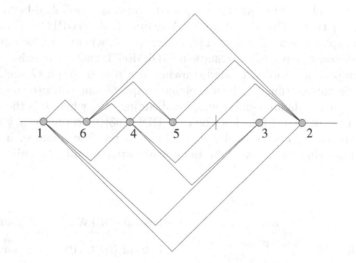

Figure 8.13 A 1-spine, 2-bend drawing of the graph of Figure 8.12, obtained by using the 1-radial, 1-bend drawing shown in Figure 8.12. Figure taken from [DDLW05].

In [DDLM05] k-radial, 0-bend drawings have been studied, for $k \geq 2$. The existence of a k-radial, 0-bend drawing of a planar graph G is related to the outerplanarity of G. The outerplanarity is defined as follows. A 1-*outerplanar embedded graph* (or simply *outerplanar embedded graph*) is an embedded planar graph where all vertices are on the external face. An embedded graph is a k-*outerplanar embedded graph* ($k > 1$) if the embedded graph obtained by removing all vertices of the external face is a $(k-1)$-outerplanar embedded graph. The planar embedding of a k-outerplanar embedded graph is called a k-*outerplanar embedding*. A graph is k-*outerplanar* if it admits a k-outerplanar embedding. A planar graph G has *outerplanarity* k (for an integer $k > 0$) if it is k-outerplanar but not $(k-1)$-outerplanar. In [DDLM05], it is proved that if a planar graph G admits a k-radial, 0-bend drawing, then its outerplanarity is at most k. The proof is by induction on the number of circles k. If G has a 1-radial, 0-bend drawing, then it is outerplanar by Theorem 8.6. Let Γ be a planar k-radial, 0-bend drawing of G. All the vertices that are on the most external circle in Γ are vertices of the external face because the drawing is planar and straight-line. Therefore, removing the vertices of the external face we are left with a $(k-1)$-radial, 0-

bend drawing and, by induction, with an embedded $(k-1)$-outerplanar graph. Therefore, G is an embedded k-outerplanar graph and its outerplanarity is at most k. In the same paper [DDLM05], an algorithm is presented to compute a k-radial, 0-bend drawing of a k-outerplanar embedded graph G. Figure 8.14 shows an example of a 2-radial, 0-bend drawing of a 2-outerplanar embedded graph. A consequence of these two results in [DDLM05] is that the class of graphs that are k-radial, 0-bend drawable, is the class of graphs with outerplanarity at most k.

Theorem 8.8 *A planar graph is k-radial, 0-bend drawable (k > 1) if and only if its outerplanarity is at most k.*

Theorem 8.8 implies that, in order to test whether a planar graph is k-radial, 0-bend drawable, one has to compute the outerplanarity of a planar graph G. In [DDLM05], it is stated that this can be done in $O(n^5 \log n)$ time based on a result by Bienstock and Monma [BM90]. Recently, this result has been improved to $O(n^4)$ by Angelini et al. [ADP11]; as a consequence the problem of deciding whether a planar graph is k-radial, 0-bend drawable can be solved in $O(n^4)$ time. The algorithm by Angelini et al. [ADP11] can also be used to compute a k-outerplanar embedding of a planar graph G, where k is the outerplanarity of G. Thus, another consequence of the results in [DDLM05] is that there exists an $O(n^4)$-time algorithm to compute a k-radial, 0-bend drawing of a planar graph G such that k is the minimum possible value. Namely, given a planar graph G, one can use the Angelini et al. algorithm to compute a planar k-outerplanar embedding of G where k is the outerplanarity of G and then use the algorithm described in [DDLM05] to compute a k-radial, 0-bend drawing. The number of circles used is the minimum possible because, if G admitted a h-radial, 0-bend drawing for $h < k$, then its outerplanarity would be smaller than k.

	0 bends	1 bend	2 bends
1 circle	outerplanar	all planar [DDLW05]	all planar [DDLW05]
$k \geq 2$ circles	outerplanarity $\leq k$ [DDLM05]	all planar [DDLW05]	all planar [DDLW05]

Table 8.3 Summary of the results about the Characterization Problem for different numbers of circles and bends.

	0 bends	1 bend	2 bends
1 circle	$O(n)$	always true	always true
$k \geq 2$ circles	$O(n^4)$ [ADP11]	always true	always true

Table 8.4 Summary of the results about the Drawability Testing Problem for different numbers of circles and bends.

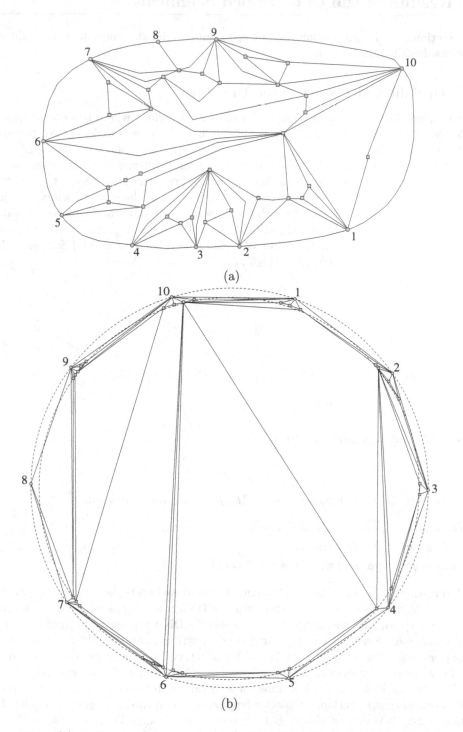

Figure 8.14 (a) A 2-outerplanar embedded graph G. (b) A 2-radial, 0-bend drawing of G. Figure taken from [DDLM05].

8.4   Results in the Constrained Scenarios

In this section, we describe results about spine and radial drawings with the additional constraints described in Section 8.2.

8.4.1   Upright and Proper Spine Drawings

We start by considering upright spine drawings, i.e., drawings where intra-layer edges are not allowed. This constraint implies that the number of layers is at least two, because on a single layer only isolated vertices can be represented. The characterization of upright 2-spine, 0-bend drawable graphs can be stated in several different but equivalent ways. A graph is a *caterpillar* if it consists of a simple path and degree-one vertices attached to this path. A 2-*claw* is a graph consisting of one vertex of degree 3 (the *center*), which is adjacent to three degree-two vertices, each of which is adjacent to the center and to a vertex of degree one. These definitions are illustrated in Figure 8.15. The following characterizations can be found in the works of Eades, McKay and Wormald [EMW86], Harary and Schwenk [HS72], and Tomii, Kambayashi, and Yajima [TKY77].

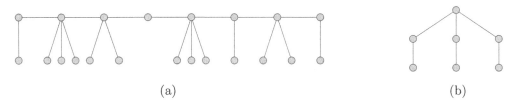

(a)   (b)

Figure 8.15   (a) A caterpillar. (b) A 2-claw.

Theorem 8.9   *Let G be a planar graph. The following are equivalent.*

1. *G is upright 2-spine, 0-bend drawable.*
2. *G is a forest of caterpillars.*
3. *G is acyclic and does not contain a 2-claw.*

An interesting work about upright 2-spine, 0-bend drawings is the one by Waterman and Griggs [WG86]. In this paper, the authors study a DNA mapping problem with applications in biology. Very roughly speaking, we have a specific DNA sequence that can be "cut" by means of enzymes. Each cut can be modeled as a partition of a straight line into intervals. Different enzymes give different cuts, i.e., different intervals. Biologists are interested in the order of the "pieces" (intervals) in the sequence, but they cannot directly observe this order. Instead, they can easily establish if different intervals of different cuts (i.e., cuts produced by different enzymes) overlap. This overlapping between intervals can be modeled as a bipartite graph. Namely, let A and B be two cuts of the same DNA sequence. We define a vertex for each interval $a_i \in A$, a vertex for each interval $b_i \in B$, and an edge (a_i, b_j) with $a_i \in A$ and $b_j \in B$ iff a_i and b_j overlap. The problem of reconstructing the two orders of the intervals in A and B can be modeled as the problem of finding an ordering of the vertices in A and an ordering of the vertices in B such that they are "consistent" with the given overlaps. But this means to find a layout of the bipartite graph on two straight lines such that there is no edge crossing. In other words, the problem of reconstructing the two orders of the intervals in A and B is equivalent to the problem of computing an upright 2-spine, 0-bend drawing of the bipartite graph representing the overlaps. Waterman and Griggs study the properties of this bipartite graph, prove that it is a caterpillar and give a linear-time algorithm to compute an upright 2-spine, 0-bend drawing.

Remaining in the case of upright drawings, when the number of spines is greater than two, the problem is different depending on whether one admits long edges (i.e., edges that span more than one level) or not.

In the case when long edges are not allowed, i.e., the case of upright proper drawings[3], Heath and Rosenberg [HR92] show that the drawability testing problem is $\mathcal{NP}$-complete if the number of spines is not fixed. By using the theory of the parametrized complexity, Dujmović et al. [DFK+08] prove that it is possible to decide whether a planar graph G admits an upright proper k-spine, 0-bend drawing in $O(f(k) \cdot n)$. This implies that, for a fixed number of layers k, k-spine, 0-bend drawable graphs can be recognized in linear time. However, the dependency of time complexity from k is given by $f(k) = 2^{32 \cdot k^3}$, which gives impractical large constants also for small values of k.

Fößmeier and Kaufmann [FK97] studied upright proper 3-spine, 0-bend drawable graphs, gave a characterization of them, and presented a linear-time algorithm to recognize them. Recently, Suderman [Sud05] pointed out some errors in the work by Fößmeier and Kaufmann and, based on the ideas found there, presented a new characterization and a new linear-time algorithm to recognize upright proper 3-spine, 0-bend drawable graphs. The characterization presented by Suderman consists of constraints on vertices and biconnected components. For example, it is not difficult to see that if C is a biconnected component of an upright proper 3-spine, 0-bend drawable graph, then $G - C$ contains at most two connected components that are not upright proper 2-spine, 0-bend drawable. However, this in itself is not sufficient to guarantee upright proper 3-spine, 0-bend drawability. Consequently, additional constraints must be defined. Suderman describes constraints on vertices and biconnected components that guarantee upright proper 3-spine, 0-bend drawability. Such constraints cannot be easily summarized. The interested reader is referred to the original work by Suderman [Sud05].

Upright spine drawings (proper or not) have also been studied by Suderman [Sud04] in the case of trees with pathwidth p. Suderman proves that every tree with pathwidth p admits an upright k-spine, 0-bend drawing with $p \le k \le \lceil 3p/2 \rceil$ and an upright proper k-spine, 0-bend drawing with $p \le k \le \lceil 3p - 3 \rceil$. Suderman also proves that these bounds are optimal and present linear-time algorithms that, given a tree with pathwidth p, compute an upright k-spine, 0-bend drawing where $k = \lceil 3p/2 \rceil$ and an upright proper k-spine, 0-bend drawing where $k = \lceil 3p - 3 \rceil$. In the same paper [Sud04], Suderman studies proper (non-upright) spine drawings of trees with pathwidth p. In this case, a lower bound of p and an upper bound of $2p - 1$ on the number of spines in a proper spine drawings of a tree with pathwidth p are given. Also in this case the bounds are optimal and a linear-time algorithm exists to compute a proper k-spine, 0-bend drawing with $k = 2p - 1$ of a tree with pathwidth p.

---

[3]These drawings are usually called simply *proper layered drawings*.

8.4.2 Partitioned Spine Drawings

As explained in Subsection 8.2.2, in the partitioned layered drawing problem the input graph is partitioned into subsets of vertices, and all vertices in the same set must be drawn on the same layer.

The special case of partitions into two sets have been studied in the literature with two different assumptions: (i) vertices of a same set are never adjacent; (ii) vertices of a same set can be adjacent. Observe that partitioned k-spine, 0-bend drawings of a bipartite planar graph with $k \in \{2,3\}$ can be regarded as upright proper k-spine, 0-bend drawings of (non-bipartite) planar graphs. Namely, if a planar graph admits an upright 2-spine, 0-bend drawing, then the vertices on each spine are not adjacent and therefore the graph is bipartite. Analogously, if a planar graph admits an upright proper 3-spine, 0-bend drawing then the vertices on the middle spine are adjacent to the vertices on the top spine and to the vertices on the bottom spine and there is no edge on each spine. This means that the vertices in the middle spine form a set of the bipartition and the vertices in the top and bottom spines form the other set. Thus, the results about upright 2-spine, 0-bend drawings and upright proper 3-spine, 0-bend drawings can also be regarded as results for bipartite graphs.

Biedl [Bie98] characterizes the family of planar graphs that admit a partitioned 2-spine, 0-bend drawing, where vertices in the same set (layer) can be adjacent. Starting from a partitioned planar graph $G = (A \cup B, E)$ Biedl constructs a graph G' whose vertex set is $A \cup B \cup \{v_a, v_b\}$. Vertex v_a is connected to all the edges in A, vertex v_b is connected to all the edges in B, and v_a and v_b are adjacent. Graph G admits a partitioned 2-spine, 0-bend drawing if and only if G' is planar and there exists a planar embedding of G' such that any triangle containing v_a or v_b is a face (see Figure 8.16).

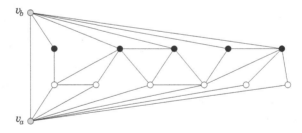

Figure 8.16   The graph G' constructed by Biedl [Bie98] in order to compute a partitioned 2-spine, 0-bend drawing of a partitioned planar graph G.

Partitioned layered drawings on three layers have been studied by Cornelsen et al. [CSW04] who considered partitioned (non-bipartite) planar graphs with the additional property that every B-vertex of degree one is adjacent to an A-vertex. Cornelsen et al. derive a graph G' from the input graph G by means of a suitable transformation and prove that G admits a partitioned 3-spine, 0-bend drawing if and only if G' admits a 2-spine, 0-bend drawing. Since G' can be computed in linear time and 2-spine, 0-bend drawability can be tested in linear time (see Section 8.3.1), we have that partitioned 3-spine, 0-bend drawable graphs can be recognized in linear time.

We remark that several other models have been introduced in the literature to draw partitioned planar graphs. We recall, for example, the LH-drawings, where only one set of the partition is required to be on a straight line while the other is drawn in one of the two half-space defined by the line itself, and the HH-drawings, where each set is drawn in one of the two half-planes defined by a straight line. These drawings, however, are not layered drawings, and therefore, we do not describe the results about them here. The interested reader is referred to the literature [Bie98, BKM98].

8.4.3 Radial Drawings with Assigned Layers

As discussed in Subsection 8.3.2 for the general scenario, a planar graph is k-radial, 0-bend drawable ($k > 1$) if and only if it has outerplanarity at most k. The algorithm that computes a k-radial, 0-bend drawing strongly relies on the possibility of choosing the radius of each circle, and therefore the distance between every two consecutive layers. This often leads to consecutive layers that are very close to each other, and the angular resolution of the drawing becomes very poor. To improve the readability of radial drawings, consecutive layers should be at least at a given distance that can be specified as part of the input. However, if the layers are given the drawability problem cannot be tackled with the technique described in [DDLM05]. Providing a complete characterization in this case is still an open problem. Partial results are given in [DD03] and in [DGL08]. In [DD03] it is proven that the family of 2-outerplanar embedded graphs whose internal vertices induce a biconnected graph are 2-radial, 0-bend drawable. The drawing can be computed in linear time in such a way that the internal vertices are placed on the internal circle and the external vertices are placed on the external circle. The idea of the drawing technique is as follows (refer to Figure 8.17).

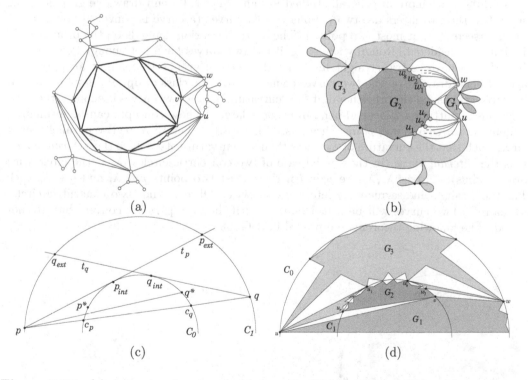

Figure 8.17 (a) A 2-outerplanar embedded graph G where the internal vertices induce a biconnected graph. (b) The structure of G decomposed into three edge-disjoint outerplanar embedded graphs. (c) Notation used in the description of the drawing algorithm. (d) A 2-radial, 0-bend drawing of G on any two given given circles. Figure taken from [DD03].

Let $G = (V, E)$ be the 2-outerplanar embedded graph given as input, let C_0 and C_1 be the external and the internal circles given in input, and let V_0 and V_1 be the external vertices and the internal vertices of G. The algorithm places all the vertices on two parallel semi-circles of C_0 and C_1. First, it chooses two distinct points, p and q, of C_0 such that: (i) the x- and y-coordinate of p is less than the x- and the y-coordinate of q, respectively; (ii) segment $\overline{pq}$ is a chord of C_0 that has two intersection points c_p, c_q with C_1, where c_p is the first point encountered while walking on $\overline{pq}$ from p to q; (iii) there are two lines t_p and t_q passing for p and q, respectively, that are tangent to C_1, and intersecting in a point lying in the portion of the annulus delimited by C_1 and C_0. Denote by $p_{ext} \neq p$ and p_{int} the points where t_p intersects C_0 and C_1, respectively. Similarly, let $q_{ext} \neq q$ and q_{int} be the points where t_q intersects C_0 and C_1, respectively. Also denote by q^* any point of C_1 between q_{int} and c_q, and by p^* any point of C_1 between p_{int} and the point $\overline{pq^*} \cap C_1$.

Then the algorithm maps all the vertices of V_1 to points of C_1, according to the clockwise order they appear on the external boundary of $G(V_1)$, in such a way that: (i) u_r and u_1 are mapped to p^* and p_{int}, respectively; (ii) w_s is mapped to q_{int}; (iii) v is mapped to q^*.

Also, it maps all vertices of V_0 to points of C_0, according to the clockwise order they appear on the external boundary of G, in such a way that: (i) u and w are mapped to p and q, respectively; (ii) all vertices from u to w are mapped to points between q_{ext} and p_{ext} (iii) all vertices from w to u are mapped to points below q.

A characterization of upright 2-radial, 0-bend drawable graphs is given by Di Giacomo et al. [DGL08] who, more in general, studied upright 2-layer, 0-bend drawable graphs in the case when the two layers are two parallel convex curves (a curve is convex if any straight line intersects it in at most two points). The characterization depends on the properties of the curves considered. Roughly speaking, if the two curves have not enough "curvature," then they behave as two straight lines and the class of graphs that admit an upright 2-layer, 0-bend drawing on the two curves coincides with the class of upright 2-spine, 0-bend drawable graphs; on the other hand, if the "curvature" of the two curves is enough, the class of graphs admitting a 2-layer, 0-bend drawing is larger. These concepts can be formalized by defining *paired* and *non-paired* curves (see Figure 8.18). Let λ_e, λ_i be two parallel convex curves such that the curvature of λ_e is less than the curvature of λ_i; λ_e is the *external curve*, λ_i is the *internal curve* (in the special case of two concentric circles, λ_e is the circle with larger radius). Curves λ_e, λ_i are *paired* if there exist two points $p \in \lambda_i$ and $q \in \lambda_e$ such that the straight-line segment $\overline{pq}$ intersects λ_i twice. Observe that two concentric circles are paired. Two curves will be called *non-paired* if they are parallel, convex, but are not paired. The following theorems are proved in [DGL08].

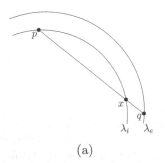
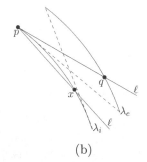

(a)   (b)

Figure 8.18   (a) Two paired curves. (b) Two non-paired curves. Figure taken from [DGL08].

Theorem 8.10 *Let C be a set of layers consisting of two non-paired curves and let G be a planar graph. G admits an upright 2-layer, 0-bend drawing on C if and only if G is a forest of caterpillars.*

Theorem 8.11 *Let C be a set of layers consisting of two paired curves and let G be a planar graph. G admits an upright 2-layer, 0-bend drawing on C if and only if G is bipartite and admits a planar embedding such that all vertices of one partite set belong to the external face.*

The proof of Theorem 8.10 is an easy adaptation of the proof of Theorem 8.9. The necessity of Theorem 8.11 can be easily proved as follows. Since the drawing is upright the graph must be bipartite with each partite set defined by the vertices drawn on each curve. Also, since the drawing is straight-line and planar, it defines a planar embedding in which all vertices of the external curve are on the external face. As for the sufficiency, Di Giacomo et al. describe a drawing algorithm based on a suitable decomposition of the graph called *bipartite fan decomposition*. A bipartite fan is a biconnected bipartite planar graph having a vertex u, called *apex*, that is shared by all its faces (including the external one). Let $u, v_0, v_1, \ldots, v_{n-2}$ be the vertices of a fan G in the counterclockwise order they have on the external face. Any three vertices $v_{2j}, v_{2j+1}, v_{2j+2}$ ($0 \leq j \leq \frac{n-4}{2}$) form a *fan triplet* of G. Notice that v_{2j+1} belongs to the same partite set as u. See Figure 8.19 (a) for an example of a bipartite fan.

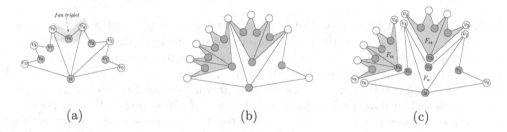

(a) (b) (c)

Figure 8.19 (a) A bipartite fan. (b) A bipartite graph G embedded with all vertices of one partition set on the external face. (c) A bipartite fan decomposition of G. Figure taken from [DGL08].

Given a biconnected bipartite graph embedded with all vertices of one partition set on the external face,[4] it is possible to decompose it into bipartite fans as follows. A first bipartite fan F_u is computed; the two edges of each fan triplet of F_u either belong to the external face, or they are a cut-set for G and they identify a subgraph that can be recursively decomposed (see Figure 8.19 (b) for an example of bipartite fan decomposition). Once G has been decomposed, a wedge W_u is defined on the paired curves; a wedge is a portion of plane delimited by the external curve λ_e and by two segments having and endpoint on each curve, one of which has two intersections with the internal curve λ_i (see Figure 8.20 (a) for an example). Fan F_u is drawn inside W_u as shown in Figure 8.20 (b). Notice that the drawing of F_u is such that each fan triplet defines a new wedge where the subgraph

---

[4]If the input graph is not biconnected, it can be augmented with vertex and edge addition to became biconnected while maintaining all the vertices of one partition set on the external face. For details, see [DGL08]

identified by the fan triplet can be recursively drawn. We conclude by mentioning that based on Theorem 8.11 upright 2-layer, 0-bend drawable graphs can be recognized in linear time and that, when the two paired curves are two circles, an upright 2-radial, 0-bend drawing can be computed in linear time [DGL08].

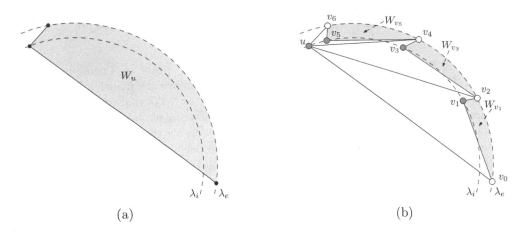

(a)   (b)

Figure 8.20   (a) A wedge W_u defined on two paired curves. (b) A 2-layer, 0-bend drawing of the fan F_u of Figure 8.19 inside W_u.

Di Giacomo et al. [DDL08b] studied k-radial drawings of graphs with assigned layers and a prescribed assignment of vertices to the layers. More precisely, the layers are concentric circles such that the difference between the radii of any two consecutive circles is constant and equal to the radius of the smallest circle. Also, a function $\phi : V \to \{0, 1, \dots, k-1\}$ is given and it is required that each vertex $v \in V$ is drawn as a point of circle $C_{\phi(v)}$. A planar graph G equipped with such a function is called a *layered planar graph*. We observe that the assignment of vertices to layers described by the function ϕ represents a stronger constraint than assigning a vertex partition. In [DDL08b], k-radial drawings with different trade-offs between the maximum number of bends along an edge and the *angular distance ratio* are studied. The angular distance ratio measures how uniform is the angular distribution of the vertices. More precisely, let $\rho_0, \rho_1, \dots, \rho_{h-1}$ ($h \geq 1$) be the distinct rays passing through the vertices in the order they are encountered in a radial sweep of the drawing. If $h > 1$, define $\alpha_i = (\angle \rho_{i+1} - \angle \rho_i)$ (the indices are taken modulo h and the angles are measured modulo 2π), $\alpha_{min} = \min_i\{\alpha_i\}$ and $\alpha_{max} = \max_i\{\alpha_i\}$. If $h = 1$, we define $\alpha_{min} = 0$ and $\alpha_{max} = 2\pi$. The angular distance ratio is defined as $ADR = \frac{\alpha_{max}}{\alpha_{min}}$. Notice that, when $h = 1$ we have $ADR = +\infty$.

In [DDL08b], it is first proved that there exist layered graphs that do not admit a k-radial, 0-bend drawing satisfying the vertex assignment that have optimal angular distance ratio (i.e., $ADR = 1$). The graph $G = (V, E, \phi)$ is defined as follows (refer to Figure 8.21). The set of vertices is $V = \{u_0, u_1, \dots, u_{h-1}\} \cup \{v_0, v_1, \dots, v_{h-1}\} \cup \{w_0, w_1, \dots, w_{h-1}\}$ with $h \geq 3$; the set of edges is $E = \{(u_i, u_{i+1}), (v_i, u_i), (v_i, u_{i+1}), (w_i, u_i), (w_i, u_{i+1}), (w_i, v_i) \mid 0 \leq i \leq h-1\}$ (indices are taken modulo h), $\phi(u_i) = 0$, $\phi(v_i) = 0$, and $\phi(w_i) = 1$ ($i = 0, \dots, h-1$). Consider now a 3-cycle u_i, u_{i+1}, v_i ($i = 0, \dots, h-2$). All the vertices of the cycle must be drawn on circle C_0 and if we want $ADR = 1$ the angle between the two rays passing through u_i and u_{i+1} must be $\frac{2\pi}{h}$. Vertex w_i must be drawn on circle C_1 and, in order to guarantee planarity, w_1 must be inside the triangle representing the 3-cycle u_i, u_{i+1}, v_i. It follows that circle C_1 must cross the segment representing the edge (u_i, u_{i+1}); thus, it must be $r_1 \geq r_0 \cos(\frac{\pi}{h})$, but this is possible only if $h < 3$ because $r_1 = \frac{1}{2}r_0$.

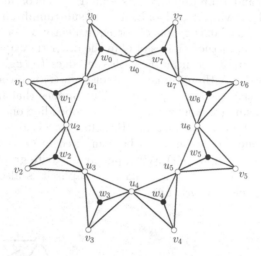

Figure 8.21 A layered planar graph that does not admit a k-radial, 0-bend drawing with optimal angular distance ratio if the white vertices are assigned to layer 0 and the black vertices are assigned to layer 1. Figure taken from [DDL08b].

The negative result above motivates the study of k-radial drawings with bends. Di Giacomo et al. [DDL08b] prove that every layered planar graph G admits a k-radial, 3-bend drawing consistent with the assignment of the vertices the layers having optimal angular distance ratio. Such a drawing can be computed in linear time. It is interesting to note that the drawing algorithm exploits the connection between 1-spine, 2-bend drawings, topological book embeddings, and Hamiltonicity observed in Section 8.3.1 and that will be explained in detail in Section 8.5.1. By using the Hamiltonian augmentation technique described in [DDLW05], a Hamiltonian augmentation $\mathsf{Ham}(G)$ of G and an augmenting Hamiltonian cycle H of G are computed. The cycle H is drawn with straight-line edges (and each vertex v drawn on circle $C_{\phi(v)}$). All the remaining edges are either inside H or outside it in the planar embedding of $\mathsf{Ham}(G)$. The edges that are outside H are drawn as a 2-bend polyline outside the polygon representing H; the edges that are inside H are drawn as a 1-bend polyline inside the polygon representing H. The properties of the cycle H computed with the augmentation technique of [DDLW05] guarantees that edges subdivided with a division vertex have at most three bends.

In [DDL08b], a drawing algorithm to compute a k-radial, 2-bend drawing consistent with the assignment of the vertices to the layers is also presented. In this case, however, the angular distance ratio is not optimal.

8.5 Related Problems

In this section, we present two applications of the results described in Subsections 8.3.1 and 8.3.2. The first application is in the field of graph theory and the second one is in computational geometry.

8.5.1   Hamiltonicity

We have already seen in the description of Section 8.3.1 that there is a connection between
1-spine, 1-bend drawings and Hamiltonicity. As stated by Theorem 8.1, a planar graph
admits a 1-spine, 1-bend drawing if and only if it is sub-Hamiltonian. Given a planar
1-spine, 1-bend drawing of a planar graph G (or equivalently a book embedding on two
pages), denote by $v_0, v_1, \ldots v_{n-1}$ the vertices of G in the order they appear along the spine.
$\mathsf{Ham}(G)$ can be computed by augmenting G with the edges (v_i, v_{i+1}) (indices are taken
modulo n) that are not in G. A Hamiltonian cycle of $\mathsf{Ham}(G)$ is given by the sequence of
edges $(v_0, v_1), (v_1, v_2), \ldots, (v_{n-2}, v_{n-1}), (v_{n-1}, v_0)$. This implies that if one can compute a
planar 1-spine, 1-bend drawing (or equivalently a book embedding on two pages) efficiently,
then it is also possible to find an augmenting Hamiltonian cycle of G efficiently. Since
a book embedding on at most two pages can be computed in $O(n)$ time for outerplanar
graphs [BK79], series-parallel graphs [DDLW06], planar bipartite graphs [ddMP95], square
grids and X-trees [CLR87], for all these families of graphs it is also possible to find an
augmenting Hamiltonian cycle in $O(n)$ time.

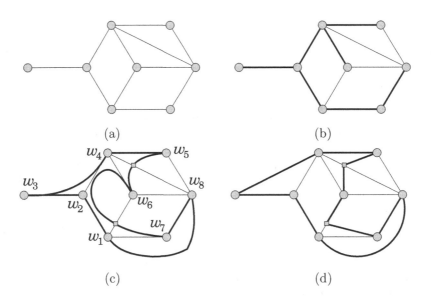

Figure 8.22   The PW Augmentation Technique [PW01]. (a) A planar graph G. (b) A
spanning tree S of G. (c) Visit of S. (d) The resulting Hamiltonian augmentation of G.
Figure taken from [PW01].

In the general case of planar graphs, there are different techniques to compute $\mathsf{Ham}(G)$
and a Hamiltonian cycle of $\mathsf{Ham}(G)$, i.e., an augmenting Hamiltonian cycle of G. A first
technique is the one described by Pach and Wenger [PW01] (see also Figure 8.22), which
we will call the *PW Augmentation Technique*. Let S be a spanning tree of G and let Γ be a
planar drawing of G. Starting at any vertex, walk clockwise around S, visiting its vertices
in order. Note that the internal vertices of S will be visited more than once. Label the
vertices with $w_1, w_2, \ldots, w_n$ by the order in which they are first visited. If w_i and w_{i+1} are
connected by an edge, then let this edge belong to the Hamiltonian cycle ($1 \leq i \leq n$ and
assume that the indices are taken modulo n). If not, connect w_i to w_{i+1} by a simple curve

clockwise around the boundary of S, passing very close to it. Wherever this curve intersects an edge of G, introduce a new vertex. This curve becomes a path whose pieces are added as edges to the graph and to its Hamiltonian cycle. Multiple edges (if any) are merged and the resulting graph is Ham(G). It can be proved that, using the PW Augmentation Technique, each edge is split at most twice. Since G has at most $3n - 6$ edges and the edges of S are not split, Ham(G) has at most $5n - 10$ vertices.

An alternative technique to compute a Hamiltonian augmentation of G is the one described in the work by Kaufman and Wiese [KW02], which we will denote as the *KW Augmentation Technique*. This technique is based on the fact that every 4-connected graph is Hamiltonian and a Hamiltonian cycle can be found in $O(n)$ time [CN89]. Thus, the idea of Kaufman and Wiese is to make a graph 4-connected. Assume that the input graph is maximal planar (if not it can be augmented in linear time to a maximal planar graph). By using an algorithm by Chiba and Nishizeki [CN85] one can find the separating triangles of G in $O(n)$ time. Each separating triangle can be removed by using the following approach. Let $e = (u, v)$ be and edge of a separating triangle. Since G is maximal planar, e is shared by two triangular faces u, v, w and u, v, z. Edge e is replaced by a chain consisting of two edges (u, d), (d, v) and a division vertex d. Furthermore, edges (d, w) and (d, z) are added to the graph (see Figure 8.23). By applying this transformation the separating triangle has been removed and no other separating triangle is created. Thus repeatedly applying this technique for every separating triangle we eventually obtain a 4-connected graph, which therefore is a Hamiltonian augmentation Ham(G) of G. The algorithm by Chiba and Nishizeki [CN89] can then be applied to find a Hamiltonian cycle in Ham(G). The KW Augmentation Technique splits each edge with at most one division vertex, therefore Ham(G) has at most $4n - 6$ vertices.

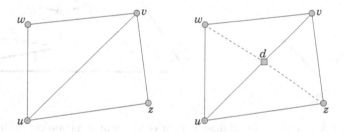

Figure 8.23 The augmentation described by Kaufmann and Wiese [KW02] to make a planar graph 4-connected.

The curve embedding defined in [DDLW05] can be used to define another alternative technique to compute a Hamiltonian augmentation of G, which will be called in the following the *DDLW Augmentation Technique*. As explained in Section 8.3.2, a curve embedding of a planar graph is a linear ordering L of the vertices of G such that G admits a planar 1-bend drawing on any concave curve Λ where the vertices appear along Λ in the same order as L. In particular, such an ordering can be computed by drawing G on a semi-circle with at most 1 bend per edge according to the technique described in Section 8.3.2. As already explained in Section 8.3.2, by using the 1-bend drawing on a semi-circle we can obtain a topological book embedding of G on two pages where each edge crosses the spine at most once. If we consider the crossings between the edges and the spine as division vertices of the edges, we have a book embedding on two pages of a subdivision sub(G) of G. Graph sub(G) has at most one division vertex per edge. Since sub(G) admits a book embedding on two pages, it is sub-Hamiltonian and we can augment it with edge addition so to make it Hamiltonian. As explained above this can be done by adding edges between non adjacent vertices that are consecutive along the spine of the book embedding and between the first and the last vertex on the spine if such an edge does not exist (see Figure 8.24). With the DDLW Augmentation Technique, each edge is split at most once and therefore, like in the case of the KW Augmentation Technique, $\mathsf{Ham}(G)$ has at most $4n-6$ vertices. However, the DDLW Augmentation Technique does not require to preliminarily make G 4-connected. The augmenting Hamiltonian cycle H of G computed by the DDLW Augmentation Technique has another interesting property. Let d be a division vertex that subdivide the edge (u,v), and consider the linear ordering of both the real vertices and the division vertices defined by the topological book embedding of $\mathsf{sub}(G)$ used to compute $\mathsf{Ham}(G)$. The division vertex d is encountered after u and before v in the considered order. This is a consequence of the fact that, according to the algorithm described in [DDLW05], the crossing between an edge and the spine always falls between the end-vertices of the edge. We say that all the division vertices of H are *flat* with respect to the considered order. The flatness of the division vertices will be used in the next application.

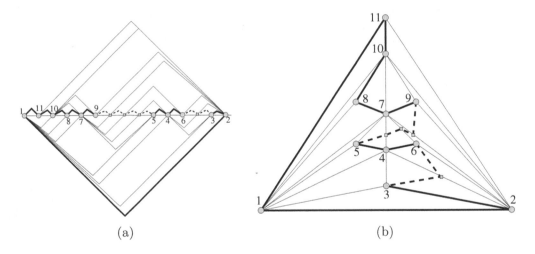

(a)   (b)

Figure 8.24   (a) A 1-spine, 2-bend drawing (or equivalently a topological book embedding on two pages) of the non-Hamiltonian graph G of Figure 8.4 obtained by using the curve embedding. (b) A Hamiltonian augmentation of G.

8.5.2   Point-Set Embeddability

The results described in Subsections 8.3.1 and 8.3.2 can be applied to the point-set embedding problem, which is widely investigated both in graph drawing and in computational geometry. Let G be a planar graph with n vertices and let S be a set of n points in the plane, a *point-set embedding* of G onto S is a planar drawing of G such that each vertex of G is represented by a point of S. Observe that there are two main variants of this problem, depending on whether the mapping between the vertices and the points is given as a part of the input or not. If the mapping is not given and the points are in general position, then every outerplanar graph admits a point-set embedding on any given set of points and straight-line edges [Bos02]. In [Bos02], an $O(n \log^3 n)$-time algorithm is also presented to compute a straight-line point-set embedding of an outerplanar graph G on a given set of points. For trees, an optimal $\Theta(n \log n)$-time algorithm is given by Bose et al. [BMS97], who improve previous results by Ikebe et al. [IPTT94] and Pach and Törőcsik [PT93].

The problem of deciding whether there exists a point set embedding with straight-line edges of a planar graph on a given set of points is, in general, $\mathcal{NP}$-hard [Cab06]. Since outerplanar graphs are the largest class of graphs admitting a straight-line point-set embedding on *every* set of points [GMPP91] in general position, Kaufmann and Wiese [KW02] investigate the problem of computing a point-set embedding of a planar graph with a small number of bends per edge. They show that any planar graph admits a point-set embedding with at most two bends per edge on any given set of points, and that two bends are required in some cases. Pach and Wenger [PW01] show that, if the mapping of the vertices of G to the points of P is given, then a planar drawing of G exists with $O(n)$ bends per edge and that $\Omega(n)$ bends per edge may be necessary even for paths. Recently, the two main variants (with or without mapping) have been unified and generalized by introducing the concept of coloured point-set embedding where the set of vertices and the set of points are coloured with k colours and it is required that each vertex is drawn on a point with the same colour [BDL08, DDL+08a, DLT06, DGLT10]. Badent et al. [BDL08] generalized the result by Pach and Wenger by proving that, for every $k \geq 2$, a k-coloured planar graph admits a k-coloured point-set embedding on every k-coloured set of points with $O(n)$ bends per edge. They also show that $\Omega(n)$ bends may be necessary.

We briefly recall here the technique of Kaufmann and Wiese [KW02] and highlight connections between this technique and spine drawings. Assume first that the input graph G is (sub)-Hamiltonian. Let $H = v_1, v_2, \ldots, v_n$ be a (augmenting) Hamiltonian cycle in G, and let Ψ be a planar embedding of G such that edge (v_1, v_n) lies on the external face. Let $p_1, p_2, \ldots, p_n$ be the sequence of points in S ordered by increasing x-coordinates (we can assume that all the points have distinct x-coordinates because, if not, we can rotate the plane to achieve this condition). Assign each vertex v_i to point p_i in P and draw the edges of path $P = H \setminus \{(v_1, v_n)\}$ as straight-line segments. Draw each remaining edge e using two segments, one with slope $\sigma > 0$ and the other with slope $-\sigma$. In order to prevent e from crossing the previously drawn edges, the slope σ is chosen to be greater than the absolute value of the slope of each edge in P. With segments of slope $\pm\sigma$, it is possible to draw e above or below P. Edge e is drawn above P if e is on the left-hand side when walking from v_1 to v_n in G, and below P otherwise. The resulting drawing is planar except that edges outside P incident to the same vertex may contain overlapping segments. To eliminate overlapping, perturb overlapping edges by decreasing the absolute value of their segment slopes by slightly different amounts (see [KW02] for details).

When the input graph G is not Hamiltonian, Kaufmann and Wiese compute a Hamiltonian augmentation $\mathsf{Ham}(G)$ of G by using the KW Augmentation Technique described in Section 8.5.1. Since $\mathsf{Ham}(G)$ has more vertices than G, the set of points S is also enriched with extra points at suitable positions. $\mathsf{Ham}(G)$ can be point-set embedded as described above. Some edges of G are split into two pieces in $\mathsf{Ham}(G)$. Let $e = (u, v)$ be an edge of G split by a division vertex d in G'. The edge e is replaced by the two edges (u, d) and (d, v); each of these two edges may have one bend. Furthermore, the two segments incident to d can have different slopes, thus creating a third bend at d. Hence, each edge of G is drawn with at most three bends. In order to remove the third bend, Kaufmann and Wiese rotate the segments incident to d and make them both vertical. Note that this may imply to rotate other segments that are "above" or "below" the rotating segments. An example of a point-set embedding of the non-Hamiltonian graph G of Figure 8.4 computed by the Kaufmann and Wiese technique is shown in Figure 8.25.

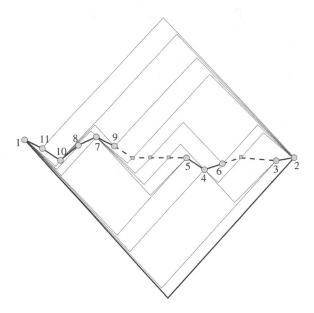

Figure 8.25   A point-set embedding of the non-Hamiltonian graph G of Figure 8.4. The drawing is created with the Kaufmann and Wiese technique [KW02] and using the Hamiltonian cycle (highlighted in the picture) shown in Figure 8.24.

As one can see from the description above, computing an augmenting Hamiltonian cycle of the input graph G plays an important role in the Kaufmann and Wiese technique. As discussed in Subsection 8.5.1, Hamiltonicity is related to spine and radial drawings. A first consequence of this fact is that one can compute a point-set embedding on any set of points with at most 1 bend per edge for all those families of (sub)-Hamiltonian graphs for which a (augmenting) Hamiltonian cycle can be found efficiently. In Section 8.5.1, we have seen that among these families we have outerplanar graphs, series-parallel graphs, planar bipartite graphs, square grids, and X-trees.

Another connection between spine and radial drawings, Hamiltonicity, and point-set embeddings, is given by the fact that one can use the DDLW Augmentation Technique (see Section 8.5.1) as an alternative to the KW Augmentation Technique to compute a Hamiltonian augmentation of the input graph G. The DDLW Augmentation Technique has the advantage that the rotation needed to avoid the third bend is not required. If $e = (u, v)$ is an edge of G split by a division vertex d, the rotation is needed only when the x-coordinate of d is not between the x-coordinates of u and v, i.e., only if d is not flat with respect to the left to right order of the vertices (see Figure 8.26). As pointed out in Section 8.5.1, the technique based on curve embeddings guarantees that d is always flat, and thus no rotation is required. To avoid the final rotation not only simplifies the drawing algorithm, but it also has impact on the area of the final drawing. Namely, Kaufmann and Wiese prove that the drawing before the rotation has area $O(W^3)$, where W is the *size* of S, i.e., the length of the side of the smallest axis parallel square containing S. The rotation may cause an exponential growth of the area of the drawing. Thus, avoiding the rotation keeps the drawing in a polynomial area.

We conclude this section by mentioning that the DDLW Augmentation Technique has been used to investigate other problems related to point-set embeddability, such as the study of universal point sets. A set S of m points is *h-bend universal* for a family of planar

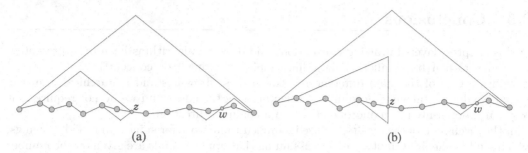

Figure 8.26 An illustration of the segments rotation performed in the technique by Kaufmann and Wiese [KW02] in order to remove a third bend. The division vertex z requires the rotation, the division vertex w does not require the rotation. Figure taken from [DDLW05].

graphs with n vertices ($n \leq m$) if each graph in the family admits a point-set embedding on a subset of S that has at most h bends per edge.

Many results about point-set embeddings can be regarded as results about universal point sets. For example, the results about the point-set embeddability of outerplanar graphs on every set of points in general position [Bos02] imply that every set of points in general position is 0-bend universal for the class of outerplanar graphs with n vertices. Analogously, the result about Kaufmann and Wiese implies that every set of points is 2-bend universal for the class of planar graphs.

De Fraysseix, Pach, and Pollack [dPP90] and independently Schnyder [Sch90] proved that a grid with $O(n^2)$ points is 0-bend universal for all planar graphs with n vertices. De Fraysseix et al. [dPP90] also showed that a 0-bend universal set of points for all planar graphs having n vertices cannot have $n + o(\sqrt{n})$ points. This last lower bound was improved by Chrobak and Karloff [CK89] and later by Kurowski [Kur04] who showed that linearly many extra points are necessary for a 0-bend universal set of points for all planar graphs having n vertices.

Since 0-bend universal point sets for planar graphs must have more that n points [Kur04], while every set of n points is 2-bend universal for planar graphs [KW02], Everett et al. [ELLW10] investigated 1-bend universal point sets and proved that there exists a set of n distinct points in the plane in general position that is 1-bend universal for all planar graphs with n vertices. The proof of the latter result is constructive. A set S of n points is defined and a point-set embedding of a planar graph G on this set of points is constructed by exploiting the DDLW Augmentation Technique. Namely, the points are chosen to be in convex position and an augmenting Hamiltonian cycle H of the input graph G is drawn as the convex hull CH of S suitably enriched with extra points that represent the division vertices. The edges of $\mathsf{Ham}(G)$ that are not in H are either inside H or outside it. Those inside are drawn as chords inside CH, the others are drawn with one bend outside CH. The choice of points and the property of H that all division vertices are flat guarantee that no additional bend is required when the division vertices are removed.

Dujmović et al. [DEL+13] study 0-bends universal point sets for sub-classes of planar graphs. They prove that there exist sets of n points that are 0-bend universal for maximum degree 3 series-parallel lattices with n vertices. They also study h-bend universal point sets with the additional requirement that bends are also constrained to be represented by points in the set. They prove that, if 1, 2, or 3 bends per edge are allowed then universal point sets exist of size $O(n^2/\log n)$, $O(n\log n)$, and $O(n)$, respectively. All these results use as a basic tool the DDLW Augmentation Technique.

8.6   Conclusions

In this chapter, layered drawing conventions and drawing algorithms have been presented, where layers can be parallel straight lines (spine drawings) or concentric circles (radial drawings). One of the main differences between these drawings and hierarchical drawings is that we do not take into account the orientation of the edges and we do not require that edges are represented as monotone curves in a common direction.

In the discussion of the results, we used a unified framework for spine and radial drawings, which studies the drawability problem assuming that upper bounds are given on the number of layers and on the number of bends along each edge. We summarized the literature by providing characterization and time-complexity results for each specific drawability problem, and we also presented variations of the problem and related results for some constrained scenarios.

Some theoretical connections between spine drawings, radial drawings, and well-studied problems in graph theory and computational geometry were also pointed-out.

References

[ADP11] Patrizio Angelini, Giuseppe Di Battista, and Maurizio Patrignani. Finding a minimum-depth embedding of a planar graph in $O(n^4)$ time. *Algorithmica*, 60:890–937, 2011.

[BDL08] Melanie Badent, Emilio Di Giacomo, and Giuseppe Liotta. Drawing colored graphs on colored points. *Theoretical Computer Science*, 408(2-3):129 – 142, 2008.

[Bie98] Therese C. Biedl. Drawing planar partitions I: LL-drawings and LH-drawings. In *Symposium on Computational Geometry*, pages 287–296, 1998.

[BK79] Frank Bernhart and Paul C. Kainen. The book thickness of a graph. *Journal Combinatorial Theory, Series B*, 27(3):320–331, 1979.

[BKM98] Therese C. Biedl, Michael Kaufmann, and Petra Mutzel. Drawing planar partitions. II. HH-drawings. In *Graph-Theoretic Concepts in Computer Science*, pages 124–136. Springer-Verlag, 1998.

[BM90] Daniel Bienstock and Clyde L. Monma. On the complexity of embedding planar graphs to minimize certain distance measures. *Algorithmica*, 5(1):93–109, 1990.

[BMS97] Prosenjit Bose, Michael McAllister, and Jack Snoeyink. Optimal algorithms to embed trees in a point set. *Journal of Graph Algorithms and Applications*, 2(1):1–15, 1997.

[Bos02] Prosenjit Bose. On embedding an outer-planar graph in a point-set. *Computational Geometry*, 23(3):303–312, 2002.

[Cab06] Sergio Cabello. Planar embeddability of the vertices of a graph using a fixed point set is np-hard. *Journal of Graph Algorithms and Applications*, 10(2):353–363, 2006.

[CK89] M. Chrobak and H. Karloff. A lower bound on the size of universal sets for planar graphs. *SIGACT News*, 20(4):83–86, 1989.

[CLR87] Fan R. K. Chung, Frank T. Leighton, and Arnold L. Rosenberg. Embedding graphs in books: A layout problem with applications to VLSI design. *SIAM Journal on Algebraic and Discrete Methods*, 8:33–58, 1987.

[CN85] Norishige Chiba and Takao Nishizeki. Arboricity and subgraph listing algorithms. *SIAM Journal on Computing*, 14:210–223, 1985.

[CN89] Norishige Chiba and Takao Nishizeki. The Hamiltonian cycle problem is linear-time solvable for 4-connected planar graphs. *J. Algorithms*, 10(2):187–211, June 1989.

[CSW04] Sabine Cornelsen, Thomas Schank, and Dorothea Wagner. Drawing graphs on two and three lines. *Journal of Graph Algorithms and Applications*, 8(2):161–177, 2004.

[DD03] Emilio Di Giacomo and Walter Didimo. Straight-line drawings of 2-outerplanar graphs on two curves. In *Graph Drawing*, volume 4912, pages 419–424, 2003.

[DDL+08a] E. Di Giacomo, W. Didimo, G. Liotta, H. Meijer, F. Trotta, and S. K. Wismath. k-colored point-set embeddability of outerplanar graphs. *Journal of Graph Algorithms and Applications*, 12(1):29–49, 2008.

[DDL08b]   Emilio Di Giacomo, Walter Didimo, and Giuseppe Liotta. Radial draw-
           ings of graphs: Geometric constraints and trade-offs. *Journal of Discrete
           Algorithms*, 6(1):109 – 124, 2008.

[DDLM05]   E. Di Giacomo, W. Didimo, G. Liotta, and H. Meijer. Computing radial
           drawings on the minimum number of circles. *Journal of Graph Algorithms
           and Applications*, 9(3):365–389, 2005.

[DDLS06]   Emilio Di Giacomo, Walter Didimo, Giuseppe Liotta, and Matthew Sud-
           erman. k-spine, 1-bend planarity. *Theoretical Computer Science*, 359(1–
           3):148–175, 2006.

[DDLW05]   Emilio Di Giacomo, Walter Didimo, Giuseppe Liotta, and Stephen K.
           Wismath. Curve-constrained drawings of planar graphs. *Computational
           Geometry: Theory and Applications*, 30:1–23, 2005.

[DDLW06]   Emilio Di Giacomo, Walter Didimo, Giuseppe Liotta, and Stephen K.
           Wismath. Book embeddability of SeriesParallel digraphs. *Algorithmica*,
           45:531–547, 2006.

[ddMP95]   Hubert de Fraysseix, Patrice Ossona de Mendez, and János Pach. A left-
           first search algorithm for planar graphs. *Discrete & Computational Geom-
           etry*, 13:459–468, 1995.

[DEL+13]   V. Dujmović, W. Evans, S. Lazard, W. Lenhart, G. Liotta, D. Rappaport,
           and S. Wismath. On point-sets that support planar graphs. *Computational
           Geometry*, 46(1):29–50, 2013.

[DETT99]   Giuseppe Di Battista, Peter Eades, Roberto Tamassia, and Ioannis G.
           Tollis. *Graph Drawing*. Prentice Hall, Upper Saddle River, NJ, 1999.

[DFK+08]   Vida Dujmović, Michael Fellows, Matthew Kitching, Giuseppe Liotta,
           Catherine McCartin, Naomi Nishimura, Prabhakar Ragde, Frances Rosa-
           mond, Sue Whitesides, and David Wood. On the parameterized complexity
           of layered graph drawing. *Algorithmica*, 52:267–292, 2008.

[DGL08]   E. Di Giacomo, L. Grilli, and G. Liotta. Drawing bipartite graphs on two
           parallel convex curves. *Journal of Graph Algorithms and Applications*,
           12(1):97–112, 2008.

[DGLT10]   Emilio Di Giacomo, Giuseppe Liotta, and Francesco Trotta. Drawing
           colored graphs with constrained vertex positions and few bends per edge.
           Algorithmica, 57:796–818, 2010.

[DLT06]   E. Di Giacomo, G. Liotta, and F. Trotta. On embedding a graph on two
           sets of points. *IJFCS, Special Issue on Graph Drawing*, 17(5):1071–1094,
           2006.

[dPP90]   Hubert de Fraysseix, János Pach, and Richard Pollack. How to draw a
           planar graph on a grid. *Combinatorica*, 10(1):41–51, 1990.

[ELLW10]   Hazel Everett, Sylvain Lazard, Giuseppe Liotta, and Stephen Wismath.
           Universal sets of n points for one-bend drawings of planar graphs with n
           vertices. *Discrete & Computational Geometry*, 43:272–288, 2010.

[EMW86]   Peter Eades, Brendan D. McKay, and Nicholas C. Wormald. On an
           edge crossing problem. In *9th Australian Computer Science Conference
           (ACSC9)*, pages 327–334, 1986.

[Fár48]   István Fáry. On straight lines representations of planar graphs. *Acta Sci.
           Math. Szeged*, 11:229–233, 1948.

[FK97]   Ulrich Fößmeier and Michael Kaufmann. Nice drawings for planar bi-partite graphs. In *3rd Italian Conference on Algorithms and Complexity (CIAC '97)*, volume 1203 of *Lecture Notes in Computer Science*, pages 122–134. Springer-Verlag, 1997.

[FLW03]   Stefan Felsner, Giuseppe Liotta, and Stephen K. Wismath. Straight-line drawings on restricted integer grids in two and three dimensions. *Journal of Graph Algorithms and Applications*, 7(4):363–398, 2003.

[GMPP91]   Peter Gritzmann, Bojan Mohar, Janos Pach, and Richard Pollack. Embedding a planar triangulation with vertices at specified points. *American Mathematical Monthly*, 98(2):165–166, 1991.

[HR92]   Lenwood S. Heath and Arnold L. Rosenberg. Laying out graphs using queues. *SIAM Journal on Computing*, 21(5):927–958, 1992.

[HS72]   Frank Harary and Allen Schwenk. A new crossing number for bipartite graphs. *Utilitas Mathematica*, 1:203–209, 1972.

[IPTT94]   Yoshiko Ikebe, Micha A. Perles, Akihisa Tamura, and Shinnichi Tokunaga. The rooted tree embedding problem into points in the plane. *Discrete Computational Geometry*, 11:51–63, 1994.

[Kur04]   Maciej Kurowski. A 1.235 lower bound on the number of points needed to draw all n-vertex planar graphs. *Inf. Process. Lett.*, 92(2):95–98, 2004.

[KW02]   Michael Kaufmann and Roland Wiese. Embedding vertices at points: Few bends suffice for planar graphs. *Journal of Graph Algorithms and Applications*, 6(1):115–129, 2002.

[PT93]   Janos Pach and Jenő Törőcsik. Layout of rooted trees. In W. T. Trotter, editor, *Planar Graphs*, volume 9 of *DIMACS*, pages 131–137. American Mathematical Society, 1993.

[PW01]   Janos Pach and Rephael Wenger. Embedding planar graphs at fixed vertex locations. *Graphs and Combinatorics*, 17:717–728, 2001.

[RM95]   S. Rengarajan and C. E. Veni Madhavan. Stack and queue number of 2-trees. In Ding-Zhu Du and Ming Li, editors, *COCOON*, volume 959 of *Lecture Notes in Computer Science*, pages 203–212. Springer, 1995.

[Sch90]   Walter Schnyder. Embedding planar graphs on the grid. In *Proc. 1st ACM-SIAM Sympos. Discrete Algorithms (SODA '90)*, pages 138–148, 1990.

[Sud04]   Matthew Suderman. Pathwidth and layered drawings of trees. *International Journal of Computational Geometry & Applications*, 14(3):203–225, 2004.

[Sud05]   Matthew J. Suderman. Proper and planar drawings of graphs on three layers. In *Graph Drawing, 13th International Symposium (GD 2005)*, volume to appear, 2005.

[Sug02]   Kozo Sugiyama. *Graph Drawing and Applications*. World Scientific, Singapore, 2002.

[TKY77]   N. Tomii, Yahiko Kambayashi, and Shuzo Yajima. On planarization algorithms of 2-level graphs. Technical Report EC77-38, Institute of Electronic and Communication Engineers of Japan (IECEJ), 1977.

[WG86]   Michael S. Waterman and Jerrold R. Griggs. Interval graphs and maps of DNA. *Bulletin of Mathematical Biology*, 48(2):189–195, 1986.

[Wig82]   Avi Wigderson. The complexity of the Hamiltonian circuit problem for maximal planar graphs. Technical Report 298, Princeton University, EECS Department, 1982.

9

Circular Drawing Algorithms

Janet M. Six
Lone Star Interaction Design

Ioannis G. Tollis
University of Crete and
Technology Hellas-FORTH

9.1 Introduction

A *circular drawing* of a graph (see Figure 9.1 for an example) is a visualization of a graph with the following characteristics:

- The graph is partitioned into clusters;
- The nodes of each cluster are placed onto the circumference of an *embedding circle*; and
- Each edge is drawn as a straight line.

There are many applications that would be strengthened by an accompanying circular graph drawing. For example, the drawing techniques could be added to tools which manipulate telecommunication [Ker93], computer [Six00], and social networks [Kre96] to show clustered views of those information structures. The partitioning of the graph into clusters can show structural information such as biconnectivity, or the clusters can highlight semantic qualities of the network such as sub-nets. Emphasizing natural group structures within the topology of the network is vital to pinpoint strengths and weaknesses within that design. It is essential that the number of edge crossings within each cluster remains low in

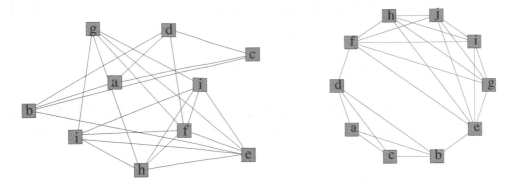

Figure 9.1   A graph with arbitrary coordinates for the nodes and a circular drawing of the same graph as produced by an implementation of Algorithm CIRCULAR. Figure taken from [ST99, ST06].

order to reduce the visual complexity of the resulting drawings. Researchers have produced several circular drawing techniques [Bra97, DMM97, KMG88, Kre96, TX95], some of which have been integrated into commercial tools. However, the resulting drawings are visually complex with respect to the number of crossings. In this chapter, we present circular drawing techniques for simple biconnected and nonbiconnected graphs which are efficient and also produce drawings with a low number of edge crossings. The first technique produces single-circle drawings of biconnected graphs. The second technique produces single-circle drawings of nonbiconnected graphs. Finally, the third technique produces multiple-circle drawings of nonbiconnected graphs.

These techniques are very useful for many applications, however, with the exception of the Graph Layout Toolkit (GLT) technique [DMM97, KMG88], these techniques do not allow the user to define which nodes should be grouped together on an embedding circle. And in the GLT technique, the layouts of the user-defined groups are themselves placed on a single embedding circle. For some graph structures, this may not be ideal. In this chapter, we also present a circular drawing algorithm that allows the user to define the node groups, draws each group of nodes efficiently and effectively, and visualizes the superstructure well. We call this approach *user-grouped circular drawing*.

An example of an application in which user-grouped circular drawing would be useful is a computer network management system in which the user needs to know the current state of the network. It would be very helpful to allow the user to group the computers by department, floor, usage rates, or other criteria. See Figure 9.2. This graph drawing could also represent a telecommunications network, social network, or even the elements of a large software project. There are, of course, many other applications which would benefit from user-grouped circular drawing: e.g., biological networks, financial market modeling, HR management, and physical science models.

The remainder of this chapter is organized as follows: Section 9.1.1 discusses previous work in this area. In Section 9.2, we present an $O(m)$ time algorithm for the circular layout of biconnected graphs. The algorithm guarantees that if a zero-crossing circular drawing exists for a biconnected graph, then it will find it. In Section 9.2.1, we discuss properties of circular drawings created by the technique in Section 9.2. In Section 9.3, we discuss an approach for reducing the number of edge crossings in circular drawings. In Section 9.4, we present an $O(m)$ time algorithm for drawing nonbiconnected graphs on a single embedding circle. In Section 9.5, we present an $O(m)$ time algorithm for drawing nonbiconnected

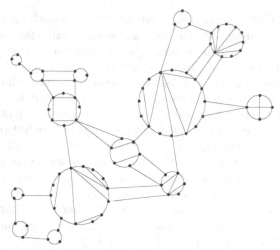

Figure 9.2 A user-grouped circular drawing. Figure taken from [ST03b].

graphs on multiple embedding circles. In Section 9.6, we introduce a framework for user-grouped circular drawing. In Section 9.7, we discuss implementation details and give results of experimental studies for these techniques. In Section 9.8, we present conclusions.

9.1.1 Other Circular Drawing Techniques

Kar, Madden, and Gilbert present a circular drawing technique and tool in [KMG88] for network management. Recognizing that a clustered view of a network can be quite helpful to its design and maintenance, the authors build a system that first partitions the network into clusters, places the clusters onto the main embedding circle, and then sets the coordinates of individual nodes. Finally, a heuristic approach is used to minimize the number of crossings. As discussed in [DMM97], an advanced version of this $O(n^2)$ technique has been implemented as part of Tom Sawyer Software's successful Graph Layout Toolkit (GLT). An early heuristic on circular drawings was presented in [Ma88].

Tollis and Xia introduced several linear time algorithms for the visualization of survivable telecommunication networks in [TX95]. Given the ring covers of a network, these algorithms create circular drawings such that the survivability of the network is clearly visible. Techniques were presented for outside (inside) drawings such that the rings are placed outside (inside) a root circle. An additional linear time algorithm produces drawings that are a combination of outside and inside drawings. This type of flexibility in a tool allows each network designer to choose the best technique given the exact application.

Citing a need for graph abstraction and reduction of today's large information structures, Brandenburg describes an approach to draw a path (or cycle) of cliques in [Bra97]. This $O(n^3)$ algorithm creates a two-level abstraction of the given graph giving the ability to project a clique on each node of the abstracted graph.

Circular drawing techniques are not limited to telecommunication and computer network applications by any means. InFlow [Kre96] is a tool to visualize human networks and produces diagrams and statistical summaries to pinpoint the strengths and weaknesses within an organization. The usually unvisualized characteristics of self-organization, emergent structures, knowledge exchange, and network dynamics can be seen in the drawings of InFlow. Resource bottlenecks, unexpected work flows, and gaps within the organization are clearly shown in these circular drawings.

In [KW02], new ideas are presented that extend the framework for circular drawings described in this chapter, in order to make the framework suitable for user interaction. They introduce the concept of hicircular drawings, a hierarchical extension of the mentioned framework replacing the circles of single vertices by circles of circular or star-like structures. Various heuristic algorithms that find an ordering of vertices that reduce the number of crossings in the corresponding circular drawing are presented in [HS04]. A two-phase heuristic for crossing reduction in circular layout is proposed in [BB05]. Their extensive experimental results indicate that they yield few crossings. Three independent, complementary techniques for lowering the density and improving the readability of circular layouts are presented in [GK07]. First, an algorithm places the nodes on the circle such that edge lengths are reduced. Second, the circular drawing style is enhanced by allowing a set of carefully selected edges to be routed around the exterior of the circle. The third technique reduces density by coupling groups of edges as bundled splines that share part of their route.

Due to lack of space, we can not describe other techniques here, but refer the reader to other works such as [BB05, GK07, HS04, KW02, Ma88].

For more information on the algorithms presented in this chapter, see [ST06, ST03b].

9.1.2   Complexity of the Circular Graph Drawing Problem

Intuitively, the problem of creating circular graph drawings while minimizing the number of edge crossings seems very hard. The general problem of placing nodes such that the number of edge crossings is minimum is the well-known NP-hard *crossing number* problem. Furthermore, the more restricted problem of finding a minimum crossing embedding such that all the nodes are placed onto the circumference of a circle and all edges are represented with straight lines is also NP-hard as proven in [MKNF87]. The authors show the NP-hardness by giving a polynomial time transformation from the NP-complete *Modified Optimal Linear Arrangement* problem.

9.2   Circular Drawings of Biconnected Graphs

In order to produce circular drawings with few crossings, the algorithm tends to place edges toward the outside of the embedding circle. This characteristic is a result of placing a few edges in the middle of the drawing to be crossed. Also, nodes are placed near their neighbors. In fact, this algorithm tries to maximize the number of edges appearing toward the periphery of the embedding circle. The algorithm achieves this improvement by selectively removing some edges and then building a depth first search (DFS) based node ordering of the resulting graph. However, the edge placement near the periphery may decrease the readability of the drawing. If this is an issue, an increase of scale will be helpful. An alternative approach where selected edges are drawn outside the embedding circle is described in [GK07].

In order to selectively remove some edges, this technique visits the nodes in a wave-like fashion. Define a *wave front node* to be adjacent to the last node processed; see Figure 9.3. A *wave center node* is adjacent to some other node that has already been processed. The algorithm starts at a lowest degree node and continues to visit wave front and wave center nodes if they are of lowest degree. If none of the current wave front or wave center nodes are of lowest degree, then some lowest degree node is chosen. The wave-like node traversal begins again from this newly chosen node and will continue from this node and the previous wave front and wave center nodes.

A *pair edge* is incident to two nodes which share at least one neighbor; see Figure 9.4. Nodes v and w are said to be *paired* by u, and u is said to *establish* the pair edge (v, w).

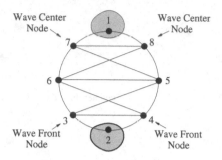

Figure 9.3 Examples of wave front and wave center nodes. The shaded region includes those nodes that have already been processed. The node labeled 2 is the most recently processed. Figure taken from [ST06].

In other words, u, v, and w form a triangle. Pair edges will be removed before the DFS step of the technique. A *triangulation* edge is a new pair edge that is inserted into the graph by the technique. The triangulation edges are also removed from the graph before the DFS portion of the algorithm. Each time a node u is visited, a list of pair edges is built. If there is an insufficient number of pair edges in the graph, the algorithm automatically inserts triangulation edges into the graph. With the ensuing removal of u, that node is inherently represented by the newly found pair edges; see Figure 9.5. The illustrations marked (a) show a degree two node u and its neighbors v and w at three different points in the algorithm. The pair edge established by u, (v, w), is shown with a bold line in the first illustration. The illustration immediately to the right shows the same graph fragment when the next node is processed. Although node u and edges (u, v) and (u, w) are not in the graph anymore, they are inherently represented by the edge (v, w). The next illustration to the right shows the same graph fragment after the pair edge (v, w) has been removed. At this point, the pair edge (v, w) is inherently represented by node u and edges (u, v) and (u, w). A similar example is shown in the illustrations labeled (b), where the current node being processed has degree three. It is this selective absorption that causes the behavior of edge placement toward the periphery of the embedding circle.

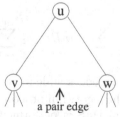

Figure 9.4 Example of a pair edge. Figure taken from [ST99, ST06].

It is important to note that we do not find all pair edges. For each node u, we visit its neighbors $v_1, v_2, \ldots, v_k$ in some order, say, the order in which they appear in the adjacency list. For example, we check to see whether (v_1, v_2) exists: if so, we add that edge to the removal list. If not, we add the triangulation edge (v_1, v_2) to the graph and to the removal list. This part of the algorithm takes $O(deg(u))$ time. Notice that a new edge is added only between two nodes that are consecutive in the adjacency list of the current node (and, of course, if such an edge does not already exist). Also note that the first and last neighbors

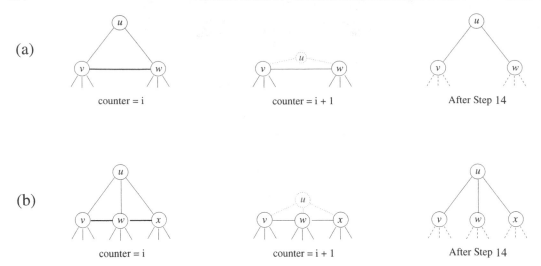

(a)

(b)

Figure 9.5　The node and edge absorption qualities of Algorithm CIRCULAR. Part (a) shows a degree-two node u and its neighbors v and w at three different points in the algorithm. First, the pair edge established by u, (v, w), is shown. Next, after node u is processed, node u and edges (u, v) and (u, w) are inherently represented by the edge (v, w). Finally, we see the same graph fragment after the pair edge (v, w) has been removed in Step 14. Part (b) shows a similar example with a degree-three node. Figure taken from [ST99, ST06].

visited cannot experience an increase in degree. For each of those nodes, the edge incident to u is removed while at most one triangulation edge is added. Next, we show that the total number of triangulation edges added is $O(m)$.

The number of triangulation edges added to G over the course of the algorithm is at most $\sum_{i=1}^{n-3} minDeg_i - 1$, where $minDeg_i$ is the minimum degree found in G at the ith iteration of the While loop. $minDeg_i \leq avgDeg$ before the ith iteration, $\forall i \geq 1$ and where $avgDeg$ is the average degree of the nodes in the original graph G. It is important to note that the visit of the neighbors starts from the lowest degree neighbor and proceeds cyclically around the adjacency list. Since we know that $minDeg_i \leq avgDeg$ before the ith iteration, $\forall i \geq 1$, we also know that

$$\sum_{i=1}^{n-3} minDeg_i - 1 < \sum_{i=1}^{n} minDeg_i \leq \sum_{i=1}^{n} avgDeg = 2m.$$

Therefore, the number of triangulation edges added is $O(m)$.

Subsequent to the edge removal, the algorithm proceeds to build an ordering of the nodes for the reduced graph. A traditional DFS is performed and then the nodes in a longest path of the DFS tree are placed around the embedding circle. Alternatively, a heuristic algorithm for finding a longest path in a graph can be used. Finally, the remaining nodes are nicely merged into the ordering. This can be accomplished by visiting each neighbor of u and asking if it is next to another neighbor of u on the embedding circle. If two neighbors of u are next to each other on the embedding circle, then we place u between those two neighbors. (If there are multiple pairs of such neighbors, we arbitrarily pick one of those pairs.) If there are no two neighbors of u next to each other on the embedding circle, then we place u next to some neighbor or u or, if there are no neighbors of u on the embedding circle yet, we pick an arbitrary position for u.

Algorithm CIRCULAR

Input: A biconnected graph, $G = (V, E)$.

Output: A circular drawing Γ of G such that each node in V lies on the periphery of a single embedding circle.

1. Bucket sort the nodes by ascending degree into a table T.
2. Set *counter* to 1.
3. While *counter* $\leq n - 3$
4.   If a wave front node u has lowest degree, then *currentNode* $= u$.
5.   Else If a wave center node v has lowest degree, then *currentNode* $= v$.
6.   Else set *currentNode* to be some node with lowest degree.
7.   Visit the adjacent nodes consecutively. For each two nodes,
8.   If a pair edge exists place the edge into *removalList*.
9.   Else place a triangulation edge between the current pair of neighbors and also into *removalList*.
10.   Update the location of *currentNode*'s neighbors in T.
11.   Remove *currentNode* and incident edges from G.
12.   Increment *counter* by 1.
13. Restore G to its original topology.
14. Remove the edges in *removalList* from G.
15. Perform a DFS (or a longest path heuristic) on G.
16. Place the resulting longest path onto the embedding circle.
17. If there are any nodes that have not been placed, then place the remaining nodes into the embedding order with the following priority: (i) between two neighbors, (ii) next to one neighbor, (iii) next to zero neighbors.

Figure 9.6 Algorithm CIRCULAR.

Figure 9.6 shows the pseudocode for Algorithm CIRCULAR. The time complexity of Algorithm CIRCULAR is $O(m)$, where m is the number of edges in G. Step 1 takes $O(m)$ time. Step 3 takes $O(m)$ time over all iterations since the use of efficient data structures (as explained in Section 6.2) allows each iteration to take only $O(deg(v_i))$ time, where v_i is the vertex chosen during the ith iteration. Notice that the number of triangulation edges added by Step 9 is $O(m)$, as shown above. Clearly, Steps 13–16 require $O(m)$ time. Finally, Step 17 also requires $O(m)$ time since at most $\sum_{i=1}^{n} deg(v_i) = O(m)$ possible placements are reviewed.

9.2.1 Properties of Algorithm CIRCULAR

In this section, we give properties of Algorithm CIRCULAR. See [ST06, ST03b] for the detailed proofs. A biconnected graph, G, is *outerplanar* if and only if G can be drawn on the plane such that all nodes lie on the boundary of a single face and no two edges cross. If the biconnected graph given to Algorithm CIRCULAR is outerplanar, then the result will be a circular visualization such that no two edges cross. In fact, the technique has been inspired by the algorithm for recognizing outerplanar graphs presented in [Mit79].

By the definition of outerplanar graphs, we know that there exists a plane circular drawing for any outerplanar graph. Also, by that same definition, we know that a graph that is not outerplanar does not admit a plane circular drawing. In fact, the set of biconnected graphs that may be drawn in a circular fashion without any crossings is exactly the set of biconnected outerplanar graphs. The requirement of placing all nodes on the periphery of some embedding circle is equivalent to placing all nodes on a single face (say, the external face) of some embedding. Furthermore, if a zero-crossing visualization exists for a biconnected graph, G, then that drawing can be found by Algorithm CIRCULAR.

Therefore, we have the following theorem:

Theorem 9.1   *Given a biconnected graph G, if G admits a circular layout with zero crossings, then Algorithm CIRCULAR produces a circular drawing with zero crossings in $O(n)$ time.*

Also, as shown in the discussion of the time requirements for Algorithm CIRCULAR, we have:

Theorem 9.2   *Algorithm CIRCULAR produces a circular drawing of any biconnected graph in $O(m)$ time.*

9.3   Further Reduction of Edge Crossings

As will be shown in the experimental results of Section 7.1, Algorithm CIRCULAR produces drawings with a low number of edge crossings and works very well in practice. We can further reduce the number of edge crossings with the technique presented in this section. As discussed in Section 9.1.2, the problem of minimizing the number of edge crossings in a circular graph drawing is NP-hard. The configuration of the nodes as determined by Algorithm CIRCULAR produces drawings with a low number of crossings, which can then be further reduced to some local minima with a monotonic crossing reduction technique. The postprocessing step visits each node v and queries whether crossings can be reduced further by moving v next to one of its neighbors.

See Figure 9.7 for Algorithm CIRCULAR-Postprocessing. The time complexity of Algorithm CIRCULAR-Postprocessing is $O(m^2)$. This order is dominated by the required time for counting the number of crossings (Steps 1 and 9). It is vitally important to the time efficiency of Algorithm CIRCULAR-Postprocessing that the number of crossings be counted in an efficient fashion. As will be shown in Lemma 9.1, Step 1 of Algorithm CIRCULAR-Postprocessing requires $O(m + \chi)$ time to find the total number of crossings, where m is the number of edges and χ is the number of crossings. The experimental study presented in Section 9.7 has shown that the loop of Step 2 needs to be iterated at most 9 times. In fact, the vast majority of drawings converged within the first two iterations. In the worst case, Step 2 requires a constant amount of time. Steps 3 and 6 require $O(n)$ time. Steps 4 and 5 require $O(m)$ time since we explore $\sum_{i=1}^{n} degree(i) = O(m)$ positions. Steps 7 and 8 require $O(m)$ time since we know there will be at most $\sum_{i=1}^{n} degree(i) = O(m)$ positions. In section 12.3.2, we will show that it takes $O(m)$ time to find the new number of crossings in Step 9. And since over the course of the algorithm, Step 9 is repeated $O(m)$ times Step 9 requires $O(m^2)$ time. Steps 10 and 11 require $O(m)$ time. So the time complexity of the entire algorithm is $O(m^2 + \chi)$. Since, each edge can cross any other edge in the drawing at most once in a circular visualization, χ is $O(\sum_{i=1}^{m} i)$, which is $O(m^2)$. Therefore, Algorithm CIRCULAR-Postprocessing has time complexity $O(m^2)$.

Algorithm CIRCULAR-Postprocessing
Input: A drawing Γ of biconnected graph $G = (V, E)$ produced by Algorithm CIRCULAR.
Output: A drawing Γ' of G with fewer or equal number of crossings.

1. *currentCrossings* = current number of crossings in the drawing.
2. For a fixed number of times
3.   For each node, u, in G
4.   Initialize $List_1$ to contain the embedding circle positions which lie between two nodes adjacent to u.
5.   If $List_1$ is empty

     (a)   Initialize $List_2$ to contain the embedding circle positions which lie next to one neighbor of u.

     (b) $PositionList = List_2$.

6.   Else $PositionList = List_1$.
7.   For each location in $PositionList$
8.   Place u at this location
9.   *newCrossings* = the new number of crossings.
10.   If *newCrossings* < *currentCrossings* then *currentCrossings* = *newCrossings*.
11.   Else Place u back into its previous position.
12. If no improvement was made during this iteration, stop.

Figure 9.7 Algorithm CIRCULAR-Postprocessing.

9.3.1   Counting All the Crossings in a Circular Drawing

Consider the straight edges e_i and e_j of Figure 9.8. The edge e_i can cross e_j if and only if one endpoint v of e_j appears between the two endpoints u and w of e_i. In this case, e_j is called an *open edge with respect to the arc uvw*. If both endpoints of e_j appear between u and w on the perimeter of the embedding circle, then e_i and e_j do not cross. So, if we order the edges as they are encountered around the embedding circle and visit their endpoints in that order, we can determine the total number of edge crossings by counting the number of open edges. Although the problem is one dimensional, this technique has some similarities to the line segment intersection algorithm presented in [PS85].

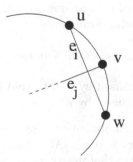

Figure 9.8 An open edge with respect to the arc *uvw*. Figure taken from [ST99, ST06].

Algorithm CountAllCrossings

Input: A single circle drawing Γ of a biconnected graph $G = (V, E)$.

Output: The number of edge crossings in Γ.

1. Order the edges as they are encountered around the circle in a clockwise order.
2. $numberOfCrossings = 0$.
3. For each edge endpoint, p_i, of edge e_i, do
4.   If p_i is the first endpoint of edge e_i append e_i to $openEdgeList$.
5.   Else

   (a)   Increase $numberOfCrossings$ by the number of open edges with respect to the arc $p_g p_h p_i$, where p_g and p_i are the endpoints of e_i and p_h is some endpoint which was visited after p_g and before p_i.

   (b)   Remove e_i from $openEdgeList$.

Figure 9.9   Algorithm CountAllCrossings.

Algorithm CountAllCrossings requires $O(m + \chi)$ time. Step 1 takes $O(m)$ time. This step can be accomplished in $O(m)$ time by visiting the incident edges of each node as they appear around the embedding circle. Steps 3, 4, and 5(b) require $O(m)$ time. Step 5(a) requires time

$$\sum_{i=1}^{2m} \chi_i = O(\chi),$$

where χ_i is the number of edge crossings caused by the edge e_i and χ is the total number of edge crossings in the embedding. We accomplish this time requirement by traversing $openEdgeList$ backward from the end of the list to the element which contains e_i. Therefore, we have the following:

LEMMA 9.1   Algorithm CountAllCrossings counts the total number of edge crossings in a single circle embedding, where m is the number of edges and χ is the number of crossings in $O(m + \chi)$ time.

9.3.2   Determining the New Number of Crossings after Moving a Node

Since we can determine the overall number of crossings at the beginning of the algorithm and then move one node at a time, it is necessary to count only the number of crossings caused by the incident edges of the current node, v, to update the number of crossings in the drawing. During each iteration of the crossing reduction, the number of crossings in the entire drawing is equal to the following formula:

$$New\ Number\ of\ Crossings = Old\ Number\ of\ Crossings - \chi_v + \chi'_v$$

where, χ_v = Number of crossings caused by v in the old location,

and χ'_v = Number of crossings caused by v in the new location.

Because we already know the old number of crossings, finding the new number of crossings is dominated by the time to find χ_v and χ'_v. Any change in the edge crossings will occur between edges incident to v and edges that have exactly one endpoint in the arc between the old and new positions of v. These *pertinent edges* are visited in order from the old toward the new position of v. A counter, *ctr*, holds the number of open edges in the arc (not including the open edges incident to v). Each time that an endpoint of an edge incident to v is encountered, the number of crossings is increased by the value in *ctr*. At the conclusion of this process, the number of crossings caused by v in the old position is known. The number of crossings caused by v in its new position is found by repeating this process from the new towards the old position of v after moving v to its new position; see Figure 9.10.

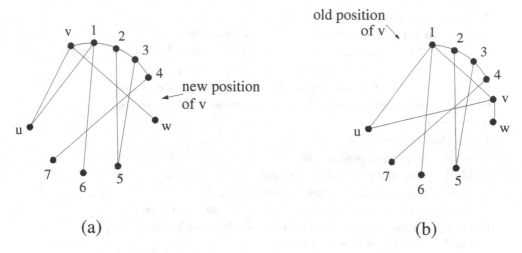

Figure 9.10 The arc created by moving node v to the position denoted with the arrow. The pertinent edges of the arc are shown. Figure taken from [ST06].

Therefore, we have the following result:

LEMMA 9.2 An $O(m)$ time algorithm exists to count the number of edge crossings gained or lost by moving a node v within a single circle embedding.

The pseudocode for Algorithm CountSingleNodeCrossings is shown in Figure 9.11. This algorithm requires $O(m)$ time. Steps 3, 4, 5, 6, 7, and 8 require $O(m)$ time since the number of pertinent edges is $O(m)$ as described above. Step 13 requires $O(m)$ time. Finally, Step 14 requires $O(m)$ time since it is a repetition of Steps 5–8.

If Algorithm CountSingleNodeCrossings is swapping the placement of two nodes which are next to each other, u and v, on the embedding circle, then Algorithm CountSingleNodeCrossings only takes $O(maxDegree)$ time, where $maxDegree$ is the maximum degree of all nodes in V. This is because the number of pertinent edges is the smaller degree of u and v, see Figure 9.12. Since a swap of these two nodes can be accomplished by moving u between v and β or moving v between α and u, we choose the move such that the number of pertinent edges (i.e., the degree of the node which is not moved) is smaller. Both of the moves produce the same node ordering, so we perform the move which requires less time. In the specific case of Figure 9.12, we choose to move node u.

Given Lemma 9.1, and Lemma 9.2, Algorithm CIRCULAR-Postprocessing produces a visualization with a reduced number of edge crossings in $O(m^2)$ time.

Algorithm CountSingleNodeCrossings

Input: A single circle drawing of a graph $G = (V, E)$,
        a node $v \in V$, and
        a new position α for v.

Output: The change in the number of edge crossings caused by moving v to α.

1. $ctr = 0$.
2. $numberOfCrossings = 0$.
3. Order the pertinent edge endpoints as they are encountered around the embedding circle.
4. Mark the pertinent edges as not seen.
5. For each pertinent edge endpoint p_i of edge e_i do
6.   If e_i is incident to v increment the $numberOfCrossings$ by ctr.
7.   Else If e_i has been seen decrement ctr by 1.
8.   Else increment ctr by 1 and mark e_i as seen.
9. $OldNumberSingleNodeCrossings = numberOfCrossings$.
10. $ctr = 0$.
11. $numberOfCrossings = 0$.
12. Move v to its new position, α.
13. Mark the pertinent edges as not seen.
14. Repeat Steps 5–8 in the opposite direction.
15. $NewNumberSingleNodeCrossings = numberOfCrossings$.
16. $changeInCrossings = NewNumberSingleNodeCrossings - OldNumberSingleNodeCrossings$.

Figure 9.11   Algorithm CountSingleNodeCrossings.

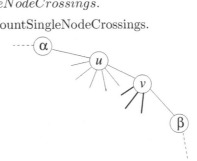

Figure 9.12   The pertinent edges for Algorithm CountSingleNodeCrossings if the two adjacent nodes u and v are being swapped. Figure taken from [ST99, ST06].

9.4   Nonbiconnected Graphs on a Single Circle

Most networks are not biconnected. Therefore, it is important for a circular drawing tool to provide a component that visualizes nonbiconnected graphs. An algorithm for producing circular drawings of nonbiconnected graphs on a single embedding circle is presented in [Six00, ST06]. Given G, a nonbiconnected graph, it can be decomposed into its biconnected components. The algorithm layouts the resulting block-cutpoint tree on a circle and then it layouts each biconnected component with a variant of Algorithm CIRCULAR.

First, we consider how to obtain a circular drawing of a tree. A DFS produces a numbering that we can use to order the nodes around the embedding circle in a crossing-free manner. From this result, we know how to order the biconnected components around the embedding circle. Next, we need to consider articulation points which are not adjacent to a bridge (*strict articulation points*). Strict articulation points appear in multiple biconnected components. In which biconnected component should a strict articulation point appear in the circular drawing? Multiple approaches to this issue are discussed in [Six00, ST99]. Due to space restrictions, we do not discuss these solutions here. A third issue to consider is how to transform the layout of each biconnected component to fit onto an arc of the embedding circle. This transformation is called *breaking*. The resulting breaks occur at an articulation point within the biconnected component.

The worst-case time requirement for the above algorithm is $O(m)$ if we use Algorithm CIRCULAR to layout each biconnected component. The resulting drawings have the property that the nodes of each biconnected component (with the exception of some strict articulation points) appear consecutively. Furthermore, the order of the biconnected components on the embedding circle are placed according to a layout of the accompanying block-cutpoint tree. Therefore, the biconnectivity structure of a graph is displayed even though all of the nodes appear on a single circle. An example drawing is shown in Figure 9.13. More details on this algorithm can be found in [Six00, ST06].

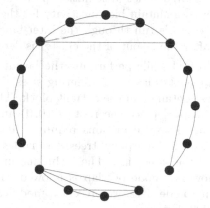

Figure 9.13 An example drawing produced by Algorithm CIRCULAR-Nonbiconnected.

9.5 Nonbiconnected Graphs on Multiple Circles

In this section, we will present a technique for producing circular drawings of graphs on multiple embedding circles. Given a nonbiconnected graph G we can decompose the structure into biconnected components in $O(m)$ time. Taking advantage of this inherent structure, we first layout the block-cutpoint tree using a radial layout technique similar to [Ber81, Ead92, Esp88], then we layout each biconnected component of the graph with a variant of Algorithm CIRCULAR. See Figure 9.14.

The algorithm addresses several issues in order to produce good quality circular drawings: 1) which biconnected component is considered to be the root of the block-cutpoint tree, 2) articulation points can appear in multiple biconnected components of the block-cutpoint tree and need to be assigned to a unique biconnected component, 3) the nodes of the block-cutpoint tree can represent biconnected components of differing size, and 4) the nodes of each biconnected component should be visualized such that the articulation points appear in good positions and also there is a low number of edge crossings. We will address each of these issues in turn.

In order to address the first issue, we can choose the root with a recursive leaf-pruning algorithm to find the "center" of the tree [DETT99]. Alternatively, we can pick the root dependent on some important metric: e.g., size of the biconnected component. Next we address the second issue. *Strict articulation points* (i.e., articulation points that are not adjacent to a bridge) are duplicated in more than one biconnected component of the block-cutpoint tree, but of course each node should appear only once in a drawing of that graph. Therefore, we offer three approaches in which each articulation point will appear only once in the drawing. The first approach assigns each strict articulation point, u, to the biconnected component which contains u and is also closest to the root in the block-cutpoint tree. This biconnected component is the parent of the other biconnected components which contain u. See Figure 9.15(a). The second approach assigns the articulation point to the biconnected component which contains the most neighbors of that articulation point, see Figure 9.15(b). The third approach assigns the articulation point to a position between its biconnected components, see Figure 9.15(c). Placing a node in this manner will highlight the fact that this node is an important articulation point. Following the assignment step, the duplicates of a strict articulation point are removed from the blocks in the block-cutpoint tree. We refer to the nodes adjacent to a removed strict articulation point in a biconnected component as *inter-block nodes*. In order to maintain biconnectivity for the method which will layout this component, a thread of edges is run through the inter-block nodes. These edges will be removed from the graph after the layout of the cluster is determined.

The third issue to be addressed while performing the layout of the block-cutpoint tree is that the biconnected components may be of differing sizes. The node sizes are proportional to the number of nodes contained in the current block. The radial layout algorithms presented in [Ber81, Ead92, Esp88] place the root at $(0,0)$ and the subtrees on concentric circles around the origin. These algorithms require linear time and produce plane drawings. However, unlike the block-cutpoint trees, the nodes of the trees laid out with [Ber81, Ead92, Esp88] are all the same size. The technique in [YFDH01] handles graphs with different node sizes; however, node overlap is allowed. In order to produce radial drawings of trees with differing node sizes, we present a modification of the classical radial layout technique [Ber81, Ead92, Esp88]:

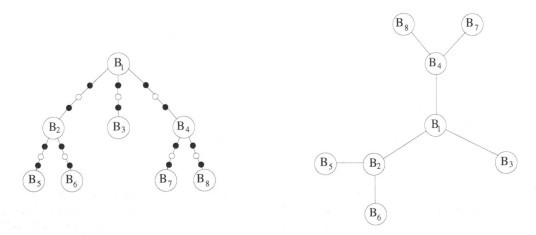

Figure 9.14 The illustration on the left shows the block-cutpoint tree of a nonbiconnected graph. The small black tree nodes represent articulation points and the small white tree nodes represent bridges. The right illustration is a drawing of the same graph where the block-cutpoint tree is laid out with a radial tree layout technique. Figure taken from [ST06].

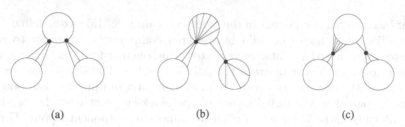

(a)   (b)   (c)

Figure 9.15 Examples of three approaches for the assignment of strict articulation points to biconnected components. The black nodes are strict articulation points. Figure taken from [ST06].

RADIAL – with Different Node Sizes: For each node, we must assign a ρ coordinate, which is the distance from point $(0,0)$ to the placement of that node and a θ coordinate which is the angle between the line from $(0,0)$ to $(\infty,0)$ and the line from $(0,0)$ to the placement of that node. The ρ coordinate of node v, $\rho(v)$, is defined to be

$$\rho(u) + \delta + \frac{d_u}{2} + \frac{max(d_1, d_2, \ldots, d_k)}{2},$$

where $\rho(u)$ is the ρ coordinate of the parent u of v, δ is the minimum distance allowed between two nodes, d_u is the diameter of u, and $max(d_1, d_2, ..., d_k)$ is the maximum of the diameters of all the children of u. It is important to note that while all descendants of a node i are placed on the same concentric circle, not all nodes in the same level of the block-cutpoint tree are placed on the same concentric circle.

In order to prevent edge crossings, each subtree must be placed inside an annulus wedge, and the width of each wedge must be restricted such that it does not overlap a wedge of any other subtree. The θ coordinate of node v depends on the widths of the descendants of v, not just the number of leaves as in [Ber81, Ead92, Esp88]. This assignment of coordinates leads to a layout of the form shown in Figure 9.16.

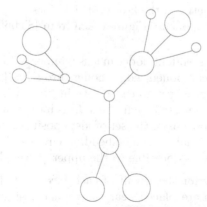

Figure 9.16 A radial layout of a tree with differing size nodes. Figure taken from [ST06].

The fourth issue to be addressed by the circular drawing technique is the visualization of each component. After performing *RADIAL – with Different Node Sizes*, we have a layout of the block-cutpoint tree and need to visualize the nodes and edges of each biconnected component. The radial layout of the block-cutpoint tree should be considered while drawing each biconnected component. See Figure 9.17. Define *ancestor nodes* to be adjacent to nodes in the parent biconnected component in the block-cutpoint tree. Likewise, define *descendant nodes* to be adjacent to nodes in child biconnected components. In order to reduce the number of crossings caused by inter-biconnected component edges, the technique tries to place ancestor nodes in the arc between the points α and β. The size of the arc from α to β is dependent on the distance between the placement of a biconnected component to the placement of its parent in the radial layout of the block-cutpoint tree. Descendant nodes are placed uniformly in the bottom half of the biconnected component layout. For example, if there are three descendant nodes, they would be placed at points γ, δ, and ϵ, as shown in Figure 9.17. These special positions for the ancestor and descendant nodes are called *ideal positions*. Because of a high number of ancestor and descendant nodes, it may not be possible to place all ancestor and descendant nodes in an ideal position; however, the algorithm places as many as possible in ideal positions.

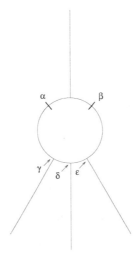

Figure 9.17   The relation between the layout of the block-cutpoint tree and the layout of an individual biconnected component. Figure taken from [ST06].

Placing the ancestor and descendant nodes in this manner reduces the number of crossings caused by inter-biconnected component edges going through a biconnected component. In fact, the only times that these edges do cause crossings are when the number of ancestor (descendant) nodes in the biconnected component B_i is more than about $\frac{n_i}{2}$, where n_i is the number of nodes in B_i. In those cases, the set of ideal positions includes all the positions in the upper (respectively lower) half of the embedding circle and also positions in the lower (upper) half which are as close as possible to the upper (lower) half.

We present two algorithms for the layout of each biconnected component such that ancestor and descendant nodes are placed near their ideal positions. The first step of each technique is to perform Algorithm CIRCULAR on the current biconnected component, B_i. This requires $O(m_i)$ time, where m_i is the number of edges in biconnected component B_i. Next, this drawing is updated so that the ancestor and descendant nodes appear near their ideal positions.

The first technique rotates the layout of the biconnected component as found by Algorithm CIRCULAR such that many ancestor and descendant nodes are placed close to their ideal positions. Then, the remaining ancestor and descendant nodes are moved to their closest ideal position. See Figure 9.18 for Algorithm LayoutCluster1. This algorithm requires $O(m_i)$ time. See Figure 9.19(b) for an example.

Algorithm LayoutCluster1
Input: A biconnected component, B_i.
Output: An circular layout of B_i such that the positions of the articulation points are placed well with respect to the ideal positions.

1. Perform Algorithm CIRCULAR on B_i and save the results in Γ_1.

2. If the number of ancestor nodes in B_i is less than the number of descendant nodes, set the block type to be descendant, otherwise, set the block type to be ancestor.

3. Loop through the nodes of B_i as they appear around the embedding circle in Γ_1 and for each node which is the same type as the block type, record the clockwise distance to the last node of that type.

4. Find the nodes that have the smallest value of the distances recorded in Step 3 and determine the median node, u, of this set.

5. If the block type is descendant, rotate the layout of B_i found in Step 1 such that u is in the middle of the lower half of the embedding circle.

6. Else rotate the layout of B_i found in Step 1 such that u is in the middle of the upper half of the embedding circle.

7. Place the remaining ancestor and descendant nodes in their closest ideal position.

Figure 9.18   Algorithm LayoutCluster1.

The second technique *LayoutCluster2* has a higher time complexity but may lead to layouts with fewer edge crossings. The first seven steps are the same as that of Algorithm LayoutCluster1. During the placement of ancestor and descendant nodes that are not in ideal positions, each such node v is placed in an ideal position, and if the number of edge crossings added exceeds a threshold T_1 or the movement of v exceeds a threshold T_2, then the size of the embedding circle is increased such that node v can be placed in an ideal position without changing the relative order between v and its neighbors on the embedding circle. See Figure 9.19(c) for an example. The thresholds are determined on a per application basis. If increasing component edge crossings or node movement is undesirable for an application, the thresholds are adjusted accordingly. The time required for Algorithm LayoutCluster2 is $O(m_i)$ if threshold T_2 (based on node movement) is used or $O(m_i * k)$, where k is the number of ancestor and descendant nodes in the cluster, if threshold T_1 (based on the number of crossings) is used.

Another technique for drawing a biconnected component would rotate the embedding circle through many positions to find a good solution.

Now that we have addressed the subproblems, we present a comprehensive technique for obtaining circular layouts of nonbiconnected graphs, called Algorithm CIRCULAR-with Radial, see Figure 9.20 for the pseudocode of the algorithm. The time complexity of Algorithm CIRCULAR-with Radial is $O(m)$ if the biconnected components are laid out with Algorithm LayoutCluster1 or $O(m * k)$, where k is the total number of ancestor and descendant nodes in the graph if Algorithm LayoutCluster2 is used. Figure 9.21 shows an example produced by this algorithm.

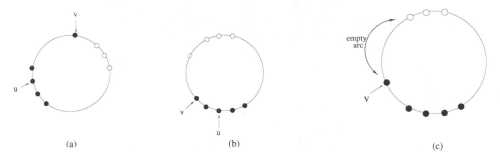

Figure 9.19 This figure demonstrates Algorithms *LayoutCluster1* and *LayoutCluster2*. The black nodes are descendant nodes and the white nodes are ancestor nodes. (a) Drawing produced by Algorithm CIRCULAR; (b) the rotated drawing of part (a) produced by Algorithm LayoutCluster1; (c) the resulting drawing of part (a) produced by Algorithm LayoutCluster2. Figure taken from [ST06].

Algorithm CIRCULAR-with Radial
Input: Any graph G.
Output: A circular drawing Γ of G.

  1. Decompose G into a block-cutpoint tree T.

  2. If G has only one biconnected component, perform Algorithm CIRCULAR on G.

  3. Else

  4.   Assign the strict articulation points to a biconnected component.

  5.   Layout the root cluster of T with Algorithm CIRCULAR.

  6.   For each subtree S of the root cluster

  7.   Perform the ρ coordinate assignment phase of *RADIAL – with Different Node Sizes* on S.

  8.   For each biconnected component, B_i, of S

  9.   Layout B_i with Algorithm LayoutCluster1, or LayoutCluster2 taking into account the radii defined for the superstructure tree in Step 7.

  10.   Considering the order of the subtrees defined during the layout of biconnected components in Step 9, perform the θ coordinate assignment phase of *RADIAL – with Different Node Sizes* on S.

  11.   Translate and rotate the clusters of S according to the radial layout of S.

Figure 9.20 Algorithm CIRCULAR-with Radial.

An extension of Algorithm CIRCULAR-with Radial to include interactive schemes has been presented by Kaufmann and Wiese in [KW02].

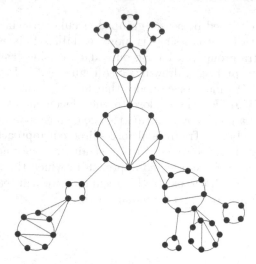

Figure 9.21 A sample drawing as produced by Algorithm CIRCULAR-with Radial. Figure taken from [ST06].

9.6 A Framework for User-Grouped Circular Drawing

The problem of producing circular drawings of graphs grouped by biconnectivity is quite different from the problem of drawing a graph whose grouping is user-defined. In the latter case, there is no known structure of either the groups or the relationship between the groups. Therefore, we must use a general method for producing this type of visualization. The four goals of a user-grouped circular drawing technique should be:

1. the user-defined groupings are highly visible,
2. each group is laid out with a low number of edge crossings,
3. the number of crossings between intra-group and inter-group edges is low, and
4. the layout technique is fast.

We know from previous work in clustered graph drawing [EFL97, EF97, EFN99, HE98] that the relationship between groups is often not very complex. We take advantage of this knowledge in this framework. Define the *superstructure* G_s of a given graph $G = (V, E, P)$, where P is the node group partition, as follows: the nodes in G_s represent the elements of P. For each edge $e \in E$ which is incident to nodes in two different node groups, place an edge between nodes representing the respective groups in G_s. The type of structure that we expect G_s to have should be visualized well with a force-directed [DETT99, Ead84] technique; therefore, the superstructure G_s will be drawn with this approach. Additionally, since G_s will likely not be a very complicated graph, it should not take much time to achieve a good drawing with a force-directed technique.

The node groups themselves will be either biconnected or not. Since Algorithms CIRCULAR and CIRCULAR-Nonbiconnected can layout biconnected and nonbiconnected graphs on a single embedding circle in linear time and have been shown to perform well in practice, we also will use those techniques here.

We have now addressed how to achieve Goals 1 and 2 with good speed. However, in order to produce good user-grouped circular graph drawings, we must successfully merge these two techniques so that we can simultaneously reach Goals 1,2, and 3. And, of course, we need a fast technique in order to achieve Goal 4. Attaining Goal 3 is very important to

the quality of drawings produced by a user-grouped circular drawing technique. As shown
in [Pur97], a drawing with fewer crossings is more readable. It is especially important to
reduce the number of intra-group and inter-group edge crossings as those can particularly
cause confusion while interpreting a drawing. See Figure 9.22. How can we achieve this
low number of crossings? We must place nodes that are adjacent to nodes in other groups
(called *outnodes* in [DMM97, KMG88]) close to the placement of those other nodes. A
force-directed approach is a good way to attain this goal since it would encourage outnodes
to be closer to their neighbors. Traditional force-directed approaches [DETT99, Ead84]
will not work here though, because we need to constrain the placement of nodes to circles.
In Section 9.6.1, we present a force-directed approach in which the nodes are restricted to
appear on circular tracks. With the use of this technique we will reach Goal 3. As will be
discussed, we can do this in a reasonable amount of time.

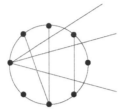

Figure 9.22   Example of intra-group and inter-group edge crossings. Figure taken from
[ST03b].

As with most force-directed techniques, the initial placement of nodes has a very significant impact on the final drawing [DETT99, Ead84]. Therefore, it is important to have
a good initial placement. This is why we should layout the superstructure and each node
group first. At the completion of those steps, we should have the almost-final drawing.
It will then be a matter of fine-tuning the drawing with the circular-track force-directed
technique. And as shown in [ST01] (see extended version in [ST03a]), once you have an
almost-final drawing, it does not take much time for a force-directed technique to converge.

9.6.1   Circular-Track Force-Directed

In order to adapt the force-directed paradigm for circular drawing, we need a way to guarantee that the nodes of a group appear on the circumference of an embedding circle, *the
circular track*. The nodes are restricted to appear on the circular track, but are allowed to
jump over each other and appear in any order, see Figure 9.23. And as in the force-directed
approach, we want to minimize the potential energy in the spring system which is modelling
the graph. In this section, we describe how this circular-track adaptation is achieved.

First, we need to look at node coordinates in a different way. Node i belongs to group α
and is located at position (x_i, y_i). Given that the center of the embedding circle on which
α is located is at (x_α, y_α) and the radius of that circle is r_α, we can restate the coordinates
of i in the following way:

$$x_i = x_\alpha + r_\alpha * cos(\theta_i) \tag{9.1}$$

$$y_i = y_\alpha + r_\alpha * sin(\theta_i) \tag{9.2}$$

Figure 9.23 Circular-track force-directed technique. Figure taken from [ST03b].

Remember that Hooke's Law [HR88] gives us the following equation for the potential energy V in a spring system:

$$V = \sum_{ij} k_{ij}[(x_i - x_j)^2 + (y_i - y_j)^2] \qquad (9.3)$$

where k_{ij} is the spring constant for the spring between nodes i and j. Equation (12.3) can be rewritten using (12.1) and (12.2):

$$V = \sum_{(i,j)\in E} k_{ij} \quad [((x_\alpha + r_\alpha * cos(\theta_i)) - (x_\beta + r_\beta * cos(\theta_j)))^2 + $$
$$((y_\alpha + r_\alpha * sin(\theta_i)) - (y_\beta + r_\beta * sin(\theta_j)))^2] \qquad (9.4)$$

where node j belongs to group β, (x_β, y_β) is the center and r_β is the radius of the embedding circle on which β appears. Thus, we have:

$$V = \sum_{(i,j)\in E} k_{ij} \quad [(x_\alpha + r_\alpha * cos(\theta_i) - x_\beta - r_\beta * cos(\theta_j))^2 + $$
$$(y_\alpha + r_\alpha * sin(\theta_i) - y_\beta - r_\beta * sin(\theta_j))^2] \qquad (9.5)$$

We can find a minimal energy solution on variables x, y, and θ. It is interesting to note that if i and j are on the same circle, then x_α and x_β are equivalent as are y_α and y_β. And, of course, $r_\alpha = r_\beta$. Now we rewrite equation (5):

$$V = \sum_{(i,j)\in E} k_{ij}[r_\alpha(cos(\theta_i) - cos(\theta_j))^2 + r_\alpha(sin(\theta_i) - sin(\theta_j))^2)] \qquad (9.6)$$

We can calculate r_α from the number of nodes in α so that means that finding the minimum V is now a one-dimensional problem based on finding the right set of θs. When we combine (12.5) or (12.6) with equations for magnetic repulsion to prevent node occlusion, we have a force-directed equation for which the nodes of a group lie on the circumference of a circle. Now we extend equation (12.5) to include repulsive forces.

$$\rho_{ij} = [(x_\alpha + r_\alpha * cos(\theta_i) - x_\beta - r_\beta * cos(\theta_j))^2 + $$
$$(y_\alpha + r_\alpha * sin(\theta_i) - y_\beta - r_\beta * sin(\theta_j))^2] \qquad (9.7)$$

$$V = \sum_{(i,j)\in E} k_{ij}\rho_{ij} + \sum_{(i,j)\in V\times V} g_{ij}\frac{1}{\rho_{ij}} \qquad (9.8)$$

where g_{ij} is the repulsive constant between nodes i and j.

Another important consideration is the set of spring constants used in the above equations. It is not necessary for the spring constant to be the same for each pair of nodes. It is also possible for these constants to change during different phases of execution.

9.6.2   A Technique for Creating User-Grouped Circular Drawings

Now that we have a force-directed technique in which the nodes are placed on circular tracks, we need to show how we will successfully merge the force-directed approach and the circular drawing techniques discussed earlier in this chapter. We present a technique for creating user-grouped circular drawings in Figure 9.24.

Algorithm CIRCULAR-with Forces
Input: A graph $G = (V, E, P)$.
Output: User-grouped circular drawing of G Γ.

1. Determine the superstructure G_s of G.

2. Layout G_s with a basic force-directed technique.

3. For each group p_i in P

    (a) If the subgraph induced by p_i, G_i, is biconnected
        layout G_i with *CIRCULAR*.

    (b) Else layout G_i with *CIRCULAR-Nonbiconnected*.

4. Place the layout of each group p_i at the respective location found in Step 2.

5. For each group p_i

    (a) rotate the layout circle and keep the position which has the lowest local potential energy.

    (b) reverse the order of the nodes around the embedding circle and repeat Step 5a.

    (c) if the result of Step 5a had a lower local potential energy than that of Step 5b revert to the result of Step 5a.

6. Apply a force-directed technique using the equations of Section 9.6.1 to G.

Figure 9.24   Algorithm CIRCULAR-with Forces.

Going back to the four goals discussed in Section 9.6, we will attain Goal 1 by using a basic force-directed technique to layout the superstructure. We will attain Goal 2 by laying out each group with either Algorithms CIRCULAR or CIRCULAR-Nonbiconnected. Attaining Goal 3 means successfully merging the results of the force-directed and circular techniques.

Once we have the layout of the superstructure and each group, we place the layout of each group at the respective location found during the layout of the superstucture. Now we have an almost-final layout: it is a matter of rotating the layouts of the groups and maybe adjusting the order of nodes around the embedding circle. Since we know that Algorithms CIRCULAR and CIRCULAR-Nonbiconnected produce good visualizations, we should change these layouts as little as possible. So first, we will fine-tune the almost-final drawing by rotating each layout and keeping the rotation that has the least local potential energy. We rotate each embedding circle through n_α positions, where n_α is the number of nodes in group α. With respect to determining local potential energy, we need to determine the lengths of inter-group edges that are incident to the nodes of α. The rotation of choice should minimize the lengths of those edges. In other words, we choose the rotation in which as many nodes as possible are close to their other-group neighbors. Since for each embedding circle we try n_α positions and examine the length of α's incident inter-group edges at each position, then the rotation step will take $O(n*m_{inter-group})$ time for the entire graph, where $m_{inter-group}$ is the number of inter-group edges. As discussed in Section 9.6, we expect $m_{inter-group} \ll m$. Then we will "flip" each layout and again rotate. We keep the rotation which has the least local potential energy. After these steps, it is still possible that some nodes will be badly placed with respect to their relationships with nodes in other groups. In other words, those placements cause intra-group and inter-group edges to cross. In order to address this problem, we will apply the force-directed technique described in Section 9.6.1. The result of this step will be the reduction of intra-group and inter-group edge crossings since nodes will be pulled to the side of the embedding circle which is closer to their other-group relatives.

Because Algorithm CIRCULAR-with Forces makes use of a force-directed technique, the worst-case time requirement is unknown. However, in practice, we expect the time requirement to be $O(n^2)$ for the following reasons: Step 1 requires $O(m)$ time. Step 2 will be on a small graph and should not require much time to reach convergence. Step 3 requires $O(m)$ time. Step 4 requires $O(n)$ time. Step 5 require $O(n * m_{inter-group})$ time. Since Step 6 is a force-directed technique, it could take $O(n^3)$ time in practice; however, the result of the previous steps will be an almost-final layout and thus should not need much time to converge. It was evidenced in [ST01] (see extended version in [ST03a]) that when a force-directed technique is applied to an almost-final layout, it does not take much more time for convergence to occur. Therefore, in practice we expect this step to require $O(n^2)$ time. Thus, we have attained Goal 4 from Section 9.6.

9.7 Implementation and Experiments

9.7.1 Experimental Analysis of Algorithm CIRCULAR

We have implemented Algorithm CIRCULAR in C++. The code runs on top of the Tom Sawyer Software Graph Layout Toolkit (GLT) version 2.3.1. We also performed an extensive experimental study to compare Algorithms CIRCULAR and CIRCULAR-Postprocessing with the circular layout component of the GLT. The circular layout technique in the GLT requires $O(n^2)$ time [DMM97, KMG88]. The results of the study show that the drawings of Algorithm CIRCULAR have about 15% fewer crossings on average than those produced by the GLT. Furthermore, the worst-case time requirement for Algorithm CIRCULAR is $O(m)$ versus the $O(n^2)$ worst-case time requirement for the GLT technique. Algorithm CIRCULAR-Postprocessing is able to significantly further reduce the number of edge crossings.

The set of input graphs for the experiments included 10,328 biconnected components of minimum size 10 extracted from the 11,399 Rome graphs [DGL+97], which have between 10 and 80 nodes. The number of edge crossings is measured for Algorithm CIRCULAR, Algorithm CIRCULAR-Postprocessing, and the circular drawing component of the GLT. As shown in the plot of Figure 9.25, the techniques produce significantly fewer crossings on average than the GLT. Specifically the drawings of Algorithm CIRCULAR have significantly fewer crossings. And as the plot shows, Algorithm CIRCULAR-Postprocessing effectively reduces the number of edge crossings even further. The percentage improvement between Algorithm CIRCULAR-Postprocessing and GLT averages is 30%. Sample drawings as produced by both GLT and the techniques are shown in Figures 9.26–9.28.

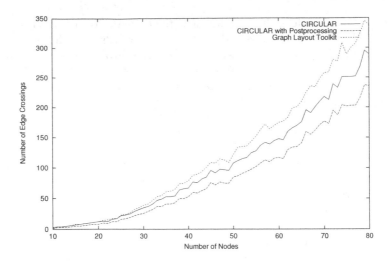

Figure 9.25   The average number of edge crossings produced by Algorithm CIRCULAR, Algorithm CIRCULAR-Postprocessing, and the Graph Layout Toolkit over 10,328 biconnected graphs. Figure taken from [ST99, ST06].

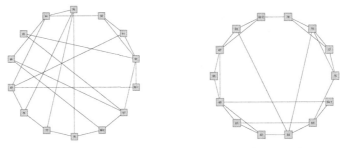

Figure 9.26   The drawing on the left is produced by the GLT. The drawing on the right is of the same graph and is produced by Algorithm CIRCULAR-Postprocessing. The drawing produced by Algorithm CIRCULAR-Postprocessing has 75% fewer crossings than the GLT drawing. Figure taken from [ST99, ST06].

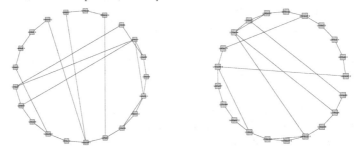

Figure 9.27   The drawing on the left is produced by the GLT. The drawing on the right is of the same graph and is produced by Algorithm CIRCULAR-Postprocessing. The drawing produced by Algorithm CIRCULAR-Postprocessing has 53% fewer crossings. Figure taken from [ST99, ST06].

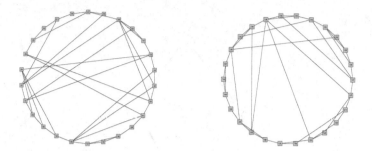

Figure 9.28 The drawing on the left is produced by the GLT. The drawing on the right is of the same graph and is produced by Algorithm CIRCULAR-Postprocessing. The drawing produced by Algorithm CIRCULAR-Postprocessing has 55% fewer crossings. Figure taken from [ST99, ST06].

9.7.2 Implementation Issues

During Step 4 of Algorithm CIRCULAR, the technique chooses a node of lowest degree with the following priority: a wave front node, a wave center node, or some lowest degree node. An efficient way to execute this is to initially sort the nodes by degree into a table of lists that reflect those categories. The table is updated as nodes and edges are removed from the graph. A bucket sort is initially used to place each node into its respective category. In order to keep the table updated, when node v, is processed, we simply move each neighbor of v into the front of its respective degree list during each iteration (similar to self-adjusting lists). This way the nodes are retrieved in the desired priority: neighbor, previous neighbor, and lowest degree node, see Figure 9.29.

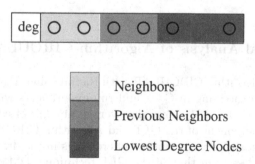

Figure 9.29 The construction of each degree list within the node table. Figure taken from [ST99, ST06].

During Step 15, the algorithm performs a DFS which will result in a DFS tree. Then we place the nodes from the longest path within that DFS tree onto the embedding circle and we merge in the nodes of the remaining DFS tree branches. See Figure 9.30. The longest path does not necessarily go through the root of the DFS tree as it does in this example.

If the input graph is outerplanar, the drawing produced by Algorithm CIRCULAR will always be plane; if not, then there might be crossings. In this case, it may be possible to further reduce the number of crossings by moving nodes to a better position on the

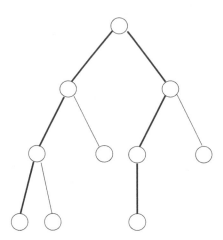

Figure 9.30   A DFS tree with the edges of the longest path designated by thick lines. Figure taken from [ST99, ST06].

embedding circle. As noted in the time complexity analysis of Algorithm CIRCULAR-Postprocessing, the order is dominated by the time required for counting the number of crossings. Therefore, it is vitally important to the time efficiency of the implementation of this algorithm that the number of crossings be counted in an effective manner. In order to lower the average time cost of counting crossings in the drawing, we ignore all edges that lie on the periphery of the embedding circle. These edges cannot possibly cause crossings. Also, in the step that determines the number of crossings caused by a single node, either the clockwise or counter-clockwise direction is first chosen dependent on which has the shorter arc.

9.7.3   Experimental Analysis of Algorithm CIRCULAR-with Radial

We have implemented Algorithm CIRCULAR-with Radial using Algorithm LayoutCluster1 and edge reduction postprocessing in C++ and run experiments with 11,399 graphs from [DGL+97]. The plot in Figure 9.31 shows the average number of edge crossings produced by the circular layout component of the GLT and Algorithm CIRCULAR-with Radial. As is shown by these results, the average number of crossings in the drawings produced by the technique is about 35% less than that of the GLT technique [DMM97, KMG88]. Sample drawings from both the GLT and Algorithm CIRCULAR-with Radial are shown in Figures 9.32 and 9.33.

The drawings produced by Algorithm CIRCULAR-with Radial clearly show the biconnectivity characteristics of networks. And although these drawings have a low number of edge crossings, they may show more details than a user would wish to see at one time. Therefore, we suggest that Algorithm CIRCULAR-with Radial can be used in an interactive environment in which the superstructure would be shown and the user would click on a node to see the details of the cluster; see Figure 9.34 for an example. Alternatively, the levels of visualization could be combined and some clusters shown in detail while others are shown with a single node.

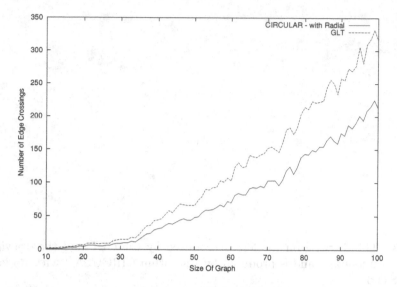

Figure 9.31 This plot shows the average number of edge crossings produced by Algorithm CIRCULAR-with Radial and the Graph Layout Toolkit when executed on 11,399 graphs from [DGL+97]. Figure taken from [ST06].

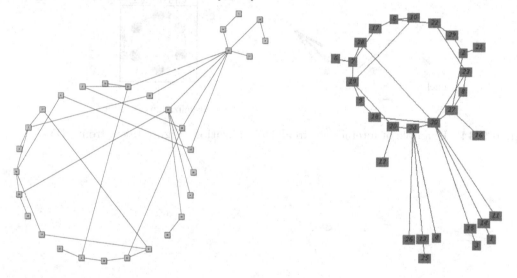

Figure 9.32 The drawing on the left is produced by the GLT and the drawing on the right is of the same graph and is produced by Algorithm CIRCULAR-with Radial. Figure taken from [ST06].

9.7.4 Implementation of Algorithm CIRCULAR-with Forces

We have implemented Algorithm CIRCULAR-with Forces so that all nodes and embedding circles are given an arbitrary initial placement. Then the force-directed equations of Section 9.6.1 are applied to the graph with the placement of group embedding circles frozen. See Figure 9.35 for a sample drawing.

An interesting behavior we noticed is that the drawing with minimal energy is not necessarily the best circular drawing. In circular drawing, a major goal is to reduce edge

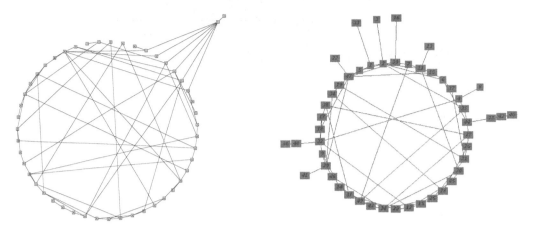

Figure 9.33 The drawing on the left is produced by the GLT and the drawing on the right is of the same graph and is produced by Algorithm CIRCULAR-with Radial. Figure taken from [ST06].

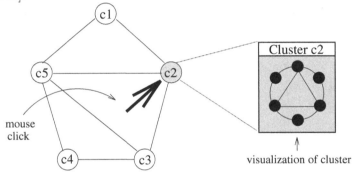

Figure 9.34 Example of interactive circular visualization. Figure taken from [ST06].

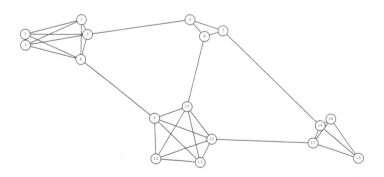

Figure 9.35 Sample user-grouped circular drawing from the preliminary implementation. Figure taken from [ST03b].

crossings. However, it is well known [DETT99] that reducing crossings sometimes means the compromise of other aesthetics, especially area, and area is related to minimum energy in spring systems. We propose adding springs from each node to its initial placement on the plane with the spring constants for these springs being high. This should keep these nodes from gravitating toward each other too much and causing extra crossings. We also suggest creating dummy nodes which are placed in the center of each embedding circle and attaching strong springs from them to every node in their respective group.

9.8 Conclusions

Circular visualizations of networks which show the inherent strengths and weaknesses of structures with clustered views are advantageous additions to many design tools.

We have presented an $O(m)$ time algorithm for drawing circular visualizations of biconnected graphs on a single embedding circle. Not only is this technique efficient, but it also produces a plane drawing of the biconnected graph if such exists. Extensive experiments show that the technique works very well in practice. We have also presented an $O(m)$ time technique which decomposes the given graph into biconnected components and visualizes each cluster on a separate embedding circle. This technique has been implemented and results of an experimental study also show this algorithm to perform very well in practice. Both techniques produce drawings that clearly show the biconnectivity structure of the given graphs and also have a low number of crossings. We have also discussed a framework for creating circular graph drawings in which the grouping is defined by the user. This framework includes the successful merging of the force-directed and circular graph drawing paradigms. Algorithm CIRCULAR-with Forces is fast and produces drawings in which the user-defined groupings are highly visible, each group is laid out with a low number of edge crossings, and the number of crossings between intra-group and inter-group edges is low.

References

[BB05]   M. Baur and U. Brandes. Crossing Reduction in Circular Layouts In *Proc. WG '04, LNCS 3353*, pages 332–343, 2005.

[Ber81]   M. A. Bernard. On the automated drawing of graphs. In *Proc. 3rd Caribbean Conf. on Combinatorics and Computing*, pages 43–55, 1981.

[Bra97]   F. J. Brandenburg. Graph clustering I: Cycles of cliques. In *Proc. GD '97, LNCS 1353*, pages 158–168, 1997.

[DETT99]   G. Di Battista, P. Eades, R. Tamassia, and I. G. Tollis. *Graph Drawing.* Prentice Hall, Upper Saddle River, NJ, 1999.

[DGL⁺97]   G. Di Battista, A. Garg, G. Liotta, R. Tamassia, E. Tassinari, and F. Vargiu. An experimental comparison of four graph drawing algorithms. *Comput. Geom. Theory Appl.*, 7:303–325, 1997.

[DMM97]   U. Dogrusoz, B. Madden, and P. Madden. Circular layout in the graph layout toolkit. In *Proc. GD '96, LNCS 1190*, pages 92–100, 1997.

[Ead84]   P. Eades. A heuristic for graph drawing. *Congr. Numer.*, 42:149–160, 1984.

[Ead92]   P. D. Eades. Drawing free trees. *Bulletin of the Institute for Combinatorics and its Applications*, 5:10–36, 1992.

[EF97]   P. Eades and Q. Feng. Multilevel visualization of clustered graphs. In *Proc. GD '96, LNCS 1190*, pages 101–112, 1997.

[EFL97]   P. Eades, Q. Feng, and X. Lin. Straight-line drawing algorithms for hier- archical graphs and clustered graphs. In *Proc. GD '96, LNCS 1190*, pages 113–128, 1997.

[EFN99]   P. Eades, Q. W. Feng, and H. Nagamochi. Drawing clustered graphs on an orthogonal grid. *Jrnl. of Graph Algorithms and Applications*, pages 3–29, 1999.

[Esp88]   C. Esposito. Graph graphics: Theory and practice. *Comput. Math. Appl.*, 15(4):247–253, 1988.

[GK07]   E. Gansner and Y. Koren. Improved Circular Layouts. In *Proc. GD '06, LNCS 4372*, pages 386–398, 2007.

[HS04]   H. He and O. Sýkora. New Circular Drawing Algorithms. Unpublished manuscript, Creative Commons License, 2004.

[HE98]   M. L. Huang and P. Eades. A fully animated interactive system for clus- tering and navigating huge graphs. In *Proc. GD '98, LNCS 1547*, pages 107–116, 1998.

[HR88]   D. Halliday and R. Resnick. *Fundamentals of Physics: 3rd Edition Ex- tended.* Wiley, New York, NY, 1988.

[KMG88]   G. Kar, B. Madden, and R. Gilbert. Heuristic layout algorithms for network presentation services. *IEEE Network*, pages 29–36, 11 1988.

[KW02]   M. Kaufmann and R. Wiese. Maintaining the Mental Map for Circular Drawings. In *Proc. GD 2002, LNCS 2528*, Pages 12–22, 2002.

[Ker93]   A. Kershenbaum. *Telecommunications Network Design Algorithms.* McGraw-Hill, 1993.

[Kre96]   V. Krebs. Visualizing human networks. *Release 1.0: Esther Dyson's Monthly Report*, pages 1–25, February 12 1996.

[Ma88]   E. Mäkinen. On Circular Layouts. In *Intl. Jrnl of Computer Mathematics*, pages 29–37, 24(1988).

[Mit79]   S. Mitchell. Linear algorithms to recognize outerplanar and maximal outerplanar graphs. *Information Processing Letters*, pages 229–232, 9(5) 1979.

[MKNF87] S. Masuda, T. Kashiwabara, K. Nakajima, and T. Fujisawa. On the NP-completeness of a computer network layout problem. In *Proc. IEEE 1987 International Symposium on Circuits and Systems, Philadelphia, PA*, pages 292–295, 1987.

[PS85]   F. P. Preparata and M. I. Shamos. *Computational Geometry: An Introduction*. Springer-Verlag, New York, NY, 1985.

[Pur97]   Helen Purchase. Which aesthetic has the greatest effect on human understanding? In *GD '97, LNCS 1353*, pages 248–261, 1997.

[Six00]   J. M. Six(Urquhart). *Vistool: A Tool For Visualizing Graphs*. PhD thesis, The University of Texas at Dallas, 2000.

[ST99]   J. M. Six and I. G. Tollis. Circular drawings of biconnected graphs. In *Proc. of ALENEX '99, LNCS 1619*, pages 57–73, 1999.

[ST01]   J. M. Six and I. G. Tollis. Effective graph visualization via node grouping. In *Proc. of IEEE InfoVis 2001*, pages 51–58, 2001. (see extended version in [ST03a])

[ST03a]   J. M. Six and I. G. Tollis. Effective graph visualization via node grouping. In K. Zhang Ed., editor, *Software Visualization: From Theory to Practice, The Kluwer Intl. Series in Engineering and Computer Science Vol. 734*. Kluwer Academic Publishers, 2003.

[ST06]   J. M. Six and I. G. Tollis. A framework and algorithms for circular drawings of graphs. *Jrnl. of Discrete Algorithms*, 4(1), pages 25–50, 2006.

[ST03b]   J. M. Six and I. G. Tollis. A framework for user-grouped circular drawings. In *Proc of GD 2003, LNCS 2912*, pages 135–146, 2003.

[TX95]   I. G. Tollis and C. Xia. Drawing telecommunication networks. In *Proc. GD '94, LNCS 894*, pages 206–217, 1995.

[YFDH01] K. Yee, D. Fisher, R. Dhamija, and M. Hearst. Animated exploration of dynamic graphs with radial layout. In *Proc. of InfoVis 2001*, pages 43–50. IEEE, 2001.

10

Rectangular Drawing Algorithms

Takao Nishizeki
Kwansei Gakuin University,
Japan

Md. Saidur Rahman
BUET, Bangladesh

10.1   Introduction

A *rectangular drawing* of a plane graph G, a planar graph G with a fixed embedding, is a drawing of G in which each vertex is drawn as a point, each edge is drawn as a horizontal or vertical line segment without edge-crossings, and each face is drawn as a rectangle. Figure 10.1(b) illustrates a rectangular drawing of the plane graph in Fig. 10.1(a).

Rectangular drawings have practical applications in VLSI floorplanning and architectural floorplanning [NR04]. In a VLSI floorplanning problem, an input is a plane graph F as illustrated in Fig. 10.2(a); F represents the functional entities of a chip, called *modules*, and interconnections among the modules; each vertex of F represents a module, and an edge between two vertices of F represents the interconnections between the two corresponding modules. An output of the problem for the input graph F is a partition of a rectangular chip area into smaller rectangles as illustrated in Fig. 10.2(d); each module is assigned to a smaller rectangle, and furthermore, if two modules have interconnections, then their corresponding rectangles must be adjacent, that is, must have a common boundary.

A similar problem arises in architectural floorplanning. When building a house, the owner may have some preference; for example, a bedroom should be adjacent to a reading room. The owner's choice of room adjacencies can be easily modeled by a plane graph F, as illustrated in Fig. 10.2(a); each vertex represents a room and an edge between two vertices represents the desired adjacency between the corresponding rooms. A rectangular drawing of a plane graph may provide a suitable solution of the floorplanning problem described above. First, obtain a plane graph F' by triangulating all inner faces of F as illustrated in Fig. 10.2(b), where dotted lines indicate new edges added to F. Then obtain a dual-like graph G of F' as illustrated in Fig. 10.2(c), where the four vertices of degree 2 drawn by white circles correspond to the four corners of the rectangular area. Finally, find a rectangular drawing of the plane graph G to obtain a possible floorplan for F as illustrated in Fig. 10.2(d).

317

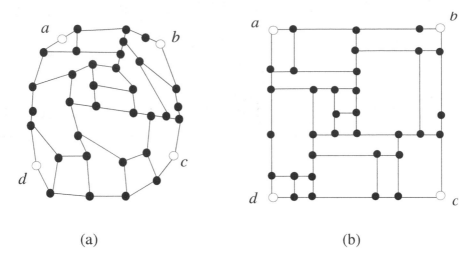

Figure 10.1   (a) Plane graph, and (b) its rectangular drawing for the designated corners a, b, c and d. (Figure taken from [NR04].)

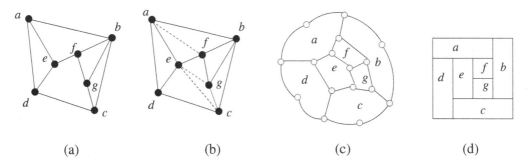

Figure 10.2   (a) Graph F, (b) triangulated graph F', (c) dual-like graph G, and (d) rectangular drawing of G. (Figure taken from [NR04].)

In a rectangular drawing of G, the outer cycle $C_o(G)$ is drawn as a rectangle and hence has four convex corners such as a, b, c and d drawn by white circles in Fig. 10.1. Such a convex corner is an outer vertex of degree two and is called a *corner of the rectangular drawing*. Not every plane graph G has a rectangular drawing. Of course, G must be 2-connected and the maximum degree Δ of G is at most four if G has a rectangular drawing. Miura et al. showed that a plane graph G with $\Delta \leq 4$ has rectangular drawing D if and only if a new bipartite graph constructed from G has a perfect matching, and D can be found in time $O(n^{1.5}/\log n)$ whenever G has D [MHN06]. In Section 10.2 we present their result on rectangular drawings of plane graphs with $\Delta \leq 4$.

Since a planar graph with $\Delta \leq 3$ often appears in many practical applications, much works are devoted to rectangular drawings of planar graphs with $\Delta \leq 3$ [BS88, Tho84, RNN98, RNN02]. In Section 10.3 we present a necessary and sufficient condition for a plane graph G with $\Delta \leq 3$ to have a rectangular drawing when four outer vertices of degree two are designated as the corners [Tho84], and also present a linear-time algorithm to obtain a rectangular drawing with the designated corners [RNN98]. The problem of examining whether a plane graph has a rectangular drawing becomes difficult when four outer vertices are not designated as the corners. We also present a necessary and sufficient condition for a plane graph with $\Delta \leq 3$ to have a rectangular drawing for some quadruplet of outer vertices appropriately chosen as the corners; the condition leads to a linear-time algorithm [RNN02, NR04].

In the floorplan described in Fig. 10.2(d), two rectangles are always adjacent if the modules corresponding to them have interconnections in F in Fig. 10.2(a). However, two rectangles may be adjacent even if the modules corresponding to them have no interconnections in F. For example, module e and module g have no interconnection in F, but their corresponding rectangles are adjacent in the floorplan in Fig. 10.2(d). Such unwanted adjacencies are not desirable in some other floorplanning problems.

In floorplanning of a MultiChip Module (MCM), two chips generating excessive heat should not be adjacent, or two chips operating on high frequency should not be adjacent to avoid malfunctioning due to their interference [She95]. Unwanted adjacencies may cause a dangerous situation in some architectural floorplanning, too [FW74]. For example, in a chemical industry, a processing unit that deals with poisonous chemicals should not be adjacent to a cafeteria. We can avoid the unwanted adjacencies if we obtain a floorplan for F by using a "box-rectangular drawing" instead of a rectangular drawing. A *box-rectangular drawing* of a plane graph G is a drawing of G such that each vertex is drawn as a rectangle, called a *box*, each edge is drawn as a straight line segment joining points on the two boxes corresponding to the ends, and the contour of each face is drawn as a rectangle, as illustrated in Fig. 10.3(c). A vertex may be drawn as a degenerate rectangle, that is, a point. A floorplan can be obtained by using a box-rectangular drawing as follows. First, without triangulating the inner faces of F, find a dual-like graph G of F as illustrated in Fig. 10.3(b). Then find a box-rectangular drawing of G to obtain a possible floorplan for F as illustrated in Fig. 10.3(c). In Fig. 10.3(c) rectangles e and g are not adjacent although there is a dead space corresponding to a vertex of G drawn by a rectangular box. Such a dead space to separate two rectangles in floorplanning is desirable for dissipating excessive heat in an MCM or for ensuring safety in a chemical industry. If G has multiple edges or a vertex of degree five or more, then G has no rectangular drawing but may have a box-rectangular drawing. However, not every plane graph has a box-rectangular drawing. Section 10.4 presents a necessary and sufficient condition for the existence of a box-rectangular drawing of a plane graph, and gives a linear algorithm to find a box-rectangular drawing if it exists.

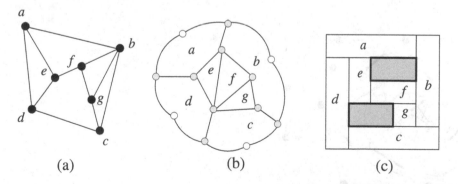

(a)   (b)   (c)

Figure 10.3 (a) F, (b) G, and (c) box-rectangular drawing of G. (Figure taken from [NR04].)

10.2   Rectangular Drawing and Matching

This section deals with rectangular drawings of plane graphs with $\Delta \leq 4$, and shows that a plane graph G with $\Delta \leq 4$ has rectangular drawing D if and only if a new bipartite graph G_d constructed from G has a perfect matching, and D can be found in time $O(n^{1.5}/\log n)$ if D exists [MHN06, NR04]. G_d is called a *decision graph*.

One may assume without loss of generality that G is 2-connected and $\Delta \leq 4$, and hence every vertex of G has degree two, three or four.

An angle formed by two edges e and e' incident to a vertex v in G is called an *angle of v* if e and e' appear consecutively around v. An angle of a vertex in G is called an *angle of G*. An angle formed by two consecutive edges on a boundary of a face F in G is called an *angle of F*. An angle of the outer face is called an *outer angle* of G, while an angle of an inner face is called an *inner angle*.

In any rectangular drawing, every inner angle is 90° or 180°, and every outer angle is 180° or 270°. Consider a labeling Θ which assigns a label 1, 2, or 3 to every angle of G, as illustrated in Fig. 10.4(b). Labels $1, 2$ and 3 correspond to angles 90°, 180° and 270°, respectively. Therefore each inner angle has label either 1 or 2, exactly four outer angles have label 3, and all other outer angles have label 2.

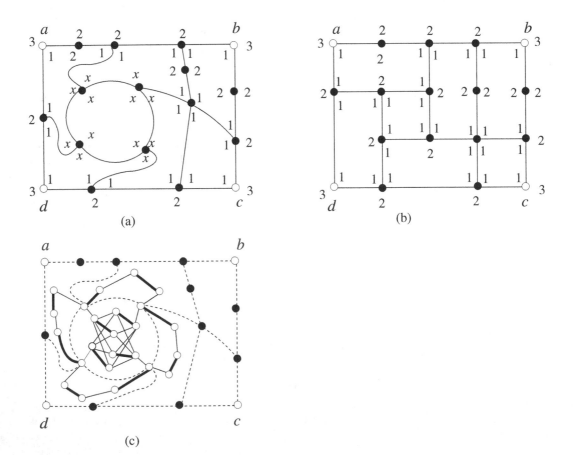

Figure 10.4   (a) Plane graph G, (b) rectangular drawing D and regular labeling Θ of G, and (c) decision graph G_d. (Figure taken from [NR04].)

We call Θ a *regular labeling* of G if Θ satisfies the following three conditions (a)–(c):

(a) For each vertex v of G, the sum of the labels of all the angles of v is equal to 4;

(b) The label of any inner angle is 1 or 2, and every inner face has exactly four angles of label 1; and

(c) The label of any outer angle is 2 or 3, and the outer face has exactly four angles of label 3;

Figure 10.4(b) depicts a regular labeling Θ of the plane graph in Fig. 10.4(a) and a rectangular drawing D corresponding to Θ. A regular labeling is a special case of an orthogonal representation of an orthogonal drawing presented in [Tam87].

Conditions (a) and (b) imply the following (i)–(iii):

(i) If a non-corner vertex v has degree two, that is, $d(v) = 2$, then the two labels of v are 2 and 2.

(ii) If $d(v) = 3$, then exactly one of the three angles of v has label 2 and the other two have label 1.

(iii) If $d(v) = 4$, then all the four angles of v have label 1.

If G has a rectangular drawing, then clearly G has a regular labeling. Conversely, if G has a regular labeling, then G has a rectangular drawing, as can be proved by means of elementary geometric considerations. We thus have the following fact.

Fact 10.1 *A plane graph G has a rectangular drawing if and only if G has a regular labeling.*

Assume now that four outer vertices a, b, c and d of degree two are designated as corners. Then the outer angles of a, b, c and d must be labeled with 3, and all the other outer angles of G must be labeled with 2, as illustrated in Fig. 10.4(a). Some of the inner angles of G can be immediately determined, as illustrated in Fig. 10.4(a). If v is a non-corner outer vertex of degree two, then the inner angle of v must be labeled with 2. The two angles of any inner vertex of degree two must be labeled with 2. If v is an outer vertex of degree three, then the outer angle of v must be labeled with 2 and both of the inner angles of v must be labeled with 1. All the four angles of each vertex of degree four must be labeled with 1. On the other hand we label all the three angles of an inner vertex of degree three with x, because one cannot determine their labels although exactly one of them must be labeled with 2 and the others with 1. Label x means that x is either 1 or 2, and exactly one of the three labels x's attached to the same vertex must be 2 and the other two must be 1. (See Figs. 10.4(a) and (b).)

We now present how to construct a decision graph G_d of G. Let all vertices of G that have been attached label x be vertices of G_d. Thus all the inner vertices of degree three are vertices of G_d, and none of the other vertices of G is a vertex of G_d. We then add to G_d a complete bipartite graph inside each inner face F of G, as illustrated in Fig. 10.5 where G_d is drawn by solid lines and G by dotted lines. Let n_x be the number of angles of F labeled with x. For example, $n_x = 3$ for the face F in Fig. 10.5. Let n_1 be the number of angles of F which have been labeled with 1. Then n_1 is the number of vertices v on F such that one of the following (i)–(iii) holds:

(i) v is a corner vertex, that is, v is an outer vertex of degree 2 and the outer angle of v is labeled with 3;

(ii) v is an outer vertex of degree 3 and the outer angle of v is labeled with 2; and

(iii) $d(v) = 4$.

Thus $n_1 = 2$ for the example in Fig. 10.5. One may assume as a trivial necessary condition that $n_1 \le 4$; otherwise, G has no rectangular drawing. Exactly $4 - n_1$ of the n_x angles of F labeled with x must be labeled with 1 by a regular labeling. Add a complete bipartite graph $K_{(4-n_1),n_x}$ in F, and join each of the n_x vertices in the second partite set with one of the n_x vertices on F whose angles are labeled with x. Repeat the operation above for each inner face F of G. The resulting graph is a *decision graph* G_d of G. The decision graph G_d of the plane graph G in Fig. 10.4(a) is drawn by solid lines in Fig. 10.4(c), where G is drawn by dotted lines.

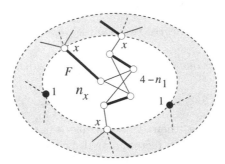

Figure 10.5   Construction of G_d for an inner face F of G. (Figure taken from [NR04].)

A *matching* of G_d is a set of pairwise non-adjacent edges in G_d. A *maximum matching* of G_d is a matching of the maximum cardinality. A matching M of G_d is called a *perfect matching* if an edge in M is incident to each vertex of G_d. A perfect matching of G_d is drawn by thick solid lines in Figs. 10.4(c) and 10.5.

Each edge e of G_d incident to a vertex v attached a label x corresponds to an angle α of v labeled with x. A fact that e is contained in a perfect matching M of G_d means that the label x of α is 2. Conversely, a fact that e is not contained in M means that the label x of α is 1. Then one can easily observe that G has a rectangular labeling if and only if G_d has a perfect matching.

Clearly, G_d is a bipartite graph, and $4-n_1 \le 4$. Obviously, n_x is no more than the number of edges on face F. Let n be the number of vertices, and let m be the number of edges in G, then we have $2m \le 4n$ since $\Delta \le 4$. Therefore the sum $2m$ of the numbers of edges on all faces is at most $4n$. One can thus know that both the number n_d of vertices in G_d and the number m_d of edges in G_d are $O(n)$. Since G_d is a bipartite graph, a maximum matching of G_d can be found either in time $O(\sqrt{n_d}m_d) = O(n^{1.5})$ by an ordinary bipartite matching algorithm [HK73, MV80, PS82] or in time $O(n^{1.5}/\log n)$ by a pseudoflow-based bipartite matching algorithm using boolean word operations on $\log n$-bit words [Hoc04, HC04]. One can find a regular labeling Θ of G from a perfect matching of G_d in linear time. It is easy to find a rectangular drawing of G from Θ in linear time. Thus the following theorem holds [MHN06].

Theorem 10.1   *Let G be a plane graph with $\Delta \le 4$ and four outer vertices a, b, c and d be designated as corners. Then G has a rectangular drawing D with the designated corners if and only if the decision graph G_d of G has a perfect matching. D can be found in time $O(n^{1.5}/\log n)$ whenever G has D.*

10.3 Linear Algorithms for Rectangular Drawing

This section presents Thomassen's theorem on a necessary and sufficient condition for a plane graph G with $\Delta \leq 3$ to have a rectangular drawing when four outer vertices of degree two are designated as the corners [Tho84], and gives a linear-time algorithm to find a rectangular drawing of G if it exists [RNN98].

10.3.1 Thomassen's Theorem

Before presenting Thomassen's theorem we recall some definitions. An edge of a plane graph G is called a *leg* of a cycle C if it is incident to exactly one vertex of C and located outside C. The vertex of C to which a leg is incident is called a *leg-vertex* of C. A cycle in G is called a *k-legged cycle* of G if C has exactly k legs in G and there is no edge which joins two vertices on C and is located outside C. Figure 10.6(a) illustrates 2-legged cycles C_1, C_2, C_3 and C_4, while Fig. 10.6(b) illustrates 3-legged cycles C_5, C_6, C_7 and C_8, where corners are drawn by white circles.

If a 2-legged cycle contains at most one corner like C_1, C_2 and C_3 in Fig. 10.6(a), then some inner face cannot be drawn as a rectangle and hence G has no rectangular drawing. Similarly, if a 3-legged cycle contains no corner like C_5 and C_8 in Fig. 10.6(b), then G has no rectangular drawing. One can thus observe the following fact.

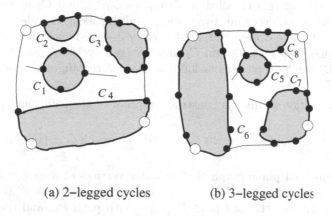

(a) 2–legged cycles   (b) 3–legged cycles

Figure 10.6 Good cycles C_4, C_6 and C_7, and bad cycles C_1, C_2, C_3, C_5 and C_8. (Figure taken from [NR04].)

Fact 10.2 *In any rectangular drawing D of G, every 2-legged cycle of G contains two or more corners, every 3-legged cycle of G contains one or more corner, and every cycle with four or more legs may contain no corner, as illustrated in Fig. 10.7.*

The necessity of the following Thomassen's theorem [Tho84] is immediate from Fact 10.2.

Theorem 10.2 *Assume that G is a 2-connected plane graph with $\Delta \leq 3$ and four outer vertices of degree two are designated as the corners a, b, c and d. Then G has a rectangular drawing if and only if*

(r1) *any 2-legged cycle contains two or more corners, and*

(r2) *any 3-legged cycle contains one or more corners.*

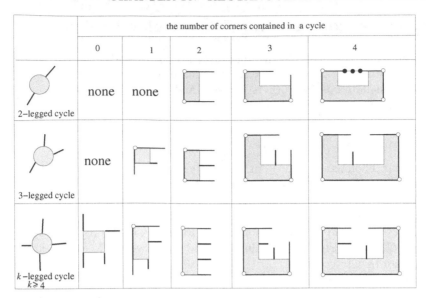

Figure 10.7 Numbers of corners in drawings of cycles. (Figure taken from [NR04].)

A cycle of type (r1) or (r2) is called *good*. Cycles C_4, C_6 and C_7 in Fig. 10.6 are good cycles; the 2-legged cycle C_4 contains two corners, and the 3-legged cycles C_6 and C_7 contain one or two corners. On the other hand, a 2-legged or 3-legged cycle is called *bad* if it is not good. Thus 2-legged cycles C_1, C_2 and C_3 and 3-legged cycles C_5 and C_8 are bad cycles. Thus Theorem 10.2 can be rephrased as follows: G has a rectangular drawing if and only if G has no bad cycle.

The rest of this section outlines a constructive proof of the sufficiency of Theorem 10.2 [RNN98].

The *union* $G = G' \cup G''$ of two graphs G' and G'' is a graph $G = (V(G') \cup V(G''), E(G') \cup E(G''))$.

In a given 2-connected plane graph G, four outer vertices of degree two are designated as the corners a, b, c and d. These four corners divide the outer cycle $C_o(G)$ of G into four paths, the north path P_N, the east path P_E, the south path P_S, and the west path P_W, as illustrated in Fig. 10.8(a). The north and south paths will be drawn as two horizontal straight line segments, and the east and west paths as two vertical line segments. Thus the embedding of $C_o(G)$ is fixed as a rectangle, which is called the *outer rectangle* of G.

A graph of a single edge, not in the outer cycle $C_o(G)$, joining two vertices in $C_o(G)$ is called a $C_o(G)$-*component* of G. A graph which consists of a connected component of $G - V(C_o(G))$ and all edges joining vertices in that component and vertices in $C_o(G)$ is also called a $C_o(G)$-*component*. The outer cycle $C_o(G)$ of the plane graph G in Fig. 10.8(a) is drawn by thick lines, and the $C_o(G)$-components J_1, J_2 and J_3 of G are depicted in Fig. 10.8(b). Clearly the following lemma holds.

LEMMA 10.1 Let $J_1, J_2, \cdots, J_p$ be the $C_o(G)$-components of a plane graph G, and let $G_i = C_o(G) \cup J_i$, $1 \le i \le p$, as illustrated in Fig. 10.9. Then G has a rectangular drawing with corners a, b, c and d if and only if, for each index i, $1 \le i \le p$, G_i has a rectangular drawing with corners a, b, c and d.

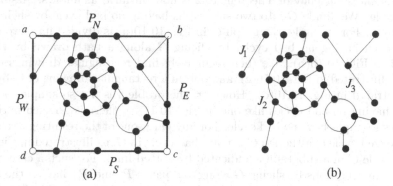

Figure 10.8 (a) Plane graph G, and (b) $C_o(G)$-components. (Figure taken from [NR04].)

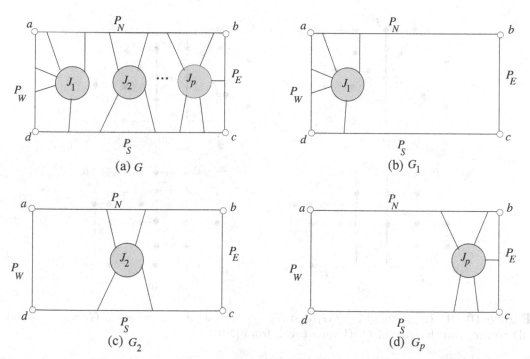

Figure 10.9 (a) G, (b) G_1, (c) G_2, and (d) G_p. (Figure taken from [NR04].)

In the remainder of this section, because of Lemma 10.1, one may assume that G has exactly one $C_o(G)$-component J.

The proof of the sufficiency of Theorem 10.2 is now outlined as follows. Assume that G has no bad cycle. We divide G into two subgraphs having no bad cycle by slicing G along one or two paths. For example, the graph G in Fig. 10.10(a) is divided into two subgraphs G_1 and G_2, each having no bad cycle, by slicing G along a path drawn by thick lines, as illustrated in Fig. 10.10(b). We then recursively find rectangular drawings of the two subgraphs as illustrated in Fig. 10.10(c), and obtain a rectangular drawing of G by patching them, as illustrated in Fig. 10.10(d). However, the problem is not so simple, because, for some graphs having no bad cycles like one in Fig. 10.11(a), there is no such path that the resulting two subgraphs have no bad cycle. For any path, one of the resulting two subgraphs has a bad 3-legged cycle C although C is not a bad cycle in G, as illustrated in Fig. 10.11(b) where a bad cycle C in a subgraph is indicated by dotted lines. For such a case, we split G into two or more subgraphs by slicing G along two paths P_c and P_{cc} having the same ends on P_N and P_S. For example, as illustrated in Fig. 10.11(c), the graph G in Fig. 10.11(a) is divided into three subgraphs G_1, G_2 and G_3, each having no bad cycle, by slicing G along path P_c indicated by dotted lines and path P_{cc} drawn by thick lines in Fig. 10.11(a). We then recursively find rectangular drawings of G_1, G_2 and G_3 as illustrated in Fig. 10.11(d), and slightly deform the drawings of G_1 and G_2, as illustrated in Fig. 10.11(e). We finally obtain a rectangular drawing of G by patching the drawings of the three subgraphs as illustrated in Fig. 10.11(f).

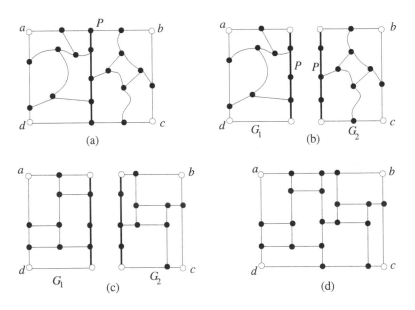

Figure 10.10   (a) G and P, (b) G_1 and G_2, (c) rectangular drawings of G_1 and G_2, and (d) rectangular drawing of G. (Figure taken from [NR04].)

We need some definitions before presenting the detail of a constructive proof. A cycle C in G like one in Fig. 10.11(a) is called "critical," because C is not a bad cycle in G but C would become a bad cycle in a subgraph obtained from G by splitting G along a path P. We now give a formal definition of a critical cycle. A cycle C in a plane graph G is *attached to a path P* if

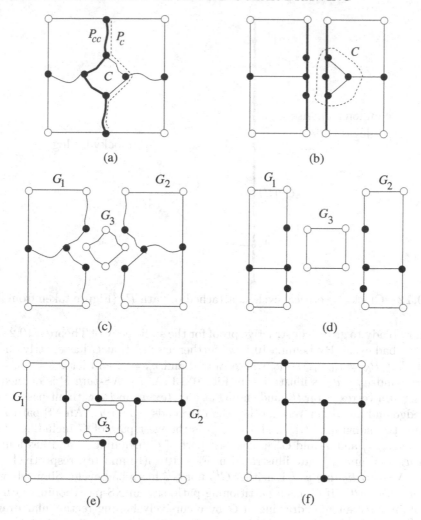

Figure 10.11   (a) G, (b) splitting G along a single path P_{cc}, (c) splitting G along two paths P_{cc} and P_c, (d) rectangular drawings of three subgraphs, (e) deformation, and (f) rectangular drawing of G. (Figure taken from [NR04].)

(i) P does not contain any vertex in the proper inside of C, and

(ii) the intersection of C and P is a single subpath of P,

as illustrated in Fig. 10.12. Let v_t be the starting vertex of the subpath, and let v_h be the ending vertex. We then call v_t the *tail vertex* of C for P, and v_h the *head vertex*. Denote by $Q_c(C)$ the path on C turning clockwise around C from v_t to v_h, and denote by $Q_{cc}(C)$ the path on C turning counterclockwise around C from v_t to v_h. A leg of C is called a *clockwise leg* for P if it is incident to a vertex in $V(Q_c(C)) - \{v_t, v_h\}$. Denote by $n_c(C)$ the number of clockwise legs of C for P. Similarly we define a *counterclockwise leg* and denote by $n_{cc}(C)$ the number of counterclockwise legs of C for P. A cycle C attached to P is called a *clockwise cycle* if $Q_{cc}(C)$ is a subpath of P, and is called a *counterclockwise cycle* if $Q_c(C)$ is a subpath of P. A cycle C is called a *critical cycle* if either C is a clockwise cycle and $n_c(C) \leq 1$ or C is a counterclockwise cycle and $n_{cc}(C) \leq 1$. Figure 10.12 illustrates a clockwise critical cycle with $n_c(C) = 1$.

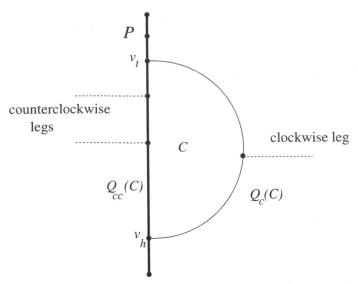

Figure 10.12   Clockwise critical cycle C attached to path P. (Figure taken from [NR04].)

We are now ready to give a constructive proof for the sufficiency of Theorem 10.2. Assume that G has no bad cycle. By Lemma 10.1, we further assume that G has exactly one $C_o(G)$-component. Let $P_N = v_0, v_1, \cdots, v_p$ where $v_0 = a$ and $v_p = b$, and let $P_S = u_0, u_1, \cdots, u_q$ where $u_0 = c$ and $u_q = d$, as illustrated in Fig. 10.13(a). An *NS-path* P is defined to be a path starting at a vertex v_i on P_N and ending at a vertex u_j on P_S without passing through any outer edge and any outer vertex other than the ends v_i and u_j. An NS-path P divides graph G into two subgraphs G_W^P and G_E^P; G_W^P is the west part of G including P and has four corners a, v_i, u_j and d, and G_E^P is the east part of G including P and has four corners v_i, b, c and u_j. G_E^P and G_W^P are illustrated in Figs. 10.13(b) and (c), respectively. We say that P is an *NS-partitioning path* if neither G_W^P nor G_E^P has a bad cycle. Similarly we define a *WE-partitioning path*. If G has a partitioning path, say an NS-partitioning path P, then one can obtain a rectangular drawing of G by recursively finding rectangular drawings of G_W^P and G_E^P and patching them together along P, as illustrated in Fig. 10.10.

An inner face of G is called a *boundary face* if its contour contains at least one outer edge. A *boundary path* is a maximal path on the contour of a boundary face connecting two outer vertices without passing through any outer edge. Note that the direction of a boundary path is the same as the contour of the face, and hence is clockwise. For $X, Y \in \{N, E, S, W\}$, a *boundary XY-path* is a boundary path starting at a vertex on path P_X and ending at a vertex on path P_Y. One can easily verify the following lemma [RNN98].

LEMMA 10.2   If G has no bad cycle, then every boundary NS-, SN-, EW- or WE-path P of G is a partitioning path, that is, G can be split along P into two subgraphs, each having no bad cycle.

Thus one may assume that G has no boundary NS-, SN-, EW- or WE-paths. Then the $C_o(G)$-component J has at least one vertex on each of the paths P_N, P_E, P_S and P_W. In this case we find a pair of partitioning paths P_c and P_{cc}, and divide G into two or more subgraphs having no bad cycles by splitting G along P_c and P_{cc}. Both P_c and P_{cc} are NS-paths which have the same ends and do not cross each other in the plane although they may share several edges. Thus, if $P_c \neq P_{cc}$, then the edge set $E(P_c) \oplus E(P_{cc}) = E(P_c) \cup E(P_{cc}) - E(P_c) \cap E(P_{cc})$

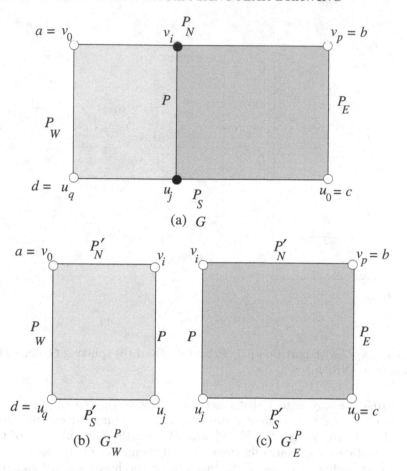

Figure 10.13 (a) Plane graph G and NS-path P, (b) G_W^P, and (c) G_E^P. (Figure taken from [NR04].)

induces vertex-disjoint cycles $C_1, C_2, \cdots, C_k$, $k \geq 1$, as illustrated in Figs. 10.14 and 10.15 where P_c and P_{cc} are indicated by dotted lines. Thus P_c and P_{cc} share $k + 1$ maximal subpaths $P_1, P_2, \cdots, P_{k+1}$, as illustrated in Fig. 10.14(a). We assume that P_c turns around cycles $C_1, C_2, \cdots, C_k$ clockwise, and P_{cc} turns around them counterclockwise. We choose P_c and P_{cc} so that each cycle C_i has exactly four legs; assuming clockwise order, the first one is contained in P_i, $1 \leq i \leq k$, the second one is a clockwise leg, the third one is contained in P_{i+1} and the fourth one is a counterclockwise leg; and the four leg-vertices of C_i will be designated as the corners of the subgraph $G(C_i)$ of G inside C_i. Thus G is divided into subgraphs $G_W^{P_{cc}}$, $G_E^{P_c}$, $G(C_1), G(C_2), \cdots, G(C_k)$, as illustrated in Figs. 10.14(b) and 10.15(b). $G_W^{P_{cc}}$ has a, d and the two ends of P_{cc} as the corners, while $G_E^{P_c}$ has b, c and the two ends of P_c as the corners. $G_i(C_i)$, $1 \leq i \leq k$, has the four leg-vertices of C_i as the corners. Then the following lemma holds [RNN98].

LEMMA 10.3 Assume that a cycle C in the $C_o(G)$-component J has exactly four legs. Then the subgraph $G(C)$ of G inside C has no bad cycle when the four leg-vertices are designated as corners of $G(C)$.

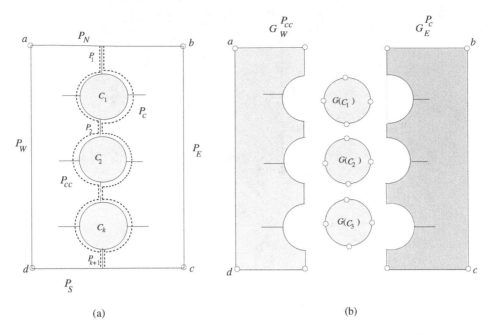

Figure 10.14   (a) G with partition-pair P_c and P_{cc}, and (b) splitting G along P_c and P_{cc}. (Figure taken from [NR04].)

By Lemma 10.3 one may assume that none of $G(C_1), G(C_2), \cdots, G(C_k)$ has a bad cycle. For each cycle C_i, $1 \le i \le k$, there are two alternative rectangular embeddings of C_i as illustrated in Fig. 10.16, where P_N', P_E', P_S' and P_W' are the four subpaths of C_i divided by the four leg-vertices. We arbitrarily choose one of them. Let G_1 be the graph obtained from $G_W^{P_{cc}}$ by contracting all edges of P_{cc} that are on the horizontal sides of rectangular embeddings of $C_1, C_2, \cdots, C_k$, as illustrated in Fig. 10.15(c). Note that every intermediate vertex on such a horizontal side has degree two in G_W^P. We denote by P_{cc}' the resulting path obtained from P_{cc} by the contraction above. Let G_1 have four corners a, d and the two ends of P_{cc}. Then one can observe that if G_1 has a rectangular drawing, in which the path P_{cc}' is drawn as a vertical straight line segment, then the rectangular drawing of G_1 can be easily modified to a drawing of $G_W^{P_{cc}}$ fitted in the area for $G_W^{P_{cc}}$ where P_{cc} is drawn as an alternating sequence of horizontal and vertical line segments, as illustrated in Figs. 10.15(b) and (c). Let G_2 be the graph obtained from $G_E^{P_c}$ by contracting all edges of P_c that are on the horizontal sides of rectangular embeddings of $C_1, C_2, \cdots, C_k$, and let P_c' be the resulting path obtained from P_c by the contraction, as illustrated in Fig. 10.15(d). Then, if G_2 has a rectangular drawing, then it can be easily modified to a drawing of $G_E^{P_c}$ fitted in the area for $G_E^{P_c}$ where P_c is drawn as an alternating sequence of horizontal and vertical line segments, as illustrated in Figs. 10.15(b) and (d). Thus if we have drawings of graphs $G_W^{P_{cc}}, G_E^{P_c}, G(C_1), G(C_2), \cdots, G(C_k)$, then we can immediately patch them to get a rectangular drawing of G. One can observe that G_1 and G_2 have no bad cycles if and only if $G_W^{P_{cc}}$ and $G_E^{P_c}$ have no bad cycles, respectively. We thus call P_c and P_{cc} a *pair of partitioning paths* or simply a *partition-pair* if neither $G_E^{P_c}$ nor $G_W^{P_{cc}}$ has a bad cycle. Especially when $P_c = P_{cc}$, it is a single partitioning path.

Thus the problem is how to prove that G has a partition-pair and to find a partition-pair efficiently. The following lemma was proved in [RNN98], and one can derive from the proof a linear algorithm to find a partition pair.

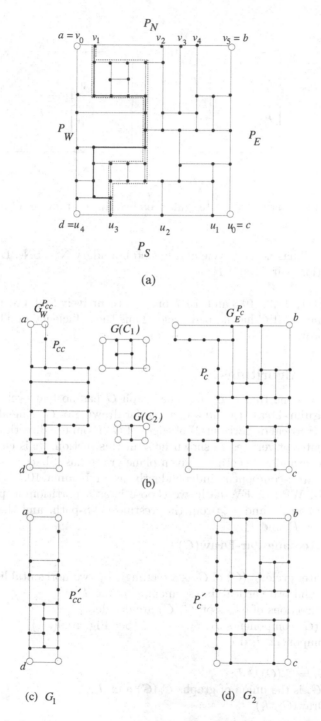

Figure 10.15   (a) G with P_c and P_{cc}, (b) splitting G along P_c and P_{cc}, (c) drawings of G_1, and (d) drawing of G_2. (Figure taken from [NR04].)

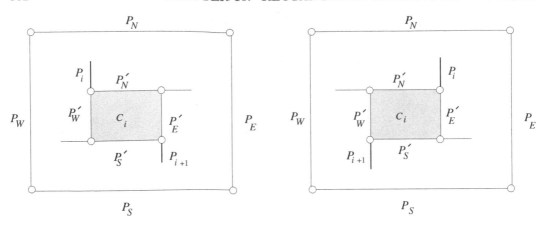

Figure 10.16 Two alternative rectangular embeddings of cycle C_i. (Figure taken from [NR04].)

LEMMA 10.4 If G has no bad cycle and has no boundary NS-, SN-, EW- or WE-path, then G has a partition-pair P_c and P_{cc}.

Using Lemmas 10.1, 10.2, 10.3 and 10.4, one can recursively find a rectangular drawing of a given plane graph G if G has no bad cycle. Thus the sufficiency of Theorem 10.2 can be constructively proved.

10.3.2   Drawing Algorithms

In this section, we assume that a given plane graph G has no bad cycle, and present an algorithm **Rectangular-Draw** to find a rectangular drawing of G. The algorithm outputs only the directions (vertical or horizontal) of edges of G. From the directions one can decide the integer coordinates of vertices as shown later in this section. It is easy to modify the algorithm so that it examines whether a given plane graph has a bad cycle or not.

We treat each $C_o(G)$-component independently as in Lemma 10.1. If there exists a boundary NS-, SN-, WE-, or EW-path, we choose it as a partitioning path. Otherwise, we find a partition-pair P_c and P_{cc} from the westmost NS-path, and then recurse to the subgraphs divided by P_c and P_{cc}.

> **Algorithm Rectangular-Draw(G)**
> **begin**
> 1   Draw the outer cycle $C_o(G)$ of G as a rectangle by two horizontal line segments
> P_N and P_S and two vertical line segments P_E and P_W;
>   {The directions of edges on $C_o(G)$ are decided.}
> 2   Find all $C_o(G)$-components $J_1, J_2, \cdots, J_p$; {See Fig. 10.9(a).}
> 3   **for** each component J_i **do**
>   **begin**
> 4 $G_i = C_o(G) \cup J_i$;
>   {G_i is the union of graphs $C_o(G)$ and J_i.}
> 5   Draw(G_i, J_i)
>   **end**
> **end.**

Procedure Draw(G, J)
begin $\{G$ has exactly one $C_o(G)$-component $J.\}$

1   **if** G has a boundary NS-, SN-, EW-, or WE-path P
   then $\{P$ is a partitioning path.$\}$
      begin $\{$See Fig. 10.17.$\}$
2   Assume without loss of generality that P is a boundary NS-path;
3   Draw all edges of P on a vertical line;
4   **if** $|E(P)| \geq 2$ **then**
            begin
5   Let $F_1, F_2, \cdots, F_q$ be the C_o-components of G_E^P;
6   **for** each component F_i, $1 \leq i \leq q$, **do**
7   Draw$(C_o(G_E^P) \cup F_i, F_i)$
            end
      end
   else $\{G$ has no boundary NS-, SN-, EW-, or WE-path. $\}$
      begin
8   Find a partition-pair P_c and P_{cc} as in the proof
         of Lemma 10.4 in [RNN98];
9   **if** $P_c = P_{cc}$ **then** $\{$See Fig. 10.10.$\}$
            begin
10   Draw all edges of P_c on a vertical line segment;
11   Let $G_1 = G_W^{P_c}$ and $G_2 = G_E^{P_c}$ be the two resulting subgraphs;
12   **for** each subgraph G_i, $i = 1, 2$, **do**
                  begin
13   Let $F_1, F_2, \cdots, F_q$ be the C_o-components of G_i;
14   **for** each component F_j, $1 \leq j \leq q$, **do**
15   Draw$(C_o(G_i) \cup F_j, F_j)$
                  end
            end
16   **else** $\{P_c \neq P_{cc}$. See Fig. 10.15.$\}$
            begin
17   Draw all edges of P_c and P_{cc} on alternating sequences
                  of horizontal and vertical line segments as in Fig. 10.15(b);
18   Let G_1 be the graph obtained from $G_W^{P_{cc}}$ by contracting
                  all edges of P_{cc} that are on horizontal sides of rectangular
                  embeddings of $C_1, C_2, \cdots, C_k$;
19   Let G_2 be the graph obtained from $G_E^{P_c}$ by contracting
                  all edges of P_c that are on horizontal sides of rectangular
                  embeddings of $C_1, C_2, \cdots, C_k$;
20   Let $G_3 = G(C_1), G_4 = G(C_2), \cdots, G_{k+2} = G(C_k)$;
21   **for** each graph G_i, $1 \leq i \leq k + 2$, **do**
                  begin
22   Let $F_1, F_2, \cdots, F_q$ be the C_o-components of G_i;
23   **for** each component F_j, $1 \leq j \leq q$, **do**
24   DRAW$(C_o(G_i) \cup F_j, F_j)$
                  end
            end
      end
   end
end

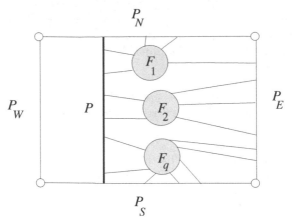

Figure 10.17 C_o-Components $F_1, F_2, \cdots, F_q$ of G_E^P. (Figure taken from [NR04].)

The algorithm **Rectangular-Draw**(G) finds only the directions of all edges in G. From the directions the integer coordinates of vertices in G can be determined in linear time as follows. From now on one may assume for simplicity that all vertices other than the four corners have degree three.

We now give a method of determining y-coordinates of the vertices in G; x-coordinates can be determined similarly. Consider a graph T_y obtained from G by deleting all upward vertical edges of three types drawn by dotted lines in Fig. 10.18. Thus any upward edge drawn by a thick line in Fig. 10.19 is not deleted. Clearly T_y is a spanning tree of G. (T_y for the graph G in Fig. 10.15(a) is drawn by thick lines in Fig. 10.20.) A rectangular drawing of G is composed of several maximal horizontal and vertical line segments. The drawing in Fig. 10.20 is composed of 16 maximal vertical line segments together with 15 maximal horizontal line segments. All these maximal horizontal line segments are contained in T_y, and every vertex of G is contained in one of them. For each maximal horizontal line segment L, we will assign an integer $y(L)$ as the y-coordinate of every vertex on L. P_S is the lowermost maximal horizontal line segment, while P_N is the topmost one. We first set $y(P_S) = 0$. We then compute $y(L)$ from bottom to top. For each vertex v in G we will assign an integer $temp(v)$ as a temporary value of the y-coordinate of v.

Figure 10.18   Deleted upward edges. (Figure taken from [NR04].)

For every vertex v on L there are two cases: either v has a neighbor u located below v or v has no neighbor u located below v. For the former case, we set $temp(v) = y(L') + 1$ where L' is the maximal horizontal line segment containing vertex u. For the latter case, we set $temp(v) = 0$. We then set $y(L) = \max_v \{temp(v)\}$ where the maximum is taken over

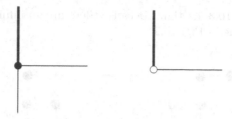

Figure 10.19 Non-deleted upward edges. (Figure taken from [NR04].)

all vertices v on L. One can easily compute $y(L)$ for all maximal horizontal line segments L from bottom to top using the counterclockwise depth-first search on T_y starting from the downward edge incident to the north-west corner a.

Thus the integer coordinates of all vertices in a rectangular grid drawing can be computed in linear time.

We now give upper bounds on the area and half perimeter of a grid for a rectangular grid drawing. Let the coordinate of the south-west corner d be $(0,0)$, and let that of the north-east corner b be (W, H). Then the grid drawing is "compact" in a sense that there is at least one vertical line segment of x-coordinate i for each integer i, $0 \le i \le W$, and there is at least one horizontal line segment of y-coordinate j for each integer j, $0 \le j \le H$. The following theorem holds on the sizes of a compact rectangular grid drawing [RNN98].

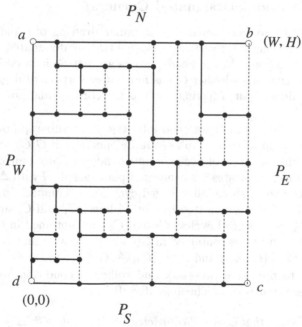

Figure 10.20 Illustration of T_y by thick lines. (Figure taken from [NR04].)

Theorem 10.3 *If all vertices of a plane graph G have degree three except the four corners, then the sizes of any compact rectangular grid drawing D of G satisfy $W + H \le \frac{n}{2}$ and $W \cdot H \le \frac{n^2}{16}$.*

The bounds in Theorem 10.3 are tight, because there are an infinite number of examples attaining the bounds, as one in Fig. 10.21.

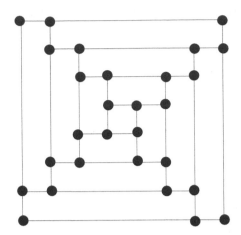

Figure 10.21   An example of a rectangular grid drawing attaining the upper bounds. (Figure taken from [NR04].)

10.3.3   Drawing without Designated Corners

In Sections 10.3.1– 10.3.2 we considered a rectangular drawing of a plane graph G with $\Delta \le 3$ for the case where four outer vertices of degree two are designated as the corners. In this section we consider a general case where corners are not designated in advance. Then our problem is how to examine whether G has four outer vertices of degree two such that there is a rectangular drawing of G having them as the corners, and how to efficiently find them if there is.

For a cycle C in a plane graph G we denote by $G(C)$ the subgraph of G inside C. We say that cycles C and C' in a plane graph G are independent if $G(C)$ and $G(C')$ have no common vertex, and that a set $\mathcal{S}$ of cycles is independent if any pair of cycles in $\mathcal{S}$ are independent. Figure 10.22 illustrates a 2-connected plane graph G with $\Delta \le 3$. Since every vertex of G has degree two or three, all 3-legged cycles are "laminar" in essence. Some of 2-legged and 3-legged cycles are indicated by dotted lines; C_1 and C_2 are 2-legged cycles, and C_3, C_4, C_5 and C_6 are 3-legged cycles. C_3 and C_4 are contained in $G(C_1)$, C_5 and C_6 are contained in $G(C_2)$, and C_5 is contained in $G(C_6)$. There are many independent sets of cycles. For example, $\mathcal{S}_1 = \{C_1, C_2\}$ and $\mathcal{S}_2 = \{C_2, C_3, C_4\}$ are independent sets of cycles.

We are now ready to present a necessary and sufficient condition for the existence of appropriate four outer vertices as in Theorem 10.4 [RNN02].

Theorem 10.4   *Assume that G is a 2-connected plane graph with $\Delta \le 3$ and has four or more outer vertices of degree two. Then four of them can be designated as the corners so that G has a rectangular drawing with the designated corners if and only if G satisfies the following three conditions:*

(a) *every 2-legged cycle in G contains at least two outer vertices of degree two;*

(b) *every 3-legged cycle in G contains at least one outer vertex of degree two; and*

(c) $2c_2 + c_3 \leq 4$ for every independent set $\mathcal{S}$ of cycles consisting of 2-legged cycles and 3-legged cycles, where c_2 and c_3 are the numbers of 2-legged cycles and 3-legged cycles in $\mathcal{S}$, respectively.

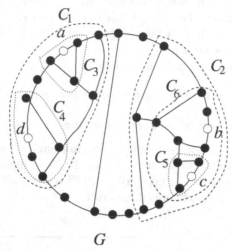

Figure 10.22 Plane graph. (Figure taken from [NR04].)

For the set $\mathcal{S}_1 = \{C_1, C_2\}$ above $c_2 = 2$, $c_3 = 0$ and hence $2c_2 + c_3 = 4$, while for $\mathcal{S}_2 = \{C_2, C_3, C_4\}$ $c_2 = 1$ and $c_3 = 2$ and hence $2c_2 + c_3 = 4$.

It is rather easy to prove the necessity of Theorem 10.4.

In order to prove the sufficiency of Theorem 10.4, it suffices to show that if the three conditions (a)–(c) in Theorem 10.4 hold then one can choose four outer vertices of degree two as the corners a, b, c and d so that the conditions (r1) and (r2) in Theorem 10.2 hold. See [RNN02] for the detail of a proof and a linear algorithm to find appropriate four outer vertices of degree two.

10.4 Box-Rectangular Drawing

A *box-rectangular drawing* of a plane graph G is a drawing of G such that each vertex is drawn as a rectangle, called a *box*, each edge is drawn as a straight line segment joining points on the two boxes corresponding to the ends, and the contour of each face is drawn as a rectangle, as illustrated in Fig. 10.24(b). A vertex may be drawn as a degenerate rectangle, that is, a point. We have seen in Section 10.1 that box-rectangular drawings have practical applications in floorplanning of MultiChip Modules (MCM) and in architectural floorplanning. If G has multiple edges or a vertex of degree five or more, then G has no rectangular drawing but may have a box-rectangular drawing. However, not every plane graph has a box-rectangular drawing. This section presents a necessary and sufficient condition for the existence of a box-rectangular drawing of a plane graph, and gives a linear algorithm to find a box-rectangular drawing if it exists [RNN00].

Before presenting the condition and the algorithm we need to present some definitions and preliminary observations regarding box-rectangular drawings.

Throughout this section we assume that a *graph* G is a so-called multigraph, which may have *multiple edges*, i.e., edges sharing both ends. If G has no multiple edges, then G is called a *simple* graph. For simplicity we assume that G has three or more vertices and is 2-connected.

We call a box-rectangular drawing D of G a *box-rectangular grid drawing* if each edge as well as each side of a box is drawn along a grid line. A vertex may be drawn as a *degenerate box*, that is, a point, in a box-rectangular drawing D. We often call a degenerate box in D a *point* and call a non-degenerate box a *real box*. We call a rectangle corresponding to an outer cycle $C_o(G)$ the *outer rectangle*, which has exactly four corners. We call a corner of the outer rectangle simply a *corner*. A box in D is called a *corner box* if it contains at least one corner. A corner box may be degenerate.

We now have the following four facts and a lemma.

Fact 10.3   *Any box-rectangular drawing has either two, three, or four corner boxes.*

Fact 10.4   *Any corner box contains either one or two corners.*

Figure 10.23(a) illustrates a box-rectangular drawing having two corner boxes; each of them is a real box and contains two corners. Figure 10.23(b) illustrates a box-rectangular drawing having three corner boxes. Figure 10.24(b) illustrates a box-rectangular drawing having four corner boxes, one of which is degenerate.

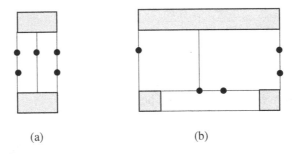

(a)   (b)

Figure 10.23   (a) Two corner boxes, and (b) three corner boxes.   (Figure taken from [NR04].)

Fact 10.5   *In a box-rectangular drawing D of G, any vertex v of degree two or three satisfies one of the following* (i), (ii) *and* (iii).

(i) *Vertex v is drawn as a point containing no corner;*

(ii) *v is drawn as a corner box containing exactly one corner; and*

(iii) *v is drawn as a corner real box containing exactly two corners.*

Fact 10.6   *In any box-rectangular drawing D of G, every vertex of degree five or more is drawn as a real box.*

LEMMA 10.5   [RNN00] If G has a box-rectangular drawing, then G has a box-rectangular drawing in which every vertex of degree four or more is drawn as a real box.

The choice of vertices as corner boxes plays an important role in finding a box-rectangular drawing. For example, the graph in Fig. 10.24(a) has a box-rectangular drawing if we choose outer vertices a, b, c and d as corner boxes as illustrated in Fig. 10.24(b). However, the graph has no box-rectangular drawing if we choose outer vertices p, q, r and s as corner boxes. If all vertices corresponding to corner boxes are designated for a drawing, then it is rather easy to examine whether G has a box-rectangular drawing with the designated corner boxes. We thus first concentrate our attention to the case where all vertices of G corresponding to corner boxes are designated.

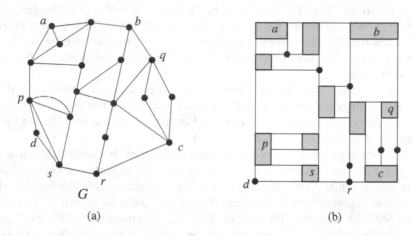

(a)   (b)

Figure 10.24 A graph G and its box-rectangular drawing with four corner boxes a, b, c and d. (Figure taken from [NR04].)

We now define two operations on a graph as follows. Let v be a vertex of degree two in G. Replace the two edges u_1v and u_2v incident to v with a single edge u_1u_2, and delete v. We call the operation above *the removal of a vertex of degree two* from G. Let v be a vertex of degree d in a plane graph, let $e_1 = vw_1, e_2 = vw_2, \cdots, e_d = vw_d$ be the edges incident to v, and assume that these edges $e_1, e_2, \cdots, e_d$ appear clockwise around v in this order as illustrated in Fig. 10.25(a). Replace v with a cycle $v_1, v_2, \cdots, v_d, v_1$, and replace edge vw_i with v_iw_i for $i = 1, 2, \cdots, d$, as illustrated in Fig. 10.25(b). We call the operation above the *replacement of a vertex by a cycle*. The cycle $v_1, v_2, \cdots, v_d, v_1$ in the resulting graph is called the *replaced cycle* corresponding to vertex v.

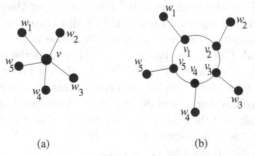

(a)   (b)

Figure 10.25 Replacement of a vertex by a cycle. (Figure taken from [NR04].)

By Fact 10.3, any box-rectangular drawing has either two, three or four corner boxes. However, we consider only box-rectangular drawings having four corner boxes for simplicity, and assume that exactly four outer vertices a, b, c and d in G are designated as the four corner boxes. We construct a new graph G'', called the *cycled graph*, from G through an intermediate graph G', and reduce the problem of finding a *box-rectangular drawing* of G with the four designated vertices to a problem of finding a *rectangular drawing* of the cycled graph G''.

We first construct G' from G as follows. If a vertex v of degree two in G, as vertex d in Fig. 10.26(a), is designated as a corner, then v must be drawn as a corner point in a box-rectangular drawing of G. On the other hand, if a vertex v of degree two is not designated as a corner, then the two edges incident to v must be drawn on a straight line segment. We thus remove all non-designated vertices of degree two one by one from G, as illustrated in Fig. 10.26(b). The resulting graph is G'. Thus all vertices of degree two in G' are designated vertices.

Clearly, G has a box-rectangular drawing with the four designated corner boxes if and only if G' has a box-rectangular drawing with the four designated corner boxes. Figure 10.26(f) illustrates a box-rectangular drawing D' of G' in Fig. 10.26(b), and Fig. 10.26(g) illustrates a box-rectangular drawing D of G in Fig. 10.26(a).

Since every vertex of degree two in G' is a designated vertex, it must be drawn as a corner point in any box-rectangular drawing of G'. Every designated vertex of degree three in G', as vertex a in Fig. 10.26(b), must be drawn as a real box since it is a corner. On the other hand, every non-designated vertex of degree three in G' must be drawn as a point. These facts together with Lemma 10.5 imply that if G' has a box-rectangular drawing then G' has a box-rectangular drawing D' in which all designated vertices of degree three and all vertices of degree four or more in G' are drawn as real boxes.

The cycled graph G'' is built from G' as follows. Replace by a cycle each of the designated vertices of degree three and the vertices of degree four or more, as illustrated in Fig. 10.26(c). The replaced cycle corresponding to a designated vertex x of degree three or more contains exactly one outer edge, say e_x, where $x = a, b, c$ or d. Put a dummy vertex x' of degree two on e_x, as shown in Fig. 10.26(d). The resulting graph is G''. We let $x' = x$ if a designated vertex x has degree two. The cycled graph G'' is a simple graph and has exactly four outer vertices a', b', c', and d' of degree two, and all the other vertices have degree three.

Then the following theorem holds.

Theorem 10.5   *Let G be a plane graph with four designated outer vertices a, b, c and d. Then G has a box-rectangular drawing with corner boxes a, b, c and d if and only if the cycled graph G'' has a rectangular drawing with designated corners a', b', c' and d'.*

Proof: The necessity is trivial, and hence it suffices to prove the sufficiency.

Assume that G'' has a rectangular drawing D'' as illustrated in Fig. 10.26(e). In D'', each replaced cycle is drawn as a rectangle, since it is a face in G''. Furthermore, the four outer vertices a', b', c' and d' of degree two in G'' are drawn as the corners of the rectangle corresponding to $C_o(G'')$. Therefore, D'' immediately gives a box-rectangular drawing D' of G' having the four vertices a, b, c and d as corner boxes, as illustrated in Fig. 10.26(f). Then, inserting non-designated vertices of degree two on horizontal or vertical line segments in D', one can immediately obtain from D' a box-rectangular drawing D of G having the designated vertices a, b, c and d as corner boxes, as illustrated in Fig. 10.26(g).   □

Furthermore the following theorem holds [RNN00].

Theorem 10.6   *Given a plane graph G of m edges and four designated outer vertices a, b, c and d, one can examine in time $O(m)$ whether G has a box-rectangular drawing D with corner boxes a, b, c and d, and if G has D, then one can find D in time $O(m)$. The half perimeter of the box-rectangular grid drawing is bounded by $m + 2$.*

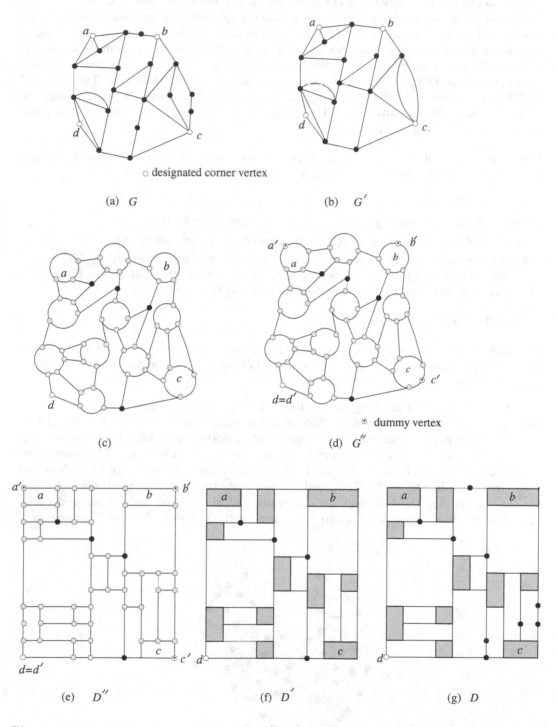

○ designated corner vertex

(a) G (b) G'

◉ dummy vertex

(c)   (d) G''

(e) D'' (f) D' (g) D

Figure 10.26 Illustration of G, G', G'', D'', D' and D. (Figure taken from [NR04].)

There are infinitely many cycles with four designated vertices for which the sum of the width and the height of any box-rectangular drawing of the cycles is $m - 2$.

The rest of this section deals with a general case where no vertices are designated as corner boxes in advance. Then our problem is how to examine whether G has some set of outer vertices such that there is a box-rectangular drawing of G having them as the corner boxes, and how to find them if there are. We first present a necessary and sufficient condition for a plane graph G with $\Delta \leq 3$ to have a box-rectangular drawing D as in Theorem 10.7 [RNN00], and then give a linear-time algorithm to find D if it exists. We then reduce the box-rectangular drawing problem of a plane graph G with $\Delta \geq 4$ to that of a new plane graph J with $\Delta \leq 3$ as in Theorem 10.9.

Theorem 10.7   *A plane graph G with $\Delta \leq 3$ has a box-rectangular drawing if and only if G satisfies the following four conditions:*

(c1) *every 2-legged or 3-legged cycle in G has an outer edge;*

(c2) *at most two 2-legged cycles of G are independent of each other;*

(c3) *at most four 3-legged cycles of G are independent of each other; and*

(c4) *if G has a pair of independent 2-legged cycles C_1 and C_2, then $\{C_1, C_2, C_3\}$ is not independent for any 3-legged cycle C_3 in G, and neither $G(C_1)$ nor $G(C_2)$ has more than two independent 3-legged cycles of G.*

Then the following theorem holds.

Theorem 10.8   *Given a plane graph with $\Delta \leq 3$, one can examine in time $O(m)$ whether G has a box-rectangular drawing D or not, and if G has D, one can find D in time $O(m)$, where m is the number of edges in G.*

Proof: One can find all 2-legged and 3-legged cycles in G, as follows. We first traverse the contour of each inner face of G containing an outer edge as illustrated in Fig. 10.27, where the traversed contours of faces are indicated by dotted lines. Clearly each outer edge is traversed exactly once, and each inner edge is traversed at most twice. The inner edges traversed exactly once form cycles, called *singly traced cycles*, the insides of which have not been traversed. In Fig. 10.27 C_4, C_8 and C_9 are singly traced cycles, the insides of which are shaded. During this traversal one can easily find all 2-legged and all 3-legged cycles that contain outer edges; C_1, C_2 and C_3 drawn by thick lines in Fig. 10.27 are some of

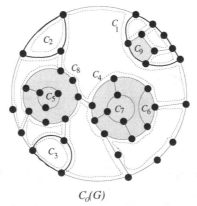

$C_o(G)$

Figure 10.27   Finding all 2-legged and 3-legged cycles. (Figure taken from [NR04].)

these cycles. (Note that a 3-legged cycle containing outer edges has two legs on $C_o(G)$ and the other leg is an inner edge which is traversed twice; if an end of a doubly traversed inner edge is an inner vertex, then it is a leg-vertex of such a 3-legged cycle.) Any of the remaining 2-legged and 3-legged cycles either is a singly traced cycle or is located inside a singly traced cycle. One can find all 2-legged and 3-legged cycles inside a singly traced cycle by recursively applying the method to the singly traced cycle. The method traverses the contour of each face by a constant number of times. Hence one can examine in time $O(m)$ whether G satisfies Condition (c1) in Theorem 10.7 or not.

One can examine Condition (c2) in Theorem 10.7 as follows. Assume that G satisfies Condition (c1). Then each 2-legged cycle must have an outer edge, and hence has the two leg-vertices on $C_o(G)$. By traversing the faces of G containing an outer edge, one can detect the leg-vertices of all 2-legged cycles of G on $C_o(G)$. While detecting the leg-vertices of 2-legged cycles, we give labels to the two leg-vertices of each 2-legged cycle; the labels indicate the name of the cycle. In Fig. 10.28, the leg-vertices of 2-legged cycles are drawn by white circles, and their labels are written next to them. It is clear that if G has k 2-legged cycles which are independent of each other then G has k minimal 2-legged cycles which are independent of each other. A 2-legged cycle C is minimal if and only if no intermediate vertex of the maximal subpath of C on $C_o(G)$ is a leg-vertex of any other 2-legged cycle. Therefore, traversing the outer vertices and checking the labels of leg-vertices, one can find all minimal 2-legged cycles, and one can also know whether two 2-legged cycles are independent or not. In Fig. 10.28 C_1, C_2 and C_3 are minimal 2-legged cycles, and they are independent. Thus one can examine Condition (c2) by traversing the edges on the contours of faces containing an outer edge by a constant number of times, and hence one can examine Condition (c2) in linear time.

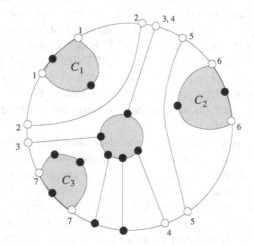

Figure 10.28   Illustration for minimal 2-legged cycles. (Figure taken from [NR04].)

One can examine Condition (c3) in linear time using a similar technique used to examine Condition (c2). One can easily examine Condition (c4) by checking the labels of the leg-vertices of minimal 2-legged cycles and minimal 3-legged cycles.

If G satisfies the conditions in Theorem 10.7, then a box-rectangular drawing of G can be found by choosing appropriate four corner boxes [RNN00, NR04]. One can find all minimal 2-legged cycles and all minimal 3-legged cycles in linear time by the technique used to examine Conditions (c2) and (c3), and hence one can choose the four designated vertices in linear time. Thus one can find a box-rectangular drawing of G in linear time.   □

We now reduce the box-rectangular drawing problem (without given corners) of a plane graph G with $\Delta \geq 4$ to that of a new plane graph J with $\Delta \leq 3$. Let G be a plane graph with $\Delta \geq 4$. We construct a new plane graph J from G by replacing each vertex v of degree four or more in G by a cycle. Figures 10.29(a) and (b) illustrate G and J, respectively. A replaced cycle corresponds to a real box in a box-rectangular drawing of G. We do not replace a vertex of degree two or three by a cycle since such a vertex may be drawn as a point by Fact 10.5. Thus $\Delta(J) \leq 3$. Then the following theorem holds.

Theorem 10.9   *Let G be a plane graph with $\Delta \geq 4$, and let J be the graph transformed from G as above. Then G has a box-rectangular drawing if and only if J has a box-rectangular drawing.*

Figures 10.29(c) and (d) illustrate D and D_J, respectively. Box f in D is a non-corner real box, and it is regarded as a face in D_J. Corner boxes a and b in D are vertices of degree three in G, and they remain as boxes in D_J. Corner boxes c and d in D are vertices of degree four or more in G, and are transformed to a drawing of a replaced cycle with one real box in D_J as illustrated in Fig. 10.29(e). We omit the proof of Theorem 10.9, which can be found in [RNN00].

10.5   Conclusions

The outer face boundary must be rectangular in a rectangular drawing, as illustrated in Fig. 10.1(b). However, the outer boundary of a VLSI chip or an architectural floor plan is not always rectangular, but is often a rectilinear polygon of L-shape, T-shape, Z-shape etc., as illustrated in Figs. 10.30(a)–(c). Such a drawing of a plane graph G is called an *inner rectangular drawing* if every inner face of G is a rectangle although the outer face boundary is not always a rectangle. Miura et al. [MHN06] reduced the problem of finding an inner rectangular drawing of a plane graph G with $\Delta \leq 4$ to a problem of finding a perfect matching of a new bipartite graph constructed from G. It immediately yields the result presented in Section 10.2 on an ordinary rectangular drawing of plane graphs with $\Delta \leq 4$.

Kozminski and Kinnen [KK84] established a necessary and sufficient condition for the existence of a "rectangular dual" of an inner triangulated plane graph, that is, a rectangular drawing of the dual graph of an inner triangulated plane graph, and gave an $O(n^2)$ algorithm to obtain it. Based on the characterization of [KK84], Bhasker and Sahni [BS88] and Xin He [He93] developed linear-time algorithms to find a rectangular dual. Kant and Xin He [KH97] presented two more linear-time algorithms. Xin He [He95] presented a parallel algorithm for finding a rectangular dual. Lai and Leinwand [LL90] reduced the problem of finding a rectangular dual of an inner triangulated plane graph G to a problem of finding a perfect matching of a new bipartite graph constructed from G. Their construction is different from that in Section 10.2, their bipartite graph has an $O(n^2)$ number of edges, and hence their method takes time $O(n^{2.5})$ to find a rectangular dual or a rectangular drawing of a plane graph with $\Delta \leq 3$.

A planar graph may have many embeddings. We say that *a planar graph G has a rectangular drawing* if at least one of the plane embeddings of G has a rectangular drawing. Since a planar graph may have an exponential number of embeddings, it is not a trivial problem to examine whether a planar graph has a rectangular drawing.

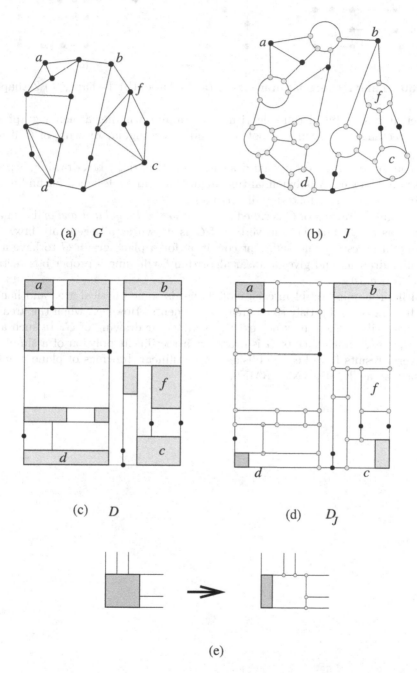

(a)   G

(b)   J

(c)   D

(d) D_J

(e)

Figure 10.29 Illustration of G, J, D_J, D and a transformation. (Figure taken from [NR04].)

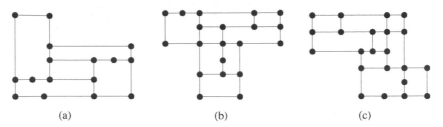

Figure 10.30 Inner rectangular drawings of (a) L-shape, (b) T-shape, (c) Z-shape.

Rahman et al. gave a linear-time algorithm to examine whether a planar graph G with $\Delta \leq 3$ has a rectangular drawing or not, and find a rectangular drawing of G if it exists [RNG04].

A similar concept of a box-rectangular drawing, called a strict 2-box drawing, is presented by Thomassen in [Tho86]. A polynomial-time algorithm can be designed for finding a strict 2-box drawing of a graph by following his method.

A box-rectangular drawing of G is called a *proper box-rectangular drawing* if every vertex of G is drawn as a real box, i.e., no vertex of G is drawn as a degenerate box. Xin He [He01] presents a necessary and sufficient condition for a plane graph G to have a proper box-rectangular drawing and gives a linear algorithm for finding a proper box-rectangular drawing of G if it exists.

In a VLSI floorplanning problem each module needs some physical area and hence each face in the drawing should satisfy some area requirements. However, when the area of each face of G is prescribed, there may not exist a rectangular drawing of G. In such a case it is desirable that each inner face of G is drawn as a rectilinear polygon of a simple shape. Recently several results have been published on rectilinear drawings of plane graphs with prescribed face areas [KN07, KN09, RMN09].

References

[BS88]   J. Bhasker and S. Sahni, A linear algorithm to find a rectangular dual of a planar triangulated graph, *Algorithmica*, 3, pp. 247-278, 1988.

[DETT99]   G. Di Battista, P. Eades, R. Tamassia, and I. G. Tollis. *Graph Drawing*. Prentice Hall, Upper Saddle River, NJ, 1999.

[FW74]   R. L. Francis and J. A. White, *Facility Layout and Location*, Prentice-Hall, New Jersey, 1974.

[He01]   X. He, A simple linear time algorithm for proper box rectangular drawings of plane graphs, *Journal of Algorithms*, 40(1), pp. 82-101, 2001.

[He93]   X. He, On finding the rectangular duals of planar triangular graphs, *SIAM J. Comput.*, 22(6), pp. 1218-1226, 1993.

[He95]   X. He, An efficient parallel algorithm for finding rectangular duals of plane triangulated graphs, *Algorithmica*, 13, pp. 553-572, 1995.

[HK73]   J. E. Hopcroft and R. M. Karp, An $n^{5/2}$ algorithm for maximum matching in bipartite graphs, *SIAM J. Comput.*, 2, pp. 225-231, 1973.

[Hoc04]   D. S. Hochbaum, Faster pseudoflow-based algorithms for the bipartite matching and the closure problems, Abstract, *CORS/SCRO-INFORMS Joint Int. Meeting*, Banff, Canada, p. 46, 2004.

[HC04]   D. S. Hochbaum and B. G. Chandran, Further below the flow decomposition barrier of maximum flow for bipartite matching and maximum closure, Working paper, 2004.

[KH97]   G. Kant and X. He, Regular edge labeling of 4-connected plane graphs and its applications in graph drawing problems, *Theoretical Computer Science*, 172, pp. 175-193, 1997.

[KK84]   K. Kozminski and E. Kinnen, An algorithm for finding a rectangular dual of a planar graph for use in area planning for VLSI integrated circuits, *Proc. Design Automation Conference*, Albuquerque, pp. 655-656, 1984.

[KN07]   A. Kawaguchi and H. Nagamochi, Orthogonal drawings of plane graphs with prescribed face areas, *Proc. of 4th International Conference on Theory and Applications of Models of Computation*, Lecture Notes in Computer Science, 4484, Springer, pp. 584-594, 2007.

[KN09]   A. Kawaguchi and H. Nagamochi, Drawing slicing graphs with face areas, *Theoretical Computer Science*, 410, pp. 1061-1072, 2009.

[LL90]   Y.-T. Lai and S. M. Leinwand, A theory of rectangular dual graphs, *Algorithmica*, 5, pp. 467-483, 1990.

[MHN06]   K. Miura, H. Haga and T. Nishizeki, Inner rectangular drawings of plane graphs, *Int. J. Computational Geometry & Applications*, 16,2 & 3, pp. 247-270, 2006.

[MV80]   S. Micali and V. V. Vazirani, An $O(\sqrt{|V|} \cdot |E|)$ algorithm for finding maximum matching in general graphs, *Proc. Symposium on Foundations of Computer Science*, pp. 17-27, 1980.

[NR04]   T. Nishizeki and M. S. Rahman, *Planar Graph Drawing*, World Scientific, Singapore, 2004.

[PS82]   C. H. Papadimitriou and K. Steiglitz, *Combinatorial Optimization*, Prentice Hall, Englewood Cliffs, New Jersey, 1982.

[RMN09]　M. S. Rahman, K. Miura and T. Nishizeki, Octagonal drawings of plane graphs with prescribed face areas, *Comp. Geom. Theo. and Appl.*, 42, pp. 214-230, 2009.

[RNG04]　M. S. Rahman, T. Nishizeki, and S. Ghosh Rectangular drawings of planar graphs, *Journal of Algorithms*, 50, pp. 62-78, 2004.

[RNN98]　M. S. Rahman, S. Nakano and T. Nishizeki, Rectangular grid drawings of plane graphs, *Comp. Geom. Theo. Appl.*, 10(3), pp. 203-220, 1998.

[RNN00]　M. S. Rahman, S. Nakano and T. Nishizeki, Box-rectangular drawings of plane graphs, *Journal of Algorithms*, 37, pp. 363-398, 2000.

[RNN02]　M. S. Rahman, S. Nakano and T. Nishizeki, Rectangular drawings of plane graphs without designated corners, *Comp. Geom. Theo. and Appl.*, 21(3), pp. 121-138, 2002.

[She95]　N. Sherwani, *Algorithms for VLSI Physical Design Automation*, 2nd edition, Kluwer Academic Publishers, Boston, 1995.

[Tam87]　R. Tamassia, On embedding a graph in the grid with the minimum number of bends, *SIAM J. Computing*, 16(3), pp. 421-444, 1987.

[Tho84]　C. Thomassen, Plane representations of graphs, (Eds.) J. A. Bondy and U. S. R. Murty, Progress in Graph Theory, Academic Press Canada, pp. 43-69, 1984.

[Tho86]　C. Thomassen, Interval representations of planar graphs, *J. of Combinat. Theory, Series B*, 40, pp. 9-20, 1986.

11

Simultaneous Embedding of Planar Graphs

11.1 Introduction

Traditional problems in graph drawing involve the layout of a single graph, whereas in simultaneous graph drawing we are concerned with the layout of multiple related graphs. In particular, consider the problem of drawing a series of graphs that share all, or parts of the same vertex set. The graphs may represent different relations between the same set of objects, or alternatively, the graphs may be the result of a single relation that changes through time.

In this chapter we survey efforts to address the following problem: Given a series of graphs that share all, or parts of the same vertex set, what is a natural way to layout and display them? The layout and display of the graphs are different aspects of the problem, but also closely related, as a particular layout algorithm is likely to be matched best with a specific visualization technique. As stated above, however, the problem is too general and it is unlikely that one particular layout algorithm will be best for all possible scenarios. Consider the case where we only have a pair of graphs in the series, and the case where we have hundreds of related graphs. The "best" way to layout and display the two series is likely going to be different. Similarly, if the graphs in the sequence are very closely related or not related at all, different layout and display techniques may be more appropriate.

For the layout of the graphs, there are two important criteria to consider: the *readability* of the individual layouts and the *mental map preservation* in the series of drawings. The readability of individual drawings depends on aesthetic criteria such as display of symmetries, uniform edge lengths, and minimal number of crossings. Preservation of the mental map can be achieved by ensuring that vertices that appear in consecutive graphs in the series, remain in the same positions. These two criteria are often contradictory. If we individually layout each graph, without regard to other graphs in the series, we may optimize readability at the expense of mental map preservation. Conversely, if we fix the vertex positions in all graphs, we are optimizing the mental map preservation but the individual layouts may be far from readable. In simultaneous graph embedding, vertices are placed in the exact same locations in all the graphs, while the layout of the edges may differ.

Visualization of related graphs, that is, graphs that are defined on the same set of vertices, arise in many different settings. Software engineering, databases, and social network analysis are all examples of areas where multiple relationships on the same set of objects are often studied. In evolutionary biology, phylogenetic trees are used to visualize the ancestral relationship among groups of species. Depending on the assumptions made, different algorithms produce different phylogenetic trees. Comparing the outputs and determining the most likely evolutionary hypothesis can be difficult if the drawings of the trees are laid out independently of each other.

While in some of the above examples the graphs are not necessarily planar, solving the planar case can provide intuition and ideas for the more general case. With this in mind, here we concentrate on the problem of simultaneous embedding of planar graphs. Simultaneous embedding of planar graphs generalizes the notion of traditional graph planarity and is motivated by its relationship with problems of graph thickness, geometric thickness, and applications such as the visualization of graphs that evolve through time.

The thickness of a graph is the minimum number of planar subgraphs into which the edges of the graph can be partitioned; see [MOS98] for a survey. Thickness is an important concept in VLSI design, since a graph of thickness k can be embedded in k *layers*, with any two edges drawn in the same layer intersecting only at a common vertex and vertices placed in the same location in all layers. A related graph property is *geometric thickness*, defined to be the minimum number of layers for which a drawing of G exists having all edges drawn as straight-line segments [DEH00]. Finally, the *book thickness* of a graph G is the minimum number of layers for which a drawing of G exists, in which edges are drawn as straight-line segments and vertices are in convex position [BK79]. It has been shown that the book thickness of planar graphs is no greater than four [Yan89].

11.1.1 Problem Definitions

This chapter is structured along three basic simultaneous embedding results for planar graphs, SIMULTANEOUS GEOMETRIC EMBEDDING (SGE), SIMULTANEOUS EMBEDDING WITH FIXED EDGES (SEFE), and SIMULTANEOUS EMBEDDING (SE), Figure 11.1 illustrates the three cases. For all three problems the input always consists of two planar graphs $G_1 = (V_1, E_1)$ and $G_2 = (V_2, E_2)$ sharing a common subgraph $G = (V, E) = (V_1 \cap V_2, E_1 \cap E_2)$.

The most strict variant is SIMULTANEOUS GEOMETRIC EMBEDDING (SGE), which asks for planar straight-line drawings of G_1 and G_2 such that common vertices have the same coordinates in both drawings. The requirements of SGE are very strict, and as we will see in Section 11.2, there exist a lot of examples that do not admit such an embedding. While the problem SIMULTANEOUS EMBEDDING WITH FIXED EDGES still requires common vertices to have the same coordinates, it relaxes the straight-line requirement by allowing arbitrary curves for representing edges. To maintain the mental map, common edges are

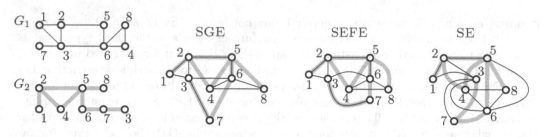

Figure 11.1 Two graphs G_1 and G_2 together with an SGE, a SEFE and an SE. In the SGE all edges are straight line segments while some edges in the SEFE are not. The SE contains common edges ($\{3,7\}$ and $\{5,6\}$) that are drawn differently with respect to G_1 and G_2.

still required to be represented by the same curves. Finally, SIMULTANEOUS EMBEDDING drops the constraints on the curves altogether and just requires common vertices to have the same coordinates.

For all these problems it is common to also use the problem name to denote a corresponding embedding, that is, we also say that G_1 and G_2 have an SGE, SEFE or SE if they admit solutions to these problems. Moreover, all these problems readily generalize to $k > 2$ input graphs $G_1, \ldots, G_k$, by requiring that the conditions hold for each pair of graphs. In this case a common restriction is to require that all input graphs share exactly the same graph G, that is, $G = G_i \cap G_j$ for $i \neq j$. We call this behavior *sunflower intersection*.

We note that simultaneous embedding problems are closely related to constrained embedding problems. For example if the planar embedding of one of the two graphs of an instance of SEFE is already fixed, the problem of finding a SEFE is equivalent to finding an embedding of the second graph respecting a prescribed embedding for a subgraph, namely the common graph. This constrained embedding problem is known as PARTIALLY EMBEDDED PLANARITY. Angelini et al. [ADF+10] show that this problem can be solved in linear time and, in the spirit of Kuratowski's theorem, Jelínek et al. [JKR11] characterize the yes-instances by forbidden substructures. A similar tie to constrained embedding problems exists in the case of SE. After fixing the drawing of one of the two input graphs it remains to draw a single graph without crossings at prescribed vertex positions. This problem is known as POINT SET EMBEDDING and Pach and Wenger show that this is always possible [PW98]. There are other, less obvious relations between simultaneous embedding and constrained embedding problems, which will be described later.

11.1.2 Overview and Outline

This chapter starts with the three simultaneous embedding problems SGE, SEFE, and SE, and we discuss each of them in one of the following sections. There are three major classes of results on simultaneous embedding problems. The first class contains algorithms that, for given graphs with certain properties, always produce a simultaneous embedding, perhaps with additional quality guarantees. These results show the existence of simultaneous embeddings for the corresponding graph classes. The second class contains counterexamples that do not admit a simultaneous embedding. The third class contains algorithms and complexity results for the problem of testing whether a given instance admits a simultaneous embedding.

We present a survey of the results on SGE in Section 11.2. Due to the strong requirements of SGE results of the first type, which identify classes of graphs that always admit a simultaneous embedding, exist only for very few and strongly restricted graph classes. For example, even a path and a tree of depth 4 may not have an SGE [AGKN12]. Moreover, it is NP-hard to decide SGE and there are no further results of the third type, that is, algorithms testing whether an instance has an SGE or not, even for restricted instances.

Section 11.3 presents the SEFE problem, which turns out to be much less restrictive than SGE. For example a tree and a path do always admit a SEFE although they do not have an SGE [Fra07]. On the other hand, examples not having a SEFE are also counterexamples for SGE. Moreover, it is still open whether SEFE can be tested in polynomial time for two graphs, whereas it is NP-complete for three or more graphs [GJP+06]. However, for two graphs, there exist several results of the third type, that is, testing algorithms, for restricted inputs. For example, it is possible to decide in linear time whether a pair of graphs admits a SEFE or not, if the common graph is biconnected [ABF+12, HJL10].

In Section 11.4, we consider the least restrictive simultaneous embedding problem, SE, which only requires common vertices to have the same coordinates in all drawings. As every planar graph can be drawn without crossings even if the position of every vertex is fixed [PW98], there are no counterexamples for SE and it is not necessary to have a testing algorithm. The results on SE focus on creating simultaneous embeddings such that edges have few bends and the resulting drawings use small area.

Sections 11.5–11.8 presents several variants of approaches to simultaneous embedding that do not quite fall into the categories of the three main problems. The problem variants discussed in Section 11.5 relax the requirement of having a fixed mapping between the vertices of G_1 and G_2. They rather ask whether a suitable mapping can be found such that a SEFE exists [BCD+07]. Colored SGEs are somewhere between and allow the mapping to identify only vertices having the same color [BEEB+11]. Section 11.6 deals with matched drawings requiring straight-line drawings of the two input graph such that each common vertex has only the same y-coordinate in both drawings. Other work, discussed in Section 11.7, deals with the problem of simultaneously representing a planar graph and its dual [Tut63] and considers different types of simultaneous representations, such as simultaneous intersection representations, as introduced by Jampani and Lubiw [JL09]. Section 11.8 presents several practical approaches to simultaneous embedding problems.

In Section 11.9 results on morphing between different planar drawings of the same graph are presented. A morph aims to preserve the mental map between different drawings of the same graph, which can be seen as the opposite to drawing different graphs such that the common part is drawn the same. Finally, in Section 11.10, we present a list of open questions. The list contains questions that have been open for several years, as well as questions that are motivated by recent research results.

11.2   Simultaneous Geometric Embedding

In this section we consider the most desirable (and most restrictive) kind of simultaneous drawings, the SGEs. Most results on that problem are summarized in Table 11.1. Figure 11.3 illustrates the relation between these results. Before we describe the results in more detail we start with a small example. While it may be tempting to say that if the union of two graphs contains a subdivision of K_5 or $K_{3,3}$ then the two graphs have no simultaneous geometric embedding, this is not the case; see Figure 11.2. In fact, while planarity testing for a single graph can be done in linear time [HT74], Estrella-Balderrama et al. [EBGJ+08] show that the decision problem SGE is NP-hard. Other results concerning the complexity of SGE (for example for restricted graph classes) are not known.

In the following we describe the results illustrated in Figure 11.3. We start with algorithms always creating an SGE when the input is restricted to special graph classes. We then

SGE Instance	Existence	Area	Ref.
G_1 & G_2 paths	✓	$n \times n$	[BCD$^+$07]
G_1 path & G_2 extended star	✓	$O(n^2) \times O(n)$	[BCD$^+$07]
G_1 caterpillar & G_2 path	✓	$n \times 2n$	[BCD$^+$07]
G_1 & G_2 caterpillar	✓	$3n \times 3n$	[BCD$^+$07]
2 stars	✓	$3 \times (n-2)$	[BCD$^+$07]
k stars	✓	$O(n) \times O(n)$	[BCD$^+$07]
G_1 & G_2 cycles	✓	$4n \times 4n$	[BCD$^+$07]
G_1 & G_2 have maximum degree 2	✓	—	[DEK04]
G_1 wheel & G_2 cycle	✓	—	[CvKL$^+$11]
G_1 tree & G_2 matching	✓	—	[CvKL$^+$11]
G_1 outerpath & G_2 matching	✓	—	[CvKL$^+$11]
G_1 tree of depth 2 & G_2 path	✓	—	[AGKN12]
G_1 level-planar w.r.t. path G_2	✓	—	[CEBFK09]
G_1 & G_2 planar	✗	—	[BCD$^+$07]
G_1 path & G_2 planar	✗	—	[BCD$^+$07, EK05a]
G_1 path & G_2 edge disjoint	✗	—	[FKK09]
three paths	✗	—	[BCD$^+$07]
G_1 matching & G_2 planar	✗	—	[CvKL$^+$11]
six matchings	✗	—	[CvKL$^+$11]
G_1 & G_2 outerplanar	✗	—	[BCD$^+$07]
G_1 & G_2 trees	✗	—	[GKV09]
G_1 tree of depth 4 & G_2 edge disjoint path	✗	—	[AGKN12]

Table 11.1 A list of classes of graphs that are either known to always have an SGE or that contain counterexamples. For the positive cases, the area consumption is given, provided that it is known.

continue with graph classes containing counterexamples. Finally, we consider the results not fitting in one of these two cases.

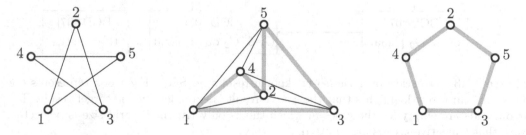

Figure 11.2 The union of the graph on the left and the graph on the right is a K_5, but the middle drawing shows a simultaneous geometric embedding of the two graphs.

11.2.1 Graph Classes with SGE

Brass et al. [BCD$^+$07] give several algorithms for different restricted graph classes always creating an SGE. In the simplest case G_1 and G_2 are both required to be paths. This result is easy to prove and also provides good intuition for most of the positive results:

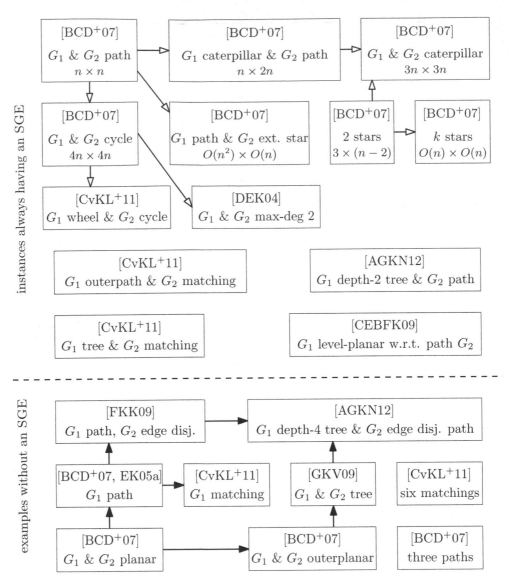

Figure 11.3 Overview over the so far known results on SGE. Each box represents one result and an arrow highlights that the source-result is extended by the target-result. The arrowheads are empty for the cases in which this is only true if the grid size is neglected. Note that transitive arrows are omitted.

Theorem 11.1 *For two paths P_1 and P_2 on the same vertex set V of size n an SGE on a grid of size $n \times n$ can be found in linear time.*

Proof: For each vertex $u \in V$, we embed u at the integer grid point (p_1, p_2), where $p_i \in \{1, 2, \ldots, n\}$ is the vertex's position in the path P_i, $i \in \{1, 2\}$. Then, P_1 is embedded as an x-monotone polygonal chain, and P_2 is embedded as a y-monotone chain. Thus, neither path is self-intersecting; see Figure 11.4 for an example. □

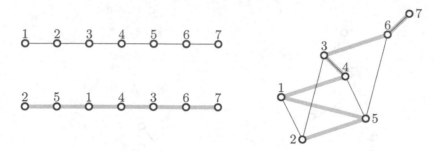

Figure 11.4 Two paths simultaneously embedded such that one path is x-monotone and the other is y-monotone.

Brass et al. [BCD$^+$07] also consider more general graph classes, such as *caterpillars* (trees being paths after the removal of all leaves), *stars* (trees with at most one inner vertex called *center*), and *extended stars* (collection of stars with an additional special root and paths from the special root to the centers of all stars). They show that a caterpillar and a path admit an SGE on a grid of size $n \times 2n$, which can be extended to two caterpillars on a grid of size $3n \times 3n$. Moreover, they can simultaneously embed two stars on a $3 \times (n-2)$ grid and extend it to the case of k stars on an $O(n) \times O(n)$-grid. Finally, the pairs path plus extended star and cycle plus cycle can be embedded on $O(n^2) \times O(n)$ and $4n \times 4n$ grids, respectively. The latter two results both extend the case of two paths (when neglecting the grid size).

The result for two cycles was further extended by Duncan et al. [DEK04] and Cabello et al. [CvKL$^+$11]. Duncan et al. [DEK04] show that a graph with maximum degree 4 has geometric thickness 2. To this end, they show that two graphs with maximum degree 2 always admit a simultaneous geometric embedding. However, their algorithm computes drawings with potentially large area.

Cabello et al. [CvKL$^+$11] show the existence of an SGE for a *wheel* (union of a star and a cycle on its leaves) and a cycle. They moreover give algorithms for the pairs tree plus *matching* (graph with maximum degree 1) and *outerpath* (outerplanar graph whose weak dual is a path) plus matching. The former algorithm uses only two slope for the matching edges, for the latter one slope suffices.

Given a planar graph and a path on the same vertices, the order of the vertices in the path induces a layering on the vertices. Cappos et al. [CEBFK09] give a linear-time algorithm that computes an SGE of a planar graph and a path if the planar graph is level-planar with respect to the layering induced by the path. Angelini et al. [AGKN12] show that every tree of depth 2 has an SGE with every path.

11.2.2 Examples without SGE

In contrast to the positive results, Brass et al. [BCD$^+$07] give several examples not admitting an SGE. They show the existence of two planar graphs without a simultaneous embedding and extended this result to two outerplanar graphs. Two results we present in more detail are the counterexample for a planar graph and a path by Brass et al. [BCD$^+$07] and Erten and Kobourov [EK05a] and the counterexample of three paths by Brass et al. [BCD$^+$07].

Theorem 11.2 *There exists a planar graph G and a path P not admitting an SGE.*

Proof Sketch: Consider the graph G and the path P as shown in Figure 11.5. Let G' be the subgraph of G induced on the vertices $\{1, 2, 3, 4, 5\}$, and let G'' be the subgraph of

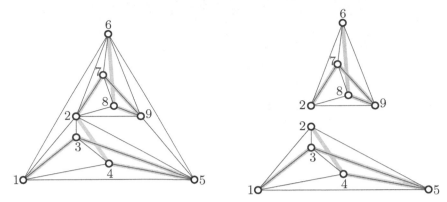

Figure 11.5   A planar graph G and a path P that do not allow an SGE.

G induced on the vertices $\{2,6,7,8,9\}$. Since G is triconnected fixing the outer face fixes an embedding for G. With the given outer face of G, the path P contains two crossings: one involving $(2,4)$, and the other one involving $(6,8)$.

Graph G' has six faces and unless we change the outer face of G' such that it contains the edge $(1,3)$ or $(3,5)$, the edge $(2,4)$ is involved in a crossing in the path. Similarly for G'', unless we change its outer face such that it contains $(2,7)$ or $(7,9)$, the edge $(6,8)$ is involved in a crossing in the path. However G' and G'' do not share any faces and removing both crossings depends on taking two different outer faces, which is impossible. Thus, regardless of the choice for the outer face of G, path P contains a crossing.   □

Theorem 11.3   *There exist three paths P_1, P_2 and P_3 not admitting an SGE.*

Proof: A path of n vertices is simply an ordered sequence of n numbers. The three paths we consider are: 714269358, 824357169 and 758261439. For example, the sequence 714269358 represents the path $(v_7, v_1, v_4, v_2, v_6, v_9, v_3, v_5, v_8)$. We will write ij for the edge connecting v_i to v_j. The union of these paths contain the following twelve edges.

$$E = \{14, 16, 17, 24, 26, 28, 34, 35, 39, 57, 58, 69\}$$

It is easy to see that the graph G consisting of these edges is a subdivision of $K_{3,3}$ and therefore non-planar: collapsing 1 and 7, 2 and 8, 3 and 9 yields the classes $\{1,2,3\}$ and $\{4,5,6\}$.

It follows that there are two nonadjacent edges of G that cross each other. It is easy to check that every pair of nonadjacent edges from E appears in at least one of the paths given above. Therefore, at least one path will cross itself which completes the proof.   □

Cabello et al. [CvKL+11] extend the counterexample for the case that G_1 is a path to the case where G_1 is a matching. Moreover, they give an example of six matchings not admitting an SGE. Note that this does not directly follow by dividing three paths without an SGE into six matchings, as the resulting matchings allow crossings that were not allowed before. Another extension of the case where G_1 is a path was given by Frati et al. [FKK09] who give a counterexample where G_1 is a path and G is a set of isolated vertices, that is, G_1 and G_2 are edge disjoint.

The question of whether two trees always admit an SGE was open for several years, before it was answered in the negative by Geyer et al. [GKV09] with a construction involving two very large trees. This of course extends the result of two outerplanar graphs not having an

SGE by Brass et al. [BCD+07]. Angelini et al. [AGKN12] further extended it to the case of a tree and a path without an SGE. More precisely, they give an example of a tree of depth 4 and an edge disjoint path not having an SGE. Recall that a tree of depth 2 does always admit a simultaneous embedding with a path, thus in this case the gap between positive and negative results is quite small.

11.2.3 Related Work

Frati et al. [FKK09] consider the restricted case where each input graph has a prescribed combinatorial embedding. They show that the pair path plus star admits an SGE even if the embedding of the star is fixed. They can extend this result to a *double-star* (tree with up to two inner vertices) if it is edge disjoint to the path. On the other hand they show that fixing the embedding of two caterpillars may lead to an counterexample, whereas they admit an SGE if the embedding is not fixed. Another counterexample is the pair outerplanar graph with fixed embedding plus edge-disjoint path.

An interesting additional restriction to SGEs was considered by Argyriou et al. [ABKS12], combining SGE with the RAC drawing convention (RAC – Right-Angular Crossing). They try to find an SGE such that crossings between exclusive edges of different graphs are restricted to right-angular crossings. Argyriou et al. consider only the case where the edge sets of both graphs are disjoint. They present one negative and one positive result for this problem. The negative result consists of a wheel and a cycle not admitting an SGE with right-angular crossings. On the other hand they show the existence of such a drawing on a small integer grid for the case that one of the graphs is a path or a cycle and the other is a matching. Moreover, they give a linear-time algorithm to compute such a drawing.

11.3 Simultaneous Embedding with Fixed Edges

In this section we drop the requirement that edges have to be straight line segments and consider the SEFE problem. Figure 11.6 shows a SEFE of the graph and the path from Figure 11.5 not admitting an SGE. Figure 11.7 and Table 11.2 illustrate the results on the problem SEFE classified in the three categories described before.

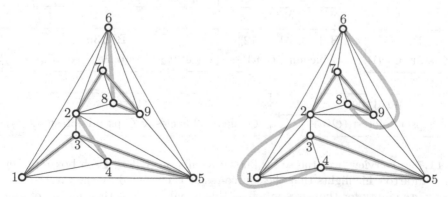

Figure 11.6 A graph and a path not admitting an SGE but a SEFE.

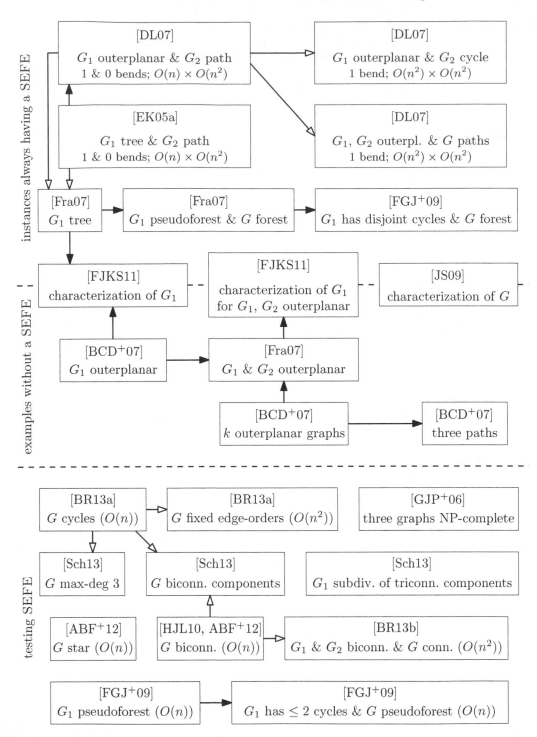

Figure 11.7 Overview over the so far known results on SEFE. Each box represents one result and an arrow highlights that the source-result is extended by the target-result. The arrowheads are empty for the cases in which this is only true, if the number of bends per edge, the consumed grid size or the necessary running time is neglected. Note that transitive arrows are omitted.

SEFE Instance	Exist.	Area	Bends	Ref.
G_1 tree & G_2 path	✓	$O(n) \times O(n^2)$	1 & 0	[EK05a]
G_1 outerplanar & G_2 path	✓	$O(n) \times O(n^2)$	1 & 0	[DL07]
G_1 outerplanar & G_2 cycle	✓	$O(n^2) \times O(n^2)$	1	[DL07]
G_1, G_2 outerplanar & G collection of paths	✓	$O(n^2) \times O(n^2)$	1	[DL07]
G_1 tree & G_2 planar	✓	—		[Fra07]
G_1 pseudoforest, G_2 planar & G forest	✓	—		[Fra07]
G_1 has disjoint cycles, G_2 planar & G forest	✓	—		[FGJ+09]
characterization of G	✓ / ✗	—		[JS09]
characterization of G_1	✓ / ✗	—		[FJKS11]
characterization of G_1 (G_1, G_2 outerplanar)	✓ / ✗	—		[FJKS11]
G_1 outerplanar & G_2 planar	✗	—		[BCD+07]
k outerplanar graphs	✗	—		[BCD+07]
three paths	✗	—		[BCD+07]
G_1 & G_2 outerplanar	✗	—		[Fra07]

SEFE Instance	Complexity	Ref.
three planar graphs	NP-complete	[GJP+06]
G_1 pseudoforest & G_2 planar	$O(n)$	[FGJ+09]
G_1 has ≤ 2 cycles, G_2 planar & G pseudoforest	$O(n)$	[FGJ+09]
G star	$O(n)$	[ABF+12]
G consists of disjoint cycles	$O(n)$	[BR13a]
G consists of components with fixed embeddings	$O(n^2)$	[BR13a]
G has maximum degree 3	polynomial	[Sch13]
G_1 subdivision of triconnected components & G_2 planar	polynomial	[Sch13]
G biconnected	$O(n)$	[HJL10]
G biconnected	$O(n)$	[ABF+12]
G consists of biconnected components	polynomial	[Sch13]
G_1, G_2 biconnected & G connected	$O(n^2)$	[BR13b]

Table 11.2 A list of graph classes that are either known to always have a SEFE or that contain counterexamples (table at the top). For the positive examples bounds on the required area and number of bends per edge are given, provided that they are known. The symbol ✓ / ✗ denotes that a complete characterization of positive and negative instances is given. The table at the bottom shows results concerning the computational complexity of SEFE.

11.3.1 Positive and Negative Examples

We start with instances that always admit a SEFE. Erten and Kobourov [EK05a] show that a tree and a path can always be embedded simultaneously. They additionally give an algorithm finding a simultaneous embedding in $O(n)$ time on a grid of size $O(n) \times O(n^2)$ such that the edges of G_1 and G_2 have at most one and zero bends per edge, respectively. Note that a grid of size $O(n^2) \times O(n^3)$ is necessary if the bends are required to be drawn on grid points. Di Giacomo and Liotta [DL07] extend this result to the case of an outerplanar graph and a path with the same grid and bend requirements. They extend it further to the case where G_1 and G_2 are outerplanar and the common graph G is a collection of paths and to the case where G_1 is outerplanar and G_2 is a cycle. However, in both cases a grid of size $O(n^2) \times O(n^2)$ and up to one bend per edge are required. If the grid and bend requirements are completely neglected, the results considering the pairs tree plus path and outerplanar graph plus path can be extended to the case where one of the two graphs is a tree.

Frati [Fra07] shows how a tree G_1 can be simultaneously embedded with an arbitrary planar graph G_2. This algorithm still works if G_1 contains one additional edge that is not a common edge, yielding the result that every graph with at most one cycle (a *pseudoforest*) can be embedded simultaneously with every other planar graph if the common graph does not contain this cycle. Fowler et al. [FGJ$^+$09] extend this result further to the case where G_1 contains only disjoint cycles and the common graph G does not contain a cycle.

Aside from instances always having a SEFE, there are also examples that cannot be simultaneously embedded. Brass et al. [BCD$^+$07] give examples for k outerplanar graphs, three paths and an outerplanar graph plus a planar graph not having a SEFE. The results concerning outerplanar graphs can be extended to the case where both graphs are outerplanar [Fra07].

In between the positive and negative results there are some characterizations stating which instances have a SEFE and which do possibly not. Fowler et al. [FJKS11] give a characterization of the graphs G_1 having a SEFE with every other planar graph. This of course extends all results concerning only G_1. In particular, the results that a tree can be simultaneously embedded with every other graph, whereas an outerplanar graph cannot, are extended. This characterization essentially requires that G_1 must not contain a subgraph homeomorphic to K_3 (a triangle) and an edge not attached to this K_3; see Figure 11.8 for an example. The considerations made for this characterization additionally yield a characterization for the biconnected outerplanar graphs G_1 having a simultaneous embedding with every other outerplanar graph G_2. This of course extends the result that two outerplanar graphs possibly do not have a SEFE.

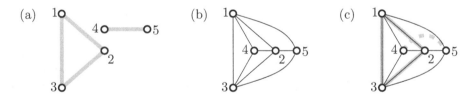

Figure 11.8 G_1 (a) and G_2 (b) do not admit a SEFE (c) as G_2 forces the vertices 4 and 5 to different sides of the triangle $\triangle 123$.

A different characterization, in terms of the common graph, is given by Jünger and Schulz [JS09]. They show that two graphs can be simultaneously embedded if the common graph G has only two embeddings, whereas in all other cases graphs G_1 and G_2 with the common graph G not having a SEFE can be constructed. They additionally show that finding a SEFE is equivalent to finding combinatorial embeddings of G_1 and G_2 inducing the same combinatorial embedding, that is the same orders of edges around vertices and the same relative positions of connected components to one another, on the common graph G [JS09, Theorem 4]. Note that it is not obvious and not even true for more than two graphs [ADF11]. As this result is heavily used in most algorithms solving the decision problem SEFE, we state it as a theorem.

Theorem 11.4   *Two graphs G_1 and G_2 with common subgraph G admit a SEFE if and only if they admit combinatorial embeddings inducing the same embedding on G.*

11.3.2   Testing SEFE

Since SEFE has positive and negative instances, it would be nice to have an algorithm deciding for given graphs, whether they can be embedded simultaneously. If more than two graphs are allowed, this problem is known to be NP-complete [GJP+06], whereas the complexity for two graphs is still open. However, there are several results solving SEFE for special cases.

Fowler et al. [FGJ+09] show how to test SEFE, if G_1 is a pseudoforest, that is, a graph with at most one cycle. Note that, as mentioned above, such an instance always has a SEFE if this single cycle is not contained in G. This result can be extended to the case where G_1 contains up to two cycles, if G does not contain the second cycle, that is, G is a pseudoforest. To achieve this result the following auxiliary problem was solved. Given a planar graph G with a designated cycle C and a partition $\mathcal{P} = \{P_1, \ldots, P_k\}$ of the vertices not contained in C, does G admit a planar embedding, such that all vertices in P_i are on the same side of the cycle for every set P_i? Note that this again is a constrained embedding problem, showing that constrained and simultaneous embedding are closely related. Despite early effort [FJKS11], testing SEFE for two outerplanar graphs remains open.

Haeupler et al. [HJL10] give a linear-time algorithm to solve SEFE for the case that the common graph is biconnected. Their solution is an extension of the planarity testing algorithm by Haeupler and Tarjan [HT08]. This planarity testing algorithm starts with a completely unembedded graph and adds vertices iteratively, such that the unembedded part is always connected, ensuring that it can be assumed to lie in the outer face of all embedded components. While inserting vertices, they keep track of the possible embeddings of the embedded parts by representing the possible orders of half-embedded edges around every component with a PQ-tree having these edges as leaves. In a *PQ-tree* every inner node is either a Q-node fixing the order of edges incident to it up to a flip or a P-node allowing arbitrary orders. In this way a PQ-tree represents a set of possible orders of its leaves.

A completely different approach is used by Angelini et al. [ABF+12] to solve SEFE in linear time if the common graph is biconnected. They choose an order for the common graph bottom up in its SPQR-tree such that the private edges can be added.

Another approach by Bläsius and Rutter [BR13b] also uses PQ-trees. They use that the possible orders of edges around every vertex of a biconnected planar graph can be represented by a PQ-tree, yielding a set of PQ-trees, one for each vertex. To obtain a planar embedding, the orders for the PQ-trees have to be chosen consistently. Bläsius and Rutter define the problem SIMULTANEOUS PQ-ORDERING asking for orders in PQ-trees that are chosen consistently, which can, among other applications, be used to represent all planar embeddings of a biconnected graph. This extends to the case of two biconnected planar graphs enforcing shared edges to be ordered the same and thus yields a quadratic time algorithm for SEFE if G_1 and G_2 are biconnected and G is connected. The latter requirement comes from the fact that only orders of edges around vertices are taken into account, relative positions of connected components to one another are neglected. Note that this result extends the case where G is biconnected for the following reason. If G is biconnected, then G is completely contained in a single block (maximal biconnected component) of G_1 and G_2. Thus, even if G_1 or G_2 are not biconnected, they contain only one block that is of interest, all other blocks can simply be attached to this block.

The result by Bläsius and Rutter can be slightly extended to the case where the graphs G_1 and G_2 contain cut-vertices incident to at most two non-trivial blocks (blocks not consisting of a single edge), including the special case where both graphs have maximum degree 5. The SIMULTANEOUS PQ-ORDERING approach again shows the strong relation between simultaneous and constrained embedding as in an instance of SEFE the two input graphs constrain the possible orders of some of the edges around vertices of one another with PQ-trees.

Angelini et al. [ABF+12] show the equivalence between SEFE and a constrained version of the PARTITIONED 2-PAGE BOOK EMBEDDING problem. An instance of PARTITIONED 2-PAGE BOOK EMBEDDING is a graph and a partition of its edges into two subsets. It asks whether all vertices can be arranged on a straight line (the *spine*) such that each of the edge partitions can be embedded without crossings in one of the two incident half-planes (*pages* of the book). PARTITIONED T-COHERENT 2-PAGE BOOK EMBEDDING additionally has a tree as input with the vertices of the graph as leaves. It is then required that the tree admits an embedding such that the order of its leaves is equal to the order of vertices on the spine. In other words, the allowed orders of vertices on the spine is constrained by a PQ-tree containing no Q-nodes. Angelini et al. [ABF+12] prove the following theorem and we sketch their proof here.

Theorem 11.5   *The problems* SEFE *for two graphs with connected intersection and* PARTITIONED T-COHERENT 2-PAGE BOOK EMBEDDING *have the same time complexity.*

Proof Sketch: Angelini et al. [ABF+12] first show that an instance of SEFE where the common graph is connected can be modified (yielding an equivalent instance) such that the common graph is a tree. Moreover, each private edge is incident to leaves of this tree. They then show the equivalence to an instance of PARTITIONED T-COHERENT 2-PAGE BOOK EMBEDDING where the common graph is the constraining tree, the leaves of this tree are the vertices that need to be placed on the spine and the private edges of each of the graphs is one of the partitions.

In the following we sketch this construction using the example in Figure 11.9. The instance in (a) having a tree T as common graph such that each private edge is incident to a leaf admits a SEFE. All private edges are embedded outside the dashed cycle around T in (b) containing all its leaves. Choosing another face as outer face and cutting the cycle at an arbitrary position yields a SEFE where all leaves of T are embedded on a straight line (c) with all private edges on the same side. This directly yields the PARTITIONED T-COHERENT 2-PAGE BOOK EMBEDDING in (d) of the private edges respecting the tree T. This shows the equivalence of SEFE and PARTITIONED T-COHERENT 2-PAGE BOOK EMBEDDING as the constructions works the same in the opposite direction.   □

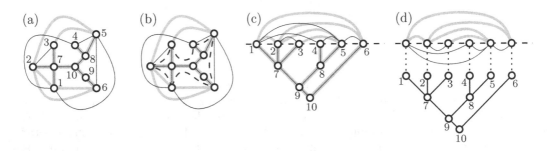

Figure 11.9   Equivalence of an instance of SEFE and the corresponding instance of PARTITIONED T-COHERENT 2-PAGE BOOK EMBEDDING.

For the restricted case that T is a star, PARTITIONED T-COHERENT 2-PAGE BOOK EMBEDDING reduces to the problem PARTITIONED 2-PAGE BOOK EMBEDDING that can be solved in linear time [HN09]. Thus the above result directly implies that SEFE can be solved in linear time if the common graph is a star.

All results mentioned thus far require G to be connected and most results also require G_1 and G_2 to be connected. Bläsius and Rutter [BR13a] consider the case where this does not hold. They show that it can be assumed without loss of generality that both graphs G_1 and G_2 are connected.

In the case that G is connected, one only has to deal with orders of edges around vertices and can neglect relative positions of connected components to one another. Bläsius and Rutter approach SEFE from the opposite direction, caring only about the relative positions, neglecting the orders of edges around vertices. More precisely, they give a linear-time algorithm solving SEFE if the common graph is a set of disjoint cycles. They can extend this result to a quadratic-time algorithm for the case where G consists of arbitrary connected components, each with a fixed planar embedding. Both results extend to an arbitrary number of graphs with sunflower intersection. Recall that sunflower intersection means that all graphs intersect in the same common subgraph. Moreover, they give a succinct representation of all simultaneous embeddings.

A completely different, algebraic approach is presented by Schaefer [Sch13]. It is based on the Hanani-Tutte theorem [Cho34, Tut70] stating that a graph is planar if and only if its *independent odd crossing number* is 0. The independent odd crossing number of a drawing is the number of pairs of non-adjacent edges whose number of crossings is odd. The independent odd crossing number of a graph is its minimum over all drawings. Thus, by the Hanani-Tutte theorem, testing planarity is equivalent to testing whether this crossing number is 0. The latter condition can be formulated as a system of linear equations over the field of two elements, leading to a simple polynomial-time planarity algorithm. Schaefer extends this result to other notions of planarity. In particular, it is shown that SEFE can be solved in polynomial time for three interesting cases, namely (1) if the common graph G consists of disjoint biconnected components and isolated vertices, (2) if the common graph has maximum degree 3, and (3) if G_1 is the disjoint union of subdivisions of triconnected graphs. When neglecting the slower running time, this extends several of the results known before; see Figure 11.7.

11.3.3 Related Work

A result not really fitting in one of the three above classes by Duncan et al. [DEK04] considers the restricted case of SEFE where each edge has to be a sequence of horizontal and vertical segments with at most one bend per edge. They show that two graphs with maximum degree 2 always admit such a SEFE on a grid of size $O(n) \times O(n)$ by adapting their linear-time algorithm computing an SGE for these types of graphs (on a larger grid).

Angelini et al. [ADF11] consider the case where the embedding of each of the input graphs is already fixed. With this restriction SEFE becomes trivial for two graphs since it remains to test whether the two graphs induce the same embedding on the common graph. They show that it can also be decided efficiently for three graphs. However, it becomes NP-hard for at least fourteen graphs. They also consider the problem SGE for the case that the embedding of each graph is fixed and show that it is NP-hard for at least thirteen graphs.

Schaefer [Sch13] shows that several other notions of planarity are related to SEFE. In particular, the well-studied cluster planarity problem reduces to SEFE, providing further incentive to study its complexity.

11.4   Simultaneous Embedding

In the most restricted version of the problem, SGE, we insist that vertices are placed in
the same position, and edges must be straight-line segments. The SEFE setting relaxes the
straight-line condition but maintains that edges common to multiple graphs are realized
the same way in each. In the least restrictive setting, SE, we allow the same edge to be
realized differently in different graphs.

It has already been mentioned that simultaneous embedding of multiple graphs can be
thought of as a generalization of the notion of planarity. A classical result about planar
graphs connects the notion of a planar graph with that of a straight-line, crossing-free
drawing thereof. Specifically, Wagner in 1936 [Wag36], Fáry in 1948 [Fár48], and Stein
in 1951 [Ste51] independently show that if a graph has a drawing without crossings, using
arbitrary curves as edges, then there exists a drawing of the graph also without crossings,
but with edges drawn as straight-line segments. For multiple graphs, however, this result
does not hold. That is, given several graphs on the same n vertices, we can surely realize
each graph without crossings, using arbitrary curves as edges and the same vertex positions
for each graph. But (except in very special circumstances such as the positive examples
in the Section 11.2) we cannot guarantee that there exist vertex positions that allow the
realization of each graph with straight-line segments and without crossings. If this were
true, then the vertex positions would be a *universal pointset* for graphs on n vertices, and
it is known that universal pointsets of linear size do not exist [dFPP90].

Pach and Wenger [PW98] show that every planar graph can be drawn without crossings
with a prespecified position for every vertex. Thus, for every pair of planar graphs an SE
can be created by drawing the first graph arbitrarily and the second graph to the vertex
positions specified by the first drawing. Thus, there are neither negative examples nor is it
necessary to have testing algorithms. However, the drawing of the second graph may have
linearly many bends per edge, thus it is of interest to find an SE with fewer bends.

Erten and Kobourov [EK05a] show that every two graphs can be drawn simultaneously
in $O(n)$ time with at most three bends per edge on an $O(n^2) \times O(n^2)$ grid ($O(n^3) \times O(n^3)$ if
bends need to be placed on grid points), where n is the number of vertices. To achieve this
result, they combine the construction of Brass et al. [BCD+07] to create an SGE of two
paths (see Theorem 11.1 in Section 11.2) with a technique by Kaufmann and Wiese [KW02],
who show that every planar graph can be drawn with at most two bends per edge if the
allowed vertex positions are restricted to a set of points. We include the main result from
this paper along with a proof sketch.

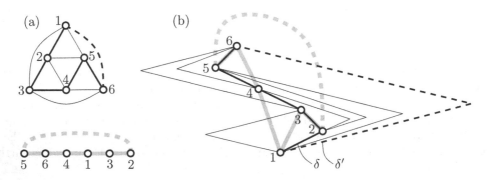

Figure 11.10   (a) The cycle H_2 (gray) with the path P_2 (not dashed) and the graph
G_1 containing the Hamiltonian cycle H_1 (bold) and the Hamiltonian path P_1 (bold, not
dashed). (b) The drawing of G_1 and P_2 according to the construction of Theorem 11.6.

Theorem 11.6 *For two planar graphs G_1 and G_2 an SE with at most three bends per edge on an $O(n^2) \times O(n^2)$ grid can be found in linear time.*

Proof Sketch: Initially, assume that G_1 and G_2 are 4-connected. This assumption is removed later using the technique of Kaufmann and Wiese.

We can compute Hamiltonian cycles H_1 and H_2 of G_1 and G_2, respectively, using the algorithm of Chiba and Nishizeki [CN89]. Let P_1 and P_2 be Hamiltonian paths contained in H_1 and H_2, respectively; see Figure 11.10(a) for an example. As in the proof of Theorem 11.1, we can construct an SGE of P_1 and P_2 such that P_1 is y-monotone, while P_2 is x-monotone. We show how to add the remaining edges of G_1 and the construction is similar for G_2.

We consider the absolute values of the slopes the edges in P_1 have and define δ to be their minimum. Let further δ' be slightly smaller. We first close the cycle H_1 by adding the missing edge using two straight-line segments with slopes δ' and $-\delta'$; see Figure 11.10(b). Similarly, all remaining edges of G_1 are drawn with two straight-line segments with slopes appropriately chosen between δ' and δ and between $-\delta$ and $-\delta'$. Dealing similarly with the remaining edges of G_2 yields an SE with at most one bend per edge on a grid of size $O(n^2) \times O(n^2)$.

For the case that G_1 and G_2 are not 4-connected, Kaufmann and Wiese [KW02] showed how they can be augmented to 4-connected planar graphs by adding new edges and subdividing every edge at most once. Drawing these augmented graphs as described above, removing the additional edges and replacing each subdivision vertex with a bend yields an SE of G_1 and G_2 with at most three bends per edge on an $O(n^2) \times O(n^2)$ grid. □

The result of Erten and Kobourov was improved by Di Giacomo and Liotta [DL05, DL07] to at most two bends per edge in general and one bend per edge, if G_1 and G_2 are both sub-Hamiltonian. That is, they can be augmented to become Hamiltonian maintaining planarity, and an augmentation together with a Hamiltonian cycle is given with the input. Similar results were obtained by Kammer [Kam06]. As series-parallel graphs [DDLW06], trees and outerplanar graphs [CLR87, BK79] are always sub-Hamiltonian and an augmentation together with a Hamiltonian cycle can be computed in linear-time this result yields a linear time algorithm to compute an SE of G_1 and G_2 with one bend per edge on a grid of size $O(n^2) \times O(n^2)$ if each of the graphs G_1 and G_2 is series-parallel, a tree or outerplanar.

Cappos et al. [CEBFK09] show that a path and an outerplanar graph can be simultaneously embedded in linear time such that edges in the outerplanar graph are straight-line segments and each edge in the path consists of a single circular arc. Alternatively, the path edges may be piecewise linear with at most two bends per edge.

11.5 Colored Simultaneous Embedding

Since SGE can be too restrictive, various relaxations have been considered. The two relaxed versions already mentioned, SEFE and SE relax the requirement of straight-line edges, and even the requirement that common edges are drawn the same way in both drawings. Another way to relax the constraints of the original SGE problem is to allow changes in vertex positions in different graphs.

Until this point we had assumed that multiple input graphs have labeled vertices and thus the mapping between the vertices of the graphs is part of the input. In *simultaneous embedding without mapping* we are interested in computing plane drawings for each of the given graphs on the same set of points, where any vertex can be placed at any of the points

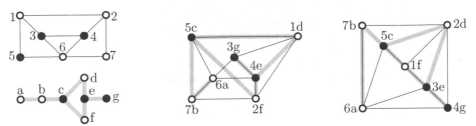

Figure 11.11   Two 2-colored graphs with two CSEs corresponding to different mappings.

in the point set. This setting of the problem was investigated in the very first paper on SGE [BCD+07] and is the source of one of the longest-standing open problems in the area.

A common generalization of the problems above is COLORED SIMULTANEOUS EMBEDDINGS (CSE), which was introduced by Brandes et al. [BEEB+11], and contains both, the version with and without mapping. Formally, the problem of CSE is defined as follows. The input is a set of planar graphs $G_1 = (V, E_1), G_2 = (V, E_2), \ldots, G_k = (V, E_k)$ on the same vertex set V and a partition of V into c classes, which we refer to as colors. The goal is to find plane straight-line drawings D_i of G_i using the same $|V|$ points in the plane for all $i = 1, \ldots, k$, where vertices mapped to the same point are required to be of the same color. We call such graphs c-colored graphs; see Figure 11.11 for an example. Given the above definition, simultaneous embeddings with and without mapping correspond to colored simultaneous embeddings with $c = |V|$ and $c = 1$, respectively. Thus, when a set of input graphs allows for a simultaneous embedding without mapping but does not allow for a simultaneous embedding with mapping, there must be a threshold for the number of colors beyond which the graphs can no longer be embedded simultaneously.

Colored simultaneous embeddings provide a way to obtain near-simultaneous embeddings, where we place corresponding vertices nearly, but not necessarily exactly, at the same locations. Relaxing the constraint on the size of the pointset allows for a way to more easily obtain near-simultaneous embeddings, where we attempt to place corresponding vertices relatively close to one another in each drawing. For example, if each cluster of points in the plane has a distinct color, then even if a red vertex v placed at a red point $p \in G_1$ has moved to another red point $q \in G_2$, the movement is limited to the area covered by the red points.

Brandes et al. [BEEB+11] show several positive and negative results about CSE. In particular they show that there exist universal pointsets of size n for 2-colored paths and spiders as well as 3-colored paths and caterpillars. It is also shown that a 2-colored tree (or even a 2-colored outerplanar graph) and any number of 2-colored paths can be simultaneously embedded. In the negative direction, there exist a 2-colored planar graph and pseudo-forest, three 3-colored outerplanar graphs, four 4-colored pseudo-forests, three 5-colored pseudo-forests, five 5-colored paths, two 6-colored biconnected outerplanar graphs, three 6-colored cycles, four 6-colored paths, and three 9-colored paths that cannot be simultaneously embedded.

Frati et al. [FKK09] continue the investigation of near-SGE's, that is, they try to find straight-line drawings of the input graphs with a small distance between every pair of common vertices in different drawings. As a negative result, they present a pair of graphs such that in every pair of drawings there exists a common vertex with distance linear in the size of the input. On the other hand, they present positive results for a sequence of paths and a sequence of trees for the case that every two consecutive graphs in the sequence are similar with respect to a parameter measuring their similarity. It can then be shown

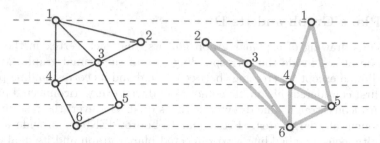

Figure 11.12 A matched drawing: corresponding vertices have the same y-coordinate.

that the distance of a common vertex in two consecutive drawings depends linearly on this parameter.

11.6 Matched Drawings

Another approach to relax requirements of SGE are the so-called *matched drawings* introduced by Di Giacomo et al. [DDvK+09]. A matched drawing of a pair of graphs is a planar straight-line drawing of each of the graphs such that each common vertex has the same y-coordinate in both drawings (instead of the same y- and x-coordinate as required for SGE); see Figure 11.12 for an example.

Di Giacomo et al. [DDvK+09] give a small counterexample consisting of two small triconnected planar graphs not admitting a matched drawing. Moreover, they give a larger example (620 vertices) of a biconnected graph and a tree not having a matched drawing.

Apart from that they also have some results on the positive side. They show that two trees are always matched drawable. Moreover, they observe that any planar graph has a matched drawing with a so-called *unlabeled level planar (ULP)* graph, that is, a graph that admits a planar straight-line drawing even if the y-coordinate of each vertex is prespecified such that no two vertices have the same y-coordinate. A characterization of ULP graphs is given by Fowler and Kobourov [FK08]. Di Giacomo et al. [DDvK+09] moreover show for a graph class containing non-ULP graphs (the *carousel graphs*) that they admit matched drawings with arbitrary planar graphs. A special case of a carousel graph is a graph consisting of a single vertex v_0 and a set of disjoint subgraphs $S_1, \ldots, S_k$, each S_i connected to v_0 over a single edge $\{v_0, v_i\}$ such that S_i is either a caterpillar with v_i on its spine, a radius-2 star with v_i as center or a cycle.

Grilli et al. [GHL+09] present further positive results on matched drawings. They show how to draw the pairs outerplane plus wheel, wheel plus wheel, outerplane plus *maximal outerpillar* (outerplane graph with triangulated inner faces and caterpillar as weak dual), and outerplane plus *generalized outerpath* (outerpath where some edges on the outer face may be replaced by some small subgraphs). Moreover, they consider matched drawings for graph triples and give algorithms creating matched drawings of three cycles, and a caterpillar and two ULP graphs.

11.7 Other Simultaneous Representations

Apart from simultaneously drawing two graphs sharing some common parts there are other ways to represent graphs simultaneously. In this section we describe how a plane graph and its dual can be represented simultaneously, and what is known about simultaneous intersection representations of (not necessarily planar) graphs.

11.7.1   A Plane Graph and Its Dual

In a simultaneous drawing of a planar graph and its dual each vertex in the dual graph
is required to be placed inside the corresponding face of the primal graph. Moreover, no
crossings are allowed except for crossings between a dual and its corresponding primal edge.
Tutte [Tut63] first considered this problem and showed that every triconnected planar graph
admits a simultaneous straight-line drawing with its dual. However, the resulting drawings
may have exponentially large area. Erten and Kobourov [EK05b] provide a linear-time
algorithm simultaneously embedding a triconnected planar graph and its dual on a grid of
size $(2n - 2) \times (2n - 2)$ such that all edges are drawn as straight-line segments. Zhang and
He [ZH06] improved this result to a grid of size $(n - 1) \times n$.

Brightwell and Scheinerman [BS93] show the existence of a simultaneous straight-line
drawing of a triconnected planar graph and its weak dual such that the crossings between
dual and the corresponding primal edges are right-angular crossings. A *circle packing*
of a planar graph represents the vertices as non-crossing circles such that two vertices are
adjacent if and only if their corresponding circles touch. Given a circle packing of a planar
graph, one obtains a planar straight-line drawing by placing each vertex at the center of
its corresponding circle. Mohar [Moh97] shows that every triconnected planar graph has
a simultaneous circle packing with its dual such that in the corresponding straight-line
drawings primal and dual edges have right-angular crossings. Argyriou et al. [ABKS12]
give a simple example of a graph that is not triconnected not admitting such a drawing.
On the positive side they give an algorithm that creates such drawings for the case that the
primal graph is outerplanar.

Another way of simultaneously representing a planar graph and its dual is the *tessellation
representation* introduced by Tamassia and Tollis [TT89]. In a tessellation representation,
every edge, every vertex and every face is represented by a (possibly degenerated) rectangle,
a so-called *tile*, such that the interiors of these tiles are pairwise disjoint, that their union
forms a rectangle, and that the incidences in the graph are represented by side contacts of
the tiles in the following way: (i) Two tiles share a horizontal line segment if and only if
they represent an edge and an incident face; and (ii) two tiles share a vertical line segment
if and only if they represent an edge and an incident vertex. Tamassia and Tollis [TT89], in
particular, showed that every biconnected planar graph admits a tessellation representation
where the tiles representing vertices and faces are vertical and horizontal line segments,
respectively.

The textbook by Di Battista et al. [DETT99, Sections 4.3 and 4.4] contains a short
description of the algorithm computing tessellation representations and of the relation to
visibility representations. Moreover, tessellation representations were also considered on
other surfaces such as the torus [MR98].

11.7.2   Intersection Representations

Jampani and Lubiw [JL09] introduce the concept of simultaneous graph representations
for other representations than drawings. An intersection representation of a graph assigns
a geometric object to each vertex such that two vertices are adjacent if and only if their
corresponding geometric objects intersect. Two graphs sharing a common subgraph are
simultaneous intersection graphs if each of them has an intersection representation such that
the common vertices are represented by the same objects. Note that every planar drawing
of a graph can be interpreted as intersection representation, each vertex is represented by
the union of its edges. This shows that deciding SEFE as a special case of recognizing
simultaneous intersection graphs.

Other popular intersection representations are the following. In an *interval representation* of a graph each vertex is represented by an interval on the real line. A graph is *chordal* if each induced cycle has length three. Gavril [Gav74] shows that chordal graphs are exactly the intersection graphs of subtrees in a tree. This shows that the class of interval graphs is contained in the class of chordal graphs. *Permutation graphs* are the intersection graphs that can be represented by a set of line segments connecting two parallel lines. Jampani and Lubiw [JL09] give $O(n^3)$-time algorithms recognizing simultaneous permutation graphs and simultaneous chordal graphs. The algorithm for simultaneous permutation graphs can be extended to more than two graphs with sunflower intersection. On the other hand, it is NP-hard to recognize simultaneous chordal graphs of this kind (for a constant number k of graphs, the complexity is still open).

In a follow-up paper Jampani and Lubiw [JL10] give an algorithm recognizing simultaneous interval graphs in $O(n^2 \log n)$ time. As interval graphs can be characterized in terms of PQ-trees, recognizing simultaneous interval graphs leads to a problem of finding orders in several PQ-trees simultaneously. Bläsius and Rutter [BR13b] consider this kind of problem in a more general leading to a $O(n)$-time algorithm recognizing simultaneous interval graphs.

Related to simultaneous intersection graphs are simultaneous comparability graphs also introduced by Jampani and Lubiw [JL09]. A *comparability graph* is a graph that can be oriented transitively where transitively means that a directed path implies the existence of a directed edge. Two graphs are *simultaneous comparability graphs* if each of them can be oriented transitively such that common edges are oriented the same in both. Jampani and Lubiw give an $O(nm)$-time algorithm recognizing simultaneous comparability graphs. It can also be used to recognize an arbitrary number of comparability graphs with sunflower intersection. Comparability graphs are related to intersection graphs as comparability graphs are exactly the graphs whose complement is a *function graph*, that is the intersection graph with respect to continuous functions on an interval [GRU83].

As for the problem SEFE, finding simultaneous representations is related to extending a representation of a subgraph to one of the whole graph. For interval graphs Klavík et al. [KKV11] give a $O(nm)$-time algorithm testing whether a partial interval representation can be extended. Bläsius and Rutter [BR13b] were able to improve the running time to $O(m)$ by constructing a second graph such that both graphs are simultaneous interval graphs if and only if the partial interval representation can be extended.

11.8 Practical Approaches to Dynamic Graph Drawing

The majority of the results reviewed above focused on the theoretical aspects of dynamic graph drawing. In this section we review practical approaches to this problem. As we have seen in the previous sections, numerous negative results show that in many of the interesting settings we cannot guarantee simultaneous embeddings. On the other hand, several efficient algorithms for different variants of the problem do exist, but they usually place additional restrictions on the number of input graphs, or limit the graphs to special sub-classes of planar graphs.

As discussed in the introduction, the problem is well motivated in practice. Of particular interest are applications to visualization of dynamic graphs and the related issues of mental map preservation and good graph readability. With this in mind we mention several more practical results here. First we focus on drawing algorithms that aim to produce simultaneous embeddings or layouts that are in some sense close to being a simultaneous embedding. Afterwards, we briefly discuss other approaches to dynamic graph drawing.

Erten et al. [EKLN05] adapt force-directed algorithms to create drawings of a series of graphs sharing subgraphs finding a tradeoff between nice drawings and similarities of common parts. Kobourov and Pitta [KP05] describe an interactive system that allows multiple users to interactively modify a pair of graphs simultaneously using a multi-user, touch-sensitive input device. While those two approaches focus on straight-line drawings (corresponding to SGE), the GraphSET system by Estrella-Balderrama et al. [EBFK10] also allows edges to have bends. GraphSET is a tool helping the user to investigate the theoretical problems SGE and SEFE and it contains implementations of several testing and drawing algorithms. Chimani et al. [CJS08] create simultaneous drawings of graphs by drawing the union of the graphs. Their objective is to minimize the number of crossings in the drawing, where crossings between edges of different graphs do not count, yielding a simultaneous embedding if and only if the number of crossings is zero.

Misue et al. [MELS95] initiated the study of drawing dynamically changing graphs and first proposed several models to capture the notion of preserving the user's mental map. In particular they suggested preservation of orthogonal orderings, proximity relations, or the topology as a formalization. Bridgeman and Tamassia [BT98] describe and evaluate difference metrics that are specialized to orthogonal graph drawings. Purchase et al. [PHG07] provide empirical evidence that preserving the user's mental map indeed assists in comprehending the evolving graph. Purchase and Samra [PS08] argue that for minimizing the node movement, finding a trade-off is worse than either keeping the exact node positions or just layouting the next graph from scratch for memorizing tasks. In a recent study, Archambault and Purchase [AP13] observed positive effects of mental map preservation for localization tasks, both in terms of speed and accuracy. Sallaberry et al. [SMM13] consider mental map preservation for large graphs and argue that restricting node movements to small distances is not sufficient for this case. They propose to cluster nodes into groups that perform the same movement in order to increase the stability of the drawing.

Bridgeman et al. [BFG$^+$97] present InteractiveGiotto, a bend-minimization algorithm for orthogonal drawings that is designed for dynamic and interactive scenarios. Their algorithm supports arbitrary graph changes and preserves the embedding, all edge crossings, and the bends of edges.

Brandes and Wagner [BW97] suggest a Bayesian framework for dynamic graph drawing that can in principle be applied to all layout styles and allows to choose a trade-off between quality and stability. Diel and Görg [DG02] introduce foresighted layouts, where the basic idea is to layout the union of the graph over all time steps and to combine vertices and edges whose life times are disjoint, in order to reduce the size of the drawing. This automatically guarantees a high stability of the layout, but possibly incurs a negative impact on the quality of individual drawings. Görg et al. [GBPD04] enhance this method by an additional step that improves the quality of the individual layouts while keeping them close to the foresighted layout.

North and Woodhull [NW02] propose a heuristic for online hierarchical graph drawing by dynamizing the classical Sugiyama algorithm [STT81]. Collberg et al. [CKN$^+$03] describe a system for visualizing the evolution of software based on force-directed methods applied to so-called time-sliced graphs. A time-sliced graph consists of disjoint copies of the graph at each point in time together with time-slice edges, which connect corresponding vertices from different points in time. The algorithm attempts to place vertices that are connected by a time-slice edge in roughly the same position. Frishman and Tal [FT08] describe an algorithm for online dynamic graph drawing that can be implemented to run on a GPU.

11.9 Morphing Planar Drawings

The main motivation for simultaneously embedding different (but related) graphs is to preserve the mental map between the unchanged parts by drawing them the same. As opposed to this, *morphing* tries to match different drawings of the same graph. More precisely, let Γ_1 and Γ_2 be two drawings of the same graph G, a morph between them is a motion of the vertices along trajectories starting at the vertex positions in Γ_1 and ending at their positions in Γ_2.

The simplest possible morph between two drawings Γ_1 and Γ_2 is the *linear morph* where each vertex moves at constant speed along a line segment from its origin in Γ_1 to its destination in Γ_2. However, the intermediate drawings of linear morphs may be pretty bad, in fact, it may even happen that the whole graph collapses to a single point. To resolve this problem Cairns [Cai44] introduced the notion of morphing planar graphs, requiring that every intermediate drawing is also planar. He showed that two planar drawings of a triangulated plane graph with an equally drawn outer face can be morphed into each other in a planar way using a sequence of linear morphs. However, this sequence of linear morphs has exponential size.

Thomassen [Tho83] extends this to drawings of general (not necessarily triangulated) planar graphs with an equally drawn outer face and convex faces by augmenting the drawings to *compatible triangulations*, that is, one must be able to add all new vertices and edges to both given drawings without violating the planarity or straight-line requirement. Compatible triangulations were further investigated by Aronov et al. [ASS93] who show that two drawings admit compatible triangulations with only $O(n^2)$ new vertices. They moreover show that $\Theta(n^2)$ new vertices are sometimes necessary. This result has the following general implication. If there exist planar morphs between drawings of triangulated graphs using $O(f(n))$ linear morphing steps, then there are morphs between drawings of arbitrary plane graphs using $O(f(n^2))$ steps.

To be able to morph with a polynomial number ($O(n^6)$) of linear steps Lubiw and Petrick [LP08] relaxed the straight-line requirement and showed how to morph between two planar drawings when edges are allowed to be bent during the morph. However, this result can also be achieved without this relaxation. Alamdari et al. [AAC+13] show that for every pair of planar straight-line drawings of a triangulated graph with an equally drawn outer face there exists a planar morph consisting of a sequence of $O(n^2)$ linear morphs. This is the first result showing that a polynomial number of morphing steps is sufficient. Using the results on compatible triangulations mentioned above [ASS93] this yields a morph with $O(n^4)$ linear steps for general plane graphs.

Floater and Gotsman [FG99] introduced a completely different approach to planar morphing of triangulations. They make use of the fact that in a planar drawing the position of each vertex is a convex combination of the neighboring vertices and that conversely fixing the coefficients of the convex combinations and fixing the outer face yields a planar drawing. This was shown by Floater [Flo97] extending the results by Tutte [Tut60, Tut63]. Floater and Gotsman [FG99] create a morph between two planar drawings by transforming the coefficients of the corresponding convex combinations into one another, yielding a sequence of coefficients and thus a sequence of planar drawings. Surazhsky and Gotsman [SG01, SG03] improve this approach further to obtain aesthetically more appealing morphs.

The approach based on convex combinations has the disadvantage that the trajectories are not explicitly computed and that it is not clear how many linear morphing steps are necessary to obtain a planar and smooth morph. Despite its theoretical shortcomings, in practice this algorithm leads to nice morphs, as shown by Erten et al. [EKP04b, EKP04a], who combine this approach with rigid motion (translation, rotation, scaling and shearing) and the triangulation algorithm by Aronov et al. [ASS93]. Moreover, they are able to morph edges with bends to straight-line edges and vice versa.

Biedl et al. [BLS06] consider a related problem of morphing so-called *parallel* straight-line drawings, that is, straight-line drawings such that for every edge e, the slope of e is the same in both drawings. Moreover, the edge slopes have to be preserved throughout the whole morph. They show that for orthogonal drawings (without bends) such a morph always exists. On the other hand, testing for the existence of such a morph becomes NP-hard if the edges are allowed to have three or more slopes. Lubiw et al. [LPS06] investigate morphs between general orthogonal drawings of planar graphs where edges may have bends. They show that for every pair of drawings there is a morph preserving planarity and orthogonality consisting of polynomial many steps, where each step is either a movement of vertices or a "twist" around a vertex that introduces new bends at the edges incident to this vertex.

Of course, problems similar to planar morphing can be considered for non-planar graphs. Examples are the results by Friedrich and Eades [FE02] and Friedrich and Houle [FH02].

11.10   Open Problems

There are many interesting problems, some of which have been open for a decade and have resisted efforts to address them. Here we list several of the current open problems.

1. Given two arbitrary planar graphs $G_1 = (V_1, E_1)$ and $G_2 = (V_2, E_2)$ with the same number of vertices, $|V_1| = |V_2|$, does there always exist a mapping from the vertex set of the first graph onto the vertex set of the second graph $V_1 \rightarrow V_2$ such that the two graphs have a SGE? That is, do pairs of planar graphs always have an SGE without mapping?

2. Given two graphs of max-degree 2, $G_1 = (V_1, E_1)$ and $G_2 = (V_2, E_2)$ with the same number of vertices, an SGE with mapping does always exist. Unlike most other results where the pair of graphs has an SGE the area of the necessary grid is not bounded. Is it possible to guarantee polynomial integer grid for the simultaneous embedding?

3. What is the complexity of SGE for two graphs with fixed planar embeddings?

4. Is it possible to decide SGE for restricted cases, for example if the common graph is highly connected?

5. What is the complexity of the decision problem SEFE for two graphs?

6. Are there interesting parameters for which SEFE or SGE are FPT? For example, tree-distance of G? What about maximum degree Δ?

7. What is the complexity of SEFE for more than two graphs with sunflower intersection?

8. What is the complexity of SEFE for four graphs, each with a fixed planar embedding?

9. What is the complexity of the optimization version of SEFE where one asks for drawings such that as many common edges as possible are drawn the same?

10. Let G_1 and G_2 be two planar graphs with given combinatorial embeddings inducing the same embedding on their intersection G, that is, a SEFE is given with the input. What is the complexity of minimizing the number of crossings in a corresponding drawing?

11. Let G_1 and G_2 be two planar graphs with given combinatorial embeddings inducing the same embedding on their intersection G, that is, a SEFE is given

with the input. Do G_1 and G_2 admit drawings with few bends on a small grid respecting the given SEFE?

12. There are many open problems in the CSE setting. A particularly interesting one concerns pairs of trees. It is known that two n-vertex trees without mapping (1-colored) have a simultaneous geometric embedding (any set of n points in convex position suffices). It is also known that at the other extreme when the mapping is given (n-colored) such geometric embedding may not exist. However, the problem is open for any number of colors $c \in \{2, \dots, n-1\}$.

13. Similarly to the previous problem, the status of the tree-path CSE problem is open for any number of colors $c \in \{3, \dots, n-1\}$.

References

[AAC+13]   Soroush Alamdari, Patrizio Angelini, Timothy M. Chan, Giuseppe Di Battista, Fabrizio Frati, Anna Lubiw, Maurizio Patrignani, Vincenzo Roselli, Sahil Singla, and Bryan T. Wilkinson. Morphing planar graph drawings with a polynomial number of steps. In *Proceedings of the Twenty-Fourth Annual ACM-SIAM Symposium on Discrete Algorithms*, SODA '13. ACM, 2013.

[ABF+12]   Patrizio Angelini, Giuseppe Di Battista, Fabrizio Frati, Maurizio Patrignani, and Ignaz Rutter. Testing the simultaneous embeddability of two graphs whose intersection is a biconnected or a connected graph. *Journal of Discrete Algorithms*, 14(0):150–172, 2012.

[ABKS12]   Evmorfia Argyriou, Michael Bekos, Michael Kaufmann, and Antonios Symvonis. Geometric RAC simultaneous drawings of graphs. In Joachim Gudmundsson, Julián Mestre, and Taso Viglas, editors, *Computing and Combinatorics*, volume 7434 of *Lecture Notes in Computer Science*, pages 287–298. Springer Berlin Heidelberg, 2012.

[ADF+10]   Patrizio Angelini, Giuseppe Di Battista, Fabrizio Frati, Vít Jelínek, Jan Kratochvíl, Maurizio Patrignani, and Ignaz Rutter. Testing planarity of partially embedded graphs. In *Proceedings of the Twenty-First Annual ACM-SIAM Symposium on Discrete Algorithms*, SODA '10, pages 202–221. Society for Industrial and Applied Mathematics, 2010.

[ADF11]   Patrizio Angelini, Giuseppe Di Battista, and Fabrizio Frati. Simultaneous embedding of embedded planar graphs. In Takao Asano, Shin-ichi Nakano, Yoshio Okamoto, and Osamu Watanabe, editors, *Algorithms and Computation*, volume 7074 of *Lecture Notes in Computer Science*, pages 271–280. Springer Berlin / Heidelberg, 2011.

[AGKN12]   Patrizio Angelini, Markus Geyer, Michael Kaufmann, and Daniel Neuwirth. On a tree and a path with no geometric simultaneous embedding. *Journal of Graph Algorithms and Applications*, 16(1):37–83, 2012.

[AP13]   D. Archambault and H. Purchase. Mental map preservation helps user orientation in dynamic graphs. In *Graph Drawing*, Lecture Notes in Computer Science. Springer Berlin / Heidelberg, 2013. To appear.

[ASS93]   Boris Aronov, Raimund Seidel, and Diane Souvaine. On compatible triangulations of simple polygons. *Computational Geometry*, 3(1):27—35, 1993.

[BCD+07]   Peter Brass, Eowyn Cenek, Christian A. Duncan, Alon Efrat, Cesim Erten, Dan Ismailescu, Stephen G. Kobourov, Anna Lubiw, and Joseph S. B. Mitchell. On simultaneous planar graph embeddings. *Computational Geometry: Theory and Applications*, 36(2):117–130, 2007.

[BEEB+11]   Ulrik Brandes, Cesim Erten, Alejandro Estrella-Balderrama, J. Joseph Fowler, Fabrizio Frati, Markus Geyer, Carsten Gutwenger, Seok-Hee Hong, Michael Kaufmann, Stephen G. Kobourov, Giuseppe Liotta, Petra Mutzel, and Antonios Symvonis. Colored simultaneous geometric embeddings and universal pointsets. *Algorithmica*, 60(3):569–592, 2011.

[BFG+97]   Stina S. Bridgeman, Jody Fanto, Ashim Garg, Roberto Tamassia, and Luca Vismara. Interactivegiotto: An algorithm for interactive orthogonal graph drawing. In Giuseppe Di Battista, editor, *Proceedings of the 5th*

International Symposium on Graph Drawing (GD'97), volume 1353 of Lecture Notes in Computer Science, pages 303–308. Springer, 1997.

[BK79]   Frank Bernhart and Paul C Kainen. The book thickness of a graph. Journal of Combinatorial Theory, Series B, 27(3):320–331, 1979.

[BLS06]   Therese Biedl, Anna Lubiw, and Michael Spriggs. Morphing planar graphs while preserving edge directions. In Patrick Healy and Nikola Nikolov, editors, Graph Drawing, volume 3843 of Lecture Notes in Computer Science, pages 13–24. Springer Berlin / Heidelberg, 2006.

[BR13a]   Thomas Bläsius and Ignaz Rutter. Disconnectivity and relative positions in simultaneous embeddings. In Graph Drawing, Lecture Notes in Computer Science. Springer Berlin / Heidelberg, 2013. To appear.

[BR13b]   Thomas Bläsius and Ignaz Rutter. Simultaneous PQ-ordering with applications to constrained embedding problems. In Proceedings of the Twenty-Fourth Annual ACM-SIAM Symposium on Discrete Algorithms, SODA '13. ACM, 2013.

[BS93]   Graham R. Brightwell and Edward R. Scheinerman. Representations of planar graphs. SIAM Journal on Discrete Mathematics, 6(2):214–229, 1993.

[BT98]   Stina Bridgeman and Roberto Tamassia. Difference metrics for interactive orthogonal graph drawing algorithms. In Sue H. Whitesides, editor, Proceedings of the 6th International Symposium on Graph Drawing (GD'98), volume 1547 of Lecture Notes in Computer Science, pages 51–71. Springer, 1998.

[BW97]   Ulrik Brandes and Dorothea Wagner. A bayesian paradigm for dynamic graph layout. In Giuseppe Di Battista, editor, Proceedings of the 5th International Symposium on Graph Drawing (GD'97), volume 1353 of Lecture Notes in Computer Science, pages 236–247. Springer, 1997.

[Cai44]   S. S. Cairns. Deformations of plane rectilinear complexes. The American Mathematical Monthly, 51(5):247–252, 1944.

[CEBFK09] Justin Cappos, Alejandro Estrella-Balderrama, J. Joseph Fowler, and Stephen G. Kobourov. Simultaneous graph embedding with bends and circular arcs. Computational Geometry: Theory and Applications, 42(2):173–182, 2009.

[Cho34]   Chaim Chojnacki (Haim Hanani). Über wesentlich unplättbare kurven im dreidimensionalen raume. Fundamenta Mathematicae, 23:135–142, 1934.

[CJS08]   Markus Chimani, Michael Jünger, and Michael Schulz. Crossing minimization meets simultaneous drawing. In IEEE Pacific Visualisation Symposium, pages 33–40, 2008.

[CKN+03] Christian Collberg, Stephen Kobourov, Jasvir Nagra, Jacob Pitts, and Kevin Wampler. A system for graph-based visualizations of the evolution of software. In Proccedings of the Symposium on Visualization, pages 77–86, 212–213. ACM, 2003.

[CLR87]   Fan R. K. Chung, Frank Thomson Leighton, and Arnold L. Rosenberg. Embedding graphs in books: a layout problem with applications to VLSI design. SIAM Journal on Algebraic and Discrete Methods, 8(1):33–58, 1987.

[CN89]   Norishige Chiba and Takao Nishizeki. The hamiltonian cycle problem is
             linear-time solvable for 4-connected planar graphs. *Journal of Algorithms*,
             10(2):187–211, 1989.

[CvKL⁺11]   Sergio Cabello, Marc J. van Kreveld, Giuseppe Liotta, Henk Meijer, Bet-
             tina Speckmann, and Kevin Verbeek. Geometric simultaneous embeddings
             of a graph and a matching. *Journal of Graph Algorithms and Applications*,
             15(1):79–96, 2011.

[DDLW06]   Emilio Di Giacomo, Walter Didimo, Giuseppe Liotta, and Stephen K.
             Wismath. Book embeddability of series-parallel digraphs. *Algorithmica*,
             45(4):531–547, 2006.

[DDvK⁺09]   Emilio Di Giacomo, Walter Didimo, Marc van Kreveld, Giuseppe Liotta,
             and Bettina Speckmann. Matched drawings of planar graphs. *Journal of
             Graph Algorithms and Applications*, 13(3):423–445, 2009.

[DEH00]   M. B. Dillencourt, D. Eppstein, and D. S. Hirschberg. Geometric thickness
             of complete graphs. *Journal of Graph Algorithms and Applications*, 4(3):5–
             17, 2000.

[DEK04]   Christian A. Duncan, David Eppstein, and Stephen G. Kobourov. The
             geometric thickness of low degree graphs. In *Proceedings of the 20th An-
             nual Symposium on Computational Geometry*, SCG '04, pages 340–346.
             ACM, 2004.

[DETT99]   G. Di Battista, P. Eades, R. Tamassia, and I. G. Tollis. *Graph Drawing:
             Algorithms for the Visualization of Graphs*. Prentice Hall, 1999.

[dFPP90]   H. de Fraysseix, J. Pach, and R. Pollack. How to draw a planar graph on
             a grid. *Combinatorica*, 10(1):41–51, 1990.

[DG02]   Stephan Diel and Carsten Görg. Graphs, they are changing – dynamic
             graph drawing for a sequence of graphs. In Michael T. Goodrich and
             Stephen G. Kobourov, editors, *Proceedings of the 10th International Sym-
             posium on Graph Drawing (GD'02)*, volume 2528 of *Lecture Notes in Com-
             puter Science*, pages 23–31. Springer, 2002.

[DL05]   Emilio Di Giacomo and Giuseppe Liotta. A note on simultaneous embed-
             ding of planar graphs. In *EuroCG*, pages 207–210, 2005.

[DL07]   Emilio Di Giacomo and Giuseppe Liotta. Simultaneous embedding of
             outerplanar graphs, paths, and cycles. *International Journal of Compu-
             tational Geometry and Applications*, 17(2):139–160, 2007.

[EBFK10]   Alejandro Estrella-Balderrama, J. Joseph Fowler, and Stephen G.
             Kobourov. GraphSET, a tool for simultaneous graph drawing. *Software:
             Practice and Experience*, 40(10):849–863, 2010.

[EBGJ⁺08]   Alejandro Estrella-Balderrama, Elisabeth Gassner, Michael Jünger, Meri-
             jam Percan, Marcus Schaefer, and Michael Schulz. Simultaneous geomet-
             ric graph embeddings. In *Graph Drawing*, volume 4875 of *Lecture Notes
             in Computer Science*, pages 280–290. Springer Berlin / Heidelberg, 2008.

[EK05a]   Cesim Erten and Stephen Kobourov. Simultaneous embedding of planar
             graphs with few bends. In Jnos Pach, editor, *Graph Drawing*, volume 3383
             of *Lecture Notes in Computer Science*, pages 195–205. Springer Berlin /
             Heidelberg, 2005.

[EK05b]   Cesim Erten and Stephen G. Kobourov. Simultaneous embedding of a planar graph and its dual on the grid. *Theory of Computing Systems*, 38(3):313–327, 2005.

[EKLN05]   Cesim Erten, Stephen G. Kobourov, Vu Le, and Armand Navabi. Simultaneous graph drawing: Layout algorithms and visualization schemes. *Journal of Graph Algorithms and Applications*, 9(1):165–182, 2005.

[EKP04a]   C. Erten, S. G. Kobourov, and C. Pitta. Morphing planar graphs. In *Proceedings of the Twentieth Annual Symposium on Computational Geometry*, SCG '04, pages 451–452. ACM, 2004.

[EKP04b]   Cesim Erten, Stephen Kobourov, and Chandan Pitta. Intersection-free morphing of planar graphs. In Giuseppe Liotta, editor, *Graph Drawing*, volume 2912 of *Lecture Notes in Computer Science*, pages 320–331. Springer Berlin / Heidelberg, 2004.

[Fár48]   I. Fáry. On straight lines representation of planar graphs. *Acta Scientiarum Mathematicarum*, 11:229–233, 1948.

[FE02]   Carsten Friedrich and Peter Eades. Graph drawing in motion. *Journal of Graph Algorithms and Applications*, 6(3):353–370, 2002.

[FG99]   Michael S. Floater and Craig Gotsman. How to morph tilings injectively. *Journal of Computational and Applied Mathematics*, 101(12):117–129, 1999.

[FGJ+09]   J. Joseph Fowler, Carsten Gutwenger, Michael Jünger, Petra Mutzel, and Michael Schulz. An SPQR-tree approach to decide special cases of simultaneous embedding with fixed edges. In *Graph Drawing*, volume 5417 of *Lecture Notes in Computer Science*, pages 157–168. Springer Berlin / Heidelberg, 2009.

[FH02]   Carsten Friedrich and Michael Houle. Graph drawing in motion ii. In Petra Mutzel, Michael Jnger, and Sebastian Leipert, editors, *Graph Drawing*, volume 2265 of *Lecture Notes in Computer Science*, pages 122–125. Springer Berlin / Heidelberg, 2002.

[FJKS11]   J. Joseph Fowler, Michael Jünger, Stephen G. Kobourov, and Michael Schulz. Characterizations of restricted pairs of planar graphs allowing simultaneous embedding with fixed edges. *Computational Geometry: Theory and Applications*, 44(8):385–398, 2011.

[FK08]   J. Fowler and Stephen Kobourov. Characterization of unlabeled level planar graphs. In Seok-Hee Hong, Takao Nishizeki, and Wu Quan, editors, *Graph Drawing*, volume 4875 of *Lecture Notes in Computer Science*, pages 37–49. Springer Berlin / Heidelberg, 2008.

[FKK09]   Fabrizio Frati, Michael Kaufmann, and Stephen G. Kobourov. Constrained simultaneous and near-simultaneous embeddings. *Journal of Graph Algorithms and Applications*, 13(3):447–465, 2009.

[Flo97]   Michael S. Floater. Parametrization and smooth approximation of surface triangulations. *Computer Aided Geometric Design*, 14:231–250, 1997.

[Fra07]   Fabrizio Frati. Embedding graphs simultaneously with fixed edges. In *Graph Drawing*, volume 4372 of *Lecture Notes in Computer Science*, pages 108–113. Springer Berlin / Heidelberg, 2007.

[FT08]   Yaniv Frishman and Ayellet Tal. Onlyne dynamic graph drawing. *IEEE Transactions on Visualizations and Computer Graphics*, 14(4):727–740, 2008.

[Gav74]   Fănică Gavril. The intersection graphs of subtrees in trees are exactly the chordal graphs. *Journal of Combinatorial Theory, Series B*, 16(1):47–56, 1974.

[GBPD04]   Carsten Görg, Peter Birke, Mathias Pohl, and Stephan Diel. Dynamic graph drawing of sequences of orthogonal and hierarchical graphs. In János Pach, editor, *Proceedings of the 12th International Symposium on Graph Drawing (GD'04)*, volume 3383 of *Lecture Notes in Computer Science*, pages 228–238. Springer, 2004.

[GHL⁺09]   Luca Grilli, Seok-Hee Hong, Giuseppe Liotta, Henk Meijer, and Stephen Wismath. Matched drawability of graph pairs and of graph triples. In *WALCOM: Algorithms and Computation*, volume 5431 of *Lecture Notes in Computer Science*, pages 322–333. Springer Berlin / Heidelberg, 2009.

[GJP⁺06]   Elisabeth Gassner, Michael Jünger, Merijam Percan, Marcus Schaefer, and Michael Schulz. Simultaneous graph embeddings with fixed edges. In Fedor Fomin, editor, *Graph-Theoretic Concepts in Computer Science*, volume 4271 of *Lecture Notes in Computer Science*, pages 325–335. Springer Berlin / Heidelberg, 2006.

[GKV09]   Markus Geyer, Michael Kaufmann, and Imrich Vrto. Two trees which are self-intersecting when drawn simultaneously. *Discrete Mathematics*, 309(7):1909–1916, 2009.

[GRU83]   Martin Charles Golumbic, Doron Rotem, and Jorge Urrutia. Comparability graphs and intersection graphs. *Discrete Mathematics*, 43(1):37–46, 1983.

[HJL10]   Bernhard Haeupler, Krishnam Jampani, and Anna Lubiw. Testing simultaneous planarity when the common graph is 2-connected. In Otfried Cheong, Kyung-Yong Chwa, and Kunsoo Park, editors, *Algorithms and Computation*, volume 6507 of *Lecture Notes in Computer Science*, pages 410–421. Springer Berlin / Heidelberg, 2010.

[HN09]   Seok-Hee Hong and Hiroshi Nagamochi. Two-page book embedding and clustered graph planarity. Technical Report 2009-004, Department of Applied Mathematics & Physics, Kyoto University, 2009.

[HT74]   J. Hopcroft and R. E. Tarjan. Efficient planarity testing. *Journal of the ACM*, 21(4):549–568, 1974.

[HT08]   Bernhard Haeupler and Robert E. Tarjan. Planarity algorithms via PQ-trees (extended abstract). *Electronic Notes in Discrete Mathematics*, 31:143–149, 2008. The International Conference on Topological and Geometric Graph Theory.

[JKR11]   Vít Jelínek, Jan Kratochvíl, and Ignaz Rutter. A Kuratowski-type theorem for planarity of partially embedded graphs. In *Proceedings of the 27th Annual ACM Symposium on Computational Geometry (SoCG'11)*, pages 107–116. ACM, 2011.

[JL09]   Krishnam Jampani and Anna Lubiw. The simultaneous representation problem for chordal, comparability and permutation graphs. In Frank Dehne, Marina Gavrilova, Jrg-Rdiger Sack, and Csaba Tth, editors, *Al-*

gorithms and Data Structures, volume 5664 of *Lecture Notes in Computer Science*, pages 387–398. Springer Berlin / Heidelberg, 2009.

[JL10]   Krishnam Jampani and Anna Lubiw. Simultaneous interval graphs. In Otfried Cheong, Kyung-Yong Chwa, and Kunsoo Park, editors, *Algorithms and Computation*, volume 6506 of *Lecture Notes in Computer Science*, pages 206–217. Springer Berlin / Heidelberg, 2010.

[JS09]   Michael Jünger and Michael Schulz. Intersection graphs in simultaneous embedding with fixed edges. *Journal of Graph Algorithms and Applications*, 13(2):205–218, 2009.

[Kam06]   Frank Kammer. Simultaneous embedding with two bends per edge in polynomial area. In Lars Arge and Rusins Freivalds, editors, *Algorithm Theory SWAT 2006*, volume 4059 of *Lecture Notes in Computer Science*, pages 255–267. Springer Berlin / Heidelberg, 2006.

[KKV11]   Pavel Klavík, Jan Kratochvíl, and Tomáš Vyskočil. Extending partial representations of interval graphs. In *Proceedings of the 8th Annual Conference on Theory and Applications of Models of Computation*, TAMC'11, pages 276–285. Springer-Verlag, 2011.

[KP05]   Stephen G. Kobourov and Chandan Pitta. An interactive multi-user system for simultaneous graph drawing. In *Graph Drawing*, volume 3383 of *Lecture Notes in Computer Science*, pages 492–501. Springer Berlin / Heidelberg, 2005.

[KW02]   Michael Kaufmann and Roland Wiese. Embedding vertices at points: Few bends suffice for planar graphs. *Journal of Graph Algorithms and Applications*, 6(1):115–129, 2002.

[LP08]   Anna Lubiw and Mark Petrick. Morphing planar graph drawings with bent edges. *Electronic Notes in Discrete Mathematics*, 31(0):45–48, 2008.

[LPS06]   Anna Lubiw, Mark Petrick, and Michael Spriggs. Morphing orthogonal planar graph drawings. In *Proceedings of the Seventeenth Annual ACM-SIAM Symposium on Discrete Algorithms*, SODA '06, pages 222–230. ACM, 2006.

[MELS95]   Kazuo Misue, Peter Eades, Wei Lai, and Kozo Sugiyama. Layout adjustment and the mental map. *Journal of Visual Languages and Computing*, 6:183–210, 1995.

[Moh97]   Bojan Mohar. Circle packings of maps in polynomial time. *European Journal of Combinatorics*, 18(7):785–805, 1997.

[MOS98]   Petra Mutzel, Thomas Odenthal, and Mark Scharbrodt. The thickness of graphs: A survey. *Graphs and Combinatorics*, 14:59–73, 1998.

[MR98]   B. Mohar and P. Rosenstiehl. Tessellation and visibility representations of maps on the torus. *Discrete & Computational Geometry*, 19:249–263, 1998.

[NW02]   Stephen C. North and Gordon Woodhall. Online hierarchical graph drawing. In Petra Mutzel, Michael Jünger, and Sebastian Leipert, editors, *Proceedings of the 9th International Symposium on Graph Drawing (GD'01)*, volume 2265 of *Lecture Notes in Computer Science*, pages 232–246. Springer, 2002.

[PHG07]   Helen C. Purchase, Eve Hoggan, and Carsten Görg. How important is the "mental map"? – an empirical investigation of a dynamic graph layout algorithm. In Michael Kaufmann and Dorothea Wagner, editors, *Proceedings of the 14th International Symposium on Graph Drawing (GD'06)*, volume 4372 of *Lecture Notes in Computer Science*, pages 184–195. Springer, 2007.

[PS08]   Helen C. Purchase and Amanjit Samra. Extremes are better: Investigating mental map preservation in dynamic graphs. In G. Stapleton, J. Howse, and J. Lee, editors, *Proceeding of the 5th International Symposium on Diagrammatic Representation and Inference (DIAGRAMS'08)*, volume 5223 of *Lecture Notes in Artificial Intelligence*, pages 60–73. Springer, 2008.

[PW98]   János Pach and Rephael Wenger. Embedding planar graphs at fixed vertex locations. In Sue Whitesides, editor, *Graph Drawing*, volume 1547 of *Lecture Notes in Computer Science*, pages 263–274. Springer Berlin / Heidelberg, 1998.

[Sch13]   Marcus Schaefer. Toward a theory of planarity: Hanani-tutte and planarity variants. In *Graph Drawing*, Lecture Notes in Computer Science. Springer Berlin / Heidelberg, 2013. To appear.

[SG01]   Vitaly Surazhsky and Craig Gotsman. Controllable morphing of compatible planar triangulations. *ACM Transactions on Graphics*, 20(4):203–231, 2001.

[SG03]   Vitaly Surazhsky and Craig Gotsman. Intrinsic morphing of compatible triangulations. *International Journal of Shape Modeling*, 9(2):191–201, 2003.

[SMM13]   A. Sallaberry, C. Muelder, and K.-L. Ma. Clustering, visualizing, and navigating for large dynamic graphs. In *Graph Drawing*, Lecture Notes in Computer Science. Springer Berlin / Heidelberg, 2013. To appear.

[Ste51]   S. K. Stein. Convex maps. *Proceedings of the American Mathematical Society*, 2(3):464–466, 1951.

[STT81]   Kozo Sugiyama, Shojiro Tagawa, and Mitsuhiko Toda. Methods for visual understanding of hierarchical system structures. *IEEE Transactions on Systems, Man and Cybernetics*, 11(2):109–125, 1981.

[Tho83]   Carsten Thomassen. Deformations of plane graphs. *Journal of Combinatorial Theory, Series B*, 34(3):244–257, 1983.

[TT89]   R. Tamassia and I. G. Tollis. Tessellation representations of planar graphs. In *Proceedings of the 27th Annual Allerton Conference on Communication, Control, and Computing*, pages 48–57, 1989.

[Tut60]   W. T. Tutte. Convex representations of graphs. *Proceedings of the London Mathematical Society*, 10:304–320, 1960.

[Tut63]   W. T. Tutte. How to draw a graph. *Proceedings of the London Mathematical Society*, 13:743–768, 1963.

[Tut70]   W. T. Tutte. Toward a theory of crossing numbers. *Journal of Combinatorial Theory*, 8(1):45–53, 1970.

[Wag36]   K. Wagner. Bemerkungen zum Vierfarbenproblem. *Jahresbericht der Deutschen Mathematiker-Vereinigung*, 46:26–32, 1936.

[Yan89]   Mihalis Yannakakis. Embedding planar graphs in four pages. *Journal of Computer and System Sciences*, 38(1):36–67, 1989.

[ZH06]   Huaming Zhang and Xin He. On simultaneous straight-line grid embedding of a planar graph and its dual. *Information Processing Letters*, 99(1):1–6, 2006.

12

Force-Directed Drawing Algorithms

Stephen G. Kobourov
University of Arizona

12.1 Introduction

Some of the most flexible algorithms for calculating layouts of simple undirected graphs belong to a class known as force-directed algorithms. Also known as spring embedders, such algorithms calculate the layout of a graph using only information contained within the structure of the graph itself, rather than relying on domain-specific knowledge. Graphs drawn with these algorithms tend to be aesthetically pleasing, exhibit symmetries, and tend to produce crossing-free layouts for planar graphs. In this chapter we will assume that the input graphs are simple, connected, undirected graphs and their layouts are straight-line drawings. Excellent surveys of this topic can be found in Di Battista *et al.* [DETT99] Chapter 10 and Brandes [Bra01].

Going back to 1963, the graph drawing algorithm of Tutte [Tut63] is one of the first *force-directed* graph drawing methods based on *barycentric representations*. More traditionally, the spring layout method of Eades [Ead84] and the algorithm of Fruchterman and Reingold [FR91] both rely on spring forces, similar to those in Hooke's law. In these methods, there are repulsive forces between all nodes, but also attractive forces between nodes that are adjacent.

Alternatively, forces between the nodes can be computed based on their graph theoretic distances, determined by the lengths of shortest paths between them. The algorithm of Kamada and Kawai [KK89] uses spring forces proportional to the graph theoretic distances. In general, force-directed methods define an objective function which maps each graph layout into a number in $\mathcal{R}^+$ representing the energy of the layout. This function is defined in such a way that low energies correspond to layouts in which adjacent nodes are near some pre-specified distance from each other, and in which non-adjacent nodes are well-spaced. A

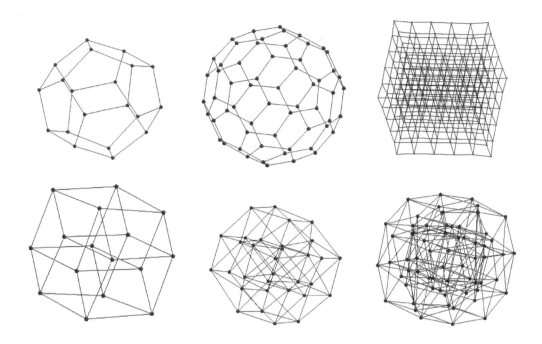

Figure 12.1 Examples of drawings obtained with force-directed algorithms. First row: small graphs: dodecahedron (20 vertices), C60 bucky ball (60 vertices), 3D cube mesh (216 vertices). Second row: Cubes in 4D, 5D and 6D [GK02].

layout for a graph is then calculated by finding a (often local) minimum of this objective function; see Figure 12.1.

The utility of the basic force-directed approach is limited to small graphs and results are poor for graphs with more than a few hundred vertices. There are multiple reasons why traditional force-directed algorithms do not perform well for large graphs. One of the main obstacles to the scalability of these approaches is the fact that the physical model typically has many local minima. Even with the help of sophisticated mechanisms for avoiding local minima the basic force-directed algorithms are not able to consistently produce good layouts for large graphs. Barycentric methods also do not perform well for large graphs mainly due to resolution problems: for large graphs the minimum vertex separation tends to be very small, leading to unreadable drawings.

The late 1990s saw the emergence of several techniques extending the functionality of force-directed methods to graphs with tens of thousands and even hundreds of thousands of vertices. One common thread in these approaches is the multi-level layout technique, where the graph is represented by a series of progressively simpler structures and laid out in reverse order: from the simplest to the most complex. These structures can be coarser graphs (as in the approach of Hadany and Harel [HH01], Harel and Koren [HK02], and Walshaw [Wal03], or vertex filtrations as in the approach of Gajer, Goodrich, and Kobourov [GGK04].

The classical force-directed algorithms are restricted to calculating a graph layout in Euclidean geometry, typically $\mathcal{R}^2$, $\mathcal{R}^3$, and, more recently, $\mathcal{R}^n$ for larger values of n. There are, however, cases where Euclidean geometry may not be the best option: Certain graphs may be known to have a structure which would be best realized in a different geometry, such as on the surface of a sphere or on a torus. In particular, 3D mesh data can be parameterized on the sphere for texture mapping or graphs of genus one can be embedded on a torus without crossings. Furthermore, it has also been noted that certain non- Euclidean geometries, specifically hyperbolic geometry, have properties that are particularly well suited to the layout and visualization of large classes of graphs [LRP95, Mun97]. With this in mind, Kobourov and Wampler describe extensions of the force-directed algorithms to Riemannian spaces [KW05].

12.2 Spring Systems and Electrical Forces

The 1984 algorithm of Eades [Ead84] targets graphs with up to 30 vertices and uses a mechanical model to produce "aesthetically pleasing" 2D layouts for plotters and CRT screens. The algorithm is succinctly summarized as follows:

> *To embed a graph we replace the vertices by steel rings and replace each edge with a spring to form a mechanical system. The vertices are placed in some initial layout and let go so that the spring forces on the rings move the system to a minimal energy state. Two practical adjustments are made to this idea: firstly, logarithmic strength springs are used; that is, the force exerted by a spring is:*

$$c_1 * \log(d/c_2),$$

> *where d is the length of the spring, and c_1 and c_2 are constants. Experience shows that Hookes Law (linear) springs are too strong when the vertices are far apart; the logarithmic force solves this problem. Note that the springs exert no force when $d = c_2$. Secondly, we make nonadjacent vertices repel each other. An inverse square law force,*

$$c_3/d^2,$$

> *where c_3 is constant and d is the distance between the vertices, is suitable. The mechanical system is simulated by the following algorithm.*

---

algorithm SPRING(G:graph);
place vertices of G in random locations;
repeat M times
    calculate the force on each vertex;
    move the vertex $c_4 * (force on vertex)$
draw graph on CRT or plotter.

> *The values $c_1 = 2$, $c_2 = 1$, $c_3 = 1$, $c_4 = 0.1$, are appropriate for most graphs. Almost all graphs achieve a minimal energy state after the simulation step is run 100 times, that is, $M = 100$.*

This excellent description encapsulates the essence of spring algorithms and their natural simplicity, elegance, and conceptual intuitiveness. The goals behind "aesthetically pleasing" layouts were initially captured by the two criteria: "all the edge lengths ought to be the same, and the layout should display as much symmetry as possible."

The 1991 algorithm of Fruchterman and Reingold added "even vertex distribution" to the earlier two criteria and treats vertices in the graph as "atomic particles or celestial bodies, exerting attractive and repulsive forces from one another." The attractive and repulsive forces are redefined to

$$f_a(d) = d^2/k, \qquad\qquad f_r(d) = -k^2/d,$$

in terms of the distance d between two vertices and the optimal distance between vertices k defined as

$$k = C\sqrt{\frac{area}{number\ of\ vertices}}.$$

This algorithm is similar to that of Eades in that both algorithms compute attractive forces between adjacent vertices and repulsive forces between all pairs of vertices. The algorithm of Fruchterman and Reingold adds the notion of "temperature" which could be used as follows: "the temperature could start at an initial value (say one tenth the width of the frame) and decay to 0 in an inverse linear fashion." The temperature controls the displacement of vertices so that as the layout becomes better, the adjustments become smaller. The use of temperature here is a special case of a general technique called *simulated annealing*, whose use in force-directed algorithms is discussed later in this chapter. The pseudo-code for the algorithm by Fruchterman and Reingold, shown in Figure 12.2 provides further insight into the workings of a spring-embedder.

Each iteration the basic algorithm computes $O(|E|)$ attractive forces and $O(|V|^2)$ repulsive forces. To reduce the quadratic complexity of the repulsive forces, Fruchterman and Reingold suggest using a grid variant of their basic algorithm, where the repulsive forces between distant vertices are ignored. For sparse graphs, and with uniform distribution of the vertices, this method allows a $O(|V|)$ time approximation to the repulsive forces calculation. This approach can be thought of as a special case of the multi-pole technique introduced in n-body simulations [Gre88] whose use in force-directed algorithms will be further discussed later in this chapter.

As in the paper by Eades [Ead84] the graphs considered by Fruchterman and Reingold are small graphs with less than 40 vertices. The number of iterations of the main loop is also similar at 50.

12.3   The Barycentric Method

Historically, Tutte's 1963 barycentric method [Tut63] is the first "force-directed" algorithm for obtaining a straight-line, crossings free drawing for a given 3-connected planar graph. Unlike almost all other force-directed methods, Tutte's guarantees that the resulting drawing is crossings-free; moreover, all faces of the drawing are convex.

The idea behind Tutte's algorithm, shown in Figure 12.3, is that if a face of the planar graph is fixed in the plane, then suitable positions for the remaining vertices can be found by solving a system of linear equations, where each vertex position is represented as a convex combination of the positions of its neighbors. This algorithm be considered a force-directed method as summarized in Di Battista *et al.* [DETT99].

In this model the force due to an edge (u, v) is proportional to the distance between vertices u and v and the springs have ideal length of zero; there are no explicit repulsive forces. Thus the force at a vertex v is described by

$$F(v) = \sum_{(u,v)\in E} (p_u - p_v),$$

where p_u and p_v are the positions of vertices u and v. As this function has a trivial minimum with all vertices placed in the same location, the vertex set is partitioned into fixed and free

area:= $W * L$; {W and L are the width and length of the frame}
$G := (V, E)$; {the vertices are assigned random initial positions}
$k := \sqrt{area/|V|}$;
function $f_a(x) := $ **begin return** x^2/k **end**;
function $f_r(x) := $ **begin return** k^2/x **end**;
for $i := 1$ **to** *iterations* **do begin**
    {calculate repulsive forces}
    for v **in** V **do begin**
        {each vertex has two vectors: *.pos* and *.disp*
        $v.disp := 0$;
        for u **in** V **do**
            if $(u \neq v)$ **then begin**
            {δ is the difference vector between the positions of the two vertices}
                $\delta := v.pos - u.pos$;
                $v.disp := v.disp + (\delta/|\delta|) * f_r(|\delta|)$
            end
    end
    {calculate attractive forces}
    for e **in** E **do begin**
        {each edges is an ordered pair of vertices *.v and .u*}
        $\delta := e.v.pos - e.u.pos$;
        $e.v.disp := e.v.disp - (\delta/|\delta|) * f_a(|\delta|)$;
        $e.u.disp := e.u.disp + (\delta/|\delta|) * f_a(|\delta|)$
    end
    {limit max displacement to temperature t and prevent from displacement
outside frame}
    for v **in** V **do begin**
        $v.pos := v.pos + (v.disp/|v.disp|) * \min(v.disp, t)$;
        $v.pos.x := \min(W/2, \max(-W/2, v.pos.x))$;
        $v.pos.y := \min(L/2, \max(-L/2, v.pos.y))$
    end
    {reduce the temperature as the layout approaches a better configuration}
    $t := cool(t)$
end

Figure 12.2 Pseudo-code for the algorithm by Fruchterman and Reingold [FR91].

vertices. Setting the partial derivatives of the force function to zero results in independent systems of linear equations for the x-coordinate and for the y-coordinate.

The equations in the for-loop are linear and the number of equations is equal to the number of the unknowns, which in turn is equal to the number of free vertices. Solving these equations results in placing each free vertex at the barycenter of its neighbors. The system of equations can be solved using the Newton-Raphson method. Moreover, the resulting solution is unique.

One significant drawback of this approach is the resulting drawing often has poor vertex resolution. In fact, for every $n > 1$, there exists a graph, such that the barycenter method computes a drawing with exponential area [EG95].

Barycenter-Draw

Input: $G = (V, E)$; a partition $V = V_0 \cup V_1$ of V into a set V_0 of at least three *fixed* vertices and a set V_1 of *free* vertices; a strictly convex polygon P with $|V_0|$ vertices

Output: a position p_v for each vertex of V, such that the fixed vertices form a convex polygon P.

1. Place each fixed vertex $u \in V_0$ at a vertex of P, and each free vertex at the origin.

2. **repeat**
   foreach free vertex $v \in V_1$ **do**

$$x_v = \frac{1}{deg(v)} \sum_{(u,v) \in E} x_u$$

$$y_v = \frac{1}{deg(v)} \sum_{(u,v) \in E} y_u$$

   until x_v and y_v converge for all free vertices v.

Figure 12.3　Tutte's barycentric method [Tut63]. Pseudo-code from [DETT99].

12.4　Graph Theoretic Distances Approach

The 1989 algorithm of Kamada and Kawai [KK89] introduced a different way of thinking about "good" graph layouts. Whereas the algorithms of Eades and Fruchterman-Reingold aim to keep adjacent vertices close to each other while ensuring that vertices are not too close to each other, Kamada and Kawai take graph theoretic approach:

> *We regard the desirable geometric (Euclidean) distance between two vertices in the drawing as the graph theoretic distance between them in the corresponding graph.*

In this model, the "perfect" drawing of a graph would be one in which the pair-wise geometric distances between the drawn vertices match the graph theoretic pairwise distances, as computed by an All-Pairs-Shortest-Path computation. As this goal cannot always be achieved for arbitrary graphs in 2D or 3D Euclidean spaces, the approach relies on setting up a spring system in such a way that minimizing the energy of the system corresponds to minimizing the difference between the geometric and graph distances. In this model there are no separate attractive and repulsive forces between pairs of vertices, but instead if a pair of vertices is (geometrically) closer/farther than their corresponding graph distance the vertices repel/attract each other. Let $d_{i,j}$ denote the shortest path distance between vertex i and vertex j in the graph. Then $l_{i,j} = L \times d_{i,j}$ is the ideal length of a spring between vertices i and j, where L is the desirable length of a single edge in the display. Kamada and Kawai suggest that $L = L_0/\max_{i<j} d_{i,j}$, where L_0 is the length of a side of the display area and $\max_{i<j} d_{i,j}$ is the diameter of the graph, i.e., the distance between the farthest pair of vertices. The strength of the spring between vertices i and j is defined as

$$k_{i,j} = K/d_{i,j}^2,$$

where K is a constant. Treating the drawing problem as localizing $|V| = n$ particles $p_1, p_2, \ldots, p_n$ in 2D Euclidean space, leads to the following overall energy function:

$$E = \sum_{i=1}^{n-1} \sum_{j=i+1}^{n} \frac{1}{2} k_{i,j} (|p_i - p_j| - l_{i,j})^2.$$

The coordinates of a particle p_i in the 2D Euclidean plane are given by x_i and y_i which allows us to rewrite the energy function as follows:

$$E = \sum_{i=1}^{n-1} \sum_{j=i+1}^{n} \frac{1}{2} k_{i,j} \left((x_i - x_j)^2 + (y_i - y_j)^2 + l_{i,j}^2 - 2l_{i,j} \sqrt{(x_i - x_j)^2 + (y_i - y_j)^2} \right).$$

The goal of the algorithm is to find values for the variables that minimize the energy function $E(x_1, x_2, \ldots, x_n, y_1, y_2, \ldots, y_n)$. In particular, at a local minimum all the partial derivatives are equal to zero, and which corresponds to solving $2n$ simultaneous non-linear equations. Therefore, Kamada and Kawai compute a stable position one particle p_m at a time. Viewing E as a function of only x_m and y_m a minimum of E can be computed using the Newton-Raphson method. At each step of the algorithm the particle p_m with the largest value of Δ_m is chosen, where

$$\Delta_m = \sqrt{\left(\frac{\partial E}{\partial x_m} \right)^2 + \left(\frac{\partial E}{\partial y_m} \right)^2}.$$

Pseudo-code for the algorithm by Kamada and Kawai is shown in Figure 12.4.

The algorithm of Kamada and Kawai is computationally expensive, requiring an All-Pair-Shortest-Path computation which can be done in $O(|V|^3)$ time using the Floyd-Warshall algorithm or in $O(|V|^2 \log |V| + |E||V|)$ using Johnson's algorithm; see the All-Pairs-Shortest-Path chapter in an algorithms textbook such as [CLRS90]. Furthermore, the algorithm requires $O(|V|^2)$ storage for the pairwise vertex distances. Despite the higher time and space complexity, the algorithm contributes a simple and intuitive definition of a "good" graph layout: A graph layout is good if the geometric distances between vertices closely correspond to the underlying graph distances.

12.5   Further Spring Refinements

Even before the 1984 algorithm of Eades, force-directed techniques were used in the context of VLSI layouts in the 1960s and 1970s [FCW67, QB79]. Yet, renewed interest in force-directed graph layout algorithms brought forth many new ideas in the 1990s. Frick, Ludwig, and Mehldau [FLM95] add new heuristics to the Fruchterman-Reingold approach. In particular, oscillation and rotations are detected and dealt with using local instead of global temperature. The following year Bruß and Frick [BF96] extended the approach to layouts directly in 3D Euclidean space. The algorithm of Cohen [Coh97] introduced the notion of an incremental layout, a precursor of the multi-scale methods described in Section 12.6.

compute $d_{i,j}$ for $1 \le i \ne j \le n$;
compute $l_{i,j}$ for $1 \le i \ne j \le n$;
compute $k_{i,j}$ for $1 \le i \ne j \le n$;
initialize $p_1, p_2, \ldots, p_n$;
while $(max_i \Delta_i > \epsilon)$
    let p_m be the particle satisfying $\Delta_m = max_i \Delta_i$;
    while $(\Delta_m > \epsilon)$
        compute δx and δy by solving the following system of equations:

$$\frac{\partial^2 E}{\partial x_m^2}(x_m^{(t)}, y_m^{(t)})\delta x + \frac{\partial^2 E}{\partial x_m \partial y_m}(x_m^{(t)}, y_m^{(t)})\delta y = -\frac{\partial E}{\partial x_m}(x_m^{(t)}, y_m^{(t)});$$

$$\frac{\partial^2 E}{\partial y_m \partial x_m}(x_m^{(t)}, y_m^{(t)})\delta x + \frac{\partial^2 E}{\partial y_m^2}(x_m^{(t)}, y_m^{(t)})\delta y = -\frac{\partial E}{\partial y_m}(x_m^{(t)}, y_m^{(t)})$$

$x_m := x_m + \delta x$;
$y_m := y_m + \delta y$;

Figure 12.4 Pseudo-code for the algorithm by Kamada and Kawai [KK89].

The 1997 algorithm of Davidson and Harel [DH96] adds additional constraints to the traditional force-directed approach in explicitly aiming to minimize the number of edge-crossings and keeping vertices from getting too close to non-adjacent edges. The algorithm uses the simulated annealing technique developed for large combinatorial optimization [KGV83]. Simulated annealing is motivated by the physical process of cooling molten materials. When molten steel is cooled too quickly it cracks and forms bubbles making it brittle. For better results, the steel must be cooled slowly and evenly and this process is known as annealing in metallurgy. With regard to force-directed algorithms, this process is simulated to find local minima of the energy function. Cruz and Twarog [CT96] extended the method by Davidson and Harel to three-dimensional drawings.

Genetic algorithms for force-directed placement have also been considered. Genetic algorithms are a commonly used search technique for finding approximate solutions to optimization and search problems. The technique is inspired by evolutionary biology in general and by inheritance, mutation, natural selection, and recombination (or crossover), in particular; see the survey by Vose [Vos99]. In the context of force-directed techniques for graph drawing, the genetic algorithms approach was introduced in 1991 by Kosak, Marks and Shieber [KMS91]. Other notable approaches in the direction include that of Branke, Bucher, and Schmeck [BBS97].

In the context of graph clustering, the *LinLog* model introduces an alternative energy model [Noa07]. Traditional energy models enforce small and uniform edge lengths, which often prevent the separation of nodes in different clusters. As a side effect, they tend to group nodes with large degree in the center of the layout, where their distance to the remaining nodes is relatively small. The node-repulsion LinLog and edge- repulsion LinLog models group nodes according to two well-known clustering criteria: the density of the cut [LR88] and the normalized cut [SM00].

12.6 Large Graphs

The first force-directed algorithms to produce good layouts for graphs with more than 1000 vertices is the 1999 algorithm of Hadany and Harel [HH01]. They introduced the multi-scale technique as a way to deal with large graphs and in the following year four related but independent force-directed algorithms for large graphs were presented at the Annual Symposium on Graph Drawing. We begin with Hadany and Harel's description on the *multi-scale method*:

> *A natural strategy for drawing a graph nicely is to first consider an abstraction, disregarding some of the graph's fine details. This abstraction is then drawn, yielding a "rough" layout in which only the general structure is revealed. Then the details are added and the layout is corrected. To employ such a strategy it is crucial that the abstraction retains essential features of the graph. Thus, one has to define the notion of coarse-scale representations of a graph, in which the combinatorial structure is significantly simplified but features important for visualization are well preserved. The drawing process will then "travel" between these representations, and introduce multi-scale corrections. Assuming we have already defined the multiple levels of coarsening, the general structure of our strategy is as follows:*
>
> 1. *Perform* fine-scale *relocations of vertices that yield a locally organized configuration.*
>
> 2. *Perform* coarse-scale *relocations (through local relocations in the coarse representations), correcting global disorders not found in stage 1.*
>
> 3. *Perform* fine-scale *relocations that correct local disorders introduced by stage 2.*

Hadany and Harel suggest computing the sequence of graphs by using edge contractions so as to preserve certain properties of the graph. In particular, the goal is to preserve three topological properties: cluster size, vertex degrees, and homotopy. For the coarse-scale relocations, the energy function for each graph in the sequence is that of Kamada and Kawai (the pairwise graph distances are compared to the geometric distances in the current layout). For the fine-scale relocations, the authors suggest using force-directed calculations as those of Eades [Ead84], Fruchterman-Reingold [FR91], or Kamada-Kawai [KK89]. While the asymptotic complexity of this algorithm is similar to that of the Kamada-Kawai algorithm, the multi-scale approach leads to good layouts for much larger graphs in reasonable time.

The algorithm of Harel and Koren [HK02] took force-directed algorithms to graphs with 15,000 vertices. This algorithm is similar to the algorithm of Hadany and Harel, yet uses a simpler coarsening process based on a k-centers approximation, and a faster fine-scale beautification. Given a graph $G = (V, E)$, the k-centers problem asks to find a subset of the vertex set $V' \subseteq V$ of size k, so as to minimize the maximum distance from a vertex to V': $\min u \in V \max_{u \in V, v \in V'} dist(u, v)$. While k-centers is an NP-hard problem, Harel and Koren use a straightforward and efficient 2-approximation algorithm that relies on Breadth-First Search [Hoc96]. The fine-scale vertex relocations are done using the Kamada-Kawai approach. The Harel and Koren algorithm is summarized in Figure 12.5.

Layout$(G(V, E))$
% Goal: Find L, a nice layout of G
% Constants:
% Rad[$= 7$] – determines radius of local neighborhoods
% Iterations[$= 4$] – determines number of iterations in local beautification
% Ratio[$= 3$] – ratio between number of vertices in two consecutive levels
% MinSize[$= 10$] – size of the coarsest graph
  Compute the all-pairs shortest path length: d_{VV}
  Set up a random layout L
  $k \leftarrow MinSize$
  while $k \leq |V|$ **do**
    $centers \leftarrow$**K-Centers**$(G(V, E), k)$
    $radius = \max_{v \in centers} \min_{u \in centers}\{d_{vu}\} * Rad$
    LocalLayout$(d_{centers \times centers}, L(centers), radius, Iterations)$
    for every $v \in V$ **do**
      $L(v) \in L(center(v)) + rand$
    $k \leftarrow kRatio$
  return L

K-Centers$(G(V, E), k)$
% Goal: Find a set $S \subseteq V$ of size k, such that $\max_{v \in V} \min_{s \in S}\{d_{sv}\}$ is minimized.
  $S \leftarrow \{v\}$ for some arbitrary $v \in V$
  for $i = 2$ to k **do**
    1. Find the vertex u farthest away from S
      (i.e., such that $\min_{s \in S}\{d_{us}\} \geq \min_{s \in S}\{d_{ws}\}, \forall w \in V$)
    2. $S \leftarrow S \cup \{u\}$
  return S

LocalLayout$(d_{V \times V}, L, k, Iterations)$
% Goal: Find a locally nice layout L by beautifying k-neighborhoods
% $d_{V \times V}$: all-pairs shortest path length
% L: initialized layout
% k: radius of neighborhoods
  for $i = 1$ to $Iterations * |V|$ **do**
    1. Choose the vertex v with the maximal Δ_v^k
    2. Compute δ_v^k as in Kamada-Kawai
    3. $L(v) \leftarrow L(v) + (\delta_v^k(x), \delta_v^k(y))$
  end

Figure 12.5 Pseudo-code for the algorithm by Harel and Koren [HK02].

MAIN ALGORITHM

create a filtration $\mathcal{V}: \ V_0 \supset V_1 \supset \ldots \supset V_k \supset \emptyset$
for $i = k$ **to** 0 **do**
    for each $v \in V_i - V_{i+1}$ **do**
        find vertex neighborhood $N_i(v), N_{i-1}(v), \ldots, N_0(v)$
        find initial position $pos[v]$ of v
    repeat *rounds* times
        for each $v \in V_i$ **do**
            compute local temperature $heat[v]$
            $disp[v] \leftarrow heat[v] \cdot \overrightarrow{F}_{N_i}(v)$
        for each $v \in V_i$ **do**
            $pos[v] \leftarrow pos[v] + disp[v]$
add all edges $e \in E$

Figure 12.6 Pseudo-code for the algorithm by Gajer *et al.* [GGK04].

The 2000 algorithm of Gajer *et al.* [GGK04], shown in Figure 12.6, is also a multi-scale force-directed algorithm but introduces several ideas to the realm of multi-scale force-directed algorithms for large graphs. Most importantly, this approach avoids the quadratic space and time complexity of previous force-directed approaches with the help of a simpler coarsening strategy. Instead of computing a series of coarser graphs from the given large graph $G = (V, E)$, Gajer *et al.* produce a *vertex filtration* $\mathcal{V}: \ V_0 \supset V_1 \supset \ldots \supset V_k \supset \emptyset$, where $V_0 = V(G)$ is the original vertex set of the given graph G. By restricting the number of vertices considered in relocating any particular vertex in the filtration and ensuring that the filtration has $O(\log|V|)$ levels an overall running time of $O(|V|\log^2|V|)$ is achieved. Filtrations based on graph centers (as in Harel and Koren [HK02]) and maximal independent sets are considered. $V = V_0 \supset V_1 \supset \ldots \supset V_k \supset \emptyset$, is a maximal independent set filtration of G if V_i is a maximal subset of V_{i-1} for which the graph distance between any pair of its elements is greater than or equal to 2^i.

In the GRIP system [GK02], Gajer *et al.* add to the filtration and neighborhood calculations of [GGK04]: they introduce the idea of realizing the graph in high-dimensional Euclidean space and obtaining 2D or 3D projections at the end. The algorithm also relies on intelligent initial placement of vertices based on graph theoretic distances, rather than on random initial placement. Finally, the notion of cooling is re-introduced in the context of multi-scale force-directed algorithms. The GRIP system produces high-quality layouts, as illustrated in Figure 12.7.

Another multilevel algorithm is that of Walshaw [Wal03]. Instead of relying on the Kamada-Kawai type force interactions, this algorithm extends the grid variant of Fruchterman-Reingold to a multilevel algorithm. The coarsening step is based on repeatedly collapsing maximally independent sets of edges, and the fine-scale refinements are based on Fruchterman-Reingold force calculations. This $O(|V|^2)$ algorithm is summarized in Figure 12.8.

The fourth 2000 multilevel force-directed algorithm is due to Quigley and Eades [QE00]. This algorithm relies on the Barnes-Hut n-body simulation method [BH86] and reduces repulsive force calculations to $O(|V|\log|V|)$ time instead of the usual $O(|V|^2)$. Similarly, the algorithm of Hu [Hu05] combines the multilevel approach with the n-bosy simulation method, and is implemented in the sfdp drawing engine of GraphViz [EGK+01].

One possible drawback to this approach is that the running time depends on the distribution of the vertices. Hachul and Jünger [HJ04] address this problem in their 2004 multilevel algorithm.

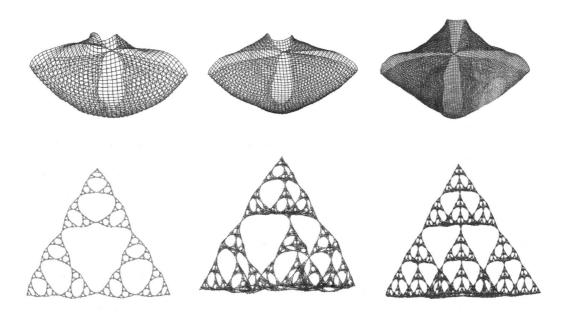

Figure 12.7 Drawings from GRIP. First row: knotted meshes of 1600, 2500, and 10000 vertices. Second row: Sierpinski graphs of order 7 (1,095 vertices), order 6 (2,050 vertices), 3D Sierpinski of order 7 (8,194 vertices) [GK02].

---

function $f_g(x, w)$:=**begin return** $-Cwk^2/x$ **end**
function $f_l(x, d, w)$:=**begin return** $\{(x - k)/d\} - f_g(x, w)$ **end**
$t := t_0$;
$Posn := NewPosn$;
while $(converged \neq 1)$ **begin**
    converged :=1;
    for $v \in V$ **begin**
        $OldPosn[v] = NewPosn[v]$
    end
    for $v \in V$ **begin**
        {initialize D, the vector of displacements of v}
        $D := 0$;
        {calculate global (repulsive) forces}
        for $u \in V, u \neq v$ **begin**
            $\Delta := Posn[u] - Posn[v]$;
            $D := D + (\Delta/|Delta|) * f_g(|\Delta|, |u|)$;
        end
        {calculate local (spring) forces }
        for $u \in \Gamma(v)$ **begin**
            $\Delta := Posn[u] - Posn[v]$;
            $D := D + (\Delta/|Delta|) * f_l(|\Delta|, |\Gamma(v)|, |u|)$;
        end
        {reposition v}
        $NewPosn[v] = NewPosn[v] + (D/|D|) * \min(t, |D|)$;
        $\Delta := NewPosn[v] - OldPosn[v]$;
        if $(|\Delta| > k \times tol)converged := 0$;
    end
    {reduce the temperature to reduce the maximum movement}
    $t := cool(t)$;
end

Figure 12.8 Pseudo-code for the algorithm by Walshaw [Wal03].

12.7   Stress Majorization

Methods that exploit fast algebraic operations offer another practical way to deal with large graphs. *Stress minimization* has been proposed and implemented in the more general setting of multidimensional scaling (MDS) [Kru64]. The function describing the stress is similar to the layout energy function of Kamada-Kawai from Section 12.4:

$$E = \sum_{i=1}^{n-1} \sum_{j=i+1}^{n} \frac{1}{2} k_{i,j} (|p_i - p_j| - l_{i,j})^2,$$

but here $k_{i,j}=1$ and $l_{i,j} = d_{i,j}$ is simply the graph theoretic distance. In their paper on graph drawing by stress minimization Gansner *et al.* [GKN04] point out that this particular formulation of the energy of the layout, or *stress function* has been already used to draw graphs as early as in 1980 [KS80]. What makes this particular stress function relevant to drawing large graphs is that it can be optimized better than with the local Newton-Raphson method or with gradient descent. Specifically, this stress function can be globally minimized via *majorization*. That is, unlike the energy function of Kamada-Kawai, the classical MDS stress function can be optimized via majorization which is guaranteed to converge.

The *strain* model, or classical scaling, is related to the stress model. In this setting a solution can be obtained via an eigen-decomposition of the adjacency matrix. Solving the full stress or strain model still requires computing all pairs shortest paths. Significant savings can be gained if we instead compute a good approximation. In PivotMDS Brandes and Pich [BP06] show that replacing the all-pairs-shortest path computation with a distance calculations from a few vertices in the graph is often sufficient, especially if combined with a solution to a sparse stress model.

When not all nodes are free to move, *constrained stress majorization* can be used to support additional constraints by, and treating the majorizing functions as a quadratic program [DKM09]. Planar graphs are of particular interest in graph drawing, and often force-directed graph drawing algorithms are used to draw them. While in theory any planar graph has a straight-line crossings-free drawing in the plane, force-directed algorithms do not guarantee such drawings.

Modifications to the basic force-directed functionality, with the aim of improving the layout quality for planar graphs, have also been considered. Harel and Sardas [HS98] improve an earlier simulated annealing drawing algorithm by Davidson and Harel [DH96]. The main idea is to obtain an initial plane embedding and then apply simulated annealing while not introducing any crossings. Overall their method significantly improved the aesthetic quality of the initial planar layouts, but at the expense of a significant increase in running time of $O(n^3)$, making it practical only for small graphs.

PrEd [Ber00] and ImPrEd [PSA11] are force-directed algorithms that improve already created drawings of a graph. PrEd [Ber00] extends the method of Fruchterman and Reingold [FR91] and can be used as a post-processing crossings-preserving optimization. In particular, PrEd takes some straight-line drawing as input and guarantees that no new edge crossings will be created (while preserving existing crossings, if any are present in the input drawing). Then the algorithm can be used to optimize a planar layout, while preserving its planarity and its embedding, or to improve a graph that has a meaningful initial set of edge crossings. To achieve this result, PrEd adds a phase where the maximal movement of each node is computed, and adds a repulsive force between (node, edge) pairs. The main aims of ImPrEd [PSA11] are to significantly reduce the running time of PrEd, achieve high aesthetics even for large and sparse graphs, and make the algorithm more stable and reliable with respect to the input parameters. This is achieved via improved spacing of the graph elements and an accelerated convergence of the drawing to its final configuration.

An alternative approach for modifying force-directed functionality is to use a preprocessing step rather than a random layout to initialize the algorithm. Experimental results indicate that combining a linear-time planar embedding step with a standard force-directed algorithm such as a Fruchterman-Reingold can lead to improved qualitative and quantitative results [FK12].

12.8 Non-Euclidean Approaches

Much of the work on non-Euclidean graph drawing has been done in hyperbolic space which offers certain advantages over Euclidean space; see Munzner [Mun97, MB96]. For example, in hyperbolic space it is possible to compute a layout for a complete tree with both uniform edge lengths and uniform distribution of nodes. Furthermore, some of the embeddings of hyperbolic space into Euclidean space naturally provide a fish-eye view of the space, which is useful for "focus+context" visualization, as shown by Lamping *et al.* [LRP95]. From a visualization point of view, spherical space offers a way to present a graph in a center-free and periphery-free fashion. That is, in traditional drawings in $\mathbb{R}^2$ there is an implicit assumption that nodes in the center are important, while nodes on the periphery are less important. This can be avoided in $\mathbb{S}^2$ space, where any part of the graph can become the center of the layout. The early approaches for calculating the layouts of graphs in hyperbolic space, however, are either restricted by their nature to the layout of trees and tree-like graphs, or to layouts on a lattice.

The hyperbolic tree layout algorithms function on the principle of hyperbolic sphere packing, and operate by making each node of a tree, starting with the root, the center of a sphere in hyperbolic space. The children of this node are then given positions on the surface of this sphere and the process recurses on these children. By carefully computing the radii of these spheres it is possible to create aesthetically pleasing layouts for the given tree.

Although some applications calculate the layout of a general graph using this method, the layout is calculated using a spanning tree of the graph and the extra edges are then added in without altering the layout [Mun98]. This method works well for tree-like and quasi-hierarchical graphs, or for graphs where domain-specific knowledge provides a way to create a meaningful spanning tree. However, for general graphs (e.g., bipartite or densely connected graphs) and without relying on domain specific knowledge, the tree-based approach may result in poor layouts.

Methods for generalizing Euclidean geometric algorithms to hyperbolic space, although not directly related to graph drawing, have also been studied. Recently, van Wijk and Nuij [vWN04] proposed a Poincaré's half-plane projection to define a model for 2D viewing and navigation. Eppstein [Epp03] shows that many algorithms that operate in Euclidean space can be extended to hyperbolic space by exploiting the properties of a Euclidean model of the space, such as the Beltrami-Klein or Poincaré.

Hyperbolic and spherical space have also been used to display self-organizing maps in the context of data visualization. Ontrup and Ritter [OR01] and Ritter [Rit99] extend the traditional use of a regular (Euclidean) grid, on which the self-organizing map is created, with a tessellation in spherical or hyperbolic space. An iterative process is then used to adjust which elements in the data-set are represented by the intersections. Although the hyperbolic space method seems to be a promising way to display high-dimensional data-sets, the restriction to a lattice is often undesirable for graph visualization.

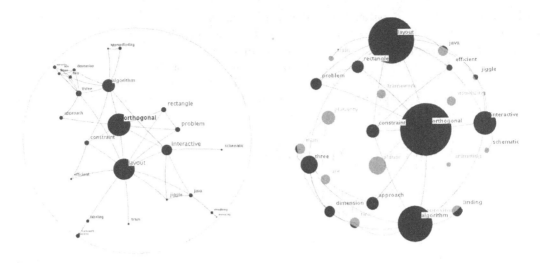

Figure 12.9 Layouts of a graph obtained from research papers' titles in hyperbolic space $\mathbb{H}^2$ and in spherical space $\mathbb{S}^2$ [KW05].

Ostry [Ost96] considers constraining force-directed algorithms to the surface of three-dimensional objects. This work is based on a differential equation formulation of the motion of the nodes in the graph, and is flexible in that it allows a layout on almost any object, even multiple objects. Since the force calculations are made in Euclidean space, however, this method is inapplicable to certain geometries (e.g., hyperbolic geometry).

Another example of graph embedding within a non-Euclidean geometry is described in the context of generating spherical parameterizations of 3D meshes. Gotsman *et al.* [GGS03] describe a method for producing such an embedding using a generalization to spherical space of planar methods for expressing convex combinations of points. The implementation of the procedure is similar to the method described in this paper, but it may not lend itself to geometries other than spherical.

Kobourov and Wampler [KW05] describe a conceptually simple approach to generalizing force-directed methods for graph layout from Euclidean geometry to Riemannian geometries. Unlike previous work on non-Euclidean force-directed methods, this approach is not limited to special classes of graphs but can be applied to arbitrary graphs; see Figure 12.9. The method relies on extending the Euclidean notions of distance, angle, and force-interactions to smooth non-Euclidean geometries via projections to and from appropriately chosen tangent spaces. Formal description of the calculations needed to extend such algorithms to hyperbolic and spherical geometries are also detailed.

In 1894 Riemann described a generalization of the geometry of surfaces, which had been studied earlier by Gauss, Bolyai, and Lobachevsky. Two well-known special cases of Riemannian geometries are the two standard non-Euclidean types, spherical geometry and hyperbolic geometry. This generalization led to the modern concept of a Riemannian manifold. Riemannian geometries have less convenient structure than Euclidean geometry, but they do retain many of the characteristics which are useful for force-directed graph layouts. A Riemannian manifold M has the property that for every point $x \in M$, the tangent space $T_x M$ is an inner product space. This means that for every point on the manifold, it is possible to define local notions of length and angle.

Using the local notions of length we can define the length of a continuous curve $\gamma : [a, b] \to M$ by

$$length(\gamma) = \int_a^b ||\gamma'(t)|| dt.$$

This leads to a natural generalization of the concept of a straight line to that of a *geodesic*, where the geodesic between two points $u, v \in M$ is defined as a continuously differentiable curve of minimal length between them. These geodesics in Euclidean geometry are straight lines, and in spherical geometry they are arcs of great circles.

We can similarly define the distance between two points, $d(x, y)$ as the length of a geodesic between them. In Euclidean space the relationship between a pair of nodes is defined along lines: the distance between the two nodes is the length of the line segment between them and forces between the two nodes act along the line through them. These notions of distance and forces can be extended to a Riemannian geometry by having these same relationships be defined in terms of the geodesics of the geometry, rather than in terms of Euclidean lines.

As Riemannian manifolds have a well-structured tangent space at every point, these tangent spaces can be used to generalize spring embedders to arbitrary Riemannian geometries. In particular, the tangent space is useful in dealing with the interaction between one point and several other points in non-Euclidean geometries. Consider three points x, y, and z in a Riemannian manifold M where there is an attractive force from x to y and z. As can be easily seen in the Euclidean case (but also true in general) the net force on x is not necessarily in the direction of y or z, and thus the natural motion of x is along neither the geodesic toward y, nor that toward z. Determining the direction in which x should move requires the notion of angle.

Since the tangent space at x, being an inner product space, has enough structure to define lengths and angles, we do the computations for calculating the forces on x in $T_x M$. In order to do this, we define two functions for every point $x \in M$ as follows:

$$\tau_x : M \to T_x M$$

$$\tau_x^{-1} : T_x M \to M$$

These two functions map points in M to and from the tangent space of M at x, respectively. We require that τ_x and τ_x^{-1} satisfy the following constraints:

1. $\tau_x^{-1}(\tau_x(y)) = y$ for all $y \in M$
2. $||\tau_x(y)|| = d(x, y)$
3. τ_x preserves angles about the origin

Using these functions it is now easy to define the way in which the nodes of a given graph $G = (V, E)$ interact with each other through forces. In the general framework for this algorithm each node is considered individually, and its new position is calculated based on the relative locations of the other nodes in the graph (repulsive forces) and on its adjacent edges (attractive forces). Then we obtain pseudo-code for a traditional Euclidean spring embedder and its corresponding non-Euclidean counterpart, as shown in Figure 12.10.

---

generic_algorithm(G)
while not done **do**
    foreach $n \in G$ **do**
        position[n] := force_directed_placement(n, G)
    end
non_Euclidean_algorithm(G)
while not done **do**
    foreach $n \in G$ **do**
        x := position[n]
        $G' := \tau_x(G)$
        $x' :=$ force_directed_placement(n, G')
        position[n] := $\tau_x^{-1}(x')$
    end
end

Figure 12.10 Pseudo-code for a traditional Euclidean spring embedder and its corresponding non-Euclidean counterpart.

12.9 Lombardi Spring Embedders

Inspired by American graphic artist Mark Lombardi, Duncan *et al.* [DEG+10a, DEG+10b] introduce the concept of a *Lombardi drawing*, which is a drawing that uses circular arcs for edges and achieves the maximum (i.e., *perfect*) amount of angular resolution possible at each vertex.

There are several force-directed graph drawing methods that use circular-arc edges or curvilinear poly-edges. Brandes and Wagner [BW00] describe a force-directed method for drawing train connections, where the vertex positions are fixed but transitive edges are drawn as Bézier curves. Finkel and Tamassia [FT05], on the other hand, describe a force-directed method for drawing graphs using curvilinear edges where vertex positions are free to move. Their method is based on adding dummy vertices that serve as control points for Bézier curve.

Chernobelskyi *et al.* [CCG+11] describe two force-directed algorithms for *Lombardi-style* (or *near-Lombardi*) drawings of graphs, where edges are drawn using circular arcs with the goal of maximizing the angular resolution at each vertex. The first approach calculates lateral and rotational forces based on the two tangents defining a circular arc between two vertices. In contrast, the second approach uses dummy vertices on each edge with repulsive forces to "push out" the circular arcs representing edges, so as to provide an aesthetic "balance". Another distinction between the two approaches is that the first one lays out the vertex positions along with the circular edges, while the second one works on graphs that are already laid out, only modifying the edges. It can be argued that Lombardi or near-Lombardi graph drawings have a certain aesthetic appeal as has been shown in recent empirical experiments [PHNK12]; see Fig. 12.11. However, another recent experimental paper on curve-based drawings [XRP+12] seems to suggest that straight-line drawings have better readability.

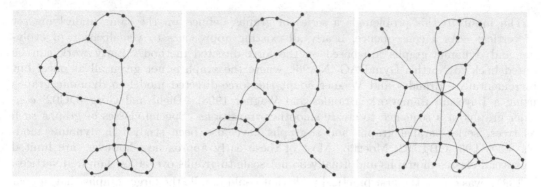

Figure 12.11 Examples of force-directed Lombardi drawings: note that every edge is a circular arc and every vertex has perfect angular resolution [CCG+11].

12.10 Dynamic Graph Drawing

While static graphs arise in many applications, dynamic processes give rise to graphs that evolve through time. Such dynamic processes can be found in software engineering, telecommunications traffic, computational biology, and social networks, among others.

Thus, dynamic graph drawing deals with the problem of effectively presenting relationships as they change over time. A related problem is that of visualizing multiple relationships on the same dataset. Traditionally, dynamic relational data is visualized with the help of graphs, in which vertices and edges fade in and out as needed, or as a time-series of graphs; see Figure 12.12.

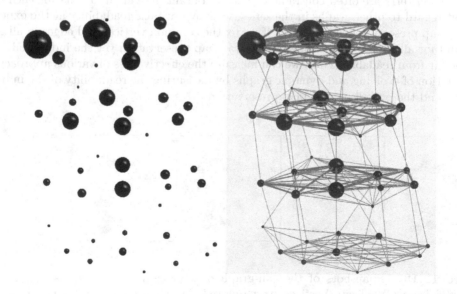

Figure 12.12 A dynamic graph can be interpreted as a larger graph made of connecting graphs in adjacent timeslices [EHK+04].

The input to this problem is a series of graphs defined on the same underlying set of vertices. As a consequence, nearly all existing approaches to visualization of evolving and dynamic graphs are based on the force-directed method. Early work can be dated back to North's DynaDAG [Nor96], where the graph is not given all at once, but incrementally. Brandes and Wagner adapt the force-directed model to dynamic graphs using a Bayesian framework [Brandes and Wagner 1998]. Diehl and Görg [DG02] consider graphs in a sequence to create smoother transitions. Special classes of graphs such as trees, series-parallel graphs and st-graphs have also been studied in dynamic models [CDTT95, CBT+92, Moe90]. Most of these early approaches, however, are limited to special classes of graphs and usually do not scale to graphs over a few hundred vertices.

TGRIP was one of the first practical tools that could handle the larger graphs that appear in the real-world. It was developed as part of a system that keeps track of the evolution of software by extracting information about the program stored within a CVS version control system [CKN+03]. Such tools allow programmers to understand the evolution of a legacy program: Why is the program structured the way it is? Which programmers were responsible for which parts of the program during which time periods? Which parts of the program appear unstable over long periods of time? TGRIP was used to visualize inheritance graphs, program call-graphs, and control-flow graphs, as they evolve over time; see Fig. 12.13.

For layout of evolving and dynamic graphs, there are two important criteria to consider:

1. *readability* of the individual layouts, which depends on aesthetic criteria such as display of symmetries, uniform edge lengths, and minimal number of crossings; and

2. *mental map preservation* in the series of layouts, which can be achieved by ensuring that vertices and edges that appear in consecutive graphs in the series, remain in the same location.

These two criteria are often contradictory. If we obtain individual layouts for each graph, without regard to other graphs in the series, we may optimize readability at the expense of mental map preservation. Conversely, if we fix the common vertices and edges in all graphs once and for all, we are optimizing the mental map preservation yet the individual layouts may be far from readable. Thus, we can measure the effectiveness of various approaches for visualization of evolving and dynamic graphs by measuring the readability of the individual layouts, and the overall mental map preservation.

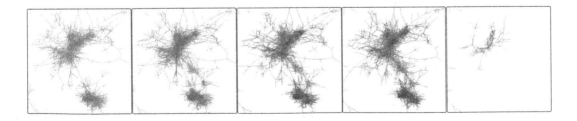

Figure 12.13 Snapshots of the call-graph of a program as it evolves through time, extracted from CVS logs. Vertices start out red. As time passes and a vertex does not change it turns purple and finally blue. When another change is affected, the vertex again becomes red. Note the number of changes between the two large clusters and the break in the build on the last image [CKN+03].

Dynamic graphs can be visualized with *aggregated views*, where all the graphs are displayed at once, *merged views*, where all the graphs are stacked above each other, and with *animations*, where only one graph is shown at a time, and morphing is used when changing between graphs (fading in/out vertices and edges that appear/disappear). When using the animation/morphing approach, it is possible to change the balance between readability of individual graphs and the overall mental map preservation, as in the system for Graph Animations with Evolving Layouts, GraphAEL [EHK+03, FKN+04]. Applications of this framework include visualizing software evolution [CKN+03], social networks analysis [MB09], and the behavior of dynamically modifiable code [DID+05].

12.11 Conclusion

Force-directed algorithms for drawing graphs have a long history and new variants are still introduced every year. Their intuitive simplicity appeals to researchers from many different fields, and this accounts for dozens of available implementations. As new relational data sets continue to be generated in many applications, force-directed algorithms will likely continue to be the method of choice. The latest scalable algorithms and algorithms that can handle large dynamic and streaming graphs are arguably of greatest utility today.

References

[BBS97]   Jürgen Branke, Frank Bucher, and Hartmut Schmeck. A genetic algorithm for drawing undirected graphs. In *Proceedings of the 3rd Nordic Workshop on Genetic Algorithms and Their Applications*, pages 193–206, 1997.

[Ber00]   Francois Bertault. A Force-Directed Algorithm that Preserves Edge Crossing Properties. *Information Processing Letters*, 74(1-2):7–13, 2000.

[BF96]   I. Bruß and A. Frick. Fast interactive 3-D graph visualization. In F. J. Brandenburg, editor, *Proceedings of the 3rd Symposium on Graph Drawing (GD)*, volume 1027 of *Lecture Notes Computer Science*, pages 99–110. Springer-Verlag, 1996.

[BH86]   Josh Barnes and Piet Hut. A hierarchical O(N log N) force calculation algorithm. *Nature*, 324:446–449, December 1986.

[BP06]   U. Brandes and C. Pich. Eigensolver methods for progressive multidimensional scaling of large data. In *Proceedings 14th Symposium on Graph Drawing (GD)*, pages 42–53, 2006.

[Bra01]   Ulrik Brandes. Drawing on physical analogies. In Michael Kaufmann and Dorothea Wagner, editors, *Drawing Graphs*, volume 2025 of *Lecture Notes in Computer Science*, pages 71–86. Springer-Verlag, 2001.

[BW00]   Ulrik Brandes and Dorothea Wagner. Using Graph Layout to Visualize Train Interconnection Data. *J. Graph Algorithms Appl.*, 4(3):135–155, 2000.

[CBT+92]   R. F. Cohen, G. Di Battista, R. Tamassia, I. G. Tollis, and P. Bertolazzi. A framework for dynamic graph drawing. In *Proceedings of the 8th Annual Symposium on Computational Geometry (SCG '92)*, pages 261–270, 1992.

[CCG+11]   R. Chernobelskiy, K. Cunningham, M. T. Goodrich, S. G. Kobourov, and L. Trott. Force-directed lombardi-style graph drawing. In *Proceedings 19th Symposium on Graph Drawing (GD)*, pages 78–90, 2011.

[CDTT95]   R. F. Cohen, G. Di Battista, R. Tamassia, and I. G. Tollis. Dynamic graph drawings: Trees, series-parallel digraphs, and planar *ST*-digraphs. *SIAM J. Comput.*, 24(5):970–1001, 1995.

[CKN+03]   C. Collberg, S. G. Kobourov, J. Nagra, J. Pitts, and K. Wampler. A system for graph-based visualization of the evolution of software. In *ACM Symposium on Software Visualization (SoftVis)*, pages 77–86, 2003.

[CLRS90]   T. H. Cormen, C. E. Leiserson, R. L. Rivest, and C. Stein. *Introduction to Algorithms*. MIT Press, Cambridge, MA, 1990.

[Coh97]   Jonathan D. Cohen. Drawing graphs to convey proximity: An incremental arrangement method. *ACM Transactions on Computer-Human Interaction*, 4(3):197–229, September 1997.

[CT96]   I. F. Cruz and J. P. Twarog. 3D graph drawing with simulated annealing. In F. J. Brandenburg, editor, *Proceedings of the 3rd Symposium on Graph Drawing (GD)*, volume 1027 of *Lecture Notes Computer Science*, pages 162–165. Springer-Verlag, 1996.

[DEG+10a]   Christian A. Duncan, David Eppstein, Michael T. Goodrich, Stephen G. Kobourov, and Martin Nöllenburg. Drawing trees with perfect angular resolution and polynomial area. In *Graph Drawing*, pages 183–194, 2010.

[DEG+10b] Christian A. Duncan, David Eppstein, Michael T. Goodrich, Stephen G. Kobourov, and Martin Nöllenburg. Lombardi drawings of graphs. In *Graph Drawing*, pages 195–207, 2010.

[DETT99] Giuseppe Di Battista, Peter Eades, Roberto Tamassia, and Ioannis G. Tollis. *Graph Drawing: Algorithms for the Visualization of Graphs*. Prentice Hall, Englewood Cliffs, NJ, 1999.

[DG02] Stephan Diehl and Carsten Görg. Graphs, they are changing. In *Proceedings of the 10th Symposium on Graph Drawing (GD)*, pages 23–30, 2002.

[DH96] Ron Davidson and David Harel. Drawing graphs nicely using simulated annealing. *ACM Transactions on Graphics*, 15(4):301–331, 1996.

[DID+05] Brad Dux, Anand Iyer, Saumya Debray, David Forrester, and Stephen G. Kobourov. Visualizing the behaviour of dynamically modifiable code. In *13th IEEE Workshop on Porgram Comprehension*, pages 337–340, 2005.

[DKM09] Tim Dwyer, Yehuda Koren, and Kim Marriott. Constrained graph layout by stress majorization and gradient projection. *Discrete Mathematics*, 309(7):1895–1908, 2009.

[Ead84] Peter Eades. A heuristic for graph drawing. *Congressus Numerantium*, 42:149–160, 1984.

[EG95] Peter Eades and Patrick Garvan. Drawing stressed planar graphs in three dimensions. In *Proceedings of the 3rd Symposium on Graph Drawing*, pages 212–223, 1995.

[EGK+01] John Ellson, Emden R. Gansner, Eleftherios Koutsofios, Stephen C. North, and Gordon Woodhull. Graphviz—open source graph drawing tools. In *Graph Drawing*, pages 483–484, 2001.

[EHK+03] C. Erten, P. J. Harding, S. G. Kobourov, K. Wampler, and G. Yee. GraphAEL: Graph animations with evolving layouts. In *11th Symposium on Graph Drawing*, pages 98–110, 2003.

[EHK+04] C. Erten, P. J. Harding, S. Kobourov, K. Wampler, and G. Yee. Exploring the computing literature using temporal graph visualization. In *Visualization and Data Analysis*, pages 45–56, 2004.

[Epp03] D. Eppstein. Hyperbolic geometry, Möbius transformations, and geometric optimization. In *MSRI Introductory Workshop on Discrete and Computational Geometry*, 2003.

[FCW67] C. Fisk, D. Caskey, and L. West. Accel: Automated circuit card etching layout. *Proceedings of the IEEE*, 55(11):1971–1982, 1967.

[FK12] Joe Fowler and Stephen G. Kobourov. Planar preprocessing for spring embedders. In *Graph Drawing*, 2012.

[FKN+04] D. Forrester, S. G. Kobourov, A. Navabi, K. Wampler, and G. Yee. graphael: A system for generalized force-directed layouts. In *12th Symposium on Graph Drawing (GD)*, 2004.

[FLM95] A. Frick, A. Ludwig, and H. Mehldau. A fast adaptive layout algorithm for undirected graphs. In R. Tamassia and I. G. Tollis, editors, *Proceedings of the 2nd Symposium on Graph Drawing (GD)*, volume 894 of *Lecture Notes in Computer Science*, pages 388–403. Springer-Verlag, 1995.

[FR91] T. Fruchterman and E. Reingold. Graph drawing by force-directed placement. *Softw. - Pract. Exp.*, 21(11):1129–1164, 1991.

[FT05]   Benjamin Finkel and Roberto Tamassia. Curvilinear Graph Drawing Using the Force-Directed Method. In *Proc. 12th Int. Symp. on Graph Drawing (GD 2004)*, pages 448–453, 2005.

[GGK04]   P. Gajer, M. T. Goodrich, and S. G. Kobourov. A fast multi-dimensional algorithm for drawing large graphs. *Computational Geometry: Theory and Applications*, 29(1):3–18, 2004.

[GGS03]   C. Gotsman, X. Gu, and A. Sheffer. Fundamentals of spherical parameterization for 3D meshes. In *ACM Transactions on Graphics, 22*, pages 358–363, 2003.

[GK02]   Pawel Gajer and Stephen G. Kobourov. GRIP: Graph dRawing with Intelligent Placement. *Journal of Graph Algorithms and Applications*, 6(3):203–224, 2002.

[GKN04]   E. Gansner, Y. Koren, and S. North. Graph drawing by stress minimization. In *Proceedings 12th Symposium on Graph Drawing (GD)*, pages 239–250, 2004.

[Gre88]   Leslie Greengard. *The Rapid Evolution of Potential Fields in Particle Systems*. MIT. Press, Cambridge, MA, 1988.

[HH01]   R. Hadany and D. Harel. A multi-scale algorithm for drawing graphs nicely. *Discrete Applied Mathematics*, 113(1):3–21, 2001.

[HJ04]   S. Hachul and M. Jünger. Drawing large graphs woth a potential-field-based multilevel algorithm. In *Proceedings of the 12th Symposium on Graph Drawing (GD)*, volume 3383 of *Lecture Notes in Computer Science*, pages 285–295. Springer-Verlag, 2004.

[HK02]   David Harel and Yehuda Koren. A fast multi-scale method for drawing large graphs. *Journal of Graph Algorithms and Applications*, 6(3):179–2002, 2002.

[Hoc96]   D. S. Hochbaum. *Approximation Algorithms for NP-Hard Problems*. PWS Publishing, 1996.

[HS98]   David Harel and Meir Sardas. An algorithm for straight-line drawing of planar graphs. *Algorithmica*, 20(2):119–135, 1998.

[Hu05]   Yifan Hu. Efficient and high quality force-directed graph drawing. *The Mathematica Journal*, 10:37–71, 2005.

[KGV83]   S. Kirkpatrick, C. D. Gelatt, and M. P. Vecchi. Optimization by simulated annealing. *Science*, 220(4598):671–680, 1983.

[KK89]   T. Kamada and S. Kawai. An algorithm for drawing general undirected graphs. *Inform. Process. Lett.*, 31:7–15, 1989.

[KMS91]   Corey Kosak, Joe Marks, and Stuart Shieber. A parallel genetic algorithm for network-diagram layout. In *Proceedings of the 4th International Conference on Genetic Algorithms*, pages 458–465, 1991.

[Kru64]   J. B. Kruskal. Multidimensional scaling by optimizing goodness of fit to a nonmetric hypothesis. *Psychometrika*, 29:1–27, 1964.

[KS80]   J. Kruskal and J. Seery. Designing network diagrams. In *Proceedings 1st General Conference on Social Graphics*, pages 22–50, 1980.

[KW05]   S. G. Kobourov and K. Wampler. Non-Euclidean spring embedders. *IEEE Transactions on Visualization and Computer Graphics*, 11(6):757–767, 2005.

[LR88]   T. Leighton and S. Rao. An approximate max-flow min-cut theorem for uniform multicommodity flow problems with applications to approximation algorithms. In *Proceedings of the 29th Annual Symposium on Foundations of Computer Science (FOCS)*, pages 422–431, 1988.

[LRP95]   John Lamping, Ramana Rao, and Peter Pirolli. A focus+context technique based on hyperbolic geometry for visualizing large hierarchies. In *Proceedings of Computer Human Interaction*, pages 401–408. ACM, 1995.

[MB96]   T. Munzner and P. Burchard. Visualizing the structure of the World Wide Web in 3D hyperbolic space. In David R. Nadeau and John L. Moreland, editors, *1995 Symposium on the Virtual Reality Modeling Language, VRML '95*, pages 33–38, 1996.

[MB09]   M. Jacomy M. Bastian, S. Heymann. Gephi: an open source software for exploring and manipulating networks. *International AAAI Conference on Weblogs and Social Media*, 2009.

[Moe90]   Sven Moen. Drawing dynamic trees. *IEEE Software*, 7(4):21–28, July 1990.

[Mun97]   Tamara Munzner. H3: Laying out large directed graphs in 3D hyperbolic space. In L. Lavagno and W. Reisig, editors, *Proceedings of IEEE Symposium on Information Visualization*, pages 2–10, 1997.

[Mun98]   T. Munzner. Drawing large graphs with H3Viewer and Site Manager. In *Proceedings of the 6th Symposium on Graph Drawing*, pages 384–393, 1998.

[Noa07]   Andreas Noack. Energy models for graph clustering. *J. Graph Algorithms Appl.*, 11(2):453–480, 2007.

[Nor96]   S. C. North. Incremental layout in DynaDAG. In *Proceedings of the 4th Symposium on Graph Drawing (GD)*, pages 409–418, 1996.

[OR01]   J. Ontrup and H. Ritter. Hyperbolic self-organizing maps for semantic navigation. In *Advances in Neural Information Processing Systems 14*, pages 1417–1424, 2001.

[Ost96]   Diethelm Ironi Ostry. Some three-dimensional graph drawing algorithms. Master's thesis, University of Newcastle, Australia, 1996.

[PHNK12]   Helen Purchase, John Hamer, Martin Nöllenburg, and Stephen G. Kobourov. On the usability of Lombardi graph drawings. In *Graph Drawing*, 2012.

[PSA11]   Daniel Archambault Paolo Simonetto and David Auber. ImPrEd: An improved force-directed algorithm that prevents nodes from crossing edges. *Computer Graphics Forum (EuroVis)*, 30(3):1071–1080, 2011.

[QB79]   N. Quinn and M. Breur. A force directed component placement procedure for printed circuit boards. *IEEE Transactions on Circuits and Systems*, CAS-26(6):377–388, 1979.

[QE00]   Aaron Quigley and Peter Eades. FADE: graph drawing, clustering, and visual abstraction. In *Proceedings of the 8th Symposium on Graph Drawing (GD)*, volume 1984 of *Lecture Notes in Computer Science*, pages 197–210. Springer-Verlag, 2000.

[Rit99]   H. Ritter. Self-organizing maps on non-euclidean spaces. In Erkki Oja and Samuel Kaski, editors, *Kohonen Maps*, pages 97–110. Elsevier, Amsterdam, 1999.

[SM00]   J. Shi and J. Malik. Normalized cuts and image segmentation. *IEEE Transaction on Pattern Analysis and Machine Intelligence*, 22(8):888–905, 2000.

[Tut63]   William T. Tutte. How to draw a graph. *Proc. London Math. Society*, 13(52):743–768, 1963.

[Vos99]   Michael D. Vose. *The Simple Genetic Algorithm: Foundations and Theory.* MIT Press, 1999.

[vWN04]   J. J. van Wijk and W. A. A. Nuij. A model for smooth viewing and navigation of large 2D information spaces. *IEEE Transactions on Visualization and Computer Graphics*, 10(4):447– 458, 2004.

[Wal03]   C. Walshaw. A multilevel algorithm for force-directed graph drawing. *Journal of Graph Algorithms and Applications*, 7(3):253–285, 2003.

[XRP⁺12]   K. Xu, C. Rooney, P. Passmore, D. H. Ham, and P. Nguyen. A user study on curved edges in graph visualization. In *IEEE InfoVis*, 2012.

13

Hierarchical Drawing Algorithms

Patrick Healy
University of Limerick

Nikola S. Nikolov
University of Limerick

13.1 Introduction

In many cases a directed graph represents a hierarchy and we want to draw it in this way. We will define a hierarchy later, but for now it is sufficient to think of a hierarchy as a cycle-free digraph where it is useful for nodes of the graph to be stratified into discrete, parallel layers. Examples of hierarchies or near-hierarchies are, among others, PERT charts for project management, object-oriented class diagrams, and function call graphs from software engineering. As usual, nodes represent entities and edges represent relationships between the entities. Closely related to hierarchically layered drawings and discussed later are radial drawings where nodes are placed on concentric circles [DDLM04, Bac07] and cyclic level drawings, where nodes are placed on "spokes" emanating from a centre-point [BBBF12].

What is common to all of the examples described above is the need to represent all the relationships graphically so that the positioning of nodes are as consistent with the transitivity of the relationship as can be achieved. That is, the edges should "flow" in a uniform direction. Whether this direction should be top-to-bottom, or left-to-right, depends on the application domain, with different disciplines having different preferences.

Di Battista *et al.* [DGL+00] conduct an experimental study of directed graph drawing algorithms. They look at two broad categories of algorithms, those that provide *layered drawings* and those that are *grid-based*. From the point of view of edge crossings, an important aspect in the readability of graph drawings [PCJ96], the hierarchical or layered approach performs better, they conclude.

For digraphs that are *almost* a hierarchy it still can be possible to take advantage of the methods we describe in this chapter. Since the methods work best for hierarchical digraphs, however, one may find the results disappointing when applied to a digraph that is fundamentally non-hierarchical. A method for computing the *hierarchical index* – that is, the amount of hierarchy in a directed graph – has been proposed [CHK02].

In addition to the applications we mentioned earlier, it is common and appropriate to represent computer file systems and social networks as a graph. Due to the possibility of symbolic links a graph representation of a computer file system may have directed cycles, and so may be more complicated than a tree. Likewise, a graph representation of relationships among a social network graph may have cycles.

In the following sections we describe the current approaches to drawing directed graphs hierarchically and consider in more depth the dominant player, the Sugiyama framework for drawing digraphs. It is perhaps a measure of its effectiveness and its success that the method has been used to draw undirected graphs, by imposing orientations on the edges. However, as we have mentioned earlier, results may be disappointing if care is not taken in how orientations of edges are fixed.

13.1.1   Current Approaches and Their Limitations

Force-directed methods, discussed elsewhere in this handbook, can be modified to take account of edge directions, and thus can be used to draw digraphs. Sugiyama and Misue [SM95b] propose modifications of Eades' spring embedder model [Ead84] to take account of the possible directedness of edges.

Far and away the most popular method of drawing directed graphs is the *Sugiyama method*, or *Sugiyama framework* [STT81], which separates the nodes into layers. The idea of layered drawings can be traced back earlier to work by Warfield [War77] and Carpano [Car80]. Systems such as *da Vinci* [FW95], *dot* [GKN02] (part of the `GraphViz` suite of tools [GN00]), *GraphLet* [Him00], the AGD graph drawing library [NPT90] and others implement this framework for drawing directed graphs. In testament to its popularity many modifications and enhancements have been proposed in the literature. However, the framework has its limitations. Figure 13.1 shows two drawings of C_4, firstly drawn using force-directed methods and secondly drawn using the Sugiyama framework. In spite of the directed edge entering node a, by imposing a leveling Figure 13.1b suggests that node d is inferior to the others.

In the following section we describe the Sugiyama framework in general terms. In subsequent sections we consider the framework in more detail, elaborating on issues specific to each step of the framework. It should be noted at the outset that this is only a framework and for many steps of the process alternative algorithms exist, each with their own merits. Equally important is the fact that the steps may interact with each other and a solution to one step can have a bearing on later steps.

13.1.2   Overview of Sugiyama's Framework

The Sugiyama framework is motivated by a number of aesthetically desirable properties that make for a more readable graph. Indeed the steps of the Sugiyama framework can be seen to address, in turn, each of the following aesthetics.

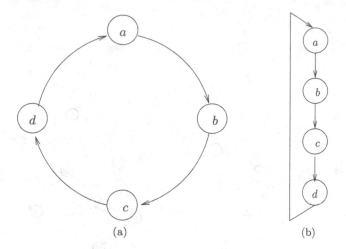

Figure 13.1 Two alternative drawings of C_4.

- Edges should point in a uniform direction
- Short edges are more readable
- Uniformly distributed nodes avoid clutter
- Edge crossings obstruct comprehension
- Straight edges are more readable

Figure 13.2 demonstrates how these aesthetics are achieved on a typical directed graph[1] G. So that all edges are directed uniformly, any directed cycles are broken by reversing a subset of edges (see Figure 13.2b). The resulting graph is then *leveled* (Figure 13.2c) where, through the introduction of *dummy vertices*, "long" edges are replaced by a series of shorter segments, after which vertices on each level are permuted in order to reduce edge crossings (Figure 13.2d). Finally, edges spanning more than one level (*long* edges) are straightened by adjusting the x-coordinates of their end vertices and by aligning the inserted dummy nodes on long edges.

As we have remarked before the Sugiyama framework draws directed graphs as layers of vertices. We have loosely described a hierarchy in terms of layers of nodes and this is part of its formal definition, also. Definition 13.1 formalizes a level graph, and the definition of a *hierarchy* follows from this.

DEFINITION 13.1 A *level graph* $G = (V, E, \lambda)$ is a directed acyclic graph with a mapping $\lambda : V \to \{1, 2, \ldots, k\}$, $k \geq 1$, that partitions the vertex set V as $V = V_1 \cup V_2 \cup \cdots \cup V_k$, $V_j = \lambda^{-1}(j)$, $V_i \cap V_j = \emptyset$ for $i \neq j$, such that $\lambda(v) = \lambda(u) + 1$ for each edge $(u, v) \in E$.

DEFINITION 13.2 A *hierarchy* is a level graph $G(V, E, \lambda)$ where for every $v \in V_j$, $j > 1$, there exists at least one edge (w, v) such that $w \in V_{j-1}$.

---

[1]The figure is due to Bachmaier *et al.* [BBBF12]; the authors' permission to reproduce the figure is gratefully acknowledged.

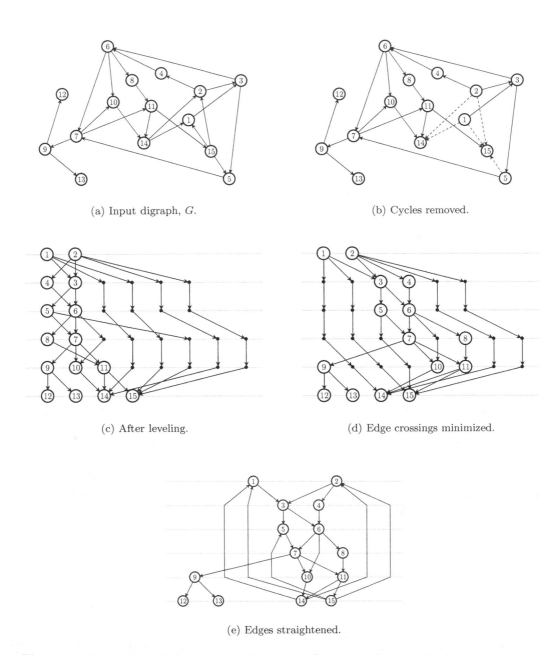

(a) Input digraph, G.

(b) Cycles removed.

(c) After leveling.

(d) Edge crossings minimized.

(e) Edges straightened.

Figure 13.2   A digraph drawn according to the Sugiyama framework.

Note that Definition 13.2 restricts all sources of the graph to appear on the first level but this may be relaxed if desired. Further, the definition implies that all edges are of unit length, a property that is necessary for the *crossing minimization* step. While an input digraph may not be a hierarchy initially, the steps described in the following subsections will transform it into an equivalent hierarchy.

13.2 Cycle Removal

The first step of the Sugiyama method is a preprocessing step that aims the reversal of the direction of some edges in order to make the input digraph acyclic. A digraph is acyclic if it does not contain any directed cycles. Note that the digraph may have undirected cycles and be acyclic. It is usually assumed that the input digraph has no *two-cycles*. A two-cycle is a cycle consisting of a pair of edges (u, v) and (v, u). If any are present then one edge of each pair can be removed before applying the Sugiyama method and reintroduced back into the final drawing.

The cycle-removal preprocessing step is necessary because the input to the the layer-assignment step must be an acyclic digraph, also called a DAG (directed acyclic graph). Once vertices are assigned to layers, the original direction of the reversed edges can be restored. These are edges which point against the flow in the final drawing. It is also possible to remove edges instead of reversing them, and introduce them back after the layer-assignment step. However, if edges are removed then the layer-assignment step will work with a subgraph of the input digraph and may have undesirable results.

A set of edges whose removal makes the digraph acyclic is commonly known as a *feedback arc set* (FAS). Following the terminology used by Di Battista *et al.* [DETT99] we call a set of edges whose reversal makes the digraph acyclic a *feedback set* (FS). Each FS is also a FAS. However, not each FAS is a FS. For example, if a digraph has only one cycle, then the set of all edges in the cycle is a FAS but not a FS.

It is always possible and easy to find a FS for a digraph. Any linear ordering of the vertices partitions the edge set into two subsets, a subset of edges whose source is before their target in the ordering, and a subset of edges edges whose source is after their target in the ordering. Each of the two subsets is a FS. However, it might be much harder to find a FS with some specific properties.

A typical requirement for a FS is to contain as few edges as possible because they are the edges against the flow in the final drawing. The problem of finding a minimum-cardinality FS is known as the *minimum FS problem*. As we mentioned above not every FAS is a FS. However, it is easy to see that every minimal cardinality FAS is also a FS. Thus, the minimum FS problem is as hard as the widely studied *minimum FAS problem* which is known to be NP-hard [Kar72, GJ79].

Any heuristic for solving the minimum FAS problem can be applied for solving the minimum FS problem as well. Consider a digraph $G = (V, E)$ and let $F \subseteq E$ be a FAS. F is a minimal FAS if for each $e \in F$ there is a cycle in $(E \setminus F) \cup \{e\}$. If F is not minimal then one by one we can remove edges from it until it becomes minimal. However, such a procedure will add additional running time.

The remainder of this section summarizes the known heuristics for solving either the minimum FS problem or the minimum FAS problem. Some of the FAS heuristics have been originally proposed as heuristic for solving its complimentary problem, i.e., the maximum acyclic subgraph problem.

13.2.1 Heuristics Based on Vertex Orderings

As we mentioned above any linear ordering of the vertices provides two FSs. Imagine the vertices placed on a horizontal line in accordance with the provided linear ordering. The first FS consists of the edges with direction from left to right, and the second FS consists of the edges with direction from right to left. It can be shown that there are digraphs and linear orderings of the vertices for which both FSs have the same size [BS90]. Thus, the simple approach is a 2-approximation heuristic for solving both the minimum FS and FAS problems.

It is clear that the outcome of this simple approach depends on the linear ordering of the vertices. Some researchers have suggested to use a linear ordering provided by a depth-first traversal [RDM+87, GKNV93]. This is based on the intuition that such an ordering may be "natural" for the digraph i.e., it may look "natural" to the viewer that the edges against the flow are drawn this way. A depth-first-traversal ordering can be computed in linear time. Here it is assumed that always the set of edges with direction to the left, i.e., against the ordering, are chosen as FSs. These are the back edges in the depth-first-traversal tree. Their number can be at most $|E| - |V| - 1$ which could be high for dense digraphs.

Eades et $al.$ proposed two alternative linear orderings [ELS89]. The first one is based on the intuition that vertices with large outdegree should appear at the top of the final drawing. First in the ordering comes vertex v with the maximum $d_G^+(v)$, next comes vertex v' with the maximum $d_{G-v}^+(v')$, etc. This linear ordering can be computed in linear time and reportedly gives smaller FSs than the ordering provided by the depth-first search.

The second linear ordering proposed by Eades at al. is the result of a divide-and-conquer approach. Assuming an input digraph $G = (V, E)$ the vertices of which have to be assigned the labels $i, i+1, \ldots, i+|V|-1$, the recursive procedure for assigning these labels works as follows. If $|V| = 1$ then the single vertex gets the label i. Otherwise V is partitioned into two subsets V_1 and V_2 and the procedure is applied recursively to $G[V_1]$ and $G[V_2]$ with sets of labels $i, i+1, \ldots, i+|V_1|-1$ and $i+|V_1|, i+|V_1|+1, \ldots, i+|V|-1$, respectively. V_1 and V_2 are such that for each pair of vertices $(v_1, v_2) \in V_1 \times V_2$ $d_G^+(v_1) \geq d_G^+(v_2)$. If $|V|$ is even then V_1 and V_2 have the same cardinality. Otherwise, V_1 contains a singe vertex, and V_2 contains the rest of the vertices. It takes $O(min((|V| + |A|)log|V|, |V|^2))$ time to compute this linear ordering. However, the the authors have observed that it regularly obtains results 20% better than the results obtained by their other ordering. They also prove performance guarantees for dense digraphs.

13.2.2 Berger-Shor Algorithm

The first polynomial-time algorithm for solving the minimum FAS problem with an approximation ratio less than 2 in the worst case is the algorithm proposed by Berger and Shor in 1990 [BS90].

Consider a digraph $G = (V, E)$ and let $E_a \subset E$ denote a set of edges such that $G[E_a]$ is acyclic. Let also $\delta(v)$ denote all edges adjacent to vertex v, and $\delta^-(v)$ and $\delta^+(v)$ denote the sets of incoming and outgoing edges of v, respectively. The algorithm starts with an empty set E_a and one by one scans all vertices of G, in an arbitrary order. For each vertex $v \in V$ if $d^+(v) \geq d^-(v)$ then $E_a \leftarrow E_a \cup \delta^+(v)$. Otherwise, $E_a \leftarrow E_a \cup \delta^-(v)$. After processing vertex v it is deleted from G (together with its adjacent edges).

The time complexity of this algorithm is $O(|V| + |E|)$. Berger and Shor prove that $G' = (V, E_a)$ is a DAG and thus $F = E \backslash E_a$ is a FAS. They also propose a modification to the algorithm for making sure the FAS is minimal. It consists of running a strongly connected components algorithm before processing each vertex, adding the edges between strongly

Figure 13.3 Greedy Cycle Removal

Require: digraph $G = (V, E)$

$S_l \leftarrow \phi$
$S_r \leftarrow \phi$
while G is not empty **do**
    while G contains a sink **do**
        Choose a sink v
        Remove v from G
        Prepend v to S_r
    end while
    while G contains a source **do**
        Choose a source u
        Remove u from G
        Append u to S_l
    end while
    if G is not empty **then**
        Choose a vertex w such that $d^+(w) - d^-(w)$ is maximum
        Remove w from G
        Append w it to S_l
    end if
end while

connected components to E_a, and removing them from E. However, this modification adds $O(|V||E|)$ to the running time.

It is easy to see that at the end F will contain at most a half of all the edges, i.e., $|F| \leq \frac{|E|}{2}$. Better approximation ratio can be achieved by processing the vertices in a special order. Berger and Shor show that if at each step the vertex to be processed is the vertex that minimizes the number of edges in the FAS over all possible orderings of the unprocessed yet vertices then at the end $|F| \leq |E|(\frac{1}{2} - \Omega(\frac{1}{\sqrt{\Delta(G)}}))$ where $\Delta(G)$ is the maximum degree of a vertex in G. It is possible to process the vertices in such an order efficiently by increasing the running time to $O(|V||E|)$ [BS90].

13.2.3 Greedy Cycle Removal

It can be observed that edges incident to either a sink or a source of the digraph cannot be a part of a cycle. By making this observation Eades *et al.* were able to improve the result of Berger and Shor [ELS93]. They proposed a linear-time algorithm with a performance guarantee at least as good as the performance guarantee of the Berger-Shor algorithm for $|E| \in O(|V|)$ and even better if $\Delta(G) \notin O(1)$ [ELS93]. The algorithm of Eades *et al.*, known as Greedy Cycle Removal, always finds a FS for the input digraph.

Similar to the approach of Berger and Shor, Algorithm 13.3 processes the vertices one by one and removes the processed vertices from the digraph. However, it builds the FAS in a different way. It computes a linear ordering of the vertices and takes the edges with direction against the ordering as an FS. Eades *et al.* show that the cardinality of the FS found by Algorithm 13.3 is at most $\frac{|E|}{2} - \frac{|V|}{6}$. For digraphs with $\Delta(G) \leq 3$ the cardinality is at most $\frac{2}{3}|E|$.

Sander has proposed a modification to Algorithm 13.3 similar to the modification which involves the computation of strongly connected components in the Berger-Shor algorithm [San96b]. It decreases the running time to $O(|V||E|)$ but reportedly leads to better results in practice. Generalizing the ideas of Sander, Eades and Lin showed how Algorithm 13.3 can be modified to guarantee that no more that one quarter of the edges will be reversed for cubic digraphs [EL95].

13.2.4   Heuristics Based on Cycle Breaking

Most of the heuristics discussed above focus on computing linear orderings of the vertices which provide FSs. Alternative point of view the minimum FAS problem (and also the minimum FS problem) is to build the FAS edge by edge choosing edges which belong to cycles.

A very simple algorithm based on this approach is the following one. Start with two empty sets S and T and scan all edges one by one. For each edge e, if $S \cup \{e\}$ is acyclic then add e to S. Otherwise add e to T. It is easy to show that at the end of this process both S and T are acyclic and the smaller of the two sets provides a FAS with at most a half of all the edges. Note that T is a minimal FAS, while S might not be.

Related to this approach is the heuristic implemented by Gansner *et al.* in their system dot [GKNV93]. It takes one non-trivial strongly connected component of the digraph at a time, in an arbitrary order. Within each component it performs a depth-first traversal and adds to the FS an edge which participates in a maximum number of cycles. This is repeated until there are no more non-trivial strongly connected components. Gansner *et al.* report that this heuristic performs well in practice. They also observed that it reverses edges whose direction against the flow is "natural" for the input digraph [GKNV93].

13.2.5   Minimum FAS in a Weighted Digraph

In some applications edges are assigned nonnegative weight. Then it might be required to reverse not the minimum number of edges but a set of edges with the minimum total weight. Demetrescu and Finocchi proposed an algorithm for the weighted minimum FAS problem that runs in $O(|V||E|)$ time. Their approach compromises between two possible approaches, i.e., greedily adding light edges to the FAS, and adding edges that belong to a large number of cycles. The latter is the approach of Gansner *et al.*, which we discussed in Section 13.2.4.

The algorithm of Demetrescu and Finocchi is presented as Algorithm 13.4. Within the while-loop heavy edges which belong to a large number of cycles become progressively more likely to be added to the FAS F. Note also that at the end edges which do not form a cycle with $E \setminus F$ are excluded from F, thus making sure F is minimal. Demetrescu and Finocchi also prove that their algorithm approximates a minimum FAS of the input digraph G within a ratio bounded by the length of a longest simple cycle of G [DF03].

13.2.6   Other Approaches

There are other approaches to the minimum FAS problem which we would like to refer the reader to. These include the heuristic of Flood [Flo90], and the best-known approximation algorithm which achieves a performance ratio $O(\log |V| \log \log |V|)$, and requires to solve a linear program [ENRS95, Sey95]. An interesting result is that all minimal solutions can be enumerated with polynomial delay [SS97].

Figure 13.4 Algorithm of Demetrescu and Finocchi

Require: digraph $G = (V, E)$, $w : E \to \mathcal{R}$

$F \leftarrow \phi$
while $V, E \setminus F)$ is not acyclic **do**
  Let C be a simple cycle in $(V, E \setminus F)$
  Let (x, y) be a minimum weight edge in C
  Let $\epsilon = w(x, y)$
  for all $(u, v) \in C$ **do**
    $w(u, v) \leftarrow w(u, v) - \epsilon$
    if $w(u, v) = 0$ **then**
      $F \leftarrow F \cup \{(u, v)\}$
    end if
  end for
end while
for all $(u, v) \in F$ **do**
  if $(V, E \setminus F \cup \{(u, v)\})$ is acyclic **then**
    $F \leftarrow F \setminus \{(u, v)\}$
  end if
end for

For an exact ILP approach to the minimum FAS problem we refer the reader to the work of Grötschel *et al.* and Rienelt *et al.* [GJR85, Rei85]. Their study of the facial structure of the acyclic subgraph polytope can be used for finding the minimum FS by a branch-and-cut algorithm.

13.3 Layer Assignment

Consider a DAG $G = (V, E)$ with a set of vertices V and a set of edges E. Let $\mathcal{L} = \{L_0, L_1, \ldots, L_h\}$ be a partition of the vertex set of G into $h \geq 1$ subsets such that if $(u, v) \in E$ with $u \in L_j$ and $v \in L_i$ then $i < j$. $\mathcal{L}$ is called a *layering* of G and the sets L_0, L_1, ..., L_h are called *layers*. A DAG with a layering is called a *layered* DAG. The problem of partitioning the vertex set of a graph into layers is known as *the layering problem* or *the layer assignment problem*.

Sometimes the term *levels* is used instead of layers. It emphasizes the usual visual representation of layers as mapped to either parallel horizontal lines or concentric circles. In this section we consider the layer assignment problem without relating it to a specific visual representation. The example drawings we give have only illustrative character. They employ the parallel horizontal levels convention, i.e., all vertices in layer L_i are placed on the horizontal level with an y-coordinate equal to i.

Let $l(u, \mathcal{L})$ be the number of the layer that contains vertex $u \in V$, i.e., $l(u, \mathcal{L}) = i$ if and only if $u \in L_i$. Sometimes $l(u, \mathcal{L})$ is called *rank* of vertex u. The *span* of edge $e = (u, v)$ in layering $\mathcal{L}$ is defined as $s(e, \mathcal{L}) = l(u, \mathcal{L}) - l(v, \mathcal{L})$. Clearly, $s(e, \mathcal{L}) \geq 1$ for each $e \in E$; edges with a span 1 are *tight* edges; edges with a span greater than 1 are *long edges*. A layering of G is *proper* if all edges are tight. The layering found by a layering algorithm might not be proper because only a small fraction of DAGs can be layered properly and also because a proper layering may not satisfy other layering requirements.

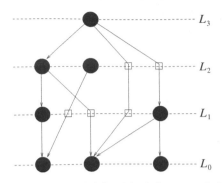
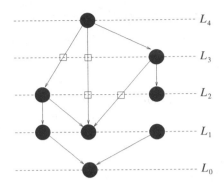

(a) A layered DAG with 4 layers and 5 dummy vertices.

(b) A layered DAG with 5 layers and 4 dummy vertices.

Figure 13.5   Two alternative layered drawings of the same DAG with introduced dummy vertices which subdivide long edges.   Dummy vertices are represented by transparent squares.

Within the Sugiyama method the vertex ordering algorithms applied after the layer assignment phase assume that their input is a DAG with a proper layering. Thus, if the layering found at the layering phase is not proper then it must be transformed into a proper one. Normally, this is done by introducing so-called *dummy vertices* which subdivide long edges (see the illustration in Figure 13.5). Formally, let $e = (u, v)$ be an edge with $l(u, \mathcal{L}) = j$ and $l(v, \mathcal{L}) = i$ and $s(e, \mathcal{L}) = j - i > 1$. Then we add dummy vertices d_e^{i+1}, d_e^{i+2}, ..., d_e^{j-1} to layers L_{i+1}, L_{i+2}, ..., L_{j-1} respectively and we replace edge e by the path $(u, d_e^{j-1}, \ldots, d_e^{i+1}, v)$. We refer to vertices which are not dummy as *original* vertices. We also denote the set of all dummy vertices introduced to a layered DAG G with a layering $\mathcal{L}$ by $\mathcal{D}(G, \mathcal{L})$. Clearly,

$$|\mathcal{D}(G, \mathcal{L})| = \sum_{e \in E} s(e, \mathcal{L}) - |E|.$$

13.3.1   Additional Criteria and Variations of the Problem

If there are no additional requirements it is not hard to find a layering of a DAG. Classical graph algorithms such as breadth-first search, depth-first search and algorithms for finding a minimum spanning tree can be easily modified to partition the vertex set of a DAG into layers. However, normally it is desirable to take into account a number of additional criteria when computing a layering [ES90].

It is desirable that $|\mathcal{D}(G, \mathcal{L})|$ is as small as possible because a large number of dummy vertices significantly slows down the vertex ordering phase of the Sugiyama method. Thus, one of the goals of a layering assignment algorithm should be to find a layering with as few as possible dummy vertices. There are also aesthetic reasons for keeping the number of dummy vertices small. A layered DAG with a small number of dummy vertices would also have a small number of undesirable long edges and edge bends. The problem of finding a layering with the minimum number of dummy vertices is in P. It can be modeled as an integer linear programming problem and safely relaxed to a linear programming problem for which there are available polynomial-time algorithms. Alternatively, it can be converted to a min-cost flow or circulation problem, for which there are polynomial-time algorithms [GKNV93]. Gansner *et al.* have introduced an integer linear programming (ILP) model of the problem and a specific network simplex algorithm for solving it which although not proven polynomial-time finds a layering with the minimum number of dummy vertices reportedly fast [GKNV93].

Other parameters of a layering that reflect on the quality of the drawing are the width and the height of a layering and the edge density between adjacent layers. The *height* of a layering is the number of layers, and the *width* is the maximum number of vertices in a layer. Usually these two parameters are used to approximate the dimensions of the final drawing. When measuring the width of a layering the contribution of the dummy vertices may or may not be taken into account. A more precise definition of the layering width takes into account both variable vertex widths and the contribution of the dummy vertices [BLME02, HN02a]. The *area* of a layering, used to approximate the area of the final drawing, is defined as the product of the layering width and the layering height.

A layering with the minimum height can be found in linear time by the longest-path algorithm [ES90]. It is also easy to find a layering with the minimum number of dummy vertices subject to an upper bound on the number of layers. The ILP model of Gansner *et al.* can be easily modified to take such an upper bound into account without making it more difficult to solve.

It is trivial to find a layering with the minimum number of original vertices per layer and no upper bound on the height. Any layering with a single vertex per layer is an optimal solution. However, it is NP-hard to find a layering with a given upper bound on the width if the width of the dummy vertices is considered greater than zero [BLME02]. A few heuristics have emerged since for layering with the minimum width and consideration of dummy vertices [BELM01, TNB04].

It is also NP-hard to find a layering with given upper bounds both on the height and on the width even without taking into account the contribution of the dummy vertices to the width and if all original vertices have the same unit width [ES90]. This variation of the layer assignment problem is equivalent to the precedence-constrained multiprocessor scheduling (PCMS) problem. The Coffman-Graham algorithm, which is an early and highly influential polynomial-time algorithm for solving PCMS approximately, has been also largely employed as a layering algorithm [CG72]. Healy and Nikolov have designed a branch-and-cut algorithm which finds layerings with the minimum number of dummy vertices and with height and width within pre-specified upper bounds [HN02a]. It is not a polynomial-time algorithm but it is an efficient way to find an optimal solution if the problem is not infeasible. The branch-and-cut algorithm of Healy and Nikolov takes into account variable vertex widths as well as the contribution of the dummy vertices to the width of the layering.

The *edge density* between layers L_i and L_j with $i < j$ is defined as the number of edges (u, v) with $u \in L_j \cup L_{j+1} \cup \ldots \cup L_h$ and $v \in L_0 \cup L_1 \cup \ldots \cup L_i$. The edge density of a layered DAG is the maximum edge density between adjacent layers. Naturally, drawings with low maximum and average edge density are clearer and easier to comprehend provided they are also compact and having not too many long edges. There has not been proposed any algorithm which find layerings specifically with consideration of edge density. A few studies have compared the edge density in layerings found by some of the algorithms mentioned above [HN02b, NT06, TNB04].

In the next section we take a closer look at the algorithms used for layer assignment.

13.3.2 Layer Assignment Algorithms

The easiest way to partition the vertex set of a DAG into layers is to take any spanning tree of the underlying undirected graph. It can be the depth first search tree or the breadth first search tree, for example. For each spanning tree a layering can be generated by picking a vertex and assigning it to an arbitrary layer, L_i. Then assign all its neighbours in the tree to either layer L_{i-1} or L_{i+1} depending on the direction of the connecting tree-edge. Then repeat the same with all neighbours of the already assigned vertices and so on until all vertices are assigned to a layer. It cannot be guaranteed that the set of layers will start with L_1 but this can be easily fixed by shifting the whole set of layers.

In general, this method does not guarantee any properties of the layering. The layer assignment algorithms, which find layerings subject to some of the criteria discussed in the previous section, broadly fall into two groups. The first group of algorithms are adopted from the area of static precedence-constrained multiprocessor scheduling. They produce layerings with either the minimum height or a specified maximum number of vertices per layer. The second group of algorithms employ network simplex and branch-and-cut techniques, respectively, for minimizing the number of dummy vertices.

List Scheduling Algorithms

The precedence-constrained multiprocessor scheduling problem is the problem of scheduling n causally related tasks (which represent a parallel program) on m processors with the goal of minimizing the completion time of the parallel program. This problem is NP-hard when $m < \infty$ [Ulm75]. It is also known as *static scheduling* because all the tasks with their causal relationship are given in advance and the schedule must be constructed prior executing any of them [KA99]. A simplified version of this problem, when all the tasks have the same computational cost and the communication time between tasks is neglected, is equivalent to the problem of finding a layering of a DAG with at most m vertices per layer and the minimum number of layers. Thus, the earliest static scheduling algorithms which deal with simplified models have also found an application as DAG layering algorithms.

Most of the static scheduling algorithms are variations of a generic list scheduling technique which consists of two main steps:

1. Build a scheduling list that contains all the tasks.
2. While the scheduling list is not empty remove the first task from it and schedule it for execution on a processor which allows earliest start time.

There are two list scheduling algorithms that have been widely employed as layering algorithms: the *longest-path algorithm* and the Coffman-Graham algorithm.

The Longest-Path Algorithm

The longest-path algorithm solves the static scheduling problem for $m = \infty$. Let π be the number of vertices in the longest directed path in a DAG. The longest path algorithm builds the scheduling list by assigning priority π to the vertices without outgoing edges. If all immediate successors of a vertex have been assigned a priority then that vertex is assigned the lowest of the priorities of its immediate successors minus one. This is repeated until all vertices are assigned a priority. The vertices with the same priority k form layer $L_{\pi-k+1}$. It has been shown that the longest-path algorithm has linear time complexity. [Meh84].

Algorithm 13.6 is a version of the longest-path algorithm where vertices are assigned to layers as soon as they are assigned priority. It employs two vertex sets U and Z which are empty in the beginning. The value of the variable *current_layer* is the label of the layer currently being built. As soon as a vertex gets assigned to a layer it is also added to the set U. Thus, U is the set of all vertices already assigned to a layer. Z is the set of all vertices assigned to a layer below the current layer. A new vertex v to be assigned to the current layer is picked among the vertices which have not been already assigned to a layer,

Figure 13.6   The Longest-Path Algorithm(G)

  Requires: DAG $G = (V, E)$

$U \leftarrow \phi$
$Z \leftarrow \phi$
$currentLayer \leftarrow 1$
while $U \neq V$ **do**
  Select vertex $v \in V \setminus U$ with $N_G^+(v) \subseteq Z$
  if v has been selected **then**
    Assign v to the layer with a number $currentLayer$
    $U \leftarrow U \cup \{v\}$
  end if
  if no vertex has been selected **then**
    $currentLayer \leftarrow currentLayer + 1$
    $Z \leftarrow Z \cup U$
  end if
end while

i.e., $v \in V \setminus U$, and which have all their immediate successors already assigned to the layers below the current one, i.e., $N_G^+(v) \subseteq Z$.

The advantages of the longest path algorithm are its simplicity and its linear time complexity. The layerings it finds have the minimum height. However, it performs very poorly in terms of drawing area, number of dummy vertices and edge density [HN02b]. The longest-path layerings tend to be very wide at the bottom layers.

The Coffman-Graham Algorithm

The second list scheduling algorithm used for DAG layering is the Coffman-Graham algorithm [CG72] which is based on an earlier algorithm by Hu [Hu61]. It approximately solves the NP-hard static scheduling problem for $m < \infty$. The technique used for building the scheduling list is more complex than the one used by the longest path algorithm. The worst-case time complexity of the Coffman-Graham algorithm is $O(|V|^2)$. It guarantees a layering with at most m original vertices per layer and in the worst case the height of the layering may become close to twice the optimal height [CG72].

The Coffman-Graham algorithm requires that the input graph $G = (V, E)$ has no transitive edges. An edge $e = (u, v) \in E$ is *transitive* if there is a directed path with a start vertex u and an end vertex v in G with length greater than 1, i.e., e is not the only directed path from u to v. Let $G_r = (V, E_r)$ and $G_c = (V, E_c)$ be also DAGs. G_c is a *transitive closure* of G if for each pair of vertices $u, v \in V$ there is a directed path from u to v in G if and only if $(u, v) \in E_c$. G_r is a *transitive reduction* of G if G_r is a DAG with a minimum number of edges among all the DAGs which have the same transitive closure as G. If a DAG contains transitive edges then the Coffman-Graham algorithm can be applied to its transitive reduction. A DAG's transitive reduction can be computed in $O(M(|V|))$ time, where $M(n)$ is the time for computing the product of two $n \times n$ matrices [AGU72].

Let $G = (V, E)$ be a DAG without transitive edges. The Coffman-Graham algorithm is the two-step Algorithm 13.7. The first step consists of computing unique labels $\lambda : V \to \mathbb{N}$ of all the vertices of G which then are used at the second step as priority for placing the vertices in layers. The computation of the labels at the first step involves the comparison

Figure 13.7   The Coffman-Graham Layering Algorithm

Require: A DAG $G = (V, E)$ without transitive edges and an integer $W > 0$

> **for all** $v \in V$ **do**
> $\lambda(u) \leftarrow \infty$
> **end for**
> **for** $i \leftarrow 1$ to $|V|$ **do**
>   Choose $v \in V$ with $\lambda(v) = \infty$ such that $N_G^-(v)$ is minimized;
> $\lambda(u) \leftarrow i$;
> **end for**
> $k \leftarrow 1, L_1 \leftarrow \emptyset, U \leftarrow \emptyset$
> **while** $U \neq V$ **do**
>   Choose $v \in V \setminus U$ such that $N_G^+(v) \subseteq U$ and $\lambda(v)$ is maximized;
>   **if** $|L_k| \leq W$ and $N_G^+(v) \subseteq L_1 \cup L_2 \cup \ldots \cup L_{k-1}$ **then**
> $L_k \leftarrow L_k \cup \{v\}$
>   **else**
> $k \leftarrow k + 1; L_k \leftarrow \{v\}$
>   **end if**
> $U \leftarrow U \cup \{v\}$
> **end while**

between vertex sets defined as follows. If U_1 and U_2 are two sets of vertices then $U_1 < U_2$ if either

- $U_1 = \emptyset$ and $U_2 \neq \emptyset$; or
- $U_1 \neq \emptyset$, $U_2 \neq \emptyset$, and $\max\{\lambda(v) : v \in U_1\} < \max\{\lambda(v) : v \in U_2\}$; or
- $U_1 \neq \emptyset$, $U_2 \neq \emptyset$, $\max\{\lambda(v) : v \in U_1\} = \max\{\lambda(v) : v \in U_2\}$, and $U_1 \setminus \{v : \lambda(v) = \max\{\lambda(u) : u \in U_1\}\} < U_2 \setminus \{v : \lambda(v) = \max\{\lambda(u) : u \in U_2\}\}$

In the second step each vertex is assigned to a layer starting from the bottom layer and going upward keeping the maximum number of vertices in a layer less than or equal to an upper bound W.

It has been observed that Coffman-Graham layerings have a large amount of dummy vertices and when they are taken into account the area of the layerings can be even worse than the area of the longest path layerings [HN02b].

Nikolov and Tarassov have proposed a vertex-promotion improvement heuristic which can be applied after either the longest-path algorithm or the Coffman-Graham algorithm for reducing the number of dummy vertices [NT06]. It is a cubic algorithm which is compensated by its simplicity. We describe the vertex-promotion heuristic in the following section.

Layering with the Minimum Width

It is NP-hard to find a layering with the minimum width if the dummy vertices are assigned non-negative width. This problem can be solved exactly by the Healy and Nikolov's branch-and-cut layering algorithm described in Section 13.3.2. In this section we describe the fast heuristic approach proposed by Tarassov *et al.* [TNB04]. Their min-width layering algorithm is the first successful attempt to design a heuristic for layering with the minimum width and consideration of dummy vertices. An earlier attempt is the heuristic developed by Branke *et al.* [BELM01].

Figure 13.8 Min-width(G, W, c)

Requires: DAG $G = (V, E)$, integers W and c

$U \leftarrow \phi$; $Z \leftarrow \phi$
$currentLayer \leftarrow 1$; $widthCurrent \leftarrow 0$; $widthUp \leftarrow 0$
while $U \neq V$ **do**
   Select vertex $v \in V \setminus U$ with $N_G^+(v) \subseteq Z$ and ConditionSelect
   if v has been selected **then**
      Assign v to the layer with a number $currentLayer$
      $U \leftarrow U \cup \{v\}$
      $widthCurrent \leftarrow widthCurrent - d^+(v) + 1$
      $widthUp \leftarrow widthup + d^-(v)$
   end if
   if no vertex has been selected OR ConditionGoUp **then**
      $currentLayer \leftarrow currentLayer + 1$
      $Z \leftarrow Z \cup U$
      $widthCurrent \leftarrow widthUp$
      $widthUp \leftarrow 0$
   end if
end while

The min-width algorithm, presented as Algorithm 13.8, is roughly based on the longest-path algorithm which is shown in detail in Algorithm 13.6. Besides the DAG G the min-width algorithm has two input parameters W and c which are explained below.

Similar to the longest-path algorithm, the min-width algorithm builds the layering layer by layer starting from layer 1. The two variables widthCurrent and widthUp are used to store the width of the current layer and the width of the layers above it, respectively. The width of the current layer, widthCurrent, is calculated as the number of original vertices already placed in that layer plus the number of potential dummy vertices along edges with a source in $V \setminus U$ and a target in Z (one dummy vertex per edge). The variable widthUp provides an estimation of the width of *any* layer above the current one. It is the number of potential dummy vertices along edges with a source in $V \setminus U$ and a target in the current layer (one dummy vertex per edge).

Vertex v is selected to be placed in a layer subject to an additional condition ConditionSelect which is true if v is the vertex with the maximum out-degree among the candidates to be placed in the current layer. Such a choice of v results in maximum reduction of widthCurrent. If either no vertex has been selected or ConditionGoUp is true then the current layer is completed and the algorithm moves to the next layer. ConditionGoUp is true if either:

- widthCurrent $\geq W$ and $d^+(v) < 1$, or
- widthUp $\geq c \times W$.

It is required that $d^+(v) < 1$ when widthCurrent $\geq W$ because the initial value of widthCurrent is determined by the dummy vertices in the current layer and it gets smaller (or at least it does not change) when a vertex with a positive out-degree gets placed in the current layer. In that case, the dummy vertices along edges with a source v are removed from the current layer and get replaced by v. If $d^+(v) \geq 1$, the condition widthCurrent $\geq W$ on its own is not a reason for moving to the upper layer because there is still a chance to add vertices to the current layer which will reduce `widthCurrent`. If $d^+(v) < 1$ then the assignment of v to the current layer increases `widthCurrent` because it does not replace any dummy vertices. This is an indication that `widthCurrent` can not be reduced further.

The min-width layering algorithm has the same time complexity as the longest-path algorithm which has been shown to run in linear time [Ulm75]. Through an extensive computational study Tarassov *et al.* have determined that the narrowest layerings are found for $1 \leq UBW \leq 4$ and $1 \leq c \leq 2$. Thus, they propose to run the algorithm for $UBW \in \{1, 2, 3, 4\}$ and $c \in \{1, 2\}$, choose the narrowest of the eight layerings and apply to it vertex-promotion heuristic described in the following section.

Improvement by Promotion of Vertices

The list-scheduling based layering algorithms described above have been shown to find layerings with a relatively large number of dummy vertices. Nikolov and Tarassov have proposed a simple vertex-promotion heuristic that can be applied to any layering for reducing its dummy vertex count [NT06].

The vertex-promotion heuristic modifies a given layering $\mathcal{L} = \{L_0, L_1, \ldots, L_h\}$ of a DAG G by promoting vertices from the layer where they are placed to the layer above. It is applied only to the original DAG vertices, not to the dummy vertices. To *promote* vertex v with $l(v, \mathcal{L}) = k$ is to move v from L_k to L_{k+1} which results in a new partition $\mathcal{L}^* = \{L_0, \ldots, L_k \setminus \{v\}, L_{k+1} \cup \{v\}, \ldots, L_h\}$. If $v \in L_h$ has to be promoted then a new empty layer L_{h+1} is added to the layering and v is promoted to it. If v has an immediate predecessor placed in layer L_{k+1} then $\mathcal{L}^*$ is *not* a layering of G. To ensure that the result of the promotion of vertex v to layer L_{k+1} is a layering all immediate predecessors of v in layer L_{k+1} (if there is any) have to be promoted to layer L_{k+2}; the same applies to their immediate predecessors and so on.

The recursive function which performs the described vertex promotion is shown in Algorithm 13.9. It takes vertex v as an input parameter and returns *dummydiff* which is the difference between the number of dummy vertices before and after the promotion v. In the **for** loop, each immediate predecessor u of v which lies in the layer above v gets promoted. The return value of its promotion is added to *dummydiff*. Then we promote v, subtract from *dummydiff* the number of immediate predecessors of v, and add to it the number of immediate successors of v. That is, we promote v one layer up, recursively promoting in advance all its immediate predecessors which need to be promoted. The time complexity of `PromoteVertex` is $O(|E|)$ because in the worst case all DAG edges might be traversed while promoting vertices recursively.

Then the vertex-promotion heuristic consists of two nested loops shown in Algorithm 13.10, an external **repeat-until** loop and an internal **for** loop. In the internal loop all vertices in a layered DAG are scanned in no particular order and each vertex with a positive in-degree gets promoted by `PromoteVertex` (see Algorithm 13.9) if its layering-preserving promotion reduces the total number of dummy vertices. The external loop goes on until the internal loop makes no promotion.

When performed after the min-width layering algorithm, described above, the vertex-promotion heuristic performs only promotions which do not increase the maximum number of vertices (original plus dummy) in a layer.

There is empirical evidence that 80 iterations of the **repeat-until** loop are enough for achieving a significant reduction of the number of dummy vertices for graphs with up to 100 vertices. If the number of iterations of the **repeat-until** loop is $O(|V|)$ then the vertex-promotion heuristic is cubic in the worst case.

Figure 13.9 `PromoteVertex(v)`

Require: A layered DAG $G = (V, E)$ with the layering information stored in a global vertex array of integers called *layering*; a vertex $v \in V$.

$dummydiff \leftarrow 0$
for all $u \in N_G^-(v)$ **do**
   if $layering[u] = layering[v] + 1$ **then**
      $dummydiff \leftarrow dummydiff + $ `PromoteVertex(u)`
   end if
end for
$layering[v] \leftarrow layering[v] + 1$
$dummydiff \leftarrow dummydiff - N_G^-(v) + N_G^+(v)$
return $dummydiff$

Figure 13.10 `Vertex-Promotion Heuristic`

Require: $G = (V, E)$ is a layered DAG; a valid layering of G is stored in a global vertex array called *layering*.

$layeringBackUp \leftarrow layering$
repeat
   $promotions \leftarrow 0$
   for all $v \in V$ **do**
      if $d^-(v) > 0$ **then**
         if `PromoteVertex(v)` < 0 **then**
            $promotions \leftarrow promotions + 1$
            $layeringBackUp \leftarrow layering$
         else
            $layering \leftarrow layeringBackUp$
         end if
      end if
   end for
until $promotions = 0$

Network-Simplex Layering Algorithm

The integer linear programming (ILP) approaches to the layering algorithm have been introduced for layering with the minimum number of dummy vertices. The first such approach is the layering technique designed by Gansner, Koutsofios, North and Vo for their system **dot**,[2] which is probably the most popular system for layered graph drawing [GKNV93].

---

[2]http://www.graphviz.org/

Figure 13.11 `Network Simplex Layer Assignment`
Require: $G = (V, E)$ is a DAG.

```
feasible_tree
while (e =leave_edge()) ≠ nil do
   f =enter_edge(e)
   exchange(e, f)
end while
normalize()
balance()
```

They model the layering problem by the following integer linear program:

$$\min \sum_{(u,v)\in E} l(u, \mathcal{L}) - l(v, \mathcal{L})$$

$$\text{subject to:}\quad l(u, \mathcal{L}) - l(v, \mathcal{L}) \geq 1, \quad \forall (u, v) \in E$$

$$l(u, \mathcal{L}) \geq 0, \quad \forall u \in V$$

$$\text{all } l(u, \mathcal{L}) \text{ are integer}$$

The linear programming relaxation of this integer program always has an integer solution because its constraint matrix is totally unimodular [NW88]. Thus, the integer program can be solved by the simplex method or any of the polynomial-time algorithms for solving linear programs. If we add an additional set of constraints of the type $l(u, \mathcal{L}) \leq H$, where H is an upper bound on the number of layers, the constraint matrix remains totally unimodular. Thus, the problem of finding a layering with the minimum number of dummy vertices subject to an upper bound on the number of layers is also in P.

Gansner *et al.* go further by introducing a network simplex algorithm for solving their ILP formulation [GKNV93]. It has not been proved to run in polynomial-time but reportedly requires a few iterations and runs fast. Its main part is presented in Algorithm 13.11.

The network simplex algorithm is based on the idea that each spanning tree of the underlying undirected graph of a DAG induces a family of equivalent (in terms of dummy vertex count) layerings. The algorithm starts with an initial spanning tree which is modified by replacing edges in order to get a spanning tree that induces a layering with the minimum number of dummy vertices. The procedure `normalize()` at the end of the algorithm is the one that makes sure the set of layers in the induced layering start from L_1.

The network simplex starts with an initial spanning tree built by the procedure `feasible_tree`. Gansner *et al.* suggest to compute a longest-path layering and then take a spanning tree of short subject to the layering edges. Then each iteration of the **while**-loop removes an edge from the spanning tree which breaks the tree into two connected components. Then a new edge is added to the tree that connects the two components into a new spanning tree. The two edges to leave and enter the tree respectively are chosen so that the new tree induces a layering with a lower dummy vertex count.

The edge to leave the tree at each iteration of the **while**-loop is chosen by the function `leave_edge()` which picks an edge with a negative *cut value*, or nil if all edges have non-negative cut value. The cut value is defined as follows. If an edge e is removed from the spanning tree, it breaks into two connected components, a *tail* and a *head*. The tail is the component that contains the source of e and head is the component that contains its target. The cut value of e is the number of all directed edges from the tail to the head, including e,

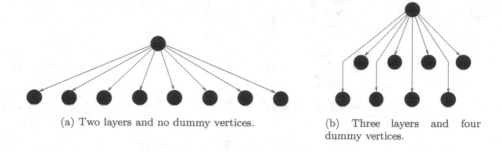

(a) Two layers and no dummy vertices.

(b) Three layers and four dummy vertices.

Figure 13.12 Two alternative layerings of the same DAG: a layering with the minimum number of dummy vertices may become too wide.

minus the number of all directed edges from the head to the tail. Typically a negative cut value of an edge means that the dummy vertex count can be reduced by lengthening that edge as much as possible, until one of the head-to-tail edges becomes tight. That tight edge is the one chosen by the function `enter_edge()`. It is the edge with the minimum span in the layering induced by the spanning tree before removing e.

After the end of the **while**-loop the spanning tree induces a layering with the minimum number of dummy vertices. The procedure `balance()`, applied at the end, moves vertices with equal in- and out-degree to a feasible layer with the fewest vertices. This is done in order to have more even distribution of vertices between layers. Gansner *et al.* also show how the network-simplex algorithm can work for graphs with weighted edges and with edges which are required to have span greater than 1 [GKNV93].

In general, layerings with the minimum number of dummy vertices lead to compact drawings. However some patterns in a DAG can result in too wide layerings with the minimum number of dummy vertices as shown in Figure 13.12.

Healy-Nikolov's Branch-and-Cut Algorithm

Another ILP approach is the branch-and-cut layering algorithm introduced by Healy and Nikolov [HN02a]. It finds layerings with the minimum number of dummy vertices subject to upper bounds on both the height and the width of the layering if there is any feasible solution. Variable vertex width and the contribution of the dummy vertices to the width of the layering can be taken into account. Since it solves exactly an NP-hard problem, this algorithm has exponential running time. It is especially designed for producing high quality layerings which satisfy exactly the pre-specified upper bounds on the width and the height.

Consider a DAG $G = (V, E)$ and let x be the incidence vector of a subset of $V \times \{1, \ldots, H\}$. Healy and Nikolov model the layering problem by the following ILP formulation.

$$\min \sum_{(u,v) \in E} \left(\sum_{k=\varphi(u)}^{\rho(u)} k x_{uk} - \sum_{k=\varphi(v)}^{\rho(v)} k x_{vk} \right) \tag{13.1}$$

$$Subject\ to \qquad \sum_{k=\varphi(v)}^{\rho(v)} x_{vk} = 1 \qquad \forall v \in V \tag{13.2}$$

$$\sum_{i=\varphi(u)}^{k} x_{ui} + \sum_{i=k}^{\rho(v)} x_{vi} \leq 1 \qquad \forall k \in LS(u) \cap LS(v),\ \forall (u,v) \in E \tag{13.3}$$

$$\sum_{v \in V_k^*} w_v x_{vk} + \mathcal{D}_k \leq W \qquad \forall k = 1, \ldots, H \tag{13.4}$$

$$\sum_{v \in V_k^*} x_{vk} \geq 1 \qquad \forall k = 1, \ldots, \pi(G) \tag{13.5}$$

all x_{vk} are binaries

W and H are upper bounds on the width and height of the layering respectively; $\pi(G)$ is the number of vertices in the longest directed path in G; $\varphi(v)$ and $\rho(v)$ are respectively the lowest and the highest layer where vertex v can be placed in; $LS(v) = \{\varphi(v), \ldots, \rho(v)\}$ and $V_k^* = \{v \in V : \varphi(v) \leq k \leq \rho(v)\}$. The objective minimizes the sum of edge spans, i.e., the number of dummy vertices. Equalities (13.2) force each vertex to be placed in exactly one layer; inequalities (13.3) force each edge to point downward; and inequalities (13.5) introduce the additional requirement of having at least one vertex in the first $\pi(G)$ layers. This reduces the number of identical layerings (but shifted vertically) if the height of the solution is less than the upper bound H.

Inequalities (13.4) restrict the width of each layer (including the dummy vertices) to be less than or equal to W: the first term on the left hand side represents the contribution of the real vertices to the width of layer V_k while $\mathcal{D}_k$ represents the contribution of the dummy vertices. We set

$$\mathcal{D}_k = \sum_{e=(u,v) \in E} w_e^d \left(\sum_{l>k}^{\rho(u)} x_{ul} - \sum_{l \geq k}^{\rho(v)} x_{vl} \right)$$

where w_e^d is the width of the dummy vertices along edge e. The difference of the two sums in the parentheses is 1 if edge $e = (u,v)$ spans layer V_k and 0 otherwise.

Healy and Nikolov propose solving their ILP formulation in a branch-and-bound framework with the employment of a cutting-plane algorithm at each vertex of the branch-and-bound tree. The cutting-plane algorithm generates valid inequalities for the constraint polytope of their formulation, some of which are facet-defining. Healy and Nikolov report that the running-time of this branch-and-cut algorithm is close to the time necessary for the ILP solver of CPLEX[3] to solve the formulation and they show some examples where the branch-and-cut algorithm is significantly faster than CPLEX.

13.3.3   The Layering Algorithms Compared

If any layering is acceptable then probably the easiest way to construct one is either to use the longest-path algorithm (Section 13.3.2) or to find any spanning tree of the underlying undirected graph and take the layering induced by it as described in the beginning of Section 13.3.2. If there is an upper bound on the the number of original vertices per layer then the Coffman-Graham algorithm, described in Section 13.3.2, is the best solution.

---

[3]http://www.ilog.com/products/cplex/

Although the longest-path algorithm finds layerings with the minimum number of layers and the Coffman-Graham algorithm finds layerings with a pre-specified maximum number of vertices in a layer, their layerings typically lead to drawings which occupy large drawing area and have too many long edges.

For a compact layering, the network simplex algorithm of Gansner *et al.*, described in Section 13.3.2, is probably the best fast solution. It finds layerings with the minimum number of dummy vertices which are also very compact in general. However, there are particular patterns in graph that may make the network simplex layering either too wide or too long. The branch-and-cut algorithm of Healy and Nikolov, outlined in Section 13.3.2, is much slower but in addition to the network simplex algorithm it guarantees that the layering's width and height will be within pre-specified bounds. It also considers variable vertex width and the contribution of the dummy vertices to the width of the layering.

The first fast heuristic for layer assignment with the minimum number of vertices per layer when both the original and the dummy vertices are considered is the min-width algorithm described in Section 13.3.2. It is not optimal but when followed by the vertex promotion heuristic, described in Section 13.3.2, it finds layerings which on average are narrower than the layerings found by any other known layering algorithm. The same vertex-promotion heuristic significantly improves the layerings found by the longest-path and the Coffman-Graham algorithms and makes them comparable to the layerings found by the network simplex algorithm. The longest-path algorithm followed by the vertex-promotion heuristic is probably the easiest to implement layering algorithm which results in good-quality layerings. Its only disadvantage is the relatively slow running time, which is cubic in the worst case.

13.3.4   Layer-Assignment with Long Vertices

The layer-assignment algorithms described above assume that all the vertices have similar height and can be aesthetically arranged on parallel horizontal levels without too much blank space between the levels. However, in some applications a few vertices in the input digraph may have large labels which can make them occupy significantly larger space in the vertical direction than the rest of the vertices. Such *long* vertices can be allowed to occupy more than one horizontal level in order to achieve aesthetically pleasant drawing.

Misue *et al.* propose to assign vertices to layers with one of the describe algorithms assuming all vertices have the same size. Then in a postprocessing step vertices can be enlarged to their original size and the eventual intersection between vertices are removed [MELS95]. However, this approach may lead to drawings which are not aesthetically acceptable.

Recently two studies have proposed layering algorithms that consider the actual size of the vertices while assigning them to layers. North and Woodhull have introduced an algorithm that assigns each vertex to two layers which correspond to the lower and the upper bounds of its height, respectively [NW01]. In a subsequent step vertices and edges which cross one or more layers are split into chains of vertices to obtain simpler layering. The second algorithm has been introduced by Friedrich and Schreiber [FS04]. It breaks the long vertices into chains of vertices while assigning them to layers.

While the special treatment of long vertices makes the final drawing compact, it also complicates the subsequent steps of the Sugiyama method. The vertex-ordering and the coordinate-assignment steps need to take the long vertices into account. It also becomes more difficult to rout edges so that they do not go through the long vertices. This may result in a large number of edge bends, which is compensated by their short length in the compact drawing.

13.4   Edge Concentration

Edge concentration is an optional step in the Sugiyama algorithm. It reduces the edge density between adjacent layers and the number of edge crossings. It can also reduce the dummy vertex count. The eventual drawbacks are that edge concentration modifies the graph and may increase the number of layers.

Consider a layered DAG $G = (V, E)$ with a layering $L = \{L_1, \ldots, L_h\}$. An *intersection* in G is a complete bipartite (biclique) subgraph I with a set of source vertices S_I and a set of target vertices T_I, such that $S_I \subseteq L_j$, and $T_I \subseteq L_i$ for some $i < j$. We use the notation $I = (S_I, T_I)$. If $|S_I| = |T_I| = 1$ then the intersection is *trivial*. The *vertex size* of I is $|S| + |T|$, and the *edge size* of I is $|S| * |T|$.

To perform *edge concentration* on a given nontrivial intersection I is to

- remove all edges between S_I and T_I from G
- add a new *edge-concentration vertex ec* to G, i.e., $V \leftarrow V \cup \{ec\}$
- add edges $\{e = (ec, u) : u \in T_I\}$ and $\{e = (u, ec) : u \in S_I\}$ to E.

That is, all edges of the intersection C are removed from the graph, a new edge-concentration vertex is added and all vertices in S_I and T_I are connected to it by an edge to form a star-like subgraph. An example is shown in Figure 13.13. If S_I and T_I occupy adjacent layers, i.e., $j - i = 1$, then a new layer is introduced between L_i and L_j and the edge-concentration vertex ec is placed in it. Otherwise ec is placed in some layer L_k with $i < k < j$.

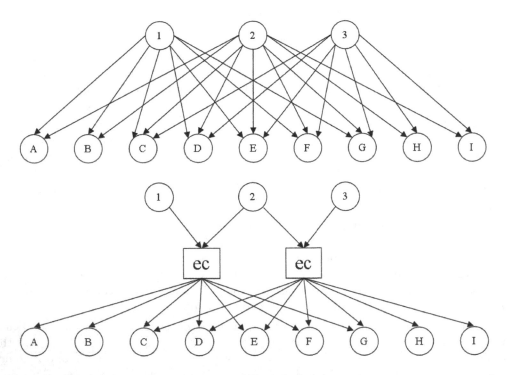

Figure 13.13   An example of a bipartite graph before and after the introduction of two edge-concentration vertices labeled "ec" [New89].

A set of intersections $\mathcal{I} = \{I_1, \ldots, I_k : 1 \leq k \leq |E|\}$ is an *intersection cover* of G if each edge of G is contained in at least one intersection in the set. The edge concentration step within the Sugiyama method consists of finding an intersection cover $\mathcal{I}$ of G and performing edge concentration on each non-trivial intersection in $\mathcal{I}$. If intersections between non-adjacent layers are considered then the dummy vertices are ignored and edge concentration is applied to intersections of original vertices and edges. After the edge concentration step dummy vertices are introduced again to subdivide long edges.

In the next section we discuss the different approaches to choosing an intersection cover for edge concentration.

13.4.1   Intersection Cover

The most important part of the edge concentration step is the choice of an intersection cover. There are two alternative approaches to this problem proposed in the graph drawing literature: choose either only intersections between adjacent layers, or only intersections between non-adjacent layers.

The only edge concentration approach with intersections between non-adjacent layers is the one employed by AT&T's **dot** [GKN02]. Only non-trivial intersections with a single target vertex are considered. This is a simple but fast solution for the edge concentration step.

Newbery as well as Eppstein *et al.* suggest a different approach to building the intersection cover [New89, EGM04]. The non-trivial intersections are only intersections between adjacent layers and the choice of intersections between two adjacent layers does not depend on the intersections between other layers. Thus, the problem of choosing an intersection cover of the whole graph is reduced to the problem of choosing a biclique (i.e., complete bipartite graph) cover of a bipartite graph.

The best biclique cover from the point of view of edge concentration is the one that will result in the fewest number of edges after applying edge concentration. This is what Newbery calls the Edge Concentration problem and it is defined as a decision problem as follows.

Edge Concentration
Instance: A bipartite graph $G = (V, E)$ and a positive number K.
Question: Is there an biclique cover $\mathcal{I} = \{I_1, \ldots, I_k : 1 \leq k \leq |E|\}$ of G with $\sum_{i=1}^{k} vs(I_i) \leq K$?

Edge Concentration is NP-complete [Lin00]. Newbery proposes a polynomial-time heuristic algorithm for solving its optimization version, i.e., to find a biclique cover $\mathcal{I} = \{I_1, \ldots, I_k : 1 \leq k \leq |E|\}$ with the minimum $\sum_{j=1}^{k} vs(I_j)$ [New89]. We describe the heuristic in detail in the following section.

Related to Edge Concentration is the problem of finding a biclique cover of a bipartite graph with no more than $K > 0$ bicliques. This problem is known as Complete Bipartite Subgraph Cover, problem GT18 of Garey and Johnson's NP-complete problems [GJ79]. In their paper on confluent layered drawings, Eppstein *et al.* propose two-layer edge concentration with a biclique cover which is an approximate solution to the optimization version of Complete Bipartite Subgraph Cover [EGM04]. Fishburn and Hammer have shown that Complete Bipartite Subgraph Cover is equivalent to a simply-restricted edge coloring problem which in turn can be transformed to a vertex coloring problem for bipartite graphs [FH96]. Thus, Eppstein *et al.* propose the biclique cover for edge concentration to be computed by one of the vertex coloring algorithms and specifically by either the

Recursive Largest First (RLF) algorithm of Leighton[Lei79] or the DSATUR algorithm of Brélaz [Bré79]. Both vertex coloring algorithms have $O(|E|^3)$ worst-case time complexity and their solutions can be transformed to a biclique cover in $O(|E|^2)$ time.

In the next section we describe in detail Newbery's heuristic for solving Edge Concentration approximately.

13.4.2　Newbery's Algorithm

Newbery's algorithm (Algorithm 13.14) finds an approximate solution to the optimization version of Edge Concentration [New89]. Consider a bipartite graph $G = (V, E)$. Two lists of bicliques, B_1 and B_2, are maintained throughout. B_1 is the list of all possible bicliques with two source vertices at all times. The bicliques in B_1 are sorted in increasing order by number of target vertices. Initially B_2 contains only the empty biclique, i.e., a biclique with an empty source and target sets. At the end B_2 is a list of the non-trivial bicliques for edge concentration.

The algorithm takes an input parameter M which is a lower bound on the edge size of a biclique. Newbery defines the edge size of a biclique x as $|S_x| * (|T_x| - 1)$ which is slightly different from the actual number of edges in the biclique. This is chosen in order to avoid bicliques with a single target vertex.

The main part of the heuristic is the **for** loop that goes through all bicliques in B_1 with two nested **for** loops that go through all bicliques in B_2. Consider a biclique $x \in B_1$ with a source set S_x and a target set T_x. If the edge size of x is less than the lower bound M then x is discarder from B_1. Otherwise x is compared to each biclique y in B_2. If x and y have the same target set then the source vertices of x are added to the source vertices of y. If there is no biclique y in B_2 with the same target set as x then x is compared once again to all bicliques in B_2 in the second nested **for** loop. Let S_y and T_y be the source and target sets of y respectively. Two cases are considered:

- **Case 1.** If $T_x \subseteq T_y$ then add $(S_y \cup S_x, T_x)$ to the front of B_2 and remove T_x from the target set of y.
- **Case 2.** If $T_y \subseteq T_x$ then add $(S_x, T_x \setminus T_y)$ to the front of B_2 and add S_x to the source set of y, i.e., $S_y \leftarrow S_y \cup S_x$.

Note that if no other condition becomes true then at the end of the second nested **for** loop x will be compared to the biclique with the empty target set in B_2 which will result in adding x to B_2. At the end all bicliques in B_2 with the same target set are merged and bicliques with size less than M are discarded. The biclique with the empty target set will also be discarded from B_2.

The heuristic has $O(n^4)$ worst-case time complexity assumed $m < n^3$. The worst case is much worse than the typical case encountered in practice.

13.5　Vertex Ordering

For readability edge crossings are one of the crucial parameters of a graph drawing [Pur97]. While edge reversal and layer assignment can have an impact on this, it is through the ordering of vertices that minimizing edge crossings between adjacent layers is mainly achieved (crossings are dependent on the relative order of vertices and not their positions). Thus, this step is often known as the *crossing minimization* or *crossing reduction* step.

Crossing minimization is of interest to VLSI-layout researchers and so has a history that predates much of the graph drawing literature. Garey and Johnson showed that the

Figure 13.14 Newbery's Biclique Cover Heuristic

Require: A bipartite graph $G = (V, E)$ and an integer $M > 0$

$B_1 \leftarrow \phi$
$B_2 \leftarrow \phi$
for all pair of source vertices (u, v) **do**
    add the largest biclique with a source set $\{u, v\}$ to B_1.
end for
Sort the bicliques in B_1 in increasing order by number of target vertices
Add a biclique with empty source and target sets to B_2
for all $x \in B_1$ **do**
    if the $S_x * (T_x - 1) < M$ **then**
    Discard x
    else
        for all $y \in B_2$ **do**
            if $T_x = T_y$ **then**
                $S_y \leftarrow S_y \cup S_x$
                Continue the external **for** loop
            end if
        end for
        for all $y \in B_2$ **do**
            if $T_x \subseteq T_y$ **then**
                Add $(S_y \cup S_x, T_x)$ to the front of B_2
                $T_y \leftarrow T_y \setminus T_x$
                Continue the external **for** loop
            end if
            if $T_y \subseteq T_x$ **then**
                Add $(S_x, T_x \setminus T_y)$ to the front of B_2
                $S_y \leftarrow S_y \cup S_x$
                Continue the external **for** loop
            end if
        end for
    end if
end for

general crossing minimization problem, where permutations for both layers that minimize the number of crossings are required, is NP-hard [GJ83]; even if one layer is fixed, the problem still remains NP-hard [EW94]. To counter these apparently negative results, several heuristic and exact methods have been proposed.

While we will mainly be concerned with algorithms that *find* a drawing with few crossings, the decision version of the problem [GJ83, EW94] also is of interest. In particular, *fixed-parameter tractable* algorithms have been developed to answer the question of whether an ordering of the vertices on one "shore" of a bipartite graph admits a k-crossing (or less) solution when the vertices of the other shore remain fixed [DFH$^+$01a, DW02, DFK03]. In the latter an $O(1.4656^k + kn^2)$-time decision algorithm is presented.

We may assume that the graph $G = (V, E)$ is properly k-layered. That is, like the level graph in Definition 13.1 $V = V_1 \cup \cdots \cup V_k$, $V_i \cap V_j = \emptyset, 1 \leq i \neq j \leq k$, although we do not insist that sources appear on the first layer. The set of edges, $E = \{(u,v) | u \in V_i, v \in V_{i+1}, 1 \leq i \leq k-1\}$. We let $n_i = |V_i|, m = |E|$ and we call $N(u) = \{v \in V | (u,v) \in E\}$ the neighborhood of vertex u. An ordering or *permutation*, π_i, of each V_i provides a solution for the crossing minimization problem since it is the relative ordering along the line $y = l_i$ that causes edges incident on that layer to cross each other. What we seek, then, is the set of permutations, $\Pi = \{\pi_i | 1 \leq i \leq k\}$ that minimizes the edge crossings $C(G, \pi_1, \pi_2)$.

In the following sections we will describe the one-layer and two-layer crossing minimization problems and solutions. We then go on to discuss techniques for handling multi-layer graphs. Finally, we discuss an alternative to crossing minimization, where the goal is to initially extract (and draw) a large planar subgraph, and then draw the remaining edges.

13.5.1   One-Sided Crossing Minimization

Few systems attempt to globally minimize edge crossings. Instead, a heuristic approach based on one-layer crossing minimization is adopted. This problem, then, is key to many of the algorithms that have been proposed for the bi- and multi-partite crossing minimization problems. The goal of the one-layered crossing minimization (OLCM) problem is to find, for a given G and π_1, the permutation π_2 that minimizes $C(G, \pi_1, \pi_2)$.

Counting Crossings

Many of the algorithms we describe below require knowledge of the exact number of crossings between two layers. Algorithms that compute crossings fall in two categories: those that simply count the crossings and those that can *report* those edge crossings, also. Since it is possible to have $\Omega(|E|^2)$ crossings, the latter class of algorithms have running time $\Omega(|E|^2)$ in the worst case. The naive algorithm that considers every pair of edges and runs in $O(|E|^2)$ time is, in this sense, optimal.

A further issue is whether the algorithm can be used to compute a *crossing number matrix* which is used by many heuristics. The crossing number matrix, with entries c_{uv}, counts the number of edge crossings between edges incident to $u, v \in V_2$, when u is to the left of v ($\pi_2(u) < \pi_2(v)$). Since it assumes that the vertices in V_1 are fixed, the matrix is only relevant for OLCM. Note that since u is to the left of v or v is to the left of u in any solution then the sum $\sum_{u,v} \min(c_{uv}, c_{vu})$ yields a simple lower bound on the optimal number of crossings. Jünger and Mutzel report that this figure is surprisingly tight on a variety of graphs [JM97]. For dense graphs, Nagamochi [Nag05] presents an upper bound on OLCM of $1.2964 + 12/(\delta - 4)$, where $\delta > 4$ is the minimum degree of a vertex.

Computing the crossing number matrix can be done naively in $O(|E|^2)$-time although Sander [San94] proposes a sweep algorithm that can compute all entries in $O(|V_1| + |V_2| +$

$|E| + C$), where C is the number of crossings. Barth *et al.* [BJM02] investigate a number of different existing algorithms for computing the bilayer crossing number, as well as proposing a new $O(|E| \log |V_s|)$-time algorithm, where V_s is the smaller of the two sets V_1 and V_2. The algorithm is based on Waddle and Malhotra's *accumulation tree* used in their earlier algorithm [WM99]. Nagamochi and Yamada [NY04] have proposed two algorithms based on dynamic programming and divide-and-conquer that run in time $O(|V_1||V_2|)$ and $O(\min\{|V_1||V_2|, |E| \log |V_s|\})$, respectively; both algorithms use $O(|E|)$-space. For dense graphs these algorithms will outperform the $O(|E| \log |V_s|)$-time algorithm by Barth *et al.* While Sander's algorithm turns out to be uncompetitive for large graphs, it does have the advantage of being able to compute the crossing matrix, which is not possible with either of the other faster-running algorithms.

Heuristic Approaches

Eades and Kelly [EK86] propose three heuristics: 1) greedy insertion, 2) greedy switching, and 3) split for the problem. All three methods require the precomputation of the crossing number matrix.

The *greedy insertion* heuristic orders vertices left to right on the "free" layer according to the one which minimizes the number of crossings that the edges incident to u make with the edges incident to vertices to u's right. That is, u is chosen to minimize $\sum_{v \in R}$, where R is the set of unchosen vertices. The algorithm runs in time $O(|V_2|^2)$. The *greedy switching* heuristic compares adjacent vertices and switches them if the change in crossing count, $c_{uv} - c_{vu} > 0$, where vertex u immediately precedes v. With a precomputed crossing number matrix the algorithm's running time is obviously $O(|V_2|^2)$. Like greedy switching, the *split* heuristic is reminiscent of sorting. Here, however, the analogy is with *quicksort* where a vertex, p, is chosen as a "pivot" and the vertices are rearranged into two consecutive sets, V_l and V_r so that $c_{up} < c_{pu}$ for all vertices in V_l and $c_{pu} \leq c_{up}$ for all vertices in V_r; the left and right subsets of vertices are then ordered recursively. As with quicksort its worst-case running time is $O(|V_2|^2)$, but its expected running time is $O(|V_2| \log |V_2|)$.

Eades and Wormald [EW94] propose the *median* heuristic that assigns to each vertex $v \in V_2$ the median position of the x-coordinates of its neighbours, $N(v) \subseteq V_1$. This heuristic has the important property that it will find an ordering with 0 crossings if such an ordering exists; in general, it guarantees an ordering, π_2, such that $C(G, \pi_1, \pi_2) \leq 3C_{\mathrm{opt}}(G, \pi_1, \pi_2)$. Computing the median position for a vertex v has time complexity $O(|N(v)|)$ and, thus, all medians can be found in $O(|E|)$-time. Sorting the set V_2 according to the computed medians provides the ordering and requires $O(|V_2| \log |V_2|)$-time.

An alternative to placing a vertex at the median of its neighbours' x-coordinates is to place it at the *average* of them. This gives rise to the *barycenter* or averaging heuristic [STT81]. It has the same running time bounds as the median heuristic and it, too, will find a crossing-free ordering if one exists, but there does not exist a general performance guarantee as exists for the median heuristic.

Some other heuristics of note are Catarci's [Cat95] *assignment heuristic* which is based on an approximation of the linear assignment problem. Catarci claims that the heuristic is more accurate than the median heuristic, especially when applied to dense graphs. Dresbach [Dre94] has proposed a *stochastic heuristic* which, having calculated *assessment* numbers for each vertex and position (the assessment number is a term borrowed from statistics), begins placing vertices in the position with the smallest assessment position, updating the remaining numbers after each placement.

Genetic algorithms have also been employed as heuristic solutions to the one-layer crossing minimization problem. Mäkinen and Sieranta [MS94] encode a permutation of V_2 as a tuple. The mutation operator is defined as removing and re-inserting a random member x_i of the tuple to a new random position, with intervening members shifted accordingly. The crossover operation on tuples generates two points $i < j$, fixes the elements in the two tuples between i and j, and reorders the remainders of the two tuples so that they are both valid permutations. Branke *et al.* [UBSE98] also use a genetic algorithm for crossing minimization. Due to its more general setting, however, we will postpone its discussion until Section 13.7.

Exact Methods

Jünger and Mutzel [JM97] present a branch-and-cut algorithm for OLCM that draws on solving the *linear ordering* problem. Letting δ_{ij}^k be a 0-1 variable that represents the ordering of vertices i and j on level $k = 1, 2$, they develop an expression for the number of crossings of a pair of permutations, π_i, which will be the objective function to minimize. That is, let $\delta_{ij}^k = 1$ if $\pi_k(i) < \pi_k(j)$ and 0 otherwise; it is clear that $\delta_{ij} = 1 - \delta_{ji}$. Then

$$C(\pi_1, \pi_2) = \sum_{i=1}^{n_2-1} \sum_{j=i+1}^{n_2} \sum_{k \in N(i)} \sum_{l \in N(j)} \delta_{kl}^1 \delta_{ji}^2 + \delta_{lk}^1 \delta_{ij}^2 \tag{13.6}$$

For OLCM we assume a fixed π_1 and seek the ordering of V_2, π_2, that minimizes the crossings. That is, we wish to minimize

$$C(\pi_2) = \sum_{i=1}^{n_2-1} \sum_{j=i+1}^{n_2} \sum_{k \in N(i)} \sum_{l \in N(j)} \delta_{kl}^1 \delta_{ji}^2 + \delta_{lk}^1 \delta_{ij}^2 \tag{13.7}$$

As currently posed $C(\pi_2)$ is problematical since it involves quadratic terms. However, using the crossing number for a pair of vertices i and j in V_2 from before

$$c_{ij} = \sum_{k \in N(i)} \sum_{k \in N(i)} \delta_{lk}^1$$

we can rewrite equation (13.7) as

$$C(\pi_2) = \sum_{i=1}^{n_2-1} \sum_{j=i+1}^{n_2} c_{ji}(1 - \delta_{ji}^2) + c_{ij}\delta_{ij}^2 \tag{13.8}$$

$$\sum_{i=1}^{n_2-1} \sum_{j=i+1}^{n_2} (c_{ij} - c_{ji})\delta_{ij}^2 + \sum_{i=1}^{n_2-1} \sum_{j=i+1}^{n_2} c_{ji} \tag{13.9}$$

The problem then reduces to finding the vector $\delta^2 \in \{0, 1\}^{\binom{n_2}{2}}$ that orders the vertices. In order for the vertices to be given a consistent ordering, we need to impose a "3-cycle" constraint that says that if vertex i precedes vertex j and vertex j precedes vertex k then vertex i must precede vertex k.

Since c_{ij} and c_{ji} can be determined beforehand and since the second double sum of equation (13.9) is a constant, the problem can be written as the following linear program, where δ_{ij}^2 is replaced by the more usual x_{ij}

$$\text{minimize} \quad \sum_{i=1}^{n_2} \sum_{j=i+1}^{n_2} a_{ij} x_{ij} \tag{13.10}$$

$$0 \le x_{ij} + x_{jk} - x_{ik} \le 1 \qquad 1 \le i < j < k \le n_2 \tag{13.11}$$

$$0 \le x_{ij} \le 1 \qquad\qquad\qquad 1 \le i < j \le n_2 \tag{13.12}$$

$$x_{ij} \text{ an integer} \tag{13.13}$$

This is a well-known formulation of the linear ordering problem and from it the number of crossings can easily be computed. As the integrality constraint (equation (13.13)) causes significant difficulty from a computational complexity consideration, a standard approach is to solve a relaxation of the formulation. This approach, called *branch-and-cut*, is achieved by dropping initially the integrality constraint and as many of the constraints as is necessary, and introducing them subsequently on an as-needed basis. In this case since there are $2\binom{n_2}{3}$ 3-cycle constraints (equation (13.11)) and $2\binom{n_2}{2}$ hypercube constraints (equation (13.12)) it has been found to be reasonable to begin simply with the hypercube constraints and to introduce constraints for violated 3-cycles until a solution is found. If the solution is non-integral a branch step takes place where a non-integral variable is selected and two subproblems are solved, one using 0 as the value of the selected variable and the second using 1 as its value.

Algorithm Performance

The most comprehensive experimental analysis of the performance of the various approaches to OLCM that we have discussed is Jünger and Mutzel's [JM97]. Their focus is in bipartite graphs and they consider the OLCM and TLCM versions of the problem using most of the heuristics we have discussed above, and others. In this section we will focus on their findings for OLCM.

Two classes of experiments are performed. First, the algorithms are applied to random instances of graphs, of varying edge density, each with 20 vertices on each side. Secondly, motivated by the sparseness that is often observable in graphs of interest, a set of sparse graphs of varying sizes were considered. It is worth noting that as graphs get denser and the exact crossing count increases all algorithms perform relatively better. Thus, the second class of experiments may tell more about the algorithms' behaviors.

For the first set of experiments it was observed that the greedy switch, greedy insertion and split heuristics had significantly higher running times as the edge densities increased. This is most certainly due to the requirement to compute the crossing number matrix. The running time of the exact method was surprisingly competitive here, although its exponential behavior became more obvious on the second class of graphs. The split algorithm performed best in terms of solution quality and also fared very well in the second class of sparse graphs.

Over both classes of graphs it would appear that barycenter is the best performer. Its running time is always among the best and its solution quality on the sparse graphs is always with 3% of optimality. In spite of their similarity, the median algorithm performs poorer than barycenter on both classes of graphs. (The barycenter algorithm is also an easier algorithm to implement.)

Finally, Mäkinen and Sieranta [MS94] claim that their genetic algorithm outperforms the barycenter method in terms of accuracy, although running time is significantly higher.

13.5.2 Multi-Layer Crossing Minimization

In the multi-layer crossing minimization problem (MLCM) we are given a k-layered graph, $G = (V_1, \ldots, V_k, E)$, with the goal of finding permutations $\pi_1, \ldots, \pi_k$ such that the edge crossings are minimized. Again, both exact and heuristic methods have been proposed.

Before we deal with the full k-layer problem, it is worth considering a number of approaches to the two-layer crossing minimization problem (TLCM). Recall that it was on this problem that first complexity results were found [GJ83].

Two-Layer Crossing Minimization

For TLCM Valls *et al.* [VML96] propose a branch-and-bound algorithm that, they claim, considers a small fraction of the total search space. They report that the average time taken to compute C_{opt} for a bipartite graph $G(U, V, E)$, $|U| = |V| = 13$ and edge-density 0.3 was 40 minutes. As the following method appears to give better results we focus on this method.

In the course of considering OLCM Jünger and Mutzel [JM97] observed that a simple lower bound on the number of crossings proved to be tight. Based on this they developed a branch-and-bound algorithm for TLCM. Given an hour of computation time on a Sun Sparcstation 10 it was possible to solve to optimality problems varying in size from vertex size $V_1 = V_2 = 11$ with 80% edge density, to vertex size $V_1 = V_2 = 16$ with 10% edge density. Of heuristic methods for TLCM iterated barycenter was the clear champion, even surpassing, surprisingly, an iterated scheme based around an exact OLCM algorithm.

Newton *et al.* [NSV02] consider two new heuristics that, they claim, outperforms the iterated barycenter method. The first method is based on a connection between the bipartite crossing number and the linear arrangement problem which seeks to order the vertices of a graph so that the absolute distance between edges is minimized [SSSV01]. Any heuristic solution to the linear arrangement problem can then be immediately used as a solution to the bipartite crossing number problem. Two methods based on computing the Fiedler eigenvector of the Laplacian of G are proposed. Another method proposed by the authors [NSV02] that also yields very good results is to repeatedly randomly choose a pair of vertices on the same side, swap their positions and keep this new solution (only) if an improvement resulted.

Heuristic Approaches

The most usual heuristic approach is to repeatedly apply a *layer-by-layer sweep* of a one-layer crossing minimization algorithm until no further improvement is possible [STT81]. That is, for a layer i, the vertices of layers $i-1$ and $i+1$ are held fixed if the layers exist and V_i is permuted. This procedure is then applied for successively increasing and decreasing i until there is no further improvement. Obviously other stopping criteria may be used instead and more sophisticated control mechanisms can replace the simple up-down strategy described here [GKNV93, San95].

An alternative to the layer-by-layer sweep approach was proposed by Matuszewski *et al.* [MSM99]. The idea, which can also be applied to OLCM, orders all vertices in $V = V_1 \cup \cdots \cup V_k$ according to their degree and finds the best position for it on its layer. It does this by *sifting* the vertex through all possible positions (all others remaining fixed), using the crossing number matrix to count the crossings that would result at each position.

A different heuristic alternative solves an exact crossing minimization problem on a subset or *window* of the multi-layer graph that matches certain criteria [EGDB02]. Some improvements over the champion iterated barycenter are presented, although running time remains a problem with the procedure.

Exact Methods

Following from their work on the linear ordering and OLCM problems Jünger *et al.* derived an integer linear programming (ILP) formulation for MLCM [JLMO97]. As in the one-layer case, quadratic terms analogous to those appearing in equation (13.7) prevent formulation as a *linear* program. In this case, however, neither layer is fixed so a different strategy is called for. The proposed solution is to introduce crossing variables, c_{ijkl}, denoting whether the edges (i, j) and (k, l) cross when $i < j$, $k < l$, $i < k$ and $j \neq l$. Using this notation along with boolean variables x_{ij} to denote $\pi_1(i) < \pi_1(j)$ and y_{ij} to denote $\pi_2(i) < \pi_2(j)$ TLCM can be formulated as

$$\text{minimize} \quad \sum_{(i,j),(k,l) \in E} c_{ijkl} \tag{13.14}$$

$$-c_{ijkl} \leq y_{jl} - x_{ik} \leq c_{ijkl} \qquad (i,j),(k,l) \in E, j < l \tag{13.15}$$

$$1 - c_{ijkl} \leq y_{lj} + x_{ik} \leq 1 + c_{ijkl} \qquad (i,j),(k,l) \in E, l > j \tag{13.16}$$

$$0 \leq x_{ij} + x_{jk} - x_{ik} \leq 1 \qquad 1 \leq i < j < k \leq n_1 \tag{13.17}$$

$$0 \leq y_{ij} + y_{jk} - y_{ik} \leq 1 \qquad 1 \leq i < j < k \leq n_2 \tag{13.18}$$

$$x_{ij}, y_{ij}, c_{ijkl} \in \{0, 1\} \tag{13.19}$$

If $c_{ijkl} = 0$ then equation (13.15) forces the ordering of vertices i and k to be the same as that of j and l; similarly, if edges (i, j) and (k, l) cross then $y_{jl} - x_{ik} = \{-1, 1\}$, forcing $c_{ijkl} = 0$. Since the variable $y_{jl}, l < j$ does not exist equation (13.16) uses the identity $y_{jl} = 1 - y_{jl}$ analogously.

The k-layer problem can then be formulated as

$$\text{minimize} \quad \sum_{r=1}^{p-1} \sum_{(i,j),(k,l) \in E_r} c_{ijkl}^r \tag{13.20}$$

$$-c_{ijkl}^r \leq x_{jl}^{r+1} - x_{ik}^r \leq c_{ijkl}^r \qquad (i,j),(k,l) \in E_r, j < l \tag{13.21}$$

$$1 - c_{ijkl}^r \leq x_{lj}^{r+1} + x_{ik}^r \leq 1 + c_{ijkl}^r \qquad (i,j),(k,l) \in E_r, l > j \tag{13.22}$$

$$0 \leq x_{ij}^r + x_{jk}^r - x_{ik}^r \leq 1 \qquad 1 \leq i < j < k \leq n_r \tag{13.23}$$

$$x_{ij}^r, c_{ijkl}^r \in \{0, 1\} \tag{13.24}$$

where

$$E_r = \{(u, v) \in E | u \in V_r, v \in V_{r+1}\}. \tag{13.25}$$

Jünger *et al.* [JLMO97] analyze the polytope associated with this ILP formulation and present some results that describe facets of the polytope. Healy and Kuusik analyze the *cycle space* of an MLCM instance [HK04] and derive additional constraints for the ILP formulation (13.20)–(13.24). In his PhD thesis Kuusik [Kuu00] uses these new inequalities. Also described is how additional constraints such as different paths that may not cross, or a group of vertices must appear consecutively can be included.

Algorithm Performance

Matuszewski *et al.* [MSM99] claim that their global sifting method leads to a 20% improvement over iterated barycenter, although the time taken is considerable. An improved version of the algorithm [GSBM01] improves the running time of the original by a factor of 10, the authors claim.

Jünger *et al.* [JLMO97] present some preliminary results of their ILP formulation of MLCM on 3-layer problems. The largest problem has $V_1 = 26, V_2 = 18, V_3 = 17, E_1 = 29, E_2 = 29$ and takes 330 seconds to find an optimal solution. Kuusik [Kuu00] presents solutions to larger problems, one measuring over 90 vertices and 112 edges spread over 9 levels (Forrester's "World Dynamics" [GKNV93]). What emerges is that the model is sensitive to the number of linear ordering variables, which is driven by the average node count per layer.

More recently, semidefinite programming (SDP) methods have begun to offer the prospect of solving larger problems to optimality. Following on from the 2-level case formulated by Buchheim *et al.* [BWZ10] general k-level solutions have been obtained. Chimani *et al.* [CHJM11] report on experiments where the method outperformed an ILP implementation. Test cases with up to 20 levels and up to 25 vertices per layer were randomly generated and while an ILP implementation was capable of solving sparse instances (inter-layer edge density, d, of up to 0.1) it was unable to solve a single case where $d \geq 0.2$, instances which their SDP solver was routinely able to solve.

13.5.3   Planarization – An Alternative

Mutzel [Mut01] observes that the understandability of a graph may be subtler than simply counting the number of crossings in the graph. It may be that if a *few* edges account for the crossings then the eye may more easily filter these crossings even if more than the minimum. Thus, an alternative to minimizing edge crossings in a k-level graph presents itself: determine the maximum level planar subgraph; draw this k-level subgraph without crossings; and, finally, reinsert the "nonplanar" edges, in the expectation that not many crossings will result. This problem has become of interest in the graph drawing community and a number of complexity results are known [DFH+01a, DFH+01b].

Mutzel and Weiskircher [MW98] adopt this approach by finding the maximum planar subgraph, which they call the two-layer planarization problem. They formulate the planarization problem as an ILP and they identify a set of facets of the associated polytope in the cases where one or both layers are fixed. In order to solve k-layer problems they solve a succession of two-layer problems, although this is clearly a sub-optimal strategy.

Kuusik uses ILP methods also for the general k-level planarization problem [Kuu00]. We state the problem as, given a k-level graph, $G = (V, E)$, find a k-level planar subgraph $G_p = (V, E_p)$ so that $|E_p|$ is maximum. An important aspect of finding the maximum k-level planar subgraph is identifying k-level subgraphs that are minimally non-planar and based on this characterization [HKL04] and the cycle space of the graph [HK04] facets of the associated polytope are identified. Some of these facets are put to use in a branch-and-cut algorithm.

An ILP formulation follows easily from that described for the minimum-crossing ILP (13.20)–(13.24). By introducing a binary variable p_{ij} for each edge where $p_{ij} = 1$ if and only if $(i, j) \in E_p$ and observing that if $(i, j) \in E_p$ and $(k, l) \in E_p$ then $c_{ijkl} = 0$, the following constraint expresses the relationship between the two

$$p_{ij} + p_{kl} + c_{ijkl} \leq 2 \tag{13.26}$$

A complete ILP may then be expressed as

$$\text{maximize} \sum_{(i,j) \in E} p_{ij} \tag{13.27}$$

subject to

$$p_{ij} + p_{kl} + c_{ijkl} \leq 2 \tag{13.28}$$
$$\text{constraints}(13.21) - (13.24) \tag{13.29}$$
$$p_{ij} \in \{0,1\} \tag{13.30}$$

Among several equal in size maximum level planar subgraph solutions it may be desirable to favor ones with that will result in fewer crossings. Objective function (13.27) can be modified to take account of crossing variables and favor solutions with fewer crossings

$$\text{maximize} \sum_{(i,j) \in E} p_{ij} + \frac{1}{k_c + 1} \sum_{(i,j),(k,l) \in E} c_{ijkl} \tag{13.31}$$

where k_c is the number of crossing variables.

Kuusik reports [Kuu00] being able to solve a set of randomly generated problems of size $k = 8$, $V_1 = \cdots = V_8 = 12$, $E = 110$ in times ranging from 62(s) to, in an extreme case, 343(s).

13.6 x-Coordinate Assignment

Having determined a relative ordering of the vertices on each level the final step requires positioning the vertices so that, insofar as is possible, the edges are straight and vertices are centered among their neighbors. Aesthetically straight edges are desirable and this can be achieved by adjusting the positioning – the x-coordinate – of each vertex; there is also some evidence that, perceptually, straight edges are preferable [HEH09]. However, it is likely that the width of the graph will increase during this step.

It will be remembered that long edges of the input graph will have been subdivided by the introduction of dummy nodes and a vertical "flow through" for these vertices is particularly desirable. Given an edge from level i to level j in the input graph it becomes the path $(v_i, v_{i+1}, \ldots, v_j)$ after the insertion of dummy nodes. It is reasonable to expect that the sub-path $(v_{i+1}, \ldots, v_{j-1})$ be drawn strictly vertical with the (at most two) bends occurring at vertices v_{i+1} and v_{j-1}, if necessary. Most algorithms in the literature make this a priority.

In their original paper Sugiyama *et al.* [STT81] propose a mathematical programming solution with a quadratic objective function that is a convex combination of two separate goals. Closeness to connected vertices and a balance, or centering, between a vertex's predecessor neighbors and its successor neighbors are the two criteria identified. Since these goals will counteract each other in general they are each weighted by parameters c and $1-c$, $0 < c \leq 1$ respectively. All dummy nodes associated with a long edge are restricted to be vertically aligned; this presupposes that if two long edges cross, the crossing occurs on either the first or last segment of both edges. Because quadratic programming was computationally expensive at the time the authors also proposed a "Priority Layout Method" that operates in a similar spirit to their crossing-minimization level-by-level sweep. Dummy nodes are assigned a high priority so that they will be aligned, though care is required to ensure that the level's vertex ordering is maintained.

Gansner *et al.*, too, propose exact and heuristic algorithms [GKNV93]. Their exact algorithm is linear programming-based again. Whereas the quadratic objective function of Sugiyama [STT81] minimizes terms (among others) such as $\sum_{(u,v \in E)}(x(u) - x(v))^2$, Gansner *et al.* replace the quadratic term by $|x(u) - x(v)|$. This can now be linearized but at the expense of introducing an additional variable and two additional constraints for each edge.

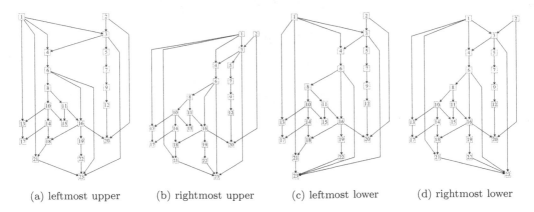

    (a) leftmost upper   (b) rightmost upper   (c) leftmost lower   (d) rightmost lower

Figure 13.15   Four "extreme" alignments computed by the Brandes-Köpf algorithm.

Thus, on the grounds that the constraint matrix grows from size VE to size $(V+E)E$, they dismiss this and propose a heuristic that operates, again, by sweeping up and down over the levels. Other heuristics have been proposed by Eades *et al.* who place nodes belonging to any layer other than the top or bottom according to a degree-weighted barycenter and sweep up and down the levels "until x converges" [ELT96], and Sander with his Pendulum heuristic, which lays out edges Manhattan style and thus may require four bends per edge [San96a].

We close this section with a more detailed look at Brandes and Köpf's algorithm [BK02], which is considered to be the algorithm of choice for assignment of x-coordinate; the algorithm's running time is linear in the number of vertices and edge segments. As we have remarked, the algorithm guarantees that, for long edges, at most two bends will occur, and if they are necessary they will occur on the external segments of an edge. Thus, provided the crossing reduction step has ensured that no two long edges cross internally, all dummy (internal) nodes of a long edge will be aligned.

Similar to Gansner *et al.* above, the algorithm uses $|x(u) - x(v)|$ as a surrogate for edge-length. Since, for a set $X = \{x_i\}$ of real numbers, $\sum_{i=1}^{k} |x - x_i|$ is minimized when x is the median of the set, the algorithm focuses on placing a vertex at the median of its neighbors.

The algorithm firstly computes four "extreme" layouts where each vertex is aligned either with its upper or lower median neighbor and, for each case, vertices on a level are considered in a left-to-right (left alignment) or right-to-left (right alignment) order. Figure 13.15 (reproduced with the authors' kind permission from their paper [BK02]) illustrates the four layouts for a given graph. Conflicts – called *type 0 conflicts* – can arise due to a pair of external segments that either cross or are incident upon the same vertex and these are resolved by the order of consideration. Because the algorithm favors vertical inner segments a crossing of an inner segment with an outer segment – a *type 1 conflict* – should give precedence to aligning the vertices of the inner segment. A preprocessing step marks such external segments as being excluded from consideration for alignment.

After each of the four extreme layouts are computed, the layout's blocks are computed. A *block* is a maximal set of vertically aligned vertices. These blocks may then be *compacted*, subject to minimum separation requirements, giving an x-coordinate for each vertex.

Having performed this procedure four times, each with a directional bias, what is surprising, perhaps, is that the four candidate positions can be combined so straightforwardly to get the final position with such satisfactory results. Of the four possibilities the algorithm chooses what is called the *average median*, which, for $x_1 \le x_2 \le \dots \le x_k$, is

$(x_{\lfloor (k+1)/2 \rfloor} + x_{\lceil (k+1)/2 \rceil})/2$. One of the arguments for this choice as opposed to the mean, say, is if a vertex is aligned twice with its upper median neighbor and unevenly positioned in the other two runs the average median will maintain verticality of the segment.

The main body of the algorithm is the four linear sweeps over the segments. The authors demonstrate how the type 1 conflicts can be detected in linear time, also. Finally, for each vertex, the average median of four can be determined in linear time.

13.7 Extensions and Alternatives to Sugiyama's Framework

There are a few attempts to extend the Sugiyama method beyond the originally proposed framework. Some extensions such as the cycle-removal step and the edge-concentration step have become standard features of the Sugiyama framework nowadays.

Do Nascimento and Eades have proposed and extensively studied methods for applying user constraints to hierarchical drawings interactively [dNE01a, dNE01b]. In particular, they allow the user to interact with an already-computed hierarchical drawing and introduce additional constraints on the relative position of some vertices. Taking the user constraints into account they show how to reapply the Sugiyama method locally to a subgraph of the digraph and efficiently recompute the coordinates of the affected nodes only. They also propose an interactive genetic algorithm with similar features [dNE02].

Although the Sugiyama method has been the most popular method for hierarchical drawing of digraphs, several researchers have proposed successful alternative methods. Sugiyama and Misue proposed a magnetic-field method for hierarchical drawing of acyclic digraphs [SM95a]. A global magnetic field is applied to the edges and they are aligned to it. As a result all edges point unidirectionally without specifically placing the vertices on parallel horizontal lines. Note that in any drawing with edges pointing into the same direction and continuous y-coordinates of the vertices levels can be easily induced from by a simple quantization procedure.

Utech *et al.* [UBSE98] proposed a genetic algorithm as an alternative to the Sugiyama method. The chromosome contains values which represent the y-coordinate (i.e., layer) and x-coordinate (i.e., position within the layer) of each original and dummy vertex. Thus, the x- and y-coordinates of the vertices evolved simultaneously. The fitness of each individual is provided by the number of edge crossings.

Carmel *et al.* proposed an energy-based alternative to the Sugiyama method [CHK04]. Based on rapidly solving a unique one-dimensional optimization problem for each of the axes their algorithm produces a clear representation of the hierarchical structure of the digraph. Vertices are not restricted to lie on horizontal levels which allows their layouts to convey the symmetries of the digraph. An interesting detail is that the algorithm can be applied without change to both cyclic or acyclic digraphs, and even to graphs containing both directed and undirected edges.

A few researchers have attempted 3D hierarchical drawing of digraphs. Ostry has proposed wrapping a 2D hierarchy around either a cylinder or a cone [Ost96]. A 3D approach intrinsically different from the Sugiyama method has been proposed within the graph drawing system GIOTTO3D [GT97]. In its first phase it applies a planarization method to draw the digraph in 2D; in the second phase vertices and edges are assigned z-coordinates so that all edges point into the same vertical direction and the total edge span is minimized; and at the third phase the shape of the vertices and the edges is determined.

Hong and Nikolov have proposed an extension to the Sugiyama framework for hierarchical drawing of digraphs in 3D [HN05b, HN05a]. They introduced the convention of drawing the digraph in a set of parallel planes, called *walls*, each containing a 2D hierarchical drawing of a subgraph of the input digraph. The partition of the vertex set into walls is done at a separate wall-assigning step applied after the layer-assignment step and before the vertex-ordering step. Hong and Nikolov proposed and evaluated various wall-assignment algorithms which partition the vertex set into walls according to different criteria.

One of the limitations of the Sugiyama framework is that decisions made at previous steps influence later steps and yet cannot be undone. We have seen that the edge of a cycle one chooses to break can have a bearing on how the graph is interpreted. Likewise, how one levels the graph impacts on the number of dummy nodes which has a bearing on the running time of the algorithm. Chimani *et al.* attempt to decouple some of these dependencies by sidestepping the three-step decomposition and instead search for a global solution with the principal goal of minimizing crossings [CGMW11, CGMW10]. The algorithm necessarily sacrifices drawing height for a lower crossing count and thus edge-length is increased. In spite of this the authors assert that their UPL system improves aspect-ratio; running time of their system suffers and is greater than their Sugiyama implementation [CGMW11].

An alternative fashion of drawing 2D hierarchies is by arranging the layers as concentric circles [RFEM88, Ead92]. Such drawings are known as *radial level drawings*. They can be very useful if the input digraph represents the structure of a website or a file system, for example. They have been used for visualizing social networks as well [BKW03]. The Sugiyama method can be applied with all its steps for making radial hierarchical drawings. However, each step may require a specific algorithm for producing aesthetically pleasing result. There are a few recent results in this area of research. Bachmaier *et al.* have proposed an algorithm for computing a radial level planar embedding if one exists along with a linear-time algorithm for the coordinate-assignment step [BFF05]. The figure below (reproduced with permission of the authors [BBF05]) illustrates one of the advantages of the drawing style: edge (18,19) of Figure 13.16a effectively can be routed "around the back of the drawing" thus avoiding the crossings, as demonstrated in Figure 13.16b.

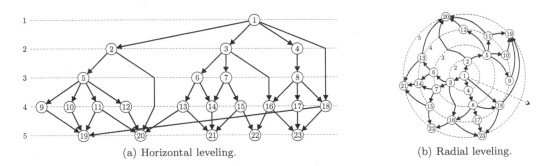

(a) Horizontal leveling.   (b) Radial leveling.

Figure 13.16   A Horizontally Leveled Drawing and its Radial Counterpart.

Much of what has gone before relies on the input (directed) graph being acyclic, or being preprocessed so that it becomes one. Yet, Sugiyama and his colleagues in their seminal paper [STT81] also addressed the drawing of a particular class of cyclic digraph they called *recurrent hierarchies*. In this situation an additional set of edges $E_n \subseteq V_n \times V_1$ connect vertices on the bottom layer to vertices on the top layer. Brunner *et al.* consider the *cyclic* style of drawing [Bru10, BBBL08] where the inevitable "implied ranking" associated with a top-to-bottom leveling (see Figure 13.1) is avoided. Vertices assigned to a level are distributed along a spoke of an imaginary wheel with spokes evenly distributed. Figure 13.17 below illustrates the technique on the example graph of Figure 13.2.

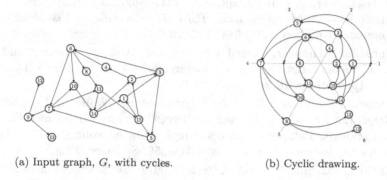

(a) Input graph, G, with cycles.   (b) Cyclic drawing.

Figure 13.17   A Cyclic Drawing of the graph G of Figure 13.2.

References

[AGU72]   A. V. Aho, M. R. Garey, and J. D. Ullman. The transitive reduction of a
           directed graph. *SIAM Journal on Computing*, 1(2):131–137, June 1972.

[Bac07]   C. Bachmaier. A radial adaptation of the Sugiyama framework for visu-
           alizing hierarchical information. *IEEE Transactions on Visualization and
           Computer Graphics*, 13(3):583 –594, 2007.

[BBBF12]   Christian Bachmaier, Franz J. Brandenburg, Wolfgang Brunner, and Ray-
           mund Fülöp. Drawing recurrent hierarchies. *Journal of Graph Algorithms
           and Applications*, 16(2):151–198, 2012.

[BBBL08]   Christian Bachmaier, Franz-Josef Brandenburg, Wolfgang Brunner, and
           Gergö Lovász. Cyclic leveling of directed graphs. In Ioannis G. Tollis
           and Maurizio Patrignani, editors, *Graph Drawing*, volume 5417 of *Lecture
           Notes in Computer Science*, pages 348–359. Springer, 2008.

[BBF05]   Christian Bachmaier, Franz J. Brandenburg, and Michael Forster. Radial
           level planarity testing and embedding in linear time. *Journal of Graph
           Algorithms and Applications*, 9(1):53–97, 2005.

[BELM01]   J. Branke, P. Eades, S. Leppert, and M. Middendorf. Width restricted
           layering of acyclic digraphs with consideration of dummy nodes. Technical
           Report No. 403, Intitute AIFB, University of Karlsruhe, 76128 Karlsruhe,
           Germany, 2001.

[BFF05]   Christian Bachmaier, Florian Fischer, and Michael Forster. Radial coor-
           dinate assignment for level graphs. In L. Wang, editor, *Proc. Computing
           and Combinatorics Conference, COCOON 2005*, volume 3595 of *Lecture
           Notes in Computer Science*, pages 401–410. Springer, 2005.

[BJM02]   Wilhelm Barth, Michael Jünger, and Petra Mutzel. Simple and efficient
           bilayer cross counting. In Kobourov and Goodrich [KG02], pages 130–141.

[BK02]   U. Brandes and B. Köpf. Fast and simple horizontal coordinate assign-
           ment. In P. Mutzel, M. Jünger, and Leipert S., editors, *Graph Drawing:
           Proceedings of 9th International Symposium, GD 2001*, volume 2265 of
           Lecture Notes in Computer Science, pages 31–44. Springer-Verlag, 2002.

[BKW03]   U. Brandes, P. Kenis, and D. Wagner. Communicating centrality in policy
           network drawings. *IEEE Transact. Vis. Comput. Graph.*, 9(2):241–253,
           2003.

[BLME02]   J. Branke, S. Leppert, M. Middendorf, and P. Eades. Width-restriced lay-
           ering of acyclic digraphs with consideration of dummy nodes. *Information
           Processing Letters*, 81(2):59–63, January 2002.

[Bré79]   D. Brélaz. New methods to color the vertices of a graph. *Communications
           of the ACM*, 22(4):251–256, 1979.

[Bru10]   Wolfgang Brunner. *Cyclic Level Drawings of Directed Graphs*. PhD thesis,
           Universitt Passau, Innstrasse 29, 94032 Passau, 2010.

[BS90]   B. Berger and P. W. Shor. Approximation algorithms for the maximum
           acyclic subgraph problem. In *Proceedings of the 1st Annual ACM-SIAM
           Symposium on Discrete Algorithms*, pages 236–243, 1990.

[BWZ10]   Christoph Buchheim, Angelika Wiegele, and Lanbo Zheng. Exact algo-
           rithms for the quadratic linear ordering problem. *INFORMS Journal on
           Computing*, 22(1):168–177, 2010.

[Car80]   M. J. Carpano. Automatic display of hierarchized graphs for computer aided decision analysis. *IEEE Transactions on Systems, Man and Cybernetics*, 10(11):705–715, 1980.

[Cat95]   T. Catarci. The assignment heuristics for crossing reduction. *IEEE Transactions on Systems, Man, and Cybernetics*, 25(3):515–521, 1995.

[CG72]   E. G. Coffman and R. L. Graham. Optimal scheduling for two processor systems. *Acta Informatica*, 1:200–213, 1972.

[CGMW10]   Markus Chimani, Carsten Gutwenger, Petra Mutzel, and Hoi-Ming Wong. Layer-free upward crossing minimization. *J. Exp. Algorithmics*, 15:2.2:2.1–2.2:2.27, March 2010.

[CGMW11]   Markus Chimani, Carsten Gutwenger, Petra Mutzel, and Hoi-Ming Wong. Upward planarization layout. *J. Graph Algorithms Appl.*, 15(1):127–155, 2011.

[CHJM11]   Markus Chimani, Philipp Hungerländer, Michael Jünger, and Petra Mutzel. An SDP approach to multi-level crossing minimization. In Matthias Müller-Hannemann and Renato F. Werneck, editors, *ALENEX*, pages 116–126. SIAM, 2011.

[CHK02]   Liran Carmel, David Harel, and Yehuda Koren. Drawing directed graphs using one-dimensional optimization. In Michael Goodrich and Stephen Koubourov, editors, *Proceedings of 10th International Symposium, GD 2002*, number 2528 in Lecture Notes in Computer Science, pages 193–206. Springer, 2002.

[CHK04]   L. Carmel, D. Harel, and Y. Koren. Combining hierarchy and energy for drawing directed graphs. *IEEE Transactions on Visualization and Computer Graphics*, 10(1):46–57, 2004.

[DDLM04]   Emilio Di Giacomo, Walter Didimo, Giuseppe Liotta, and Henk Meijer. Computing radial drawings on the minimum number of circles. In János Pach, editor, *Graph Drawing*, volume 3383 of *Lecture Notes in Computer Science*, pages 251–261. Springer, 2004.

[DETT99]   G. Di Battista, P. Eades, R. Tamassia, and I. G. Tollis. *Graph Drawing*. Prentice Hall, 1999.

[DF03]   C. Demetrescu and I. Finocchi. Combinatorial algorithms for feedback problems with directed graphs. *Information Processing Letters*, 86:129–136, 2003.

[DFH+01a]   Vida Dujmovic, Michael R. Fellows, Michael T. Hallett, Matthew Kitching, Giuseppe Liotta, Catherine McCartin, Naomi Nishimura, Prabhakar Ragde, Frances A. Rosamond, Matthew Suderman, Sue Whitesides, and David R. Wood. On the parameterized complexity of layered graph drawing. In Friedhelm Meyer auf der Heide, editor, *ESA*, volume 2161 of *Lecture Notes in Computer Science*, pages 488–499. Springer, 2001.

[DFH+01b]   Vida Dujmovic, Michael R. Fellows, Michael T. Hallett, Matthew Kitching, Giuseppe Liotta, Catherine McCartin, Naomi Nishimura, Prabhakar Ragde, Frances A. Rosamond, Matthew Suderman, Sue Whitesides, and David R. Wood. A fixed-parameter approach to two-layer planarization. In Petra Mutzel, Michael Jünger, and Sebastian Leipert, editors, *Graph Drawing*, volume 2265 of *Lecture Notes in Computer Science*, pages 1–15. Springer, 2001.

[DFK03]   Vida Dujmovic, Henning Fernau, and Michael Kaufmann. Fixed parame-
           ter algorithms for one-sided crossing minimization revisited. In Giuseppe
           Liotta, editor, *Graph Drawing*, volume 2912 of *Lecture Notes in Computer
           Science*, pages 332–344. Springer, 2003.

[DGL⁺00]   Giuseppe Di Battista, Ashim Garg, Giuseppe Liotta, Armando Parise,
           Roberto Tamassia, Emanuele Tassinari, Francesco Vargiu, and Luca Vis-
           mara. Drawing directed acyclic graphs: An experimental study. *Int. J.
           Comput. Geometry Appl.*, 10(6):623–648, 2000.

[dNE01a]   H. A. D. do Nascimento and P. Eades. A framework for human-computer
           interaction in directed graph drawing. In *InVis.au*, pages 63–69. CRPIT,
           2001.

[dNE01b]   H. A. D. do Nascimento and P. Eades. User hints for directed graph
           drawing. In P. Mutzel, M. Jünger, and S. Leipert, editors, *Graph Drawing:
           Proceedings of 9th International Symposium, GD 2001*, volume 2265 of
           Lecture Notes in Computer Science, pages 205–219. Springer-Verlag, 2001.

[dNE02]   H. A. D. do Nascimento and P. Eades. A focus and constraint-based
           genetic algorithm for interactive directed drawing. *HIS*, pages 634–643,
           2002.

[Dre94]   S. Dresbach. A new heuristic layout algorithm for directed acyclic graphs.
           In U. Derigs, A. Bachem, and A. Drexl, editors, *Operations Research Pro-
           ceedings 1994*, pages 121–126, Berlin–Heidelberg, 1994.

[DW02]   Vida Dujmovic and Sue Whitesides. An efficient fixed parameter tractable
           algorithm for 1-sided crossing minimization. In Kobourov and Goodrich
           [KG02], pages 118–129.

[Ead84]   Peter Eades. A heuristic for graph drawing. *Congressus Numerantium*,
           42:149–160, 1984.

[Ead92]   P. Eades. Drawing free trees. *Bulletin of the Institute of Combinatorics
           and its Applications*, 5:10–36, 1992.

[EGDB02]   Thomas Eschbach, Wolfgang Günther, Rolf Drechsler, and Bernd Becker.
           Crossing reduction by windows optimization. In Kobourov and Goodrich
           [KG02], pages 285–294.

[EGM04]   D. Eppstein, M. T. Goodrich, and J. Y. Meng. Confluent layered draw-
           ings. In J. Pach, editor, *Graph Drawing: Proceedings of 12th International
           Symposium, GD 2004*, volume 3383 of *Lecture Notes in Computer Science*,
           pages 184–194. Springer-Verlag, 2004.

[EK86]   P. Eades and D. Kelly. Heuristics for reducing crossings in 2-layered net-
           works. *Ars Combinatoria*, 21-A:89–98, 1986.

[EL95]   P. Eades and X. Lin. A new heuristic for the feedback arc set problem.
           Australian Journal of Combinatorics, 12:15–26, 1995.

[ELS89]   P. Eades, X. Lin, and W. F. Smyth. Heuristics for the feedback arc set
           problem. Technical Report 1, Curtin University of Technology, School of
           Computing Science, Perth, Australia, 1989.

[ELS93]   P. Eades, X. Lin, and W. F. Smyth. A fast and effective heuristic for the
           feedback arc set problem. *Information Processing Letters*, 47(6):319–323,
           1993.

[ELT96]   Peter Eades, Xue-Min Lin, and Roberto Tamassia. An algorithm for draw-
           ing hierarchical graphs. *Int. J. Comput. Geom. Appl.*, 6:145–156, 1996.

[ENRS95]   G. Even, J. Naor, S. Rao, and B. Schieber. Divide-and-conquer approximation algorithms via spreading metrics. In *Proceedings of the 36th Annual IEEE Symposium on Foundations of Computer Science*, pages 62–71, 1995.

[ES90]   P. Eades and K. Sugiyama. How to draw a directed graph. *Journal of Information Processing*, 13(4):424–437, 1990.

[EW94]   P. Eades and N. C. Wormald. Edge crossings in drawings of bipartite graphs. *Algorithmica*, 11:379–403, 1994.

[FH96]   P. C. Fishburn and P. L. Hammer. Bipartite dimensions and bipartite degree of graphs. *Discrete Mathematics*, 160:127–148, 1996.

[Flo90]   M. M. Flood. Exact and heuristic algorithms for the weighted feedback arc set problem: A special case of the skew-symetric quadratic assignment problem. *Networks*, 20:1–23, 1990.

[FS04]   C. Friedrich and F. Schreiber. Flexible layering in hierarchical drawings with nodes of arbitrary size. In *Proceedings of the 27th Conference on Australasian Computer Science*, pages 369–376, Dunedin, New Zealand, 2004.

[FW95]   M. Fröhlich and M. Werner. Demonstration of the interactive graph visualization system daVinci. In R. Tamassia and I. Tollis, editors, *Proceedings of DIMACS Workshop on Graph Drawing '94, Princeton (USA) 1994*, volume 894 of *LNCS*. Springer-Verlag, 1995.

[GJ79]   M. R. Garey and D. S. Johnson. *Computers and Intractability: A Guide to the Theory of NP-Completeness*. W. H. Freeman, New York, 1979.

[GJ83]   M. R. Garey and D. S. Johnson. Crossing number is NP-complete. *SIAM J. Algebraic and Discrete Methods*, 4(3):312–316, 1983.

[GJR85]   M. Grötschel, M. Jünger, and G. Reinelt. On the acyclic subgraph polytope. *Mathematical Programming*, 33(1):28–42, 1985.

[GKN02]   Emden Gansner, Eleftherios Koutsofios, and Stephen North. Drawing graphs with *dot*. Technical report, AT&T Labs—Research, 2002. http://www.research.att.com/sw/tools/graphviz/dotguide.pdf.

[GKNV93]   E. R. Gansner, E. Koutsofios, S. C. North, and K.-P Vo. A technique for drawing directed graphs. *IEEE Transactions on Software Engineering*, 19(3):214–230, March 1993.

[GN00]   Emden R. Gansner and Stephen C. North. An open graph visualization system and its applications to software engineering. *Softw., Pract. Exper.*, 30(11):1203–1233, 2000.

[GSBM01]   W. Günther, R. Schönfeld, B. Becker, and P. Molitor. k-layer straightline crossing minimization by speeding up sifting. In J. Marks, editor, *Graph Drawing: Proceedings of 8th International Symposium, GD 2000*, volume 1984 of *Lecture Notes in Computer Science*, pages 253–258. Springer-Verlag, 2001.

[GT97]   A. Garg and R. Tamassia. GIOTTO: A system for visualizing hierarchical structures in 3D. In S. North, editor, *Graph Drawing: Symposium on Graph Drawing, GD '96*, volume 1190 of *Lecture Notes in Computer Science*, pages 193–200. Springer-Verlag, 1997.

[HEH09]   Weidong Huang, Peter Eades, and Seok-Hee Hong. A graph reading behavior: Geodesic-path tendency. In Peter Eades, Thomas Ertl, and Han-Wei Shen, editors, *PacificVis*, pages 137–144. IEEE Computer Society, 2009.

[Him00]   Michael Himsolt. Graphlet: design and implementation of a graph editor. *Softw., Pract. Exper.*, 30(11):1303–1324, 2000.

[HK04]   Patrick Healy and Ago Kuusik. Algorithms for multi-level graph planarity testing and layout. *Theoretical Computer Science*, 320(2–3):331–344, 2004.

[HKL04]   Patrick Healy, Ago Kuusik, and Sebastian Leipert. A characterisation of level planar graphs. *Discrete Mathematics*, 280(1–3):51–63, 2004.

[HN02a]   P. Healy and N. S. Nikolov. A branch-and-cut approach to the directed acyclic graph layering problem. In M. Goodrich and S. Koburov, editors, *Graph Drawing: Proceedings of 10th International Symposium, GD 2002*, volume 2528 of *Lecture Notes in Computer Science*, pages 98–109. Springer-Verlag, 2002.

[HN02b]   P. Healy and N. S. Nikolov. How to layer a directed acyclic graph. In P. Mutzel, M. Jünger, and S. Leipert, editors, *Graph Drawing: Proceedings of 9th International Symposium, GD 2001*, volume 2265 of *Lecture Notes in Computer Science*, pages 16–30. Springer-Verlag, 2002.

[HN05a]   S.-H. Hong and N. S. Nikolov. Hierarchical layouts of directed graphs in three dimensions. In P. Healy and N. S. Nikolov, editors, *Graph Drawing: Proceedings of 13th International Symposium, GD 2005*, volume 3843 of *LNCS*. Springer-Verlag, 2005.

[HN05b]   S.-H. Hong and N. S. Nikolov. Layered drawings of directed graphs in three dimensions. In S.-H. Hong, editor, *Information Visualisation 2005: Asia-Pacific Symposium on Information Visualisation (APVIS2005)*, volume 45, pages 69–74. CRPIT, 2005.

[Hu61]   T. Hu. Parallel sequencing and assembly line problems. *Operations Research*, 9:841–848, November 1961.

[JLMO97] M. Jünger, E. K. Lee, P. Mutzel, and T. Odenthal. A polyhedral approach to the multi-layer crossing minimization problem. In G. Di Battista, editor, *Graph Drawing: 5th International Symposium, GD '97*, volume 1353 of *Lecture Notes in Computer Science*, pages 13–24, Rome, Italy, September 1997. Springer-Verlag.

[JM97]   Michael Jünger and Petra Mutzel. 2-layer straightline crossing minimization: Performance of exact and heuristic algorithms. *J. Graph Algorithms Appl.*, 1, 1997.

[KA99]   Y.-K. Kwok and I. Ahmad. Static scheduling algorithms for allocating directed task graphs to multiprocessors. *ACM Computing Surveys*, 31(4):406–471, 1999. ACM Computing Surveys 31(4): 406-471 (1999).

[Kar72]   R. M. Karp. Reducibility among combinatorial problems. In R. E. Miller and J. W. Thatcher, editors, *Complexity of Computer Computations*, pages 85–103. Plenum Press, 1972.

[KG02]   Stephen G. Kobourov and Michael T. Goodrich, editors. *Graph Drawing, 10th International Symposium, GD 2002*, volume 2528 of *Lecture Notes in Computer Science*. Springer, 2002.

[Kra99]   Jan Kratochvíl, editor. *Graph Drawing, 7th International Symposium, GD'99, Stirín Castle, Czech Republic, September 1999, Proceedings*, volume 1731 of *Lecture Notes in Computer Science*. Springer, 1999.

[Kuu00]   Ago Kuusik. *Integer Linear Programming Approaches to Hierarchical Graph Drawing*. PhD thesis, University of Limerick, 2000.

[Lei79]   F. T. Leighton. A graph coloring algorithm for large scheduling problems. *Journal of Research of National Bureau of Standard*, 84:489–506, 1979.

[Lin00]   X. Lin. On the computational complexity of edge concentration. *Discrete Applied Mathematics*, 101:197–205, 2000.

[Meh84]   K. Mehlhorn. *Data Structures and Algorithms, Volume 2: Graph Algorithms and NP-Completeness*. Springer-Verlag, Heidelberg, Germany, 1984.

[MELS95]   K. Misue, P. Eades, W. Lai, and K. Sugiyama. Layout adjustment and the mental map. *Journal of Visual Languages and Computing*, 6:183–210, 1995.

[MS94]   E. Mäkinen and M. Sieranta. Genetic algorithms for drawing bipartite graphs. Technical Report A-1994-1, Department of Computer Science, University of Tampere, 1994.

[MSM99]   Christian Matuszewski, Robby Schönfeld, and Paul Molitor. Using sifting for k-layer straightline crossing minimization. In Kratochvíl [Kra99], pages 217–224.

[Mut01]   P. Mutzel. An alternative method to crossing minimization on hierarchical graphs. *SIAM Journal on Optimization*, 11(4):1065–1080, 2001.

[MW98]   Petra Mutzel and René Weiskircher. Two-layer planarization in graph drawing. In Kyung-Yong Chwa and Oscar H. Ibarra, editors, *ISAAC*, volume 1533 of *Lecture Notes in Computer Science*, pages 69–78. Springer, 1998.

[Nag05]   Hiroshi Nagamochi. On the one-sided crossing minimization in a bipartite graph with large degrees. *Theor. Comput. Sci.*, 332(1-3):417–446, 2005.

[New89]   F. J. Newbery. Edge concentration: A method for clustering directed graphs. In *Proceedings of the 2nd International Workshop on Software Configuration Management*, pages 76–85. ACM Press, 1989.

[NPT90]   Frances Newbery-Paulisch and Walter F. Tichy. Edge: An extendible graph editor. *Softw., Pract. Exper.*, 20(S1):S1/63–S1/88, 1990.

[NSV02]   Matthew Newton, Ondrej Sýkora, and Imrich Vrto. Two new heuristics for two-sided bipartite graph drawing. In Kobourov and Goodrich [KG02], pages 312–319.

[NT06]   N. S. Nikolov and A. Tarassov. Graph layering by promotion of nodes. *Discrete Applied Mathematics*, 154(5):848–860, 2006.

[NW88]   G. L. Nemhauser and L. A. Wolsey. *Integer and Combinatorial Optimization*. Wiley-Interscience series in discrete mathematics and optimization. John Wiley & Sons, Inc., 1988.

[NW01]   S. North and G. Woodhull. Online hierarchical graph drawing. In P. Mutzel, M. Jünger, and Leipert S., editors, *Graph Drawing: Proceedings of 9th International Symposium, GD 2001*, volume 2265 of *LNCS*, pages 232–246. Springer-Verlag, 2001.

[NY04]   Hiroshi Nagamochi and Nobuyasu Yamada. Counting edge crossings in a 2-layered drawing. *Inf. Process. Lett.*, 91(5):221–225, 2004.

[Ost96]   D. Ostry. Some three-dimensional graph drawing algorithms. Master's thesis, University of Newcastle, 1996.

[PCJ96]   H. C. Purchase, R. F. Cohen, and M. James. Validating graph drawing aesthetics. In F. J. Brandenburg, editor, *Graph Drawing: Symposium on Graph Drawing, GD '95*, volume 1027 of *Lecture Notes in Computer Science*, pages 435–446. Springer-Verlag, 1996.

[Pur97]   H. C. Purchase. Which aesthetic has the greatest effect on human understanding? In Giuseppe Di Battista, editor, *Graph Drawing. 5th International Symposium, GD '97*, volume 1353 of *Lecture Notes in Computer Science*, pages 248–261. Springer-Verlag, 1997.

[RDM⁺87]   L. A. Rowe, M. Davis, E. Messinger, C. Mayer, C.and Spirakis, and A. Tuan. A browser for directed graphs. *Software Practice and Experience*, 17(1):61–76, 1987.

[Rei85]   G. Reinelt. The linear ordering problem: Algorithms and applications. In *Research and Exposition in Mathematics*, volume 8. Heldermann, 1985.

[RFEM88]   M. G. Reggiani and F. E. F. E. Marchetti. A proposed method for representing hierarchies. *IEEE Transact. Systems, Man, and Cyb.*, 18(1):2–8, 1988.

[San94]   G. Sander. Graph layout through the VCG tool. Technical Report A03/94, Universität des Saarlandes, 1994.

[San95]   G. Sander. Graph layout through the VCG tool. In *Graph Drawing. DIMACS International Workshop GD'94*, volume 894 of *Lecture Notes in Computer Science*. Springer-Verlag, 1995.

[San96a]   G. Sander. A fast heuristic for hierarchical Manhattan layout. In Franz Brandenburg, editor, *Graph Drawing*, volume 1027 of *Lecture Notes in Computer Science*, pages 447–458. Springer Berlin / Heidelberg, 1996. 10.1007/BFb0021828.

[San96b]   G. Sander. Graph layout for applications in compiler construction. Technical Report A01-96, Universität des Saarlandes, FB 14 Informatik, 1996.

[Sey95]   P. D. Seymour. Packing directed circuits fractionally. *Combinatorica*, 15:281–288, 1995.

[SM95a]   K. Sugiyama and K. Misue. Graph drawing by the magneting spring model. *Journal of Visual Languages and Computing*, 6(3):217–231, 1995.

[SM95b]   K. Sugiyama and K. Misue. A simple and unified method for drawing graphs: Magnetic-spring algorithm. In R. Tamassia and I. G. Tollis, editors, *Graph Drawing: DIMACS International Workshop, GD '94*, volume 894 of *Lecture Notes in Computer Science*, pages 364–375. Springer-Verlag, 1995.

[SS97]   B. Schwikowski and E. Speckenmeyer. On computing all minimal solutions for feedback problems. Technical report, Universität zu Köln, 1997.

[SSSV01]   Farhad Shahrokhi, Ondrej Sýkora, László A. Székely, and Imrich Vrto. On bipartite drawings and the linear arrangement problem. *SIAM J. Computing*, 30:1773–1789, 2001.

[STT81]   K. Sugiyama, S. Tagawa, and M. Toda. Methods for visual understanding of hierarchical system structures. *IEEE Transaction on Systems, Man, and Cybernetics*, 11(2):109–125, 1981.

[TNB04]   A. Tarassov, N. S. Nikolov, and J. Branke. A heuristic for minimum-width of graph layering with consideration of dummy nodes. In C. C. Ribeiro and S. L. Martins, editors, *Experimental and Efficient Algorithms, Third International Workshop, WEA 2004*, volume 3059 of *Lecture Notes in Computer Science*, pages 570–583. Springer-Verlag, 2004.

[UBSE98] J. Utech, J. Branke, H. Schmeck, and P. Eades. An evolutionary algorithm for drawing directed graphs. In *Proceedings of the 1998 International Conference on Imaging Science, Systems, and Technology (CISST'98)*, pages 154–160, 1998.

[Ulm75]   J. Ulman. NP-complete scheduling problems. *Journal of Computer and System Sciences*, 10:384–393, 1975.

[VML96]   V. Valls, R. Marti, and P Lino. A branch and bound algorithm for minimizing the number of crossing arcs in bipartite graphs. *Eur. J. Op. Res.*, 90:303–319, 1996.

[War77]   J. N. Warfield. Crossing theory and hierarchical mapping. *IEEE Transactions on Systems, Man and Cybernetics*, 7(7):502–523, 1977.

[WM99]   Vance E. Waddle and Ashok Malhotra. An $E \log E$ line crossing algorithm for levelled graphs. In Kratochvíl [Kra99], pages 59–71.

Three-Dimensional Drawings

Vida Dujmović
Carleton University

Sue Whitesides
University of Victoria

14.1   Introduction

Two-dimensional graph drawing, that is, graph drawing in the plane, has been widely studied. While this is not yet the case for graph drawing in 3D, there is nevertheless a growing body of research on this topic, motivated in part by advances in hardware for three-dimensional graphics, by experimental evidence suggesting that displaying a graph in three dimensions has some advantages over 2D displays [WF94, WF96, WM08], and by applications in information visualization [WF94, WM08], VLSI circuit design [LR86], and software engineering [WHF93]. Furthermore, emerging technologies for the nano through micro scale may create demand for 3D layouts whose design criteria depend on, and vary with, these new technologies.

Not surprisingly, the mathematical literature is a source of results that can be regarded as early contributions to graph drawing. For example, a theorem of Steinitz states that a graph G is a skeleton of a convex polyhedron if and only if G is a simple 3-connected planar graph.

It is natural to generalize from drawing graphs in the plane to drawing graphs on other surfaces, such as the torus. Indeed, surface embeddings are the object of a vast amount of research in topological graph theory, with entire books devoted to the topic. We refer the interested reader to the book by Mohar and Thomassen [MT01] as an example.

Numerous drawing styles or conventions for 3D drawings have been studied. These styles differ from one another in the way they represent vertices and edges. We focus on the most common ones and on the algorithms with provable bounds on layout properties and running time.

In this chapter, by a *drawing* we always mean a graph representation (realization, layout, embedding) where no two vertices overlap and no vertex-edge intersections occur unless there is a corresponding vertex-edge incidence in the combinatorial graph. We say that two edges *cross* if they intersect at a point that is not the location of a shared endpoint of the edges in the combinatorial graph. A drawing is *crossing-free* if no two edges cross.

It is natural to represent each vertex by a point and each edge by a straight-line segment joining its endpoint vertices. These so-called *straight-line* drawings are one of the earliest drawing styles considered both in the plane and in 3D. Steinitz's Theorem, for example, ensures the existence of 3D straight-line crossing-free drawings of all 3-connected planar graphs. In fact, as will be seen later, all graphs have such drawings in 3D.

Regardless of the application, the placement of vertices is usually limited to points in some discretized space. For example, when a drawing is to be displayed on a computer screen, vertices must be mapped to integer grid points (pixels). This motivates the study of *grid drawings*, where vertices are required to have integer coordinates. An attractive feature of such drawings is that they ensure a minimum separation of at least one grid unit between any pair of vertices. This aids readability and is thus a desirable aesthetic in visualization applications.

straight-line crossing-free drawings whose vertices are located at points in $\mathbb{Z}^3$ are called *3D (straight-line) grid drawings*. The relaxation where edges are represented with polygonal chains with bends (if any) also at grid-points gives rise to the so-called *3D polyline grid drawings*. Here, a point where a polygonal chain changes its direction is called a *bend*. Straight-line grid drawings are thus a special case of polyline grid drawings. Polyline drawings provide great flexibility. In particular, they allow 3D drawings with smaller volume than is possible in the straight-line model. The number of bends, however, should be kept as small as possible, since bends typically reduce the readability of a drawing.

If each segment of each edge in a polyline drawing is parallel to one of the three coordinate axes, then we say the drawing is an *orthogonal drawing*. Orthogonal drawings are thus special cases of polyline drawings. Since the orthogonal style guarantees very good angular resolution, it is commonly chosen for VLSI design and data-flow diagrams. However, since each vertex is represented by a point, for a graph to admit a 3D orthogonal drawing, each vertex must have degree at most six. To overcome this difficulty, *orthogonal box drawings* were introduced, where each vertex is represented by an axis-aligned box. In such drawings, in addition to the volume and number of bends, various aspects of the sizes and shapes of the boxes are taken as quality measures for the drawing.

Different drawing styles may be subject to different measures of quality. More often than not, however, the measure of a good drawing, regardless of its purpose, rewards having few edge crossings. When a drawing is to be displayed on a page or a computer screen, or is to be used for VLSI design, it is important to keep the volume small to avoid wasting space. On the other hand, a bend on an edge increases the difficulty for the eye to follow the course of the edge. For this reason, it is desirable to keep the edges straight, or at least to keep small the total number of bends and the maximum number of bends per edge.

Since by definition 3D grid drawings have straight edges and no crossings, volume is the main aesthetic criterion for this drawing style. The convention for measuring the volume of a drawing is to multiply together the number of grid points on each of three mutually orthogonal sides of the axis-aligned bounding box of the drawing. In polyline and orthogonal 3D drawings, in addition to the volume, the number of bends is a measure of the quality of the drawing.

In the last decade, this topic has been extensively studied by the graph drawing community. Hence much of the following chapter, in particular Sections 14.2 and 14.3, is dedicated to reviewing the results obtained for 3D (polyline) grid drawings and 3D orthogonal drawings with the volume and the number of bends as the main aesthetic criteria.

Other measures of quality for 3D drawings include: *angular resolution*, defined as the size of the smallest angle between any pair of edges incident to the same vertex; *aspect ratio*, which is the ratio of the length of the longest side to the length of the shortest side of the bounding box of the drawing; and *edge resolution*, which is the minimum distance between a pair of edges not incident to the same vertex. When the underlying combinatorial graph has non-trivial automorphisms, displaying some of the symmetries of the graph can produce beautiful drawings. The display of symmetry in a 3D drawing is one of the various topics covered in Section 14.5. Another one concerns 3D crossing-free straight-line drawings where vertices have real coordinates, that is, they are not restricted to lie on the integer grid.

Suppose edge crossings are permitted for graphs drawn in the plane, but that the edges must then be colored so that no two edges that cross each other have the same color. The minimum number of colors, taken over all possible drawings of that graph, is the classical graph parameter known as *thickness*. If the edges are required to be straight, then this parameter is called the *geometric thickness*. If, in addition, the vertices are required to lie in convex position (i.e., the convex hull of the vertices contains no vertices in its interior), then the parameter is called the *book thickness*.

These three extensively studied graph parameters have a natural interpretation in 3D graph drawing that is important for multilayered VLSI design. Undesired crossings of uninsulated wires are avoided by having wires placed onto several different physical layers, making each layer crossing-free. The graph drawing convention associated with this application area represents each vertex as a line-segment parallel to the Z-axis. Each vertex is intersected by all layers (that is, by planes orthogonal to the Z-axis). Each edge is confined to one of the layers and is drawn between its endpoints in its layer. Edges in the same layer are not allowed to cross. Associating layers, and the edges placed in them, with colors, clearly two edges with the same color do not cross. Thus the minimum possible number of layers corresponds to the thickness parameter. Motivated by the fact that only a limited but increasing number of layers is possible in VLSI technology and also noting that a small number of layers is easier for humans to understand visually, the number of layers of a drawing, that is, its thickness, is the main criterion for the quality for such drawings. The thickness parameters are the subject of Section 14.4.

Graph theory notation used in this chapter: In what follows, all graphs are simple unless stated otherwise. A multigraph is a graph with no loops but it may have multiple copies of edges. A graph G with $n = |V(G)|$ vertices, $m = |E(G)|$ edges, maximum degree at most Δ, and *chromatic number* c is referred to as an n-vertex m-edge degree-Δ c-colorable graph. The complete graph on n vertices is denoted by K_n.

A graph H is a *minor* of a graph G if H is isomorphic to a graph obtained from a subgraph of G by contracting edges. A class of graphs is *minor-closed* if for any graph in the class, all its minors are also in the class. For example, the class of all planar graphs is minor-closed since contracting and/or deleting an edge in a planar graph results in another planar graph. On the contrary, contracting an edge in a 4-regular graph may result in a vertex of degree higher than 4, thus the class of all 4-regular graphs is not minor-closed. A minor-closed class of graphs is *proper* if it is not the class of all graphs.

14.2 Straight-Line and Polyline Grid Drawings

14.2.1 Straight-Line Grid Drawings

A *three-dimensional straight-line grid drawing* (sometimes called a *three-dimensional Fáry grid drawing*) of a graph, henceforth called a *3D grid drawing*, represents the vertices by distinct points in $\mathbb{Z}^3$ (called *grid-points*), and represents each edge as a line-segment between its endpoints, such that edges only intersect at common endpoints, and an edge intersects only the two vertices that are its endpoints (see Figure 14.1). In contrast to the case for the plane, every graph has a 3D grid drawing, by a folklore construction. It is therefore of interest to optimize certain quality measures of such drawings. The most commonly studied measure for 3D grid drawings is their volume, measured as follows.

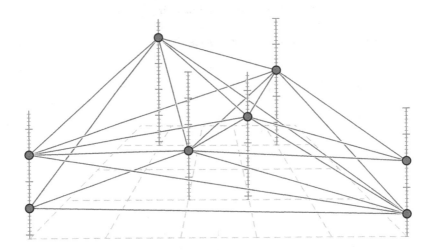

Figure 14.1   A 3D grid drawing of a graph.

The *bounding box* of a 3D grid drawing is the minimum axis-aligned box containing the drawing. If the bounding box has side lengths $X - 1$, $Y - 1$ and $Z - 1$, then we speak of an $X \times Y \times Z$ grid drawing with *volume* $X \cdot Y \cdot Z$. That is, the volume of a 3D grid drawing is the number of gridpoints in the bounding box. This definition is formulated so that two-dimensional straight-line grid drawings have positive volume.

A starting point for many results on 3D grid drawings is the following simple fact.

Fact 14.1   *A straight-line drawing of a graph (on $n > 3$ vertices) such that no four vertices are coplanar has no crossings.*

This fact is key to the folklore construction that proves that every graph has a 3D grid drawing. In particular, a *moment curve* M is a curve defined by parameters (q, q^2, q^3). It is not difficult to prove that no four distinct points on this curve are coplanar. Thus given a graph G on n vertices, a 3D grid drawing of G can be obtained by placing each vertex $v_i \in V(G)$, $1 \le i \le n$, at (i, i^2, i^3). This construction gives an $n \times n^2 \times n^3$ 3D grid drawing with $\mathcal{O}(n^6)$ volume. Cohen *et al.* [CELR96] improved this bound by placing each vertex v_i at the grid-point $(i, i^2 \bmod p, i^3 \bmod p)$, where p is a prime such that $n < p \le 2n$. The resulting drawing is an $n \times 2n \times 2n$ 3D grid drawing with $\mathcal{O}(n^3)$ volume. This construction is a generalization of an analogous two-dimensional technique due to Erdös [Erd51]. Furthermore, Cohen *et al.* [CELR96] proved that the $\Omega(n) \times \Omega(n) \times \Omega(n)$ bounding box and thus the $\Theta(n^3)$ volume bound is asymptotically optimal in the case of the complete graph K_n. The proof of this lower bound is based on the fact that in any 3D grid drawing of K_n, no five vertices can be coplanar, so each side of the bounding box has size at least $n/4$.

Theorem 14.1   [CELR96] *Every n-vertex graph has a 3D grid drawing with $\mathcal{O}(n^3)$ volume. Moreover, the bounding box of every 3D grid drawing of K_n, the complete graph on n vertices, is at least $\frac{n}{4} \times \frac{n}{4} \times \frac{n}{4}$, and thus has $\Omega(n^3)$ volume.*

Since complete graphs require cubic volume, it is of interest to identify fixed graph parameters that allow for 3D grid drawings with smaller volume. The first such parameter to be studied was the chromatic number [CS97, PTT99]. Calamoneri and Sterbini [CS97] proved that each 4-colorable graph has a 3D grid drawing with $\mathcal{O}(n^2)$ volume. Generalizing this result, Pach *et al.* [PTT99] proved the following theorem.

Theorem 14.2 [PTT99] *Every n-vertex graph with chromatic number χ has a 3D grid drawing with $\mathcal{O}(\chi^2 n^2)$ volume. This bound is asymptotically optimal for the complete bipartite graphs with equal sized bipartitions.*

The main idea behind this result is similar to the one for general graphs. In case of complete graphs, crossings are avoided by ensuring that no four vertices are coplanar. That restriction, however, necessarily leads to cubic volume 3D grid drawings and is overly cautious for graphs that have small chromatic number. In particular, vertices that belong to the same color class may all be coplanar, as there are no edges between them. To avoid crossings, it suffices to ensure that if two edges share an endpoint, that they are not collinear and otherwise, that they are not coplanar. The construction in [PTT99] does exactly that. All the vertices that belong to the same color class have the same x-coordinate; in particular, they all belong to some plane orthogonal to the X-axis. Edge crossings are then avoided by appropriate choice of y- and z-coordinates for the vertices. Specifically, if $p \in \mathcal{O}(n)$ is a suitably chosen prime, the main step of this algorithm represents the vertices in the i-th color class by grid-points in the set $\{(i, t, it) : t \equiv i^2 \pmod{p}\}$. It follows that the volume bound is $\mathcal{O}(c^2 n^2)$ for c-colorable graphs.

Many interesting graph families have bounded chromatic number, including planar graphs, bounded genus graphs, and bounded treewidth graphs. In fact all proper minor-closed families have bounded chromatic number. By the above result, all such families have 3D grid drawings with quadratic volume. This naturally gives rise to the question of which graph families admit 3D grid drawings with subquadratic, or even linear volume for each member of a class. Since n distinct points on the 3D integer grid cannot fit in a sublinear volume bounding box, linear volume grid drawings are the best possible for any graph. Pach *et al.* [PTT99] proved that the quadratic volume bound is asymptotically optimal for the complete bipartite graph with equal sized bipartitions. This was generalized by Bose *et al.* [BCMW04] for all graphs.

Theorem 14.3 [BCMW04] *Every 3D grid drawing with n vertices and m edges has volume at least $\frac{1}{8}(n + m)$. In particular, the maximum number of edges in an $X \times Y \times Z$ drawing is exactly $(2X - 1)(2Y - 1)(2Z - 1) - XYZ$.*

For example, graphs admitting 3D grid drawings with $\mathcal{O}(n)$ volume have $\mathcal{O}(n)$ edges.

Planar graphs are one natural class to consider as a candidate for admitting 3D grid drawings with small volume. They have chromatic number at most four, and thus, by the above results [CS97, PTT99], they admit $\mathcal{O}(n^2)$ volume 3D grid drawings. More strongly, the classical result of de Fraysseix *et al.* [dFPP90] and Schnyder [Sch89] states that every planar graph has a $1 \times \mathcal{O}(n) \times \mathcal{O}(n)$ 3D grid drawing, that is, planar graphs admit 2D grid drawings in $\mathcal{O}(n^2)$ area. In 2D this is the best possible, as there are planar graphs that require quadratic area. Intuition suggests, however, that in 3D one should be able to do better. The following open problem has been first suggested by Felsner *et al.* [FLW01].

Open Problem 14.1 [FLW01] *Do planar graphs admit linear volume 3D grid drawings?*

Although the problem is still open, in a recent breakthrough, Di Battista *et al.* [DFP10] showed that planar graphs admit $\mathcal{O}(n \log^{16} n)$ volume 3D grid drawings. Some progress has also been made for more general classes of graphs. In particular, all proper minor-closed families of graphs have been proved to admit $\mathcal{O}(n^{\frac{3}{2}})$ volume 3D grid drawings [DW04c]. Refer to Table 14.1 for exact bounds.

Most, if not all, of the successful attempts to derive linear volume bounds have been done by constructing 3D grid drawings that fit in a bounding box with dimensions $\mathcal{O}(1) \times \mathcal{O}(1) \times \mathcal{O}(n)$. In such a drawing all the vertices lie on $\mathcal{O}(1)$ parallel lines. Thus not only does such a drawing have many quadruples of vertices that are coplanar, but in fact a constant fraction of all vertices are collinear.

Consider a drawing of a graph where all vertices lie on t lines parallel to the Z-axis, such that no three lines are coplanar and no two vertices on the same line are adjacent. Suppose there is a pair of edges that cross in such a drawing and that we would like to remove just that one crossing. If the four endpoints of the edges belong to four distinct parallel lines, as illustrated in Figure 14.2, then, for example, increasing the z-coordinate of the highest vertex removes the crossing. Whenever four endpoints belong to three distinct lines, the two edges do not cross in the projection to the XY-plane and thus cannot cross in the drawing. If, however, the endpoints belong to two parallel lines, then the only way to remove the crossing is to change the ordering of the vertices on one of the two lines, as illustrated in Figure 14.2. These are the difficult crossings to handle, as they arise from a combinatorial situation of "bad" vertex orderings. Having that in mind, Dujmović *et al.* [DMW02] introduced track layouts of graphs, although similar structures are implicit in much previous work [FLW01, HLR92, HR92, RVM95].

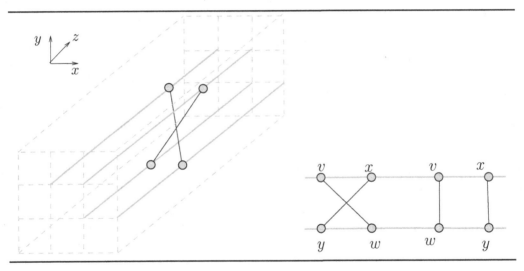

Figure 14.2   Removing a crossing when the edge endpoints are on parallel lines.

Let $\{V_i : i \in I\}$ be a proper vertex t-coloring of a graph G. Let $<_i$ be a total order on each color class V_i. Then $\{(V_i, <_i) : i \in I\}$ is a *t-track assignment* of G. An *X-crossing* in a track assignment consists of two edges vw and xy such that $v <_i x$ and $y <_j w$, for distinct colors i and j. A *t-track layout* of G is a t-track assignment of G with no X-crossing. The *track-number* of G, denoted by $\mathsf{tn}(G)$, is the minimum integer t such that G has a t-track layout. Some authors [DLMW05, Di 03, DLW02, DM03] use a slightly different definition of track layout (called *improper*), in which *intra-track* edges are allowed between consecutive vertices in a track.

Track layouts, which are a purely combinatorial structure, and 3D grid drawings are intrinsically related. In particular, a graph G has a $\mathcal{O}(1) \times \mathcal{O}(1) \times \mathcal{O}(n)$ 3D grid drawing if and only if G has $\mathcal{O}(1)$ track number [DMW05]. More precisely:

Theorem 14.4 [DMW05, DW04c] *Let G be an n-vertex graph with chromatic number $\chi(G) = c$ and track-number $\mathrm{tn}(G) = t$. Then:*

(a) G has an $\mathcal{O}(t) \times \mathcal{O}(t) \times \mathcal{O}(n)$ 3D grid drawing with $\mathcal{O}(t^2 n)$ volume, and

(b) G has an $\mathcal{O}(c) \times \mathcal{O}(c^2 t) \times \mathcal{O}(c^4 n)$ 3D grid drawing with $\mathcal{O}(c^7 tn)$ volume.

Conversely, if a graph G has an $X \times Y \times Z$ 3D grid drawing, then G has track-number $\mathrm{tn}(G) \leq 2XY$.

The key to proving part (a) of the theorem is knowing that there are no bad orderings, that is, no X-crossings; the rest is a generalization of the number theoretic teachings of Erdös that assigns appropriate z-coordinates to vertices such that crossings between edges whose endpoints belong to four distinct tracks are avoided. Proving part (b) of this theorem is much more involved.

Theorem 14.4 (a) says that graphs that have bounded track number admit linear volume 3D grid drawings. Part (b) says that graphs that have bounded chromatic number and sublinear track number have sub-quadratic 3D grid drawings. This provides a strong motivation for studying track layouts of different graph families. Consider first a few simple examples. A *caterpillar* is a tree such that deleting the leaves gives a path. It is simple to verify that a graph has track-number two if and only if it is a caterpillar. Trees have track number at most three. That can be verified by starting with a natural 2D crossing-free drawing of a tree, then wrapping it around a triangular prism, as illustrated in Figure 14.3.

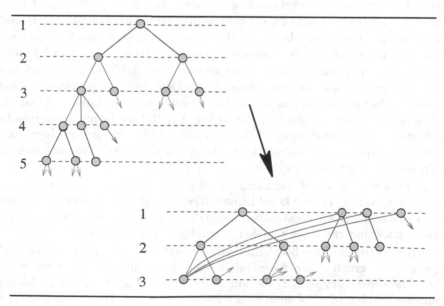

Figure 14.3 3-track layout of trees.

For track layouts such that no two adjacent vertices are allowed to be in the same track, the chromatic number of a graph is a lower bound for its track number. For example,

$\text{tn}(K_n) = n$. However, that lower bound is very weak. Observe, for example, that the complete bipartite graph $K_{n,n}$, although 2-colorable, has track number $n+1$: if two vertices from the same bipartition belong to the same track, then no pair of vertices from the other bipartition can lie on the same track, as otherwise that would imply that $K_{4,4}$ has track number two.

The concept of track layouts, in the case of three tracks, is implicit in the work of Felsner *et al.* [FLW01]. They established the first non-trivial $\mathcal{O}(n)$ volume bound for outerplanar graphs. Their algorithm "wraps" a two-dimensional drawing around a triangular prism. They proved that outerplanar graphs have improper track number at most three.

Dujmović *et al.* [DMW05] proved that graphs of bounded treewidth have bounded track number and therefore have linear volume 3D grid drawings. Many graphs arising in applications of graph drawing have small tree-width. Outerplanar and series-parallel graphs are the obvious examples. They have treewidth at most two. Another example arises in software engineering applications. Thorup [Tho98] proved that the control-flow graphs of go-to free programs in many programming languages have treewidth bounded by a small constant: in particular, 3 for Pascal and 6 for C. Other families of graphs having bounded tree-width (for constant k) include: almost trees with parameter k, graphs with a feedback vertex set of size k, band-width k graphs, cut-width k graphs, planar graphs of radius k, and k-outerplanar graphs. If the size of a maximum clique is a constant k then chordal, interval and circular arc graphs also have bounded tree-width.

Note that bounded tree-width is not necessary for a graph to have a 3D grid drawing with $\mathcal{O}(n)$ volume. The $\sqrt{n} \times \sqrt{n}$ plane grid graph has $\Theta(\sqrt{n})$ tree-width and has a $\sqrt{n} \times \sqrt{n} \times 1$ grid drawing with n volume. It also has a 3-track layout (simply wrap the grid graph, along its diagonals, around a triangular prism,) and thus has a $\mathcal{O}(1) \times \mathcal{O}(1) \times \mathcal{O}(n)$ 3D grid drawing.

The track number of a graph is at most its pathwidth plus one [DMW02]. Many interesting graph families have bounded chromatic number and pathwidth at most $\mathcal{O}(\sqrt{n})$. Thus by Theorem 14.4 (b) they have $\mathcal{O}(n^{\frac{3}{2}})$ volume 3D grid drawings [DW04c]. Included in this family are planar graphs, graphs of bounded genus, graphs with no K_h-minor where h is a constant, and in fact all proper minor-closed families. Refer to Table 14.1 for details.

A vertex coloring is said to be a *strong star coloring* [DW04c] if, for each pair of color classes, all edges (if any) between them are incident to a single vertex. That is, each bichromatic subgraph consists of a star and possibly some isolated vertices. The *strong star chromatic number* of a graph G, denoted by $\chi_{\text{sst}}(G)$, is the minimum possible number of colors in a strong star coloring of G. No matter what ordering on the vertices in each color class in a strong star coloring, there is no X-crossing. Thus the track-number $\text{tn}(G) \leq \chi_{\text{sst}}(G)$, as observed in [DW04c].

Every graph with m edges and maximum degree Δ has track number at most $14\sqrt{\Delta m}$. The proof relies on the Lovász Local Lemma [DW04c]. It is well known that the chromatic number χ of a graph G is at most its maximum degree plus one. Together with Theorem 14.4 (b), this implies that graphs of bounded degree have 3D grid drawings with $\mathcal{O}(n^{\frac{3}{2}})$ volume.

Recently these results have been improved by essentially replacing Δ by the weaker notion of degeneracy. A graph G is *d-degenerate* if every subgraph of G has a vertex of degree at most d. The *degeneracy* of G is the minimum integer d such that G is d-degenerate. A d-degenerate graph is $(d+1)$-colorable by a greedy algorithm. For example, every forest is 1-degenerate, every outerplanar graph is 2-degenerate, and every planar graph is 5-degenerate. Dujmović and Wood proved that every m-edge d-degenerate graph G satisfies $(\text{tn}(G) \leq) \chi_{\text{sst}}(G) \leq 5\sqrt{2dm}$ and $(\text{tn}(G) \leq) \chi_{\text{sst}}(G) \leq (4 + 2\sqrt{2})m^{2/3}$. Again, Theorem 14.4 (b) implies that graphs of bounded degeneracy have 3D grid drawings with $\mathcal{O}(n^{\frac{3}{2}})$ volume.

The family of graphs with bounded degeneracy is vast. It includes all proper minor-closed families, such as, for example, planar graphs. In fact the family is strictly larger than that, since there are graph classes with bounded degeneracy but with unbounded clique minors. For example, the graph K'_n obtained from K_n by subdividing every edge once has degeneracy two, yet contains a K_n minor.

An affirmative answer to the following open problem would imply linear volume 3D grid drawings for planar graphs and thus an affirmative answer to Open Problem 14.1.

Open Problem 14.2 [DMW05] *Do planar graphs have* $\mathcal{O}(1)$ *track-number?*

A tight relationship between track layout and another well-studied type of graph drawing called queue layout has been established in [DPW04]. Queue layouts were introduced by Heath *et al.* [HLR92, HR92] and are defined as follows. A *queue layout* of a graph $G = (V, E)$ consists of a total order $<$ on the vertices $V(G)$, and a partition of the edges $E(G)$ into *queues*, such that no two edges in the same queue are *nested* with respect to $<$: two edges vw and xy are nested with respect to $<$ if $v < x < y < w$. The minimum number of queues in a queue layout of G is called the *queue-number* of G, and is denoted by $\mathsf{qn}(G)$.

It has been established in [DPW04] that a graph has a bounded track number if and only if it has a bounded queue number. Thus Open Problem 14.2 is equivalent to following open problem from 1992 due to Heath *et al.* [HLR92, HR92].

Open Problem 14.3 [HLR92, HR92] *Do planar graphs have* $\mathcal{O}(1)$ *queue-number?*

The best-known upper bound for the queue-number of planar graph is $\mathcal{O}(log^4 n)$, due to Di Battista *et al.* [DFP10]. Unfortunately, for more general proper minor closed families, the best-known bound for both the track number and the queue number is $\mathcal{O}(\sqrt{n})$. The bound follows easily from the fact that proper minor closed families have pathwidth bounded by $\mathcal{O}(\sqrt{n})$.

The best-known bounds on the volume of 3D grid drawings for different graph families are summarized in Table 14.1.

Although almost all of the results on 3D grid drawings focus on the volume of such drawings, some results about aspect ratio of 3D grid drawings were reported in [DMW02].

3D grid drawings have been generalized in a number of ways.

Crossings allowed: Pór and Wood [PW04] considered a variation of 3D grid drawings where edges are allowed to cross. Specifically, they considered 3D drawings where each vertex is represented by a distinct grid point in $\mathbb{Z}^3$ such that the line-segment representing each edge does not intersect any vertex, except the two at the endpoints of the edge. Let such drawings be called 3D *straight-line grid drawings*. With that relaxation, better volume bounds are possible. For instance, a 3D straight-line grid drawing of the complete graph K_n is nothing more than a set of n gridpoints with no three collinear, and such a set can be found with grid volume $\Theta(n^{\frac{3}{2}})$ [PW04]. Generalizing this construction, Pór and Wood [PW04] proved that if edge crossings are allowed, every c-colorable graph has a 3D straight-line grid drawing with $\mathcal{O}(n\sqrt{c})$ volume. That bound is optimal for the c-partite Turán graph.

Dujmović *et al.* [DMS13] studied the crossing number of graphs that have linear volume 3D straight-line grid drawings. In particular, they showed that in every 3D straight-line grid drawing of volume N of a graph with $m \geq 16N$ edges, there are at least $\Omega(\frac{m^2}{N} \log \log \frac{m}{N})$ crossings. They also showed that this bound cannot be much bigger, namely for all $m \leq N^2/4$, there is a graph with m edges that has a 3D straight-line grid drawing of volume

N and $\mathcal{O}(\frac{m^2}{N} \log \frac{m}{N})$ crossings. One such graph is the complete bipartite graph, $K_{N/2,N/2}$. They showed similar results in higher dimensions.

14.2.2   Upward

Another straight-line graph drawing model for the 3D integer grid is the upward 3D grid drawing. A 3D grid drawing of a directed graph G is *upward* if $\mathsf{z}(v) < \mathsf{z}(w)$ for every arc vw of G. Obviously an upward 3D grid drawing can only exist if G is a directed acyclic graph (a *dag*). Upward two-dimensional drawings have been widely studied.

Poranen [Por00] proved that series-parallel digraphs have upward 3D grid drawings with $\mathcal{O}(n^3)$ volume, and that this bound can be improved to $\mathcal{O}(n^2)$ and $\mathcal{O}(n)$ in certain special cases.

Di Giacomo *et al.* [DLMW05] extended the definition of track layouts to dags as follows. An *upward track layout* of a dag G is a track layout of the underlying undirected graph of G, such that if G^+ is the directed graph obtained from G by adding an arc from each vertex v to the successor vertex in the track that contains v (if it exists), then G^+ is still acyclic. The upward track number of G, denoted by $\mathsf{utn}(G)$, is the minimum integer t such that G has an upward t-track layout. Di Giacomo *et al.* [DLMW05] proved the following analogue of Theorem 14.4 (a).

Theorem 14.5   [DLMW05] *Let G be an n-vertex graph with upward track-number $\mathsf{utn}(G) \le t$. Then G has an $\mathcal{O}(t) \times \mathcal{O}(t) \times \mathcal{O}(tn)$ upward 3D grid drawing with $\mathcal{O}(t^3 n)$ volume. Conversely, if a dag G has an $X \times Y \times Z$ upward 3D drawing then G has upward track-number $\mathsf{utn}(G) \le 2XY$.*

This theorem provides motivation for studying upward track layouts of dags. Di Giacomo *et al.* [DLMW05] proved that directed trees have upward track number at least four and at most seven. The upper bound was subsequently improved to five [DW06]. Together with the above theorem, that implies that all directed trees have upward 3D grid drawings with linear volume [DLMW05]. Although undirected outerplanar graphs (and all bounded treewidth graphs) have bounded track number and linear volume 3D grid drawings, the situation is much different in the case of dags. In particular, Di Giacomo *et al.* [DLMW05] proved that there is an outerplanar dag that requires $\Omega(n^{3/2})$ volume in every upward 3D grid drawing. In particular, as illustrated in Figure 14.4, let G_n be the dag with vertex set $\{u_i : 1 \le i \le 2n\}$ and arc set $\{\overrightarrow{u_i u_{i+1}} : 1 \le i \le 2n - 1\} \cup \{\overrightarrow{u_i u_{2n-i+1}} : 1 \le i \le n\}$.

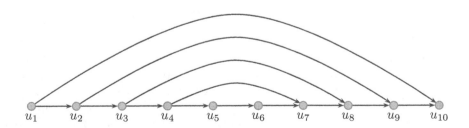

Figure 14.4   Illustration of G_5.

Suppose that G_n has an $X \times Y \times Z$ upward 3D grid drawing. Observe that G_n is outerplanar and has a Hamiltonian directed path $(u_1, u_2, \ldots, u_{2n})$. Thus $(u_1, u_2, \ldots, u_{2n})$ is the only topological ordering of G_n. Thus $Z \ge 2n$. Di Giacomo *et al.* [DLMW05] proved that $\mathrm{utn}(G_n) \geq \sqrt{2n}$. Theorem 14.5 implies that $2XY \geq \mathrm{utn}(G_n) \geq \sqrt{2n}$. Hence the volume is $\Omega(n^{3/2})$ [DLMW05].

This result highlights a substantial difference between 3D grid drawings of undirected graphs and upward 3D grid drawings of dags, since every (undirected) outerplanar graph has a 3D grid drawing with linear volume [FLW01]. In the full version of their paper, Di Giacomo *et al.* [DLMW05] constructed an upward 3D grid drawing of G_n with $\mathcal{O}(n^{3/2})$ volume. It is unknown whether every n-vertex outerplanar dag has an upward 3D grid drawing with $\mathcal{O}(n^{3/2})$ volume.

The proof that every graph has a 3D grid drawing with $\mathcal{O}(n^3)$ volume [CELR96] generalizes to upward 3D grid drawings. In particular,

Theorem 14.6 [DW06] *Every dag G on n vertices has a $2n \times 2n \times n$ upward 3D grid drawing with $4n^3$ volume. Moreover, the bounding box of every upward 3D grid drawing of the complete dag on n vertices is at least $\frac{n}{4} \times \frac{n}{4} \times n$, and thus has $\Omega(n^3)$ volume.*

As already stated, Pach *et al.* [PTT99] proved that every c-colorable graph has an $\mathcal{O}(c) \times \mathcal{O}(n) \times \mathcal{O}(cn)$ drawing with $\mathcal{O}(c^2n^2)$ volume. The result generalizes to upward 3D grid drawings as follows.

Theorem 14.7 [DW06] *Every n-vertex c-colorable dag G has a $c \times 4c^2n \times 4cn$ upward 3D grid drawing with volume $\mathcal{O}(c^4n^2)$.*

Every acyclic orientation of $K_{n,n}$ requires $\mathcal{O}(n^2)$ volume in every upward 3D grid drawing [PTT99]. Hence Theorem 14.7 is tight for constant c. The theorem implies the quadratic volume upper bound for numerous families of dags, including series-parallel dags, planar dags, dags of constant treewidth, all proper minor-closed dags, dags with bounded degeneracy, and so on.

14.2.3 Polyline

Consider a relaxation of 3D straight-line grid drawings where edges are allowed to have bends. In particular, a *three-dimensional polyline grid drawing* of a graph, henceforth called a *3D polyline drawing*, represents the vertices by distinct gridpoints, and represents each edge as a polygonal chain between its endpoints with bends (if any) also at gridpoints, such that distinct edges only intersect at common endpoints, and each edge only intersects a vertex that is an endpoint of that edge. Here a point where a polygonal chain changes its direction is called a *bend*. A 3D polyline drawing with at most b bends per edge is called a *3D b-bend drawing*. Thus 0-bend drawings are 3D grid drawings.

As discussed in the next section, the volume and number of bends in 3D polyline drawings where edges are restricted to be axis-aligned have been studied extensively. The study of 3D polyline drawings has only recently been initiated [DW04b]. Tools developed for 3D (straight-line) grid drawings, such as track layouts, turned out to be useful for the polyline drawings as well. That is simply because a 3D b-bend drawing of a graph G is precisely a 3D straight-line drawing of a subdivision of G with at most b division vertices per edge. This provides a motivation for a study of track layouts of graph subdivisions. Recall that a *subdivision* of a graph G is a graph D obtained from G by replacing each edge $vw \in E(G)$ by a path having v and w as endpoints and having at least one edge. Internal vertices on this path are called *division* vertices.

Dujmović and Wood [DW04b] proved that every n-vertex m-edge graph G has a subdivision D with at most $\log n$ division vertices per edge and such that the track number of D is at most four. Thus by the aforementioned relationship to the 3D grid drawings, D has a (straight-line) 3D grid drawing with $\mathcal{O}(|V(D)|)$ volume. Since $|V(D)| = m \log n$, it follows that every graph G has a 3D polyline drawing with $\mathcal{O}(m \log n)$ volume and at most $\log n$ bends per edge. These results are further generalized [DW04b] as indicated in Table 14.1. For example, complete graphs admit 2-bend 3D polyline grid drawings in $\mathcal{O}(n^2)$ volume. That bound is best possible if the number of bends per edge is restricted to be at most two. If only one bend per edge is allowed, then the complete graphs admit 1-bend 3D polyline grid drawings with $\mathcal{O}(n^{5/2})$ [DEL+05] volume. The best-known lower bound in this case is $\Omega(n^2)$.

Table 14.1 summarizes the best-known upper bounds on the volume and bends per edge in 3D grid drawings and 3D polyline drawings. In general, there is a trade-off between few bends and small volume in such drawings, which is evident in Table 14.1.

graph family	bends per edge	volume	reference
straight-line			
arbitrary	0	$\mathcal{O}(n^3)$	[CELR96]
arbitrary	0	$\mathcal{O}(m^{4/3}n)$	[DW04c]
maximum degree Δ	0	$\mathcal{O}(\Delta mn)$	[DW04c]
maximum degree Δ	0	$\mathcal{O}(\Delta^{15/2}m^{1/2}n)$	[DW06]
d-degenerate	0	$\mathcal{O}(dmn)$	[DW06]
d-degenerate	0	$\mathcal{O}(d^{15/2}m^{1/2}n)$	[DW04c]
c-colorable	0	$\mathcal{O}(c^2n^2)$	[PTT99]
c-colorable	0	$\mathcal{O}(c^6m^{2/3}n)$	[DW04c]
proper minor-closed	0	$\mathcal{O}(n^{3/2})$	[DW04c]
planar	0	$\mathcal{O}(n \log^{16} n)$	[DFP10]
outerplanar	0	$\mathcal{O}(n)$	[FLW01]
bounded treewidth	0	$\mathcal{O}(n)$	[DMW05]
polyline			
c-colorable q-queue	1	$\mathcal{O}(cqm)$	[DW04b]
arbitrary	1	$\mathcal{O}(nm)$	[DW04b]
arbitrary	1	$\mathcal{O}(n^{5/2})$	[DEL+05]
q-queue	2	$\mathcal{O}(qn)$	[DW04b]
q-queue (constant $\epsilon > 0$)	$\mathcal{O}(1)$	$\mathcal{O}(mq^\epsilon)$	[DW04b]
q-queue	$\mathcal{O}(\log q)$	$\mathcal{O}(m \log q)$	[DW04b]

Table 14.1 Volume of 3D straight-line and polyline drawings of graphs with n vertices and $m \geq n$ edges.

In the case of dags, upward variants of 3D polyline grid drawings have also been considered. For instance, with two bends per edge allowed, every n-vertex dag G has an upward 2-bend $n \times 2 \times 2n$ 3D grid drawing with volume $4n^2$ [DW06].

14.3   Orthogonal Grid Drawings

3D polyline (b-bend) drawings where all edge segments are restricted to be parallel to one of the three axes are called *3D orthogonal (b-bend) point-drawings*. This restriction implies that only graphs with maximum degree at most six have such drawings. For that reason the notion is generalized to *3D orthogonal (b-bend) (box-)drawings*, where vertices of the graph are represented by pairwise non-intersecting boxes. A *box* is a rectanguloid with all of its corners at grid points. A 3D orthogonal (*b*-bend) (box)-drawing where all boxes degenerate to cubes, line-segments, or points is called, respectively, a *3D orthogonal (b-bend) cube-, line-,* or *point-drawing*.

The 3D orthogonal drawings have very good angular resolution, which makes them suitable for numerous applications. Minimum edge separation and minimum vertex separation are also guaranteed in such drawings. Notice that neither good angular resolution nor good edge separation is a feature of 3D (straight-line) grid drawings. The main quality measures for 3D orthogonal drawings are the volume and the number of bends (per edge). Other criteria of importance include the length of the edges, and, in the case of 3D orthogonal box-drawings, the size and the shape of the boxes. While the focus of this section is orthogonal drawings in 3D, degree-4 graphs admit 3D polyline drawings with angular resolution even better than 90 degrees. Study of such drawings with small number of bends and good volume bounds has recently been initiated by Eppstein *et al.* [ELMN11].

It is $\mathcal{NP}$-hard to optimize most of these aesthetic criteria for 3D orthogonal drawings. Using straightforward extensions of known two-dimensional hardness results, Eades *et al.* [ESW96] showed that it is $\mathcal{NP}$-hard to find a 3D orthogonal point-drawing of a graph that minimizes any one of the following aesthetic criteria: the volume, the number of bends per edge, the total number of bends, and the total edge length.

Not surprisingly, the 3D orthogonal point-drawings were the first to be studied; we consider them in the next section, followed by a review of 3D orthogonal box-drawings in Section 14.3.2.

14.3.1   Point-Drawings

In a 3D orthogonal point-drawing a vertex can have at most six neighbors. Thus only graphs of degree at most six may admit such drawings. In fact a graph has a 3D orthogonal point-drawing if and only if its maximum degree is at most six. This result will be discussed shortly (Theorem 14.8 below). The drawings used in establishing this result have many bends. This is unavoidable, since every 3D orthogonal point-drawing of the triangle (that is, K_3) obviously has at least one bend. Moreover, to draw an edge between any pair of vertices not on the same grid line, at least one bend is required, and to draw and edge between a pair not on the same grid plane, at least two bends are required. This sheds light on the fact that no nontrivial class of graphs (excluding trees) is known to admit 3D orthogonal point-drawings with zero bends. Less obvious is the well-known result that any 3D orthogonal point-drawing of a multi-graph comprising of two vertices and six edges has an edge with at least three bends. For simple graphs, K_5 requires an edge with at least two bends [Woo03a]. This provides the best-known lower bound on the number of bends per edge for 3D orthogonal point-drawings of degree-6 graphs.

Volume $\Theta(n^{3/2})$:

One of the earliest results concerning 3D orthogonal point-drawings is due to Kolmogorov and Barzdin [KB67] and established a lower bound of $\Omega(n^{3/2})$ for the volume of degree-6 graphs. This lower bound was matched with an upper bound by Eades *et al.* [ESW96] to establish the following theorem.

Theorem 14.8   [ESW96, KB67] *Every n-vertex degree-6 graph has a 3D orthogonal point-drawing in $\mathcal{O}(n^{3/2})$ volume, and that bound is best possible for some degree-6 graphs.*

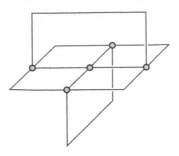

Figure 14.5 3D orthogonal 2-bend point-drawing of K_5 (in coplanar model).

To obtain the upper bound, Eades *et al.* [ESW96] developed an $\mathcal{O}(n)$-time algorithm[1] that produces a 3D orthogonal point-drawing for a degree-6 graph G. Their algorithm is a modification of the method developed by Kolmogorov and Barzdin [KB67] for a similar problem. The algorithm places all the vertices of G on an $\mathcal{O}(n) \times \mathcal{O}(n)$ grid in the $\mathbb{Z} = 0$ plane and draws each edge with at most sixteen bends. This model of drawing where all the vertices intersect one grid plane is known as the *coplanar* model. Figure 14.5 illustrates a 2-bend orthogonal point-drawing of K_5 in the coplanar model.

2 and 3 Bends:

Theorem 14.8 states that for the point-drawings, the optimal volume for degree-6 graphs is known (at least asymptotically). The situation is different for the number of bends per edge. As noted above two bends per edge may be necessary. The best-known upper bound is three. This result was first proved by Eades *et al.* [ESW00].

Theorem 14.9   [ESW00] *Every degree-6 graph has a 3D 3-bend orthogonal point-drawing.*

We now overview the most commonly used approach for producing 3D orthogonal point-drawings. The approach was first taken by Eades *et al.* [ESW00] in their 3-bend algorithm that establishes Theorem 14.9.

A *cycle cover* of a graph G, also called a 2-*factor*, is a 2-regular spanning subgraph of G, that is, a spanning subgraph that consists of cycles. If the graph is directed, then the cycles in the cover are required to be directed as well. Eades *et al.* [ESW00] gave an algorithmic proof that the edges of every degree-6 graph G can be oriented in such a way that G is a subgraph of some directed graph G' (possibly with loops) such that the edges of G' can be colored with three colors each of which induces a directed cycle cover of G'. The proof can be viewed as a repeated application of the classical result of Petersen that every regular graph of even degree has a 2-factor. The cycle covers can be computed in $\mathcal{O}(n)$ time for n-vertex graphs.

Having this in mind, most algorithms for producing 3D orthogonal point-drawings start off with the decomposition of G' into three cycle covers, denoted, say, by $\mathcal{C}_{red}$, $\mathcal{C}_{blue}$, and $\mathcal{C}_{green}$. In the second step vertices of G' are positioned on the 3D grid in some way that makes drawing the red cycles easy. For example, in the coplanar model, vertices can be placed in the $\mathbb{Z} = 0$ plane and all red edges can be drawn in that plane. The remaining

---

[1]The running time in the conference paper is $\mathcal{O}(n^{3/2})$. This was later reduced in [ESW00].

edges $\mathcal{C}_{blue}$ and $\mathcal{C}_{green}$ are then routed above and below the $\mathbb{Z} = 0$ plane, respectively. In general, the third step involves finding drawings for the edges in $\mathcal{C}_{blue}$ and $\mathcal{C}_{green}$.

The 3-bend algorithm of Eades *et al.* [ESW00] positions each vertex v_i of G' at $(3i, 3i, 3i)$ for some arbitrary vertex ordering $(v_1, v_2, \ldots, v_n)$ of $V(G')$. This model of 3D orthogonal point-drawings, where vertices are place along the 3D diagonal of a cube, is called the *diagonal model*. The resulting drawings have volume at most $8n^3$ after all the grid planes not containing a vertex or a bend are deleted. Wood [Woo04] modifies the 3-bend algorithm of Eades *et al.* [ESW00] to produce 3-bend drawings in the diagonal model with $n^3 + o(n^3)$ volume, which is to date the best volume bound on 3D orthogonal 3-bend drawings. To achieve this, Wood places each vertex v_i of G' at (i, i, i) in a particular vertex ordering $(v_1, v_2, \ldots, v_n)$ stemming from book embeddings. For more on book embeddings, refer to the next section on graph thickness. While the algorithm of Eades *et al.* runs in $\mathcal{O}(n)$ time, the algorithm of Wood runs in $\mathcal{O}(n^{5/2})$ time due to the book embedding computation. The diagonal model was also used in the incremental algorithm of Papakostas and Tollis [PT99]. Their algorithm, which runs in $\mathcal{O}(n)$ time, supports on-line insertion of vertices in constant time. The resulting 3D orthogonal 3-bends point-drawings have volume at most $4.63n^3$.

The upper bound from Theorem 14.9 and the lower bound of two on the number of bends per edge leave the following open problem.

Open Problem 14.4 [ESW00] *Does every degree-6 graph have a 3D 2-bend orthogonal point-drawing?*

This problem is considered to be the most important open problem concerning 3D orthogonal point-drawings. The answer to the question remains unknown even when attention is restricted to more specific classes of graphs, including degree-6 planar graphs, degree-6 series-parallel graphs, and degree-6 outerplanar graphs. It is easy to observe that every degree-6 tree has a 3D orthogonal point-drawing with no bends.

A natural candidate for answering Open Problem 14.4 in the negative was K_7, as conjectured in the conference version of [ESW00]. The counterexample to that conjecture was discovered by Wood [Woo03a]. His construction is illustrated in Figures 14.6 and 14.7 (courtesy of David R. Wood). Moreover, Wood exhibited 3D 2-bend point-drawings for other small multipartite 6-regular graphs: $K_{6,6}$, $K_{3,3,3}$ and $K_{2,2,2,2}$.

For degree-5 graphs, Wood [Woo03b] answered Open Problem 14.4 in the affirmative.

Theorem 14.10 [Woo03b] *Every degree-5 graph has a 3D 2-bend orthogonal point-drawing.*

The $\mathcal{O}(n^2)$-time algorithm of Wood that establishes this result produces 3D orthogonal point-drawings of degree-6 graphs in the so-called *general position model*, where no pair of vertices belongs to the same grid plane. (Note, for example, that a drawing in the diagonal model is also in the general position model.) In the case of degree-5 graphs, the algorithm outputs 2-bend drawings in the general position model. While this model allows for 2-bend drawings for degree-5 graphs, the same is not the case for degree-6 graphs. In particular, Wood [Woo03a] constructed an infinite family of degree-6 graphs that have an edge with at least 3 bends in every 3D orthogonal point-drawing in the general position model.

Tradeoffs and more bounds:

Tradeoff issues between the maximum number of bends per edge and the volume of 3D orthogonal point-drawings were first studied by Eades *et al.* [ESW00]. They began with an algorithm to draw a degree-6 graph in the coplanar model with $\mathcal{O}(n^{3/2})$ volume and at most 7 bends per edge. By successive refinements of this algorithm, they obtained 3D orthogonal point-drawings of degree-6 graphs with the following bounds: volume $\mathcal{O}(n^2)$ with at most 6

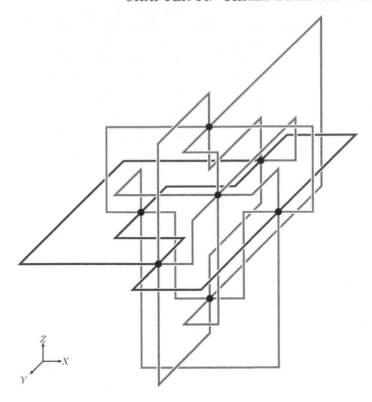

Figure 14.6 A 3D orthogonal 2-bend point-drawing of K_7. (Figure taken from [Woo03a].)

bends per edge, and volume $\mathcal{O}(n^{5/2})$ with at most 5 bends per edge. For drawings in $\mathcal{O}(n^2)$ volume, Biedl [BJSW01] reduced the number of bends per edge to 4.

Numerous refinements of these results have appeared in the literature. Table 14.2 summarizes the best-known bounds on 3D orthogonal point-drawings. Some of the algorithms associated with the bounds in Table 14.2 are dynamic, supporting operations such as vertex insertion [PT99, CGJW01] and deletion, as well as edge deletion and insertion [CGJW01]. See also [DPV00].

In addition to the number of bends per edge, the total number of bends in 3D orthogonal point-drawings has also been investigated. Wood [Woo03a] showed that every 3D orthogonal point-drawing of K_7 has at least 20 bends, which implies the lower bounds of $20m/21$ bends for simple m-edge graphs. The algorithm of Wood [Woo03b] that establishes Theorem 14.10 also produces 3D orthogonal point-drawings for simple m-edge degree-6 graphs with at most $16m/7$ bends, thus having an average of $2\frac{2}{7}$ bends per edge. The drawings are in the general position model, for which the bound is optimal since K_7 requires $\frac{16}{7}|E(K_7)|$ bends in that model, as established in [Woo03a].

14.3.2 Box-Drawings

Only degree-6 graphs admit 3D orthogonal point-drawings. Hence it was only natural to consider the extension to box-drawings for general graphs. For point-drawings, it was enough to consider K_3 to realize that there are degree-6 graphs that do not admit such drawings with straight-line edges. It is less obvious that not all graphs admit 3D orthogonal box-drawings with straight-line edges (that is, with zero bends). In a straight-line orthogo-

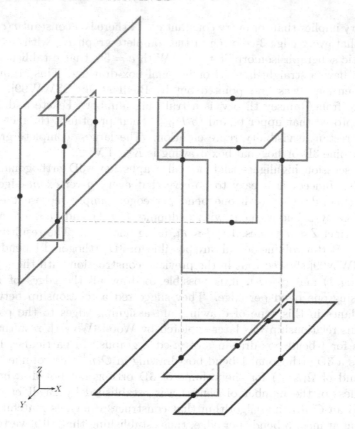

Figure 14.7 Breakaway view of the 3D orthogonal 2-bend point-drawing of K_7. (Figure taken from [Woo03a].)

graph family	max. (avg.) bends per edge	volume	reference
multigraph	7	$\Theta(n^{3/2})$	[ESW00]
multigraph (dynamic)	14	$\Theta(n^{3/2})$	[BJSW01]
multigraph	4	$\mathcal{O}(n^2)$	[BJSW01]
multigraph (dynamic)	5	$\mathcal{O}(n^2)$	[CGJW01]
multigraph $\Delta \leq 4$	3	$\mathcal{O}(n^2)$	[ESW00]
simple	$4\ (2\frac{2}{7})$	$2.13n^3$	[Woo03b]
multigraph (dynamic)	3	$4.63n^3$	[PT99]
multigraph	3	$n^3 + o(n^3)$	[Woo04]
simple $\Delta \leq 5$	2	n^3	[Woo03b]

Table 14.2 The volume and the number of bends per edge in 3D orthogonal point-drawings of n-vertex graphs with maximum degree $\Delta \leq 6$.

nal box-drawing of a graph G, each edge is a line segment parallel to one of the three axes. This defines an associated coloring of the edges with three colors, where a subgraph of G induced by each color class has a visibility representation by rectangles. (Refer to the last section, page 478, for the definition of a visibility representation.) Bose *et al.* [BEF+98] proved that K_n does not have such a representation for $n \geq 56$.

Ramsey theory implies that for every constant $c \in \mathbb{N}$ there is a constant $r(c)$ (the Ramsey number) such that every edge 3-coloring of the complete graph K_n with $n \geq r(c)$ contains a monochromatic subgraph isomorphic to K_c. With $c = 56$, that establishes the fact that $K_{r(56)}$ does not have a straight-line 3D orthogonal box-drawing. This argument (in three and higher dimensions) was first pointed out by Biedl *et al.* [BSWW99]. The constant $r(56)$, stemming from Ramsey theory, is a truly big number. Fekete and Meijer [FM99] significantly improved that upper bound to K_{184}. Their proof uses the fact that K_{56} does not have a 3D rectangle visibility representation. The largest complete graph known to admit a straight-line 3D orthogonal box-drawing is K_{56} [FM99].

The above discussion highlights that not all graphs have 3D orthogonal box-drawings with zero bends. Indeed, it is easy to observe that every n-vertex m-edge graph G has an orthogonal (line)-drawing with one bend per edge: simply represent each vertex v_i, $1 \leq i \leq n$, of G by a line-segment with endpoints $(i, i, 1)$ and (i, i, m), and then draw each edge in distinct $\mathsf{Z} = j$ planes, $1 \leq j \leq m$, using one bend. The resulting drawing has $\mathcal{O}(n^2 m)$ volume. Better volume bounds are possible for 3D orthogonal 1-bend box-drawings. Biedl *et al.* [BSWW99] showed that in the previous construction with the segments having endpoints at $(i, i, 1)$ and (i, i, n), it is possible to draw all the edges of K_n in $\mathsf{Z} = j$, $1 \leq j \leq m$, using one bend per edge. They suggested a relationship between assigning edges to the planes in this type of drawing and assigning edges to the pages of a book embedding. This relationship was later explored by Wood [Woo01], resulting in improved volume bounds for 1-bend box-drawings of m-edge graphs. In particular, he proved that every graph has a 3D orthogonal 1-bend box-drawing in $\mathcal{O}(n^{3/2} m)$ volume.

A lower bound of $\Omega(n^{5/2})$ for the volume of 3D orthogonal box-drawings of n-vertex graphs (regardless of the number of bends) was established by Biedl *et al.* [BSWW99]. They developed an $\mathcal{O}(m)$-time algorithm that constructs drawings matching that volume bound and using at most 3 bends per edge, thus establishing that all n-vertex graphs have 3D orthogonal 3-bend box-drawings in $\Theta(n^{5/2})$ volume. Closing the gap between the $\mathcal{O}(n^3)$ upper bound and the $\Omega(n^{5/2})$ lower bound for 3D orthogonal 1-bend box-drawings of K_n remains an interesting open problem.

The lower bound of Biedl *et al.* [BSWW99] was established using the complete graph K_n. The proof relies critically on the fact that between any two disjoint vertex sets of size $\Omega(n)$ in K_n, there are $\Theta(n^2)$ edges. To generalize this lower bound to sparse graphs and to be able to express it in terms of the number of edges, Biedl *et al.* [BTW06] exhibited graphs such that between any two disjoint vertex sets of size $\Omega(n)$ there are $\Theta(m)$ edges. That allowed them to extend the arguments of [BSWW99] to establish the lower bound of $\Omega(m\sqrt{n})$ on the volume of 3D orthogonal box-drawings of m-edge n-vertex graphs. They developed an $\mathcal{O}(m^2/\sqrt{n})$-time algorithm that constructs drawings matching that volume bound and using at most 4 bends per edge, thus establishing that all graphs have 3D orthogonal 4-bend box-drawings in $\Theta(m\sqrt{n})$ volume. It is unknown whether all m-edge graphs admit 3D orthogonal box-drawings with such volume and at most 3 bends per edge, as is the case for K_n.

The discussion above pertains to drawings where the volume and the number of bends per edge are the only concerns. The shapes and the sizes of boxes used to represent vertices are unrestricted. However, for box-drawings the size and the shape of a vertex with respect to its degree are also important aesthetic criteria. For a vertex v in a 3D orthogonal box-drawing the *surface* of v is the number of grid lines intersecting the box-representing v times two. The surface of v indicates the number of grid lines available for drawing edges incident to v. In point-drawings, for example, the surface of each vertex is six. Generally, in any 3D orthogonal box-drawing, the surface of each vertex v is at least the degree of v. Ideally, the surface of v should also not be much bigger than the degree of v. Biedl *et al.* [BTW06] defined a 3D orthogonal box-drawing of a graph G to be *degree-restricted* if there exists some constant $\alpha \geq 1$ such that for every vertex v in G, surface(v) $\leq \alpha \cdot$degree(v).

Degree-restricted drawings do not, however, impose any aesthetic restriction on the shape of the boxes used to represent vertices. The *aspect ratio* of a vertex in a 3D orthogonal box-drawing is the ratio of the length (measured in the number of grid points) of the longest side of the box representing that vertex to the shortest side of that box. 3D orthogonal box-drawings have a *bounded vertex-aspect ratio* if there exists a constant r such that all vertices have aspect ratios at most r. Note that $r \geq 1$, and for the case of 3D orthogonal point-drawings and cube-drawings, it is one. Also note that degree-restricted drawings may have unbounded vertex-aspect ratio; consider, for example, a drawing in which each vertex is represented by a segment with length equal to its degree.

The discussion at the beginning of this subsection pertains to 3D orthogonal box-drawings with (possibly) unbounded vertex-aspect ratios and with no degree-restrictions. The best-known upper bounds on the volume and the number of bends per edge in such unrestricted 3D orthogonal box-drawings are summarized in the top part of Table 14.3. The upper bounds can be compared to the best-known lower bound on the volume of such drawings which, as discussed above, is $\Omega(m\sqrt{n})$ regardless of the number of bends [BTW06]. The table exhibits the tradeoff between the number of bends per edge and the volume of such drawings.

Biedl *et al.* [BTW06] derived lower bounds for the volume of 3D orthogonal box-drawings that are required to be degree-restricted and/or have bounded vertex-aspect ratio. In particular, they proved an $\Omega(m^{3/2}/\alpha)$ lower bound on the volume of 3D orthogonal box-drawings that are degree-restricted for some $\alpha \geq 1$, as well as an $\Omega(m^{3/2}/\sqrt{r})$ lower bound on the volume of 3D orthogonal box-drawings for which each vertex has aspect ratio at most r. For bounded α and bounded r, both bounds become $\Omega(m^{3/2})$. The discussion pertaining to the proof technique of Biedl *et al.* [BTW06] used to derive the $\Omega(m\sqrt{n})$ volume bound for unrestricted drawings applies to these two lower bounds as well.

Biedl *et al.* [BTW06] also developed an algorithm that constructs the corresponding 3D orthogonal box-drawings matching the volume lower-bound and using at most 6 bends per edge, thus establishing that all m-edge graphs have 3D orthogonal 6-bend box-drawings with volume $\Theta(m^{3/2})$ such that the drawings are degree-restricted and have bounded aspect ratio.

The best-known upper bounds on the volume and the number of bends per edge in degree-restricted 3D orthogonal box-drawings are summarized in the middle part of Table 14.3, while drawings that are both degree-restricted and have bounded vertex-aspect ratio are addressed at the bottom of the table. These upper bounds on the volume can be compared to the best-known lower bound of $\Omega(m^{3/2})$.

The table reveals that no further asymptotic improvements are possible for the volume of drawings in all three aesthetic models discussed. There is room for improvement, however, with regard to the number of bends per edge, as suggested by some of the open problems mentioned in this subsection.

14.4 Thickness

Thickness is a classical graph parameter that has been studied since the early 1960s. It was first defined by Tutte [Tut63]. The *thickness* of a graph G, denoted by $\theta(G)$, is the minimum $k \in \mathbb{N}$ such that the edge set of G can be partitioned into k planar subgraphs.

For ease of exposition in this section, we express the concept of thickness in terms of drawings in the plane. The *thickness* of a drawing in the plane with vertices represented

graphs	bends	volume	reference
unbounded vertex-aspect ratio / not degree-restricted			
simple	1	$\mathcal{O}(n^3)$	[BSWW99]
simple	1	$\mathcal{O}(n^{3/2}m)$	[Woo01]
simple	2	$\mathcal{O}(nm)$	[Woo01]
simple	3	$\mathcal{O}(n^{5/2})$	[BSWW99]
multigraphs	3	$\mathcal{O}(nm)$	[BTW06]
simple	4	$\Theta(m\sqrt{n})$	[BTW06]
unbounded vertex-aspect ratio / degree-restricted			
simple	2	$\mathcal{O}(n^2m)$	[Bie98, Woo99]
simple	2	$\mathcal{O}(n^2\Delta)$	[Bie98]
multigraphs	5	$\mathcal{O}(m^2)$	[BTW06]
multigraphs	6	$\Theta(m^{3/2})$	[BTW06]
bounded vertex-aspect ratio / degree-restricted			
simple	2	$\mathcal{O}((nm)^{3/2})$	[Bie98, Woo99]
simple	2	$\mathcal{O}(nm\sqrt{\Delta})$	[Bie98]
multigraphs	5	$\mathcal{O}(m^2)$	[BTW06]
simple	10	$\mathcal{O}((n\Delta)^{3/2})$	[HTS83]
multigraphs	6	$\Theta(m^{3/2})$	[BTW06]

Table 14.3 Volume and the maximum number of bends in 3D orthogonal (box)-drawings of n-vertex m-edge degree-Δ graphs for various aesthetic criteria.

as points and edges represented as simple curves is the minimum $k \in \mathbb{N}$ such that the edges of the drawing can be partitioned into k subgraphs such that each subgraph has no crossings in the drawing; that is, each edge is assigned one of k colors such that no pair of like-colored edges of the drawing cross. Since any planar graph can be drawn with its vertices at prespecified points in the plane (see, for example, [PW01]), a graph has thickness k if and only if it has a drawing in the plane with thickness k [Hal91]. However, in such a drawing the edges may be highly curved and thus unsuitable for most applications. For instance, when the edges are represented by polygonal chains, then $\Omega(n)$ bends per edge may be needed [PW01]. This motivates the notion of geometric thickness.

A drawing of a graph in the plane is *geometric* if every edge is represented by a straight-line segment. The *geometric thickness* of a graph G, denoted by $\bar{\theta}(G)$, is the minimum $k \in \mathbb{N}$ such that there is a geometric drawing of G with thickness k. Kainen [Kai73] first defined geometric thickness under the name of *real linear thickness*, and it has also been called *rectilinear thickness*. By the Fáry-Wagner theorem, a graph has geometric thickness one if and only if it is planar. Graphs of geometric thickness two, the so-called *doubly linear* graphs, were studied by Hutchinson *et al.* [HSV99] in the context of rectangle-visibility graphs.

Another parameter closely related to geometric thickness is book thickness. A geometric drawing in which the vertices are in convex position is called a *book embedding*. The *book thickness* of a graph G, denoted by $\mathsf{bt}(G)$, is the minimum $k \in \mathbb{N}$ such that there is book embedding of G with thickness k. The book embeddings have also been called *stack layouts*, and book thickness is also called *stacknumber*, *pagenumber* and *fixed outerthickness*.

Whether two edges cross in a book embedding is simply determined by the relative positions of their endpoints in the cyclic order of the vertices around the convex hull. One can think of the vertices as being ordered on the spine of a book and each plane subgraph being drawn without crossings on a single page. A graph has book thickness one if and only if it is outerplanar [BK79]. Bernhart and Kainen [BK79] proved that a graph has book thickness at most two if and only if it is a subgraph of a Hamiltonian planar graph. Unlike thickness, being able to partition the edge set of a graph G into k outerplanar subgraphs does not imply that G has book thickness at most k. For example, the edge set of K_5 can be partitioned into two cycles, yet K_5 has book thickness more than two, since it is not a subgraph of a Hamiltonian planar graph. The situation is similar for geometric thickness as will soon become clear.

Book embeddings, first defined by Ollmann [Oll73], are ubiquitous structures with a variety of applications; see [DW04a] for a survey with over 50 references. These applications include sorting permutations, fault-tolerant VLSI design, and compact graph encodings as well as graph drawing. In general, drawings arising from the study of thickness have applications in graph visualization (where each plane subgraph is colored by a distinct color), and in multilayer VLSI (where each plane subgraph corresponds to a set of wires that can be routed without crossings in a single layer).

First we consider the relationship between the three thickness parameters. By definition, for every graph G

$$\theta(G) \le \overline{\theta}(G) \le \mathsf{bt}(G). \tag{14.1}$$

These inequalities have been shown to be strict for certain graphs [DEH00]. In the other direction, no such relationship is possible for any bounding function. Eppstein [Epp01] proved that geometric thickness is not bounded by any function of book thickness. In particular, the graph obtained by subdividing each edge of K_n once has geometric thickness at most two. On the other hand, a Ramsey-theoretic argument shows that the book thickness of that graph is not bounded by any constant.

Using a more elaborate Ramsey-theoretic argument applied to graphs formed by starting with n points and adding a new point adjacent to each triple of the n points, Eppstein [Epp04a] proved that geometric thickness is not bounded by any function of thickness. In particular, for every t there exists a graph with thickness three and geometric thickness at least t. This leaves an interesting open problem.

Open Problem 14.5 [Epp04a] *Do graphs with thickness two have bounded geometric thickness?*

Complete graphs: The thickness of the complete graph K_n was intensely studied in the 1960s and 1970s. Results by a number of authors [AG76, Bei67, BH65, May72] together prove that $\theta(K_n) = \lceil (n+2)/6 \rceil$, unless $n = 9$ or 10, in which case $\theta(K_9) = \theta(K_{10}) = 3$.

Bernhart and Kainen [BK79] proved that $\mathsf{bt}(K_n) = \lceil n/2 \rceil$. In fact, they proved that every convex drawing of K_n can be partitioned into $\lceil n/2 \rceil$ plane spanning paths.

Bose *et al.* [BHRCW06] proved that every geometric drawing of K_n has thickness at most $n - \sqrt{n/12}$. It is unknown whether every geometric drawing of K_n has thickness at most $(1 - \epsilon)n$. Dillencourt *et al.* [DEH00] studied the geometric thickness of K_n, and proved that

$$\lceil (n/5.646) + 0.342 \rceil \le \overline{\theta}(K_n) \le \lceil n/4 \rceil . \tag{14.2}$$

Their upper bound construction generalizes to show that for any n, $\overline{\theta}(K_n) \le \lceil n/4 \rceil$. What is $\overline{\theta}(K_n)$? It seems likely that the answer is closer to $\lceil n/4 \rceil$ rather than to the above lower bound.

Maximum degree: Next, we consider the relationships among the three thickness parameters and the maximum degree. Recall that, a graph with maximum degree Δ is called a *degree-Δ graph*. Wessel [Wes84] and Halton [Hal91] proved independently that the thickness of a degree-Δ graph is at most $\lceil \Delta/2 \rceil$. The proof is based on the classical result of Petersen that every regular graph of even degree has a 2-factor, that is, a set of vertex disjoint cycles that together cover all the vertices. The theorem implies that the edges of a Δ-regular graph for even Δ can be partitioned into $\Delta/2$ sets of vertex disjoint cycles. Vertex disjoint cycles are planar, and thus the upper bound follows by proving that every degree-Δ graph is a subgraph of some Δ-regular graph. Sýkora *et al.* [SSV04] proved that this bound is tight.

Malitz [Mal94b] proved that there exist Δ-regular n-vertex graphs with book thickness at least $\Omega(\sqrt{\Delta}n^{1/2-1/\Delta})$. Thus, unlike thickness, book thickness is not bounded by any function of maximum degree. The proof is based on a probabilistic construction. Malitz [Mal94b] also derived an upper bound of $\mathcal{O}(\sqrt{m}) \in \mathcal{O}(\sqrt{\Delta n})$ for the book thickness, and thus the geometric thickness, of m-edge graphs.

Eppstein [Epp04a] asked whether bounded degree graphs have bounded geometric thickness. Duncan *et al.* [DEK04] gave an affirmative answer for degree-4 graphs. By Petersen's theorem, the edges of a degree-4 graph G can be partitioned into two sets each of which induces a subgraph comprised of vertex disjoint paths and cycles in G. Duncan *et al.* [DEK04] proved that two such subgraphs can be drawn simultaneously on some planar point set using straight-line edges, thus proving that G has a geometric drawing with thickness at most two. Moreover, they provided a linear-time algorithm to produce such thickness-2 geometric drawings for degree-4 graphs. In the case of degree-3 graphs, the resulting drawings fit in the $n \times n$ grid.

In a recent development, the above-mentioned question of Eppstein has been answered in the negative. Barát *et al.* [BMWR3] have shown that bounded degree graphs may have unbounded geometric thickness, even approaching the square root of the number of vertices. In particular, for all $\Delta \geq 9$ there exists a Δ-regular n-vertex graph with geometric thickness $\Omega(\sqrt{\Delta}n^{1/2-4/\Delta-\epsilon})$. The proof is non-constructive and based on counting arguments. The authors have shown that there are more graphs with bounded degree than with bounded geometric thickness. To count the number of n-vertex graphs of thickness k, they considered the number of order types of n points and all the ways of connecting the points in an order type into a geometric drawing of thickness k.

Open Problem 14.6 [BMWR3] *Do degree-Δ graphs with $\Delta \in \{5, 6, 7, 8\}$ have bounded geometric thickness?*

Proper minor-closed families: Blankenship and Oporowski [Bla03, BO01] proved that all proper minor-closed families have bounded book thickness and therefore, by Equation 14.1, bounded thickness and geometric thickness. Proper minor-closed families include, for example, planar graphs, bounded genus graphs, and bounded treewidth graphs. The proof depends on Robertson and Seymour's deep structural characterization of the graphs excluding a fixed minor. As a result, the obtained bound on book thickness for graphs excluding a K_ℓ-minor is a truly huge function of ℓ.

A much better bound is known for the thickness of such families. Kostochka [Kos82] and Thomason [Tho84] proved independently that graphs excluding a K_ℓ-minor have thickness at most $\mathcal{O}(\ell \log \ell)$. Better bounds on book thickness (and thus geometric thickness) are also known for many minor-closed families. The question of book thickness of planar graphs was settled by Yannakakis [Yan86] in 1986: he proved that the book thickness of planar graphs is at most four and that there are planar graphs with book thickness matching that bound. There is some dispute over this lower bound. The construction is given in the conference version of the paper only [Yan86], where the proof is far from complete.

Endo [End97] determined that the book thickness of toroidal graphs, that is, graphs with genus one, is at most seven. Malitz [Mal94a] proved by a probabilistic argument that the book thickness of graphs with genus γ is at most $\mathcal{O}(\sqrt{\gamma})$.

Exact bounds are known for all three thickness parameters in relation to treewidth. In particular, for graphs of treewidth k the maximum thickness and the maximum geometric thickness both equal $\lceil k/2 \rceil$ [DW05]. This says that the lower bound for thickness can be matched by an upper bound, even in the more restrictive geometric setting. For graphs of treewidth k, the maximum book thickness equals k if $k \leq 2$ and equals $k+1$ if $k \geq 3$. While the lower bounds are proved in [DW05], the upper bounds on book thickness are due to Ganley and Heath [GH01].

Computational complexity: The graphs with book thickness one are precisely the outerplanar graphs [BK79], and thus can be recognized in linear time. The graphs with book thickness two are characterized as the subgraphs of planar Hamiltonian graphs [BK79], which implies that it is $\mathcal{NP}$-complete to test if $\mathrm{bt}(G) \leq 2$ [Wig82]. In fact, even determining thickness of a given book embedding is hard. Specifically, a book embedding with k pairwise crossing edges has thickness at least k, since each edge must receive a distinct color. However, the converse is not true. There exist book embeddings with no $(k+1)$ pairwise crossing edges for graphs that have thickness at least $\Omega(k \log k)$ [KK97]. Moreover, it is $\mathcal{NP}$-complete to test if a given book embedding of a graph has thickness k [GJMP80].

Testing whether a graph has thickness k is $\mathcal{NP}$-hard [Man83] even for $k = 2$. Eppstein [Epp04b] considered the problem of testing if a given geometric drawing has thickness k. For $k = 2$ the problem can be solved in polynomial time but becomes $\mathcal{NP}$-complete for $k \geq 3$. Dillencourt *et al.* [DEH00] asked what the complexity is for determining the geometric thickness of a given graph.

Open Problem 14.7 [DEH00] *Is it $\mathcal{NP}$-hard to test if the geometric thickness of a graph is k?*

We close this section with an open problem that relates book thickness and 3D grid drawings.

Open Problem 14.8 [DW04b] *Do all bipartite graphs that have book thickness three have bounded track-number?*

By studying book thickness of graph subdivisions Dujmović and Wood [DW04b] proved that an affirmative answer to this question would imply an affirmative answer to Open Problems 14.1, 14.2, and 14.3. More generally, it would imply that the queue-number is bounded by book-thickness, which is a long standing open problem [HLR92]. Since all proper minor-closed graph families have bounded book thickness [BO01], an affirmative answer to this question would further imply that all proper minor-closed graph families have linear volume 3D grid drawings.

14.5 Other (Non-Grid) 3D Drawing Conventions

3D crossing-free straight-line drawings with real coordinates: Three dimensional straight-line crossing-free graph drawings in which the vertices are allowed real coordinates have also been studied. Naturally, having a less restrictive model allows for drawings with better bounds, for example better volume bounds, in comparison to the grid model. One disadvantage to using real coordinates, however, becomes evident when a drawing is to be displayed, on a computer screen for example. Then the real vertex coordinates must be converted into integer coordinates. There are no guarantees that rounding off will maintain the correctness of the embedding.

As in the grid model, the main criterion for measuring the quality of a drawing is its volume. To make a discussion about volume meaningful, that is, to disallow arbitrary scaling, the vertices are required to lie at least unit distance apart. As noted in the introduction, a classical result of Steintz states that the triconnected planar graphs are exactly the 1-skeletons of convex polyhedra in 3D, that is, they admit 3D convex drawings. This may be considered as one of the first results in the real coordinates model. The construction, however, seems to require exponential volume in the number of vertices of a graph. The same is true for the number of bits needed to represent the coordinates of the vertices. This outlook has been greatly improved by Chrobak *et al.* [CGT96]. The technique they used to derive their results falls under the category of so-called *force directed* methods.

Force directed methods model the graph as a physical system. For example, edges can be modeled as springs and vertices as charged particles that repel each other. A configuration where the sum of the forces on each particle is zero, that is, a local minimum of the system, gives a straight-line drawing of the graph. The famous *barycenter method* developed by Tutte [Tut60] is an example of the force directed approach. Specifically, the barycenter method takes a 3-connected plane graph G and fixes the vertices of the outer face in a convex position in the plane. The remaining vertices of G are then added one by one at the barycenter of their neighbors. The resulting system of linear equations gives coordinates for the internal vertices, and results in a 3D drawing of G where all internal faces are convex. This method can be extended to 3D.

As noted above, the best-known bounds are due to Chrobak *et al.* [CGT96]. They developed a force-directed algorithm that, given an n-vertex triconnected planar graph G, outputs a 3D drawing of G with $\mathcal{O}(n)$ volume. Moreover, the vertex coordinates in the drawing can be represented by $\mathcal{O}(n \log n)$-bit rational numbers. The algorithm runs in $\mathcal{O}(M(n^{1/2}))$ time, where $M(n)$ is the time needed to multiply two $n \times n$ matrices. They also showed that if the minimum angle between two edges incident to the same vertex is required to be some fixed function of the maximum degree, then there are bounded-degree triconnected planar graphs that require $2^{\Omega(n)}$ volume in any 3D convex drawing.

In other results in the real coordinate model, Garg *et al.* [GTV96] proved that all graphs with bounded chromatic number can be drawn in $\mathcal{O}(n^{3/2})$ volume with constant aspect ratio and using $\mathcal{O}(\log n)$-bit rational numbers for vertex coordinates. If the number of bits is increased to $\mathcal{O}(n \log n)$, they showed that all graphs have 3D straight-line crossing-free drawings in $\mathcal{O}(n)$ volume. Their algorithms run in $\mathcal{O}(n)$ time provided that the graph coloring is given as a part of the input.

Simulated annealing techniques for generating 3D straight-line drawings of general graphs have also been considered [CT96].

3D graph representations: In a *graph representation*, vertices are depicted as some set of objects and edges indicate a relationship between the objects. In the case of visibility representations, for example, there is an edge between two vertices in the graph if and only if there is a line-segment that joins the objects representing the vertices and that does not intersect any other object, that is, if the two objects are (mutually) visible. Typically, these line-segments may be required to align with an axis. In two dimensions, popular visibility representations studied are *bar-* and *rectangle visibility*. Both models are related to orthogonal drawings in the plane. Only thickness-2 graphs have such two-dimensional visibility representations, which motivates the study of 3D counterparts.

The concept generalizes naturally to three dimensions. The vertices may be disjoint 2D objects parallel to the XY-plane, and the edges may be line-segments parallel to Z-axis connecting pairs of visible objects. It is easy to see that all graphs have such a representation if the objects may be arbitrary non-convex polygons. Attention has therefore been restricted to convex polygons. For instance, K_7 has a representation with unit squares and K_8 does not, and every graph has a representation with unit disks. Bose *et al.* [BEF+98] proved that K_n has a representation with arbitrary rectangles for $n \leq 22$, while for $n \geq 56$ it does not. They also showed that all planar graphs and all complete bipartite graphs have a representation with arbitrary rectangles, but that the family of representable graphs is not closed under graph minors.

Alt *et al.* [AGW98] considered representations with arbitrary convex polygons and showed that there is no convex polygon P that would allow every complete graph to have a visibility representation by shifted copies of P. In particular, for $n > 2^{2^k}$, K_n cannot be represented by a convex k-gon. This bound has been improved by Štola [Što04], who proved that the maximum size of a complete graph with a visibility representation by copies of regular k-gon is between $k+1$ and 2^{6k}. Visibility representations with boxes have also been considered [FM99].

Kotlov *et al.* [KLV97] discovered a relationship between graph representations by touching spheres in 3D and the algebraic graph invariant μ introduced by Colin de Verdière.

Surfaces and the theory of graph minors: The field of topological graph theory studies geometric realizations of graphs in 3-space and embeddings on surfaces. Embeddings of graphs on higher surfaces are a natural generalization of embeddings in the plane.

The celebrated *graph minors theorem* of Robertson and Seymour [RS] implies that there is a finite number of forbidden minors for graphs embeddable on any given fixed surface. The Kuratowski theorem identifies the forbidden minors for the plane. The projective plane is the only other surface for which all the forbidden minors (35 of them) are known. Mohar [Moh99] gave a linear-time algorithm that for any graph and any fixed surface S, either finds an embedding of the given graph in S or identifies a subgraph homeomorphic to a forbidden minor for S.

The power of the graph minors theorem can be nicely illustrated by means of the following 3D graph drawing problem. A graph is *knotless* if it has an embedding in 3D that does not contain a non-trivial knot, that is, if it has an embedding such that every cycle in the embedding bounds a disk. For example K_7 is known not to have a knotless embedding. It is easy to observe that the class of all knotless graphs is minor-closed. One algorithmic consequence of the graph minors theory is that there is a cubic time algorithm to test membership of a graph in any proper minor-closed family. Thus, remarkably, there exists a cubic time algorithm to test if a graph is knotless. This problem was not even known to be decidable before the advent of the graph minors theory. At present, however, no explicit algorithm is known, let alone a polynomial-time one, as the theory only guarantees the existence of such an algorithm.

A related concept is that of a linkless embedding. A graph is *linkless* if it has an embedding in 3D that does not contain a pair of linked cycles, that is, two cycles in the embedding that cannot be separated by a 2-sphere embedded in 3D. For example, K_6 is known not to be linkless. Unlike the case for knotless graphs, the full characterization of linkless graphs is known. In particular, a graph is linkless if and only if it does not contain a minor one of the six members of the Peterson family of graphs. A ΔY-*exchange* in a graph replaces a triangle by a 3-star, while a $Y\Delta$-*exchange* replaces a 3-star by a triangle. The Peterson family is comprised of the six graphs that can be obtained from K_6 by a sequence of ΔY- and $Y\Delta$-exchanges. It is also known that a graph G is linkless if and only if its Colin de Verdière invariant $\mu(G)$ is at most four. Whether knotless graphs are precisely those graphs whose Colin de Verdière invariant is at most five is an interesting open problem.

Good viewpoints: In most visualization applications, a 3D drawing of a graph will eventually be displayed as an image on some kind of 2D medium, such as a computer screen or a sheet of paper. This can be achieved by using projections. In computer graphics the most commonly used projections are the parallel and perspective projections. A 2D image, by its very nature, will necessarily contain less information than the original 3D drawing. It is therefore desirable to find *viewpoints* (the position and the direction the viewer is facing) that result in "nice" 2D images, that is, projections that preserve as much information about the 3D drawing as possible. Having an edge of the 3D drawing map to one point in the projection is lossy in that context, as is having two vertices project to the same point.

Bose *et al.* [BGRT99] developed an algorithm that, given a 3D straight-line drawing, computes an arrangement of curves that describe all bad viewpoints for that drawing. A viewpoint is bad if it maps three 3D points to the same point in the projection (vertices count as two points). Their algorithm runs in $\mathcal{O}(m^4 \log m + k)$ time, where m is the number of edges of the graph and k may be $\mathcal{O}(m^6)$ in the worst case.

The arrangement above distinguishes between bad and good viewpoints. Eades *et al.* studied a model with a continuous measure of goodness for a viewpoint [EHW97]. In particular the goodness of a viewpoint increases with distance from its nearest bad point. They also considered different definitions of bad points and developed an algorithm to compute them based on techniques of Bose *et al.* [BGRT99].

3D symmetry: Connections between symmetry and aesthetics have long been recognized. Thus displaying automorphisms of a graph as symmetries in its drawing is a very desirable feature. Drawing graphs symmetrically involves solving at least two problems. The first is to determine the symmetries (automorphisms) of a graph. The second problem is, given the graph automorphisms, to display as many of them as possible as geometric symmetries of a drawing of the graph. Symmetries in 3D can be displayed by, for example, rotation, reflection, and inversion. For a detailed account on symmetric drawings, including 3D symmetric drawings, the reader is referred to Chapter 3.

Higher dimensions: One of the basic problems in discrete geometry is determining when a graph can be realized with prescribed edge lengths in $\mathbb{R}^d$. An interesting graph invariant related to that concept is the *dimension* of a graph, introduced by Erdős *et al.* [EHT65]. It is defined as the minimum d such that the graph has a drawing in $\mathbb{R}^d$ with straight-line edges all of unit length (with possible crossings). They show, among other results, that the dimension of the complete graph K_n is $n - 1$ and that the dimension of the complete bipartite graph is at most four.

A concept related to the dimensionality of a graph is that of realizability. A *realization* of a graph is a straight-line "drawing" with vertices represented as points, where there is no restriction on how vertices and edges may intersect. A graph G is *d-realizable* if, given any realization of G in $\mathbb{R}^t$, there exists a realization of G with the same edge-lengths in $\mathbb{R}^d$. For example, a path is 1-realizable since its vertices can be arranged on a line with any desired edge-lengths. A tree is also 1-realizable. On the other hand, the triangle is not 1-realizable, since it has a realization in R^2 with unit distance edges but no such realization is possible in R^1. Connelly and Sloughter [BC07] proved that a graph is 1-realizable if and only if it is a forest. It is 2-realizable if and only if it has treewidth at most two, that is, if it is a series-parallel graph. They showed that a graph is 3-realizable if and only if it does not contain K_5 or an octahedral graph as a minor.

A relationship between the connectivity of graphs and higher dimensional drawings has been established [LLW88]. In particular, k-connected graphs were characterized in terms of particular convex drawings in $\mathbb{R}^{k-1}$. A force directed method was used to derive these results.

Dujmović *et al.* [DMS13] studied higher dimensional straight-line grid drawings (with possible crossings). They showed that in every d-dimensional ($d \geq 4$) straight-line grid drawing of volume N of a graph with $m \geq (2^2 + 1)N$ edges, there are at least $\Omega(\frac{m^2}{N})$ crossings. They also showed that there are graphs for which this bound is tight.

Some other directions explored include the idea of producing 2D drawings by starting with a "nice" higher dimensional drawing of a graph and then projecting it to a plane. Higher-dimensional visibility representations with hyper-rectangles [CDH+96] have also been considered.

Applications and information visualization: This chapter was mainly focused on theory and foundations of 3D Graph Drawing, that is, results with provable bounds on properties on drawings and provable bounds on the running times of drawing algorithms. An important theme outside the scope of this chapter is that of development of software packages for 3D graph drawing (see, for example, [GT97, PV97]) as well as information visualization in 3D. Graph drawing in 3D relates to this area particularly because graphs model hierarchies and networks. Understanding large social and biological trees and networks requires the support of visualization tools [BvLH+11, LLB+12, NJBJ09]. A substantial body of research literature explores the possibility of combining 3D graphics and interactive animation technology with an understanding of human perception for the purpose of conveying information, including graph models, to humans [XRP+12]. Classic work of Robertson *et al.* [RMC91] proposed to visualize organizational hierarchies with 3D animations of trees. Work on visualization of graphs is found not only in the information visualization literature, but also domain specific literature such as that of biology and bioinformatics. Important key words include information visualization, human computer interaction, computer graphics, animation, human perception.

References

[AG76]   V. B. Alekseev and V. S. Gonchakov. Thickness of arbitrary complete graphs. *Mat. Sbornik*, 101:212–230, 1976.

[AGW98]   H. Alt, M. Godau, and S. Whitesides. Universal 3-dimensional visibility representations for graphs. *Comput. Geom.*, 9:111–125, 1998.

[BC07]   M. Belk and R. Connelly. Realizability of graphs. *Discrete and Computational Geometry*, 37(2):125–137, 2007.

[BCMW04]   P. Bose, J. Czyzowicz, P. Morin, and D. R. Wood. The maximum number of edges in a three-dimensional grid-drawing. *J. of Graph Algorithms and Appl.*, 8(1):21–26, 2004.

[BEF+98]   P. Bose, H. Everett, S. Fekete, M. E. Houle, A. Lubiw, H. Meijer, K. Romanik, G. Rote, T. C. Shermer, S. Whitesides, and C. Zelle. A visibility representation for graphs in three dimensions. *J. of Graph Algorithms and Appl.*, 2(3):1–16, 1998.

[Bei67]   L. W. Beineke. The decomposition of complete graphs into planar subgraphs. In *Graph Theory and Theoretical Physics*, pages 139–154. 1967.

[BGRT99]   P. Bose, F. Gómez, P. A. Ramos, and G. T. Toussaint. Drawing nice projections of objects in space. *J. of Visual Communic. and Image Representation*, 10:155–172, 1999.

[BH65]   L. W. Beineke and F. Harary. The thickness of the complete graph. *Canad. J. Math.*, 17:850–859, 1965.

[BHRCW06] P. Bose, F. Hurtado, E. Rivera-Campo, and D. R. Wood. Partitions of complete geometric graphs into plane trees. *Computational Geometry: Theory and Applications*, 34(2):116–125, 2006.

[Bie98]   T. C. Biedl. Three approaches to 3D-orthogonal box-drawings. In *Proc. 6th Int. Symp. on Graph Drawing (GD'98)*, volume 1547 of *LNCS*, pages 30–43. Springer, 1998.

[BJSW01]   T. Biedl, J. R. Johansen, T. C. Shermer, and D. R. Wood. Orthogonal drawings with few layers. In *Proc. 9th Int. Symp. on Graph Drawing (GD'01)*, volume 2265 of *LNCS*, pages 297–311. Springer, 2001.

[BK79]   F. Bernhart and P. C. Kainen. The book thickness of a graph. *J. Combin. Theory Ser. B*, 27(3):320–331, 1979.

[Bla03]   Robin Blankenship. *Book Embeddings of Graphs*. PhD thesis, Dept. of Mathematics, Louisiana State University, U.S.A., 2003.

[BMWR3]   J. Bárat, J. Matoušek, and D. R. Wood. Bounded-degree graphs have arbitrarily large geometric thickness. *The Electronic Journal of Combinatorics*, 13, 2006, R3.

[BO01]   R. Blankenship and B. Oporowski. Book embeddings of graphs and minor-closed classes. In *Proc. 32nd Southeastern Int. Conf. on Combinatorics, Graph Theory and Comp.* Dept. of Math., Louisiana State University, 2001.

[BSWW99]   T. C. Biedl, T. C. Shermer, S. Whitesides, and S. K. Wismath. Bounds for orthogonal 3-D graph drawing. *J. of Graph Algorithms and Appl.*, 3(4):63–79, 1999.

[BTW06]   T. Biedl, T. Thiele, and D. R. Wood. Three-dimensional orthogonal graph drawing with optimal volume. *Algorithmica*, 44(3):233–255, 2006.

[BvLH+11]   S. Bremm, T. von Landesberger, M. Hess, T. Schreck, P. Weil, and K. Hamacherk. Interactive visual comparison of multiple trees. In *Visual Analytics Science and Technology (VAST), 2011 IEEE Conference on*, pages 31–40, 2011.

[CDH+96]   F. Cobos, J. Dana, F. Hurtado, A. Márquez, and F. Mateos. On a visibility representation of graphs. In *Proc. Int. Symp. on Graph Drawing (GD'95)*, volume 1027 of *LNCS*, pages 152–161. Springer, 1996.

[CELR96]   R. F. Cohen, P. Eades, T. Lin, and F. Ruskey. Three-dimensional graph drawing. *Algorithmica*, 17(2):199–208, 1996.

[CGJW01]   M. Closson, S. Gartshore, J. Johansen, and S. K. Wismath. Fully dynamic 3-dimensional orthogonal graph drawing. *J. of Graph Algorithms and Appl.*, 5(2):1–34, 2001.

[CGT96]   M. Chrobak, M. T. Goodrich, and R. Tamassia. Convex drawings of graphs in two and three dimensions (preliminary version). In *Proc. 12th Annual Symposium on Computational Geometry*, pages 319–328, 1996.

[CS97]   T. Calamoneri and A. Sterbini. 3D straight-line grid drawing of 4-colorable graphs. *Inform. Process. Lett.*, 63(2):97–102, 1997.

[CT96]   I. F. Cruz and J. P. Twarog. 3D graph drawing with simulated annealing. In *Proceedings of the Symposium on Graph Drawing*, GD '95, pages 162–165, London, UK, UK, 1996. Springer-Verlag.

[DEH00]   M. B. Dillencourt, D. Eppstein, and D.S. Hirschberg. Geometric thickness of complete graphs. *J. of Graph Algorithms and Appl.*, 4(3):5–17, 2000.

[DEK04]   C. A. Duncan, D. Eppstein, and S. G. Kobourov. The geometric thickness of low degree graphs. In *Proc. of the 20th Annual Symp. on Computational Geometry (SoCG'04)*, pages 340–346, 2004.

[DEL+05]   O. Devillers, H. Everett, S. Lazard, M. Pentcheva, and S. Wismath. Drawing K_n in three dimensions with one bend per edge. In *Proc. 13th Int. Symp. on Graph Drawing (GD'05)*, volume 3843 of *LNCS*, pages 83–88. Springer, 2005. Also in, *J. Graph Algorithms Appl.*, 10(2): 287-295 (2006).

[DFP10]   G. Di Battista, F. Frati, and J. Pach. On the queue number of planar graphs. In *Proc. 51st Annual IEEE Symposium on Foundations of Computer Science (FOCS)*, pages 365 –374, 2010.

[dFPP90]   H. de Fraysseix, J. Pach, and R. Pollack. How to draw a planar graph on a grid. *Combinatorica*, 10(1):41–51, 1990.

[Di 03]   E. Di Giacomo. Drawing series-parallel graphs on restricted integer 3D grids. In *Proc. 11th Int. Symp. on Graph Drawing (GD'03)*, volume 2912 of *LNCS*, pages 238–246. Springer, 2003.

[DLMW05]   E. Di Giacomo, G. Liotta, H. Meijer, and S. K. Wismath. Volume requirements of 3D upward drawings. In *Proc. 13th Int. Symp. on Graph Drawing (GD'05)*, volume 3843 of *LNCS*, pages 101–110. Springer, 2005. Aslo in *Discrete Mathematics*, 309(7):1824–1837 (2009).

[DLW02]   E. Di Giacomo, G. Liotta, and S. Wismath. Drawing series-parallel graphs on a box. In *Proc. 14th Canadian Conf. on Computational Geometry (CCCG'02)*, pages 149–153. The University of Lethbridge, Canada, 2002.

[DM03]   E. Di Giacomo and H. Meijer. Track drawings of graphs with constant queue number. In *Proc. 11th Int. Symp. on Graph Drawing (GD'03)*, volume 2912 of *LNCS*, pages 214–225. Springer, 2003.

[DMS13]   V. Dujmović, P. Morin, and A. Sheffer. Crossings in grid drawings. *Arxiv preprint*, January 2013. http://arxiv.org/abs/1301.0303.

[DMW02]   V. Dujmović, P. Morin, and D. R. Wood. Path-width and three-dimensional straight-line grid drawings of graphs. In *Proc. 10th Int. Symp. on Graph Drawing (GD'02)*, volume 2528 of *LNCS*, pages 42–53. Springer, 2002.

[DMW05]   V. Dujmović, P. Morin, and D. R. Wood. Layout of graphs with bounded tree-width. *SIAM J. of Computing*, 34(3):553–579, 2005.

[DPV00]   G. Di Battista, M. Patrignani, and F. Vargiu. A split & push approach to 3D orthogonal drawing. *J. Graph Algorithms Appl.*, 4(3):105–133, 2000.

[DPW04]   V. Dujmović, A. Pór, and D. R. Wood. Track layouts of graphs. *Discrete Mathematics and Theoretical Computer Sci.*, 6(2):497–522, 2004.

[DW04a]   V. Dujmović and D. R. Wood. On linear layouts of graphs. *Discrete Mathematics and Theoretical Computer Sci.*, 6(2):339–358, 2004.

[DW04b]   V. Dujmović and D. R. Wood. Stacks, queues and tracks: layouts of graph subdivisions. In *Proc. 12th Int. Symp. on Graph Drawing (GD'04)*, volume 3383 of *LNCS*, pages 133–143. Springer, 2004. Also in *Discrete Mathematics and Theoretical Computer Sci.*, DMTCS, 7:155–202, 2005.

[DW04c]   V. Dujmović and D. R. Wood. Three-dimensional grid drawings with sub-quadratic volume. In János Pach, editor, *Towards a Theory of Geometric Graphs*, volume 342 of *Contemporary Mathematics*, pages 55–66. Amer. Math. Soc., 2004.

[DW05]   V. Dujmović and D. R. Wood. Graph treewidth and geometric thickness parameters. In *Proc. 13th Int. Symp. on Graph Drawing (GD'05)*, volume 3843 of *LNCS*, pages 129–140. Springer, 2005. Also in, *Discrete and Computational Geometry*, 37(4): 641-670 (2007).

[DW06]   V. Dujmović and D. R. Wood. Upward three-dimensional grid drawings of graphs. *Order*, 23(1):1–20, 2006.

[EHT65]   P. Erdős, F. Harary, and W. T. Tutte. On the dimension of a graph. *Mathematika*, 12:118–122, 1965.

[EHW97]   P. Eades, M. E. Houle, and R. Webber. Finding the best viewpoints for three-dimensional graph drawings. In *Proc. Int. Workshop on Graph Drawing*, volume 1353 of *LNCS*, pages 87–98. Springer, 1997.

[ELMN11]   D. Eppstein, M. Löffler, E. Mumford, and M. Nöllenburg. Optimal 3D angular resolution for low-degree graphs. In *Proc. 18th Symposium on Graph Drawing (GD'10)*, volume 6502, pages 208–219, 2011.

[End97]   T. Endo. The pagenumber of toroidal graphs is at most seven. *Discrete Math.*, 175(1-3):87–96, 1997.

[Epp01]   D. Eppstein. Separating geometric thickness from book thickness. arXiv.org math.CO/0109195, Sept. 2001.

[Epp04a]   D. Eppstein. Separating thickness from geometric thickness. In János Pach, editor, *Towards a Theory of Geometric Graphs*, number 342 in Contemporary Mathematics, pages 75–86. Amer. Math. Soc., 2004.

[Epp04b]   D. Eppstein. Testing bipartiteness of geometric intersection graphs.
            In *Proc. 15th Annual ACM-SIAM Symp. on Discrete Algorithms
            (SODA'04)*, pages 860–868, 2004. Also in *ACM Transactions on Al-
            gorithms*, 5(2), Article No. 15, 2009.

[Erd51]   P. Erdös. Appendix. In K. F. ROTH, On a problem of Heilbronn. *J.
            London Math. Soc.*, 26:198–204, 1951.

[ESW96]   P. Eades, C. Stirk, and S. Whitesides. The techniques of Kolmogorov
            and Barzdin for three dimensional orthogonal graph drawings. *Inform.
            Proc. Lett.*, 60(2):97–103, 1996.

[ESW00]   P. Eades, A. Symvonis, and S. Whitesides. Three dimensional orthogonal
            graph drawing algorithms. *Discrete Applied Math.*, 103:55–87, 2000.

[FLW01]   S. Felsner, G. Liotta, and S. Wismath. Straight-line drawings on re-
            stricted integer grids in two and three dimensions. In *Proc. 9th Int.
            Symp. on Graph Drawing (GD'01)*, volume 2265 of *LNCS*, pages 328–342.
            Springer, 2001. Also in *J. of Graph Algorithms and Appl.*, 7(4):363–398,
            2003.

[FM99]   S. Fekete and H. Meijer. Rectangle and box visibility graphs in 3D.
            Internat. J. Comput. Geom. Appl., 9(1):1–27, 1999.

[GH01]   J. L. Ganley and L. S. Heath. The pagenumber of k-trees is $O(k)$. *Discrete
            Appl. Math.*, 109(3):215–221, 2001.

[GJMP80]   M. R. Garey, D. S. Johnson, G. L. Miller, and C. H. Papadimitriou.
            The complexity of coloring circular arcs and chords. *SIAM J. Algebraic
            Discrete Methods*, 1(2):216–227, 1980.

[GT97]   A. Garg and R. Tamassia. GIOTTO3D: A system for visualizing hi-
            erarchical structures in 3D. In *Proc. Symposium on of Graph Drawing
            (GD'96)*, volume 1190, pages 193–200. Springer-Verlag, 1997.

[GTV96]   A. Garg, R. Tamassia, and P. Vocca. Drawing with colors. In *Proc. 4th
            Annual European Symp. on Algorithms (ESA'96)*, volume 1136 of *LNCS*,
            pages 12–26. Springer, 1996.

[Hal91]   J. H. Halton. On the thickness of graphs of given degree. *Inform. Sci.*,
            54(3):219–238, 1991.

[HLR92]   L. S. Heath, F. T. Leighton, and A. L. Rosenberg. Comparing queues
            and stacks as mechanisms for laying out graphs. *SIAM J. Discrete Math.*,
            5(3):398–412, 1992.

[HR92]   L. S. Heath and A. L. Rosenberg. Laying out graphs using queues. *SIAM
            J. Comput.*, 21(5):927–958, 1992.

[HSV99]   J. P. Hutchinson, T. Shermer, and A. Vince. On representations of some
            thickness-two graphs. *Comput. Geom.*, 13(3):161–171, 1999.

[HTS83]   K. Hagihara, N. Tokura, and N. Suzuki. Graph embedding on a three-
            dimensional model. *Systems-Comput.-Controls*, 14(6):58–66, 1983.

[Kai73]   P. C. Kainen. Thickness and coarseness of graphs. *Abh. Math. Sem.
            Univ. Hamburg*, 39:88–95, 1973.

[KB67]   A. N. Kolmogorov and Ya. M. Barzdin. On the realization of nets in
            3-dimensional space. *Problems in Cybernetics*, 8:261–268, 1967.

[KK97]   A. V. Kostochka and J. Kratochvíl. Covering and coloring polygon-circle
            graphs. *Discrete Math.*, 163(1-3):299–305, 1997.

[KLV97]   A. Kotlov, L. Lovász, and S. Vempala. The Colin de Verdière number
           and sphere representations of a graph. *Combinatorica*, 17(4):483–521,
           1997.

[Kos82]   A. V. Kostochka. The minimum Hadwiger number for graphs with a
           given mean degree of vertices. *Metody Diskret. Analiz.*, 38:37–58, 1982.

[LLB⁺12]   A. G. Landge, J. A. Levine, A. Bhatele, K. E. Isaacs, T. Gamblin,
           M. Schulz, S. H. Langer, P.-T. Bremer, and V. Pascucci. Visualizing
           network traffic to understand the performance of massively parallel sim-
           ulations. *IEEE Trans. Vis. Comput. Graph.*, 18(12):2467–2476, 2012.

[LLW88]   N. Linial, L. Lovász, and A. Wigderson. Rubber bands, convex embed-
           dings and graph connectivity. *Combinatorica*, 8(1):91–102, 1988.

[LR86]   F. T. Leighton and A. L. Rosenberg. Three-dimensional circuit layouts.
           SIAM J. Comput., 15(3):793–813, 1986.

[Mal94a]   S. M. Malitz. Genus g graphs have pagenumber $O(\sqrt{g})$. *J. Algorithms*,
           17(1):85–109, 1994.

[Mal94b]   S. M. Malitz. Graphs with E edges have pagenumber $O(\sqrt{E})$. *J. Algo-
           rithms*, 17(1):71–84, 1994.

[Man83]   A. Mansfield. Determining the thickness of graphs is NP-hard. *Math.
           Proc. Cambridge Philos. Soc.*, 93(1):9–23, 1983.

[May72]   J. Mayer. Décomposition de K_{16} en trois graphes planaires. *J. Combi-
           natorial Theory Ser. B*, 13:71, 1972.

[Moh99]   B. Mohar. A linear time algorithm for embedding graphs in an arbitrary
           surface. *SIAM J. Discrete Math.*, 12(1):6–26, 1999.

[MT01]   B. Mohar and C. Thomassen. *Graphs on Surfaces*. Johns Hopkins Uni-
           versity Press, 2001.

[NJBJ09]   C. B. Nielsen, S. D. Jackman, I. Birol, and S. J. M. Jones. ABySS-
           Explorer: Visualizing genome sequence assemblies. *IEEE Trans. Vis.
           Comput. Graph.*, 15(6):881–888, 2009.

[Oll73]   L. T. Ollmann. On the book thicknesses of various graphs. In *Proc. 4th
           Southeastern Conference on Combinatorics, Graph Theory and Comput-
           ing*, volume VIII of *Congressus Numerantium*, page 459, 1973.

[Por00]   T. Poranen. A new algorithm for drawing series-parallel digraphs in
           3D. Technical Report A-2000-16, Dept. of Computer and Information
           Sciences, University of Tampere, Finland, 2000.

[PT99]   A. Papakostas and I. G. Tollis. Algorithms for incremental orthogonal
           graph drawing in three dimensions. *J. of Graph Algorithms and Appl.*,
           3(4):81–115, 1999.

[PTT99]   J. Pach, T. Thiele, and G. Tóth. Three-dimensional grid drawings of
           graphs. In Bernard Chazelle, Jacob E. Goodman, and Richard Pollack,
           editors, *Advances in Discrete and Computational Geometry*, volume 223
           of *Contemporary Mathematics*, pages 251–255. Amer. Math. Soc., 1999.

[PV97]   M. Patrignani and F. Vargiu. 3DCube: a tool for three dimensional graph
           drawing. In *Proc. Symposium on Graph Drawing (GD'97)*, volume 1353,
           pages 284–290. Springer, 1997.

[PW01]   J. Pach and R. Wenger. Embedding planar graphs at fixed vertex loca-
           tions. *Graphs and Combinatorics*, 17:717–728, 2001.

[PW04]   A. Pór and D. R. Wood. No-three-in-line-in-3D. In *Proc. 12th Int. Symp. on Graph Drawing (GD'04)*, volume 3383 of *LNCS*, pages 395–402. Springer, 2004. Also in *Algorithmica*, 47(4): 481–488, 2007.

[RMC91]   G. G. Robertson, J. D. Mackinlay, and S. K. Card. Cone trees: animated 3D visualizations of hierarchical information. In *Proceedings of the SIGCHI Conference on Human Factors in Computing Systems*, CHI '91, pages 189–194, New York, NY, USA, 1991. ACM.

[RS]   N. Robertson and P. D. Seymour. Graph minors *I–XX*. *J. Combin. Theory Ser. B*. 1983–2004.

[RVM95]   S. Rengarajan and C. E. Veni Madhavan. Stack and queue number of 2-trees. In *Proc. 1st Annual Int. Conf. on Computing and Combinatorics (COCOON'95)*, volume 959 of *LNCS*, pages 203–212. Springer, 1995.

[Sch89]   W. Schnyder. Planar graphs and poset dimension. *Order*, 5(4):323–343, 1989.

[SSV04]   O. Sýkora, L. A. Székely, and I. Vrto. A note on Halton's conjecture. *Information Sci.*, 164(1-4):61–64, 2004.

[Što04]   J. Štola. 3D visibility representations of complete graphs. In *Proc. 12th Int. Symp. on Graph Drawing (GD'04)*, volume 3383 of *LNCS*, pages 226–237. Springer, 2004.

[Tho84]   A. Thomason. An extremal function for contractions of graphs. *Math. Proc. Cambridge Philos. Soc.*, 95(2):261–265, 1984.

[Tho98]   M. Thorup. All structured programs have small tree-width and good register allocation. *Information and Computation*, 142(2):159–181, 1998.

[Tut60]   W. T. Tutte. Convex representations of graphs. *Proc. London Math. Soc.*, 10(3):304–320, 1960.

[Tut63]   W. T. Tutte. The thickness of a graph. *Nederl. Akad. Wetensch. Proc. Ser. A 66=Indag. Math.*, 25:567–577, 1963.

[Wes84]   W. Wessel. Über die abhängigkeit der dicke eines graphen von seinen knotenpunktvalenzen. In *Proc. of the Geometrie und Kombinatorik '83*, volume 2, pages 235–238, 1984.

[WF94]   C. Ware and G. Franck. Viewing a graph in a virtual reality display is three times as good as a 2D diagram. In *Proc. IEEE Symp. Visual Languages (VL'94)*, pages 182–183. IEEE, 1994.

[WF96]   C. Ware and G. Franck. Evaluating stereo and motion cues for visualizing information nets in three dimensions. *ACM Trans. Graphics*, 15(2):121–140, 1996.

[WHF93]   C. Ware, D. Hui, and G. Franck. Visualizing object oriented software in three dimensions. In *Proc. IBM Centre for Advanced Studies Conf. (CASCON'93)*, pages 1–11, 1993.

[Wig82]   A. Wigderson. The complexity of the hamiltonian circuit problem for maximal planar graphs. Technical Report EECS 198, Princeton University, USA, 1982.

[WM08]   Colin Ware and Peter Mitchell. Visualizing graphs in three dimensions. *ACM Trans. Appl. Percept.*, 5(1), 2008.

[Woo99]   D. R. Wood. Multi-dimensional orthogonal graph drawing with small boxes. In *Proc. 7th Int. Symp. on Graph Drawing (GD'99)*, volume 1731 of *LNCS*, pages 311–222. Springer, 1999.

[Woo01]   D. R. Wood. Bounded degree book embeddings and three-dimensional orthogonal graph drawing. In *Proc. 9th Int. Symp. on Graph Drawing (GD'01)*, volume 2265 of *LNCS*, pages 312–327. Springer, 2001.

[Woo03a]   D. R. Wood. Lower bounds for the number of bends in three-dimensional orthogonal graph drawings. *J. of Graph Algorithms and Appl.*, 7(1):33–77, 2003.

[Woo03b]   D. R. Wood. Optimal three-dimensional orthogonal graph drawing in the general position model. *Theoret. Comput. Sci.*, 299(1-3):151–178, 2003.

[Woo04]   D. R. Wood. Minimising the number of bends and volume in three-dimensional orthogonal graph drawings with a diagonal vertex layout. *Algorithmica*, 39(3):235–253, 2004.

[XRP+12]   K. Xu, C. Rooney, P. Passmore, D.-H. Ham, and P. Nguyen. A user study on curved edges in graph visualization. *IEEE Trans. Vis. Comput. Graph.*, 18(12):2449–2456, 2012.

[Yan86]   M. Yannakakis. Four pages are necessary and sufficient. In *Proc. 18th ACM Symp. on Theory of Comput. (STOC'86)*, pages 104–108, 1986.

15

Labeling Algorithms

Konstantinos G. Kakoulis

T.E.I. of West Macedonia, Greece

Ioannis G. Tollis

University of Crete, Greece

15.1 Introduction

An important aspect of information visualization is the automatic placement of text or symbol labels corresponding to graphical features of drawings and maps. Labels are textual descriptions that convey information or clarify the meaning of complex structures presented in a graphical form. The automatic label placement problem is identified as an important research area by the ACM Computational Geometry Task Force [C+99]. It has applications in many areas including cartography [RMM+95], geographic information systems [Fre91], and graph drawing [DETT99].

Because the labeling process is a monotonous and very demanding task, its automation is very desirable. It is very difficult to quantify all the characteristics of a good label placement since they reflect human visual perception, intuition, and experience, which have been perfected through the centuries by cartographers who have elevated the placement of labels into an art. Hence, it is unlikely that computer-based systems will be able to deliver fully automated placement of labels in maps of a sufficient quality to be comparable to those produced manually by experienced cartographers. Nevertheless, there are many areas where the requirements for high aesthetic quality are not as strict and automatic labeling techniques may be applied. For example, these techniques may be used for real time name placement in the context of on-line geographic information systems or internet-based map search, and special-purpose maps such as those used to display census [EG90], oil exploration [Zor90] or soil survey data [FMC96]. Additionally, semi-automated interactive name placement systems may be the most practical approach at the present time. Labeling systems may produce an initial label placement that could be improved manually by cartographers to produce desirable results. Furthermore, the whole concept of map labeling may change depending on computer capabilities [RMM+95]. Maps may be viewed in an electronic format allowing the user to interact and display information on demand as opposed to viewing all the map information at once.

In the following sections, we study the labeling problem not only in its traditional form (i.e., cartography), but also in the context of information visualization, specifically as it relates to graph drawing. In Section 15.2 we present a model for the labeling problem: we discuss the qualities of good label assignment and give a formal definition of the problem. In Section 15.3, we present a variety of algorithms for the labeling problem. Finally, we discuss how one can modify a drawing to accommodate the placement of labels in Section 15.3.5.

15.2   The Labeling Problem

15.2.1   Searching for a Good Label Assignment

Let Γ be a drawing and F be the set of graphical features of Γ to be labeled. A solution to the labeling problem for drawing Γ assigns text or symbol labels to each graphical feature f of F such that the relevant information is communicated in the best possible way. This can be achieved by positioning the labels in the most appropriate places. For each graphical feature there is a large number of potential label positions, and the most preferable among them must be assigned.

Good label placement aids in conveying information and enhances the aesthetics of the input drawing. It is difficult to quantify all the characteristics of a good label placement, because they reflect human visual perception and intuition. It is trivial to place a label when its associated object is isolated. The real difficulty arises when the freedom to place a label is restricted by the presence (in close proximity) of other objects of the drawing. In this common scenario, we must consider not only the position of a label with respect to its associated object, but also how it relates to other labels and objects in the surrounding area.

In a successful label assignment, labels must be positioned such that they are legible and follow basic aesthetic quality criteria. According to cartographers like Imhof [Imh75] and Yoeli [Yoe72], who have extensively studied this subject, labels must be placed in the best position available following some basic rules: Labels must be easily read, quickly located, a label and the object to which it belongs should be easily recognized, labels must be placed very close to the objects they belong to, labels must not obscure other labels or objects, a label must be placed in the most preferred position, among all legible positions. We summarize the labeling quality evaluation in the following three basic rules:

- No overlaps of a label with other labels or other graphical features of the drawing are allowed.
- Each label can be easily identified with exactly one graphical feature of the drawing.
- Each label must be placed in the best possible position (among all acceptable positions).

The order of preference among possible label positions varies depending on the specific application.

In the production of geographical maps, we rank label positions according to rules developed through years of experience with manual placement, which typically capture the aesthetic quality of label positions. A typical rule when labeling points (nodes) is that labels must be placed to the right and above the point. For example, in Figure 15.1(a) the number of each label position reveals the rank (priority) of the label. In addition, a point label is allowed to touch but not overlap its associated point or any other graphical feature in the drawing. In the case of labeling lines (edges), a label is allowed to touch the edge that

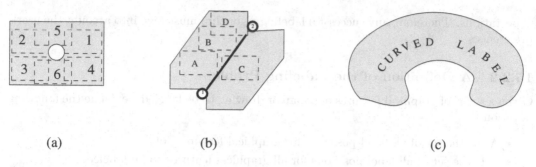

(a)   (b)   (c)

Figure 15.1 (a) Labeling space of a node. (b) Labeling space of an edge. (c) Labeling space of an area. Figure taken from [KT03].

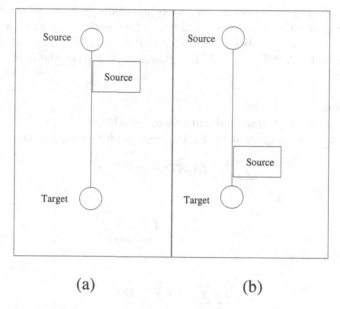

(a)   (b)

Figure 15.2 (a) A good label assignment. (b) A misleading label assignment. Figure taken from [KT03].

it belongs to, but it should not overlap any other graphical feature in a drawing. In Figure 15.1(b), where the graphical feature to be labeled is an edge, labels like A, B and D are preferable but certainly a label like C, which overlaps its associated edge, can be acceptable with some appropriate cost assigned to it. The accepted practice for placing a label associated with an area is to have the label span the entire area and conform to its shape, as shown in Figure 15.1(c). For more details on name placement rules for geographical maps, see [FA87, Imh75, vR89, Yoe72].

When the graphical objects to be labeled belong to a technical map or drawing, then, usually a different set of rules govern the preferred label positions. These rules depend on the particular application, and must follow user specifications. For example, if the graphical feature is an edge of a graph drawing, the user must be able to specify that the preferred position for an edge label is closer to the source or destination node. For example, a label of a single edge that is relevant to its source node must be placed close to the source node (see Figure 15.2(a)) to avoid ambiguity (see Figure 15.2(b)). It is important to emphasize that a user must be able to customize the rules of label quality to meet specific needs and/or expectations. Therefore, any successful labeling algorithm must take into account the user's preferences.

15.2.2   A Definition of the Labeling Problem

Given a set F of graphical features of a map or drawing to be labeled we define the following notation:

- Λ_f is the set of all label positions for graphical feature f of F.
- Λ is the set of all label positions for all graphical features to be labeled.
- $\lambda : F \to \Lambda$ is a function that assigns a label position from Λ to graphical feature f in F, that is $\lambda(f) = \lambda_f \in \Lambda_f$.

The labeling problem can be viewed as an optimization problem where the objective is to find a label assignment of minimum total cost where each graphical feature has a label position assigned to it. Each label position λ_f that is part of a final label assignment is associated with a cost. $COST : \Lambda \to \mathcal{N}$ is a function that gives the cost of label λ_f with respect to quality.

Labeling Problem
Instance: Let F be a set of graphical features to be labeled.
Question: Find a label assignment that minimizes the following function:

$$\sum_{i \in F} \sum_{j \in \Lambda_i} COST(\lambda(i)) P(i,j)$$

Where:

$$P(i,j) = \begin{cases} 1, & if\ \lambda(i) = j, \\ 0, & otherwise \end{cases}$$

and

$$\sum_{i \in F} \sum_{j \in \Lambda_i} P(i,j) = |F|$$

Where:

$$\sum_{j \in \Lambda_i} P(i,j) = 1, \qquad i \in F.$$

$\square$

15.3   Solving the Labeling Problem

Most of the research addressing the labeling problem has been focused on labeling graphical features of geographical and technical maps. The label placement problem is typically partitioned into three tasks: (a) labeling points (e.g., cities), (b) labeling lines (e.g., roads or rivers), and (c) labeling areas (e.g., lakes or oceans).

Progress has been made in solving the problem of assigning labels to a set of points or nodes, the *Node Label Placement* (NLP) problem [CMS95, DMM⁺97, FW91, Hir82, WW95, Zor90]. The problem of assigning labels to a set of lines or edges, also known as the *Edge Label Placement* (ELP) problem, has been addressed in [DKMT07, KT98, vR89, Zor90]. The general labeling problem, the *Graphical Feature Label Placement* (GFLP) problem (where a graphical feature can be a node, edge, or area), has been addressed primarily in the context of cartography; however, it has direct application in the area of graph drawing [AF84, DF92, ECMS97, EG90, FA87, KT03].

In many practical applications, each graphical feature may have more than one label. The need for assigning multiple labels is necessary not only when objects are large or long, but also when it is necessary to display different attributes of an object. This problem is known as the *Multiple Label Placement* (MLP) problem and has been addressed in [FA87, KT06].

The labeling process is not allowed to modify the underlying geometry of geographical and technical maps which is fixed. However, one can modify a graph drawing in order to accommodate the placement of labels. In [Hu09, KT11], algorithms that modify an existing layout of a graph drawing to make room for the placement of labels are presented.

In [BDLN05, DDPP99, KM99], algorithms that combine the layout and labeling process of orthogonal drawings of graphs are presented.

An alternative approach for displaying edge labels is presented in [WMP+05]. Each edge is replaced by its corresponding edge label in the drawing. The label font starts out larger from the source node and shrinks gradually until it reaches the destination node. The tapered label also indicates the direction of the edge.

It is worth noting that both the NLP [FW91, KI88, MS91] and ELP [KT01] problems are NP-hard. Because automatic labeling is a very difficult problem we rely on heuristics to provide practical solutions for real world problems.

A variety of types of algorithms have been used in order to solve the labeling problem: greedy algorithms [CMS95, Hir82], exhaustive search algorithms [DF92, EG90, FA87], algorithms that simulate physical models (i.e., Simulated Annealing [CMS95]), algorithms that reduce the labeling problem to a variant of 0-1 integer programming [Zor90], algorithms that restrict the labeling problem to a variant of the 2-SAT problem [FW91, WW95], and algorithms that transform the labeling problem into a matching problem [KT98, KT03, KT06].

15.3.1 The GFLP Problem

Most labeling algorithms that address the general labeling problem are based on local and exhaustive search algorithms [DF92, EG90, FA87]. These algorithms perform well for small problems. These methods use inferior optimization techniques, as pointed out in [Zor90] and verified in [CMS95]. Actually, these methods use a rule based approach to evaluate good label placement and variants of depth-first search to explore different labeling configurations. The approach in [ECMS97] uses simulated annealing to find solutions for the general labeling problem, and it separates the cartographic knowledge needed to recognize the best label positions from the optimization procedure needed to find them.

All of the above techniques for the general labeling problem first create an initial label assignment in which conflicts between labels are allowed. Then conflicts are resolved by repositioning assigned labels until all conflicts are resolved, or no further improvement can be achieved. Furthermore, they start with a rather small initial set of potential label positions from which they derive a final label assignment. The performance of these techniques decreases when the number of potential label positions increases.

In [KT03] the labeling problem is transformed into a matching problem. The general framework of this technique is flexible and can be adjusted for particular labeling requirements. In the next section this technique is presented in more detail.

A practical matching algorithm for the GFLP problem

The placement of labels is a post-layout operation (i.e., performed on a fixed geometry of nodes and edges). The basic idea behind this labeling technique is the following: a set of discrete potential label solutions for each object is carefully selected. This set of labels is reduced by removing heavily overlapping labels. Finally, an assignment of labels is performed by solving a variant of the matching problem. This method is shown in Figure 15.3. An example of the resulting label placement is given in Figure 15.4.

Basic Labeling Algorithm

INPUT: A drawing Γ and a set F of objects to be labeled.
OUTPUT: A label assignment free of overlaps.
1. A set of discrete potential label solutions for each object in F is carefully selected.
2. This set of labels is reduced by removing heavily overlapping labels. The remaining labels are assigned to groups, such that, if two labels overlap then they belong to the same group.
3. Labels are assigned by solving a variant of the matching problem, where at most one label position from each group is part of the solution.

Figure 15.3   Basic labeling algorithm.

Next, the three basic steps of the basic labeling algorithm are presented in detail.

Selecting labels

To find a set of discrete label positions for each graphical feature, a number of heuristics can be used. For points, a number of label positions that touch their corresponding point is defined. In most algorithms a finite set of potential label positions are associated with each point, typically the size of this set is four or eight as shown in Figure 15.5 (see also [CMS95]).

It is generally accepted, especially in the framework of cartography, that area labels must follow the general shape of their corresponding area, and that they must be inside the boundaries of the area. For each area, a number of potential label positions is defined according to the techniques described in [FA87, Fre88, PF96, vR89].

Next, a simple heuristic for finding a set of label positions corresponding to edges of graph drawings is presented. As Figure 15.6 illustrates, a number of equally spaced points on each edge is defined. Each assigned label position λ_i is associated with exactly one of these points i, such that one of the corners of label λ_i coincides with point i. In addition, label λ_i does not overlap its corresponding graphical feature or any other graphical feature (except other label positions). A global approach for finding an initial set of label positions for non-horizontal edges can be found in Section 15.3.2 and in [KT98].

Reducing labels

The size of the initial set of label positions must be kept reasonably small since it affects the performance of any labeling algorithm.

In order to reduce the set of label positions an intersection graph is first created, where each label position is a node and if two label positions intersect then there is an edge

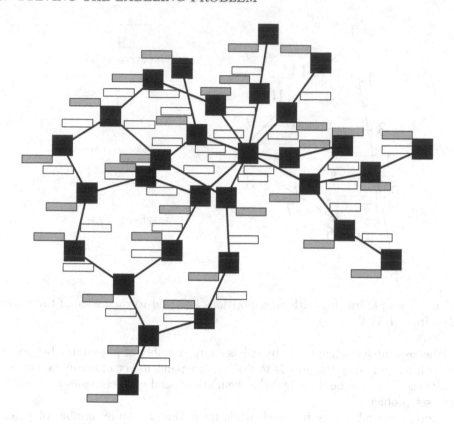

Figure 15.4 A force-directed drawing where labels are positioned by the matching technique for the GFLP problem. The labels are parallel to the horizontal axis. The grey boxes are node labels and the white boxes are edge labels. Figure taken from [KT03].

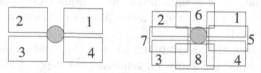

Figure 15.5 Potential label positions for a point. Figure taken from [KT03].

connecting their corresponding nodes. If label positions are parallel to the axis then overlaps can be detected using the techniques for detecting overlaps among isothetic rectangles in $O(n \log n + K)$ time (n is the number of rectangles and K the number of intersections) [Ede83a, Ede83b]. Otherwise, in order to detect overlaps of labels with arbitrary orientation, the techniques of [GJS96] can be used to detect intersections between convex polygons in $O(n^{4/3+\epsilon} + K)$ time (n is the total number of vertices of the polygons, K is the number of pairs of polygons that intersect, and ϵ is any constant greater than zero).

Then, heavily overlapping labels are removed. The remaining labels are assigned to groups, such that, if two labels overlap then they belong to the same group. The goal of the third step of the algorithm is to select at most one label from each group as part of the solution. This way, the algorithm will produce a label assignment free of overlaps.

The optimal solution would be one with the maximum number of minimum size complete subgraphs (groups) of the intersection graph, with the additional constraint that each object has a large number of label positions as part of some groups. It is most likely to have a

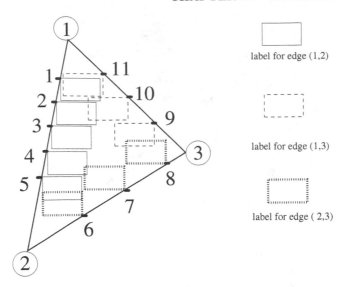

Figure 15.6 A graph drawing with label positions assigned to each edge of the drawing. Figure taken from [KT03].

successful label assignment when each object has a large number of potential label positions associated with it. In reality the goal is to find an independent set of complete subgraphs of label positions. This can be done by using heuristics based on techniques that solve the independent set problem.

Heavily overlapping labels are removed, while maintaining a large number of potential labels for each object f by keeping track of the number of labels associated with f. Our aim is to reduce the intersection graph into a set of disconnected subgraphs.

First, in order to make this process more efficient a preprocessing step is applied that eliminates unnecessary labels or assigns labels in obvious cases. For example, if a label position l of an object f is free of overlaps, then all label positions for f with lower ranking than l can be safely removed.

Next, an appropriate number of overlapping labels is removed. A simple and very successful (according to experiments) technique for removing overlapping labels is the following: If a subgraph c must be split, then the node with the highest degree is removed from c, unless that node corresponds to a label position of some object with very few label positions. In that case the next highest degree node from c is removed. This process is repeated until either c is split into at least two disjoint subgraphs, or c is complete.

Matching labels to objects

To further clarify the main idea of this technique the *matching* graph is introduced:

DEFINITION 15.1   Given a drawing Γ, a set F of graphical features to be labeled, and a set Λ of label positions for F, the matching graph $G_m(V_f, V_c, E_m)$ is defined as follows:

- Each node in V_f corresponds to a graphical feature in F.
- Each node in V_c corresponds to a group of overlapping labels.
- Each edge (i, j) in E_m connects a node i in V_f, to a node j in V_c, if and only if the graphical feature that corresponds to i has a label position that is a member of the group that corresponds to j.

Notice that G_m is a bipartite graph and the cost of assigning label l to graphical feature f is the weight of edge (f, l) in G_m. Therefore, a maximum cardinality minimum weight matching for graph G_m will give us an optimal (maximum number of labels with minimum cost) label assignment with no overlaps with respect to the *reduced* set of label positions.

By representing the labeling problem as a bipartite graph, the inherent hardness of the problem is revealed. According to [KR92], the labeling problem is closely related to the *independent set* problem. Indeed, consider the very simple case where there is only one potential label position for each graphical feature in the drawing, the problem of assigning labels to the maximum number of graphical features is equivalent to finding a maximum size independent set.

Once the set of groups is found, the construction of the matching graph is trivial. A final label assignment can be found by solving the maximum cardinality minimum weight matching problem (see [GK95, Tar83] for efficient algorithms) for graph G_m. The size of the matching graph depends not only on the size of the input drawing, but also on the size of set Λ of labels and the density of overlaps. Notice that at most one label position from each group may be part of a label assignment. Thus, a matching of graph G_m produces an assignment free of overlaps. Because the label assignment is free of overlaps, the cost of each label position will depend only on the ranking of that label. This implies that the cost of each label position can be computed by a preprocessing step.

15.3.2 The ELP Problem

The problem of assigning labels to a set of lines or edges, also known as the *Edge Label Placement* (ELP) problem, has been addressed in the context of geographical and technical maps [AH95, ECMS97, vR89, WKvK+00, Zor90] and graph drawing [KT98, DKMT07]. Furthermore, any of the techniques for solving the general labeling problem (see section 15.3.1) can be applied in solving the ELP problem.

In the context of geographical and technical maps edges are linear features. Labels should be placed alongside and parallel to rivers, boundaries, roads or linear features. If the linear feature is curved, the shape of the label must follow the curvature of the linear feature. The positioning of linear labels has the greatest degree of freedom since labels can be placed almost anywhere along the linear feature, thus cartographers have been focusing their efforts on finding the right shape of the linear label.

However, in the context of graph drawing, placing labels to edges is a more complicated process. Edges are not necessarily long, they are usually straight lines or polygonal chains and they have to follow user preferences and specifications. For example an edge label might be related to the source node of the edge, thus it must be placed closer to the source node rather than the target node to avoid a misleading label assignment (see Figure 15.2).

In [DKMT07] a labeling system is presented that includes a very functional interface and labeling engine that addresses the ELP problem in the context of a graph drawing editor. It is noteworthy that the interface of that system allows the user to set the labeling preferences interactively.

In the following section a fast and simple technique, first proposed in [KT98], is presented for solving the problem of positioning text or symbol labels corresponding to edges of a graph drawing.

A fast and simple algorithm for labeling edges of graph drawings

This technique is based on the matching technique for solving the general labeling problem presented in section 15.3.1.

The technique works for labels that are parallel to the horizontal axis, and have approximately equal height and arbitrary width. In order to simplify the discussion the following assumptions are made:

- All labels have the same size.
- Each edge has only one label associated with it.

The goal of this technique is to assign to each edge a label position that is free of overlaps and touches only its associated edge. The main idea of this technique is the following:

First, a set Λ of label positions is produced. Next, label positions are grouped such that each label position that is part of a group overlaps every other label position that belongs to the same group. Then, edges to label positions are matched by allowing at most one label position from each group to be part of a label assignment by using a fast matching heuristic. The key to restricting the ELP problem to a matching problem is to create a suitable initial set of label positions.

The initial set of label positions is created in the following way. The input drawing is divided into consecutive horizontal strips of equal height. The height of each strip is equal to the height of the labels. Next, a set of label positions Λ_e for each edge e is found. Each label position must be inside a horizontal strip. Labels are slided inside each horizontal strip until a label touches its edge, say e. That label position is included into set Λ_e if it does not overlap any other graphical feature or only overlaps label positions of some edge other than e, as shown in Figure 15.7. Label positions that overlap nodes or edges of the layout are not considered. Also label positions are not allowed to intersect their associated edges. Label positions lie entirely inside horizontal strips. Thus, label positions can only overlap other labels that belong to the same horizontal strip. Hence, the following are true:

- A label position of an edge e does not overlap any other label position of e.
- If two label positions overlap then they are inside the same horizontal strip.
- Each label position overlaps at most one other label position.

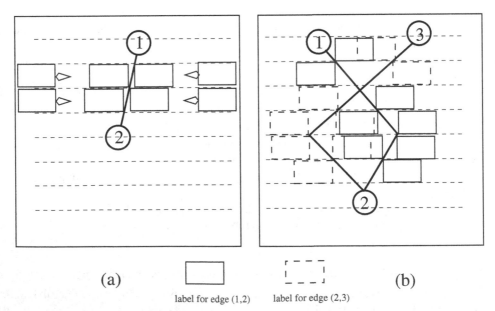

(a)   (b)

label for edge (1,2)   label for edge (2,3)

Figure 15.7 Assigning potential labels to edges of a drawing. Figure taken from [KT98].

If two label positions overlap then they belong to the same group. If a label position is free of overlaps then it belongs to a single member group.

The size of the initial set of label positions must be kept reasonably small since it affects the performance of any labeling algorithm. The above method of defining a set of potential label positions is very practical and effective because it partitions the solution space and identifies the areas of the drawing where conflicts of label assignment may occur. In addition, it significantly reduces the search space for potential conflicts (overlaps).

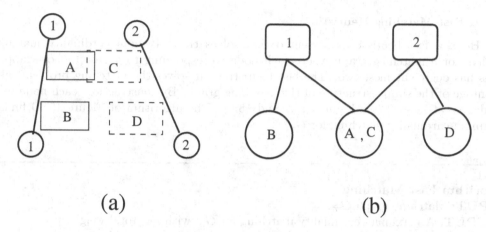

(a)   (b)

Figure 15.8 (a) A simple drawing with label positions for each edge. (b) The corresponding matching graph. Figure taken from [KT03].

In Figure 15.8 a simple example of how to construct the matching graph (see Def. 15.1) is presented. Figure 15.8(a) represents a simple drawing with two edges and two label positions for each edge. Figure 15.8(b) shows its corresponding matching graph. Label positions that overlap belong to the same node in the matching graph, in this example label position A of edge 1 overlaps label position C of edge 2, thus they are represented by a single node in the matching graph.

Since at most one label position from each group may be part of a label assignment, a matching of graph G_m produces an assignment free of overlaps. A maximum cardinality matching of graph G_m assigns labels to the maximum number of edges.

A description of the labeling technique is given in the algorithm of Figure 15.9.

Algorithm ELP
INPUT: A drawing Γ of graph $G(V, E)$.
OUTPUT: A label assignment free of overlaps.
1. Split Γ into horizontal strips.
2. Find all label positions for each edge and construct the groups of overlapping labels.
3. Create the matching graph G_m for Γ.
4. Match label positions to edges, by finding a maximum cardinality minimum weight matching of G_m.

Figure 15.9 Algorithm ELP.

The size of the matching graph depends on the size of set Λ of label positions. Unfortunately the size of Λ can be large with respect to the size of the original graph G. This implies that a typical matching algorithm might take a long time. The best algorithms for finding a maximum cardinality minimum weight matching of G_m take more than quadratic time with respect to the size of G_m [GK95, Tar83]. In order to reduce the time complexity of the matching, in the next section a heuristic is presented that finds a maximum cardinality matching with low total weight in linear time with respect to the size of G_m, by taking advantage of the structure and properties of graph G_m.

A Fast Matching Heuristic

Here, a fast heuristic is presented that solves the maximum cardinality matching problem for a matching graph where each node corresponding to a group of overlapping labels has degree at most two. The fast heuristic that solves the matching problem takes advantage of the simple structure of the matching graph. By construction, each node in V_c has degree at most 2 (see Figures 15.7 and 15.8). The algorithm of Figure 15.10 finds a maximum cardinality matching for G_m.

Algorithm Fast Matching
 INPUT: Matching graph G_m.
OUTPUT: A maximum cardinality matching for G_m with low total weight.
  1. *If* the minimum weight incident edge of a node in V_f connects this node to a node in V_c of degree 1 *then*:
    1.1. Assign this edge as a matched edge.
    1.2. Update G_m.
  2. *If* a node in V_f has degree 1 *then*:
    2.1. Assign its incident edge as a matched edge.
    2.2. Update graph G_m.
  3. *Repeat* Steps 1 and 2 until no new edge can be matched.
  4. Delete all nodes of degree 0 from G_m.
  5. *For* each node f in V_f do
    5.1. Remove all but the two incident edges of f with the least weight.
  6. The remaining graph consists of simple cycles and/or paths.
    6.1. Find the only two maximum cardinality matchings for each component.
    6.2. Choose the matching of minimum weight.

Figure 15.10 Algorithm Fast Matching.

Note: The *Update* G_m operation removes the two nodes incident to a new matched edge and stores that edge and its incident nodes as part of the matching. Also removes all incident edges from the two nodes.

In Step 1 matched edges are found that are part of any optimal solution. In Step 2 edges are matched to those nodes in V_f that are of degree 1. If two nodes of degree 1 in V_f are connected to the same node in V_c, as matched edge is chosen the edge with minimum weight. This implies that one of the edges will have no label. In Step 4 nodes are removed from G_m that correspond to either edges that have no potential labels assigned to them or have potential labels that will not be part of a final labeling assignment. In Step 5, for each node in V_f of degree more than 2, only its two incident edges of least weight are kept and the rest of the edges are removed. The remaining bipartite graph has a simple structure: It consists of simple cycles or simple paths, because each node in V_f has degree 2 and each node in V_c has degree at most 2. Each path or cycle has exactly two maximum cardinality matchings. It is trivial to find both of them by simply traversing the cycle or path and picking as part of the matching only the even or odd numbered edges.

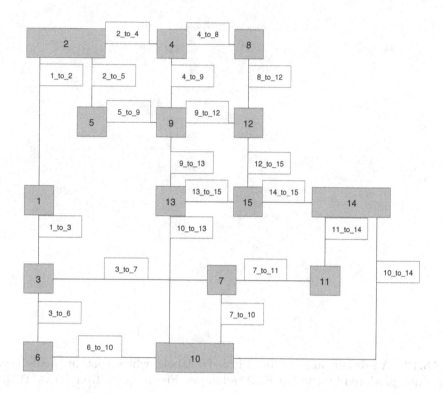

Figure 15.11 An orthogonal drawing with edge labels produced by the fast ELP technique. Figure taken from [KT03].

It is trivial to see that Algorithm Fast Matching runs in linear time. Notice that it also finds a maximum cardinality matching with low total weight because in the last step it considers only the two incident edges of nodes in V_e with the lowest weight. Figure 15.11 shows a label assignment produced by the fast ELP technique.

Further Improvements

For the above fast and simple technique for labeling edges of graph drawings it is clear that the longer and the more vertical the edges are, the more potential label positions are associated with each edge. Thus, the greater the possibility for the labeling algorithm to assign a label to each of these edges. Therefore, hierarchical drawings are particularly suitable for this algorithm since edges are usually long and almost vertical. This technique performs very well also for straight-line drawings, such as ones produced by force-directed and circular techniques. One weakness of this labeling technique is that it ignores horizontal edges or edge segments. Thus, as presented, this technique is not suitable for orthogonal drawings. However, one can use the general technique by dividing an orthogonal drawing into horizontal and vertical strips in order to find a set of label positions, followed by the assignment of labels to edges. Figure 15.12 shows the results with an example.

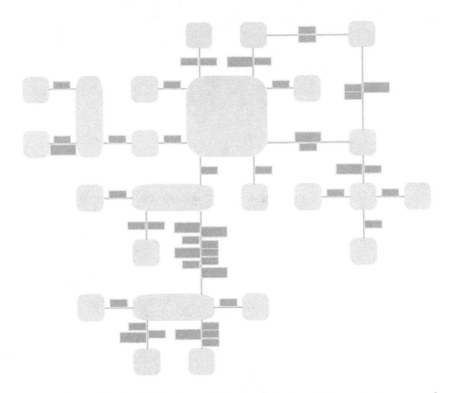

Figure 15.12 An orthogonal drawing with edge labels, which contains many horizontal edge segments, produced by the fast ELP technique. Figure taken from [DKMT07].

In addition, when drawings are very dense or there is a large number of oversized labels, the default label assignment produced by the labeling system might not be satisfactory. In such cases, the user can fine tune the algorithm by relaxing the labeling quality constraints by allowing overlaps (see Figure 15.13).

15.3.3 The NLP Problem

The problem of assigning labels to a set of points or nodes, also known as the *Node Label Placement* (NLP) problem, has been extensively studied in the context of automated cartography and many successful algorithmic approaches have been introduced [CMS95, DMM+97, FW91, Hir82, WW95, Zor90]. Also, any of the techniques for solving the general labeling problem (see section 15.3.1) can be applied to the NLP problem.

Algorithms based on local and exhaustive search [DF92, EG90, FA87] and simulated annealing [ECMS97] are well suited for solving the NLP problem. Experimental results [CMS95] have shown that simulated annealing outperforms all algorithms based on local and exhaustive search. In addition simulated annealing is one of the easiest algorithms to implement.

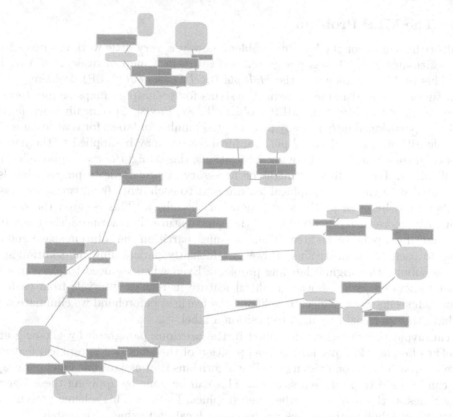

Figure 15.13 A circular drawing with edge labels, where labels are allowed to overlap other graph objects, produced by the fast ELP technique. Figure taken from [DKMT07].

These algorithms start with a rather small initial set of potential label positions from which they derive a final label assignment. This is because the size of the initial set of label positions plays a critical role in the performance of these algorithms. This precondition works well when solving the NLP problem. For example, each point is given at most four or eight potential label positions (see Figure 15.5).

Approximation algorithms for restricted versions of the NLP problem are presented in [DMM+97, FW91]. Specifically, the approach of [FW91] assigns labels of *equal* size to all points while attempting to maximize the size of the assigned labels. The work in [WW95] improves the results in [FW91] by using heuristics. A similar approach has been taken in [DMM+97]. In effect, finding the maximum label size is equivalent to finding the smallest factor by which the map has to be zoomed out such that each point has a label assigned to it. However, it is not clear how these techniques can be modified to solve real-world problems, including the labeling of graphical features of graph drawing, where the label size is usually predefined and labels are not necessarily of equal size.

Another approach to solve the NLP problem is based on the sliding model, where sliding labels can be attached to the point they label anywhere on their boundary. This model was first introduced in [Hir82] who gave an iterative algorithm that uses repelling forces between labels in order to eventually find a placement of labels. A polynomial-time approximation scheme and a fast factor-2 approximation algorithm for maximizing the number of points that are labeled by axis-parallel sliding rectangular labels of common height, based on the sliding model, is presented in [vKSW99].

15.3.4   The MLP Problem

Many algorithms exist for the labeling problem; however, very little work has been directed toward positioning many labels per graphical feature in a map or drawing [FA87, Fre88, KT06]. This problem is known as the *Multiple Label Placement* (MLP) problem.

In existing automated name placement systems for geographic maps simple techniques have been utilized to address the MLP problem [FA87, Fre88]. Specifically, each feature to be labeled is partitioned into as many pieces as the number of labels for that feature. Then, labeling algorithms for single label per graphical feature may be applied to the new set of partitioned graphical features. In many applications, this straightforward approach presents some difficulties. For example, it might be necessary or preferable to position labels that are associated with the same graphical feature next to each other (e.g., two labels assigned to an edge must be close to the source node of the edge). This is often the case when labels describe more than one attribute of the same feature. Furthermore, the feature to be labeled might be a point or an area. Then, we must partition the solution space and assign one label to each of the partitions. However, efficiently partitioning the solution space is as hard as solving the original labeling problem. Even when we need to place more than one label associated with a linear graphical feature in regular intervals from each other, this approach seems weak. Since, by splitting the features beforehand we eliminate solution space that otherwise could be used to position a label.

One can avoid the situations described in the previous paragraph by allowing each of the labels to be placed in any legible label position of the associated graphical feature. An iterative approach based on existing labeling algorithms that assigns one label per graphical feature can be used to produce a solution. This can be done by applying these algorithms as many times as the number of labels per graphical feature. This scheme presents a new challenge: most labeling algorithms are based on local and exhaustive search. Thus, their performance (running time and quality of solutions) is sensitive to the size of the graphical features to be labeled and to the density of the drawing. Clearly, if each graphical feature in a drawing is associated with i labels, then the size of the problem is i times larger. Therefore, the above techniques might be slow even for small instances.

In [KT06] the MLP problem is treated in the context of graph drawing. A framework for evaluating the quality of label positions is presented. In addition, two algorithmic schemes are presented: (i)A simple and practical iterative technique and (ii) A flow-based technique which is an extension of the matching technique presented in section 15.3.1. In the following sections these techniques will be presented in detail.

Labeling quality rules for the MLP problem

Multiple labels per graphical feature are needed not only when objects are very long (i.e., long edges) and repetition is necessary, but also when more than one attribute per graph object must be displayed. Therefore, some additional considerations have to be taken into account with respect to the quality of a label assignment, when graphical features have many labels. Specifically, we must take into account how labels for the same graphical feature influence each other. For example, many times each of the labels corresponds to some attribute of a graphical feature and the relative position of a label with respect to other labels of the same graphical feature reveals that attribute.

Next, we present some constraints that may be used to ensure that each label is unambiguous, easily read and recognized, when more than one label is associated with a graphical feature. These constraints can be divided into three general categories: (i) *proximity*, (ii) *partial order*, and (iii) *priority*. In order to illustrate the three different sets of constraints we will use as an example the labeling of a single edge (s,t) with two labels l_s and l_t. Label

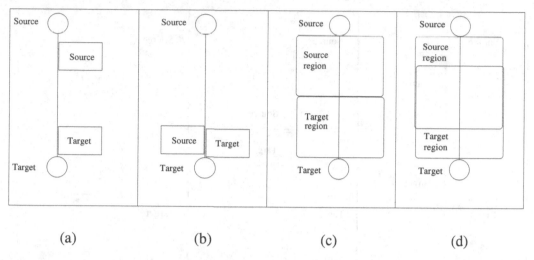

(a)   (b)   (c)   (d)

Figure 15.14   (a) A preferable label assignment. (b) A misleading label assignment. (c) Defining strict proximity constraints. (d) Defining relaxed proximity constraints. Figure taken from [KT06].

l_s is associated with the source node and label l_t is associated with the target node, as shown in Figure15.14(a).

Proximity:

Label l_s (resp. l_t) must be in close proximity with the source (resp. target) node to avoid ambiguity. Therefore, it is necessary to define a maximum distance from the source (resp. target) node that label l_s (resp. l_t) may be positioned. When edge (s,t) is associated with exactly one label, then that label may be located anywhere inside the solution space. If there are more than one label associated with (s,t), then each label must be positioned inside an area that is a subset of the solution space.

In Figure 15.14 we illustrate the importance of the proximity constraints. For example the label assignment in Figure 15.14(a) is a preferable assignment. The assignment in Figure 15.14(b) does not convey clearly the meaning of the labels, because they are very close to the target node; hence by observing the picture we cannot establish with certainty that the source label is associated with the source node. In Figure 15.14(c) the proximity constraint is that the distance between the source (resp. target) node and its label must be at most half the length of the edge. This implies that the source (resp. target) label must be inside the source (resp. target) region. The defined proximity constraints in Figure 15.14(c) are too restrictive, since the defined regions do not intersect. One could define more relaxed proximity constraints, as shown in Figure 15.14(d), where intersecting of different regions is allowed. In practice the latter is preferable since it increases the labeling solution space and improves the possibility for finding a labeling assignment, especially in cases where the drawing is crowded.

Partial Order:

A label associated with the source (resp. target) node must be closer to the source (resp. target) node than any other label to avoid ambiguity. Thus, in many cases, it is appropriate to define a partial order between labels of the same graphical feature according to some invariant (e.g., x or y axis, distance from a fixed point).

In Figure 15.15(c) we present an example where the absence of a partial order rule produces a misleading label assignment, since by simply looking at the picture we associate the target (resp. source) label to the source (resp. target) node. In Figures 15.15(a) and

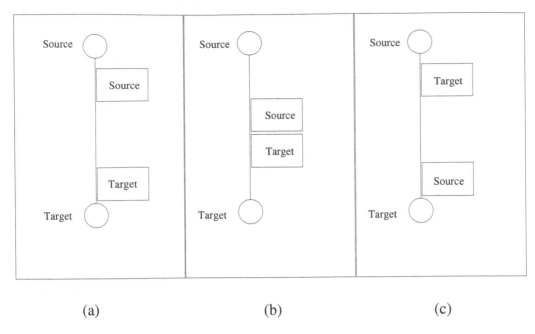

(a)   (b)   (c)

Figure 15.15   (a) A preferable label assignment. (b) An acceptable label assignment. (c) A misleading label assignment. Figure taken from [KT06].

15.15(*b*) the additional condition that a label associated with the source node must be closer to the source node than the label associated with the target node ensures the correct interpretation of the label assignment. If we define restrictive proximity constraints, as shown in Figure 15.14(*c*), then a partial order constraint is not necessary. However, if we relax the proximity constraints, as shown in Figure 15.14(*d*), then we need to define a partial order constraint in order to avoid misleading labeling assignments.

Priority:

In many cases, it is impossible to assign all labels associated with a graphical feature, due to the density of the drawing. Then, the user might prefer to have the important labels assigned first, and then assign the rest of the labels if there is available space.

These three sets of constraints present a succinct framework for a good label assignment with respect to the MLP problem.

In the following sections we focus on two sets of heuristics, iterative and flow-based, to solve the MLP problem.

An iterative algorithm for the MLP problem

First a simple iterative approach to solve the problem of assigning multiple labels to each graphical feature of a drawing is presented. For simplicity, let us assume that each graphical feature is associated with the same number of label positions. The main idea is the following: existing algorithms solve the labeling problem for single label per graphical feature. Therefore, one could solve the MLP problem by applying these algorithms as many times as the number of labels per graphical feature. This method consists of a main loop, and we execute the loop as many times as the number of labels per graphical feature. In particular, at the *i-th* execution of the loop, we assign the *i-th* label to each graphical feature.

This technique can take into account all three sets of constraints: (a) proximity (by considering only the label positions that respect the proximity rules), (b) partial order (by eliminating from the set of potential label positions, after each execution of the loop, the label positions that do not respect the partial order) and (c) priority (by selecting, if possible, the label position of highest priority among the available label positions). One can refine this technique by first finding a set of label positions before entering the loop, and then executing inside the loop only the step of positioning labels. The refinement works because the cited labeling algorithms produce a label assignment from an initial finite set of discrete potential label positions. The refined algorithm is shown in Figure 15.16.

Iterative Algorithm

INPUT: A drawing Γ, a set of graphical features F in Γ to be labeled,
  a number N of labels for each graphical feature f in F.

OUTPUT: A label assignment.

1. Find an initial set of label positions L.
2. *For $i = 1$ to N do*:
   2.1. Assign the *i-th* label to each graphical feature in F from the set L of potential labels using existing labeling algorithms.
   2.2. Remove potential label positions from L that overlap already assigned labels.

Figure 15.16   Iterative algorithm.

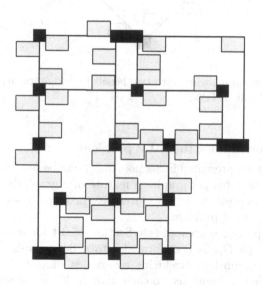

Figure 15.17   An orthogonal drawing with two labels per edge, positioned by the Iterative algorithm. Figure taken from [KT06].

Even though this technique is very attractive, especially because it can be realized by using existing labeling algorithms, it presents some challenges that have to be addressed. Labeling techniques based on local or exhaustive search first create an initial label assignment where conflicts between labels are allowed. Then conflicts are resolved by repositioning assigned labels until all conflicts are resolved, or no further improvement can be achieved. When applying these techniques in the context of the iterative algorithm one can either apply repositioning only for labels assigned in the current run of the loop or for any assigned label (even in previous runs of the loop). In either case such techniques are slow.

This iterative approach is especially suited for the labeling algorithms presented in [KT98, KT03], because they first find a set of label positions, and then they produce a label assignment in a single step without any repositioning of labels (see Figures 15.17 and 15.18).

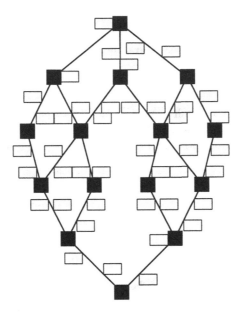

Figure 15.18   A hierarchical drawing with two labels per edge, positioned by the Iterative algorithm. Figure taken from [KT06].

A Flow-Based algorithm for the MLP problem

The matching technique presented in Section 15.3.1 can be further extended to support placement of more than one label per graphical feature of a graph drawing. The algorithm presented here assigns label positions in a non-iterative fashion. It solves the MLP problem by reducing it to an assignment problem.

First the matching graph G_m is created (see Section 15.3.1 for more details).

Next, the matching graph G_m is transformed into a flow graph $G_{flow}(s, t, V_f, V_c, E_f)$. G_m is converted to an st-graph by introducing two nodes s and t. Node s is connected to each node in V_f, and node t is connected to each node in V_c, as shown in Figure 15.19.

Finally capacities to each edge of the flow graph G_{flow} are assigned in the following way:

- Each edge of the original matching graph has capacity one.
- Each edge (c, t) of G_{flow} incident to the target node has capacity one.
- Each edge (s, v) incident to the source node has capacity equal to the number of labels associated with the graphical feature of the input graph that is represented by node v in G_m.

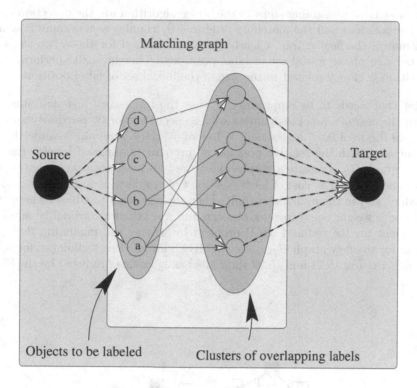

Figure 15.19 The flow graph. Figure taken from [KT06].

Clearly a maximum flow of graph G_{flow} will produce a maximum cardinality label assignment with respect to the set of labels encoded in the matching graph. Sophisticated techniques can solve the maximum flow problem in $O(nm \log n)$ time [AMO93], where n is the number of vertices and m is the number of edges of the flow graph.

This technique is summarized in the algorithm of Figure 15.20.

Flow-based Algorithm
INPUT: A drawing Γ, a set of graphical features F in Γ to be labeled,
    a number $M(f)$ of labels for each graphical feature f in F.
OUTPUT: A label assignment free of overlaps.
1. Find a set of label positions for each graphical feature in the drawing.
2. Arrange overlapping label positions into groups.
3. Create the matching graph G_m.
4. Augment graph G_m to a flow graph G_{flow}.
5. Assign capacities to each edge of G_{flow}.
6. Assign cost to edges of G_{flow}.
7. Find the maximum flow minimum cost of graph G_{flow}.
8. Assign labels according to the results of Step 7.

Figure 15.20 Flow-based algorithm.

The two most time consuming steps of the above algorithm are the detection of overlaps between label positions and the matching produced by running a maximum flow minimum cost algorithm on the flow graph. Clearly the time required for those two steps depends highly on the size of the initial set of label positions. Therefore, the performance of the above algorithm is closely related to the size of the initial set of label positions.

One point that needs to be emphasized is that the framework just described can take into account the cost of a label assignment with respect to priority, proximity and aesthetic criteria. Since the final label assignment is free of overlaps, one may assume that there is no cost associated with the relative position of any pair of assigned labels. Each edge in the bipartite graph G_m connects a graphical feature to a label position of that feature that belongs to some group. The cost of label position l of graphical feature f is included as the weight of edge (f, l) in the matching graph. Then, by assigning to edges incident to source and target nodes weight equal to zero, one can find a maximum cardinality minimum cost label assignment for the reduced MLP problem by solving the maximum flow minimum cost problem for the flow graph G_{flow} (see [AMO93] for efficient techniques for solving the flow problem). Figures 15.21 and 15.22 show label assignments produced by the Flow-based algorithm.

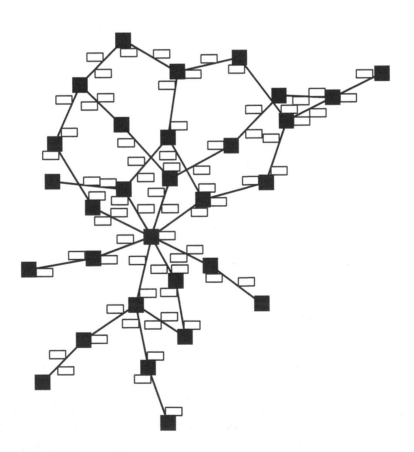

Figure 15.21 A force-directed drawing with two labels per edge, positioned by the Flow-based algorithm. Figure taken from [KT06].

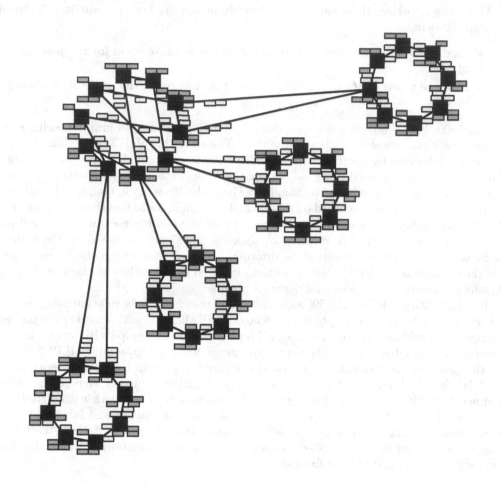

Figure 15.22 A circular drawing with three labels for each edge and node positioned by the Flow-based algorithm. The white boxes are edge labels and the dark boxes are node labels. Figure taken from [KT06].

15.3.5 Placing Labels by Modifying the Drawing

Automatic labeling is a very difficult problem, and because we rely on heuristics to solve it, there are cases where the best methods available do not always produce an acceptable or legible solution even if one exists. Furthermore, there are cases where no feasible solution exists. Given a specific drawing and labels of fixed size, then it might be impossible to assign labels without violating any of the basic rules of a good label assignment (e.g., label to label overlap, legibility, unambiguous assignment). These cases appear often in practical applications when drawings are dense, labels are oversized, or the label assignment must meet minimum requirements set by the user (e.g., font size or preference of placing labels).

To solve the labeling problem where the best solution we can have is either incomplete or not acceptable one must modify the drawing. This approach cannot be applied in drawings that represent geographical or technical maps where the underlying geometry is fixed by definition. However, the layout of a given graph drawing can be changed since it is the result of the algorithm used to draw the graph.

Generally speaking, there can be two algorithmic approaches in modifying the layout of a graph drawing:

- Modify the existing layout of a graph drawing to make room for the placement of labels.
- Produce a new layout of a graph drawing that integrates the layout and labeling process.

In [Hu09, KT11] algorithms that modify an existing layout of a graph drawing to make room for the placement of labels are presented. The algorithm of [KT11] modifies an existing orthogonal drawing by inserting extra space in order to accommodate the placement of edge labels that are free of overlaps. First, an edge label assignment is computed, where overlaps are allowed, by using existing techniques. Then, the drawing is modified by applying a polynomial time algorithm based on minimum flow techniques to find the extra space needed to eliminate label overlaps, while preserving the orthogonal representation of the drawing. In [Hu09] label overlaps are resolved by applying an algorithm based on the techniques used to produce force-directed layout drawings. It iteratively moves the labels to remove overlaps, while keeping the relative positions between them as close to those in the original layout as possible, and edges as straight as possible.

In [BDLN05, DDPP99, KM99] algorithms that combine the layout and labeling process of orthogonal drawings of graphs are presented. In [KM99] the authors study the problem of computing a grid drawing of an orthogonal representation of a graph with labeled nodes and minimum total edge length. They show an integer linear programming (ILP) formulation of the problem and present a branch-and-cut based algorithm that combines compaction and labeling techniques. The work in [BDLN05] makes a further step in the direction defined in [KM99] by integrating the topology-shape-metrics approach with algorithms for edge labeling. In [DDPP99] an approach to combining the layout and labeling process of orthogonal drawings is presented. Labels are modeled as dummy nodes and the topology-shape-metrics approach is applied to compute an orthogonal drawing where the dummy nodes are constrained to have fixed size.

References

[AF84]   J. Ahn and H. Freeman. A program for automatic name placement. *Cartographica*, 21(2 & 3):101–109, 1984.

[AH95]   D. H. Alexander and C. S. Hantman. Automating Linear Text Placement Within Dense Feature Networks. In *Proc. Auto-Carto 12*, pages 311–320. ACSM/ASPRS, Bethesda, 1995.

[AMO93]   R. K. Ahuja, T. L. Magnanti, and J. B. Orlin. *Network Flows: Theory, Algorithms, and Applications*. Prentice Hall, Englewood Cliffs, NJ, 1993.

[BDLN05]   C. Binucci, W. Didimo, G. Liotta, and M. Nonato. Orthogonal Drawings of Graphs with Vertex and Edge Labels. *CGTA*, 32(2):71–114, 2005.

[C⁺99]   Bernard Chazelle et al. Application challenges to computational geometry: CG impact task force report. In B. Chazelle, J. E. Goodman, and R. Pollack, editors, *Advances in Discrete and Computational Geometry*, volume 223 of *Contemporary Mathematics*, pages 407–463. American Mathematical Society, Providence, 1999.

[CMS95]   J. Christensen, J. Marks, and S. Shieber. An empirical study of algorithms for Point Feature Label Placement. *ACM Trans. on Graphics*, 14(3):203–232, 1995.

[DDPP99]   G. Di Battista, W. Didimo, M. Patrignani, and M. Pizzonia. Orthogonal and Quasi-upward Drawings with Vertices of Prescribed Size. In J. Kratochvil, editor, *Graph Drawing (Proc. GD '99)*, volume 1731 of *Lecture Notes in Computer Science*, pages 297–310. Springer-Verlag, 1999.

[DETT99]   G. Di Battista, P. Eades, R. Tamassia, and I. G. Tollis. *Graph Drawing*. Prentice Hall, Upper Saddle River, NJ, 1999.

[DF92]   J. S. Doerschler and H. Freeman. A rule based system for dense map name placement. *Communications of ACM*, 35(1):68–79, 1992.

[DKMT07]   U. Doğrusöz, K. G. Kakoulis, B. Madden, and I. G. Tollis. On Labeling in Graph Visualization. Special Issue on Graph Theory and Applications, Information Sciences Journal, vol. 177/12, pp. 2459-2472, 2007.

[DMM⁺97]   S. Doddi, M. V. Marathe, A. Mirzaian, B. M. Moret, and B. Zhu. Map Labeling and Its Generalizations. In *Proc. 8th ACM-SIAM Sympos. Discrete Algorithms*, pages 148–157, 1997.

[ECMS97]   S. Edmondson, J. Christensen, J. Marks, and S. M. Schieber. A General Cartographic Labeling Algorithm. *Cartographica*, 33(4):321–342, 1997.

[Ede83a]   H. Edelsbrunner. A new approach to rectangle intersections, Part I. *Internat. J. Comput. Math.*, 13:209–219, 1983.

[Ede83b]   H. Edelsbrunner. A new approach to rectangle intersections, Part II. *Internat. J. Comput. Math.*, 13:221–229, 1983.

[EG90]   L. R. Ebinger and A. M. Goulete. Noninteractive automated names placement for the 1990 decennial census. *Cartography and Geographic Information Systems*, 17(1):69–78, 1990.

[FA87]   H. Freeman and J. Ahn. On the problem of placing names in a geographical map. *Int. J. of Pattern Rec. and Artificial Intelligence*, 1(1):121–140, 1987.

[FMC96]   H. Freeman, S. Marrinan, and H. Chitalia. Automated Labeling of Soil Survey Maps. In *Proc. 8th Canadian Conference on Computational Ge-*

ometry, volume 1 of *Proceedings of the ASPRS-ACSM Annual Convention*, pages 51–59, 1996.

[Fre88] H. Freeman. An Expert System for the Automatic Placement of Names on a Geographic Map. *Information Sciences*, 45:367–378, 1988.

[Fre91] H. Freeman. Computer name placement. In D. J. Maguire, M. F. Goodchild, and D. W. Rhind, editors, *Geographical Information Systems: Principles and Applications*, pages 445–456. Longman, London, 1991.

[FW91] M. Formann and F. Wagner. A packing problem with applications to lettering of maps. In *Proc. 7th Annu. ACM Sympos. Comput. Geom.*, pages 281–288, 1991.

[GJS96] P. Gupta, R. Janardan, and M. Smid. Efficient Algorithms for Counting and Reporting Pairwise Intersections between Convex Polygons. In *Proc. 8th Canadian Conference on Computational Geometry*, pages 8–13. Carleton University Press, 1996.

[GK95] A. V. Goldberg and R. Kennedy. An Efficient Cost Scaling Algorithm for the Assignment Problem. *Mathematical Programming*, 71:153–178, 1995.

[Hir82] S. A. Hirsch. An algorithm for automatic name placement around point data. *The American Cartographer*, 9(1):5–17, 1982.

[Hu09] Y. Hu. Visualizing Graphs with Node and Edge Labels. *CoRR*, abs/0911.0626, 2009.

[Imh75] E. Imhof. Positioning names on maps. *The American Cartographer*, 2(2):128–144, 1975.

[KI88] T. Kato and H. Imai. The NP-completeness of the character placement problem of 2 or 3 degrees of freedom. In *Record of Joint Conference of Electrical and Electronic Engineers in Kyushu*, pages 11–18, 1988. In Japanese.

[KM99] G. W. Klau and P. Mutzel. Combining Graph Labeling and Compaction. In J. Kratochvil, editor, *Graph Drawing (Proc. GD '99)*, volume 1731 of *Lecture Notes in Computer Science*, pages 27–37. Springer-Verlag, 1999.

[KR92] D. Knuth and A. Raghunathan. The problem of compatible representatives. *SIAM J. Disc. Math.*, 5:36–47, 1992.

[KT98] K. G. Kakoulis and I. G. Tollis. An Algorithm for Labeling Edges of Hierarchical Drawings. In G. Di Battista, editor, *Graph Drawing (Proc. GD '97)*, volume 1353 of *Lecture Notes in Computer Science*, pages 169–180. Springer-Verlag, 1998.

[KT01] K. G. Kakoulis and I. G. Tollis. On the Complexity of the Edge Label Placement Problem. *Computational Geometry*, 18(1):1–17, 2001.

[KT03] K. G. Kakoulis and I. G. Tollis. A Unified Approach to Automatic Label Placement. *International Journal of Computational Geometry and Applications*, 13(1):23–60, 2003.

[KT06] K. G. Kakoulis and I. G. Tollis. Algorithms for the Multiple Label Placement Problem. *Computational Geometry*, 35(3):143–161, 2006.

[KT11] K. G. Kakoulis and I. G. Tollis. Placing Edge Labels by Modifying an Orthogonal Graph Drawing. In U. Brandes and S. Cornelsen, editors, *Graph Drawing (Proc. GD 2010)*, volume 6502 of *Lecture Notes in Computer Science*, pages 395–396. Springer-Verlag, 2011.

[MS91]   J. Marks and S. Shieber. The computational complexity of cartographic label placement. Technical Report 05-91, Harvard University, 1991.

[PF96]   I. Pinto and H. Freeman. The Feedback Approach to Cartographic Area Text Placement. In P. Perner, P. Wang, and A. Rosenfeld, editors, *Advances in Structural and Syntactical Pattern Recognition*, volume 1121 of *Lecture Notes in Computer Science*, pages 341–350. Springer-Verlag, 1996.

[RMM+95]   A. H. Robinson, J. L. Morrison, P. C. Muehrcke, A. J. Kimerling, and S. C. Guptill. *Elements of Cartography*. John Wiley & Sons, Inc., 6th edition, 1995.

[Tar83]   R. E. Tarjan. *Data Structures and Network Algorithms*, volume 44 of *CBMS-NSF Regional Conference Series in Applied Mathematics*. Society for Industrial and Applied Mathematics, Philadelphia, PA, 1983.

[vKSW99]   M. van Kreveld, T. Strijk, and A. Wolff. Point Labeling with Sliding Labels. *CGTA*, 13:21–47, 1999.

[vR89]   J. W. van Roessel. An algorithm for locating candidate labeling boxes within a polygon. *The American Cartographer*, 16(3):201–209, 1989.

[WKvK+00]   A. Wolff, L. Knipping, M. van Kreveld, T. Strijk, and P. K. Agarwal. A Simple and Efficient Algorithm for High-Quality Line Labeling. In P. M. Atkinson and D. J. Martin, editors, *Innovations in GIS VII: GeoComputation*, chapter 11, pages 147–159. Taylor and Francis, 2000.

[WMP+05]   P. C. Wong, P. Mackey, K. Perrine, J. Eagan, H. Foote, and J. Thomas. Dynamic Visualization of Graphs with Extended Labels. In *Proceedings of the 2005 IEEE Symposium on Information Visualization*, INFOVIS '05, pages 73–80, 2005.

[WW95]   F. Wagner and A. Wolff. Map Labeling Heuristics: Provably Good and Practically Useful. In *Proc. 11th Annu. ACM Sympos. Comput. Geom.*, pages 109–118, 1995.

[Yoe72]   P. Yoeli. The logic of automated map lettering. *The Cartographic Journal*, 9(2):99–108, 12 1972.

[Zor90]   S. Zoraster. The solution of large 0-1 integer programming problems encountered in automated cartography. *Operation Research*, 38(5):752–759, September-October 1990.

16

Graph Markup Language (GraphML)

Ulrik Brandes
University of Konstanz

Markus Eiglsperger

Jürgen Lerner
University of Konstanz

Christian Pich
Swiss Re

16.1   Introduction

Graph drawing tools, like all other tools dealing with relational data, need to store and exchange graphs and associated data. Despite several earlier attempts to define a standard, no agreed-upon format is widely accepted and, indeed, many tools support only a limited number of custom formats which are typically restricted in their expressibility and specific to an area of application.

Motivated by the goals of tool interoperability, access to benchmark data sets, and data exchange over the Web, the Steering Committee of the Graph Drawing Symposium started a new initiative with an informal workshop held in conjunction with the 8th Symposium on Graph Drawing (GD 2000) [BMN01]. As a consequence, an informal task group was formed to propose a modern graph exchange format suitable in particular for data transfer between graph drawing tools and other applications.

Thanks to its XML syntax, GraphML can be used in combination with other XML based formats. On the one hand, its own extension mechanism allows to attach `<data>` labels with complex content (possibly required to comply with other XML content models) to GraphML elements. Examples of such complex data labels are Scalable Vector Graphics [W3Ca] describing the appearance of the nodes and edges in a drawing. On the other hand, GraphML can be integrated into other applications, e.g., in SOAP messages [W3Cb].

A modern graph exchange format cannot be defined in a monolithic way, since graph drawing services are used as components in larger systems and Web-based services are emerging. Graph data may need to be exchanged between such services, or stages of a service, and between graph drawing services and systems specific to areas of applications.

The typical usage scenarios that we envision for the format are centered around systems designed for arbitrary applications dealing with graphs and other data associated with them. Such systems will contain or call graph drawing services that add or modify layout and graphics information. Moreover, such services may compute only partial information or intermediate representations, for instance because they instantiate only part of a staged layout approach such as the topology-shape-metrics or Sugiyama frameworks [DBETT99, STT81]. We hence aimed to satisfy the following key goal.

> The graph exchange format should be able to represent arbitrary graphs with arbitrary additional data, including layout and graphics information. The additional data should be stored in a format appropriate for the specific application, but should not complicate or interfere with the representation of data from other applications.

GraphML is designed with this and the following more pragmatic goals in mind:

- *Simplicity*: The format should be easy to parse and interpret for both humans and machines. As a general principle, there should be no ambiguities and thus a single well-defined interpretation for each valid GraphML document.

- *Generality*: There should be no limitation with respect to the graph model, i.e., hypergraphs, hierarchical graphs, etc. should be expressible within the same basic format.

- *Extensibility*: It should be possible to extend the format in a well-defined way to represent additional data required by arbitrary applications or more sophisticated use (e.g., sending a layout algorithm together with the graph).

- *Robustness*: Systems not capable of handling the full range of graph models or added information should be able to easily recognize and extract the subset they can handle.

16.1.1   Related Formats

Besides GraphML there is a multitude of file formats for serializing graphs. Among the simplest ones are direct ASCII-based codings of tables (matrices) or lists, such as tab-separated value files. Specific instances of these include UCINET's `*.dl` files [BEF99] and Pajek's `*.net` files [DMB05]. XML-based formats to represent graphs include GXL [Win02], and DyNetML [TRC03].

16.2   Basic Concepts

In this section, we describe how graphs and simple graph data are represented in GraphML. The graph model used in this section is a *labeled mixed multigraph*, i.e., a tuple

$$G = (V, E, \mathcal{D}),$$

where V is a set of *nodes*, E a *multi-set* containing *directed* and *undirected edges*, and $\mathcal{D}$ a set of *data labels* that are partial functions from $\{G\} \cup V \cup E$ into some specified range of values. The data labels can encode, e.g., properties of nodes and edges such as graphical variables or, if nodes correspond to social actors, demographic characteristics such as gender or age. Thus, our graph model includes graphs that can contain both directed and undirected edges, loops, and multi-edges. This graph model will be extended in Section 16.3, where advanced concepts for the graph topology, like nested graphs, hypergraphs, and ports, are introduced. As an example, consider the document fragment and the graph it describes in Figure 16.1.

```
<graphml>
  <graph edgedefault="directed">
    <node id="v1"/>
    <node id="v2"/>
    <node id="v3"/>
    <node id="v4"/>

    <edge source="v1" target="v2"/>
    <edge source="v1" target="v3"/>
    <edge source="v2" target="v4"/>
    <edge source="v2" target="v4" directed="false"/>
  </graph>
</graphml>
```

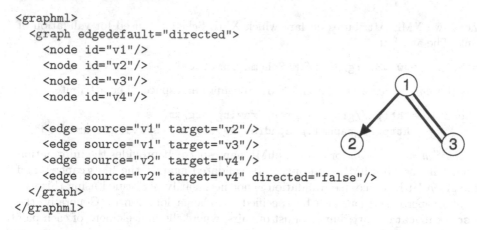

Figure 16.1 A graph and its representation in GraphML.

16.2.1 Header

The document fragment shown in Figure 16.1 is not yet a *valid* XML document. Valid XML documents must declare in their header either a DTD (*document type definition*) or an XML schema. Both DTDs or schemas define a subset of all XML documents that forms a certain language. The GraphML language has been defined by a schema. Although a DTD is provided to support parsers that cannot handle schema definitions, the only normative specification is the GraphML schema located at

http://graphml.graphdrawing.org/xmlns/1.1/graphml.xsd

The document shown in Figure 16.2 is minimal to be a GraphML document that can be validated against the above schema. Actually, it defines an empty set of graphs. Areas starting with <!-- and ending with --> are comments.

```
<?xml version="1.0" encoding="UTF-8"?>
<graphml xmlns="http://graphml.graphdrawing.org/xmlns"
    xmlns:xsi="http://www.w3.org/2001/XMLSchema-instance"
    xsi:schemaLocation="http://graphml.graphdrawing.org/xmlns
      http://graphml.graphdrawing.org/xmlns/1.1/graphml.xsd">
    <!--Content: List of graphs and data-->
</graphml>
```

Figure 16.2 A minimal valid GraphML document.

The first line of the GraphML document in Figure 16.2 is an XML process instruction which defines that the document adheres to the XML 1.0 standard and that the encoding of the document is UTF-8, the standard encoding for XML documents. Of course other encodings can be chosen for GraphML documents.

The second line contains the *root-element*XS of a GraphML document: the <graphml> element. The <graphml> element, like all other GraphML elements, belongs to the namespace http://graphml.graphdrawing.org/xmlns. For this reason we define this namespace as the *default namespace* in the document by adding the XML Attribute

xmlns="http://graphml.graphdrawing.org/xmlns"

to it. The next two XML Attributes declare which XML Schema is used for validation of this document. The attribute

```
xmlns:xsi="http://www.w3.org/2001/XMLSchema-instance"
```

defines `xsi` as the *namespace prefix* for the XML Schema namespace. The attribute,

```
xsi:schemaLocation="http://graphml.graphdrawing.org/xmlns
                    http://graphml.graphdrawing.org/xmlns/1.1/graphml.xsd"
```

defines the *XML Schema location* for the GraphML namespace. It provides the information that all elements in the GraphML namespace are validated against the file `graphml.xsd` located at the given URL. Of course, validation is not necessarily performed using this file. Local copies of `graphml.xsd` can also be specified as schema locations. (Generally, the value of the `schemaLocation` attribute is a list of pairs, where the first element of each pair denotes a namespace and the second points to a file where elements of this namespace are defined.)

The XML Schema reference provides means to validate the document and is therefore strongly recommended. If validation is not considered necessary, the schema location declaration can be omitted. A minimal GraphML document without Schema reference is shown in Figure 16.3. Note that this file is not a valid document according to the XML specifica-

```
<?xml version="1.0" encoding="UTF-8"?>
<graphml xmlns="http://graphml.graphdrawing.org/xmlns" >
        <!--Content: List of graphs and data-->
</graphml>
```

Figure 16.3   A minimal GraphML document without a schema reference.

tion.

16.2.2   Topology

In this section, we describe how the basic graph-topology (nodes and edges) are represented in GraphML.

Remind the document fragment shown in Figure 16.1. A graph is represented in GraphML by a `<graph>` element. The `<graphml>` element can contain any number of `<graph>`s. The nodes of a graph are represented by a list of `<node>` elements. Each node must have an `id` attribute. The edge set is represented by a list of `<edge>` elements. Edges and nodes may be ordered arbitrarily and it is not required that all nodes are listed before all edges. Clearly, the space requirement for storing a graph with n nodes and m edges in GraphML is in $\mathcal{O}(n + m)$.

Edges point to source- and target-nodes by the values of their attributes `source` and `target`, respectively. It is ensured in the GraphML Schema specification that node-ids are unique within the enclosing `<graph>` and that the attribute values of the `source` and `target` attributes match the `id` of some `<node>` within the enclosing `<graph>`. The possibility of enforcing this constraint already in the definition of the GraphML language is one of the advantages of using XML schema instead of a DTD.

The `edgedefault` attribute of `<graph>` declares whether edges are understood as directed or undirected per default. Individual `<edge>`s can overwrite this default by setting the value of their `directed` attribute to `true` or `false`, respectively.

```
<?xml version="1.0" encoding="UTF-8"?>
<graphml>
  <key id="d0" for="node"
        attr.name="color" attr.type="string">
    <default>yellow</default>
  </key>
  <key id="d1" for="edge"
        attr.name="weight" attr.type="double"/>
  <graph id="G" edgedefault="undirected">
    <node id="n0">
      <data key="d0">green</data>
    </node>
    <node id="n1"/>
    <node id="n2">
      <data key="d0">blue</data>
    </node>
    <node id="n3">
      <data key="d0">red</data>
    </node>
    <node id="n4"/>
    <node id="n5">
      <data key="d0">turquoise</data>
    </node>
    <edge id="e0" source="n0" target="n2">
      <data key="d1">1.0</data>
    </edge>
    <edge id="e1" source="n0" target="n1">
      <data key="d1">1.0</data>
    </edge>
    <edge id="e2" source="n1" target="n3">
      <data key="d1">2.0</data>
    </edge>
    <edge id="e3" source="n3" target="n2"/>
    <edge id="e4" source="n2" target="n4"/>
    <edge id="e5" source="n3" target="n5"/>
    <edge id="e6" source="n5" target="n4">
      <data key="d1">1.1</data>
    </edge>
  </graph>
</graphml>
```

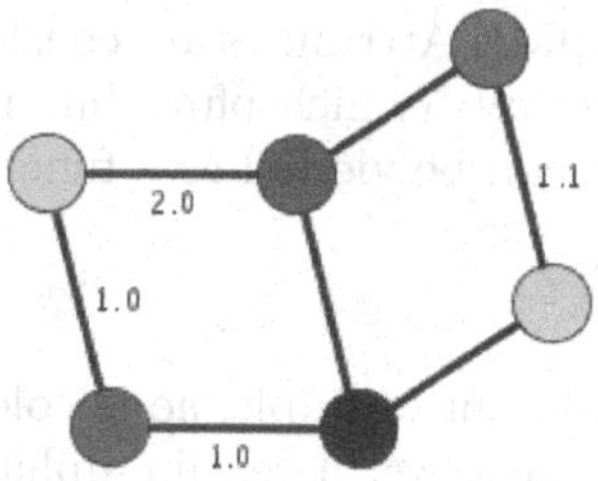

Figure 16.4 Graph with attributes. Edges have weights and nodes have colors. (For readability, the namespace declarations and schema location information has been left out.)

16.2.3 Attributes

In the previous section we discussed how to describe the topology of a graph in GraphML. While pure topological information may be sufficient for some applications of GraphML, for most of the time additional information is needed. With the help of the extension *GraphML-Attributes* one can specify additional information of simple type for the elements of the graph. Simple type means that the information is restricted to scalar values, e.g., numerical values and strings. The GraphML-Attributes extension is already included in the file

```
http://graphml.graphdrawing.org/xmlns/1.1/graphml.xsd
```

thus the header of the following example file may look like the one in Section 16.2.1. GraphML-Attributes must not be confused with XML-attributes which are a different concept (putting it in a simple way, GraphML-Attributes add information to graphs, sets of graphs, or parts of graphs and XML-attributes add information to XML elements).

In most cases, additional information can and should be attached to GraphML elements by usage of GraphML-Attributes as described in this section. This ensures readability for other GraphML parsers. If a custom data-format is necessary, then the GraphML language can be extended to include arbitrary data in well-defined places. How extensions can be defined is described in Section 16.4.

GraphML-Attributes are considered to be partial functions that assign values to elements of the graph (which often but not necessarily have the same type). For example edges weights can be viewed as a function from the set of edges E to the real numbers.

$$\text{weight:} \, E \to \mathbf{R}.$$

As a different example, node colors can be represented by a function from the set of nodes V to strings over a certain alphabet Σ.

$$\text{color:} \, V \to \Sigma^*.$$

To add data functions to graph elements, the GraphML *key/data* mechanism has to be used. A `<key>` element, at the beginning of the document, *declares* a new data function; more precisely, the `<key>` element specifies the function's id, name, domain, and range of values. The values of the function are *defined* by `<data>` elements.

The declaration of all data functions right at the beginning of the document has the benefit that parsers can build up appropriate data structures at the beginning of the parsing process. Likewise, parsers can recognize if some required data is missing. The GraphML document shown in Figure 16.4 is an example illustrating the key/data mechanism. The *weight* function is declared in the line

```
<key id="d1" for="edge" attr.name="weight" attr.type="double"/>
```

A `<key>` has an XML attribute called `for` that specifies the *domain* of the data function. The attribute `for` may assume values like `graph`, `node`, `edge`, `graphml` and names of other graph element types introduced later in Section 16.3. The XML attribute `for` may also assume the value `all` having the meaning that these data labels can be attached to all graph elements. The attribute `for` as well as a unique `id` are mandatory for `<key>` elements. The GraphML-Attributes extension provides two more attributes for `<key>`: the attribute `attr.name`, which defines the name of the data function and is used by parsers to recognize "their" data, and the attribute `attr.type`, which specifies the range of the data values.

Possible values for `attr.type` are `boolean`, `int`, `long`, `float`, `double`, and `string` having the obvious meaning.

A parser that handles edge weights will typically, after parsing the above line, initialize some internal data structure that stores doubles for each edge. Conversely, a parser that does not know or does not need a function for edges with the name "weight" will simply ignore the associated `<data>` elements. Values for the data functions are defined in `<data>` elements. For example, the code fragment

```
<edge id="e0" source="n0" target="n2">
  <data key="d1">1.0</data>
</edge>
```

defines a value of 1.0 as weight for the enclosing `<edge>`. The `<data>` elements point to `<key>`s by their `key` attribute. It is ensured in the GraphML schema that the value of the `key` attribute must match the `id` of some `<key>` element within the same document.

Since in general data labels are only partial functions, `<data>` elements need not be present for all edges. For example the edge

```
<edge id="e3" source="n3" target="n2"/>
```

does not define a value for the weight function. However, `<key>`s can define default values for the associated data function. For example

```
<key id="d0" for="node" attr.name="color" attr.type="string">
  <default>yellow</default>
</key>
```

declares a function named `color` on the set of nodes and defines `yellow` as the default node color. Thus, the node

```
<node id="n4"/>
```

is understood as being colored yellow. Nodes can overwrite the default by their `<data>` element. For instance, the node

```
<node id="n0">
  <data key="d0">green</data>
</node>
```

is colored green. The default mechanism serves to save space if many elements assume the same value.

16.2.4 Parseinfo

There is one more extension, called *GraphML-Parseinfo*, to the core structural part of GraphML. GraphML-Parseinfo makes it possible to write simple parsers that rely on additional information in the GraphML files. The GraphML-Parseinfo extension is already included in the file

http://graphml.graphdrawing.org/xmlns/1.1/graphml.xsd

thus the header of the example file in Figure 16.5 may look like the one in Section 16.2.1.

```xml
<?xml version="1.0" encoding="UTF-8"?>
<graphml>
  <graph id="G" edgedefault="directed"
         parse.nodes="11" parse.edges="12"
         parse.maxindegree="2"
         parse.maxoutdegree="3"
         parse.nodeids="canonical"
         parse.edgeids="free"
         parse.order="nodesfirst">
    <node id="n0" parse.indegree="0" parse.outdegree="1"/>
    <node id="n1" parse.indegree="0" parse.outdegree="1"/>
    <node id="n2" parse.indegree="2" parse.outdegree="1"/>
    <node id="n3" parse.indegree="1" parse.outdegree="2"/>
    <node id="n4" parse.indegree="1" parse.outdegree="1"/>
    <node id="n5" parse.indegree="2" parse.outdegree="1"/>
    <node id="n6" parse.indegree="1" parse.outdegree="2"/>
    <node id="n7" parse.indegree="2" parse.outdegree="0"/>
    <node id="n8" parse.indegree="1" parse.outdegree="3"/>
    <node id="n9" parse.indegree="1" parse.outdegree="0"/>
    <node id="n10" parse.indegree="1" parse.outdegree="0"/>
    <edge id="edge0001" source="n0" target="n2"/>
    <edge id="edge0002" source="n1" target="n2"/>
    <edge id="edge0003" source="n2" target="n3"/>
    <edge id="edge0004" source="n3" target="n5"/>
    <edge id="edge0005" source="n3" target="n4"/>
    <edge id="edge0006" source="n4" target="n6"/>
    <edge id="edge0007" source="n6" target="n5"/>
    <edge id="edge0008" source="n5" target="n7"/>
    <edge id="edge0009" source="n6" target="n8"/>
    <edge id="edge0010" source="n8" target="n7"/>
    <edge id="edge0011" source="n8" target="n9"/>
    <edge id="edge0012" source="n8" target="n10"/>
  </graph>
</graphml>
```

Figure 16.5 Example demonstrating the use of *GraphML-Parseinfo* meta data.

To make it possible to implement optimized parsers for GraphML documents, meta-data can be attached as XML-attributes to some GraphML elements. There are two kinds of meta-data intended for parsers: information about the number of elements and information about how specific data is encoded in the document. For instance, a parser that stores nodes and incident edges in (non-extensible) arrays can profit from information about the number of nodes in the graph and the nodes' degrees, respectively. All XML-attributes denoting meta-data for parsers are prefixed with `parse`.

For the first kind, information about the number of elements, the following XML-attributes for the `<graph>` element are defined. The value of the attribute `parse.nodes` gives the number of `<node>`s in the `<graph>`. Likewise, the value of `parse.edges` gives the number of `<edge>`s, `parse.maxindegree` is for the maximum indegree of the all `<node>`s in the `<graph>`, and `parse.maxoutdegree` for the maximum outdegree. For `<node>` elements the value of the attribute `parse.indegree` gives the indegree and `parse.outdegree` the outdegree of `<node>`s, respectively.

For the second kind, information about element encoding, the following XML-attributes for the `<graph>` element are defined. If the attribute `parse.nodeids` has the value `canonical`, all `<node>`s have identifiers following the pattern `nX`, where `X` denotes the number of occurrences of `<node>` elements before the current element. Otherwise the value of `parse.nodeids` equals `free`. The same holds for `<edge>`s for which the corresponding XML-attribute `parse.edgeids` is defined, with the only difference that the identifiers of `<edge>`s follow the pattern `eX`. The XML-attribute `parse.order` of `<graph>` gives information about the order in which `<node>` and `<edge>` elements occur in the `<graph>`. If `parse.order` assumes the value `nodesfirst`, all `<node>` elements appear the first occurrence of an `<edge>`. If `parse.order` assumes the value `adjacencylist`, the declaration of a `<node>` is followed by the declaration of its adjacent `<edge>`s. If `parse.order` assumes the value `free`, no order is imposed. The example in Figure 16.5 demonstrates the use of parse info meta-data.

16.3 Advanced Concepts

In this section we discuss advanced topological features for graphs. The graph model from Section 16.2 is extended to include a *nesting hierarchy*, *hyperedges* and *ports*. Since many graph applications do not support these extended graph models, we describe at the end of each subsection the specified fall-back behavior.

The GraphML elements that are introduced in this section can be specified as the domain of data-functions, i.e., as the value of the `for` attributes of `<key>`s (compare Section 16.2.3).

16.3.1 Nested Graphs

GraphML supports *nested graphs*, i.e., graphs in which the nodes are hierarchically ordered. The hierarchy tree is encoded in the GraphML document tree. A `<node>` in a GraphML document may contain a `<graph>` element which itself contains the `<node>`s which are in the hierarchy below this `<node>`.

Figure 16.6 is an example of a document describing a nested graph. Note that in the drawing of the graph the hierarchy is expressed by containment, i.e., a node u is below a node v in the hierarchy if and only if the graphical representation of u is entirely inside the graphical representation of v.

The edges between two nodes in a nested graph have to be declared in a graph that is an ancestor of both nodes in the hierarchy. Note that this is true for our example. Declaring the edge between node `n3` and node `n2` inside graph `G1` would be wrong while declaring it

```
<graphml>
  <graph id="G0" edgedefault="undirected">
    <node id="n1">
      <graph id="G1" edgedefault="undirected">
        <node id="n3"/>
        <node id="n4"/>
        <node id="n5"/>
        <edge source="n3" target="n4"/>
        <edge source="n4" target="n5"/>
      </graph>
    </node>
    <node id="n2">
      <graph id="G2" edgedefault="undirected">
        <node id="n6"/>
      </graph>
    </node>
    <edge source="n1" target="n2"/>
    <edge source="n3" target="n2"/>
    <edge source="n3" target="n6"/>
  </graph>
</graphml>
```

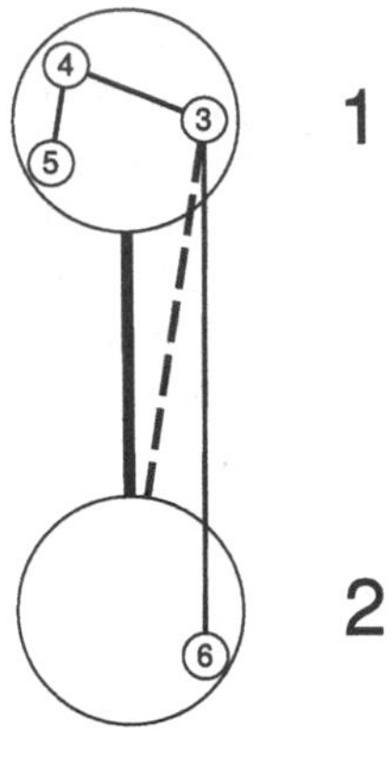

Figure 16.6   A nested graph.

in graph G0 is correct. A good policy is to place the edges at the least common ancestor of the nodes in the hierarchy.

The GraphML language includes an element called `<locator>` which makes it possible to define some of the document content in another file. More specifically, the elements `<graph>` and `<node>` can contain a `<locator>` element whose attribute `xlink:href` points to a file in which the content of this `<graph>`, respectively `<node>` is defined. If a particular `<graph>` or `<node>` element contains a `<locator>`, then this `<graph>`, respectively `<node>` does not contain any other element. For instance, the document fragment

```
<graph id="G0" edgedefault="undirected">
  <node id="n1">
    <graph id="G1" edgedefault="undirected">
      <locator xlink:href="content_of_G1.graphml"/>
    </graph>
  </node>
  ...
</graph>
```

(which is a modified version of the document in Figure 16.6) tells the parser that the content of the `<graph>` with `id="G1"` is defined in the file `content_of_G1.graphml`. Likewise, the content of `<node>`s can be outsourced to another file with the help of `<locator>` elements.

For applications that cannot handle nested graphs, the fall-back behavior is to ignore nodes that are not contained in the top-level graph and to ignore edges that do not have both endpoints in the top-level graph.

16.3.2 Hypergraphs

Hyperedges are a generalization of edges in the sense that they do not only relate two endpoints to each other but rather express a relation between an arbitrary number of endpoints. Hyperedges are declared by a `<hyperedge>` element in GraphML. For each endpoint of the hyperedge, this `<hyperedge>` element contains an `<endpoint>` element. The `<endpoint>` element must have an XML-attribute `node`, which contains the `id` of a `<node>` in the document. The example in Figure 16.7 contains two hyperedges and two edges. The hyperedges are illustrated by joining arcs, the edges by straight lines.

```
<?xml version="1.0" encoding="UTF-8"?>
<graphml>
  <graph id="G" edgedefault="undirected">
    <node id="n0"/>
    <node id="n1"/>
    <node id="n2"/>
    <node id="n3"/>
    <node id="n4"/>
    <node id="n5"/>
    <node id="n6"/>
    <hyperedge>
       <endpoint node="n0"/>
       <endpoint node="n1"/>
       <endpoint node="n2"/>
    </hyperedge>
    <hyperedge>
       <endpoint node="n3"/>
       <endpoint node="n4"/>
       <endpoint node="n5"/>
       <endpoint node="n6"/>
    </hyperedge>
    <hyperedge>
       <endpoint node="n1"/>
       <endpoint node="n3"/>
    </hyperedge>
    <edge source="n0" target="n4"/>
  </graph>
</graphml>
```

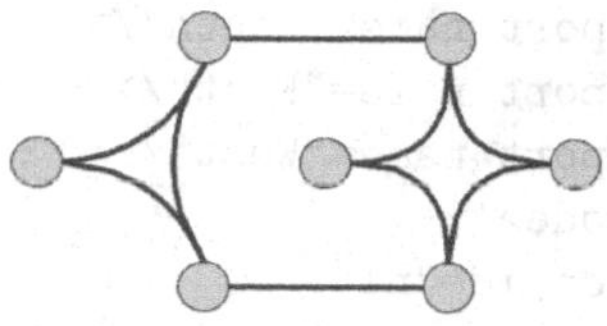

Figure 16.7 A hypergraph.

Note that edges can be either specified by an `<edge>` element or by a `<hyperedge>` element containing exactly two `<endpoint>` elements. Obviously, the latter option is only recommendable for applications that can handle hyperedges. The `<endpoint>` elements have an optional attribute called `type` which may assume the values `in`, `out`, and `undir` and is set to `undir` by default. The fall-back behavior for applications that cannot handle hyperedges is simply to ignore them.

16.3.3 Ports

A node may specify different logical locations for edges and hyperedges to connect. The logical locations are called *ports*. As an analogy, think of the graph as a motherboard, the nodes as integrated circuits and the edges as connecting wires. Then the pins on the integrated circuits correspond to ports of a node.

The ports of a node are declared by `<port>` elements as children of the corresponding `<node>` element. `<port>` elements may be nested, i.e., they may contain `<port>` elements themselves. Each `<port>` element must have an XML-attribute `name`, which is an identifier for this port. Port names are unique only within the enclosing `<node>` (see the example in Figure 16.8). The `<edge>` element has optional XML-attributes `sourceport` and `targetport` with which an edge may specify the port on the source, resp. target, node. Correspondingly, the `<endpoint>` element has an optional XML-attribute `port`. An example of a GraphML document with ports is shown in Figure 16.8. The fall-back behavior for applications that can not handle ports is simply to ignore them.

```xml
<?xml version="1.0" encoding="UTF-8"?>
<graphml>
  <graph id="G" edgedefault="directed">
    <node id="n0">
      <port name="North"/>
      <port name="South"/>
      <port name="East"/>
      <port name="West"/>
    </node>
    <node id="n1">
      <port name="North"/>
      <port name="South"/>
      <port name="East"/>
      <port name="West"/>
    </node>
    <node id="n2">
      <port name="NorthWest"/>
      <port name="SouthEast"/>
    </node>
    <node id="n3">
      <port name="NorthEast"/>
      <port name="SouthWest"/>
    </node>
    <edge source="n0" target="n3"
          sourceport="North" targetport="NorthEast"/>
    <hyperedge>
        <endpoint node="n0" port="North"/>
        <endpoint node="n1" port="East"/>
        <endpoint node="n2" port="SouthEast"/>
    </hyperedge>
  </graph>
</graphml>
```

Figure 16.8 Document of a graph with ports.

16.4 Extending GraphML

GraphML is designed to be easily extensible. With GraphML the topology of a graph and simple attributes of graph elements (see Section 16.2.3) can be serialized. To store more complex application data one has to extend GraphML which will be discussed in this section.

GraphML can be extended in two different ways: adding additional attributes to GraphML elements (discussed in Section 16.4.1) and extending the content of the `<data>` elements by allowing them to contain elements from other XML languages (discussed in Section 16.4.2).

Extensions of GraphML should be defined by an XML Schema (the other possibility, extending the DTD, is not described here). The Schema which defines the extension can be derived from the GraphML Schema documents by using a standard mechanism similar to the one used by XHTML.

```xml
<?xml version="1.0" encoding="UTF-8"?>
<xs:schema
    targetNamespace="http://graphml.graphdrawing.org/xmlns"
    xmlns="http://graphml.graphdrawing.org/xmlns"
    xmlns:xlink="http://www.w3.org/1999/xlink"
    xmlns:xs="http://www.w3.org/2001/XMLSchema"
    elementFormDefault="qualified"
    attributeFormDefault="unqualified">

<xs:import namespace="http://www.w3.org/1999/xlink"
           schemaLocation="xlink.xsd"/>

<xs:redefine
   schemaLocation="http://graphml.graphdrawing.org/xmlns/1.1/graphml.xsd">
   <xs:attributeGroup name="node.extra.attrib">
     <xs:attributeGroup ref="node.extra.attrib"/>
     <xs:attribute ref="xlink:href" use="optional"/>
   </xs:attributeGroup>
</xs:redefine>

</xs:schema>
```

Figure 16.9 File `graphml+xlink.xsd` : an XML Schema Definition that extends the GraphML language by adding attribute `xlink:href` to element `<node>`.

16.4.1 Adding XML-Attributes

In most cases, additional information can and should be attached to GraphML elements by usage of GraphML-Attributes (see Section 16.2.3). This assures readability for other GraphML parsers. However, sometimes it might be more convenient to use specific XML attributes. Suppose a graph whose nodes model WWW pages should be stored in GraphML. A node could then point to the associated page by storing the URL in an `xlink:href` attribute within the `<node>` element:

```
<node id="n0" xlink:href="http://graphml.graphdrawing.org"/>
```

The string `http://graphml.graphdrawing.org` could as well be stored within a `<data>` element contained in the node n0. However, when storing this string as the value of the `xlink:href` attribute, then its semantic (being a URL) becomes more obvious.

The element `<node>` as written above would not be valid for the core GraphML, since there is no `xlink:href` attribute defined for `<node>`. To add XML attributes to GraphML elements one has to extend GraphML. This extension can be defined by an XML Schema. The document in Figure 16.9 is an *XML Schema Definition* that extends the GraphML language by adding the `xlink:href` attribute to `<node>`.

The document in Figure 16.9 has a `<schema>` element as its root element (every XML Schema Definition does so). The element `<schema>` has a couple of attributes:

```
targetNamespace="http://graphml.graphdrawing.org/xmlns"
```

specifies that the language defined by this document is GraphML. The next three lines specify the default namespace (identified by the GraphML URL) and the namespace prefixes for XLink and XMLSchema. The attributes `elementFormDefault` and `attributeFormDefault` are of no importance for this example.

The import instruction

```
<xs:import namespace="http://www.w3.org/1999/xlink"
           schemaLocation="xlink.xsd"/>
```

gives access to the XLink namespace (assumed that the Schema Definition for XLink is located at the file `xlink.xsd`).

The extension is done in the `<redefine>` element. The attribute

```
    schemaLocation="http://graphml.graphdrawing.org/xmlns/1.1/graphml.xsd"
```

of `<redefine>` specifies the file (part of) which is being redefined. The document fragment

```
    <xs:attributeGroup name="node.extra.attrib">
      <xs:attributeGroup ref="node.extra.attrib"/>
      <xs:attribute ref="xlink:href" use="optional"/>
    </xs:attributeGroup>
```

extends the *attribute group* called `node.extra.attrib` which (by the core GraphML specification) is an empty set, but included in the attribute-list of the element `<node>`. After redefinition, this attribute group has its old content plus one more attribute, namely `xlink:href`. This attribute is declared as being optional for `<node>`. It is a good policy to always add the old content to the newly defined attribute groups, as there might be more than one Schema definitions extending the same attribute group.

As there is the attribute group `node.extra.attrib` for the element `<node>`, there are corresponding attribute groups for all GraphML elements. These attribute groups are empty in the core GraphML definition but can be extended as illustrated above.

The schema `graphml+xlink.xsd` can be used to validate the document shown in Figure 16.10.

Storing additional information directly in the attributes of GraphML elements, as illustrated in this section, may seem to be preferable to storing them within a `<data>` element, as explained in Section 16.2.3 (at least it can be observed that less characters are necessary). However, such a user-specified extension comes at a price: since these non-standard attributes are not declared by `<key>` elements, GraphML parsers might not be able to handle them.

```
<?xml version="1.0" encoding="UTF-8"?>
<graphml xmlns="http://graphml.graphdrawing.org/xmlns"
            xmlns:xlink="http://www.w3.org/1999/xlink"
            xmlns:xsi="http://www.w3.org/2001/XMLSchema-instance"
            xsi:schemaLocation="http://graphml.graphdrawing.org/xmlns
                                graphml+xlink.xsd">
  <graph edgedefault="directed">
    <node id="n0" xlink:href="http://graphml.graphdrawing.org"/>
    <node id="n1" />
    <edge source="n0" target="n1"/>
  </graph>
</graphml>
```

Figure 16.10 A document that can be validated with the XSD shown in Figure 16.9. Note that the `schemaLocation` attribute of `<graphml>` points to the file `graphml+xlink.xsd`.

16.4.2   Adding Structured Content

In some cases it might be convenient to use other XML languages to represent data in GraphML. For example a user wants to store images for nodes, written in SVG, as in the following document fragment.

```
    ...
xmlns:svg="http://www.w3.org/2000/svg"
    ...
<node id="n0" >
  <data key="k0">
    <svg:svg width="4cm" height="8cm" version="1.1">
      <svg:ellipse cx="2cm" cy="4cm" rx="2cm" ry="1cm" />
    </svg:svg>
  </data>
</node>
    ...
```

The attributes of `<svg>` and `<ellipse>` could also be stored in data functions as described in Section 16.2.3. However, the representation above is much more convenient, since applications can use existing parsers or viewers for SVG images.

GraphML can be extended to validate such a document. Arbitrary elements can be added to the content of `<data>`—but only to `<data>`—while the core GraphML cannot be changed. This decision has been made to ensure that parsers can understand at least the structural part and ignore possibly unknown content of `<data>`.

Figure 16.11 shows the XML Schema Definition that adds SVG elements to the content of `<data>`.

The schema in Figure 16.11 is similar to the one in Figure 16.9. First the namespace declarations are made. Then the SVG namespace is imported. As before, the extension is done in the `<redefine>` element. Within this element the *complex type* `data-extension.type` is extended by the SVG element `<svg>`. `data-extension.type` is the *base-type* for the content of the elements `<data>` and `<default>`. This type has empty content in the core GraphML definition, but can be extended by arbitrary XML elements.

Documents that are validated against the Schema in Figure 16.11 can thus have `<data>` elements that contain `<svg>`. An example is shown in Figure 16.12. The node with id

```xml
<?xml version="1.0" encoding="UTF-8"?>
<xs:schema
    targetNamespace="http://graphml.graphdrawing.org/xmlns"
    xmlns="http://graphml.graphdrawing.org/xmlns"
    xmlns:svg="http://www.w3.org/2000/svg"
    xmlns:xs="http://www.w3.org/2001/XMLSchema"
    elementFormDefault="qualified"
    attributeFormDefault="unqualified"
>

<xs:import namespace="http://www.w3.org/2000/svg"
           schemaLocation="svg.xsd"/>

<xs:redefine
    schemaLocation="http://graphml.graphdrawing.org/xmlns/1.1/graphml.xsd">
  <xs:complexType name="data-extension.type">
    <xs:complexContent>
      <xs:extension base="data-extension.type">
        <xs:sequence>
          <xs:element ref="svg:svg"/>
        </xs:sequence>
      </xs:extension>
    </xs:complexContent>
  </xs:complexType>
</xs:redefine>

</xs:schema>
```

Figure 16.11 File `graphml+svg.xsd` : an XML Schema Definition that extends the GraphML language by adding element `<svg:svg>` to the content of `<data>`.

```xml
<?xml version="1.0" encoding="UTF-8"?>
<graphml xmlns="http://graphml.graphdrawing.org/xmlns"
         xmlns:svg="http://www.w3.org/2000/svg"
         xmlns:xsi="http://www.w3.org/2001/XMLSchema-instance"
         xsi:schemaLocation="http://graphml.graphdrawing.org/xmlns
                             graphml+svg.xsd">
  <key id="k0" for="node">
    <default>
      <svg:svg width="5cm" height="4cm" version="1.1">
        <svg:desc>Default graphical representation for nodes
        </svg:desc>
        <svg:rect x="0.5cm" y="0.5cm" width="2cm" height="1cm"/>
      </svg:svg>
    </default>
  </key>
  <key id="k1" for="edge">
    <desc>Graphical representation for edges
    </desc>
  </key>
  <graph edgedefault="directed">
    <node id="n0">
      <data key="k0">
        <svg:svg width="4cm" height="8cm" version="1.1">
          <svg:ellipse cx="2cm" cy="4cm" rx="2cm" ry="1cm" />
        </svg:svg>
      </data>
    </node>
    <node id="n1" />
    <edge source="n0" target="n1">
      <data key="k1">
        <svg:svg width="12cm" height="4cm" viewBox="0 0 1200 400">
          <svg:line x1="100" y1="300" x2="300" y2="100"
          stroke-width="5"  />
        </svg:svg>
      </data>
    </edge>
  </graph>
</graphml>
```

Figure 16.12 A document that can be validated with the XSD shown in Figure 16.11. Note that the schemaLocation attribute of <graphml> points to graphml+svg.xsd.

n1 admits the default graphical representation given within key k0. The above example shows also the usefulness of XML Namespaces. There are two different <desc> elements, one in the GraphML namespace and one in the SVG namespace. Possible conflicts, due to elements from different XML languages that happen to have identical names, are resolved by different namespaces.

We note that it is not only possible to use other XML languages (like SVG) within GraphML. GraphML can also be used to represent graph data within extensible XML languages like SVG or XHTML. The possibility to combine modularly built XML languages ensures the reusability of parsers and other software. For example, SVG viewers could call graphdrawing software to layout graphs that are stored in GraphML within an SVG file.

16.5   Transforming GraphML

It is straightforward to provide access to graphs represented in GraphML by adding input and output filters to an existing software application. However, we find that *Extensible Stylesheet Language Transformations (XSLT)* [W3Cc] offer a more natural way of exploiting XML data, in particular when the resulting format of a computation is again based on XML. The mappings that transform input GraphML documents to output documents are defined in XSLT style sheets and can be used stand-alone, as components of larger systems, or in, say, web services [BP04].

Basically, the transformations are defined in style sheets (sometimes also called transformation sheets), which specify how an input XML document gets transformed into an output XML document in a recursive pattern matching process. The underlying data model for XML documents is the Document Object Model (DOM), a tree of DOM nodes representing the elements, attributes, text etc., which is held completely in memory. Figure 16.13 shows the basic workflow of a transformation.

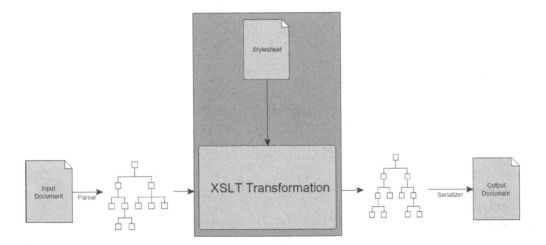

Figure 16.13 Workflow of an XSLT transformation. First, XML data is converted to a tree representation, which is then used to build the result tree as specified in the style sheet. Eventually, the result tree is serialized as XML. Taken from [BP04].

DOM trees can be navigated with the *XPath* language, a sublanguage of XSLT: It expresses paths in the document tree seen from a particular context node (similar to a directory tree of a file system) and serves to address sets of its nodes that satisfy given conditions. For example, if the context node is a `<graph>` element, all node identifiers can be addressed by `child::node/attribute::id`, or `node/@id` as shorthand. Predicates can be used to specify more precisely which parts of the DOM tree to select; for example, the XPath expression `edge[@source='n0']/data` selects only those `<data>` children of `<edge>`s starting from the `<node>` with the given identifier.

The transformation process can be roughly described as follows: A style sheet consists of a list of templates, each having an associated pattern and a template body containing the actions to be executed and the content to be written to the output. Beginning with the root, the processor performs a depth-first traversal (in document order) through the DOM tree. For each DOM node it encounters, it checks whether there is a template whose pattern it satisfies; if so, it selects one of the templates and executes the actions given in that template body (potentially with further recursive pattern matching for the subtrees), and does not do any further depth-first traversal for the DOM subtree rooted at that DOM node; else, it automatically continues the depth-first traversal recursively at each of its children. See Figure 16.14 for an example of an XSLT transformation sheet.

16.5.1 Means of Transformation

The expressivity and usefulness of XSLT transformations goes beyond their original purpose of adding some style to the input. The following is an overview of some important basic concepts of XSLT and how these concepts can particularly be employed in order to formulate advanced GraphML transformations that also take into account the underlying combinatorial structure of the graph instead of only the DOM tree.

16.5.2 Transformation Types

Since GraphML is designed as a general format not bound to a particular area of application, an abundance of XSLT use cases exist. However, we found that transformations can be filed into three major categories, depending on the actual purpose of transformation. Note that there may of course be transformations that belong to more than one of these categories.

Internal While one of GraphML's design goals is to require a well-defined interpretation for all GraphML files, there is no uniqueness the other way round, i.e., there are various GraphML representations for a graph; for example, its `<node>`s and `<edge>`s may appear in arbitrary order. However, applications may require their GraphML input to satisfy certain preconditions, such as the appearance of all `<node>`s before any `<edge>` in order to set up a graph in memory on-the-fly while reading the input stream.

Generally, some frequently arising transformations include

- pre- and postprocessing the GraphML file to make it satisfy given conditions, such as rearranging the markup elements or generating unique identifiers,
- inserting default values where there is no explicit entry, e.g., edge directions or default values for `<data>` tags,
- resolving XLink references in distributed graphs,
- filtering out unneeded `<data>` tags that are not relevant for further processing and can be dropped to reduce communication or memory cost, and
- converting between graph classes, for example eliminating hyperedges, expanding nested graphs, or removing multiedges.

```
<xsl:stylesheet version="2.0"
    xmlns:xsl="http://www.w3.org/1999/XSL/Transform">
  <xsl:output method="xml" indent="yes" encoding="iso-8859-1"/>

  <xsl:template match="data|desc|key|default"/> <!-- empty template-->

  <xsl:template match="/graphml">
    <graphml>
      <xsl:copy-of select="key|desc|@*"/>
      <xsl:apply-templates match="graph"/> <!-- process graph(s) -->
    </graphml>
  </xsl:template>

  <xsl:template match="graph">  <!-- override template -->
    <graph>
      <xsl:copy-of select="key|desc|@*"/>
      <xsl:copy-of select="node"/> <!-- nodes first -->
      <xsl:copy-of select="edge"/> <!-- then edges -->
    </graph>
  </xsl:template>
</xsl:stylesheet>
```

Figure 16.14 Example of an XSLT transformation sheet removing the elements `<data>`, `<desc>`, `<key>`, and `<default>` from the document and reorders nodes and edges such that all `<node>` elements appear before any `<edge>` element.

Format Conversion   Although in recent years, GraphML and similar formats like GXL [Win02] and GML [GML] have become increasingly used in various areas of interest, there are still many applications and services not (yet) capable of processing them. To be compatible, formats need to be translatable to each other, preserving as much information as possible.

In doing so, it is essential to take into account possible structural mismatch in terms of both the graph models and concepts that can be expressed by the involved formats, and their support for additional data. Of course, the closer the conceptual relatedness between source and target format is, the simpler the style sheets typically are.

While conversion will be necessary in various settings, two use cases appear to be of particular importance:

- *Conversion into another graph format:* We expect GraphML to be used in many applications to archive attributed graph data and in Web services to transmit aspects of a graph. While it is easy to output GraphML, style sheets can be used to convert GraphML into other graph formats [BLP05] and can thus be used in translation services like GraphEx [Bri04].
- *Export to some graphics format:* Of course, graph-based tools in general and graph drawing tools in particular will have to export graphs in graphics formats for visualization purposes.

The transformation need not be applied to a filed document, but can also be carried out in memory by applications that ought to be able to export in some target format. Note that, even though XSLT is typically used for mapping between XML documents, it can also be utilized to generate non-XML output.

Algorithmic Algorithmic style sheets appear in transformations which create fragments in the output document that do not directly correspond to fragments in the input document, i.e., when there is structure in the source document that is not explicit in the markup. This is typical for GraphML data: For example, it is not possible to determine whether a given `<graph>` contains cycles by just looking at the markup; some algorithm has to be applied to the represented graph.

To get a feel for the potential of algorithmic style sheets, we implemented some basic graph algorithms using XSLT, and with recursive templates, it proved powerful enough to formulate even more advanced algorithms. For example, a style sheet can be used to compute the distances from a single source to all other nodes or execute a layout algorithm, and then attach the results to `<node>`s in `<data>` labels.

16.5.3 Language Binding

We found that pure XSLT functionality is expressive enough to solve even more advanced GraphML related problems. However, it suffers from some general drawbacks:

- With growing problem complexity, the style sheets tend to become disproportionately verbose.
- Algorithms must be reformulated in terms of recursive templates, and there is no way to use existing implementations.
- Computations may perform poorly, especially for large input. This is often due to excessive DOM tree traversal and overhead generated by template instantiation internal to the XSLT processor.
- There is no direct way of accessing system services, such as date functions or database connectivity.

Therefore, most XSLT processors allow the integration of extension functions implemented in XSLT or some other programming language. Usually, they support at least their native language. For example, Saxon [Sax] can access and use external Java classes since itself is written entirely in Java. In this case, extension functions are methods of Java classes available on the class path when the transformation is being executed, and get invoked within XPath expressions. Usually, they are static methods, thus staying compliant with XSLT's design idea of declarative style and freeness of side effects. However, XSLT allows to create objects and to call their instance-level methods by binding the created objects to XPath variables.

The architecture shown in Figure 16.15 consists of three layers:

- The style sheet that instantiates the wrapper and communicates with it
- A wrapper class (the actual XSLT extension) that converts GraphML markup to a wrapped graph object, and provides computation results
- Java classes for graph data structures and algorithms

Thus, the wrapper acts as a mediator between the graph object and the style sheet. The wrapper instantiates a graph object corresponding to the GraphML markup, and, for instance, applies a graph drawing algorithm to it. In turn, it provides the resulting coordinates and other layout data in order for the style sheet to insert it into the XML (probably GraphML) result of the transformation, or to do further computations.

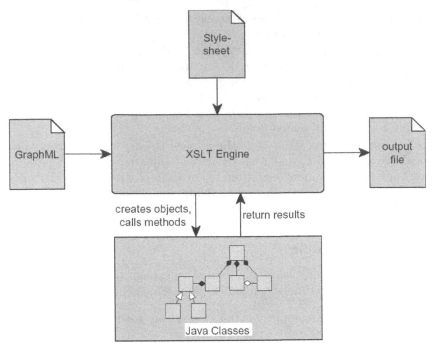

Figure 16.15   Using extension functions in XSLT. Taken from [BP04].

The approach presented here is only one of many ways of mapping an external graph description file to an internal graph representation. A stand-alone application could integrate a GraphML parser, build up its graph representation in memory apart from XSLT, execute a transformation, and serialize the result as GraphML output. However, the intrinsic advantage of using XSLT is that it generates output in a natural and embedded way, and that the output generation process can be customized easily.

XSL transformations are a simple, lightweight approach to processing graphs represented in GraphML. They have proven to be useful in various areas of application, when the target format of a transformation is GraphML again, or another format with a similar purpose, and the output structure does not vary too much from input.

They are even powerful enough to specify advanced transformations that go beyond mapping XML elements directly to other XML elements or other simple text units. However, advanced transformations may result in long-winded style sheets that are intricate to maintain, and most likely to be inefficient. Extension functions appear to be the natural way out of such difficulties.

We found that, as rule-of-thumb, XSLT should be used primarily to do the structural parts of a transformation, such as creating new elements or attributes, whereas specialized extensions are better for complex computations that are difficult to express or inefficient to run using pure XSLT.

16.6 Using GraphML

The easiest way to read and write GraphML files is to use a graph-processing software that can handle this format. GraphML is the principal I/O format of visone [BBB⁺02] and of the graph editor yEd from yWorks.[1] Besides these there are several software tools or libraries that can either import or export (or both) GraphML, including Pajek [DMB05], ORA [CR04], and JUNG [OFS⁺05]. If a customary GraphML reader has to be implemented it is convenient to make use of one of many available XML parsers and adapt it to the purpose at hand.

---

[1]http://www.yworks.com/

References

[BBB+02]   Michael Baur, Marc Benkert, Ulrik Brandes, Sabine Cornelsen, Marco
           Gaertler, Boris Köpf, Jürgen Lerner, and Dorothea Wagner. visone –
           software for visual social network analysis. In *Proc. 9th Intl. Symp. Graph
           Drawing (GD '01)*, pages 463–464, 2002.

[BEF99]   Stephen P. Borgatti, Martin G. Everett, and Linton C. Freeman. *UCINET
           6.0.* Analytic Technologies, 1999.

[BLP05]   Ulrik Brandes, Jürgen Lerner, and Christian Pich. GXL to GraphML and
           vice versa with XSLT. *Electronic Notes in Theoretical Computer Science*,
           127(1):113–125, 2005.

[BMN01]   Ulrik Brandes, M. Scott Marshall, and Stephen C. North. Graph data
           format workshop report. In Joe Marks, editor, *Proceedings of the 8th
           International Symposium on Graph Drawing (GD 2000)*, volume 1984 of
           Lecture Notes in Computer Science, pages 410–418. Springer, 2001.

[BP04]   Ulrik Brandes and Christian Pich. GraphML transformation. In János
           Pach, editor, *Proceedings of the 11th International Symposium on Graph
           Drawing (GD '04)*, volume 3383 of *Lecture Notes in Computer Science*,
           pages 89–99. Springer, 2004.

[Bri04]   Stina Bridgeman. GraphEx: An improved graph translation service. In
           Giuseppe Liotta, editor, *Proceedings of the 11th International Symposium
           on Graph Drawing (GD '03)*, volume 2912 of *Lecture Notes in Computer
           Science*, pages 307–313. Springer, 2004.

[CR04]   Kathleen Carley and Jeffrey Reminga. ORA: Organization risk analyzer.
           Technical Report CMU-ISRI-04-106, Carnegie Mellon University, 2004.

[DBETT99] Giuseppe Di Battista, Peter Eades, Roberto Tamassia, and Ioannis G.
           Tollis. *Graph Drawing: Algorithms for the Visualization of Graphs.* Pren-
           tice Hall, 1999.

[DMB05]   Wouter De Nooy, Andrej Mrvar, and Vladimir Batagelj. *Exploratory
           social network analysis with Pajek.* Cambridge University Press, 2005.

[GML]   GML.   *The   Graph   Modeling   Language   File   Format.*
           http://www.infosun.fmi.uni-passau.de/Graphlet/GML/.

[OFS+05]   Joshua O'Madadhain, Danyel Fisher, Padhraic Smyth, Scott White, and
           Yan-Biao Boey. Analysis and visualization of network data using JUNG.
           Journal of Statistical Software, 2005.

[Sax]   Saxon   Open   Source   Project.   *Saxon   home   page.*
           http://saxon.sourceforge.net/.

[STT81]   Kozo Sugiyama, Shojiro Tagawa, and Mitsuhiko Toda. Methods for visual
           understanding of hierarchical system structures. *IEEE Transactions on
           Systems, Man and Cybernetics*, 11(2):109–125, February 1981.

[TRC03]   Max Tsvetovat, Jeffrey Reminga, and Kathleen Carley. DyNetML: In-
           terchange format for rich social network data. In *NAACSOS Conference,
           Pittsburgh, PA*, 2003.

[W3Ca]   W3C. *Scalable Vector Graphics.* http://www.w3.org/TR/SVG/.

[W3Cb]   W3C. *SOAP.* http://www.w3.org/TR/soap12-part0/.

[W3Cc]   W3C. *XSL Transformations.* http://www.w3.org/TR/xslt/.

[Win02]   Andreas Winter. Exchanging graphs with GXL. In Petra Mutzel, Michael Jünger, and Sebastian Leipert, editors, *Proceedings of the 9th International Symposium on Graph Drawing (GD '01)*, volume 2265 of *Lecture Notes in Computer Science*, pages 485–500. Springer, 2002.

17

The Open Graph Drawing Framework (OGDF)

Markus Chimani
Friedrich-Schiller-Universität Jena

Carsten Gutwenger
TU Dortmund

Michael Jünger
University of Cologne

Gunnar W. Klau
Centrum Wiskunde & Informatica

Karsten Klein
TU Dortmund

Petra Mutzel
TU Dortmund

17.1 Introduction

We present the Open Graph Drawing Framework (OGDF), a C++ library of algorithms and data structures for graph drawing. The ultimate goal of the OGDF is to help bridge the gap between theory and practice in the field of automatic graph drawing. The library offers a wide variety of algorithms and data structures, some of them requiring complex and involved implementations, e.g., algorithms for planarity testing and planarization, or data structures for graph decomposition. A substantial part of these algorithms and data structures are building blocks of graph drawing algorithms, and the OGDF aims at providing such functionality in a reusable form, thus also providing a powerful platform for implementing new algorithms.

The OGDF can be obtained from its website at:

http://www.ogdf.net

This website also provides further information like tutorials, examples, contact information, and links to related projects. The source code is available under the GNU General Public License (GPL v2 and v3).

17.1.1   The History of the OGDF

Back in 1996, the development of the AGD library [AGMN97] (Algorithms for Graph Drawing) started at the Max-Planck Institute for Computer Science in Saarbrücken, Germany, originating from the DFG-funded project *Design, Analysis, Implementation, and Evaluation of Graph Drawing Algorithms*. The project later moved to Vienna University of Technology. AGD was designed as a C++ library of algorithms for graph drawing, based on the LEDA library [MN99] of efficient data structures and algorithms.

In 1999, a new branch of the library was developed as an internal project at the oreas GmbH and the research center caesar. The main goal was to have a code basis that could be built independently of any other libraries. This resulted in a new design and complete rewrite by starting from scratch, thereby concentrating on the strengths of the library, i.e., planarity and orthogonal layout, and implementing a wealth of required basic data structures and algorithms.

Later on, this internal project was renamed to OGDF and made open source under the GPL. The OGDF is currently maintained and further developed by researchers at the Universities of Dortmund, Cologne, and Jena.

17.1.2   Outline

After introducing the major design concepts and goals in Section 17.2, we dedicate two sections to the algorithms and data structures contained in the library. Section 17.3 introduces general graph algorithms and related data structures, and Section 17.4 focuses on drawing algorithms and layout styles. Finally, we conclude this chapter with selected success stories in Section 17.5.

17.2   Major Design Concepts

Many sophisticated graph drawing algorithms build upon complex data structures and algorithms, thus making new implementations from scratch cumbersome and time-consuming. Obviously, graph drawing libraries can ease the implementation of new algorithms a lot. E.g., the AGD library was very popular in the past, since it covered a wide range of graph drawing algorithms and—together with the LEDA library—data structures. However, the lack of publicly available source-code restricted the portability and extendability, not to mention the understanding of the particular implementations. Other currently available graph drawing libraries suffer from similar problems, or are even only commercially available or limited to some particular graph layout methods.

Our goals for the OGDF were to transfer essential design concepts of AGD and to overcome AGD's main deficiencies for use in academic research. Our main design concepts and goals are the following:

- Provide a wide range of graph drawing algorithms that allow a user to reuse and replace particular algorithm phases by using a dedicated *module mechanism*.
- Include *sophisticated data structures* that are commonly used in graph drawing, equipped with rich public interfaces.
- A *self-contained* source code that does not require additional libraries (except for some optional LP-/ILP-based algorithms).
- *Portable* C++-code that supports the most important compilers for the major operating systems (Linux, MacOS, and Windows) and that is available under an *open source license* (GPL).

17.2.1 Modularization

In the OGDF, an algorithm (e.g., a graph drawing algorithm or an algorithm that can be used as building block for graph drawing algorithms) is represented as a class derived from a base class defining its interface. Such algorithm classes are also called *modules* and their base classes *module types*. E.g., general graph layout algorithms are derived from the module type `LayoutModule`, which defines as interface a `call` method whose parameters provide all the relevant information for the layout algorithm: the graph structure (`Graph`) and its graphical representation like node sizes and coordinates (`GraphAttributes`).[1] The algorithm then obtains this information and stores the computed layout in the `GraphAttributes`.

Using common interface classes for algorithms allows us to make algorithms exchangeable. We can write an implementation that utilizes several modules, but each module is used only through the interface defined by its module type. Then, we can exchange a module by a different module implementing the same module type. The OGDF provides a mechanism called module options that even makes it possible to exchange modules at runtime. Suppose an algorithm A defines a module option M of a certain type T representing a particular phase of the algorithm, and adds a set-method for this option. A *module option* is simply a pointer to an instance of type T, which is set to a useful default value in A's constructor and called for executing this particular phase of the algorithm. Using the set-method, this implementation can be changed to *any* implementation implementing the module type T, even new implementations not contained in the OGDF itself.

Module options are the key concept for modularizing *algorithm frameworks*, thus allowing users to experiment with different implementations for particular phases of the algorithm, or to evaluate new implementations for phases without having to implement the whole framework from scratch. Figure 17.1 shows how module options are used in `Sugiyama-Layout`. In this case, `SugiyamaLayout` is a framework with three customizable phases (ranking, 2-layer crossing minimization, and layout), and the constructor takes care of setting useful initial implementations for each phase. Using a different implementation of, e.g., the crossing minimization step is simple:

```
SugiyamaLayout sugi;
sugi.setCrossMin(new MedianHeuristic);
```

In Section 17.4, we will illustrate the main drawing frameworks available in the OGDF using class diagrams, thereby showing the interconnections between the various classes in the OGDF.

17.2.2 Self-Contained and Portable Source Code

It was important for us to create a library that runs on all important systems, and whose core part can be built without installing any further libraries. Therefore, all required basic data structures are contained in the library, and only a few modules based on linear programming require additional libraries: COIN-OR [Mar10] as LP-solver and ABACUS [JT00] as branch-and-cut framework.

For reasons of portability and generality, the library provides only the drawing algorithms themselves and not any graphical display elements. Such graphical display would force us

---

[1] More precisely, the `call` method has only one parameter, the `GraphAttributes`, which allows us to get a reference to the `Graph` itself.

```
class SugiyamaLayout : public LayoutModule
{
protected:
    ModuleOption<RankingModule>        m_ranking;
    ModuleOption<TwoLayerCrossMin>     m_crossMin;
    ModuleOption<HierarchyLayoutModule> m_layout;
    ...
public:
    SugiyamaLayout() {
        m_ranking .set(new LongestPathRanking);
        m_crossMin.set(new BarycenterHeuristic);
        m_layout  .set(new FastHierarchyLayout);
        ...
    }
    void setRanking(RankingModule *pRanking)    { m_ranking.set(pRanking); }
    void setCrossMin(TwoLayerCrossMin *pCrossMin) { m_crossMin.set(pCrossMin); }
    void setLayout(HierarchyLayoutModule *pLayout) { m_layout.set(pLayout); }
    ...
};
```

Figure 17.1 Excerpt from the declaration of `SugiyamaLayout` demonstrating the use of module options.

to use very system-dependent GUI or drawing frameworks, or to have the whole library based on some cross-platform toolkit. Instead of this, the OGDF simply computes basic layout information like coordinates of nodes or bend points, and an application that uses the OGDF can create the required graphical display by using the GUI framework of its choice.

For creating graphics in common image formats, the OGDF project provides the command line utility `gml2pic`.[2] This utility converts graph layouts stored in GML or OGML file formats into images in PNG, JPEG, TIFF, SVG, EPS, or PDF format. We recommend to use the new OGML (*Open Graph Markup Language*) file format, since it offers a wide range of clearly specified formatting options. Hence, it is easy to save graph layouts in OGML format using the OGDF, and then apply `gml2pic` for creating high-quality graphics. All graph layouts in this chapter have been created with `gml2pic`. Figures 17.9 and 17.11, e.g., demonstrate the automatic creation of Bézier curves from ordinary polylines.

17.3 General Algorithms and Data Structures

The OGDF contains many basic data structures like arrays, lists, hashing tables, and priority queues, as well as fundamental data structures for the representation of graphs (`Graph`, `ClusterGraph`, and associative arrays for nodes, edges, etc.). Many basic graph algorithms can be found in `basic/simple_graph_alg.h`, e.g., functions dealing with parallel edges, connectivity, biconnectivity, and acyclicity. In this section, we focus on the more sophisticated algorithms and data structures in the OGDF.

---

[2]available at http://www.ogdf.net/doku.php/project:gml2pic

17.3.1 Augmentation and Subgraph Algorithms

Augmentation Algorithms. Several augmentation modules are currently available in the library for adding edges to a graph to achieve biconnectivity. This can be done either by disregarding the planarity of the graph or by taking care not to introduce non-planar subgraphs.

Augmenting a planar graph to a planar biconnected graph by adding the minimum number of edges is an NP-hard optimization problem. It has been introduced by Kant and Bodlaender [KB91], who also presented a simple 2-approximation algorithm for the problem. They also claimed to have a 3/2-approximation algorithm, but Fialko and Mutzel [FM98] have shown that this algorithm is erroneous and cannot be corrected. However, their suggested 5/3-approximation algorithm was shown to approximate the optimal solution by a factor of 2, only (see [GMZ09a]). Experiments show that the Fialko-Mutzel algorithm performs very good in practice. The module `PlanarAugmentation` implements the Fialko-Mutzel algorithm, which proceeds roughly as follows. The biconnected components of a graph induce a so-called *block tree* whose nodes are the cut vertices and blocks of the graph. The algorithm first constructs a block tree T from the given graph and then iteratively adds edges between blocks of degree one in T. Experiments on a set of benchmark graphs have shown that in about 96% of all the cases the approximation algorithm finds the optimal solution to the planar augmentation problem [FM98].

In addition, the OGDF contains the module `DfsMakeBiconnected`. The underlying algorithm uses depth-first search and adds a new edge whenever a cut vertex is discovered. If the input graph is planar, the augmented graph also remains planar. However, in general, this approach adds a significantly higher number of edges than the `PlanarAugmentation` module.

A special variant of the planar augmentation problem is solved by the `PlanarAugmentationFix` module. Here, a planar graph with a fixed planar embedding is given, and this embedding shall be extended such that the graph becomes biconnected. `PlanarAugmentationFix` implements the optimal, linear-time algorithm by Gutwenger, Mutzel, and Zey [GMZ09b].

Acyclic Subgraphs. Two modules are available to compute acyclic subgraphs of a digraph $G = (V, A)$. These modules determine a feedback arc set $F \subset A$ of G, i.e., if G contains no self-loops, an acyclic digraph is obtained by reversing all the arcs in F. `DfsAcyclicSubgraph` computes an acyclic subgraph in linear time by removing all back arcs in a depth-first-search tree of G. On the other hand, `GreedyCycleRemoval` implements the linear-time greedy algorithm by Eades and Lin [EL95]. If G is connected and has no two-cycles, the algorithm guarantees that the number of non-feedback arcs is at least $|A|/2 - |V|/6$.

The OGDF provides further modules for the computation of planar subgraphs. These are covered in the context of graph planarization; see Section 17.3.3.

17.3.2 Graph Decomposition

Besides the basic algorithm for computing the biconnected components of a graph [Tar72, HT73b] (function `biconnectedComponents` in `basic/simple_gaph_alg.h`), the OGDF provides further powerful data structures for graph decomposition. `BCTree` represents the decomposition of a graph into its biconnected components as a BC-tree and `StaticSPQRTree` represents the decomposition of a biconnected graph into its triconnected components as an SPQR-tree [DT89, DT96]. An *SPQR-tree* is a tree whose nodes are associated with the triconnected components (called the *skeletons* of the tree nodes) of the graph: *S-nodes* correspond to serial structures, *P-nodes* to parallel structures, and *R-nodes* to simple, triconnected structures; *Q-nodes* simply correspond to the edges in the graph and are hence not required by an implementation. Both data structures can be build in linear time; the latter constructs the SPQR-tree by applying the corrected version [GM01] of Hopcroft and Tarjan's algorithm [HT73a] for decomposing a graph into its triconnected components. The OGDF is one of the few places where one can find a correct implementation of this complex and involved algorithm (to the best of our knowledge, AGD was the first library providing such an implementation).

BC- and SPQR-trees are important data structures for many graph algorithms dealing with planar graphs, since they efficiently encode all planar embeddings of a planar graph. Notice that a planar graph might have exponentially many planar embeddings. The embeddings of the skeleton graphs of an SPQR-tree induce a unique embedding of the original graph. `StaticPlanarSPQRTree` is a specialized version of `StaticSPQRTree` with additional support for planar graphs. It provides basic operations for changing the currently represented embedding of the graph, like flipping the skeleton of an R-node and permuting the order of the edges in the skeleton of a P-node, and a method for embedding the graph according to the embeddings of the skeletons.

In addition to the static versions of BC- and SPQR-trees, the OGDF also contains efficient implementations of dynamic BC- and SPQR-trees. The supported update operations are insertion of nodes and edges. `DynamicBCTree` implements the update operations as described by Westbrook and Tarjan [WT92], and `DynamicSPQRTree` as described by Di Battista and Tamassia [DT96].

17.3.3   Planarity and Planarization

The OGDF provides a unique collection of algorithms for planar graphs, including algorithms for planarity testing and planar embedding, computation of planar subgraphs, and edge reinsertion. These algorithms can be combined using the planarization approach, yielding excellent crossing minimization heuristics. The planarization approach for crossing minimization is realized by the module `SubgraphPlanarizer`, and the two layout algorithms `PlanarizationLayout` and `PlanarizationGridLayout` implement a complete framework for planarization and layout. Figure 17.2 gives an overview of the OGDF's planarization framework for graph layout, illustrating the interconnection between the modules involved; the various implementations for `EmbedderModule` are shown in Figure 17.3. An in-depth description of this framework can be found in [Gut10].

Planarity Testing and Embedding.   The OGDF provides two algorithms for planarity testing. `PlanarModule` implements the node-addition algorithm [LEC67, BL76, CNAO85]) based on PQ-trees, and `BoyerMyrvold` implements the edge-addition algorithm by Boyer and Myrvold [BM04] which is based on depth-first search. Both modules can also compute a planar embedding of the graph.

However, for many graphs it is highly beneficial for a graph layout algorithm not to use just any embedding but an embedding that optimizes certain criteria. The OGDF contains such embedding algorithms which optimize criteria like a large external face or a small *block-nesting depth* (which is a measure for the topological nesting of the biconnected components in the embedding). `EmbedderMinDepthPiTa` implements the algorithm by Pizzonia and Tamassia [PT00], which minimizes the block-nesting depth for fixed embeddings of the blocks. On the other hand, `EmbedderMinDepth` minimizes the block-nesting depth without any restrictions, and `EmbedderMaxFace` maximizes the size of the external face; these

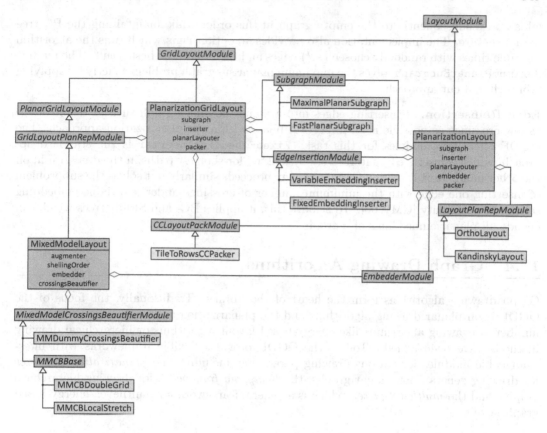

Figure 17.2 The Planarization framework for graph layout in the OGDF library.

two modules implement algorithms presented by Gutwenger and Mutzel [GM03]. Notice that just maximizing the external face still leaves a lot of freedom for embedding inner faces. Therefore, Kerkhof [Ker07] developed an extension of `EmbedderMaxFace`, realized by `EmbedderMaxFaceLayers`, which considers the *layers* of the embedding and the sizes of their *boundaries*. Here, layer i is formed by the faces with distance i to the external face in the dual graph, and the boundary B_i of layer i is roughly given by the edges shared by layers i and $i + 1$. Then, the algorithm computes an embedding such that $|B_0|, |B_1|, \ldots$ is lexicographically maximal. There are also combinations of these algorithms, realized by `EmbedderMinDepthMaxFace` and `EmbedderMinDepthMaxFaceLayers`.

Upward Planarity Testing. Although the general upward planarity testing problem is NP-complete [GT01], the problem can be solved efficiently for digraphs with only a single source, also called sT-digraphs. The OGDF provides a linear-time implementation of the sophisticated algorithm by Bertolazzi et al. [BDMT98], which is based on decomposing the underlying undirected graph using SPQR-trees.

Planar Subgraphs. The module `FastPlanarSubgraph` computes a planar subgraph of an input graph G by deleting a set of edges using the PQ-tree data structure [JLM98]. The algorithm is similar to the one by Jayakumar et al. [JTS89] and is one of the best heuristics for the NP-hard *maximum planar subgraph problem* that asks for the smallest set of edges whose removal leads to a planar graph. The heuristic proceeds in a similar manner as PQ-tree-based planarity testing: First, it constructs an s-t-numbering and then adds nodes subsequently to the empty graph in this order while maintaining the PQ-tree data structure. The implementation also provides an option `runs` which runs the algorithm multiple times with randomly chosen (s, t)-edges and then takes the best result. The module `MaximumPlanarSubgraph` solves the maximum planar subgraph problem exactly by applying a branch-and-cut approach.

Edge Reinsertion. Reinserting edges into a planar auxiliary graph such as to introduce as few crossings as possible is an important phase within the planarization approach. The OGDF offers two modules for this task: `FixedEmbeddingInserter` is the standard approach, which reinserts edges iteratively by looking for shortest paths in the dual graph; on the other hand, `VariableEmbeddingInserter` proceeds similarly but solves the subproblem of inserting one edge with the minimum number of crossings under a variable embedding setting to optimality [GMW05]. To achieve this, it applies BC- and SPQR-trees which can encode all planar embeddings of a graph.

17.4   Graph Drawing Algorithms

Graph drawing algorithms form the heart of the library. Traditionally, the focus of the OGDF is on planar drawing algorithms and the planarization approach. However, a large number of drawing algorithms like energy-based layout algorithms or hierarchical drawing methods have been added. Today, the OGDF provides flexible frameworks with interchangeable modules for various drawing paradigms, including the *planarization approach* for drawing general, non-planar graphs, the *Sugiyama framework* for drawing hierarchical graphs, and the *multilevel-mixer*, which is a general framework for multilevel, energy-based graph layout.

17.4.1   Planar Drawing Algorithms

The planar layout algorithms can be divided into those that compute straight-line layouts and those that produce drawings with bends along the edges, in particular orthogonal layouts. Figure 17.3 gives an overview of the available layout algorithms and their module options; the orthogonal layouts (`OrthoLayout` and `KandinskyLayout`) are not covered in this figure, since they are only used within the planarization framework (see Figure 17.2).

Straight-Line Layouts. The class `PlanarStraightLayout` implements planar straight-line drawing algorithms based on a shelling (or canonical) order of the nodes. This order determines the order in which the nodes are placed by the algorithm. `PlanarStraight-Layout` provides a module option `shellingOrder` for selecting the shelling order used by the algorithm. There are two implementations in the OGDF: Using the `Triconnected-ShellingOrder` realizes the algorithm by Kant [Kan96], which draws triconnected planar graphs such that all internal faces are represented as convex polygons; using the `Bi-connectedShellingOrder` realizes the relaxed variant by Gutwenger and Mutzel [GM97] for biconnected graphs. To make the drawing algorithm more generally applicable, it provides the additional module option `augmenter` for setting an augmentation module that is called as a preprocessing step. This augmentation module must ensure that the graph has the required connectivity when computing the shelling order. In all cases, the algorithm guarantees to produce a drawing on a $(2n - 4) \times (n - 2)$ grid, where $n \geq 3$ is the number of nodes in the graph.

An improved version of `PlanarStraightLayout` is `PlanarDrawLayout`. It provides the same module options but implements a slightly modified drawing algorithm, which guaran-

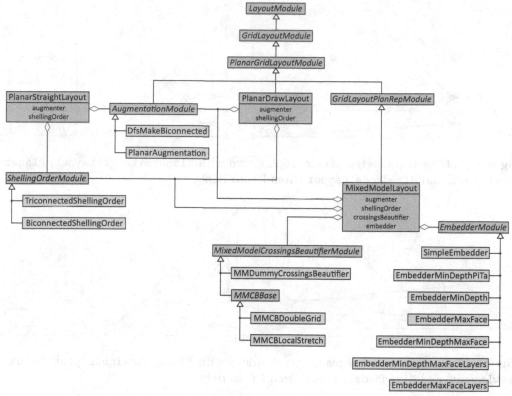

Figure 17.3 Planar graph drawing in the OGDF library.

tees a smaller grid size of $(n-2) \times (n-2)$. Some sample drawings of a tri- and a biconnected graph are shown in Figures 17.4 and 17.5.

Mixed-Model Layouts. In mixed-model layouts, each edge is drawn in an orthogonal fashion, except for a small area around its endpoints. The class `MixedModelLayout` represents the layout algorithm by Gutwenger and Mutzel [GM97], which is based upon ideas by Kant [Kan96]. In particular, Kant's algorithm has been changed concerning the placement phase and the node boxes, which determine the routing of the incident edges around a node. It has also been generalized to work for connected planar graphs.

This algorithm draws a d-planar graph G on a grid such that every edge has at most three bends and the minimum angle between two edges is at least $\frac{2}{d}$ radians. The grid size is at most $(2n - 6) \times \left(\frac{3}{2}n - \frac{7}{2}\right)$, the number of bends is at most $5n - 15$, and every edge has length $O(n)$ if G has n nodes.

Similar to the planar straight-line drawing algorithms, `MixedModelLayout` is based on a shelling order (`shellingOrder` module option) and an augmentation module is used to ensure the required connectivity. It also performs an enhancement for the placement of degree-one nodes, which are temporarily removed in a preprocessing step and later considered again when computing the node boxes. A further enhancement improves the drawing of edge crossings when using `MixedModelLayout` within the planarization approach (`PlanarizationGridLayout`). In this case, nodes representing crossings are drawn with four 90° angles, which is not the case for the original version. Figures 17.4 and 17.5 also show the corresponding mixed-model drawings of the graphs drawn with the planar straight-line methods.

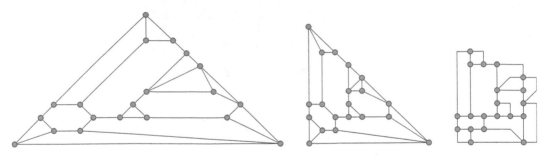

Figure 17.4   A triconnected planar graph drawn with `PlanarStraightLayout`, `Planar-DrawLayout`, and `MixedModelLayout` (from left to right).

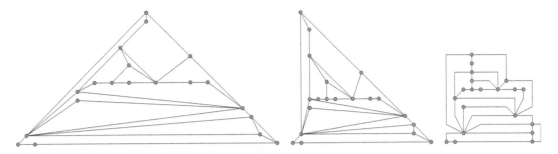

Figure 17.5   A biconnected planar graph drawn with `PlanarStraightLayout`, `Planar-DrawLayout`, and `MixedModelLayout` (from left to right).

Orthogonal Layouts. Orthogonal drawings represent edges as sequences of horizontal and vertical line segments. Bends occur where these segments change directions. The OGDF provides orthogonal layout algorithms for graphs without degree restrictions; these are embedded in the planarization approach realized by `PlanarizationLayout`. Thereby, the orthogonal layout algorithm receives as input a planarized representation of the possibly non-planar input graph, i.e., a planar graph in which some nodes represent edge crossings.

By default, `PlanarizationLayout` uses `OrthoLayout` as layout algorithm. This is a variation of Tamassia's bend minimizing algorithm [Tam87], generalized to work with graphs of arbitrary node degrees. Tamassia's algorithm requires a planar graph G of maximal node degree four and a planar embedding Γ of G. Notice that pure orthogonal drawings in which the nodes are mapped to points in the grid are only admissible for this class of planar graphs. The computation of the layout follows the so-called *topology-shape-metrics* approach, see, e.g., [DETT99]. According to the given planar embedding Γ the algorithm constructs a network in which a minimum-cost flow determines a bend-minimal representation of the orthogonal shape of G. In a last phase, a compaction module assigns lengths to this representation and thus fixes the coordinates of the drawing.

The OGDF contains two implementations for orthogonal compaction. `LongestPathCompaction` relies on computing longest paths in the so-called *constraint graphs*, an underlying pair of directed acyclic graphs that code placement relationships. `FlowCompaction` computes a minimum-cost flow in a pair of dual graphs and results in shorter edge lengths. Both algorithms rely on a dissection of the original face structures into rectangular faces. In addition, a branch-and-cut approach that produces provably optimal solutions for the two-dimensional compaction problem [KM99, Kla01] is in preparation.

In order to extend Tamassia's algorithm to graphs of arbitrary node degree, `OrthoLayout` uses ideas from quasi-orthogonal drawings [KM98] and Giotto layout [TDB88], combined

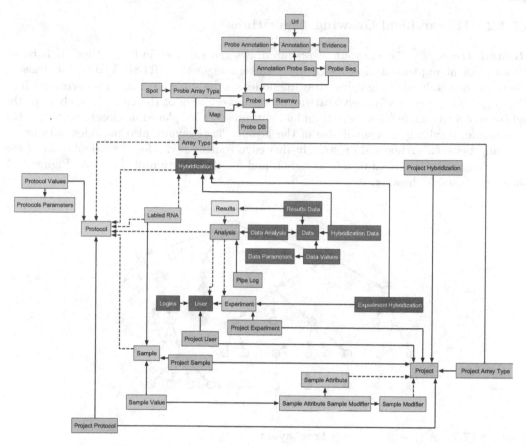

Figure 17.6 An entity-relationship diagram drawn with the planarization approach and OrthoLayout.

with a local orthogonal edge routing algorithm. The common idea is to replace high-degree nodes by artificial faces which will be drawn as larger boxes in an intermediate drawing. The node is then placed within this boxes and its incident edges are routed orthogonally to the corresponding connection points on the surrounding box. An ER-diagram drawn by using OrthoLayout with PlanarizationLayout is shown in Figure 17.6.

An alternative to OrthoLayout is KandinskyLayout which extends the basic approach to graphs of arbitrary degree by allowing 0° angles between two successive edges adjacent to a node. Nodes in a graph are modeled as square boxes of unified size placed on a coarse grid, whereas edges are routed on a finer grid. Feasibility is achieved by maintaining the so-called *bend-or-end* property: Let e_1 and e_2 be the two edges incident to the same side of a node v in a Kandinsky drawing, e_1 following e_2 in the given embedding, and let f be the face to which e_1 and e_2 are adjacent. Then either e_1 must have a last bend with a 270° angle in f or e_2 must have a first bend with 270° angle in f. See [FK96] for a detailed description of the Kandinsky drawing model. The KandinskyLayout implementation does not use an extension of the original bend-minimization flow network as described in [FK96] to compute a shape for the input graph since this network has a flaw that may lead to suboptimal solutions or not a feasible solution at all [Eig03]. Instead, an ILP formulation is used, and hence KandinskyLayout requires COIN-OR.

17.4.2   Hierarchical Drawing Algorithms

Rooted Trees. The `TreeLayout` algorithm draws general trees in linear time. It is based on an efficient implementation [BJL06] of Walker's algorithm [RT81, Wal90] for drawing trees. In the resulting straight-line drawing nodes on the same level lie on a horizontal line. The algorithm works recursively starting on the lowest level of the tree. In each step, the subtrees of a tree node (that have been laid out already) are placed as closely to each other as possible, resulting in a small size of the layout. `TreeLayout` also provides options for choosing between orthogonal or straight-line edge routing style, for the orientation of the layout (e.g., top to bottom or left to right), and for the selection of the root. Figure 17.7 shows an example drawing.

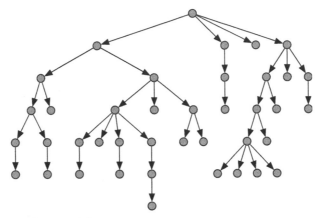

Figure 17.7   A tree drawn with `TreeLayout`.

Sugiyama Framework. The OGDF provides a flexible implementation of Sugiyama's framework [STT81] for drawing directed graphs in a hierarchical fashion. This framework basically consists of three phases, and for each phase various methods and variations have been proposed in the literature. The corresponding OGDF implementation `SugiyamaLayout` provides a module option for each of the three phases; optionally, a packing module can be used to pack multiple connected components of the graph. The available OGDF modules and their dependencies are shown in Figure 17.8.

The three phases in Sugiyama's framework and their implementations are:

1. *Rank assignment:* In the first phase (realized by a `RankingModule`), the nodes of the input digraph G are assigned to layers. If G is not acyclic, then we compute a preferably large acyclic subgraph and reverse the edges not contained in the subgraph by one of the modules described in Section 17.3.1. Currently, the OGDF contains three algorithms for computing a layer assignment for an acyclic digraph in which the edges are directed from nodes on a lower level to nodes on a higher level. `LongestPathRanking` is based on the computation of longest paths and minimizes the number of layers (height of the drawing), `OptNodeRanking` minimizes the total edge length [GKNV93] (here the length of an edge is the number of layers it spans), and `CoffmanGrahamRanking` computes a layer assignment with a predefined maximum number of nodes on a layer (width of the drawing) [CG72]. If edges span several layers, they are split by inserting additional artificial nodes such that edges connect only nodes on neighboring layers.

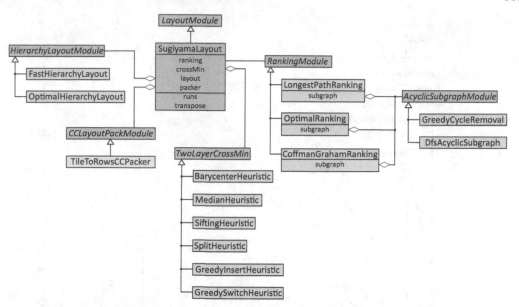

Figure 17.8 Sugiyama's framework for hierarchical graph layout in the OGDF.

2. *k-layer crossing minimization:* The second phase determines permutations of the nodes on each layer such that the number of edge crossings is small. The corresponding optimization problem is NP-hard. A reasonable method consists of visiting the layers from top to bottom, fixing the order of the nodes on the layer and trying to find a permutation of the nodes on the lower next layer that minimizes the number of crossings between edges connecting the two adjacent layers, also referred to as two-layer crossing minimization (realized by a `TwoLayerCrossMin` module). Then, the algorithm proceeds from bottom to top and so on until the total number of crossings does not decrease anymore. `SugiyamaLayout` contains a sophisticated implementation that uses further improvements like calling the crossing minimization several times (controlled by parameter `runs`) with different starting permutations, or applying the `transpose` heuristic [GKNV93].
   Several heuristics for two-layer crossing minimization have been proposed. The library offers the choice between the barycenter heuristic [STT81], the weighted median heuristic [GKNV93], the sifting heuristic [MSM00], as well as the split, the greedy insert, and the greedy switch heuristics presented in [EK86].

3. *Coordinates assignment:* The third phase (realized by a `HierarchyLayout-Module`) computes the final coordinates of the nodes and bend points of the edges, respecting the layer assignment and ordering of the nodes on each layer. The OGDF contains two implementations for the final coordinate assignment phase. The first, `OptimalHierarchyLayout`, tries to let edges run as vertical as possible by solving a linear program; the second, `FastHierarchyLayout`, proposed by Buchheim, Jünger, and Leipert [BJL00] guarantees at most two bends per edge and draws the whole part between these bends vertically.

Figure 17.9 shows a layered drawing produced by `SugiyamaLayout` using the LP-based coordinate assignment method. The digraph displays the history of the UNIX operating system and the layers correspond to the time line depicted on the right side.

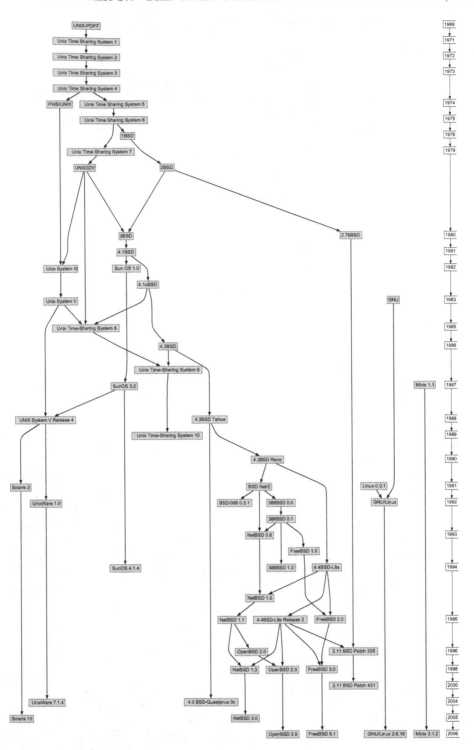

Figure 17.9 A layered digraph illustrating the history of UNIX; each layer represents a point in time. Drawn by applying Sugiyama layout and the LP-based coordinate assignment with angle optimization and special node balancing. Source: http://en.wikipedia.org/wiki/File:Unix_history-simple.svg.

Upward Planarization. Though the commonly applied approach for hierarchical graph drawing is based on the Sugiyama framework, there is a much better alternative that produces substantially less edge crossings. This alternative adapts the crossing minimization procedure known from the planarization approach and is thus called *upward planarization* [CGMW10]. Like the traditional planarization approach for undirected graphs, the algorithm consists of two steps: In the first step, a *feasible upward planar subgraph U* is constructed; in the second step, the arcs not yet contained in *U* are inserted one-by-one so that few crossings arise. These crossings are replaced by dummy nodes so that the digraph in which arcs are inserted can always be considered upward planar. The final outcome of this crossing minimization procedure is an *upward planar representation*; it can be turned into a drawing of the original digraph by replacing the dummy nodes with arc crossings.

The upward planarization framework in the OGDF follows the presentations in [CGMW09] and [CGMW10]; see Figure 17.10. The class `UpwardPlanarizationLayout` represents the layout algorithm, which is implemented in two phases: The first phase realizes the upward crossing minimization procedure and computes an upward planarized representation of the input digraph; the second phase is realized by a `UPRLayoutModule` and computes the final layout. Currently, the layout computation is implemented by reusing modules from Sugiyama's framework, namely the rank assignment and hierarchy layout modules.

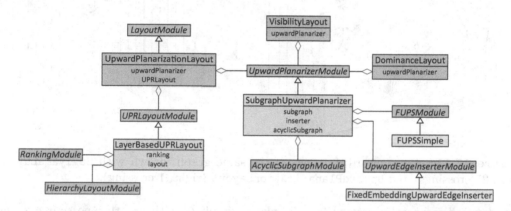

Figure 17.10 The upward planarization framework for hierarchical graph layout in the OGDF; modules for ranking, hierarchy layout, and acyclic subgraphs are omitted and can be found in Figure 17.8.

The crossing minimization step, realized by an `UpwardPlanarizerModule`, is the heart of the upward planarization. The OGDF modularizes this step similarly as for the planarization approach. First, a feasible upward planar subgraph is computed by a `FUPSModule`, which is implemented by `FUPSSimple`, and then the remaining edges are inserted by an `UpwardEdgeInserterModule`, implemented by applying a fixed embedding approach (`FixedEmbeddingUpwardEdgeInserter`). Figure 17.11 compares two upward drawings of the same digraph, one produced by the Sugiyama approach and the other one by applying upward planarization. Typically, Sugiyama drawings tend to become quite flat, thus enforcing many crossings, whereas the upward planarization approach unfolds the digraph well, thereby saving a lot of crossings and revealing the true structure of the digraph.

The crossing minimization step is also used by two further algorithms: `Visibility-Layout` based on the computation of a visibility representation by Rosenstiehl and Tarjan [RT86] and `DominanceLayout` based on dominance drawings of *s-t*-planar digraphs. An

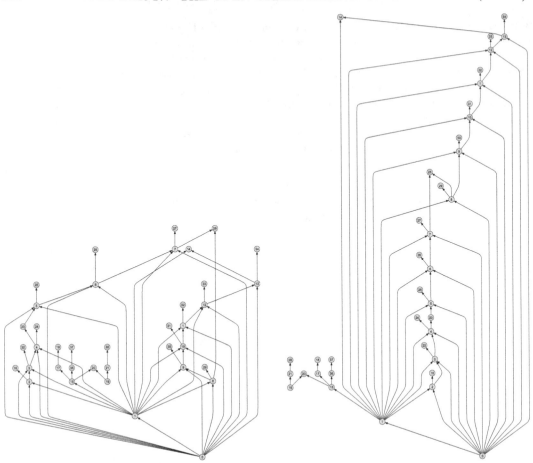

Figure 17.11 Two upward drawings of the same graph, drawn with `SugiyamaLayout` (left, 27 crossings) and `UpwardPlanarizationLayout` (right, 1 crossing).

s-t-planar digraph is a directed, acyclic planar graph G with exactly one source s and exactly one sink t. `DominanceLayout` applies the layout algorithm for *s-t*-planar digraphs by Di Battista, Tamassia, and Tollis [DTT92]. If the input digraph G contains no transitive edges, the algorithm computes a planar dominance grid drawing of G, i.e., a straight-line embedding such that, for any two nodes u and v, there is a directed path from u to v if and only if $x(u) \leq x(v)$ and $y(u) \leq y(v)$. Dominance drawings characterize the transitive closure of a digraph by means of the geometric dominance relation among the nodes [DTT92]. If G does contain transitive edges, the algorithm splits these edges by introducing artificial nodes and computes a dominance drawing for the resulting digraph in which the artificial nodes represent bend points.

17.4.3 Energy-Based Drawing Algorithms

Energy-based drawing algorithms constitute the most common drawing approach for undirected graphs. They are reasonably fast for medium sized graphs, intuitive to understand, and easy to implement—at least in their basic versions. The fundamental underlying idea of energy-based methods is to model the graph as a system of interacting objects that contribute to the overall energy of the system, such that an energy-minimized state of the system corresponds to a nice drawing of the graph. In order to achieve such an optimum, an energy or cost function is minimized. There are various models and realizations for this approach, and the flexibility in the definition of both the energy model and the objective function enables a wide range of optimization methods and applications. There is a wealth of publications concerning energy-based layout methods; see the overview in [DETT99, KW01] and the comprehensive discussion in Chapter 12.

Single-level Algorithms. The OGDF provides implementations for several classical algorithms, such as the force-directed spring embedder algorithm [Ead84], the grid-variant of Fruchterman and Reingold [FR91] (`SpringEmbedderFR`), and the simulated annealing approach by Davidson and Harel [DH96] (`DavidsonHarelLayout`). We also implemented the energy-based approach by Kamada and Kawai [KK89] (`SpringEmbedderKK`), which uses the shortest graph-theoretic distances as ideal pairwise distance values and subsequently tries to obtain a drawing that minimizes the overall difference between ideal and current distances. Further implementations include the GEM algorithm [FLM95] (`GEMLayout`) and Tutte's barycenter method [Tut63] (`TutteLayout`). All implementations of energy-based drawing algorithms are directly derived from the class `LayoutModule`.

An important advantage of energy-based methods—based on the iterative nature of the numerical methods for computing the layout—is that they provide an animation of the change from a given layout to a new one, thus allowing us to use a given drawing as input. In addition, these algorithms support stopping the computation when either the improvement of successive steps falls under a certain threshold or as soon as a prespecified energy value is reached. Our implementations therefore provide the corresponding interfaces to adjust the respective parameters.

Multi-level Algorithms. In addition to these single level algorithms, the OGDF provides a generic framework for the implementation of multilevel algorithms, realized by the class `ModularMultilevelMixer`. Multilevel approaches can help to overcome local minima and slow convergence problems of single level algorithms. Their result does not depend on the quality of an initial layout, and they are well suited also for large graphs with up to tens or even hundreds of thousands of nodes.

The multilevel framework allows us to obtain results similar to those of many different multilevel layout realizations [Wal03, GK02, HJ04]. Instead of implementing these versions from scratch, only the main algorithmic phases— coarsening, placement, and single level layout—have to be implemented or reused from existing realizations. The module concept allows us to plug in these implementations into the framework, enabling also a comparison of different combinations as demonstrated in [BGKM10]. Figure 17.12 shows two example drawings of large graphs.

On the one hand, the multilevel framework provides high flexibility for composing multilevel approaches out of a variety of realizations for the different layout steps. On the other hand, this modularity prohibits fine-tuning of specific combinations by adjusting the different phases to each other. Therefore the OGDF also contains a dedicated implementation of the fast multipole multilevel method (`FMMMLayout`) by Hachul and Jünger [HJ04], as well as an engineered and optimized version of this algorithm supporting multicore hardware [Gro09] (`FastMultipoleMultilevelEmbedder`). Figure 17.13 shows a drawing of a very large graph with 143,437 nodes, which was obtained—using this engineered version—in just 2.1 seconds on an Intel Xeon E5430 (2.66GHz) quadcore machine.

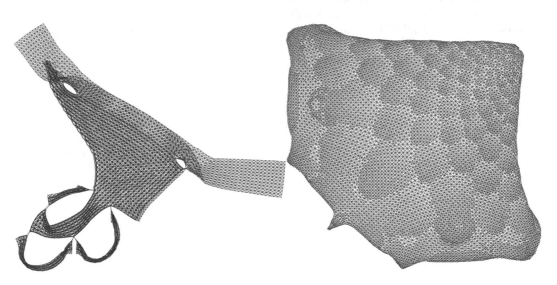

Figure 17.12   Two drawings obtained with OGDF's multilevel framework: graph `data` (left; 2,851 nodes; 15,093 edges) and graph `crack` (right; 10,240 nodes; 30,380 edges).

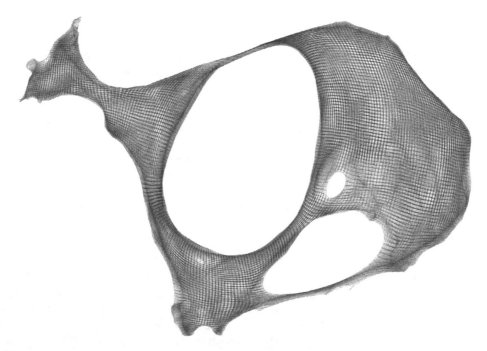

Figure 17.13   The graph `fe_ocean` (143,437 nodes; 409,593 edges) drawn with the `Fast-MultipoleMultilevelEmbedder` in 2.1 seconds.

17.4.4 Drawing Clustered Graphs

A clustered graph $C = (G, T)$ is a tuple consisting of a graph $G = (V, E)$ and a hierarchical structuring T called *cluster tree*. Every node of V is assigned to exactly one inner node of T, which is the cluster to which it belongs.

In the OGDF, a clustered graph C is represented by an instance of class `ClusterGraph`, which stores the necessary information together with a reference to the underlying graph G. The OGDF provides methods to test c-planarity of arbitrary clustered graphs and to draw such graphs in either orthogonal or hierarchical style.

Orthogonal Layout. Similar to the planarization approach for general graphs, we implemented the planarization approach for clustered graphs based on the method by Di Battista et al. [DDM01], which is realized by the class `ClusterPlanarizationLayout`. This method uses the topology-shape-metrics approach and is suitable only for *c-connected* clustered graphs, i.e., clustered graphs with the property that every subgraph induced by the nodes of some cluster c and its subclusters is connected.

The cluster planarization algorithm works as follows: First, we calculate a minimum spanning tree for each cluster, thereby treating its subclusters as simple nodes. Afterward these trees are joined together. We start to insert the remaining edges one after another if this is possible without introducing any crossings. We call the resulting graph the *maximal planar cluster subgraph G'*. Hence, after this step there generally remains a set of edges which could not yet be inserted.

We apply the following steps for every edge $e = (u, v)$ that still has to be inserted: We generate a dual graph D with the following properties: faces within the same cluster are joined by bidirectional arcs. Arcs between faces of different clusters are only generated if these clusters are on the path between u and v in T. The direction of such arcs is chosen accordingly. Finally we add arcs from u to all of its incident faces, and arcs from all faces incident to v, to v. Then we search for the shortest path between u and v in D and generate edge crossings according to the used arcs.

Thanks to the OGDF's modularity we can easily reuse all available code concerning bend minimization and compaction—originally only intended for planar graphs—without a single change.

In order to also cope with non-c-connected clustered graphs, the OGDF provides two quite different approaches: As a simple and fast heuristic, non-connected clusters are made connected by temporarily adding single dummy edges between the components of the induced subgraphs. A much more sophisticated approach is implemented by the class `Maximum-CPlanarSubgraph`: This class applies a branch-and-cut approach that computes the maximum c-planar subgraph of a clustered graph [CGJ+08]. The result can be used for the first step of the cluster planarization approach and at the same time also constitutes the first practical c-planarity testing algorithm for arbitrary (i.e., also non-c-connected) clustered graphs. The branch-and-cut approach is based on the result that every c-planar clustered graph can be augmented to a completely connected clustered graph, i.e., where for each cluster both the cluster and its complement are connected [CW06]. As the call function also returns the set of edges that is eventually added to achieve c-connectivity, these edges can be used in order to make the input graph c-connected, allowing us to apply the planarization and drawing approach without adding unnecessary crossings.

Figure 17.14 shows an example drawing of a c-planar (but not c-connected) clustered graph, obtained with `ClusterPlanarizationLayout` and the simple heuristic for making the clustered graph c-connected.

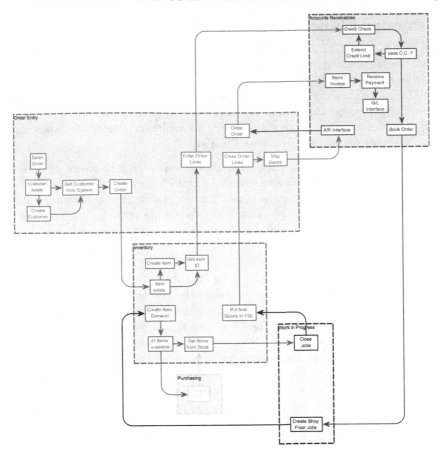

Figure 17.14   A clustered graph drawn in orthogonal style with `ClusterPlanarization-Layout`.

Hierarchical Layout.   For drawing directed graphs with an additional cluster structure in a hierarchical style, `SugiyamaLayout` provides an additional call method. This methods implements a cluster hierarchical layout algorithm that is based on Sugiyama's framework as described by Sanders [San96a, San96b] and applies improvements for the crossing reduction strategy proposed by Schreiber [Sch01] and Forster [For02].

Though this approach would also allow us to draw *compound graphs*, i.e., a generalization of clustered graphs where edges can also be attached at clusters, the implementation currently supports only clustered graphs. The reason is that there is yet no representation of compound graphs in the OGDF; it will be added in a future release.

17.5   Success Stories

This section showcases some outstanding implementations in the OGDF and the story behind their development. These are also good examples for demonstrating design decisions and engineering aspect.

17.5.1 SPQR-Trees

In the early 1970s, Hopcroft and Tarjan [HT73a] published the first linear-time algorithm for computing the triconnected components of a graph. This decomposition is in particular important in graph drawing—here usually known as the data structure SPQR-tree—, as it allows us to encode all planar embedding of a graph. Hence, this algorithm has been cited over and over again for showing that SPQR-trees can be constructed in linear time. However, for a long time nobody was able to come up with an implementation of this algorithm, since the paper was hard to understand and contained various flaws, thus preventing a straightforward implementation.

This situation is a classical example for showing the need of publicly available reference implementations, revealing all the algorithmic details and simplifying the application of the algorithm. A breakthrough was achieved in the early 2000s—almost 20 years after the publication of the algorithm—by Gutwenger and Mutzel [GM01]. They described how to fix the flaws in the Hopcroft and Tarjan algorithm, and also provided a stable implementation of SPQR-trees in the AGD library. This implementation is now part of the OGDF and thus open source, allowing everybody to study and understand it. Since this implementation is publicly available, we observed a lot of interest in it, ranging from applications that just apply the data structure to reimplementations, e.g., in other programming languages.

17.5.2 Exact Crossing Minimization

One of the most challenging problems in graph drawing is the *crossing number problem*, i.e.: What is the minimum number of edge crossings required when drawing a given graph? See Chapter 2. for an extensive introduction to this topic. For a long time, no exact algorithms existed that could compute the crossing number for at least some interesting graphs in practice. The classical benchmark instances for evaluating crossing minimization algorithms are the Rome graphs, a benchmark set of quite sparse graphs with up to 100 nodes. The first approach that could compute the exact solutions for a handful of interesting Rome graphs was based on an ILP formulation presented by Buchheim et al. [BEJ+05] and implemented using AGD and CPLEX. In the following years, this approach was revised and reimplemented using the OGDF by Chimani et al. [BCE+08, CGM09], and the resulting implementation was able to solve many more graphs. The key ideas were to use branch-and-cut-and-price (by applying the ABACUS framework) and better primal heuristics (the planarization approach provided by the OGDF). The currently best algorithm for exact crossing minimization [Chi08, CMB08] is also implemented using the OGDF, and now allows us to solve the majority of Rome graphs exactly.

This example demonstrates how the OGDF's modular design supports the development of new algorithms. Here, the planarization approach (see Section 17.3.3) is used as primal heuristic, as well as other important components like testing planarity, extraction of Kuratowski subdivisions [CMS08], and the non-planar core reduction [CG09] as a preprocessing strategy. It also shows that exact algorithms based on ILP formulations require additional frameworks providing an LP-solver and support for the design and implementation of branch-and-cut(-and-price) algorithms. Hence, we decided to (optionally) use the libraries COIN-OR as LP-solver interface (which also allows us to choose CPLEX as LP-solver) and ABACUS in the OGDF.

17.5.3   Upward Graph Drawing

The classical approach for upward drawing of acyclic digraphs is Sugiyama's framework, which was already proposed in the early 1980s. The first step of this framework layers the graph, and the subsequent steps ensure that each node is finally placed on its assigned layer. It is well known that such a fix layer assignment forces unnecessary crossings in the drawing, but the Sugiyama framework is still widely used. A breakthrough in upward crossing minimization was recently achieved by Chimani et al. [CGMW08, CGMW09, CGMW10]. They propose a new method for upward crossing minimization that does not need to layer the graph and report on substantial reduction in crossings (compared to Sugiyama's framework and other approaches) for commonly used benchmark graphs.

They developed this new approach using the OGDF and made use of the OGDF's modular design by reusing some of the modules from Sugiyama's framework (see also Figure 17.10), as well as sophisticated algorithms like upward planarity testing for sT-digraphs. Within their experimental study, they could apply the OGDF's Sugiyama layout algorithm providing state-of-the-art crossing minimization heuristics for the layered approach. The resulting implementation is also modularized and thus allows us to easily replace particular phases of the algorithm with alternative implementations.

This example demonstrates how the OGDF helps in developing alternative approaches, and how new frameworks can be established such that other users can easily experiment with it and modify some of the phases.

Acknowledgments

Markus Chimani was funded via a junior professorship by the Carl-Zeiss-Foundation.

The OGDF, as it is today, is by far not only the product of the authors of this chapter. It benefits from contributions of many additional supporters, in alphabetical order:[3] *Dino Ahr, Gereon Bartel, Christoph Buchheim, Tobias Dehling, Martin Gronemann, Stefan Hachul, Mathias Jansen, Thorsten Kerkhof, Joachim Kupke, Sebastian Leipert, Daniel Lückerath, Jan Papenfuß, Gerhard Reinelt, Till Schäfer, Jens Schmidt, Michael Schulz, Andrea Wagner, René Weiskircher, Hoi-Ming Wong, and Bernd Zey.*

---

[3]See also `http://www.ogdf.net/doku.php/team:about` for an up-to-date list.

References

[AGMN97] D. Alberts, C. Gutwenger, P. Mutzel, and S. Näher. AGD-library: A library of algorithms for graph drawing. In *Proc. WAE '97*, pages 112–123, 1997.

[BCE+08] C. Buchheim, M. Chimani, D. Ebner, C. Gutwenger, M. Jünger, G. W. Klau, P. Mutzel, and R. Weiskircher. A branch-and-cut approach to the crossing number problem. *Discrete Optimization, Special Issue in Memory of George B. Dantzig*, 5(2):373–388, 2008.

[BDMT98] P. Bertolazzi, G. Di Battista, C. Mannino, and R. Tamassia. Optimal upward planarity testing of single-source digraphs. *SIAM J. Comput.*, 27(1):132–169, 1998.

[BEJ+05] C. Buchheim, D. Ebner, M. Jünger, P. Mutzel, and R. Weiskircher. Exact crossing minimization. In P. Eades and P. Healy, editors, *Graph Drawing (Proc. GD '05)*, Lecture Notes Comput. Sci. Springer-Verlag, 2005. To appear.

[BGKM10] G. Bartel, C. Gutwenger, K. Klein, and P. Mutzel. An experimental evaluation of multilevel layout methods. In *18th International Symposium on Graph Drawing 2010 (GD10)*, number 6502 in Lecture Notes Comput. Sci., pages 80–91. Springer-Verlag, 2010.

[BJL00] C. Buchheim, M. Jünger, and S. Leipert. Fast layout algorithm for k-level graphs. In J. Marks, editor, *Proc. Graph Drawing 2000*, volume 1984 of *Lecture Notes Comput. Sci.*, pages 229–240. Springer-Verlag, 2000.

[BJL06] C. Buchheim, M. Jünger, and S. Leipert. Drawing rooted trees in linear time. *Software: Practice and Experience*, 36(6):651–665, 2006.

[BL76] K. Booth and G. Lueker. Testing for the consecutive ones property interval graphs and graph planarity using PQ-tree algorithms. *J. Comput. Syst. Sci.*, 13:335–379, 1976.

[BM04] J. M. Boyer and W. Myrvold. On the cutting edge: Simplified $o(n)$ planarity by edge addition. *J. Graph Algorithms Appl.*, 8(3):241–273, 2004.

[CG72] E. G. Coffman and R. L. Graham. Optimal scheduling for two processor systems. *Acta Informatica*, 1:200–213, 1972.

[CG09] M. Chimani and C. Gutwenger. Non-planar core reduction of graphs. *Discrete Mathematics*, 309(7):1838–1855, 2009.

[CGJ+08] M. Chimani, C. Gutwenger, M. Jansen, K. Klein, and P. Mutzel. Computing maximum c-planar subgraphs. In I. G. Tollis and M. Patrignani, editors, *Graph Drawing*, volume 5417 of *Lecture Notes Comput. Sci.*, pages 114–120. Springer-Verlag, 2008.

[CGM09] M. Chimani, C. Gutwenger, and P. Mutzel. Experiments on exact crossing minimization using column generation. *ACM Journal of Experimental Algorithmics*, 14(3):4.1–4.18, 2009.

[CGMW08] M. Chimani, C. Gutwenger, P. Mutzel, and H.-M. Wong. Layer-free upward crossing minimization. In C. McGeoch, editor, *Experimental Algorithms, 7th International Workshop, WEA 2008*, volume 5038 of *Lecture Notes Comput. Sci.*, pages 55–68. Springer-Verlag, 2008.

[CGMW09] M. Chimani, C. Gutwenger, P. Mutzel, and H.-M. Wong. Upward planarization layout. In D. Eppstein and E. Gansner, editors, *Proceedings of the 17th Symposium on Graph Drawing 2009 (GD 2009)*, volume 5849 of *Lecture Notes Comput. Sci.*, pages 94–106. Springer-Verlag, 2009.

[CGMW10] M. Chimani, C. Gutwenger, P. Mutzel, and H.-M. Wong. Layer-free upward crossing minimization. *ACM J. Exp. Algorithmics*, 15:Article No. 2.2, 2010.

[Chi08] M. Chimani. *Computing crossing numbers.* PhD thesis, TU Dortmund, 2008. `http://hdl.handle.net/2003/25955`.

[CMB08] M. Chimani, P. Mutzel, and I. Bomze. A new approach to exact crossing minimization. In *Proc. ESA '08*, volume 5193 of *Lecture Notes Comput. Sci.*, pages 284–296. Springer-Verlag, 2008.

[CMS08] M. Chimani, P. Mutzel, and J. M. Schmidt. Efficient extraction of multiple Kuratowski subdivisions. In *Proc. GD '07*, volume 4875 of *Lecture Notes Comput. Sci.*, pages 159–170. Springer-Verlag, 2008.

[CNAO85] N. Chiba, T. Nishizeki, S. Abe, and T. Ozawa. A linear algorithm for embedding planar graphs using PQ-trees. *J. Comput. Syst. Sci.*, 30(1):54–76, 1985.

[CW06] S. Cornelsen and D. Wagner. Completely connected clustered graphs. *J. Discrete Algorithms*, 4(2):313–323, 2006.

[DDM01] G. Di Battista, W. Didimo, and A. Marcandalli. Planarization of clustered graphs. In P. Mutzel, M. Jünger, and S. Leipert, editors, *Graph Drawing*, volume 2265 of *Lecture Notes Comput. Sci.*, pages 60–74. Springer-Verlag, 2001.

[DETT99] G. Di Battista, P. Eades, R. Tamassia, and I. G. Tollis. *Graph Drawing.* Prentice Hall, Upper Saddle River, NJ, 1999.

[DH96] R. Davidson and D. Harel. Drawing graphs nicely using simulated annealing. *ACM Trans. Graph.*, 15(4):301–331, 1996.

[DT89] G. Di Battista and R. Tamassia. Incremental planarity testing. In *Proc. 30th Annu. IEEE Sympos. Found. Comput. Sci.*, pages 436–441, 1989.

[DT96] G. Di Battista and R. Tamassia. On-line maintenance of triconnected components with SPQR-trees. *Algorithmica*, 15:302–318, 1996.

[DTT92] G. Di Battista, R. Tamassia, and I. G. Tollis. Constrained visibility representations of graphs. *Inform. Process. Lett.*, 41:1–7, 1992.

[Ead84] P. A. Eades. A heuristic for graph drawing. In *Congressus Numerantium*, volume 42, pages 149–160, 1984.

[Eig03] M. Eiglsperger. *Automatic Layout of UML Class Diagrams: A Topology-Shape-Metrics Approach.* PhD thesis, Eberhardt-Karl-Universität (Tübingen), 2003.

[EK86] P. Eades and D. Kelly. Heuristics for reducing crossings in 2-layered networks. *Ars Combinatoria*, 21(A):89–98, 1986.

[EL95] P. Eades and X. Lin. A new heuristic for the feedback arc set problem. *Australian J. Combin.*, 12:15–26, 1995.

[FK96] U. Fössmeier and M. Kaufmann. Drawing high degree graphs with low bend number. In *Graph Drawing (Proc. GD '95)*, Lecture Notes Comput. Sci. Springer-Verlag, 1996.

[FLM95] A. Frick, A. Ludwig, and H. Mehldau. A fast adaptive layout algorithm for undirected graphs. In *GD '94: Proceedings of the DIMACS International*

Workshop on Graph Drawing, pages 388–403, London, UK, 1995. Springer-Verlag.

[FM98]   S. Fialko and P. Mutzel. A new approximation algorithm for the planar augmentation problem. In *Proceedings of the Ninth Annual ACM-SIAM Symposium on Discrete Algorithms (SODA '98)*, pages 260–269, San Francisco, California, 1998. ACM Press.

[For02]   M. Forster. Applying crossing reduction strategies to layered compound graphs. In S. G. Kobourov and M. T. Goodrich, editors, *Graph Drawing*, volume 2528 of *Lecture Notes Comput. Sci.*, pages 276–284. Springer-Verlag, 2002.

[FR91]   T. M. J. Fruchterman and E. M. Reingold. Graph drawing by force-directed placement. *Softw. Pract. Exper.*, 21(11):1129–1164, 1991.

[GK02]   P. Gajer and S. G. Kobourov. GRIP: Graph drawing with intelligent placement. *J. Graph Algorithms Appl.*, 6(3):203–224, 2002.

[GKNV93]   E. R. Gansner, E. Koutsofios, S. C. North, and K. P. Vo. A technique for drawing directed graphs. *IEEE Trans. Softw. Eng.*, 19:214–230, 1993.

[GM97]   C. Gutwenger and P. Mutzel. Grid embedding of biconnected planar graphs. Extended Abstract, Max-Planck-Institut für Informatik, Saarbrücken, Germany, 1997.

[GM01]   C. Gutwenger and P. Mutzel. A linear time implementation of SPQR trees. In J. Marks, editor, *Proceedings of the 8th International Symposium on Graph Drawing (GD 2000)*, volume 1984 of *Lecture Notes Comput. Sci.*, pages 77–90. Springer-Verlag, 2001.

[GM03]   C. Gutwenger and P. Mutzel. Graph embedding with minimum depth and maximum external face. In G. Liotta, editor, *11th Symposium on Graph Drawing 2003, Perugia*, volume 2912 of *Lecture Notes Comput. Sci.*, pages 259–272. Springer-Verlag, 2003.

[GMW05]   C. Gutwenger, P. Mutzel, and R. Weiskircher. Inserting an edge into a planar graph. *Algorithmica*, 41(4):289–308, 2005.

[GMZ09a]   C. Gutwenger, P. Mutzel, and B. Zey. On the hardness and approximability of planar biconnectivity augmentation. In H. Q. Ngo, editor, *Proceedings of the 15th Annual International Computing and Combinatorics Conference 2009*, volume 5609 of *Lecture Notes Comput. Sci.*, pages 249–257. Springer-Verlag, 2009.

[GMZ09b]   C. Gutwenger, P. Mutzel, and B. Zey. Planar biconnectivity augmentation with fixed embedding. In J. Fiala, J. Kratochvl, and M. Miller, editors, *Proceedings of the 20th International Workshop on Combinatorial Algorithms 2009*, volume 5874 of *Lecture Notes Comput. Sci.*, pages 289–300. Springer-Verlag, 2009.

[Gro09]   M. Gronemann. Engineering the fast-multipole-multilevel method for multicore and SIMD architectures. Master's thesis, Technische Universität Dortmund, 2009.

[GT01]   A. Garg and R. Tamassia. On the computational complexity of upward and rectilinear planarity testing. *SIAM J. Comput.*, 31(2):601–625, 2001.

[Gut10]   C. Gutwenger. *Application of SPQR-Trees in the Planarization Approach for Drawing Graphs*. PhD thesis, Technische Universität Dortmund, Germany, Fakultät für Informatik, 2010.

[HJ04]   S. Hachul and M. Jünger. Drawing large graphs with a potential-field-based multilevel algorithm. In Janos Pach, editor, *Proc. Graph Drawing 2004*, volume 3383 of *Lecture Notes Comput. Sci.*, pages 285–295. Springer-Verlag, 2004.

[HT73a]   J. Hopcroft and R. E. Tarjan. Dividing a graph into triconnected components. *SIAM J. Comput.*, 2(3):135–158, 1973.

[HT73b]   J. E. Hopcroft and R. E. Tarjan. Efficient algorithms for graph manipulation. *Communications of the ACM*, 16(6):372–378, 1973.

[JLM98]   M. Jünger, S. Leipert, and P. Mutzel. A note on computing a maximal planar subgraph using PQ-trees. *IEEE Transactions on Computer-Aided Design*, 17(7):609–612, 1998.

[JT00]   M. Jünger and S. Thienel. The ABACUS system for branch-and-cut-and-price algorithms in integer programming and combinatorial optimization. *Software: Practice and Experience*, 30:1325–1352, 2000. See also http://www.informatik.uni-koeln.de/abacus/.

[JTS89]   R. Jayakumar, K. Thulasiraman, and M. N. S. Swamy. $O(n^2)$ algorithms for graph planarization. *IEEE Trans. Comp.-Aided Design*, 8:257–267, 1989.

[Kan96]   G. Kant. Drawing planar graphs using the canonical ordering. *Algorithmica*, 16:4–32, 1996. (special issue on Graph Drawing, edited by G. Di Battista and R. Tamassia).

[KB91]   G. Kant and H. L. Bodlaender. Planar graph augmentation problems. In *Proc. WADS '91*, volume 519 of *Lecture Notes Comput. Sci.*, pages 286–298. Springer-Verlag, 1991.

[Ker07]   T. Kerkhof. Algorithmen zur Bestimmung von guten Graph-Einbettungen für orthogonale Zeichnungen. Master's thesis, University of Dortmund, 2007.

[KK89]   T. Kamada and S. Kawai. An algorithm for drawing general undirected graphs. *Inform. Process. Lett.*, 31(1):7–15, 1989.

[Kla01]   G. W. Klau. *A Combinatorial Approach to Orthogonal Placement Problems*. PhD thesis, Universität des Saarlandes, Saarbrücken, Germany, Fachbereich Informatik, Technische Fakultät I, 2001.

[KM98]   G. W. Klau and P. Mutzel. Quasi-orthogonal drawing of planar graphs. Technical Report MPI-I-98-1-013, Max Planck Institut für Informatik, Saarbrücken, Germany, 1998.

[KM99]   G. W. Klau and P. Mutzel. Optimal compaction of orthogonal grid drawings. In G. Cornuejols, R. E. Burkard, and G. J. Woeginger, editors, *Integer Programming and Combinatorial Optimization*, volume 1610 of *Lecture Notes Comput. Sci.*, pages 304–319. Springer-Verlag, 1999.

[KW01]   M. Kaufmann and D. Wagner, editors. *Drawing Graphs*, volume 2025 of *Lecture Notes Comput. Sci.* Springer-Verlag, 2001.

[LEC67]   A. Lempel, S. Even, and I. Cederbaum. An algorithm for planarity testing of graphs. In *Theory of Graphs: Internat. Symposium (Rome 1966)*, pages 215–232, New York, 1967. Gordon and Breach.

[Mar10]   K. Martin. Tutorial: COIN-OR: Software for the OR community. *Interfaces*, 40(6):465–476, 2010. See also http://www.coin-or.org.

[MN99]   K. Mehlhorn and S. Näher. *LEDA: A Platform for Combinatorial and Geometric Computing.* Cambridge University Press, 1999.

[MSM00]   C. Matuszewski, R. Schönfeld, and P. Molitor. Using sifting for k-layer crossing minimization. In *Graph Drawing (Proc. GD '99)*, Lecture Notes Comput. Sci. Springer-Verlag, 2000. to appear.

[PT00]   M. Pizzonia and R. Tamassia. Minimum depth graph embedding. In M. Paterson, editor, *Proceedings of the 8th Annual European Symposium on Algorithms (ESA 2000)*, volume 1879 of *Lecture Notes Comput. Sci.*, pages 356–367. Springer-Verlag, 2000.

[RT81]   E. Reingold and J. Tilford. Tidier drawing of trees. *IEEE Trans. Softw. Eng.*, SE-7(2):223–228, 1981.

[RT86]   P. Rosenstiehl and R. E. Tarjan. Rectilinear planar layouts and bipolar orientations of planar graphs. *Discrete Comput. Geom.*, 1(4):343–353, 1986.

[San96a]   G. Sander. Layout of compound directed graphs. Technical Report A/03/96, Universität Saarbrücken, 1996.

[San96b]   G. Sander. *Visualisierungstechniken für den Compilerbau.* PhD thesis, Universität Saarbrücken, Germany, 1996.

[Sch01]   F. Schreiber. *Visualisierung biochemischer Reaktionsnetze.* PhD thesis, Universität Passau, Germany, 2001.

[STT81]   K. Sugiyama, S. Tagawa, and M. Toda. Methods for visual understanding of hierarchical systems. *IEEE Trans. Syst. Man Cybern.*, SMC-11(2):109–125, 1981.

[Tam87]   R. Tamassia. On embedding a graph in the grid with the minimum number of bends. *SIAM J. Comput.*, 16(3):421–444, 1987.

[Tar72]   R. E. Tarjan. Depth-first search and linear graph algorithms. *SIAM J. Comput.*, 1(2):146–160, 1972.

[TDB88]   R. Tamassia, G. Di Battista, and C. Batini. Automatic graph drawing and readability of diagrams. *IEEE Trans. Syst. Man Cybern.*, SMC-18(1):61–79, 1988.

[Tut63]   W. T. Tutte. How to draw a graph. *Proc Lond Math Soc*, 13:743–767, 1963.

[Wal90]   J. Q. Walker II. A node-positioning algorithm for general trees. *Softw. - Pract. Exp.*, 20(7):685–705, 1990.

[Wal03]   C. Walshaw. A multilevel algorithm for force-directed graph-drawing. *J. Graph Algorithms Appl.*, 7(3):253–285, 2003.

[WT92]   J. Westbrook and R. E. Tarjan. Maintaining bridge-connected and biconnected components on-line. *Algorithmica*, 7:433–464, 1992.

18

GDToolkit

Giuseppe Di Battista
University "Roma Tre"

Walter Didimo
University of Perugia

18.1 Introduction

GDToolkit (available at http://www.dia.uniroma3.it/~gdt) is an object-oriented graph drawing library, written in the C++ programming language. It provides many facilities that support users to develop specific graph visualization interfaces that can be used in real-world domains.

The computation of a drawing is typically decomposed into a sequence of logical steps, and several algorithms can be chosen for each step, which offer different compromises in terms of efficiency and effectiveness. Developers can tune the ratio between the performance of their applications and the quality of the computed drawings, by combining the different algorithms available for each step. Generic drawing algorithms and drawing conventions can be customized and tailored for a specific application context by means of different types of constraints that the developer can apply on the drawings.

The design of GDToolkit started in 1996, as a part of the ALCOM-IT European Project. For the use of basic data-structures like vectors, lists, maps, and sets, GDToolkit was originally strongly based on the LEDA library [MN95, MN00]. After several years, the current version of GDToolkit is now completely LEDA free, since the basic data-structures have been totally re-implemented. GDToolkit is now under a commercial license; detailed information about license terms and conditions can be found at the official web page of the project.

This chapter describes the main functionalities and architectural aspects of GDToolkit and it is structured as follows. The key features and the design principles of GDToolkit are first described (Section 18.2). Specific architectural aspects concerned with the design and the use of class constructors are then examined (Section 18.4). The constraint management system of the library is discussed in Section 18.5. Finally, some examples of real-world applications developed using GDToolkit are illustrated (Section 18.6).

18.2   Key Features of GDToolkit

Several key features have been taken into account in the design of GDToolkit. They are listed and discussed below:

A specific class for each type of graph. In GDToolkit each type of graph is modeled as a specific class, called a *graph-class*. Graph-classes are organized into a hierarchy that reflects different levels of abstraction, ranging from graph topology to graph geometry (see Section 18.3 for a detailed description of the hierarchy). A similar architecture has been previously proposed in other projects of graph drawing libraries [BBDL91, DGST90]. There are basic graph-classes to model graphs with different topological properties, like general multi-graphs, directed graphs, planar graphs, flow networks, trees. In addition, there are graph-classes for representing graphs with associated some drawing information. For example, there exist intermediate graph-classes that model orthogonal drawings and upward drawings only in terms of drawing "shape" (see, e.g., [DETT99]), and there are graph-classes that model drawings of graphs in terms of vertex and edge-bend coordinates.

All the graph algorithms implemented in GDToolkit are encapsulated as methods of the topmost graph-class in which they are safely applicable. Derived graph-classes inherit methods from the ancestor ones, optionally refining or hiding them when unsafe. Inheritance and encapsulation effectively help the application developer in dealing with the intrinsic complexity of graph algorithms and data-structures.

A graph drawing algorithm is viewed as a sequence of steps. Each step maps an object of a graph-class to an object of another graph-class. A drawing is typically the result of a sequence of constructors; each time a constructor is applied to a graph-object g, a new graph-object g' is created and equipped with additional drawing features with respect to g. For example, the code in Figure 18.1 shows how to create an orthogonal drawing of a graph as a simple sequence of constructors.

```
/* creates a graph ug, loading it from file "my-graph" */
undi_graph ug;
ug.read ("my-graph");
/* computes a planar embedding for ug, with possible crossing nodes */
plan_undi_graph pug (ug);
/* computes an orthogonal shape for the planar embedded graph */
orth_plan_undi_graph opug (pug);
/* compacts the orthogonal shape to create the final drawing */
draw_undi_graph dug (opug);
```

Figure 18.1   A fragment of code that computes an orthogonal drawing of the graph described by the graph-object ug. The graph is loaded from a file and the drawing is computed according to the topology-shape-metrics approach [Tam87]. Each computation step is performed by a different constructor.

Efficient object constructors. Suppose that B is a graph-class that inherits from A. Invoking a constructor of B that takes in input a graph-object g of A has the effect of creating a new graph-object g' of B that contains additional drawing information (attributes) with respect to g. Typically, in the construction process, all the structures of g are first copied in the state of g', and then the state of g' is enriched with additional data computed by some algorithms. Sometimes however, once g' has been created, g is no longer needed in the program. In these cases one may wish to "promote" g to become an object of class B, avoiding to duplicate data for the structures of g. Such a promoting mechanism makes it possible to save computational time and space resources, especially when g describes a large graph. The graph-classes of GDToolkit are designed to allow that. Details about the promoting mechanisms of GDToolkit are given in Section 18.4.

Management of Constraints. As many other graph drawing libraries, GDToolkit is not devoted to a specific application field. It is mainly thought as a general purpose graph drawing collection of objects and algorithms, which can be used in several real-world contexts. However, different application domains may need to deal with different variants of a generic graph drawing convention, depending on the specificity of the domain itself. These variants often reflect into a set of drawing constraints, and therefore it is crucial that the drawing algorithm is able to deal with these constraints. For example, some applications might require that a subset of edges is not allowed to cross, or that some vertices should have a prescribed dimension.

GDToolkit makes it possible to customize its drawing conventions and its drawing algorithms by means of an effective constraint management system. The graph-objects of GDToolkit can be equipped with constraints that can be viewed as additional properties for vertices, edges, or faces. Constraints can be added or removed at each time of the life-cycle of a graph-object. If a constraint is added to a graph-object, this constraint remains consistent even if the object is updated. Also, GDToolkit constructors automatically preserve (and in case enforce) the constraints when a new graph-object is created as a refinement or as a copy of an existing graph-object. The constraint management system of GDToolkit is described in detail in Section 18.5

Extensibility. In order to make the extensibility of the library easy, the definition of new classes, constructors, and constraints is done according to specific patterns, which should be taken into account by programmers that wish to extend the library. The principles of these patterns are described in Sections 18.3, 18.4, and 18.5.

In the next section the architecture of GDToolkit is described, focusing on the key features discussed above. Several code and drawing examples are provided in order to better illustrate the use of the library.

18.3 Graph-classes and their Hierarchy

The graph-classes of GDToolkit are structured into a hierarchy, and provide objects for each specific type of graph. The design of the hierarchy is mainly driven by the well-known *topology-shape-metrics approach* [DETT99, Tam87] for orthogonal drawings. According to this approach, a drawing is computed into three phases:

Topology: A *planar embedding* of the input graph is computed, by possibly adding dummy vertices to replace crossings if the graph is not planar; the planar embedding is described by the circular lists of edges incident to each vertex, or equivalently by the set of faces.

Shape: An *orthogonal shape* is computed within the planar embedding found in the previous phase, where one of the faces is chosen as the external face; the shape describes the sequence of left and right bends along the edges, and the angles formed by two consecutive edges incident to the same vertex in a circular order.

Metrics: The final position of the vertices and bends is computed, while preserving the shape determined in the previous phase. Then, dummy vertices are removed.

GDToolkit applies and extends this approach to other drawing conventions. Indeed, more in general, a drawing is described (and constructed) at three different levels of abstractions, where each level adds drawing information to the parent level. The first level describes the topology (embedding) of the drawing, the second level its "shape", and the third level the final geometry of the drawing in terms of vertex and edge-bend coordinates. The shape of a drawing can be regarded as a partial description of the drawing that typically determines the relative position of vertices and edges, without deciding their final placement. For some drawing conventions the concept of shape does not make sense, and in this case it is possible to skip an abstraction level in the construction of the drawing. The hierarchy of the main graph-classes of GDToolkit is depicted in Figure 18.2.

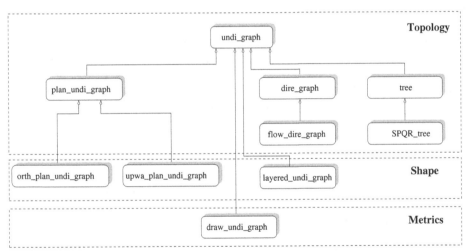

Figure 18.2   The hierarchy of the main graph-classes of GDToolkit.

18.3.1   Topology level

The root of the graph classes hierarchy is the *undi_graph* class, whose objects represent generic graphs that can be connected or not, and that can have multiple edges and self-loops. Also, any edge of an *undi_graph* object can be optionally oriented, i.e, an *undi_graph* can have both undirected and directed edges at the same time.

Each node and each edge of an *undi_graph* object is associated with a non-negative integer identifier. No duplication of identifiers is allowed in the same class of elements. Methods for automatically generating identifiers are provided by the library, but identifiers can also be manually set or changed by the programmer. When a graph object is copied into another graph object, nodes and edges are duplicated while preserving their identifiers. Therefore, identifiers can be used to keep a one-to-one correspondence between the set of nodes and edges of the two graphs.

An *undi_graph* stores information about its embedding, i.e., the circular ordering of the edges incident to every node. This embedding is preserved during any copy operation of the graph. Also, class *undi_graph* contains a large set of basic methods to access and update the topology of the graph, and advanced methods to deal with its embedding, orientation, and connectivity. For example, there are methods that test the existence of planar embeddings for the graph, or of planar *bimodal embeddings* in the case the graph has only directed edges. We recall that a bimodal embedding for a planar digraph is such that, for each vertex, all the incoming edges (as well as all the outgoing edges) are consecutive around the vertex. There are methods to compute *st*-orientations, methods to connect the graph by adding a minimal set of extra edges, and methods to perform different kinds of traversal of the graph. Figure 18.3 shows a fragment of code that creates a new *undi_graph* object ug reading the structure of the graph from a file, executes two copies ug1 and ug2 of ug, and updates (if possible) the embedding of ug1 into a planar one, and the embedding of ug2 into a planar bimodal one, after an orientation for ug2 is found. Note that, if ug2 is biconnected, the program computes an *st*-orientation for it.

```
/* creates an object ug, loading it from file "my_graph",
 * and makes two copies of ug */
undi_graph ug;
ug.read("my_graph");
undi_graph ug1 (ug);
undi_graph ug2 (ug);

/* makes the embedding of ug1 planar, if possible*/
if (!ug1.make_embedding_planar ())
    cout << "\nThe graph is not planar" << flush;

/* if graph ug2 is biconnected, makes it st-oriented,
 * else makes it randomly oriented*/
if (ug2.is_biconnected())
    ug2.make_directed(ug2.first_node(),ug2.last_node());
else ug2.make_directed(true);

/* makes the embedding of ug2 planar bimodal, if possible */
if (!ug2.make_embedding_cand_planar())
    cout << "\nThe oriented graph is not planar bimodal" << flush;
```

Figure 18.3 A fragment of code that illustrates how to use some methods of the *undi_graph* class.

Most graph algorithms implemented as methods of an undi_graph object runs in linear time. For example, an *st*-orientation of a graph with 200,000 vertices and 600,000 edges is executed in about 14 seconds under Linux on a typical machine with i5-540M Intel processor and 4 GB RAM.

Embedded planar graphs are modeled by the class *plan_undi_graph*, which enriches the basic topological structure of an *undi_graph* with the description of a set of faces. Following the philosophy of the library, a *plan_undi_graph* object can be created using a constructor that takes as a constant parameter an *undi_graph* object. This constructor applies a planarity testing algorithm and, if the graph is not planar, a planarization algorithm that replaces crossings with "dummy" nodes, called *crossing nodes*.

The planarity testing algorithm implemented in GDToolkit is the one described by Boyer et al. [BCPD04]; it has been shown to be faster than the one implemented in the LEDA library [MN95, MN00]. The planarization algorithm is based on a technique that inserts an edge per time by following a shortest path in the dual graph of the planar embedded graph computed so far [DETT99]. While planarizing sparse graphs is rather fast, executing the planarization algorithm on dense graphs might require a significant computational effort, due to the high number of crossings. For example, a graph with 500 vertices and 750 edges is planarized in about 13 seconds under Linux on a computer with i5-540M Intel processor and 4 GB RAM. A much smaller but much denser graph consisting of 100 vertices and 500 edges is planarized in about 42 seconds.

Class *tree* offers methods to perform standard operations on ordered rooted trees, like visits in different orders, re-rooting, and so on. The *SPQR_tree* class inherits from class **tree**, and models the data structure introduced by Di Battista and Tamassia [DT96] to represent the triconnected components and the different embeddings of a biconnected graph. It is possible to use *SPQR*-tree objects to enumerate and change the embeddings of a graph, although the current implementation of SPQR-trees in GDToolkit only deals with planar graphs. In GDToolkit, *SPQR*-trees are extensively used to implement branch-and-bound algorithms that compute drawing with the minimum number of bends in the variable embedding setting [BDD00, BDD02].

Class *dire_graph* defines specialized methods performing on directed graphs. A special subclass of *dire_graph* is the *flow_dire_graph* class, which represents flow networks [AMO93] with *capacities*, *costs*, and *flow values* for the arcs. Every node of a flow network may also have a certain *balance* value that can be either negative or positive, depending on the fact that the node demands or supplies flow (by default, the balance value of a node is zero, which means that its total entering flow equals its total leaving flow). The class provides methods to compute feasible flows in a given network while optimizing some function, like for example the total cost. Several drawing algorithms in the library extensively use a *flow_dire_graph* object to compute a feasible flow with prescribed value and minimum cost. The flow is computed in a network that is typically constructed from the topology of the graph to be drawn.

18.3.2   Shape Level

At the shape level, GDToolkit offers three classes: *orth_plan_undi_graph*, which models orthogonal representations; *upwa_plan_undi_graph*, which models upward and quasi-upward representations, and *layered_undi_graph*, which models layered graphs. In the following we give some details about orthogonal, upward, and quasi-upward representations in GDToolkit, by also recalling some basic concepts related to these drawing conventions.

Orthogonal representations in GDToolkit are modeled according to the *simple-podevsnef* drawing convention[BDD00], a simplified and pretty robust version of the *podevsnef* convention (also called *Kandinsky*) defined in [FK96]. Vertices are represented as small rectangles, all having the same size, and any vertex can have any number of incident edges.

The library offers different algorithmic choices to compute orthogonal representations, with different compromises between efficiency and effectiveness. Figure 18.4 shows two

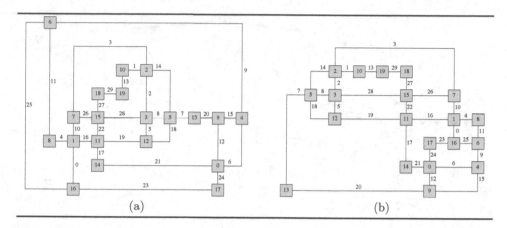

Figure 18.4 Two orthogonal representations of the same planar graph computed by GDToolkit. (a) An orthogonal representation with the minimum number of bends for a given planar embedding; (b) An orthogonal representation with the minimum number of bends over all possible planar embeddings of the graph. Node and edge identifiers are shown in the drawing.

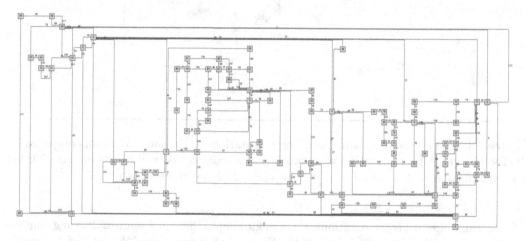

Figure 18.5 An orthogonal representation of a graph with 100 vertices.

examples of orthogonal representations of a planar graph G, one having the minimum number of bends within the given planar embedding of G, and the other having the minimum number of bends over all planar embeddings of G. The representation in (a) has been computed with an $O(n^2 \log n)$-time algorithm based on a flow technique that extends the one described in [Tam87]. The representation in (b) has been computed with an exponential-time algorithm based on a branch-and-bound technique, which enumerates and explores the embeddings of the graph using $SPQR$-trees. Both the polynomial-time algorithm and the exponential-time algorithm are described in [BDD00].Figure 18.5 shows an orthogonal representation of a graph with 100 vertices. The computation of a minimum-bend orthogonal drawing for an embedded planar graph with 100 vertices and 200 edges takes about 0.2 seconds under Linux on a machine with i5-540M Intel processor and 4 GB RAM. Computing a bend-minimum orthogonal drawing over all planar embeddings for a graph with 30 vertices and 50 edges takes about 10 seconds.

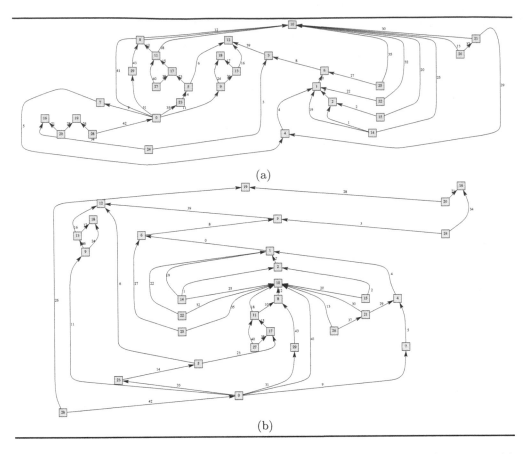

Figure 18.6 Two quasi-upward planar representations of the same digraph, computed by GDToolkit. (a) A representation with two bends on edge 5 and two bends on edge 29; (b) A representation with no bend, i.e., it is an upward planar representation.

Concerning upward representations, GDToolkit adopts the *quasi-upward* drawing convention defined by Bertolazzi et al. [BDD02]. We recall that an *upward drawing* of a directed graph is a drawing such that each vertex is represented as a distinct point of the plane and each edge is drawn as a simple curve monotonically increasing in the upward direction (i.e., from bottom to top), according to its orientation. An upward planar drawing is a drawing that is planar and upward a the same time. An upward planar drawing can exist only if the digraph is acyclic and admits a bimodal embedding. An *upward planar representation* is a partial description of an upward planar drawing, which defines the two linear lists of outgoing and incoming edges for each vertex, without fixing the final positions of the vertices. Unfortunately, acyclicity and bimodality are not sufficient conditions for the existence of an upward planar drawing, and in practice most digraphs do not admit such a layout. A quasi-upward drawing is a generalization of an upward drawing, which allows *bends* along the edges. A bend is a point in which the edge inverts its vertical direction, switching from upward to downward or vice-versa (if the edge is drawn as a smoothed curve, a bend along the edge is a point with horizontal tangent for the edge). The only requirement of a quasi-upward drawing is that for each directed edge (u, v), the edge enters v from below and leaves u from above. This implies that each edge has an even number of bends (possibly zero bends). Every digraph admits a quasi-upward drawing (even if it is acyclic) and a planar digraph admits a quasi-upward planar drawing if and only if it admits a bimodal embedding. Note that, an upward drawing can be regarded as a quasi-upward drawing with no bends along the edges. A *quasi-upward planar representation* is a partial description of a quasi-upward planar drawing; it defines the two linear lists of incoming and outgoing edges for each vertex and the sequence of bends along the edges.

As for orthogonal representations, GDToolkit provides different methods to compute a quasi-upward planar representation of a digraph. Figure 18.6 shows two examples of quasi-upward planar representations of the same digraph; the first representation is computed by using a flow-based $O(n^2 \log n)$-time algorithm that minimizes the number of bends within a given planar bimodal embedding; the second one is computed with a branch-and-bound exponential-time algorithm that minimizes the number of bends over all planar bimodal embeddings of the digraph. The algorithms for computing quasi-upward representations are those described in [BDD02]. In practice, the computation of a quasi-planar representation is very fast and takes less time then computing orthogonal drawings. For example, a quasi-planar representation of a bimodal planar digraph with 200 vertices and 240 edges is computed in about 0.05 seconds under Linux on a computer with i5-540M Intel processor and 4 GB RAM.

```
/* creates a graph ug, loading it from file "my-graph" */
undi_graph ug;
ug.read("my_graph");
/* computes a planar embedding for ug, with possible crossing nodes */
plan_undi_graph pug (ug);
/* computes an orthogonal shape for the planar embedded graph,
 * specifying the external face and the desired algorithm */
orth_plan_undi_graph opug (pug,pug.last_face(),PLAN_ORTH_OPTIMAL);
```

Figure 18.7 A fragment of code that creates an orthogonal shape of a graph.

GDToolkit also offers the possibility of orienting an undirected embedded planar graph in such a way that the number of its sources and sinks is minimized and it has an upward planar representation. As described in [DP03], this helps in the implementation of drawing algorithms for visibility representations in case the graph is not biconnected. Observe that, for a biconnected graph an upward orientation with the minimum number of sources and sinks always coincides with an *st*-orientation of the graph.

Objects of classes *orth_plan_undi_graph* and *upwa_plan_undi_graph* are usually constructed from *plan_undi_graph* objects, by specifying the wanted layout algorithm. It is also possible to specify an external face if the selected algorithm preserves the planar embedding. To give an example, consider the simple code in Figure 18.7. It constructs an *orth_plan_undi_graph* object opug by the *plan_undi_graph* object pug. When opug's constructor is invoked, a face of pug is chosen to be the external face; if such a face is not specified, it is chosen as the first in the list of faces of pug. The algorithm PLAN_ORTH_OPTIMAL selected to construct opug corresponds to the algorithm that computes an orthogonal representation of the graph in the simple-podevsnef model, with the minimum number of bends within the given planar embedding.

18.3.3   Metrics Level

At the bottom level of the hierarchy GDToolkit provides the *draw_undi_graph* class, which is very easy to use. Indeed, an object of this class is an *undi_graph* object with additional basic geometric information, like vertex-coordinates and bend-coordinates; *draw_undi_graph* objects are also equipped with some attributes to define colors and labels for vertices and edges.

The basic philosophy of the *draw_undi_graph* class is to provide one or more constructors from each other graph-class of the library. Often, GDToolkit provides different algorithms to compute a drawing in a specific convention; each algorithm has a different trade-off between drawing aesthetics and time performance. For instance, an orthogonal drawing can be computed from an *orth_plan_undi_graph* object by selecting an algorithm in a wide set of compaction algorithms, obtained by combining different alternatives like:

- Decomposing the faces of the orthogonal representation into rectangles [Tam87] or into *regular faces* [BBD+00].

- Computing the coordinates of vertices and bends with a linear-time algorithm based on topological numbering or with an $O(n^2 \log n)$-time algorithm based on flow-techniques [DETT99].

- Applying or not a one-dimensional compaction post-processing to further reduce the area and the total edge length of the drawing, if possible.

The code in Figure 18.8 computes two different orthogonal drawings with the same shape. The first drawing, **dug1**, is computed by applying the fastest compaction algorithm in the library, while the second one, **dug2**, is constructed by using the slowest compaction algorithm. The resulting drawings, **dug1** and **dug2**, are depicted in Figure 18.9; observe that **dug2** is much more compact in terms of area and total edge length.

```
/* creates a graph ug, loading it from file "my-graph" */
undi_graph ug;
ug.read("my_graph");
/* computes a planar embedding for ug, with possible crossing nodes */
plan_undi_graph pug (ug);
/* computes an orthogonal shape for the embedded graph */
orth_plan_undi_graph opug (pug);

/* computes two drawings of the orthogonal shape,
 * using different compaction algorithms */
draw_undi_graph dug1 (opug, FAST_COMPACTION);
draw_undi_graph dug2 (opug, SLOW_REGULAR_COMPACTION_2_REFINED);
```

Figure 18.8   A fragment of code that computes two different orthogonal drawings with the same shape.

As another example, visibility and polyline drawings can be directly computed from an object **pug** of class *plan_undi_graph*, by choosing between a linear-time compaction algorithm or a polynomial-time compaction algorithm based on flow techniques [Did00]. Indeed, for these kind of drawing conventions the concept of shape is not defined. Figure 18.10 shows two visibility drawings and two polyline drawings of the same embedded planar graph. The

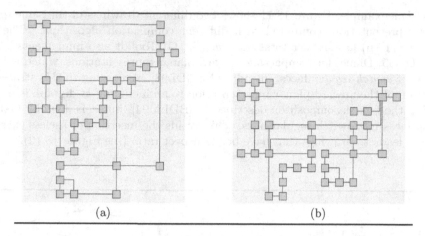

Figure 18.9 (a) Drawing `dug1` and (b) drawing `dug2`, computed with the code of Figure 18.8. The two drawings have the same shape but different geometry. The drawing in (b) is much more compact, both in terms of area and in terms of total edge length.

drawings in Figures 18.10 (a)-(b) are obtained by executing a linear-time drawing algorithm, while the drawings in Figures 18.10 (c)-(d) are obtained by applying an $O(n^2 \log n)$-time compaction algorithm that reduces the total edge length.

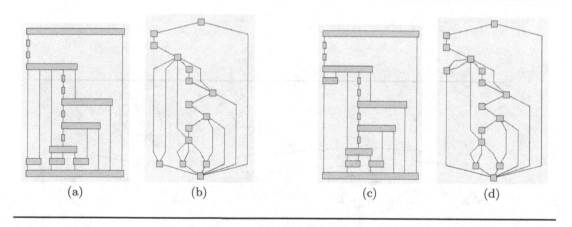

Figure 18.10 Two visibility drawings and two polyline drawings of the same embedded planar graph. The total edge length of the drawings (c) and (d) is smaller than the one of the drawing (a) and (b).

It is also interesting to observe that, since a quasi-upward drawing can be computed by using a visibility representation as intermediate step, the same compaction algorithms applied above can be used for computing a quasi-upward drawing of a quasi-upward representation. For example, Figure 18.11 shows two different drawings of the same quasi-upward planar representation, computed with different compaction algorithms. The drawing in Figure 18.11 (b) has smaller total edge length. GDToolkit also implements a recent algorithm [Did05, Did06] for compacting upward planar representations, which is based on the concept of *switch-regular* faces, introduced in [DL98]. According to this strategy, the augmentation of the upward planar representation to an including *st*-digraph is not performed by using the face decomposition described in [BDLM94], but it is done by first decomposing the faces into switch-regular ones. This avoids the insertion of useless extra edges and typically leads to drawings that have better aspect ratio (see Figure 18.12).

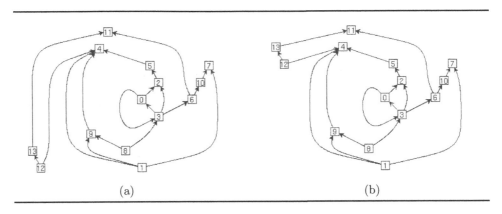

Figure 18.11   Two different quasi-upward drawings of the same quasi-upward representation. The total edge length of the drawing in (b) is smaller that the total edge length of the drawing in (a).

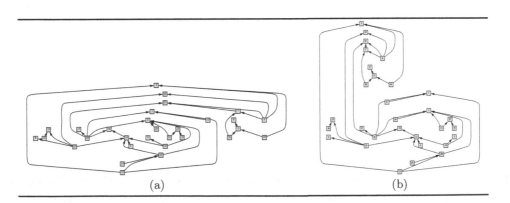

Figure 18.12   Two upward drawings of the same upward representation. (a) The drawing has been computed using the standard augmentation technique described in [BDLM94]. (b) The drawing has been computed with the new algorithm described in [Did05].

18.4 Constructors

As observed in the previous sections, a drawing algorithm in GDToolkit typically reflects in a path of constructors. For this reason, constructors play a crucial role in the library and they are written following a common pattern, which is depicted in Figure 18.13.

Suppose that a graph-class B inherits a graph-class A. According to the pattern of Figure 18.13, a constructor of B first invokes a constructor or a copy operator of A to transfer the inherited information; then, the constructor of B invokes a private method, local_new, that allocates memory for the local structures that are needed to store additional information, and finally it calls another private method, loca_init, that computes and stores the data in the new local structures.

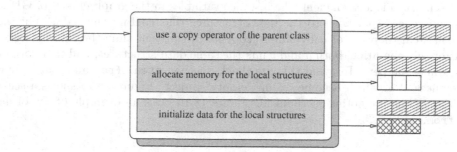

Figure 18.13 A schematic illustration of the design pattern of GDToolkit constructors.

As a concrete example, imagine that a *plan_undi_graph* object pug is created from an *undi_graph* object ug. Object pug must have the same nodes and edges as ug and, additionally, it defines a set of faces and possible extra nodes that replace crossings. The construction of the planar embedding of pug is done by applying a planarization algorithm on the topology of ug, possibly subject to some planarization constraints (see Section 18.5). Figure 18.14 shows the code of a constructor of class *plan_undi_graph*. Parameter po specifies if the new graph object must have the same embedding as ug or not. Parameter err_mess enables/disables an error-handler in the case some planarization constraints can not be satisfied. Method local_new allocates memory for the list of faces, while method local_init executes the planarization algorithm.

```
    plan_undi_graph::
plan_undi_graph (const undi_graph& ug, planarize_options po, bool err_mess)
  {
  /* copies the basic structure of the graph (nodes and edges) */
  undi_graph::operator=(ug);
  /* creates the additional data structures required by
   * a plan_undi_graph object */
  local_new();
  /* executes a planarization algorithm to computes faces
   * and related objects */
  local_init(po,err_mess);
  }
```

Figure 18.14 A constructor of class *plan_undi_graph*. The code reflects the pattern illustrated in Figure 18.13.

As mentioned in Section 18.2, another key aspect of GDToolkit is the possibility of constructing a new graph-object by means of a *promoting* mechanism. Suppose for example that *a* is an *undi_graph* object and suppose we want to construct a *plan_undi_graph* object *b* with the same set of vertices and edges as *a*. As explained above, *b* enriches the information stored in *a* with a set of faces, which defines a planar embedding for *a*. Suppose also that *a* is no longer needed in the program after the construction of *b*; indeed, *b* contains a super-set of information of *a*. In this situation it could be useful to get *b* as the result of a promoting procedure applied to *a* that avoids duplication of data, so saving computational time and memory space. The graph-classes of GDToolkit support such a promoting mechanism by means of a public method, called `steal_from`. Referring to the example above, method `steal_from` invoked on *b* "steals" the data-structures of *a* and then constructs a set of new data-structures to store the additional information of *b* (in the specific example a set of faces). To make this idea efficient, the instance variables in the graph-classes of GDToolkit are just references (pointers) to the data-structures that contain the data. This implies that *b* can steal the data of *a* by simply copying in constant time the internal references of *a*. After this operation, both *a* and *b* link the same data-structures, and therefore update collisions may happen. To avoid this drawback, method `steal_from` automatically cleans the references of *a*, which becomes as an "empty" object. Figure 18.15 shows a schematic description of the promoting mechanism. Figure 18.16 gives an example of use of method `steal_from`.

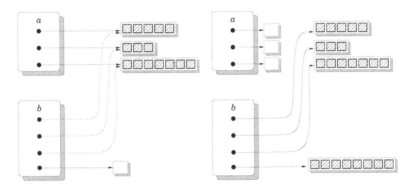

Figure 18.15 A schematic description of the promoting mechanism. Object *b* is the result of the promoting, and object *a* is made useless after the promoting process.

```
/* creats an undi_graph object (*ug), loading it from file "my-graph" */
undi_graph *ug = new undi_graph();
(*ug).read ("my-graph");
/* computes an empty planar embedded graph */
plan_undi_graph pug();
/* initializes pug with the nodes and edges of (*ug);
 * object (*ug) will be useless from now on */
pug.steal_from(*ug);
/* (*ug) is deallocated from the main memory */
delete(ug);
```

Figure 18.16 A *plan_undi_graph* object is constructed promoting an *undi_graph* object.

18.5   Management of Constraints

GDToolkit is equipped with a flexible architecture for managing constraints. Different types of constraint can be concurrently applied on the graph, which are taken into account by the involved layout algorithms. A *constraint* type is a reference to an object as well as types *node* and *edge*, and each constraint still has a unique identifier. The *undi_graph* class provides a set of methods for adding, removing, and copying constraints.

Constraints in GDToolkit have a special "intelligent" management system, which is explained in the following points.

- Each constraint is described by a specific set of parameters that depends on the type of the constraint itself. For example, a constraint that makes an edge *e* not crossable is described by the only parameter *e*; a constraint that forces a vertex *v* to have height *h* and width *w* is described by the triple (v, h, w). In addition, each constraint has an internal read-only parameter that specifies the type of constraint. This type can be accessed by means of a public method.

- A graph G' that is obtained as a copy or by inheritance of a graph G, also inherits all constraints of G. Furthermore, constraints react according to their type to all the relevant events occurring on their node and edge parameters. More precisely, each type of constraint is represented by a specific class that encapsulates its behavior with respect to changes of the nodes and edges involved in the constraint. An abstract class provides the set of virtual reaction methods common to all the derived constraint classes, and each derived constraint class provides its own implementation for each reaction method.

- The main events with a potential impact on a constraint applied on a given node/edge are the deletion, the split, and the merge of that node/edge. For each of these events, each constraint class defines a reacting method. For example, if an edge *e* is split into two edges e_1 and e_2, a reaction method called `update_after_edge_split()` is automatically invoked on all the constraints applied on *e*, so that each constraint executes its specific implementation of this method.

- Theoretically, any number of constraints can be set on a graph at any time. However, each algorithm decides its own policy about each kind of constraint. This means that the programmer can decide to implement an algorithm that takes into account or not a specific type of constraint. Also, some constraints could be not compatible to each other; in this case, an algorithm that takes them into account, typically causes an error.

GDToolkit currently offers several types of predefined constraints involving both topology, shape, and metrics. The use of constraints in the topology-shape-metrics framework have been addressed in several papers, including [BDLN05, CGM+10, DDLP10, EFK00, GKM08, Tam98]. GDToolkit implements some of the constraints described in the literature or their variants. However, any programmer can define a new constraint by extending the base abstract class and by providing an implementation for each reaction method. In the following the main predefined constraints of GDToolkit are described.

18.5.1   Topology Constraints

Concerning the topology of a graph, GDToolkit provides three different types of constraints; all of them are taken into account by the planarization algorithm.

The first type of constraint imposes that an edge e is not allowed to cross any other edge. If edge e is split into two edges e_1 and e_2, the constraint is propagated on both e_1 and e_2. Symmetrically, if an edge e is obtained by merging two edges e_1, e_2, and at least one of them is not crossable, then e will become not crossable too. If the planarization algorithm encounters an edge e that is not crossable, it omits to insert its dual edge in the dual graph of the planar embedded graph computed so far. This implies that a shortest path in the dual graph never intersects e.

The second type of constraint imposes that a specified set of vertices $\{v_1, v_2, \ldots v_k\}$ belongs to the same face. In order to maintain this property, the planarization algorithm temporarily adds to the graph a *star gadget*, consisting of a dummy vertex u and dummy edges $(u, v_1), (u, v_2), \ldots, (u, v_k)$, where the dummy edges of the star are made not crossable, applying on them the previous type of constraint. The star gadget is removed at the end of the planarization process. Figure 18.17 shows an example of application of this constraint.

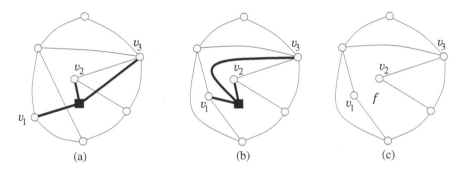

(a)   (b)   (c)

Figure 18.17   Illustration of the star gadget used to force a set of vertices to stay in the same face. In this example, the vertices are v_1, v_2, v_3. (a) A star gadget is added; it consists of the square black vertex and of the bold edges. (b) A planar embedding of the enhanced graph is computed; (c) The final planar embedding for the input graph. At the end of the planarization process, v_1, v_2, v_3 belong to the same face f.

The third type of topological constraint is a variant of the previous one. It imposes that a certain set of vertices $\{v_1, v_2, \ldots, v_k\}$ belongs to the same face f and that these vertices circularly occur on the boundary of f in the specified order. To satisfy this constraint, the planarization algorithm uses the star gadget shown above, with the additional property that the circular sequence of edges incident to the dummy vertex of the star gadget is fixed.

Figure 18.18 shows two orthogonal drawings: The drawing in (a) has been obtained without any topological constraint. The drawing in (b) has been computed imposing that vertices $7, 12, 1, 0$ belong to the same face (the face is highlighted). In order to satisfy this constraint, the planarization algorithm introduced some edge-crossings.

18.5.2   Shape Constraints

At the shape level, GDToolkit provides several predefined constraints that are taken into account by the flow-based algorithms that compute orthogonal and quasi-upward representations. These constraints are listed and discussed below.

- *Number of bends per edge.* This constraint can be applied on an edge e, in order to establish a certain policy in bending e. Two different policies can be applied:

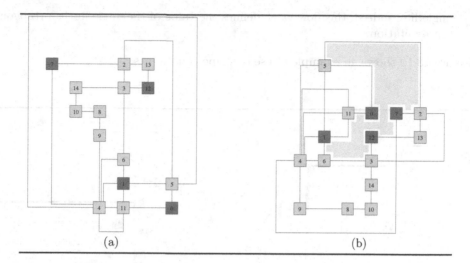

(a)   (b)

Figure 18.18 Two orthogonal drawings of the same graph. (a) The drawing has been computed with no constraint. (b) The drawing has been computed forcing vertices $7, 12, 1, 0$ to stay in the same face.

> – Edge e must have zero bends, i.e., it must be a straight-line edge.
>
> – Edge e can have any number of bends. This means that the algorithm will assign a zero cost to each bend of e, and therefore e will turn any time this avoids to bend other edges.

Each of the two policies is translated into a suitable constraint in the flow network associated with the orthogonal or quasi-upward representation. We recall that in such a flow network (see also [BDD02, DETT99, Tam87]) there is a node v_f for each face f of the graph and there is a pair of directed arcs $e_{fg} = (v_f, v_g), e_{gf} = (v_g, v_f)$ for each edge e shared by two (possibly coincident) faces f and g. The flow along the arcs e_{fg}, e_{gf} determines the right bends and the left bends along e in the final representation. In order to guarantee that e has no bend in the representation, it is sufficient to set an infinite cost (or zero upper capacity) on e_{fg}, e_{gf}. Conversely, in order to assign the highest turn priority to e, one can assign cost zero and infinite upper capacity to both e_{fg} and e_{gf}.

- *Turn direction.* This constraint forces an edge $e = (u, v)$ to turn only in a specified direction. This means that e can be forced to have only right bends or only left bends, while moving on it either from u or from v. To implement this constraint, we just remove in the flow network one of the two arcs e_{fg}, e_{gf}, where f and g are the two (possibly coincident) faces shared by e.

- *Angle type.* This last constraint allows the programmer to decide the value that a specified angle must have in an orthogonal representation. Possible angle values in degrees are $\{0, 90, 180, 270, 360\}$. The constraint is specified by a triple (e, v, a), where e is an edge incident to v, and a is the value of the angle formed at v between edge e and its successive edge in clockwise order around v. This type of constraint is still translated into a suitable constraint in the flow network associated with the orthogonal representation. More precisely, it is sufficient to fix the value of the flow along the arc of the network that connects v to the face in which the angle lies; indeed, this flow value defines the value of the angle in the orthogonal representation.

Figure 18.19 shows an example of use of shape constraints.

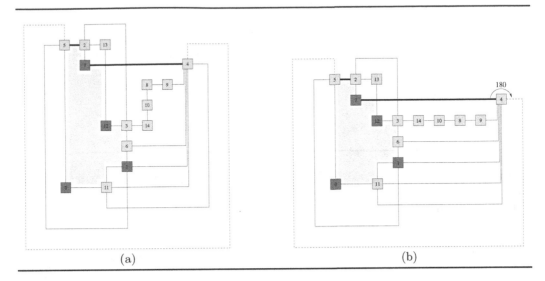

| (a) | (b) |

Figure 18.19   (a) The orthogonal drawing has been computed from the graph of Figure 18.18, with the constraint that edges $(5, 2), (4, 7)$ cannot bend (the edges are in bold), while edge $(5, 2)$ can have any number of bends (the edge is dashed). (b) An orthogonal drawing computed by adding the further constraint that the angle at vertex 4 to the right of edge $(4, 7)$ is a 180 degrees angle.

18.5.3   Metrics Constraints

Concerning the metrics of a drawing, GDToolkit currently offers two predefined constraints for orthogonal drawings.

The first constraint allows the programmer to customize the size of each vertex, independently to each other. More in details, every vertex v can be drawn has a rectangle having a predefined width w and a predefined height h in terms of units of an integer coordinate grid. In absence of constraints, v will be drawn as a small rectangle that occupies only a grid unit, that is, v will have width and height equal to zero. The constraint on the dimension of the nodes is handled in the compaction step of the topology-shape-metrics approach, by using the flow-based technique described in [DDPP99]. Figure 18.20 shows two orthogonal drawings of the same embedded planar graph. In the drawing of Figure 18.20 (a) all vertices have dimensions zero, while in the drawing of Figure 18.20 (b) some vertices have been expanded imposing constraint dimensions. Observe that the shape of the two drawings is the same.

The second constraint makes it possible to specify the points where an edge will be incident to a side of a vertex. More precisely, assume that an edge e is incident to a vertex v. An orthogonal representation fixes the side s of v on which e will be incident. If on v a

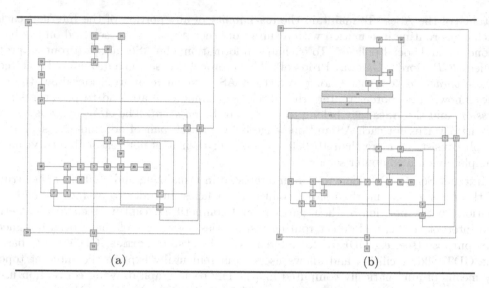

(a)   (b)

Figure 18.20 Two orthogonal drawings with the same shape: (a) The drawing has no constraint; (b) The dimensions of some vertices have been preassigned.

dimension constraint has been fixed so that s has length l, the programmer can impose any distance $d \leq l$ between the incidence point of e on s and a corner of s (see Figure 18.21).

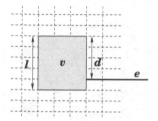

Figure 18.21 Illustration of the constraint that makes it possible to fix the incidence point of an edge on the side of a vertex in an orthogonal drawing.

18.6 Examples of Applications

The GDToolkit library has been effectively used to develop several applications in different real-world domains, which is a proof of its flexibility. In the following we briefly discuss some of these applications.

18.6.1 Internet Analysis

At a high level of abstraction, the Internet can be seen as a network of so called *Autonomous Systems*. An Autonomous System (AS in the following) is a group of sub-networks under the same administrative authority, and is identified by a unique integer number. In this sense, an AS can be seen as a portion of the Internet, and the Internet can be seen as the totality of the ASes. To maintain the reachability of any portion of the Internet, each AS exchanges routing information with a subset of other ASes, mainly selected on the basis of economic and social policies. To exchange information, the ASes adopt a routing protocol called *BGP* (Border Gateway Protocol). This protocol is based on a distributed architecture where *border routers* that belong to distinct ASes exchange information about the routes they know. Two border routers that directly communicate are said to perform a *peering session*, and the ASes they belong to are said to be *adjacent*. The *ASes graph* is the graph having a vertex for each AS and one edge between each pair of adjacent ASes. The ASes graph consists of more than 10,000 vertices and then it is not reasonable to visualize it completely on a computer screen.

Internet Service Providers are often interested in visualizing and analyzing the structure of the ASes graph and the related connection policies, in order to extract valuable information on the position of their partners and competitors, capture recurrent patterns in the Internet traffic, and detect routing instabilities. Several tools have been designed for this purpose (see, e.g., [DK01] for references). The system Hermes [CDD+02] is based on the GDToolkit facilities, and allows users to incrementally explore the Internet topology by means of automatically computed maps. The basic graph drawing convention used to represent the maps is the Kandinsky model for orthogonal drawings. However, since the handled graphs often have many vertices of degree one connected to the same vertex, the Kandinsky model is enriched with new features for effectively representing such vertices.

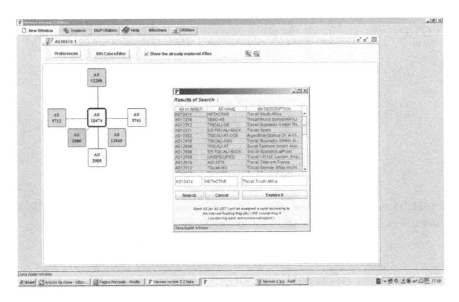

Figure 18.22   A map showing the ASes adjacent to AS10474, NETACTIVE, Tiscali South Africa. (Figure taken from [DL07].)

The graphical user interface of Hermes offers several exploration facilities. The user can search for a specific AS and can start the exploration of the Internet from that AS. At each successive step, the user can display information about the routing policies of the ASes contained in the current map, or she can expand the map by exploring one of these ASes. For example, Figure 18.22 shows a snapshot of the system where the AS10474 (NETACTIVE, Tiscali South Africa) is searched and selected by the user for exploration; a first map that consists of the ASes adjacent to AS10474 is then automatically computed

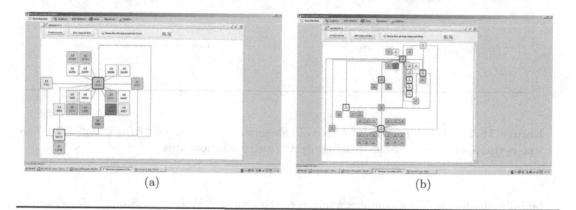

(a) (b)

Figure 18.23 (a) A new map obtained from the previous map by exploring AS11845. (b) A more complex map obtained by performing several exploration steps. (Figure taken from [DL07].)

and displayed by the system. Figure 18.23 shows how the map is expanded when the user decides to explore other ASes.

18.6.2 Web Searching

The output of a classical Web search engine consists of an ordered list of links (URLs) that are selected and ranked according to the user's query, the documents content, and (in some cases, like `Google`) the popularity of the links in the World Wide Web. The returned list can however consist of several hundreds of URLs and users may omit to check some URLs that might be relevant for them just because these links do not appear in the first positions of the list.

A *Web meta-search clustering engine* is a system conceived to support the user in retrieving data from the Web by overcoming some of the limitations of traditional search engines. A Web meta-search clustering engine provides a visual interface to the user who submits a query; it forwards the query to (one or more) traditional search engines, and returns a set of clusters, also called *categories*, which are typically organized into a hierarchy. Each category contains URLs of documents that are semantically related to each other and is labeled with a string that describes its content. As a consequence, the user of a meta-search clustering engine has a global view of the different semantic areas involved by her query and can more easily retrieve the Web data relative to those topics in which she is interested.

Although an effective representation of the categories and of their semantic relationships is essential for efficiently retrieving the wanted information, most Web meta-search clustering engines (see, e.g., `Vivísimo`, `iBoogie`[1], `SnakeT`[2] [FG04, FG05]) have a GUI in which the hierarchy of clusters is displayed as a tree. However, this type of representation may not

---

[1]http://www.iboogie.com/
[2]http://snaket.di.unipi.it/

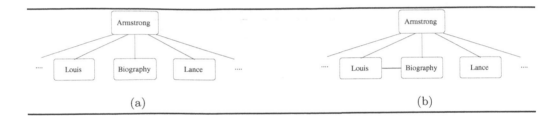

Figure 18.24   (a) A portion of a tree of categories for the query "Armstrong". (b) The tree is equipped with an edge that highlights cluster relationships.

be fully satisfactory for a complex analysis of the returned Web data. Suppose for example that the user's query is "Armstrong" and that the clusters hierarchy returned by a Web meta-search clustering engine is the tree depicted in Figure 18.24 (a). Is the category "Biography" related to "Louis" or to "Lance" or to both (or to no one of them but to the astronaut Neil Armstrong?). If instead of a tree, the systems returned a graph as the one in Figure 18.24 (b), the user would be facilitated in deciding whether the category "Biography" is of her interest.

WhatsOnWeb [DDGL05, DDGL06, DDGL07] is a meta-search clustering engine that makes it possible to retrieve data from the Web by using drawings of graphs. The nodes represent categories of semantically coherent URLs and the edges describe relationships between pairs of categories. The graphical environment of WhatsOnWeb consists of two frames (see, e.g., Figure 18.25). In the left hand side frame the hierarchy of categories is represented as a classical directories tree. In the right hand side frame the user interacts with the drawing of a *clustered graph* [FCE95], where each cluster coincides with a semantic category.

The drawing is computed using the orthogonal drawing algorithms of GDToolkit. The user can expand/contract clusters in the graph and the drawing changes accordingly. Using the constraint dimensions described in Section 18.5.3, each cluster is drawn as a box having the minimum size required to host just its label (if the cluster is contracted) or a drawing of its sub-clusters (if the cluster is expanded). The map in Figure 18.25 (a) shows a snapshot of the interface, where the results for the query "Armstrong" are presented; in the figure, the category "Louis Armstrong" has been expanded by the user. In order to preserve the user mental map during the browsing, WhatsOnWeb preserves the orthogonal shape of the drawing during after every expansion or contraction operation. For example, Figure 18.25 (b) shows the map obtained by expanding the categories "Jazz", "School", and "Louis Armstrong Stamp" in the first map.

18.6.3   Database Analysis

The third example of real-world application based on GDToolkit is focused on the analysis of a relational database. The *logical schema* of a relational database (also called *relational schema*) describes the database as a set of *tables*, where each table consists of a set of *attributes*. Links between tables might be present. A link between two tables A and B represents either an *integrity constraint* or a *join* between an attribute of A and an attribute of B; these two attributes are called the *attributes of the link*.

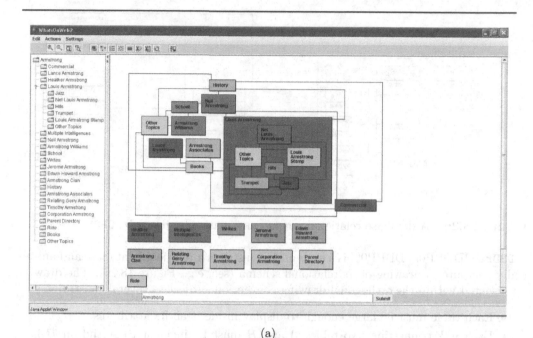

(a)

(b)

Figure 18.25 Snapshots of the user interface of WhatsOnWeb. (a) A map for the query "Armstrong"; in the map the user performed the expansion of the category "Louis Armstrong". (b) A subsequent map obtained by expanding the categories "Jazz", "School", and "Louis Armstrong Stamp"; this last category contains two URLs, described by reporting their titles. (Figure taken from [DL07].)

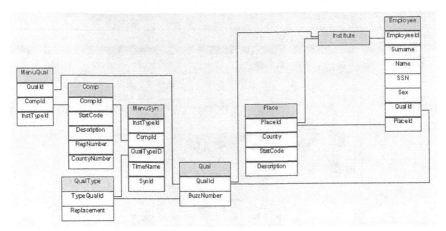

Figure 18.26   A database relational schema automatically drawn by `DBDraw`.

`DBDraw` [DDPP03, DDPP02] is a system that inspects a relational database and automatically computes a drawing of its relational schema (see, e.g., Figure 18.26). The drawing is represented within the orthogonal drawing convention subject to several constraints:

- Each table must be large enough to display inside it all its attributes.
- Each link connecting two tables A and B must be incident to A and on B in correspondence of the attributes of the link.
- Links cannot be incident to a table from north or from south.

The three constraints above are enforced by using the topology constraints and the metrics constraints described in Section 18.5.1 and Section 18.5.3.

Acknowledgements

Many people contributed to the development of GDToolkit, other than the authors of this chapter. We wish to warmly tank some of them whose contribution has been crucial for the success of the project. In alphabetic order, thanks to: Pier Francesco Cortese, Antonio Leonforte, Alessandro Marcandalli, Francesco Matera, Maurizio Patrignani, and Maurizio Pizzonia.

References

[AMO93]   R. K. Ahuja, T. L. Magnanti, and J. B. Orlin. *Network Flows: Theory, Algorithms, and Applications*. Prentice Hall, Englewood Cliffs, NJ, 1993.

[BBD+00]   S.S. Bridgeman, G. Di Battista, W. Didimo, G. Liotta, R. Tamassia, and L. Vismara. Turn-regularity and optimal area drawings of orthogonal representations. *Computational Geometry: Theory and Applications*, 16:53–93, 2000.

[BBDL91]   M. Beccaria, P. Bertolazzi, G. Di Battista, and G. Liotta. A tailorable and extensible automatic layout facility. In *Proc. IEEE Workshop on Visual Languages (VL '91)*, pages 68–73, 1991.

[BCPD04]   J. M. Boyer, P. F. Cortese, M. Patrignani, and G. Di Battista. Stop minding your P's and Q's: Implementing a fast and simple dfs-based planarity testing and embedding algorithm. In *Proc. 11th Symposium on Graph Drawing, LNCS*, volume 2912, pages 25–36, 2004.

[BDD00]   P. Bertolazzi, G. Di Battista, and W. Didimo. Computing orthogonal drawings with the minimum number of bends. *IEEE Trans. on Computers*, 49(8):826–840, 2000.

[BDD02]   P. Bertolazzi, G. Di Battista, and W. Didimo. Quasi-upward planarity. *Algorithmica*, 32(3):474–506, 2002.

[BDLM94]   P. Bertolazzi, G. Di Battista, G. Liotta, and C. Mannino. Upward drawings of triconnected digraphs. *Algorithmica*, 6:476–497, 1994.

[BDLN05]   Carla Binucci, Walter Didimo, Giuseppe Liotta, and Maddalena Nonato. Orthogonal drawings of graphs with vertex and edge labels. *Comput. Geom.*, 32(2):71–114, 2005.

[CDD+02]   A. Carmignani, G. Di Battista, W. Didimo, F. Matera, and M. Pizzonia. Visualization of the high level structure of the internet with HERMES. *Journal of Graph Algorithms and Applications*, 6(3):281–311, 2002.

[CGM+10]   Markus Chimani, Carsten Gutwenger, Petra Mutzel, Miro Spönemann, and Hoi-Ming Wong. Crossing minimization and layouts of directed hypergraphs with port constraints. In *Graph Drawing*, volume 6502 of *Lecture Notes in Computer Science*, pages 141–152, 2010.

[DDGL05]   E. Di Giacomo, W. Didimo, L. Grilli, and G. Liotta. A topology-driven approach to the design of web meta-search clustering engines. In *Theory and Practice of Computer Science (SOFSEM '05)*, volume 3381 of *Lecture Notes in Computer Science*, pages 106–116, 2005.

[DDGL06]   E. Di Giacomo, W. Didimo, L. Grilli, and G. Liotta. Using graph drawing to search the web. In *13th International Symposium on Graph Drawing, GD 2005*, volume 3843 of *Lecture Notes in Computer Science*, pages 480–491, 2006.

[DDGL07]   Emilio Di Giacomo, Walter Didimo, Luca Grilli, and Giuseppe Liotta. Graph visualization techniques for web clustering engines. *IEEE Trans. Vis. Comput. Graph.*, 13(2):294–304, 2007.

[DDLP10]   Emilio Di Giacomo, Walter Didimo, Giuseppe Liotta, and Pietro Palladino. Visual analysis of one-to-many matched graphs. *J. Graph Algorithms Appl.*, 14(1):97–119, 2010.

[DDPP99] G. Di Battista, W. Didimo, M. Patrignani, and M. Pizzonia. Orthogonal and quasi-upward drawings with vertices of prescribed size. In *Symposium on Graph Drawing (GD'99)*, volume 1731 of *LNCS*, pages 297–310, 1999.

[DDPP02] G. Di Battista, W. Didimo, M. Patrignani, and M. Pizzonia. Drawing database schemas. *Software - Practice and Experience*, (32):1065–1098, 2002.

[DDPP03] G. Di Battista, W. Didimo, M. Patrignani, and M. Pizzonia. DBDraw - automatic layout of relational database schemas. In M. Jünger and P. Mutzel, editors, *Graph Drawing Software*, pages 237–256. Springer-Verlag, 2003.

[DETT99] G. Di Battista, P. Eades, R. Tamassia, and I. G. Tollis. *Graph Drawing*. Prentice Hall, Upper Saddle River, NJ, 1999.

[DGST90] G. Di Battista, A. Giammarco, G. Santucci, and R. Tamassia. The architecture of Diagram Server. In *Proc. IEEE Workshop on Visual Languages (VL '90)*, pages 60–65, 1990.

[Did00]   W. Didimo. *Flow Techniques and Optimal Drawing of Graphs*. PhD thesis, Dipartimento di Informatica e Sistemistica, Univeristà di Roma"La Sapienza", 2000.

[Did05]   W. Didimo. Computing upward planar drawings using switch-regularity heuristics. In *Theory and Practice of Computer Science (SOFSEM '05)*, volume 3381 of *LNCS*, pages 117–126, 2005.

[Did06]   Walter Didimo. Upward planar drawings and switch-regularity heuristics. *J. Graph Algorithms Appl.*, 10(2):259–285, 2006.

[DK01]   M. Dodge and R. Kitchin. *Atlas of Cyberspace*. Addison Wesley, 2001.

[DL98]   G. Di Battista and G. Liotta. Upward planarity checking: "faces are more than polygons". In *Symposium on Graph Drawing (GD'98)*, volume 1547 of *LNCS*, pages 72–86, 1998.

[DL07]   W. Didimo and G. Liotta. *Mining Graph Data*, chapter Graph Visualization and Data Mining, pages 35–64. Wiley, 2007.

[DP03]   W. Didimo and M. Pizzonia. Upward embeddings and orientations of undirected planar graphs. *Journal of Graph Algorithms and Applications*, 7(2):221–241, 2003.

[DT96]   G. Di Battista and R. Tamassia. On-line planarity testing. *SIAM Journal on Computing*, 25:956–997, 1996.

[EFK00]   Markus Eiglsperger, Ulrich Fößmeier, and Michael Kaufmann. Orthogonal graph drawing with constraints. In *SODA*, pages 3–11, 2000.

[FCE95]   Q. Feng, R. F. Choen, and P. Eades. How to draw a planar clustered graph. In *COCOON'95*, volume 959 of *LNCS*, pages 21–31, 1995.

[FG04]   P. Ferragina and A. Gullí. The anatomy of a hierarchical clustering engine for web-page, news and book snippets. In *Fourth IEEE International Conference on Data Mining (ICDM'04)*, pages 395–398, 2004.

[FG05]   P. Ferragina and A. Gullí. A personalized search engine based on web-snippet hierarchical clustering. In *14th international conference on World Wide Web*, pages 801–8106, 2005.

[FK96]   U. Fößmeier and M. Kaufmann. Drawing high degree graphs with low bend numbers. In *Symposium on Graph Drawing (GD'95)*, volume 1027 of *LNCS*, pages 254–266, 1996.

[GKM08]   Carsten Gutwenger, Karsten Klein, and Petra Mutzel. Planarity testing and optimal edge insertion with embedding constraints. *J. Graph Algorithms Appl.*, 12(1):73–95, 2008.

[MN95]   K. Mehlhorn and S. Näher. LEDA: A platform for combinatorial and geometric computing. *Commun. ACM*, 38(1):96–102, 1995.

[MN00]   K. Mehlhorn and S. Näher. *LEDA: A Platform for Combinatorial and Geometric Computing.* Cambridge University Press, Cambridge, UK, 2000.

[Tam87]   R. Tamassia. On embedding a graph in the grid with the minimum number of bends. *SIAM J. Comput.*, 16(3):421–444, 1987.

[Tam98]   Roberto Tamassia. Constraints in graph drawing algorithms. *Constraints*, 3(1):87–120, 1998.

19

PIGALE

Hubert de Fraysseix
CNRS UMR 8557. Paris

Patrice Ossona de Mendez
CNRS UMR 8557. Paris

19.1 Introduction

This chapter gives an overview of the *Public Implementation of a Graph Algorithm Library and Editor* (Pigale). Pigale integrates a graph algorithm library written in C++ and a graph editor based on the Qt© and OpenGL™ libraries. This program runs under Linux, Mac OS X™ and Windows™ platforms. It is particularly intended for academic researchers working on topological graph theory.

Pigale is available under GPL[1] license and may be downloaded on sourceforge.net at http://pigale.sourceforge.net. Pigale may be used as a library, as a graph editor or as a multi-threaded graph algorithm server.

The GNU General Public License is a free, copyleft license for software and other kinds of works. the GNU General Public License is intended to guarantee your freedom to share

---

and change all versions of a program — to make sure it remains free software for all its users (see http://www.gnu.org/licenses/gpl.html).

The library is built on an original data structure. This data structure optimizes operations performed on static graphs.

19.1.1   Why GPL?

Free software has the following advantages, which we believe are essential for academic software:

- It increases the exchanges between research centers and facilitates the integration of algorithms originating from several contexts into a coherent framework, thus inducing *de facto* new standards in the concerned field.

- It increases the visibility of the laboratories's skills, thus offering a showcase toward potential industrial partners and allows the development of industrial software based on well-designed license-free libraries.

- It allows to reduce the economic gap between rich and poor countries and contributes to the competitiveness of local laboratories and companies by reducing the cost linked to the acquisition of foreign licenses.

- It allows the users to control the source code of sometimes strategic modules of their projects and suppresses the dramatic dependence on a single software provider, which ties the users to the perennity and the goodwill of a particular actor.

19.1.2   Chapter Organization

The rest of this chapter is organized as follows. Section 19.2 discusses data structures for representing graphs and their embeddings. In Section 19.3, we describe fundamental graph algorithms provided by Pigale. The map generators available in Pigale are outlined in Section 19.4. In Section 19.5, we present the drawing algorithms supported by Pigale. The implementation of Pigale, including the graphical interface for creating graphs in Pigale is illustrated in Section 19.6. Finally, in Section 19.7, we show an example of use of Pigale as a software libary.

19.2   Data Structures

In this section, we present the graph model and data structures we have developed in Pigale.

19.2.1   The Topological Quasi-Static Model

Pigale provides two main graph data structures, depending on whether one considers dense graphs or sparse ones:

- For dense graphs, a matrix is used, which represents the adjacency relation among vertices or the vertex-edge incidence relation;

- For sparse graphs, either a list of incidences (i.e., a list of all edges with vertex incidences) or lists of adjacencies for the vertices are used.

Although the matrix encoding allows constant-time adjacency testing, it does not allow to list the edges incident to a vertex in constant time per incident edge. Also, this encoding needs space quadratic in the number of vertices. As the `Pigale` software is mainly concerned with topological graph algorithms, particularly traversal-based algorithms, it has been a natural choice to consider list encodings of graphs. On the one hand, we shall allow to input graphs encoded as a list of edge incidences in order to simplify the interface to other software (see Figure 19.1). On the other hand, the internal representation of graphs is tailored to fit the types of topological graph algorithms we mainly consider.

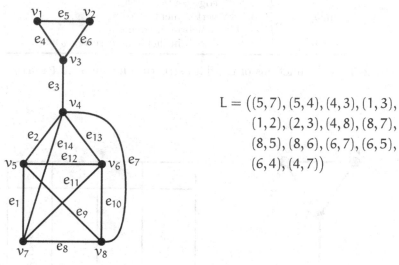

$$L = \big((5,7),(5,4),(4,3),(1,3),$$
$$(1,2),(2,3),(4,8),(8,7),$$
$$(8,5),(8,6),(6,7),(6,5),$$
$$(6,4),(4,7)\big)$$

Figure 19.1 Encoding of a graph by a list of incidences

Internal graph representation is a major issue for the efficiency of graph algorithms. Although most of the data structures used by graph algorithm libraries are oriented to fully dynamic graphs, thus offering constant-time insertion and deletion operations, the `Pigale` data structure is oriented to *quasi-static* graphs, that is, graphs on which only few modifications are done. Moreover, these modifications mainly correspond to a sequence of additions and (after some computations) of deletions of the added elements. In such a context, it is of particular interest to index vertices by consecutive integer values from 1 to n (where n is the order of the graph) and edges by consecutive integer values from 1 to m (where m is the size of the graph).

Since `Pigale` is designed to ease the writing of topological graph algorithms, the data structure is based on the mathematical notion of combinatorial map. A *combinatorial map* is a triple (B, τ, σ), where B is a set of half-edges, each called a *brin* (also sometimes called *flag* or *dart*), τ is a fixed point free involution of B whose orbits are the edges of the map, and σ is a permutation of B whose orbits are the vertices of the map.

This combinatorial structure is particularly efficient for map traversals. However, edges and vertices only have an implicit description in this model. This is the reason why `Pigale`'s graph description slightly differs from the one of the combinatorial map. The structure describing maps in `Pigale` is based on the functions shown in Table 19.1, where $V = \{1, \ldots, n\}$ is the index set of the vertices, $E = \{1, \ldots, m\}$ is the index set of the edges, and $B = \{-m, \ldots, -1, 1, \ldots, m\}$ is the index set of the brins.

Note that for technical reasons, the vertex set, the edge set and the brin set are actually $\{0,\ldots,n\}$, $\{0,\ldots,m\}$ and $\{-m,\ldots,m\}$. The operators are extended to 0 with reserved values $\mathrm{cir}[0] = \mathrm{acir}[0] = \mathrm{vin}[0] = \mathrm{pbrin}[0] = 0$. (see Figure 19.2).

Operator	Domain	Description
$-b$	$B \to B$	brin opposite to b ($\tau(b)$)
cir[b]	$B \to B$	brin next to b in circular order ($\sigma(b)$)
acir[b]	$B \to B$	brin before b in circular order ($\sigma^{-1}(b)$)
$\lvert b \rvert$	$B \to E$	edge containing b
vin[b]	$B \to V$	vertex incident to b
e	$E \to B$	first brin of edge e
pbrin[v]	$V \to B$	first brin incident to vertex v

Table 19.1 Functions of the data structure for maps in `Pigale`.

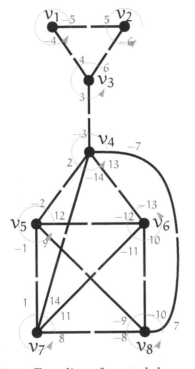

brin #	Cir	Acir	Vin
−14	13	2	4
−13	−12	10	6
−12	−11	−13	6
−11	10	−12	6
−10	−9	7	8
−9	−8	−10	8
−8	7	−9	8
−7	−3	13	4
−6	5	5	2
−5	−4	−4	1
−4	−5	−5	1
−3	2	−7	4
−2	−1	12	5
−1	9	−2	5
1	8	14	7
2	−14	−3	4
3	6	4	3
4	3	6	3
5	−6	−6	2
6	4	3	3
7	−10	−8	8
8	11	1	7
9	12	−1	5
10	−13	−11	6
11	14	8	7
12	−2	9	5
13	−7	−14	4
14	1	11	7

Figure 19.2 Encoding of a graph by a combinatorial map.

19.2.2 Graph Properties

Since in our model vertices, edges and brins are represented by integer values, most of the properties attached to the elements of the graph will be scalar. In order to reduce the slow down of calls to constructors and destructors of complex types, it has been decided to favor scalar properties.

Since the graph structure is a very general abstract one, most algorithms and applications need more or less specific properties to be added to vertices, edges or brins. It appears that class derivation, which is suitable in contexts where a limited number of distinct sets of properties are meaningful, does not work well in our case. This is the reason why we have opted for a more flexible framework in which properties may be added or suppressed dynamically. Then only a few subsets of properties have to be distinguished, the subsets corresponding to coherent views of a graph as a mere *graph* (i.e., a list of edge to vertex incidences), a *topological graph* (where circular orders around the vertices are defined) or a *geometric graph* (where vertices have coordinates, labels, colors, ...), leading to three *logical views* in `Pigale`, namely: `Graph`, `TopologicalGraph`, and `GeometricGraph`, of the set of graph properties stored in a `GraphContainer` data structure.

19.3   Basic Graph Algorithms

In this section, we describe the implementation of several basic graph algorithms in `Pigale`.

19.3.1   Depth-First Search

Depth-First Search (DFS) is central to the planarity algorithm implemented in `Pigale`. It is responsible for a sensible percentage of the execution time. Thus, the optimization of this particular algorithm has strong consequences on the efficiency of other important algorithms.

One of the main characteristics of DFS is that the DFS-tree it builds is traversed several times and that the tree/cotree partition it induces is intensively used in the planarity testing algorithm. For these reasons, it appeared that an efficient optimization stands in the renumbering of the vertices and the edges of the graph using the following scheme (see Figure 19.3):

- the vertices are numbered $1, \dots, n$ in the order of first discovery by the DFS;
- the tree edges are numbered $1, \dots, n - 1$ in the order of first traversal by the DFS. Precisely, brin i is adjacent to the parent of vertex $(i + 1)$ and brin $-i$ is adjacent to vertex $(i + 1)$;
- the cotree edges are numbered $n, \dots, m$ in order opposite to the order in which their low incidences are met by the DFS. The positive brin is incident to the lower vertex according to tree order.

From the above numbering, it follows that a traversal of the edges in DFS order may be simulated using a simple `for(e=1; e<n; e++)` loop. Also, testing if an edge belongs to the tree is performed by a simple `(e<n)` test.

19.3.2   Planarity and Nonplanar Subgraph Exhibition

The linear-time planarity testing algorithm implemented in `Pigale` is based on the characterization by de Fraysseix and Rosenstiehl [FR85, FR82, FR83a, FR83b] and its improvement [FOdMR06, FOdM12, Fra08]. This algorithm is currently the fastest-implemented planarity testing algorithm [BCPD04].

A linear-time algorithm to find a Kuratowski subdivision in a nonplanar graph (see Figure 19.4) has been implemented in `Pigale`, based on a theoretical characterization of DFS cotree-critical graphs [FOdM01a, FOdM02, FOdM03].

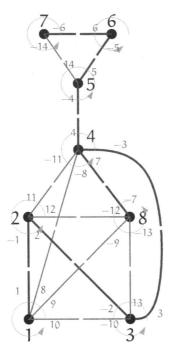

Figure 19.3 DFS numbering of a combinatorial map.

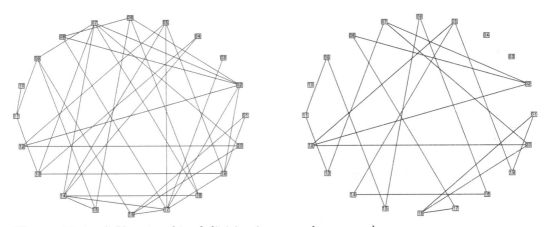

Figure 19.4 A Kuratowski subdivision in a nonplanar graph.

This algorithm relies on the concept of DFS cotree-critical graphs, which is a by-product of our planarity testing algorithms. Roughly speaking, a DFS cotree-critical graph is a simple graph of minimum degree 3 having a DFS tree, such that any nontree (i.e., cotree) edge is critical, in the sense that its deletion would lead to a planar graph. A first study of DFS cotree-critical graphs appeared in [FR83a], where it is proved that a DFS cotree-critical graph is either isomorphic to K_5 or includes a subdivision of $K_{3,3}$ and no subdivision of K_5.

The algorithm consists of two steps:

1. Extraction of a DFS cotree-critical subgraph by a case analysis algorithm; and
2. Extraction of a Kuratowski subdivision from the DFS cotree-critical subgraph.

Step 2 is performed by an algorithm whose simplicity contrasts with the complexity of its theoretical justification (which relies on the full characterization of DFS cotree-critical graphs proved in [FOdM03]. This algorithm roughly works as follows:

- It first computes the set of the critical edges of a graph, using the property that a tree edge is critical if and only if it belongs to a fundamental cycle of length 4 of some cotree edge to which it is not adjacent.

- Then, three pairwise non-adjacent non-critical edges are found to complete a Kuratowski subdivision isomorphic to $K_{3,3}$.

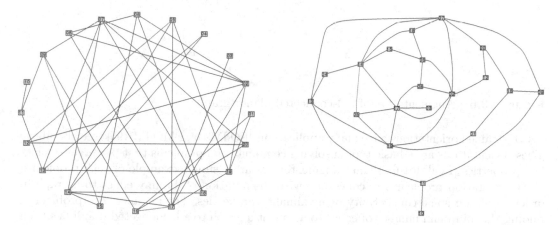

Figure 19.5 Finding a minimal subset of nonplanar edges. On the right is a Bezier drawing of the planar graph obtained after deletion of the computed set of nonplanar edges.

This algorithm is the central routine of a heuristic for exhibiting an inclusion-minimal set of edges whose deletion ensures the planarity of the graph (see Figure 19.5).

19.3.3 Connectivity Tests

Based on properties of regular orientations of planar graphs, `Pigale` offers a linear-time algorithm to test whether a planar graph is 3-connected and a linear-time algorithm to test whether a maximal planar graph is 4-connected [FOdM01b, FOdM04, FOdM01d]. The study of graphs by means of special orientations is relatively recent. For instance, bipolar orientations have become a basic tool in many graph drawing problems [OdM94, FOdMR95].

Constrained orientations (i.e., orientations with bounded indegrees) lead to new characterizations of connectivity for planar undirected graphs. Although standard 3-connectivity testing algorithms for planar graphs are heavily related to planarity testing algorithms (see [HT73, Tar74] and PQ-tree algorithms), the algorithm in `Pigale` assumes that the input graph is already embedded in the plane so the problem drastically reduces to the acyclicity testing of a particular orientation. Concerning the 4-connectivity testing of a maximal planar graph, the use of an indegree bounded orientation was already used in [CE91] to enumerate triangles. In `Pigale`, the use of a specific orientation allows to further simplify the algorithm. The 4-connectivity test itself also reduces to an acyclicity test. It should be noted that no special data structure is used for these algorithms since in the planar case the acyclicity of an orientation can be efficiently tested using a dual topological sort.

19.3.4   Augmentation of Planar Graphs

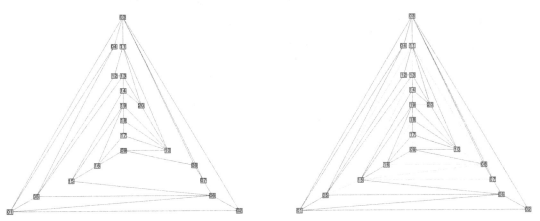

Figure 19.6   Augmentation of a 3-connected planar graph.

Constrained orientations have many applications [FOdM94a, FMOdMR95] (see also above). These orientations are a basic tool in solving combinatorial problems that preserve topological properties [FOdM01d]. Planar augmentations are a simple example of such problems.

Augmentation problems are concerned with the addition of dummy edges to a graph in order to obtain some connectivity or maximality properties. For instance, the problem of finding the minimum number of edges to augment a graph to a biconnected graph has been solved in [ET76]. If the original graph is planar and if it is required to preserve the planarity, the problem is NP-complete [Kan93]. Triangulating a biconnected graph while minimizing the maximum degree has also been proved to be an NP-complete problem.

`Pigale` offers several optimal augmentation algorithms, including a linear-time algorithm for augmenting a 3-connected planar graph to a maximal planar graph (see Figure 19.6) that increases the degree of any vertex of the graph by no more than 6 (which is optimal) [FOdM95, FOdM94b].

19.3.5   Graph Symmetry and Clustering

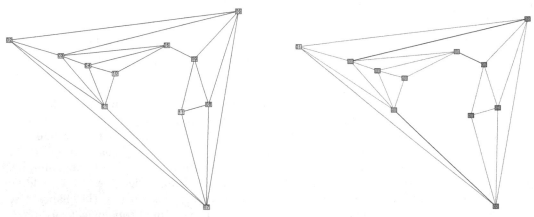

Figure 19.7   Finding a symmetry of a graph.

Based on spectral analysis [FK92], `Pigale` offers a heuristic to find symmetries in a general simple graph (planar or not) [Fra99, FOdM06] (see Figure 19.7). These symmetries

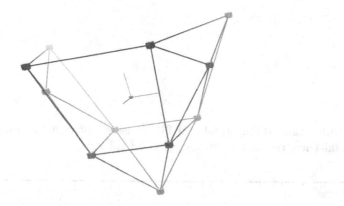

Figure 19.8 3D view of the graph displaying the symmetry.

may actually be viewed in the 3D drawing built from the spectral analysis of the graph (see Figure 19.8).

Using spectral analysis, Babai proved in 1978 that the abstract automorphism group of any multigraph G having s distinct eigenvalues with respective multiplicities $m_1, m_2, \ldots, m_s$ is a subgroup of $\omega(m_1) \oplus \omega(m_2) \oplus \ldots \oplus \omega(m_s)$, where $\omega(m)$ denotes the real orthogonal group of dimension m [Bab78]. As a consequence, if all the eigenvalues of G are simple, the only automorphisms of G are involutions.

Some years before, Mani proved that every triconnected planar graph G can be realized as the 1-skeleton of a convex polytope P in $\mathbb{R}^3$ such that all automorphisms of G are induced by isometries of P [Man71]. One non trivial consequence of this result is that the automorphism group $\mathrm{Aut}(G)$ of a triconnected planar graph G has a chain of normal subgroups $\mathrm{Aut}(G) = G_0 \triangleright G_1 \triangleright \ldots \triangleright G_m = 1$, where each quotient G_i/G_{i-1} is either cyclic, or isomorphic to a symmetric group or A_5.

The result of Mani may be expressed in a weaker form: any triconnected planar graph has an embedding f into $\mathbb{R}^3$, such that $\mathrm{Aut}(G)$ is the group of isometries of $\mathbb{R}^3$ globally preserving the point set $P = f(V(G))$, that we shall denote by $\omega(3, P)$.

These two results are generalized in [FOdM06], where it is proved that every twin-free loopless multigraph G has some *regular embedding*, that is, some embedding $f : V(G) \to \mathbb{R}^k$ such that $\mathrm{Aut}(G)$ is isomorphic to the group $\omega(k, f(V(G)))$ of isometries of $\mathbb{R}^k$ globally preserving $f(V(G))$, and that this group might be expressed as a subgroup of a group sum relying on spectral considerations. This result is proved using techniques similar to those used in the symmetry detection heuristic presented in [Fra99]. The problem of finding regular embeddings is reduced to the one of finding metrics on the vertex set of the multigraph that define *Euclidean, reconstructing*, and *commuting* distance matrices, which may be built from particular symmetric real matrices with 0 on the diagonal (the commuting reconstructing predistances).

Several such distances have been implemented in `Pigale` (see Table 19.2 and Figure 19.9).

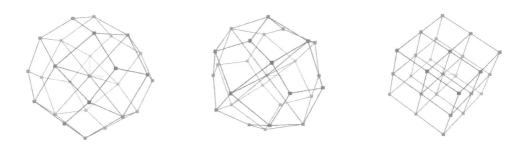

Figure 19.9 Embedding a cube in $\mathbb{R}^{n-1}$ using different distances (from left to right): Czekanovski-Dice, translated adjacency, and Laplacian.

Czekanovski-Dice distance	$\mathrm{dist}^2(i,j) = 1 - \dfrac{|N(i) \cap N(j)|}{|N(i)| + |N(j)|}$
Oriented distance	$\mathrm{dist}^2(i,j) = 1 - \dfrac{|N^-(i) \cap N^-(j)|}{|N^-(i)| + |N^-(j)|} - \dfrac{|N^+(i) \cap N^+(j)|}{|N^+(i)| + |N^+(j)|}$
Adjacency distance (not Euclidean)	$\mathrm{dist}^2(i,j) = \begin{cases} 0, & \text{if } i = j \text{ or } i \text{ and } j \text{ are adjacent} \\ 1, & \text{otherwise} \end{cases}$
Translated adjacency distance	$\mathrm{dist}^2(i,j) = \begin{cases} 0, & \text{if } i = j \\ 1 - \frac{2}{n}, & \text{if } i \text{ and } j \text{ are adjacent} \\ 1, & \text{otherwise} \end{cases}$
Bisection distance	$\mathrm{dist}^2(i,j) = \begin{cases} 0, & \text{if } i = j \\ 1 - \frac{2}{d(i)+d(j)+2}, & \text{if } i \text{ and } j \text{ are adjacent} \\ 1, & \text{otherwise} \end{cases}$
$\mathbb{R}^2$ distance	$\mathrm{dist}^2(i,j) = (x(i) - x(j))^2 + (y(i) - y(j))^2$
Laplacian distance	$\mathrm{dist}^2(i,j) = \begin{cases} 0, & \text{if } i = j \\ 2n - d(i) - d(j), & \text{if } i \text{ and } j \text{ are adjacent} \\ 2n - d(i) - d(j) + 2, & \text{otherwise} \end{cases}$
Q distance	$\mathrm{dist}^2(i,j) = \begin{cases} 0, & \text{if } i = j \\ 1, & \text{if } i \text{ and } j \text{ are non adjacent} \\ 1 - \frac{1}{\sqrt{d(i)d(j)}}, & \text{otherwise} \end{cases}$

Table 19.2 Choice of distances for the spectral analysis/embedding in `Pigale`; $N(i)$ (resp. $N^-(i), N^+(i)$) denotes the set of the neighbors (resp. in-neighbors, out-neighbors) of vertex i and $d(i) = |N(i)|$ denotes the degree of vertex i.

19.4   Random Map Generators

Several polynomial-time random planar map uniform generators have been implemented by Gilles Schaeffer in `Pigale` [Sch99]:

- planar maps (connected, 2-connected, or 3-connected),
- planar cubic maps (2-connected, 2-connected bipartite, 3-connected, 3-connected bipartite, or dual-4-connected),
- planar 4-regular maps (2-connected, 3-connected, or bipartite),
- planar bipartite maps.

Also, linear-time uniform generators of outerplanar maps have been implemented by Nicolas Bonichon [BGH03].

The implementation of these algorithms in `Pigale` uses the uniform pseudo-random number generator of Matsumoto and Nishimura [MN98]. This pseudo-random number generator is also used to generate random graphs where edges are independently included with fixed probabilities (Erdős-Rényi model).

19.5   Graph Drawing Algorithms

This section is devoted to the graph drawing algorithms provided by `Pigale`.

19.5.1   Planar Straight-Line Grid Drawings

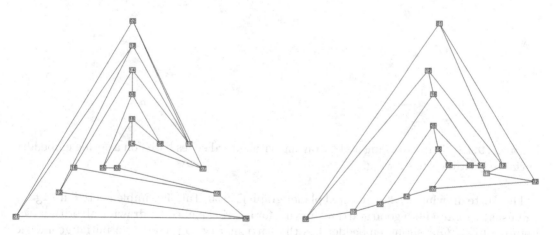

Figure 19.10 Fraysseix Pach Pollack (with edge augmentation) and Schnyder (using vertex augmentation).

`Pigale` includes several linear-time planar straight-line drawing algorithms for simple planar graphs, including the Fraysseix-Pach-Pollack algorithm [FPP88, FPP90] and Schnyder's algorithm [Sch89, Sch90] (see Figure 19.10). Some bounds and conjectures on the size of straight-line drawings may be found in [FOdM01c].

A linear-time compact convex drawing algorithm for 3-connected planar graphs [BFM04], as well as a compact polyline drawing for simple planar graphs [BLSM02], have been added by Nicolas Bonichon (see Figure 19.11).

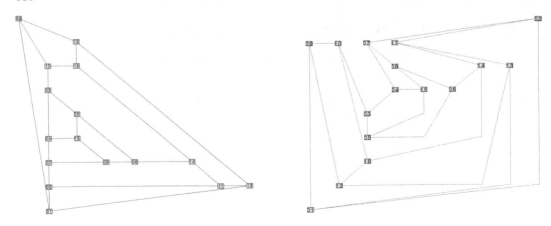

Figure 19.11   Compact convex drawing and compact polyline drawing.

19.5.2   Spring Embedders

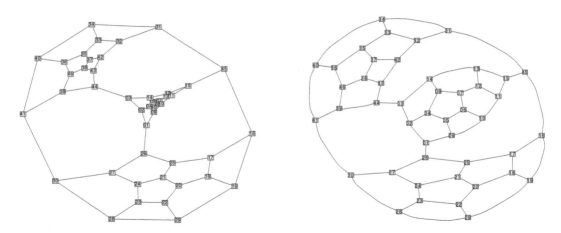

Figure 19.12   Tutte drawing and a drawing with curved edges based on a spring embedder initialized with Tutte drawing.

The Tutte drawing of a 3-connected planar graph [Tut60, Tut63] is implemented in `Pigale` and usually represents a good starting drawing for a spring embedder drawing algorithm (see Figure 19.12). Our spring embedder has the particularity to preserve an initial geometric map (relative positions and crossings) of a (nonplanar) graph.

19.5.3   Visibility Drawing and Variants

Visibility and rectilinear drawings [RT86, TT86] have received much attention because of their good readability (see Figure 19.13). All the algorithms mentioned in this section are linear-time algorithms. With the exception of the Polrec algorithm, all the representations described in this section concern simple planar graphs. The area of the drawing may be further reduced by allowing horizontal and vertical visibility, as in an algorithm proposed by de Fray", Pach, and Pollack (see Figure 19.14).

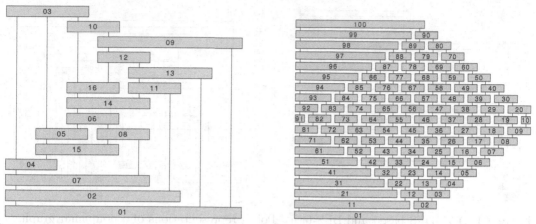

Figure 19.13 Visibility drawings. The drawing on the right is within a 10×10 grid.

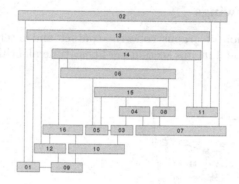

Figure 19.14 Rectilinear drawing constructed by an algorithm by de Fraysseix, Pach, and Pollack.

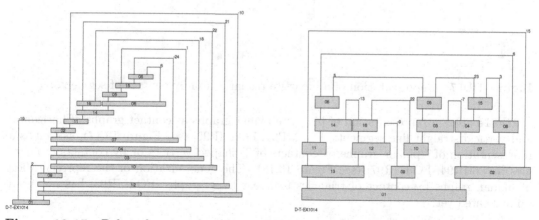

Figure 19.15 Polrec drawings based on a DFS-tree and a BFS-tree, respectively.

The Polrec algorithm produces a drawing where vertices are represented by boxes, a tree is represented using straight-line vertical segments and cotree-edges are represented by U-shaped polylines (see Figure 19.15). Such a representation can be used for non-simple nonplanar graphs with loops (see Figure 19.16).

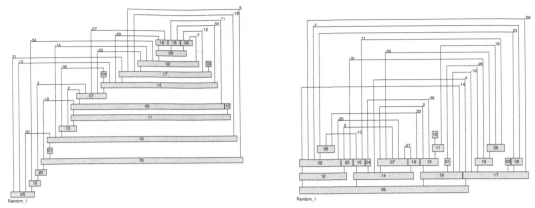

Figure 19.16   DFS-based and BFS-based Polrec representations of a nonplanar graph.

19.5.4   Contact Drawings

An emerging representation of graphs concerns contact and intersection representations. All the algorithms mentioned in this section are linear-time algorithms and concern simple planar graphs.

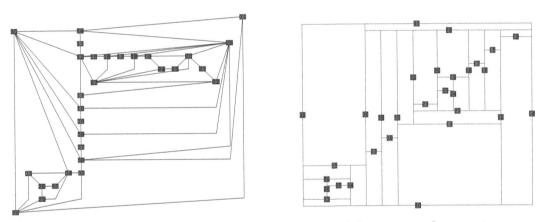

Figure 19.17   Representation of a bipartite planar graph by contact of segments.

Pigale offers a representation of bipartite planar graphs as contact graphs of horizontal and vertical straight line segments [FOdMP91, FOdMP95] (see Figure 19.17), as well as a representation of   planar graphs by contacts of T-shaped vertices or by contacts of triangles [FOdMR94, FOdMR97] (see Figure 19.18). The generalization of the representation of planar graphs by contact of triangles to linear hypergraphs [FOdMR08] has not been implemented yet.

19.5.5   Spectral Drawings in $\mathbb{R}^n$

As mentioned in Section 19.3.5, spectral analysis may be used to generate 3D visualizations of (nonplanar) graphs in polynomial time (see Figure 19.19). The time complexity of the algorithm derives from the complexity of the computation of the eigenvalues of an $n \times n$ matrix, where n is the number of vertices of the represented graph.

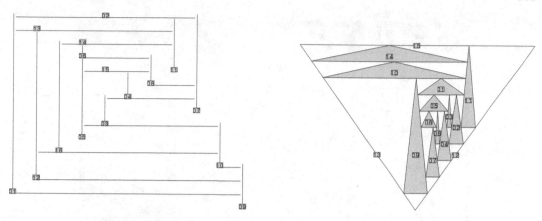

Figure 19.18 Contacts of Ts and contacts of triangles.

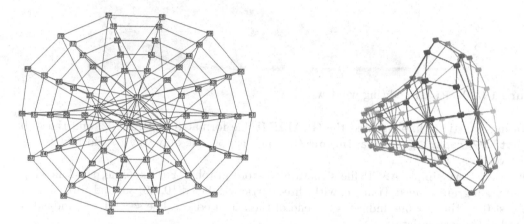

Figure 19.19 3D embedding of a nonplanar graph with 3D symmetries.

19.6   Implementation

19.6.1   User Interface

Pigale provides a graph editor that allows the user to load, save or generate graphs, to edit them, to check the properties of the graph (automatically displayed by the program), to perform several transformations (augmentations, orientations, computation of duals, etc.), and to compute representations of the graph.

While mouse-editing a graph, a user can add, delete, contract, bisect, orient, reorient, unorient, and color edges, and can set their width; the user can also add, move, delete, and color vertices, and can put numerical labels on them.

19.6.2   File Storage

We use a general proprietary format, called TGF. A TGF file contains records, here corresponding to graphs. Each record consisting of a variable number of fields. One of its main advantage is that we can write and read any complex data structure. But it is dependent of the processor type (e.g., big-endian or little-endian).

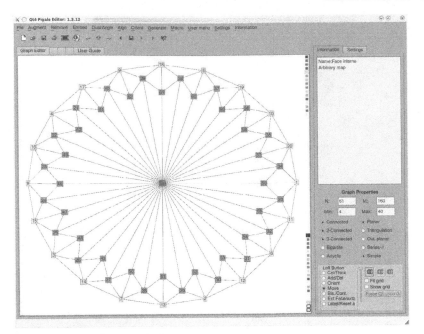

Figure 19.20 `Pigale` editing window.

We have partially implemented the GRAPHML file format (cf. [BEH+02] and the web site `http://graphml.graphdrawing.org/`), which is now the only way to add text labels to the vertices.

We use a very simple ASCII file structure to store graphs. For example, the following file defines a graph, called Triangle, with three vertices, labeled 10, 20, 30, and three edges. The first 0 on the last line indicates the end of the list of edges. The second zero indicates the end of the graph data.

```
PIG:0 Triangle
10 20
20 30
30 10
0 0
```

19.6.3   Macro Recording

One can record any number of functions from the menus into macros, which can be saved as text files.

A macro can be repeated any number of times (possibly until the user will press the ESC key). If the first record of the macro is not a call to a graph generator, the macro will start loading the next graph of the current file.

Macros can be used to develop and benchmark algorithms and to test conjectures.

19.6.4   Multi-Threaded Server

The `Pigale` editor may be put in server mode, which allows the editor to be controlled by a client application. A simple program `client` is provided as an example of how to communicate with the server. The client reads its instructions from `stdin` so that it should not be difficult for applications to communicate with the server. However, it is not difficult to write, for instance, a web server that acts as a front end to Pigale.

19.7 Interfacing with PIGALE

As mentioned in its name, Pigale is not only an editor, but also a library. Nearly all the algorithms in Pigale may be run in a non-graphic context through a library call.

An example of a simple C++ program using Pigale library is given below.

```cpp
#include <Pigale.h>

int main ()
{
  GraphContainer GC; // defined in TAXI/graph.h
  // GC is the object that will contain all the information of a graph.
  int n = 4; // n =  number of vertices [1,n]
  int m = 5; // m = number of edges      [1,m]

  GC.setsize(n,m);  // defines the size of the container
  /*
   - a tvertex v is an integer v(): 1 <= v() <= n = GC.nv()
   - a tedge   e is an integer e(): 1 <= e() <= m = GC.ne()
   - a tedge   e is composed of 2 tbrin b0,b1 equal to e() and -e()
   tvertex, tedge, tbrin behave like integers in many respects
  */

  Prop<tvertex> vin(GC.Set(tbrin()),PROP_VIN);
  // vin is an array  of tbrin whose values are tvertex.

  // Create the edges: each edge (tedge) is incident to 2 vertices (tvertex)
  vin[1] = 1; vin[-1] = 2;  // edge 1 is incident to vertices 1 and 2
  vin[2] = 1; vin[-2] = 3;
  vin[3] = 2; vin[-3] = 3;
  vin[4] = 3; vin[-4] = 4;
  vin[5] = 2; vin[-5] = 4;

  // create a topological graph access
  TopologicalGraph G(GC); // defined in TAXI/graphs.h

  // print the number of vertices and edges
  cout << "Nodes: " << G.nv() << "\tEdges: " << G.ne()<< endl;

  // print the edges (if e is a tedge, e() is the int that represents it)
  cout << "Edges:" << endl;
  for(tedge e = 1; e <= G.ne();e++)
    cout << e() << " = [" << G.vin[e] << "," << G.vin[-e] << "]" <<endl;

  // For planarity test, graphs should be LOOPLESS. You can remove loops:
  // int nloops = RemoveLoops();

  // Compute a planar embedding or return -1
  if(G.Planarity() == 0)
    {cout << "not planar" << endl; return -1;}
```

```
// At each vertex v  there is a tbrin G.pbrin[v] incident to it:
//          G.vin[G.pbrin[v]] = v;
// So we can print the planar map, that is the cirular order of
// half edges around each  vertex.
cout << "Map (half edges):"<<endl;
for(tvertex v = 1; v <= G.nv() ; v++)
  {cout << v() <<"  -> ";
    tbrin first = G.pbrin[v];
    tbrin b = first;
    do
      {cout << b() << " ";
      }
    while((b = G.cir[b]) != first);
    cout << endl;
  }
// Or you could print the circular order of vertices aroud each vertex
cout << "Map (vertices):"<<endl;
for(tvertex v = 1; v <= G.nv() ; v++)
  {cout << v() <<"  -> ";
    tbrin first = G.pbrin[v];
    tbrin b = first;
    do
      {cout << G.vin[-b]() << " ";
      }
    while((b = G.cir[b]) != first);
    cout << endl;
  }
return 0;
}
```

References

[Bab78]   L. Babai. Automorphism group and category of cospectral graphs. *Acta Math. Acad. Sci. Hung.*, 31:295–306, 1978.

[BCPD04]   J. M. Boyer, P. F. Cortese, M. Patrignani, and G. Di Battista. Stop minding your P's and Q's: implementing fast and simple DFS-based planarity and embedding algorithm. In *Graph Drawing*, volume 2912 of *Lecture Notes in Computer Science*, pages 25–36. Springer, 2004.

[BEH+02]   U. Brandes, M. Eiglsperger, I. Herman, M. Himsolt, and M. S. Marshall. GraphML progress report: Structural layer proposal. In Springer-Verlag, editor, *Proc. 9th Intl. Symp. Graph Drawing (GD '01)*, volume 2265, pages 501–512, 2002.

[BFM04]   N. Bonichon, S. Felsner, and M. Mosbah. Convex drawings of 3-connected planar graphs (extended abstract). In J. Pach, editor, *Graph Drawing 2004*, volume 3383 of *Lecture Notes in Computer Science*, pages 60–70. Springer Verlag, 2004.

[BGH03]   N. Bonichon, C. Gavoille, and N. Hanusse. Canonical decomposition of outerplanar maps and application to enumeration, coding and generation. In Springer-Verlag, editor, *29th International Workshop, Graph-Theoretic Concepts in Computer Science (WG)*, volume 2880 of *Lecture Notes in Computer Science*, pages 81–92, 2003.

[BLSM02]   N. Bonichon, B. Le Saëc, and M. Mosbah. Optimal area algorithm for planar polyline drawings. In Springer-Verlag, editor, *28th International Workshop, Graph-Theoretic Concepts in Computer Science (WG)*, volume 2573 of *Lecture Notes in Computer Science*, pages 35–46, 2002.

[CE91]   M. Chrobak and D. Eppstein. Planar orientations with low out-degree and compaction of adjacency matrices. *Theoret. Comput. Sci.*, 86:243–266, 1991.

[ET76]   K. P. Eswaran and R. E. Tarjan. Augmentation problems. *SIAM J. Comput.*, 5:653–665, 1976.

[FK92]   H. de Fraysseix and P. Kuntz. Pagination of large scale networks. *Algorithms review*, 2(3):105–112, 1992.

[FMOdMR95]   H. de Fraysseix, T. Matsumoto, P. Ossona de Mendez, and P. Rosenstiehl. Regular Orientations and Graph Drawing. In *Third Slovenian International Conference in Graph Theory*, pages 12–13, 1995. abstract.

[FOdM94a]   H. de Fraysseix and P. Ossona de Mendez. On regular orientations. In *Prague Midsummer Combinatorial Workshop*, pages 9–13, 1994. Abstract.

[FOdM94b]   H. de Fraysseix and P. Ossona de Mendez. Some augmentation problems. In *Effiziente Algorithmen*, volume 34/1994, page 11, 1994. Abstract.

[FOdM95]   H. de Fraysseix and P. Ossona de Mendez. Regular orientations, arboricity and augmentation. In *DIMACS International Workshop, Graph Drawing 94*, volume 894 of *Lecture Notes in Computer Science*, pages 111–118, 1995.

[FOdM01a]   H. de Fraysseix and P. Ossona de Mendez. An algorithm to find a Kuratowski subdivision in DFS cotree critical graphs. In Edy Try Baskoro, editor, *Proceedings of the Twelfth Australasian Workshop on Combinatorial Algorithms (AWOCA 2000)*, pages 98–105, Indonesia, 2001. Institut Teknologi Bandung.

[FOdM01b]   H. de Fraysseix and P. Ossona de Mendez. Connectivity of planar graphs. *Journal of Graph Algorithms and Applications*, 5(5):93–105, 2001.

[FOdM01c]   H. de Fraysseix and P. Ossona de Mendez. Lower bounds on sets supporting Fáry drawings. In O. Pangrac, editor, *Graph Theory Day V*, volume 2001-539 of *KAM Series*, pages 35–37, 2001.

[FOdM01d]   H. de Fraysseix and P. Ossona de Mendez. On topological aspects of orientations. *Discrete Mathematics*, 229(1-3):57–72, 2001.

[FOdM02]   H. de Fraysseix and P. Ossona de Mendez. A characterization of DFS cotree critical graphs. In *Graph Drawing*, volume 2265 of *Lecture notes in Computer Science*, pages 84–95, 2002.

[FOdM03]   H. de Fraysseix and P. Ossona de Mendez. On cotree-critical and DFS cotree-critical graphs. *Journal of Graph Algorithms and Applications*, 7(4):411–427, 2003.

[FOdM04]   H. de Fraysseix and P. Ossona de Mendez. Connectivity of planar graphs. In *Graphs Algorithms and Applications 2*. World Scientific, 2004.

[FOdM06]   H. de Fraysseix and P. Ossona de Mendez. Regular embeddings of multigraphs. In M. Klazar, J. Kratochvil, M. Loebl, J. Matousek, R. Thomas, and P. Valtr, editors, *Topics in Discrete Mathematics*, volume 26 of *Algorithms and Combinatorics*, pages 553–563. Springer-Verlag, 2006. Dedicated to Jarik Nešetřil on the occasion of his 60th birthday.

[FOdM12]   H. de Fraysseix and P. Ossona de Mendez. Planarity and Trémaux trees. *European Journal of Combinatorics*, 33(3):279–293, 2012.

[FOdMP91]   H. de Fraysseix, P. Ossona de Mendez, and J. Pach. Representation of planar graphs by segments. *Intuitive Geometry*, 63:109–117, 1991.

[FOdMP95]   H. de Fraysseix, P. Ossona de Mendez, and J. Pach. A left-first search algorithm for planar graphs. *Discrete Computational Geometry*, 13:459–468, 1995.

[FOdMR94]   H. de Fraysseix, P. Ossona de Mendez, and P. Rosenstiehl. On triangle contact graphs. *Combinatorics, Probability and Computing*, 3:233–246, 1994.

[FOdMR95]   H. de Fraysseix, P. Ossona de Mendez, and P. Rosenstiehl. Bipolar orientations revisited. *Discrete Applied Mathematics*, 56:157–179, 1995.

[FOdMR97]   H. de Fraysseix, P. Ossona de Mendez, and P. Rosenstiehl. On triangle contact graphs. In *Combinatorics, Geometry and Probability: A Tribute to Paul Erdős*, pages 165–178. Cambridge University Press, 1997.

[FOdMR06]   H. de Fraysseix, P. Ossona de Mendez, and P. Rosenstiehl. Depth-first search and planarity. *International Journal of Foundations of Computer Science*, 17(5):1017–1029, 2006. Special Issue on Graph Drawing.

[FOdMR08]   H. de Fraysseix, P. Ossona de Mendez, and P. Rosenstiehl. Representation of planar hypergraphs by contacts of triangles. In *Proceedings of Graph Drawing 2007*, volume 4875/2008 of *Lecture Notes in Computer Science*, pages 125–136. Springer, 2008.

[FPP88]   H. de Fraysseix, J. Pach, and R. Pollack. Small sets supporting Fary embeddings of planar graphs. In *Twentieth Annual ACM Symposium on Theory of Computing*, pages 426–433, 1988.

[FPP90]   H. de Fraysseix, J. Pach, and R. Pollack. How to draw a planar graph on a grid. *Combinatorica*, 10:41–51, 1990.

[FR82]   H. de Fraysseix and P. Rosenstiehl. A depth-first search characterization of planarity. *Annals of Discrete Mathematics*, 13:75–80, 1982.

[FR83a]   H. de Fraysseix and P. Rosenstiehl. A discriminatory theorem of Kuratowski subgraphs. In J. W. Kennedy, M. Borowiecki, and M. M. Sysło, editors, *Graph Theory, Łagów 1981*, volume 1018 of *Lecture Notes in Mathematics*, pages 214–222. Springer-Verlag, 1983. Conference dedicated to the memory of Kazimierz Kuratowski.

[FR83b]   H. de Fraysseix and P. Rosenstiehl. Système de référence de Trémaux d'une représentation plane d'un graphe planaire. *Annals of Discrete Mathematics*, 17:293–302, 1983.

[FR85]   H. de Fraysseix and P. Rosenstiehl. A characterization of planar graphs by Trémaux orders. *Combinatorica*, 5(2):127–135, 1985.

[Fra99]   H. de Fraysseix. An heuristic for graph symmetry detection. In J. Kratochvíl, editor, *Graph Drawing*, volume 1731 of *Lecture Notes in Computer Science*, pages 276–285. Springer, 1999.

[Fra08]   H. de Fraysseix. Trémaux trees and planarity. In P. Ossona de Mendez, M. Pocchiola, D. Poulalhon, J.L. Ramírez Alfonsín, and G. Schaeffer, editors, *The International Conference on Topological and Geometric Graph Theory*, volume 31 of *Electronic Notes in Discrete Mathematics*, pages 169–180. Elsevier, 2008.

[HT73]   J. Hopcroft and R. E. Tarjan. Dividing a graph into triconnected components. *SIAM J. Comput.*, 2(3):135–158, 1973.

[Kan93]   G. Kant. *Algorithms for Drawing Planar Graphs*. PhD thesis, Dept. Comput. Sci., Univ. Utrecht, Utrecht, Netherlands, 1993.

[Man71]   P. Mani. Automorphismen von polyedrischen Graphen. *Mathematische Annalen*, 192:279–303, 1971.

[MN98]   M. Matsumoto and T. Nishimura. Mersenne twister: A 623-dimensionally equidistributed uniform pseudo-random number generator. *ACM Transactions on Modeling and Computer Simulation*, 8(1):3–30, 1998.

[OdM94]   P. Ossona de Mendez. *Orientations bipolaires*. PhD thesis, Ecole des Hautes Etudes en Sciences Sociales, Paris, 1994.

[RT86]   P. Rosenstiehl and R. E. Tarjan. Rectilinear planar layout and bipolar orientation of planar graphs. *Discrete and Computational Geometry*, 1:343–353, 1986.

[Sch89]   W. Schnyder. Planar graphs and poset dimension. *Order*, 5:323–343, 1989.

[Sch90]   W. Schnyder. Embedding planar graphs in the grid. In *First ACM-SIAM Symposium on Discrete Algorithms*, pages 138–147, 1990.

[Sch99]   G. Schaeffer. Random sampling of large planar maps and convex polyhedra. In ACM, editor, *Annual ACM Symposium on Theory of Computing (Atlanta, GA, 1999)*, pages 760–769 (electronic), New-York, 1999.

[Tar74]   R. E. Tarjan. Testing graph connectivity. In *Conference Record of Sixth Annual ACM Symposium on Theory of Computing (Seattle, Washington)*, pages 185–193, 1974.

[TT86]   R. Tamassia and I. G. Tollis. A unified approach to visibility representations of planar graphs. *Discrete Comput. Geom.*, 1(4):321–341, 1986.

[Tut60]   W. T. Tutte. Convex representations of graphs. In *Proc. London Math. Society*, volume 10, pages 304–320, 1960.

[Tut63]   W. T. Tutte. How to draw a graph. In *Proc. London Math. Society*, volume 13, pages 743–768, 1963.

20

Biological Networks

Christian Bachmaier
University of Passau

Ulrik Brandes
University of Konstanz

Falk Schreiber
IPK Gatersleben and
University of Halle-Wittenberg

20.1 Introduction

Biological processes are often represented in the form of networks such as protein-protein interaction networks and metabolic pathways. The study of biological networks , their modeling, analysis, and visualization are important tasks in life science today. An understanding of these networks is essential to make biological sense of much of the complex data that is now being generated. This increasing importance of biological networks is also evidenced by the rapid increase in publications about network-related topics and the growing number of research groups dealing with this area. Most biological networks are still far from being complete and they are usually difficult to interpret due to the complexity of the relationships and the peculiarities of the data. Network visualization is a fundamental method that helps scientists in understanding biological networks and in uncovering important properties of the underlying biochemical processes. This chapter therefore deals with major biological networks, their visualization requirements and useful layout methods. We start with some basic biology and important biological networks.

20.1.1 Molecular Biological Foundations

A cell consists of many different (bio-)chemical compounds. A crucial macromolecule in organisms is DNA (deoxyribonucleic acid), which is the carrier of genetic information. But DNA itself is not able to provide the structure of a cell, to act as a catalyst for chemical reactions or to sense changes in the cell's environment. Such functions are carried out by proteins, large molecules which are built according to information stored in DNA sequences. The central dogma of molecular biology deals with the information transfer from DNA to proteins. It states that proteins do not code for the production of other proteins, DNA or RNA (ribonucleic acid), i.e., that information cannot be transferred from one protein to another protein directly or from a protein back to nucleic acid. Instead, the standard pathway of information flow is from DNA to RNA to protein. Genes represented by DNA sequences are transcribed into RNA sequences which are then translated into proteins, see Figure 20.1. These proteins have different types such as structural components (which give cells their shape and help them move), transport proteins (which carry substances such as oxygen), enzymes (which catalyze most chemical processes in cells and help change metabolites into each other) and regulatory proteins (which regulate the expression of other genes). Crick summarized the standard pathway of information flow as "DNA makes RNA, RNA makes protein and proteins make us" [Kel00].

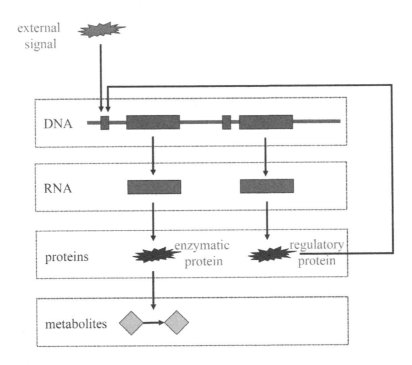

Figure 20.1 The standard pathway of information flow: DNA→RNA→protein. Two kinds of proteins (enzymatic and regulatory proteins) are shown as well as two types of gene regulation (via regulatory protein and external signal).

20.1.2   Biological Networks

Several highly important biological networks are related to molecules such as DNA, RNA, proteins and metabolites and to interactions between them. Gene regulatory and signal transduction networks describe how genes can be activated or repressed and therefore which proteins are produced in a cell at a particular time. Such regulation can be caused by regulatory proteins or external signals. The related networks are considered in Section 20.2. Protein-protein interaction networks represent the interaction between proteins such as the building of protein complexes and the activation of one protein by another protein. Section 20.3 deals with these networks and their visualization in detail. Metabolic networks show how metabolites are transformed, for example to produce energy or synthesize specific substances. Metabolic and closely related networks are studied in Section 20.4. In Section 20.5 we consider phylogenetic trees, special networks or hierarchies which are often built on information from molecular biology such as DNA or protein sequences. Phylogenetic trees represent the ancestral relationships between different species. They are used to study evolution, which describes and explains the history of species, i.e., their origins, how they change, survive, or become extinct. Finally, signal transduction, gene regulatory, protein-protein interaction and metabolic networks interact with each other and build a complex network of interactions; furthermore these networks are not universal but species-specific, i.e., the same network differs between different species. These topics are discussed in Section 20.6.

Often established layout methods as described in the previous chapters are used to visualize biological networks. Sometimes these methods are slightly modified, e.g., by adding extra forces to force-directed approaches. We will not discuss all these modifications in detail for each network, instead we focus on two topics: metabolic networks and phylogenetic trees. Metabolic networks have been studied for a long time in biology and biochemistry, and specific visualization requirements are given, e.g., by established drawing styles. We present some algorithmic extensions of the hierarchical layout approach which aim to fulfil these requirements. Phylogenetic tree visualizations are quite different to usual tree drawings. Therefore we discuss specific algorithms which have been developed to produce information-rich layouts of phylogenetic trees.

20.2   Signal Transduction and Gene Regulatory Networks

A key issue in biology is the response of a cell to internal and external stimuli and the subsequent regulation of its genetic activity. Signal transduction and gene regulatory pathways and networks describe processes to coordinate the cell's response to such stimuli. Here we consider both networks together as the underlying mechanisms have many similarities, the networks share some common elements and both often result in the regulation of gene expression. Consequently, similar visualization approaches are used for signal transduction and gene regulatory pathways and networks.

20.2.1   Definition

Signal transduction is a communication process within a cell to coordinate its responses to an environmental change. The *stimulus* comes from the cell's environment, e.g., molecules such as hormones. The *response* is a reaction of the cell, e.g., the activation of a gene or the production of energy. A *signal transduction pathway* is a directed network of chemical reactions in a cell from a stimulus (an external molecule which binds to a receptor on the cell membrane) to the response (e.g., the activation of a gene). Here we focus on signal transduction pathways that aim at transcription factors and thus alter the expression of genes in a cell. The *signal transduction network* of a cell is the complete network of all signal transduction pathways. A *signaling cascade* is a process where signal transduction involves an increasing number of molecules in the steps from the stimulus to the response.

Gene regulation is a general term for cellular control of the synthesis of proteins at the transcription step. Gene regulation can also be seen as the response of a cell to an internal stimulus. Often one gene is regulated by another gene via the corresponding protein (called transcription factor), thus gene regulation is coordinated in a *gene regulatory network*. This network directs the level of expression for each gene in the cell by controlling whether and how often that gene will be transcribed into RNA. Similar to signaling cascades in signal transduction networks a gene can activate more genes in turn and an initial stimulus can trigger the expression of large sets of genes.

As mentioned above we study signal transduction and gene regulation together. Figure 20.1 sketches both processes with signal transduction going from an external signal via several steps to the activation of a gene as one possible response and gene regulation going from a gene via a protein to another gene.

Events of signal transduction and gene regulatory processes occur in different parts of a cell (cellular compartments). To represent compartments these networks can be modeled as clustered graphs. A clustered graph $C = (G, T)$ consists of a directed graph $G = (V, E)$ and a rooted tree T, such that the leaves of T are exactly the nodes of G. The nodes $v \in V$ of the graph are chemical and biochemical compounds (ranging from ions, to small molecules, macromolecules and genes) and the edges $e \in E$ are biochemical events (e.g., binding, transportation and reaction). The occurrence of signal transduction and gene regulatory events in different cellular compartments can be modeled be the tree T. Each node $t \in T$ represents a cluster of nodes of G consisting of the leaves of the subtree rooted at t. The modeling of such networks based on clustered graphs can be used for cluster-preserving layout algorithms [EH00]. However, as it is only partly known in which compartment an event occurs, signal transduction and gene regulatory processes are usually modeled by graphs. The pathways and networks can be derived from databases such as KEGG [KGKN02, KGH+06] and TransPath [KVC+03] (for an overview of biological databases see, for example, [CG10]).

20.2.2   Visualization Requirements

Important goals of the visualizations of signal transduction and gene regulatory pathways are the understanding of the regulation of cellular processes by external and internal signals, the flow of information through the pathways and networks, the interconnection of genes, the discovering of master-genes responsible for the regulation of larger sets of genes, and the identification of main and alternative regulatory paths.

The main visualization requirements are:

- *Pathways*: The main direction of the processes (e.g., from top to bottom) should be clearly visible to express the temporal order of the events.

- *Compartments*: Events of signal transduction and gene regulation occur in different cellular compartments and this information should be visually represented.

- *Complexes*: Especially during signal transduction one event occurring frequently is the building of molecular complexes. Their structure and how they are built by interacting molecules should be displayed.

Signal transduction and gene regulatory pathways often contain metabolic reactions, therefore the visualization requirements discussed in Section 20.4 are also of interest. However, there is no need for the consideration of open and closed cycles (see Section 20.4.2) and usually co-substances are not considered.

20.2.3   Layout Methods

There are two established approaches to visualize signal transduction and gene regulatory pathways and networks: force-directed and hierarchical layout methods. It should be noted that some visualizations of gene regulatory networks in books and articles also use orthogonal or grid-based drawing styles.

Figure 20.2 A hierarchical layout of a part of the gene regulatory network of *E. coli.*

There are some systems supporting force-directed layouts for the visualization of signal transduction and gene regulatory pathways and networks. These tools are either based on re-implementations of well-known algorithms or on existing layout libraries. Usually the visualizations do not meet the main requirements, especially the main direction and the consideration of compartments. There are a few approaches to improve the general force-directed method. Examples are the PATIKA system [DBD+02, GD06] where the force-directed layout has been extended to deal with several application specific requirements, e.g., cellular compartments, and the approach presented in [SDMW09] where placement, directional, compartmental and other constraints are considered.

Another common approach for the visualization of signal transduction and gene regulatory networks are graph drawing solutions based on hierarchical layout methods, see Figure 20.2. There exist several systems which use hierarchical layouts for the visualization of these networks, e.g., TransPath [KVC+03]. Most are based on existing layout libraries such as dot [KN95] and Pajek [BM02]. These approaches meet some visualization requirements such as the main direction of pathways.

20.3 Protein-Protein Interaction Networks

Proteins are one of the most important molecule groups for living cells. For example, they serve as enzymes for catalysis of metabolic processes, signaling substances (hormones), structural or mechanical material (hair), or transporters for other substances (oxygen). The primary structure of a protein is a long sequence out of essentially twenty different *amino acids* connected by *peptide bonds*.

20.3.1 Definition

A protein can interact with another protein, e.g., to build a protein complex or to activate it. *Protein-protein interactions* form large networks. Their visualization aids biologists in pinpointing the role of proteins and in gaining new insights about the processes within and across cellular processes and compartments, e.g., for formulating and experimentally testing specific hypotheses about gene function.

Often only the existence of an interaction between two proteins is known, but the interaction type, such as activation, binding to, or phosphorylation, remains unknown. However, for the understanding of biological processes, information about the interaction type is crucial, although up to now databases contain little information about that. Therefore we define a protein-protein interaction network as a directed graph $G = (V, E, \tau)$ where V is the set of proteins, E the set of directed interactions (the initiator defines the source), and $\tau: E \to T$ defines the type of each edge (interaction type). Protein-protein interaction networks can be derived from databases such as BIND [BDH03] and DIP [XFS+01].

20.3.2   Visualization Requirements

Important goals of the visualization of protein-protein interaction networks are the under-standing of the overall structure of the interactions, the interactions between two proteins, and the functions of proteins by investigating the functions of their neighbors or of all proteins within a cluster the protein belongs to. These networks are inherently complex: large, non-planar with many edge crossings, many separate components, and nodes of a wide range of degrees [HJP02]. Thus, the main visualization requirements are the common aesthetic criteria for graph layouts such as even node distribution, symmetry, uniform edge lengths, or Euclidian distances reflecting graph-theoretic distances.

20.3.3   Layout Methods

The established approach for the visualization of protein-protein interaction networks is the force-directed layout method. For drawing networks where interactions are not typed or not of interest accelerated force-directed methods are used: Basalaj and Eilbeck [BE99] use an incremental multidimensional scaling heuristic [Bas99] and Han, Ju and Park [HJP02] use Walshaw's algorithm [Wal02], which is a multi level variant of the original algorithm of Fruchterman and Reingold [FR91]. Both algorithms can generate two and three dimensional drawings. For example, Figure 20.3 shows a force-directed layout of interactions in yeast (Saccharomyces cerevisiae).

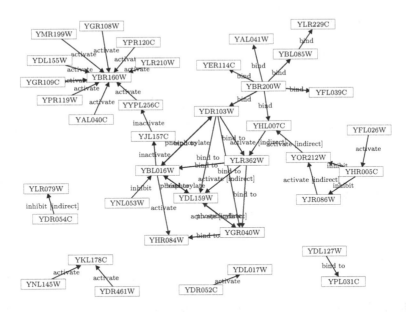

Figure 20.3   A force-directed 2D layout of protein-protein interactions in yeast (redrawn from [FS03]).

However, the general methods cannot cope well with the complexity of protein-protein interaction networks containing typed interactions. In those networks it is not only necessary to show the interactions, but also to explore their different type. For computing visual representations of a network depending on the type of interaction a combination of circular and force-directed algorithms has been suggested [FS03]: Proteins not supporting a selected type of interaction $t \in T$ are placed on an outer circle, whereas proteins that support that type, i.e., to which an edge of type t is incident, are clustered inside the circle, see Figure 20.4. Thereby the radius of the circle is chosen as big as possible while still fitting in the drawing canvas. As the node labels have a font and thus a fixed height, the circular placement is done with constant vertical distance between them rather than with equal distribution. In the second phase, the positions of the nodes which are involved in the selected interaction are recomputed. Let $G' = (V', E')$ with $E' = \{\, e \in E \mid \tau(e) = t \,\}$ and $V' \subseteq V$ the set of vertices adjacent to an edge in E' be the subgraph representing the interaction t. Based on a variation of the force-directed GEM layout [FLM95] the drawing of G' is generated. GEM optimizes minimal node distances and constant edge lengths while it also tends to display symmetries. However, the gravity force to attract nodes to the center is not suitable to keep all nodes in V' inside the circle. Either the gravity force has to be set so high that it distorts the drawing, or it is not strong enough to prevent nodes from escaping the circle. Thus, a reflective barrier at 80% of the circle radius is introduced. Any node which is about to leave this perimeter is reflected toward the interior of the circle while the energy acting on it is slightly dampened.

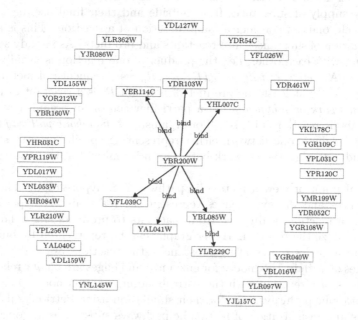

Figure 20.4 The graph of Figure 20.3 with focus on interaction "bind" (redrawn from [FS03]).

While working with a visualization focusing on a special type of interaction, users build a mental map of the picture. Thus, when working with a dynamic visualization tool which allows frequent changes of the interaction type of interest, it is important to help the user in maintaining the mental map. In the described method [FS03] animations are used to provide smooth transitions between different visualizations and ensure that the position of the nodes on the outer circle are fixed over all types of interactions. After computing the new drawing, the nodes are moved on straight lines from their initial positions to their final positions. Thereby the node speed is increased in the beginning and decreased toward the end to allow an easy perception. Edges which have been visible in the initial drawing fade into the background while newly active edges fade from background to foreground color.

20.4   Metabolic Networks

Metabolic reactions are fundamental to life processes, e.g., for the production of energy and the synthesis of substances. A huge number of reactions occur at any time in living cells and the product of one reaction is usually used by another reaction, thus metabolic reactions are strongly interconnected and form metabolic pathways and networks.

20.4.1   Definition

A *metabolic reaction* R is a transformation of chemical substances or metabolites (*reactants*) into other substances (*products*) usually catalyzed by *enzymes*. In general metabolic reactions are reversible, that is, they occur in both directions. Such reactions are characterized by a steady state, i.e., if occurring isolated they reach a state where the amount of change in both directions is equal. A cell is in a constant exchange of substances with its environment. Furthermore, many reactions are regulated, i.e., they are suppressed or enhanced by other factors (allosteric control). This shifts the steady state and together with the steady supply of substances from outside and their final use, e.g., by exporting them from the cell, one can consider a main direction of a reaction. This is also expressed by the differentiation of substances into reactants and products. As already seen, metabolic reactions interact with each other, i.e., the product of one reaction is usually a reactant of another reaction. A *metabolic path* $P = (R_1, \ldots, R_n)$ is a sequence of metabolic reactions where for all $1 \leq i < n$ at least one product of reaction R_i is a reactant of reaction R_{i+1}. The *metabolic network* or *metabolism* of a particular cell or an organism is the complete network of metabolic reactions of this cell or organism. A *metabolic pathway* is a connected sub-network of the metabolic network either representing specific processes or defined by functional boundaries, e.g., the network between an initial and a final substance as shown in Figure 20.5.

From a formal point of view a metabolic pathway is a hyper-graph. The nodes represent the substances and the hyper-edges represent the reactions. A hyper-edge connects all substances of a reaction, is directed from reactants to products and is labeled with the enzymes that catalyze the reaction. Hyper-graphs can be represented by bipartite graphs. Additionally to the nodes representing substances, the reactions are nodes (either labeled with the enzymes or with further nodes for enzymes) and edges are binary relations connecting the substances of a reaction with the corresponding reaction node. This is a common modeling of metabolic pathways, e.g., for their simulation using Petri-nets [HT98, RML93]. For the analysis and visualization of metabolic pathways substances are often divided into two types [MZ03]: *main* substances and *co*-substances. Co-substances are usually small or current metabolites, e.g., ATP, ADP, H_2O, NH_3 and NADH. These substances normally transfer electrons or functional groups such as phosphate and amino groups [NIS90]. Main substances are all other metabolites. However, this is not a global property but is given according to the reaction [MZ03], and a small metabolite such as ATP may be considered as main substance in a particular reaction. For visualization purposes this distinction is important as main substances and co-substances are often differently visually represented.

Here a metabolic pathway is modeled as directed bipartite graph $G = (V_S, V_R, E)$ with nodes $u_1, \ldots, u_n, w_1, \ldots, w_m \in V_S$ representing substances, nodes $v \in V_R$ representing reactions (including the enzyme(s) catalyzing the reaction) and directed edges $(u_1, v), \ldots, (u_n, v)$, $(v, w_1), \ldots, (v, w_m) \in E$ representing the transformation of substances $u_1, \ldots, u_n$ to substances $w_1, \ldots, w_m$ by the reaction v. A reversible reaction does not contain backward edges as in some models for simulation purposes, instead this property of an reaction is represented by an attribute. Another attribute is used to mark main and co-substances.

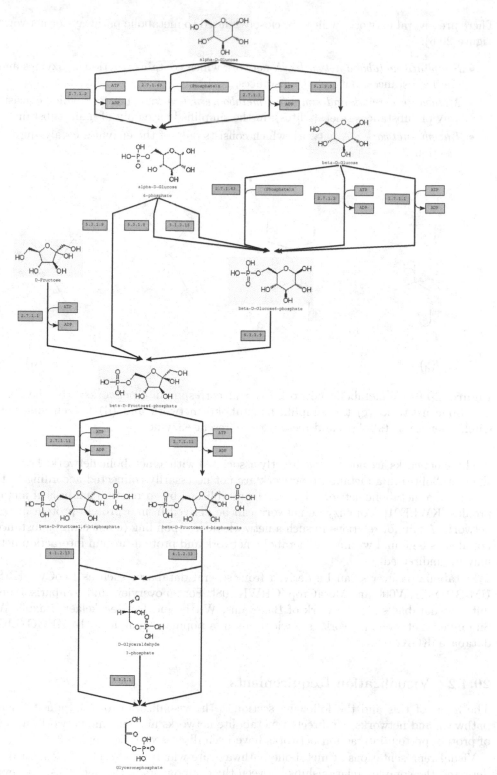

Figure 20.5 An example of a metabolic pathway.

There are several networks which are closely related to metabolic pathways or networks (see Figure 20.6):

- *Simplified metabolic network* : A network which contains reactions, enzymes and main substances, but no co-substances.
- *Metabolite network and simplified metabolite network*: A network which consists only of substances (metabolites); in the simplified case only of main substances.
- *Enzyme network* : A network which consists only of the enzymes catalyzing the reactions.

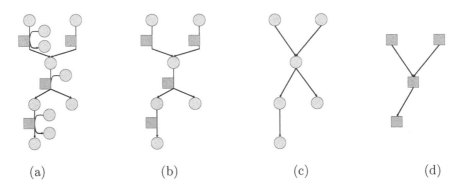

(a)   (b)   (c)   (d)

Figure 20.6 A metabolic network (a) and corresponding networks: (b) the simplified metabolic network, (c) the simplified metabolite network and (d) the enzyme network. Circles denote metabolites and rectangles represent enzymes

These networks are not always directly associated with a metabolic network. For example, the metabolites in a metabolite network are not necessarily connected according to the reactions of a metabolic network, but can be established by correlation analysis of metabolite profiles [KWLF01]. An enzyme network can be derived from a protein-protein interaction network. Again for relations in such a network a corresponding (connecting) substance cannot always be found within the metabolic network and protein-protein interaction networks may be undirected.

Metabolic pathways can be derived from several databases such as EcoCyc [KRS+00], UM-BBD [EHW00], and MetaCrop [GBWK+08]. For an overview and comparison between different databases see the work of Baxevanis, Wittig and De Beuckelaer [Bax03, WB01]. Simplified metabolic networks are widely used, a popular example is the KEGG/LIGAND database [KGKN02].

20.4.2   Visualization Requirements

The focus of this and the following section is the visualization of (simplified) metabolic pathways and networks. Undirected metabolite networks and enzyme networks as a subset of protein-protein interaction networks have been discussed in Section 20.3.

Visual representations of metabolic pathways are widely used and help scientists to understand the complex relationships between the components of the networks. However, the style of pathway visualizations varies significantly [Mic98]. Examples are biochemical and biological textbooks [Cam96, LNC93, Mic99], pathway posters [Mic93, Nic97] and electronic databases [ABH94, KGKN02, OLP⁺00]. Visualizations of metabolic pathways should help understanding the interconnections between metabolites, analyzing the flow of substances through the network and identifying main and alternative paths. The established presentation styles and discussions with users result in several visualization requirements [Sch02]:

1. *Parts of reactions*: The display of substances and enzymes is application and user-specific. Usually for main substances their name, structural formula or both should be shown. Co-substances should be displayed using their name or abbreviation and enzymes should be represented by their name or EC-number [Int92].

2. *Reactions*: The reaction arrow(s) should be shown from the reactants to the products with enzymes placed on one side of the reaction arrow and co-substances on the opposite side. The reversibility of a specific reaction should be clearly visible. For co-substances their temporal order, which depends on the reaction mechanism, is important, and they should be placed according to this order.

3. *Pathways*: The main direction of reactions (e.g., from top to bottom) should be clearly visible to express the temporal order of reactions. There are important exceptions to the main direction used for the visualization of specific pathways, e.g., the citrate acid cycle or the fatty acid synthesis. The structure of these cyclic reaction chains should be emphasized. Such pathways are characterized by the continuous repetition of a reaction sequence in which the product of the sequence re-enters in the next loop as a reactant. There are two mechanisms. First, the reactant and the product of the reaction sequence are identical from loop to loop (e.g., citrate acid cycle)— a mechanism called a *closed cycle*. Second, the reactant of the reaction sequence varies slightly from the product (e.g., fatty acid cycle) - this is called an *open cycle*.

Besides usual quality criteria, e.g., low number of edge crossings, these visualization requirements result in some specific layout criteria: the hierarchical placement of nodes depending on the structure of the network, the treatment of nodes of varying sizes and the consideration of layout constraints for the order of co-substances and the visualization of specific pathways. Often closed and open cycles are displayed as circles and spirals, respectively. In a spiral related reaction steps from different loops and corresponding substances are placed side by side to emphasis the cyclic structure. As this drawing style needs much space and makes it difficult for a user to trace the reaction sequence of long pathways, an alternative visualization would be to unravel the spiral and align related reactions and substances horizontally.

20.4.3 Layout Methods

There are two established approaches to visualizing metabolic pathways and networks: force-directed and hierarchical layout methods.

Force-directed methods are often used and several pathway analysis tools support such layout. Frequently they visualize not only metabolic and metabolite pathways, but different types of biochemical pathways and networks. Examples are PathwayAssist [NEDM03], PathDB [MBF⁺00] and pathSCOUT [MdRW03]. These tools use either their own implementations of well-known algorithms or are based on existing layout libraries. For example, VisANT [HMWD04] contains an algorithm based on the layout method of Eades [Ead84], and the method described by Rojdestvenski [Roj03] is based on the force-directed method of Kamada and Kawai [KK89]. On the other hand Cytoscape [SMO⁺03] uses the yFiles li-

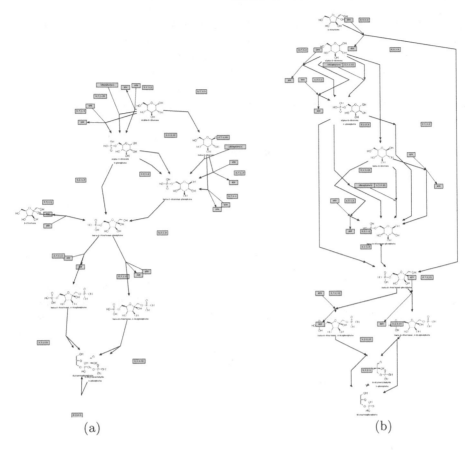

(a)   (b)

Figure 20.7   Visualizations of the metabolic pathway shown in Figure 20.5 using (a) a force-directed algorithm [KK89] and (b) a hierarchical approach [STT81].

brary [WEK01] and the layout of BioJAKE [SMKS99], a tool for the creation, visualization and manipulation of metabolic pathways, which is based on Graphviz [EGK+01].

Force-directed approaches do not meet the visualization requirements described in the previous section and visualizations based on this method are very different to the diagrams in posters and books, see Figure 20.7 (a). Different node sizes, the special placement of co-substances and enzymes, the partitioning of substances into reactants and products as well as the general direction of pathways are not considered. A few approaches extend this layout method to deal with application specific requirements. Advanced approaches are the algorithms described in [DBD+02, GD06] where directional and rectangular regional constrains are considered which can be used to enforce different node types (e.g., main and co-substances), layout directions and subcellular locations (cellular compartments), and in [SDMW09] where placement, directional, compartmental and other constraints are considered.

The second layout method for (simplified) metabolic pathways is hierarchical layout. Tools supporting this layout are largely based on existing libraries. Such solutions show the main direction of reactions and are sometimes able to deal with different node sizes. However, there is no specific placement of co-substances, furthermore, open and closed cycles are not emphasized. Figure 20.7 (b) shows a typical example of such a visualization. For example, PathFinder [GHM+02] is restricted to acyclic pathways which are modeled as directed acyclic graphs and drawn using the VCG library [San95]. The hierarchical layout of BioMiner [SSE$^+$02] is based on yFiles [WEK01]. Some improved approaches consider cyclic structures within the network or depict pathways of different topology using different layouts, e.g., linear, circular and tree structured. Becker and Rojas [BR01] present a graph layout algorithm for drawing metabolic pathways which emphasizes cyclic structures. However, these cycles are computed based on the topology of the network and not on biological knowledge. Therefore pathways may be shown as circles even if they are not closed cycles and closed cycles may not be emphasized by this method, e.g., if they contain shortcuts within the cycle. Furthermore, open cycles are not considered. PathDB contains a component for the visualization of metabolic pathways based on hierarchical layout which allows co-substances to be represented in a smaller font on the side of the reaction arrow [Men00, MBF$^+$00].

The most advanced algorithms try to consider all the visualization requirements discussed in Section 20.4.2. The approach of Karp et al. [KP94, KPR02] based on the Grasper-CL system [KLSW94] depicts pathways of different topology using different layout algorithms (linear, circular, tree, hierarchical). It places co-substances and enzymes beside reaction arrows, but has restrictions concerning the order of co-substances or the layout of open cycles. Another approach [Sch02] extends the hierarchical layout for different node sizes; consideration of co-substances and enzymes and special layout of open and closed cycles is implemented in the BioPath system [BFP$^+$04]. The algorithm temporarily builds larger nodes containing the layout of co-substances and enzymes for each reaction, extends the layering step of hierarchical layouts by a local layering [FS04] and the crossing reduction step by constraint crossing reduction [For04]. A drawing produced with this method is shown in Figure 20.5.

The extensions of layering and crossing reduction are of interest also for other graph drawing applications. Usually the layering step of hierarchical layouts computes a global layering, i.e., a layering where nodes belong to a particular layer depending on the topological sorting of the graph. Global layering of graphs tends to produce large drawings as the distance between two layers is determined by the highest node of the layer. An algorithm to compute a local layering, i.e., a layering where each node may be assigned to its own layer depending only on the layers of its direct predecessors and their particular heights is shown in Figure 20.8. It computes the layers from top to bottom. The y-coordinate of a node, i.e., the upper boundary of the rectangle representing the node, and its layer are computed together. Nodes can be split such that a high node may belong to a number of consecutive layers. To reduce the number of layers and dummy nodes layers are joined together if they are situated in an area starting from the current layer with depth y_d. For local and global layering the final part is the replacement of each edge-layer crossing by a dummy node in order to compute a so called *proper* layering. This part is not shown in the algorithm, but takes $\mathcal{O}(|V| * |E|)$ in both the global and the local layering method. This is also the overall running time for these algorithms.

For constraint crossing reduction Forster [For04] presents a heuristic shown in Algorithm 20.9 which extends the well known barycenter heuristic [DETT99]. It starts with partitioning the node set V_2 into ordered node lists with one singleton list $L(v) = \langle v \rangle$ for each node v. Later these lists are pairwise concatenated according to violated constraints. Each violated constraint $c = (s, t)$, i.e., a constraint that node s should be placed left of node t, is removed. The lists containing s and t are concatenated in the required order and treated as a cluster of vertices. The nodes s and t are replaced by a node v_c to represent the concatenated list $L(v_c) = L(s) \circ L(t)$. This node has a barycenter value which is computed as if all edges incident to a node in $L(v_c)$ were incident to v_c. After all violated constraints have been removed the remaining nodes/node lists are sorted according to their barycenter

Input: $G = (V, E)$, height of nodes ($h: V \to \mathbb{R}$), minimum node distance d, depth of area
  where layers are joint y_d
Output: Coordinates $y: V \to \mathbb{R}$ and layers $l: V \to \mathbb{N}$
Data: Min-heap H, counter $c: V \to \mathbb{N}$ for the nodes
  $y \leftarrow y_{next} \leftarrow 0;\ l \leftarrow 0$
  for all $v \in V$ **do**
    $c(v) \leftarrow$ indegree$(v);\ h(v) \leftarrow h(v) + d$
    if $c(v) = 0$ **then**
      H.insert(v)
    end if
  end for
  while !H.isEmpty() **do**
    {Place nodes on current and consecutive layers within y_d in one layer}
    $l \leftarrow l + 1;\ y \leftarrow y_{next}$
    $v \leftarrow H$.delmin();$\ l(v) \leftarrow l;\ y(v) \leftarrow y$
    $y_{next} \leftarrow y + h(v)$
    while $(y + h(H.\text{top}())) \leq (y_{next} + y_d)$ **do**
      $v \leftarrow H$.delmin();
      $l(v) \leftarrow l;\ y(v) \leftarrow y$
      for all $u \in$ children(v) **do**
        $c(u) \leftarrow c(u) - 1;$
      end for
    end while
    $y_{next} := y + h(v);$
    for all $v \in H$ **do**
      {Split large nodes (on this and next layer)}
      In $G = (V, E)$ replace v by v_1, v_2 and the edge (v_1, v_2);
      $l(v_1) \leftarrow l;\ y(v_1) \leftarrow y;\ h(v_1) \leftarrow y_{next} - y;\ h(v_2) \leftarrow h(v) - h(v_1)$
      Replace in heap H node v by node v_2
    end for
    for all $v \in V$ **do**
      if $v \notin H$ and v not already placed and $c(v) = 0$ **then**
        H.insert(v)
      end if
    end for
  end while

Figure 20.8   Computing a local layering of the nodes

value. The result is a vertex permutation that satisfies all constraints and has few cross-ings. During the algorithm the violated constraints have to be considered in an order which avoids the generation of constraint cycles. This is done by the procedure FIND-VIOLATED-CONSTRAINT(V, C) and with the $\mathcal{O}(|C|)$ algorithm for this procedure [For04] the running time of the complete algorithm is $\mathcal{O}(|V_2| \log |V_2| + |E| + |C|^2)$.

20.5   Phylogenetic Trees

A fundamental issue in biology is the hierarchical classification of organisms in an evolu-tionary context, i.e., reconstruction of ancestral relationships between different *taxons*, e.g.,

Input: A two-level graph $G = (V_1, V_2, E)$, acyclic constraints $C \subseteq V_2 \times V_2$
Output: A permutation of V_2 (result in L)
Data: singleton lists L and barycenter $b \colon V \to \mathbb{Q}_0^+$ for all nodes
    for all $v \in V_2$ **do**
        $b(v) \leftarrow \sum_{u \in V} \text{position}(u) / \text{degree}(v)$
        $L(v) \leftarrow \langle v \rangle$
    end for
    $V \leftarrow \{ s, t \mid (s, t) \in C \}$ {constrained vertices}
    $V' \leftarrow V_2 - V$
    while $(s, t) \leftarrow$ FIND-VIOLATED-CONSTRAINT$(V, C) \neq \perp$ **do**
      create new vertex v_c
      $\text{degree}(v_c) \leftarrow \text{degree}(s) + \text{degree}(t)$ {update barycenter value}
      $b(v_c) \leftarrow \big(b(s) \cdot \text{degree}(s) + b(t) \cdot \text{degree}(t) \big) / \text{degree}(v_c)$
      $L(v_c) \leftarrow L(s) \circ L(t)$
      for all $c \in C$ **do**
        if c is incident to s or t **then**
          make c incident to v_c instead of s or t
        end if
      end for
      $C \leftarrow C - \{(v_c, v_c)\}$ {remove self loops}
      $V \leftarrow V - \{s, t\}$
      if v_c has incident constraints **then**
        $V \leftarrow V \cup \{v_c\}$
      else
        $V' \leftarrow V' \cup \{v_c\}$
      end if
    end while
    $V'' \leftarrow V \cup V'$
    sort V'' by $b()$
    $L \leftarrow \langle \rangle$ {concatenate vertex lists}
    for all $v \in V''$ **do**
      $L \leftarrow L \circ L(v)$
    end for

Figure 20.9   Computing a constrained crossing reduction

species, genes, or DNA sequences. The common approach for determining such relations is the construction of a phylogenetic tree.

20.5.1   Definition

For hierarchical classification of a set of taxons A there are two common types of approaches: The first are the *phenetic* methods, which have an $|A| \times |A|$ *distance matrix* Δ assigning each pair of taxons a quantitative difference as input. The goal is to group (commonly two) most similar taxons/ancestors and thus to find out how an ancestor of theirs may look like according to the principle of *minimum evolution*. This is done recursively until a common ancestor is reached and a phylogenetic tree is obtained. All these methods are based on clustering and thus explicitly do not consider evolutionary history. The second type of approach is the *cladistic* methods, which have an $|A| \times |M|$ *characteristic matrix* Γ assigning each taxon $|M|$ characteristics like number of legs, ability to fly, or color of skin as input. These methods try to find out the actual genealogy according to a model of the real evolutionary development assuming that identical characteristics of different taxons indicate a common ancestry.

A *phylogenetic tree* (in literature also called *evolutionary tree*) $T = (V, E, \delta)$ is a tree consisting of nodes V (taxons) and edges E (links). Leave nodes, i.e., nodes with exactly one link, represent species, sequences, or similar entities; they are called *operational taxonomical units* (and are represented by $A \subseteq V$). Internal nodes represent (hypothetical) ancestors generated from phylogenetic analysis; they are called *hypothetical taxonomic units*. The lengths of the edges $\delta : E \rightarrow \mathbb{R}_0^+$ quantify the biological divergence between the incident nodes, e.g., biological time or genetic distance. Phylogenetic trees are often stored in the Newick file format [Fel95], which makes use of the correspondence between trees and nested parentheses.

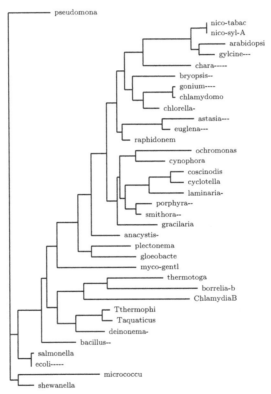

Figure 20.10   An example of a phylogenetic tree (phylogram, redrawn from [DS04]).

A simple phenetic representative for creating a phylogenetic tree $T = (V, E, \delta)$ is the $\mathcal{O}\left(|A|^2 \log |A|^2\right)$ time "Unweighted Pair Group Method with Arithmetic Mean" (UPGMA) [MS57]: Initially define clusters $C \leftarrow \{ c_i \mid 1 \leq i \leq |A| \}$, each containing one taxon of A, set the cluster sizes $s(c_i) \leftarrow 1$, and let $V \leftarrow C$. Then iterate until there is only one cluster left: Find two closest clusters $c_i \neq c_j$ according to Δ (with the help of a priority queue over the $|A|^2$ elements of Δ). Join the clusters c_i and c_j to a new cluster c_p by $C \leftarrow C \cup \{c_p\} - \{c_i, c_j\}$ with $s(c_p) \leftarrow s(c_i) + s(c_j)$, and add it to T with $V \leftarrow V \cup \{c_p\}$. Introduce new edges $E \leftarrow E \cup \{(c_p, c_i), (c_p, c_j)\}$ with $\delta\left((c_p, c_i)\right) \leftarrow \delta\left((c_p, c_j)\right) \leftarrow \frac{\Delta_{ij}}{2}$. Then compute the distances from c_p to all clusters $c_k \in C$ with $k \neq i, j$:

$$\Delta_{pk} \leftarrow \Delta_{kp} \leftarrow \frac{s(c_i)}{s(c_p)} \cdot \Delta_{ik} + \frac{s(c_j)}{s(c_p)} \cdot \Delta_{jk} \tag{20.1}$$

At the end of the iteration delete the two columns i and j and the two rows i and j in Δ.

If Δ is an ultrametric matrix, then UPGMA guarantees for the unique way W between any two nodes $v_i, v_j \in V : \sum_{e \in W} \delta(e) = \Delta_{ij}$ and T is said to be ultrametric, too. Otherwise, UPGMA is a heuristic.

Another common phenetic approach is the $\mathcal{O}\left(|A|^3\right)$ time "Neighbor-Joining" (NJ) method [SN87] which is an enhancement of UPGMA especially for protein and nucleotide data (DNA does not evolute by accident, but follows some constraints which can be included in the computation of NJ). The idea of NJ is to join clusters which are not only close to each other, but also far from the rest. The initialization is the same as in UPGMA, whereas the iteration for $|C| > 2$ is the following: For each cluster c_i compute the mean distance to an arbitrary other cluster $c_k \in C$ by $d(c_i) \leftarrow \sum_{k \neq i} \frac{\Delta_{ik}}{|C|-2}$. Find two closest clusters $c_i \neq c_j$ with least $\Delta_{ij} - (d(c_i) + d(c_j))$. Join the clusters c_i and c_j to a new cluster c_p by $C \leftarrow C \cup \{c_p\} - \{c_i, c_j\}$, and add it to T with $V \leftarrow V \cup \{c_p\}$. Introduce new edges $E \leftarrow E \cup \{(c_p, c_i), (c_p, c_j)\}$ with lengths as shown in (20.2) and compute the distances from c_p to all clusters $c_k \in C$ with $k \neq i, j$ with (20.3).

$$\delta\left((c_p, c_i)\right) \leftarrow \frac{1}{2}\Delta_{ij} + \frac{1}{2}\left(d(c_i) - d(c_j)\right), \qquad \delta\left((c_p, c_j)\right) \leftarrow \frac{1}{2}\Delta_{ij} + \frac{1}{2}\left(d(c_j) - d(c_i)\right) \tag{20.2}$$

$$\Delta_{pk} \leftarrow \Delta_{kp} \leftarrow \frac{\Delta_{ik} + \Delta_{jk} - \Delta_{ij}}{2} \tag{20.3}$$

Delete the two columns and the two rows i and j in Δ. If $|C| = 2$, i.e., $C = \{c_s, c_t\}$, then connect $c_s, c_t \in V$ by $E \leftarrow E \cup \{(c_s, c_t)\}$ with $\delta\left((c_s, c_t)\right) \leftarrow \Delta_{st}$ and stop.

A typical representative of the cladistic category is the "Maximum Parsimony" (MP) method. The idea is to define the (non-unique) tree T as optimal, which posits fewest mutations as possible. For the "Small Parsimony" problem the topology of T is already given and only the labels $l(v) = \bigcup_{1 \leq j \leq |M|} l_j(v)$ of the inner nodes $v \in V$, i.e., the position $l_j(v)$ of each characteristic $m_j \in M$ has to be determined. It can be solved in $\mathcal{O}\left(|A||M| \cdot \max\{|\text{dom}(m_j)| \mid m_j \in M\}\right)$ time [Fit71], where $\text{dom}(m_j)$ is the set of all possible values which a taxon can adopt for m_j. A solution is the following algorithm: Assign each $v_i \in V$ for each $m_j \in M$ in a postorder traversal of T a set $S_j(v_i) \subseteq \text{dom}(m_j)$ with (20.4), where $w_1, w_2 \in V$ are the children of v_i.

$$S_j(v_i) \leftarrow \begin{cases} \Gamma_{ij}, & \text{if } v_i \text{ is a leaf,} \\ S_j(w_1) \cap S_j(w_2), & \text{if } S_j(v_1) \cap S_j(v_2) \neq \emptyset, \\ S_j(w_1) \cup S_j(w_2), & \text{otherwise.} \end{cases} \tag{20.4}$$

In a subsequent preorder traversal of T for each node $v \in V$ which has a parent u with $l_j(u) \in S_j(v)$ set $l_j(v) \leftarrow l_j(u)$. If no such u exists or v is a leaf set $l_j(v)$ to an arbitrary element of $S_j(v)$. The number of (independent) mutations in T is equal to how many times the third item of (20.4) was used.

In the "Weighted Small Parsimony" version the probability of different mutations is not unique, i.e., $p_j(a, b)$ defines the "price" of a change for a characteristic $m_j \in M$ from state $a \in \text{dom}(m_j)$ to $b \in \text{dom}(m_j)$. The goal is not to minimize the number of mutations, but the sum of their prizes while the topology of T again is given. For that we present the $\mathcal{O}\left(|A||M| \cdot \max\{|\text{dom}(m_j)| \mid m_j \in M\}\right)$ time algorithm [San75], which is a generalization of [Fit71]: Assign in a postorder traversal of T to each $v_i \in V$ quantities $S_j(v_i, t_k(m_j))$ for each m_j and all values $t_k(m_j) \in \text{dom}(m_j)$ with (20.5) for a leaf v_i and (20.6) for an internal node v_i, where $w_1, w_2 \in V$ are the children of v_i. Considering only mutations of characteristic m_j, then $S_j(v, t_k(m_j))$ is the minimum total cost for the subtree rooted at v_i if $l_j(v_i)$ was set to $t_k(m_j)$.

$$S_j(v_i, t_k(m_j)) \leftarrow \begin{cases} 0, & \text{if } \Gamma_{ij} = t_k(m_j), \\ \infty, & \text{otherwise.} \end{cases} \tag{20.5}$$

$$\begin{aligned} S_j(v_i, t_k(m_j)) \leftarrow \quad & \min\{p_j(t_k(m_j), t) + S_j(w_1, t) \mid t_k(m_j) \neq t \in \text{dom}(m_j)\} \\ & + \min\{p_j(t_k(m_j), t) + S_j(w_2, t) \mid t_k(m_j) \neq t \in \text{dom}(m_j)\} \end{aligned} \tag{20.6}$$

The minimum total cost of T with root r is $\sum_{m_j \in M} \min\{S_j(r, t) \mid t \in \text{dom}(m_j)\}$. In a subsequent preorder traversal of T update the labels of each $v_i \in V$, where u is the parent of v_i:

$$l_j(v_i) \leftarrow \begin{cases} \arg\min\{S_j(r, t) \mid t \in \text{dom}(m_j)\}, & \text{if } v_i = r, \\ \arg\min\{p_j(l_j(u), t) + S_j(v_i, t) \mid t \in \text{dom}(m_j)\}, & \text{otherwise.} \end{cases} \tag{20.7}$$

In contrast to the above, the "Large Parsimony" problem, where the topology of T is not given, is $\mathcal{NP}$-hard, regardless if discrete or weighted. However, there are some heuristics, e.g., [HP82] which uses branch&bound to find the cheapest tree T among all trees. This approach guarantees to find T, but its time complexity is in the worst case exponential in $|A|$ (exhaustive search). Another heuristic is "Nearest Neighbor Interchange" (NNI) [MGB73], which defines a relation between each pair of trees and then uses well-known concepts like greedy algorithms or simulated annealing to find a (local) optimum.

Given a tree T with known edge lengths δ, the likelihood of T is $P(M|T)$. It is a statistical measure of how well it describes the biological data. Let $P_{a \to b}(\delta(e))$ be the probability that character $a \in \text{dom}(m_j)$ will transform to $b \in \text{dom}(m_j)$ within the time $\delta(e)$, $P(a)$ be character frequency of $a \in \text{dom}(m_j)$ fixed throughout biological history, L be the set of all reconstructions of T, i.e., all full labelings of internal nodes, and $r \in V$ be the root of T. Then [Fel73]:

$$P(M|T) = \prod_{j \in M} \left(\sum_{l \in L} \left((P(l_j(r)) \cdot \prod_{(u,v) in E} P_{l_j(u) \to l_j(v)}(\delta((u,v))) \right) \right) \tag{20.8}$$

If the character substitution is reversible, i.e., $P_{a \to b}(\delta(e)) = P_{b \to a}(\delta(e))$, then T is unrooted and r can be chosen arbitrarily without changing $P(M|T)$. The "Maximum Likelihood" method (ML) [Fel73] computes the likelihood of a tree with dynamic programming in $\mathcal{O}(|A||M| \cdot \max\{|\text{dom}(m_j)| \mid m_j \in M\})$ time, i.e., it computes the likelihood of each bifurcation and declares the tree with the greatest sum of likelihoods as the best. There are also statistical methods for computing the optimum edge lengths δ for a given tree T with regard to a maximum tree likelihood [SL99].

The topology of T is fixed. However, there is in most cases the freedom of permutation of each node's children and thus there are $2^{|V|-1}$ possible linear leaf orderings consistent with the structure of a binary T. From a biological view it makes sense to order the leaves such that similar leaves are close together. Remember, the dissimilarity of each pair of leaves is stored in the distance matrix Δ. Therefore, the goal is to minimize the sum of the lengths of the ways from each leaf to each other. In an optimal tree the lengths of all ways correspond exactly to the entries in Δ. Since in the general case no such optimal tree exists (Δ represents a complete graph and not only a tree), leaf ordering makes sense. It can be done, e.g., with the dynamic programming approach [BJDG$^+$03] which needs $\mathcal{O}\left(4^k|V|^3\right)$ time for a k-ary T. There, an optimal leaf ordering consistent with a binary tree T is determined by a bottom-up computation of subintervals. Define $M(u, w_l, w_r)$ to be the cost of the best linear order of the leaves in the subtree $T(u)$ induced by $u \in V$ that begins with leaf w_l and ends with leaf w_r. If u is a leaf, then $M(u, u, u) \leftarrow 0$. Otherwise, let v_1 and v_2 be the children of u such that $w_l \in T(v_1)$ and $w_r \in T(v_2)$. Then the optimality criterion of (20.9) holds. For a k-ary tree, denote the children of u by $v_1, \ldots, v_p$, $1 \leq p \leq k$. If $w_l \in T(v_1)$ and $w_r \in T(v_p)$, any ordering of $v_2, \ldots, v_{p-1}$ is possible. Thus for each of the $p!$ orderings $M(u, w_l, w_r)$ is computed in the same way as for binary trees by inserting $k-1$ internal binary dummy nodes while maintaining the current order.

$$
M(u, w_l, w_r) \leftarrow
$$
$$
\min \left\{ M(v_1, w_l, a_i) + \Delta_{ij} + M(v_2, b_j, w_r) \mid \text{leaf } a_i \in T(v_1), \text{leaf } b_j \in T(v_2) \right\} \quad (20.9)
$$

20.5.2   Visualization Requirements

As seen earlier, the graphs to visualize are directed (and thus rooted) or undirected trees $T = (V, E, \delta)$ with given edge lengths δ. T is either a binary tree or very similar to a binary tree, i.e., there are view nodes with a degree higher than three. Irrespectively of edge direction, T should be laid out hierarchically to visualize the ancestral relationships between taxons. Since the sum over the edge lengths on the unique path from one taxon to another is the evolutionary distance, it is desirable to reflect this in the lengths of the curves drawn for the edges. This means in the most simple case that $\delta(e)$ is the curve length of $e \in E$. Traditional algorithms for drawing trees explicitly do not consider given edge lengths. They follow aesthetic criteria as edges should have the same length and nodes of the same depth should be drawn on the same y-coordinate [RT81, Wal90, WS79] or radius [Ead92]. In most cases the nodes as well as the edges contain labels, which should be drawn non-overlapping. Further a good layout follows common criteria for graph/tree layout like no unnecessary edge crossings, compactness, and use of the entire available drawing area.

As we will see in the next section, some layout methods will use the freedom of permuting children to generate nice drawings. However, if not especially mentioned, we assume to have already a fixed leave ordering given.

Although there is need to edit layouts dynamically [Car04a], e.g., collapsing and expanding subtrees or editing annotations, for an easy understanding of large trees, we restrict ourselves to static layouts for the sake of simplicity. Since there is an ongoing trend to larger trees, which may contain several hundred thousand of nodes, a layout algorithm must be efficient.

20.5.3   Layout Methods

The most common layouts for phylogenetic trees are vertical or circular *dendrograms* or radial drawings [Car04b]. The typical representatives of the first group are the orthogonal *phylograms* (see Figure 20.10), where the tree is drawn hierarchically and from left to right and thus the vertices vertically from top to bottom. Each edge $e = (u, v)$ has exactly one bend b at the x-coordinate of u and at the y-coordinate v. The length of the horizontal edge segment (b, v) represents $\delta(e)$. A parent node is vertically placed, e.g., in the middle between its extremal children or in the arithmetic mean of all its children. Since the topology of the tree, the horizontal edge lengths, and the leave ordering (and thus the y-coordinates of the leaves) are already fixed, the layout is already fixed and can be computed by the $\mathcal{O}\left(|V|\right)$ time algorithm in Figure 20.11. Phylograms are easy to interpret and leave space for edge annotations [Car04b]. *Cladograms* and *curvograms* drawing edges as straight lines or splines are subtypes of phylograms and thus are not treated separately.

Input: $T = (V, E, \delta)$, y-coordinates of leaves
Output: Coordinates $x, y \colon V \to \mathbb{R}$ for the nodes and $x_b, y_b \colon E \to \mathbb{R}$ for the bends
Data: Stack S

    $r \leftarrow \mathsf{root}(T)$
    $S.\mathsf{push}(r)$
    $x(r) \leftarrow 0$
    while $!S.\mathsf{isEmpty}()$ **do**
        $v \leftarrow S.\mathsf{top}()$
        if v has an unmarked child w **then**
            mark w; $S.\mathsf{push}(w)$
            $x_b\left((v, w)\right) \leftarrow x(v)$
            $x(w) \leftarrow x(v) + \delta\left((v, w)\right)$
        else
            $S.\mathsf{pop}()$
            if v is an internal node **then**
                $y(v) \leftarrow \frac{1}{2}\left(\min\left\{ y(w) \mid w \text{ is a child of } v \right\} + \max\left\{ y(w) \mid w \text{ is a child of } v \right\}\right)$
            end if
            if $v \neq r$ **then**
                $u \leftarrow S.\mathsf{top}()$ {the parent of v}
                $y_b\left((u, v)\right) \leftarrow y(v)$
            end if
        end if
    end while

Figure 20.11   Computing coordinates for drawing a phylogram.

Another style of dendrograms is the *circle layout*, which draws the trees concentric around the root with an unique radius for the leaves. Again, each edge $e = (u, v)$ bends exactly once at the radius of the parent u. The "vertical" segment is drawn as a segment of a circle, whereas the "horizontal" one is an interval of a straight line from the root through the child v, see Figure 20.13. The algorithm for computing a circle layout is similar to Algorithm 20.11 if treating x as levels $(x, x_b \colon V \to \{0, 1, \ldots, \mathsf{height}(T)\})$ with $x(r) = 0$ and y as angles $(y, y_b \colon V \to [0, \ldots, 2\pi])$. Instead of the Cartesian coordinates, the algorithm needs the polar angles of the leaves distributed uniformly on a circle as input. Since the radius now is unique for all leaves, we set $x(w) \leftarrow x(v) + 1$ instead of $x(w) \leftarrow x(v) + \delta\left((v, w)\right)$ for each edge (v, w). This ignores edge lengths δ, however. Another approach [BBS05] which considers edge lengths is to distribute the leaves uniformly on a circle, to set each inner node v on the weighted Cartesian barycenter of its parent u and its children W as shown in (20.10), and to draw each edge as a straight line. See Figure 20.13 for an example. The arising equation system can be solved in $\mathcal{O}\left(|V|\right)$ time. Algorithm 20.12 shows the computation in a unit circle. If reordering of the leaves is acceptable, the postorder traversal of the children w of each node v can be ordered according to ascending height of $T(w)$ (in terms of δ) plus

$\delta\left((v,w)\right)$. This should support the algorithm to draw edges with their desired length, but raises the running time to $\mathcal{O}\left(|V|\log|V|\right)$, however. Since even this cannot guarantee exact lengths, the edges are colored, i.e., blue color means too short and red color too large, such that the color saturation reflects the multiplicative failure.

$$((x(v), y(v)) \leftarrow \frac{(x(u), y(u))}{\delta\left((u,v)\right)} + \sum_{w\in W} \frac{(x(w), y(w))}{\delta\left((v,w)\right)\cdot|W|} \qquad (20.10)$$

Input: $T = (V, E, \delta)$ with $\delta(e) > 0$ for all edges e
Output: Coordinates $x, y\colon V \to \mathbb{R}$ for the nodes
Data: Coefficients $c\colon V \to \mathbb{R}$, offsets $d\colon V \to \mathbb{R}^2$, and edge weights $s\colon E \to \mathbb{R}$
  for each $v \in V$ **if** $\deg(v) = 1$ **then** $l \leftarrow l + 1$
  $i \leftarrow 0$
  postorder_traversal(root(T))
  preorder_traversal(root(T))

  procedure postorder_traversal(node v)
    for each child w of v **do** postorder_traversal(w) {optionally ordered}
    if v is a leaf or ($v = $ root(T) and $\deg($root(T)$) = 1$) **then**
      $c(v) \leftarrow 0;\ d(v) \leftarrow \left(\cos\left(\frac{2\pi i}{l}\right), \sin\left(\frac{2\pi i}{l}\right)\right)$ {fix vertex on circle}
      $i \leftarrow i + 1$
    else
      $s \leftarrow 0$
      for each adjacent edge $e \leftarrow \{u, v\}$ **do**
        if $v = $ root(T) or w is the parent of v **then** $s(e) \leftarrow \frac{1}{\delta(e)}$
        else $s(e) \leftarrow \frac{1}{\delta(e)\cdot(\deg(v)-1)}$
        $s \leftarrow s + s(e)$
      end for
      $t \leftarrow t' \leftarrow 0$
      for each outgoing edge $e \leftarrow (v, w)$ **do** $t \leftarrow t + \frac{s(e)}{s}\cdot c(w);\ t' \leftarrow t' + \frac{s(e)}{s}\cdot d(w)$
      if $v \neq $ root(T) **then** let e be the incoming edge of v; $c(v) \leftarrow \frac{s(e)}{s\cdot(1-t)}$
      $d(v) \leftarrow \frac{t'}{1-t}$
    end if
  end procedure

  procedure preorder_traversal(node v)
    if $v = $ root(T) **do** $x(v) \leftarrow d(v)$
    else let u be the parent of v; $x(v) \leftarrow c(v)\cdot x(u) + d(v)$
    for each child w of v **do** preorder_traversal(w)
  end procedure

Figure 20.12 Cartesian barycenter method for generating a circle layout.

Circle layouts provide the best use of the available space for trees with more than 100 leaves [Car04b]. Dendrograms in general are a good choice to visualize the leaf ordering.

The second type of drawings are the radial tree drawings [BBS05], which are preferred for visualizing unrooted trees. Their edges are drawn as straight lines. To obtain coordinates for the vertices, Algorithm 20.14 traverses T in preorder (here, breadth first search) from a given root to the leaves. Thereby it assigns each subtree a wedge according to its size,

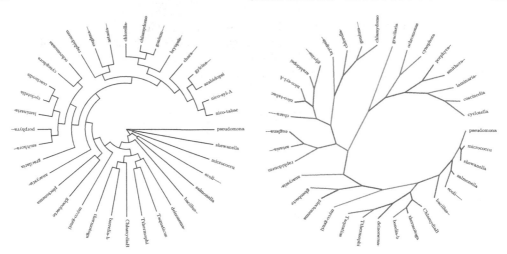

Figure 20.13   Circle layouts with levels and weighted Cartesian barycenter.

i.e., according to its number of leaves (leafcount). Note that here all degree one vertices are treated as leaves. Since the wedge sizes are independent of the root, rerooting the tree only results in a different ordering of the children of the new root.

Input: $T = (V, E, \delta)$
Output: Coordinates $x, y \colon V \to \mathbb{R}$ for the nodes
Data: Queue Q, leafcount$\colon V \to \mathbb{N}^+$ {from a previous postorder traversal}

  $r \leftarrow \mathsf{root}(T)$
  $Q.\mathsf{insert}(r)$
  rightborder$(r) \leftarrow 0$
  wedgesize$(r) \leftarrow 2\pi$
  $x(r) \leftarrow y(r) \leftarrow 0$
  while $!Q.\mathsf{isEmpty}()$ **do**
    $v \leftarrow Q.\mathsf{delete\_first}()$
    $\eta \leftarrow$ rightborder(v)
    for each child w of v **do**
      $Q.\mathsf{insert}(w)$
      rightborder$(w) \leftarrow \eta$
      wedgesize$(w) \leftarrow \frac{2\pi \cdot \mathrm{leafcount}(w)}{\mathrm{leafcount}(r)}$
      $\alpha \leftarrow$ rightborder$(w) + \frac{\mathrm{wedgesize}(w)}{2}$
      $x(w) \leftarrow x(v) + \cos(\alpha) \cdot \delta\left((v, w)\right);\ y(w) \leftarrow y(v) + \sin(\alpha) \cdot \delta\left((v, w)\right)$
      $\eta \leftarrow \eta + \mathrm{wedgesize}(w)$
    end for
  end while

Figure 20.14   Computing coordinates for drawing of radial tree drawings.

Clearly, Algorithm 20.14 has an $\mathcal{O}\left(|V|\right)$ running time if newly discovered children are distributed in random order around their parent, e.g., as they occur in the adjacency list. Advanced versions use the freedom of reordering the children. The first aims to reach a symmetric layout: For each child v the metric of (20.11) is computed with a postorder traversal of T. It is a measure of how far the biological development goes on in the induced subtree of v. Alternating, depending on the depth of the parent node, the child with higher value is drawn on the left or on the right side of the corresponding wedge. If the parent has more than two children, then the child with highest value is drawn in the middle and the other children on its left and right side according to descending m. The second method is to put evolutionary closely related children on near positions. For this (20.12) is used to order the children ascending according to average distance of the leaves in the induced subtree to the parent. However, in both cases the running time raises to $\mathcal{O}\left(|V|\log|V|\right)$ and ordering of children makes no sense for UPGMA-trees, since each child will have the same m-value.

$$m(v) \leftarrow \begin{cases} \delta\left((u,v)\right), & \text{if } v \text{ is a leaf,} \\ \delta\left((u,v)\right) + \max\left\{\, m(w) \mid w \text{ is a child of } v \,\right\}, & \text{otherwise.} \end{cases} \qquad (20.11)$$

$$m(v) \leftarrow \begin{cases} \delta\left((u,v)\right), & \text{if } v \text{ is a leaf,} \\ \dfrac{\sum_{(v,w)}\left(\delta((u,v))+m(w)\right)}{|\{\,w\mid w \text{ is a child of } v\,\}|}, & \text{otherwise.} \end{cases} \qquad (20.12)$$

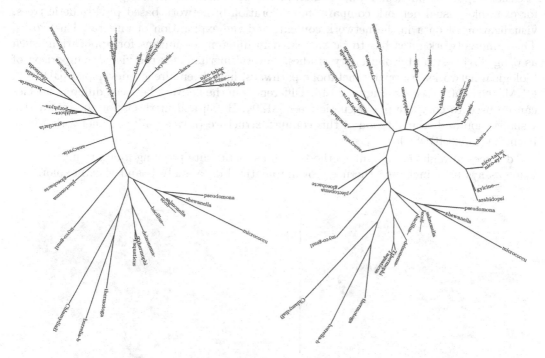

Figure 20.15 Radial tree layout with the same root as in Figure 20.10 and leaf reordering for drawing those closely related near. The right drawing is with spreading.

A lot of space is wasted by simply giving the wedge for a child v from the parent u to v, i.e., the area between the pairwise parallel wedge borders. This can be avoided by spreading (the subtrees induced by) the children w of v to use the full wedge of v originated at u and not at v except of a small buffer. Spreading is done in a postprocessing step and needs $\mathcal{O}\left(|V|^2\right)$ time. Each label is drawn as an extension of the incoming edge of the corresponding leaf, i.e., in the corresponding wedge. To leave space for labels in spreaded layouts, the lengths of the labels are added to the δ values of the respective incoming edges, for computation only. Another more simple solution is to draw the labels with an angle of a ray from the root through the leaves. Figure 20.15 shows a standard and a spreaded layout of our running example. To overcome the problem of zero edge lengths, e.g., incoming edges of ecoli----- or nico-tabac and nico-syl-A, a user definable minimum edge length is useful to indicate edges and to simplify the labeling.

20.6   Discussion

In this chapter we discussed the visualization of biological networks. We focused on important networks closely related to molecular biology: gene regulatory, signal transduction, protein-protein interaction and metabolic networks. Furthermore, we studied the visualization of phylogenetic trees, hierarchies which are often built on information from molecular biology such as DNA or protein sequences. However, there are many more networks in biology: ecological networks such as food-webs, biological data analysis networks such as correlation networks, and neuronal networks to name just a few. Moreover, even for the networks discussed we presented only some visualization aspects.

Other topics of particular importance in the visualization of biological networks are, for example, visual network comparison, exploration of network based phylogenetic trees, visualization of data in the network context, and the exploration of integrated networks. The same network often has to be compared in different organisms for applications such as drug discovery and evolutionary studies. Several methods for the visual comparison of biological networks, especially metabolic pathways, have been already developed [BDS04b, GHM+02, Sch03], see also Figure 20.16. Differences in the network between different species can be used to compute phylogenetic trees [MZ04, HS03] and methods for the interactive visualization and triangulation of this complex structure (a tree built over networks) have been developed [BDS04a].

Advances in high-throughput methods such as metabolite profiling and automatized enzyme assays have increased the need for automatized data analysis and visual exploration

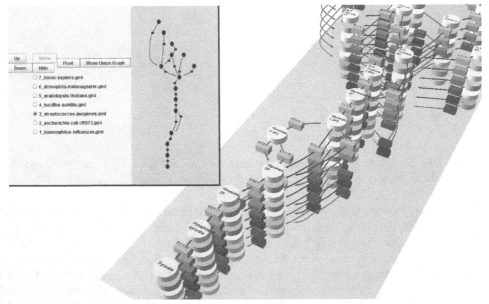

Figure 20.16   Visual comparison of metabolic pathways in $2\frac{1}{2}$ dimensions.

techniques to deduct biologically meaningful interpretations from the large amount of experimental data. The visualization of these data-rich networks provides new challenges for algorithms such as the consideration of complex graphical elements and of different node sizes. There is an increasing amount of approaches which look into this area, early approaches were, for example, [BHK+05, DRS04, JKS06, TSS+05], and a comparison is given in [KAO+09]. Also, the integration of different networks is increasingly important. Elements of one biological network often belong to several networks. For example, a protein of a protein-protein interaction network may be an enzyme of a metabolic network, an element of a gene regulatory network, or a leaf of a phylogenetic tree. This complex structure of interwoven networks requires new visualization and exploration methods which are the topic of current research. Finally, the standardization of the visual representation of elements of biological networks has been the focus of recent developments. The Systems Biology Graphical Notation (SBGN) [LHM+09] provides a set of standards for graphically representing biological information. It can be considered as the biology equivalent of the circuit diagram in electronics. The standard also contains layout requirements for SBGN maps.

A detailed presentation of the above-mentioned and newly emerging topics would easily fill not only another chapter, but a book. Biological network visualization is growing at an extremely fast pace. However, our sole intention in this chapter was to raise awareness of the relevance of graph drawing for the area of biological networks and provide an introduction to this topic. The interested reader is referred to journals such as *Bioinformatics* and *BMC Bioinformatics* as well as newly founded conferences such as *VIZBI* (since 2010) or *IEEE BioVis* (since 2011) for ongoing developments.

References

[ABH94]   R. D. Appel, A. Bairoch, and D. F. Hochstrasser. A new generation of information retrieval tools for biologists: The example of the ExPASy WWW server. *Trends Biochemical Sciences*, 19:258–260, 1994.

[Bas99]   W. Basalaj. Incremental multidimensional scaling method for database visualization. In R. F. Erbacher, P. C. Chen, and C. M. Wittenbrink, editors, *Visual Data Exploration and Analysis VI (Proc. SPIE)*, volume 3643 of *Proceedings of SPIE*, pages 149–158, 1999.

[Bax03]   A. D. Baxevanis. The molecular biology database collection: 2003 update. *Nucleic Acids Research*, 31(1):1–12, 2003.

[BBS05]   C. Bachmaier, U. Brandes, and B. Schlieper. Drawings of phylogenetic trees (extended abstract). In X. Deng and D. Du, editors, *Algorithms and Computation, Proc. ISAAC 2005*, volume 3827 of *LNCS*, pages 1110–1121. Springer, 2005.

[BDH03]   G. D. Bader, D. Betel D, and C. W. Hogue. BIND: the biomolecular interaction network database. *Nucleic Acids Research*, 31(1):248–250, 2003.

[BDS04a]   U. Brandes, T. Dwyer, and F. Schreiber. Visual triangulation of network-based phylogenetic trees. In O. Deussen, C. Hansen, D. Keim, and D. Saupe, editors, *Data Visualization (Proc. VisSym'04)*, pages 75–84. Eurographics Association, 2004.

[BDS04b]   U. Brandes, T. Dwyer, and F. Schreiber. Visual understanding of metabolic pathways across organisms using layout in two and a half dimensions. *Journal of Integrative Bioinformatics*, 1:2 (EPub), 2004.

[BE99]   W. Basalaj and K. Eilbeck. Straight-line drawings of protein interactions (system demonstration). In J. Kratochvíl, editor, *Graph Drawing (Proc. GD '99)*, volume 1731 of *Lecture Notes Comput. Sci.*, pages 259–266. Springer-Verlag, 1999.

[BFP+04]   F. J. Brandenburg, M. Forster, A. Pick, M. Raitner, and F. Schreiber. *Graph Drawing Software*, chapter BioPath – Exploration and Visualization of Biochemical Pathways, pages 215–236. Springer Mathematics and Visualization Series, 2004.

[BHK+05]   L. Borisjuk, M.-R. Hajirezaei, C. Klukas, H. Rolletschek, and F. Schreiber. Integrating data from biological experiments into metabolic networks with the DBE information system. *In Silico Biology*, 5(2):93–102, 2005.

[BJDG+03] Z. Bar-Joseph, E. D. Demaine, D. K. Gifford, A. M. Hamel, T. S. Jaakkola, and N. Srebro. K-ary clustering with optimal leaf ordering for gene expression data. *Bioinformatics*, 19(9):1070–1078, 2003.

[BM02]   V. Batagelj and A. Mrvar. Pajek – analysis and visualization of large networks. In P. Mutzel, M. Jünger, and S. Leipert, editors, *Graph Drawing (Proc. GD '01)*, volume 2265 of *Lecture Notes Comput. Sci.*, pages 477–478, 2002.

[BR01]   M. Y. Becker and I. Rojas. A graph layout algorithm for drawing metabolic pathways. *Bioinformatics*, 17(5):461–467, 2001.

[Cam96]   N. A. Campbell. *Biology*. The Benjamin-Cummings Publishing Company, 1996.

[Car04a]   S. F. Carrizo. Phylogenetic trees: An information visualization perspective. In Y.-P. Phoebe Chen, editor, *Bioinformatics (Proc. APBC 2004)*, volume 29 of *Conf. Res. Pract. Inform. Techn.*, pages 315–320, 2004.

[Car04b]   S. F. Carrizo. A survey of phylogenetic researchers: Results. `http://www.cs.usyd.edu.au/~scarrizo/Carrizo_PhylogeneticsSurveyResults.doc`, January 2004.

[CG10]   G. R. Cochrane and M. Y. Galperin. The 2010 Nucleic Acids Research database issue and online database collection: a community of data resources. *Nucleic Acids Research*, 38:D1–D4, 2010.

[DBD⁺02]   E. Demir, O. Babur, U. Dogrusöz, A. Gürsoy, G. Nisanci, R. Çetin Atalay, and M. Ozturk. PATIKA: an integrated visual environment for collaborative construction and analysis of cellular pathways. *Bioinformatics*, 18(7):996–1003, 2002.

[DETT99]   G. Di Battista, P. Eades, R. Tamassia, and I. G. Tollis. *Graph Drawing*. Prentice Hall, Upper Saddle River, NJ, 1999.

[DRS04]   T. Dwyer, H. Rolletschek, and F. Schreiber. Representing experimental biological data in metabolic networks. In Y. P. Chen, editor, *Bioinformatics (Proc. APBC'04)*, volume 29 of *Conf. Res. Pract. Inform. Techn.*, pages 13–20, 2004.

[DS04]   T. Dwyer and F. Schreiber. Optimal leaf ordering for two and a half dimensional phylogenetic tree visualization. In N. Churcher and C. Churcher, editors, *Information Visualisation (Proc. invis.au 2004)*, volume 35 of *Conf. Res. Pract. Inform. Techn.*, pages 109–115, 2004.

[Ead84]   P. Eades. A heuristic for graph drawing. *Congr. Numer.*, 42:149–160, 1984.

[Ead92]   P. D. Eades. Drawing free trees. *Bulletin of the Institute for Combinatorics and its Applications*, 5:10–36, 1992.

[EGK⁺01]   J. Ellson, E. R. Gansner, E. Koutsofios, S. C. North, and G. Woodhull. Graphviz – open source graph drawing tools. In P. Mutzel, M. Jünger, and S. Leipert, editors, *Graph Drawing (Proc. GD'01)*, volume 2265 of *Lecture Notes Comput. Sci.*, pages 483–484, 2001.

[EH00]   P. Eades and M. L. Huang. Navigating clustered graphs using force-directed methods. *Journal of Graph Algorithms Applications*, 4(3):157–181, 2000.

[EHW00]   L. B. Ellis, C. D. Hershberger, and L. P. Wackett. The university of minnesota biocatalysis/biodegradation database: Microorganisms, genomics and prediction. *Nucleic Acids Research*, 28(1):377–379, 2000.

[Fel73]   J. Felsenstein. Maximum likelihood and minimum-steps methods for estimating evolutionary trees from data on discrete characters. *Systematic Zoology*, 22:240–249, 1973.

[Fel95]   J. Felsenstein. The newick tree format. `http://evolution.gs.washington.edu/phylip/newicktree.html`, 1995.

[Fit71]   W. M. Fitch. Toward defining the course of evolution: Minimum change for a specified tree topology. *Systematic Zoology*, 20:406–416, 1971.

[FLM95]   A. Frick, A. Ludwig, and H. Mehldau. A fast adaptive layout algorithm
           for undirected graphs. In R. Tamassia and I. G. Tollis, editors, *Graph
           Drawing (Proc. GD '94)*, volume 894 of *Lecture Notes Comput. Sci.*,
           pages 388–403. Springer-Verlag, 1995.

[For04]   M. Forster. A fast and simple heuristic for constrained two-level crossing
           reduction. In *Graph Drawing (Proc. GD'04)*, volume 3383 of *Lecture
           Notes Comput. Sci.*, pages 206–216, 2004.

[FR91]   T. Fruchterman and E. Reingold. Graph drawing by force-directed place-
           ment. *Softw. – Pract. Exp.*, 21(11):1129–1164, 1991.

[FS03]   C. Friedrich and F. Schreiber. Visualisation and navigation methods
           for typed protein-protein interaction networks. *Applied Bioinformatics*,
           2(S3):19–24, 2003.

[FS04]   C. Friedrich and F. Schreiber. Flexible layering in hierarchical drawings
           with nodes of arbitrary size. In V. Estivill-Castro, editor, *Computer
           Science (Proc. ACSC 2004)*, volume 26 of *Conf. Res. Pract. Inform.
           Techn.*, pages 369–376, 2004.

[GBWK⁺08] E. Grafahrend-Belau, S. Weise, D. Koschützki, U. Scholz, B. H. Junker,
           and F. Schreiber. MetaCrop – a detailed database of crop plant
           metabolism. *Nucleic Acids Research*, 36:D954–D958, 2008.

[GD06]   B. Genc and U. Dogrusöz. A layout algorithm for signaling pathways.
           Information Sciences, 176:135–149, 2006.

[GHM⁺02]   A. Goesmann, M. Haubrock, F. Meyer, J. Kalinowski, and R. Giegerich.
           PathFinder: reconstruction and dynamic visualization of metabolic path-
           ways. *Bioinformatics*, 18(1):124–129, 2002.

[HJP02]   K. Han, B.-H. Ju, and J. H. Park. InterViewer: Dynamic visualization
           of protein-protein interactions. In M. T. Goodrich and S. G. Kobourov,
           editors, *Graph Drawing (Proc. GD '02)*, volume 2528 of *Lecture Notes
           Comput. Sci.*, pages 364–365. Springer-Verlag, 2002.

[HMWD04]   Z. Hu, J. Mellor, J. Wu, and C. DeLisi. VisANT: an online visualization
           and analysis tool for biological interaction data. *BMC Bioinformatics*,
           5(1):17 (EPub), 2004.

[HP82]   M. D. Hendy and D. Penny. Branch and bound algorithms to determine
           minimal evolutionary trees. *Mathematical Bioscience*, 60:133–142, 1982.

[HS03]   M. Heymans and A. K. Singh. Deriving phylogenetic trees from the simi-
           larity analysis of metabolic pathways. *Bioinformatics*, 19(Suppl. 1):138–
           146, 2003.

[HT98]   R. Hofestädt and S. Thelen. Qualitative modeling of biochemical net-
           works. *In Silico Biology*, 1(1):39–53, 1998.

[Int92]   International Union of Biochemistry and Moleculare Biology, Nomencla-
           ture Commitee. *Enzyme Nomenclature*. Academic Press, 1992.

[JKS06]   B. H. Junker, C. Klukas, and F. Schreiber. VANTED: A system for
           advanced data analysis and visualization in the context of biological net-
           works. *BMC Bioinformatics*, 7:109, 2006.

[KAO⁺09]   N. Kono, K. Arakawa, R. Ogawa, N. Kido, K. Oshita, K. Ikegami,
           S. Tamaki, and M. Tomita. Pathway Projector: Web-based zoomable
           pathway browser using KEGG atlas and Google Maps API. *PLoS ONE*,
           4(11):e7710, 2009.

[Kel00]   E. F. Keller. *The Century of the Gene*. Harvard University Press, 2000.

[KGH⁺06]   M. Kanehisa, S. Goto, M. Hattori, K. F. Aoki-Kinoshita, M. Itoh, S. Kawashima, T. Katayama, M. Araki, and M. Hirakawa. From genomics to chemical genomics: new developments in KEGG. *Nucleic Acids Research*, 34:D354–357, 2006.

[KGKN02]   M. Kanehisa, S. Goto, S. Kawashima, and A. Nakaya. The KEGG databases at GenomeNet. *Nucleic Acids Research*, 30(1):42–46, 2002.

[KK89]   T. Kamada and S. Kawai. An algorithm for drawing general undirected graphs. *Inform. Process. Lett.*, 31:7–15, 1989.

[KLSW94]   P. D. Karp, J. Lowrance, T. Strat, and D. Wilkins. The Grasper-CL graph management system. *LISP and Symbolic Computation*, 7:245–282, 1994.

[KN95]   E. Koutsofios and S. North. Drawing graphs with *dot*. Technical report, AT&T Bell Laboratories, Murray Hill, NJ., 1995. Available from `http://www.research.bell-labs.com/dist/drawdag`.

[KP94]   P. D. Karp and S. M. Paley. Automated drawing of metabolic pathways. In H. Lim, C. Cantor, and R. Bobbins, editors, *Proc. of the 3rd International Conference on Bioinformatics and Genome Research*, pages 225–238, 1994.

[KPR02]   P. D. Karp, S. M. Paley, and P. Romero. The pathway tools software. *Bioinformatics*, 18(Suppl. 1):S225–S232, 2002.

[KRS⁺00]   P. D. Karp, M. Riley, M. Saier, I. Paulsen, S. M. Paley, and A. Pellegrini-Toole. The EcoCyc and MetaCyc database. *Nucleic Acids Research*, 28(1):56–59, 2000.

[KVC⁺03]   M. Krull, N. Voss, C. Choi, S. Pistor, A. Potapov, and E. Wingender. TRANSPATH: an integrated database on signal transduction and a tool for array analysis. *Nucleic Acids Research*, 31:97–100, 2003.

[KWLF01]   F. Kose, W. Weckwerth, T. Linke, and O. Fiehn. Visualizing plant metabolomic correlation networks using clique-metabolite matrices. *Bioinformatics*, 17(12):1198–1208, 2001.

[LHM⁺09]   N. Le Novère, M. Hucka, H. Mi, S. Moodie, F. Schreiber, A. Sorokin, E. Demir, K. Wegner, M. Aladjem, S. M. Wimalaratne, F. T. Bergman, R. Gauges, P. Ghazal, K. Hideya, L. Li, Y. Matsuoka, A. Villéger, S. E. Boyd, L. Calzone, M. Courtot, U. Dogrusoz, T. Freeman, A. Funahashi, S. Ghosh, A. Jouraku, S. Kim, F. Kolpakov, A. Luna, S. Sahle, E. Schmidt, S. Watterson, G. Wu, I. Goryanin, D. B. Kell, C. Sander, H. Sauro, J. L. Snoep, K. Kohn, and H. Kitano. The Systems Biology Graphical Notation. *Nature Biotechnology*, 27:735–741, 2009.

[LNC93]   A. L. Lehninger, D. L. Nelson, and M. M. Cox. *Principles of Biochemistry*. Worth Publisher, 1993.

[MBF⁺00]   P. Mendes, D. L. Bulmore, A. D. Farmer, P. A. Steadman, M. E. Waugh, and S. T. Wlodek. PathDB: a second generation metabolic database. In J.-H. S. Hofmeyr, J. M. Rohwer, and J. L. Snoep, editors, *Proc. of the 9th International BioThermoKinetics Meeting*, pages 207–212. Stellenbosch University Press, 2000.

[MdRW03]   E. Minch, M. de Rinaldis, and S. Weiss. pathSCOUT: exploration and analysis of biochemical pathways. *Bioinformatics*, 19(3):431–432, 2003.

[Men00]   P. Mendes. Advanced visualization of metabolic pathways in PathDB. In *Proc. of the 8th Conference on Plant and Animal Genome*, 2000.

[MGB73]   G. W. Moore, M. Goodman, and J. Barnabas. A method for constructing maximum parsimony ancestral amino acid sequences on a given network. *Journal of Theoretical Biology*, 38(3):459–483, 1973.

[Mic93]   G. Michal. *Biochemical Pathways (Poster)*. Boehringer Mannheim, 1993.

[Mic98]   G. Michal. On representation of metabolic pathways. *BioSystems*, 47:1–7, 1998.

[Mic99]   G. Michal. *Biochemical Pathways*. Spektrum Akademischer Verlag, 1999.

[MS57]   C. D. Michener and R. R. Sokal. A quantitative approach to a problem in classification. *Evolution*, 11:130–162, 1957.

[MZ03]   H. Ma and A.-P. Zeng. Reconstruction of metabolic networks from genome data and analysis of their global structure for various organisims. *Bioinformatics*, 19(2):270–277, 2003.

[MZ04]   H. W. Ma and A. P. Zeng. Phylogenetic comparison of metabolic capacities of organisms at genome level. *Molecular Phylogenetics and Evolution*, 31(1):204–213, 2004.

[NEDM03]   A. Nikitin, S. Egorov, N. Daraselia, and I. Mazo. Pathway studio – the analysis and navigation of molecular networks. *Bioinformatics*, 19(16):2155–2157, 2003.

[Nic97]   D. E. Nicholson. *Metabolic Pathways Map (Poster)*. Sigma Chemical Co., St. Louis, 1997.

[NIS90]   F. C. Neidhardt, J. L. Ingraham, and M. Schaechter. *Physiology of the Bacterial Cell: A Molecular Approach*. Sinauer Associates, 1990.

[OLP$^+$00] R. A. Overbeek, N. Larsen, G. D. Pusch, M. D'Souza, E. Selkov Jr., N. Kyrpides, M. Fonstein, N. Maltsev, and E. Selkov. WIT: integrated system for high-throughput genome sequence analysis and metabolic reconstruction. *Nucleic Acids Research*, 28(1):123–125, 2000.

[RML93]   V. N. Reddy, M. L. Mavrovouniotis, and M. N. Liebman. Petri net representations of metabolic pathways. In L. Hunter, D. Searls, and J. Shavlik, editors, *Intelligent Systems for Molecular Biology (Proc. ISMB '93)*, pages 328–336, 1993.

[Roj03]   I. Rojdestvenski. Metabolic pathways in three dimensions. *Bioinformatics*, 19(18):2436–2441, 2003.

[RT81]   E. Reingold and J. Tilford. Tidier drawing of trees. *IEEE Trans. Softw. Eng.*, SE-7(2):223–228, 1981.

[San75]   D. D. Sankoff. Minimal mutation trees of sequences. *SIAM J. Appl. Math.*, 28:35–42, 1975.

[San95]   G. Sander. Graph layout through the VCG tool. In R. Tamassia and I. G. Tollis, editors, *Graph Drawing (Proc. GD '94)*, volume 894 of *Lecture Notes Comput. Sci.*, pages 194–205. Springer-Verlag, 1995.

[Sch02]   F. Schreiber. High quality visualization of biochemical pathways in BioPath. *In Silico Biology*, 2(2):59–73, 2002.

[Sch03]   F. Schreiber. Visual comparison of metabolic pathways. *Journal of Visual Languages and Computing*, 14(4):327–340, 2003.

[SDMW09] F. Schreiber, T. Dwyer, K. Marriott, and M. Wybrow. A generic algorithm for layout of biological networks. *BMC Bioinformatics*, 10:375, 2009.

[SL99] G. Shavit and C. Linhart. Algorithms for molecular biology, chapter 9. `http://http://www.math.tau.ac.il/~rshamir/algmb/98/scribe/pdf/lec09.pdf`, January 1999.

[SMKS99] W. Salamonsen, K. Y. Mok, P. Kolatkar, and S. Subbiah. BioJAKE: A tool for the creation, visualization and manipulation of metabolic pathways. In *Proc. of the 4th Pacific Symposium on Biocomputing*, pages 392–400, 1999.

[SMO⁺03] P. Shannon, A. Markiel, O. Ozier, N. S. Baliga, J. T. Wang, D. Ramage, N. Amin, B. Schwikowski, and T. Ideker. Cytoscape: a software environment for integrated models of biomolecular interaction networks. *Genome Research*, 13(11):2498–2504, 2003.

[SN87] N. Saitou and M. Nei. The neighbor-joining method: A new method for reconstructing phylogenetic trees. *Molecular Biology and Evolution*, 4(4):406–425, 1987.

[SSE⁺02] M. Sirava, T. Schäfer, M. Eiglsperger, M. Kaufmann, O. Kohlbacher, E. Bornberg-Bauer, and H.-P. Lenhof. BioMiner – modeling, analyzing, and visualizing biochemical pathways and networks. *Bioinformatics*, 18(S2):S219–S230, 2002.

[STT81] K. Sugiyama, S. Tagawa, and M. Toda. Methods for visual understanding of hierarchical systems. *IEEE Trans. Syst. Man Cybern.*, SMC-11(2):109–125, 1981.

[TSS⁺05] T. Tokimatsu, N. Sakurai, H. Suzuki, H. Ohta, K. Nishitani, T. Koyama, T. Umezawa, N. Misawa, K. Saito, and D. Shibatanenell. KaPPA-View. a web-based analysis tool for integration of transcript and metabolite data on plant metabolic pathway maps. *Plant Physiology*, 138:1289–1300, 2005.

[Wal90] J. Q. Walker II. A node-positioning algorithm for general trees. *Softw. – Pract. Exp.*, 20(7):685–705, 1990.

[Wal02] C. Walshaw. A multilevel algorithm for force-directed graph drawing. In J. Marks, editor, *Graph Drawing (Proc. GD '01)*, volume 1984 of *Lecture Notes Comput. Sci.*, pages 171–182. Springer-Verlag, 2002.

[WB01] U. Wittig and A. De Beuckelaer. Analysis and comparison of metabolic pathway databases. *Briefings in Bioinformatics*, 2(2):126–142, 2001.

[WEK01] R. Wiese, M. Eiglsperger, and M. Kaufmann. yFiles: Visualization and automatic layout of graphs. In P. Mutzel, M. Jünger, and S. Leipert, editors, *Graph Drawing (Proc. GD'01)*, volume 2265 of *Lecture Notes Comput. Sci.*, pages 453–454, 2001.

[WS79] C. Wetherell and A. Shannon. Tidy drawing of trees. *IEEE Trans. Softw. Eng.*, SE-5(5):514–520, 1979.

[XFS⁺01] I. Xenarios, E. Fernandez, L. Salwinski, X.J. Duan, M. J. Thompson, E. M. Marcotte, and D. Eisenberg. DIP: The database of interaction proteins: 2001 update. *Nucleic Acids Research*, 29(1):239–241, 2001.

21

Computer Security

Olga Ohrimenko
Brown University

Charalampos
Papamanthou
University of California,
Berkeley

Bernardo Palazzi
Brown University and Italian
National Institute of Statistics

21.1 Introduction

As the number of devices connected to the Internet continues to grow rapidly and software systems are being increasingly deployed on the Web, security and privacy have become crucial properties for networks and applications. Because of the complexity and subtlety of cryptographic methods and protocols, software architects and developers often fail to incorporate security principles in their designs and implementations. Also, most users have minimal understanding of security threats. While several tools for developers, system administrators, and security analysts are available, these tools typically provide information in the form of textual logs or tables, which are cumbersome to analyze. Thus, in recent years, the field of security visualization has emerged to provide novel ways to display security-related information, thus making such information easier to understand.

Securing computers and cyberspace is one of today's grand challenges for science and engineering. Computers and networks are under continuous threat from attackers who want to steal credit card numbers, intellectual property, and other sensitive information. Also, massive distributed denial of service attacks can impair even the largest of companies and government organizations.

Computer security research aims at developing methods and associated protocols to analyze and defend against a growing number and variety of attacks. The development of security tools is a continuous process that keeps on reacting to newly discovered hardware and software vulnerabilities and newly deployed technologies.

21.1.1   Motivation

Both the discovery of vulnerabilities and the development of security protocols can be greatly assisted by visualization. For example, network traffic can be naturally displayed as a graph whose nodes are hosts and whose edges are associated with packets going from one host to another. Also, a visual representation of a complex multiparty security protocol can give experts better intuition of its execution and security properties. Traditionally, instead, computer security analysts read through large logs produced by applications, operating systems, and network devices. Inspecting such logs is quite cumbersome and often unwieldy, even for experts. Motivated by the growing need for automated visualization methods and tools for computer security, the field of *security visualization* has recently emerged as an interdisciplinary community of researchers with its own annual meeting (VizSec).

For basic background on computer security, see the textbook by Goodrich and Tamassia [GT11]. The book by Raffael Marty [Mar08] provides an excellent introduction to methods and tools for visualizing computer networks to analyze their security.

21.1.2   Chapter Organization

In this chapter, we give a survey of approaches to the visualization of computer security concepts that use graph drawing techniques. We consider a variety of fundamental security and privacy issues, focusing on network security, access control, and attack strategies. We show how graphs can be used as an effective modeling tool in computer security and we give examples of how several classic graph drawing techniques have been used in current security visualization prototypes. Finally, we mention an approach for privacy-preserving drawing in a cloud computing scenario.

Thanks to their versatility, graph drawing techniques are one of the main approaches employed in security visualization. Indeed, not only computer networks are naturally modeled as graphs, but also data organization (e.g., file systems) and vulnerability models (e.g., attack trees) can be effectively represented by graphs. In particular, we consider the following security visualization problems:

1. *Network monitoring.* (Section 21.2) The visualization of network traffic helps network administrators identify anomalous patterns, such as scans, worm infections, and hosts trying to gain unauthorized access to the network. Thus, it is an effective component of intrusion detection systems. Also, traffic visualization can be used to identify unusually heavy network activity and quickly track down machines that generate or receive a large volume of packets. Early detection is crucial when defending against denial of service attacks. It is also interesting to monitor and visualize the evolution of highly dynamic services on the Internet, such as the root name servers.

2. *Border gateway protocol (BGP).* (Section 21.3) This protocol manages reachability between hosts in different Autonomous Systems, i.e., networks controlled by Internet Service Providers. The visualization of BGP-related information is important to ensure that routing in the Internet has not changed and has not been tampered with. In particular, displaying the topology of Autonomous Systems and the evolution of BGP routing patterns can assist the detection of disruptions in Internet traffic caused by attacks or router configuration errors.

3. *Access control.* (Section 21.4) Access to resources on a computer system or network is regulated by organizational policies and enforced with technological mechanisms for authentication and authorization. Resources need to be protected not only from malicious activity by outside attackers but also from accidental disclosure to unauthorized legitimate users. Access control mechanisms for file systems, databases, and distributed applications are complex and tricky to configure. Visualization helps both users and administrators gain an intuitive understanding of the vast set of permissions that are in place in the system and allows them to efficiently spot sensitive resources that are insufficiently protected. Also, visualizing flows of information in a system can help keeping sensitive data private and defend against the leakage of confidential information. Access control is especially challenging in distributed environments without centralized administrative control. An aspect of access control that is gaining increasing importance is the management of privacy settings by users of a social network.

4. *Attack graphs.* (Section 21.5) Starting with a vulnerable component of a system, an attacker can compromise other components to reach the desired goal. Attack graphs are used to describe dependencies between vulnerabilities in a system. They characterize the paths through the system that can be followed by an adversary. The visualization of attack graphs helps computer security analysts identify and remedy vulnerabilities.

Sections 21.2 through 21.5 are organized around the four security topics mentioned above. For each topic, we overview visualization tools that employ graph drawing techniques. Table 21.1.2 classifies the papers surveyed in these sections according to the security topic addressed and the graph drawing method used.

	Force Directed	Layered Drawing	Bipartite Drawing	Circular Drawing	Treemap or Gmap	3D
Network Monitoring	[MMK07, TN00, GB98, MMB05, DSN12]		[YYT+04, BFN04, Con07]	[Tol]	[DSN12, BvO09]	[XMB+06]
BGP	[BMPP04, TRNC06]			[TRNC06]		[OKB06]
Access Control	[MLA12]	[MFG+06, Yee06, YSTW05]			[HPPT08]	
Attack Graphs		[NJKJ05, NJ04]			[CIL+10]	

Table 21.1 Classification of the papers on security and privacy visualization surveyed in this chapter according to the security topic addressed and the graph drawing method used.

Finally, in Section 21.6, we take a different perspective and consider the subject of privacy protection when a client outsources the task of drawing a graph to a server in the cloud. We present a technique that provides a high level of privacy, going beyond encryption, and is computationally efficient.

Chapter 24 overviews related work on the visualization of computer networks.

21.2 Network Monitoring

In this section, we overview selected papers on graph-based visualization techniques for network monitoring. Related work includes, e.g., [FMK$^+$08, MFK$^+$09].

21.2.1 Intrusion Detection

In [TN00], the authors use a combination of force-directed drawing, graph clustering, and regression-based learning in a system for intrusion detection (see Figure 21.1). Their system consists of the following components:

- a packet collecting module;
- a graph construction and clustering module;
- a visualization module; and
- an event generation module.

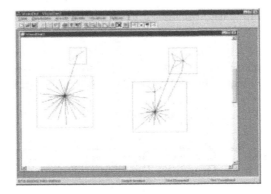

Figure 21.1 Force-directed clustered drawing for intrusion detection (thumbnail of image from [TN00]).

The authors model the computer network with a graph where the nodes are computers and the edges are communication links with weight proportional to the network traffic on that link. The clustering of the graph is performed with a simple iterative method. Initially, every node forms its own cluster. Next, nodes join clusters that already have most of their neighbors, breaking ties at random. The resulting graph is a simplified version of traffic exchanges where entities that communicate often are joined into clusters. Two clusters A and B are connected by an edge if there is at least one edge between some node of cluster A and some node of cluster B.

The classic force-directed spring embedder method [Ead84, FR91] is used to determine the layout of the graph of clusters and of the graph within each cluster. Since forces are proportional to the weights of the edges, if there is a lot of communication between two hosts, their nodes are placed close to each other.

Various features of the clustered graph (including statistics on the node degrees, number of clusters, and internal/external connectivity of clusters) are used to describe the current state of network traffic and are summarized by a feature vector. Using test traffic samples and a regression-based strategy, the system learns how to map feature vectors to intrusion detection events. The visualization of the clustered graph can help a security analyst in assessing the severity of the intrusion detection events generated by the system.

21.2.2 Traffic Analysis

A tool for visualizing the evolution over time of the volume and type of network traffic using force-directed graph drawing techniques is described in [MMK07] (see Figure 21.2). Since there are different types of traffic protocols (HTTP, FTP, SMTP, SSH, etc.) and multiple time periods, this multidimensional data set is modeled by a graph with two types of nodes: *dimension nodes* represent traffic protocols and *observation nodes* represent the state of a certain host in a given time interval. Edges are also of two types: *trace edges* link observation nodes of consecutive time intervals and *attraction edges* link observation nodes with dimension nodes and have weight proportional to the traffic of that type.

The layout of the above graph is computed starting with a fixed placement of the dimension nodes and then executing a modified version of the Fruchterman-Reingold force-directed algorithm [FR91] that aims at achieving uniform edge lengths. The authors show how intrusion detection alerts can be associated with visual patterns in the layout of the graph.

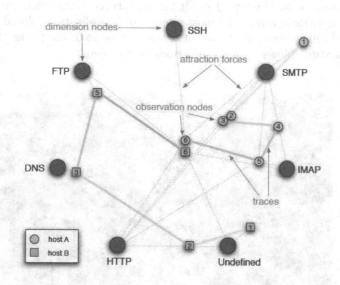

Figure 21.2 Evolution of network traffic over time (thumbnail of image from [MMK07]): dimension nodes represent types of traffic and observation nodes represent the state of a host at a given time.

EtherApe [Tol] shows traffic captured on the network via the pcap interface (Figure 21.3). A simple circular layout places the hosts around a circle and represents network traffic between hosts by straight-line edges between them. Each protocol is distinguished by a different color and the width of an edge shows the amount of traffic. This tool allows to quickly understand the role of a host in the network and the changes in traffic patterns over time. Beyond the graphical representation, it is also possible to display detailed traffic statistics of active ports.

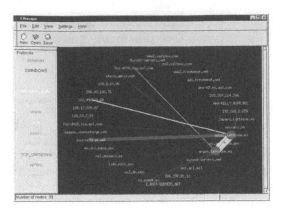

Figure 21.3 Traffic monitoring with Etherape (thumbnail of image from [Tol]). The size of the nodes and the thickness of the edges are proportional to the traffic volume. The color of an edge denotes the prevalent protocol of the associated traffic.

RUMINT [Con07] system (named after RUMor INTelligence) is a free tool for network and security visualization (Figure 21.4). It takes captured traffic as input and visualizes it in various unconventional ways. The most interesting visualization related to graph drawing is a parallel plot that allows one to see at a glance how multiple packet fields are related. An animation feature allows to analyze various trends over time.

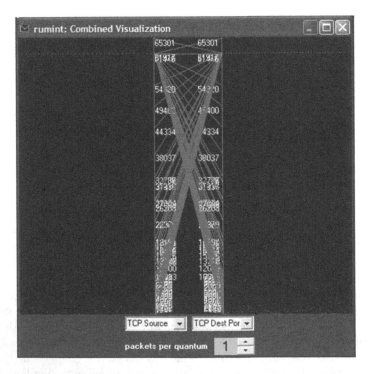

Figure 21.4 Visualization of an NMAP scan with RUMINT (thumbnail of image from [Con07]).

21.2.3 Internal vs. External Hosts

In [YYT$^+$04], the authors apply a simple bipartite drawing technique to provide a visualization solution for network monitoring and intrusion detection (see Figure 21.5). The nodes, representing internal hosts and external domains, are placed on three vertical lines. The external domains that send traffic to some internal host are placed on the left line. The domains of the internal hosts are placed on the middle line. The external domains that receive traffic from some internal host are placed on the right line. Each edge represents a network flow, which is a sequence of related packets transmitted from one host to another host (e.g., a TCP packet stream). Basically, the layout represents a tripartite graph. The vertical ordering of the domains along each line is computed by the drawing algorithm with the goal of minimizing crossings.

The tool uses a slider to display network flows at various time intervals and provides three views. In the global view, the entire tripartite graph is displayed to show all the communication between internal and external hosts. In the internal view and domain view, the tool isolates certain parts of the network, such as internal senders and internal receivers, and correspondingly displays a bipartite graph. The domain view and internal view are easier to analyze and provide more details on the network activity being visualized but on the other hand, the global view produces a high-level overview of the network flows. The authors apply the tool in various security-related scenarios, such as virus outbreaks and denial-of-service attacks.

In [BFN04], the authors use a matrix display combined with a simple graph drawing method in order to visualize the traffic between domains in network and external domains (see Figure 21.6). To visualize the internal network, the authors use a square matrix: each entry of the matrix corresponds to a host of the internal network. External hosts are represented by squares placed outside the matrix, with size proportional to the traffic sent or received.

Straight-line edges represent traffic between internal and external hosts and can be colored to denote the predominant direction of the traffic (outgoing, incoming, or bidirectional). The placement of the squares arranges hosts from subnets of the same size along the same vertical line and attempts to reduce the number of edge crossings. Further details on the type of traffic can be also displayed in this tool. For example, vertical lines inside each square indicate ports with active traffic. This system can be used to visually identify traffic patterns associated with common attacks, such as virus outbreaks and network scans.

21.2.4 Similarity Analysis for Traffic Logs and Scans

In [GB98], the authors present a technique to visualize log entries obtained by monitoring network traffic. Each log entry stores a multidimensional vector whose elements correspond to features of the network traffic, including origin IP, destination IP, and traffic volume. The authors build a weighted similarity graph for the log entries using a simple distance metric for two entries given by the sum of the differences of the respective elements. The force-directed drawing algorithm of [Cha96] is used to compute a 2D drawing of the similarity graph of the entries, which shows clusters of similar entries (see Figure 21.7). For example, this visualization allows to focus on entries associated with small clusters, which denote unusual events that could be associated with anomalous behavior of the network or a security breach.

The work by [MMB05] considers network scans, often used as the preliminary phase of an attack. The authors develop a visualization system that shows the relationships between different network scans (see Figure 21.8). The authors set up a graph where each node

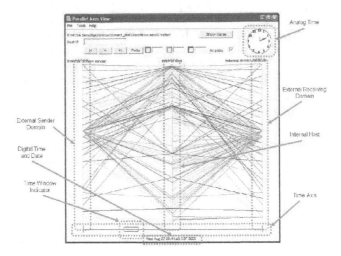

Figure 21.5 Global view of network flows using a tripartite graph layout: nodes represent external domains (on the left and right) and internal domains (in the middle) and edges represent network flows (packet streams) between domains (thumbnail of image from [YYT+04]).

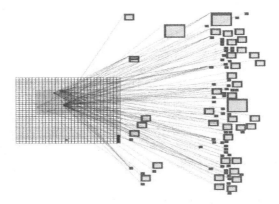

Figure 21.6 Visualization of internal vs. external hosts using a matrix combined with a straight-line drawing. Internal hosts correspond to entries of the matrix while external hosts are drawn as squares placed around the matrix. The size of the square for an external host is proportional to the amount of traffic from/to that host (thumbnail of image from [BFN04]).

represents a scan and the connection between them is weighted according to some metric (similarity measure) that is defined for the two scans. Features taken into consideration for the definition of the similarity measure include the origin IP, the destination IP, and the time of the connection. To avoid displaying a complete graph, the authors define a minimum weight threshold, below which edges are removed. The LinLog force-directed layout method [Noa04] is used for the visualization of this graph. In the drawing produced, sets of similar scans are grouped together, thus facilitating the visual identification of malicious scans.

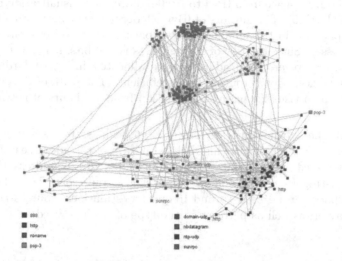

Figure 21.7 Similarity graph of traffic log entries (thumbnail of image from [GB98]).

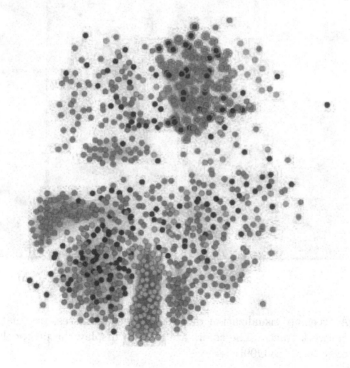

Figure 21.8 Similarity graph of network scans (thumbnail of image from [MMB05]). Nodes represents scans. Only edges with weight (similarity) above a certain threshold are displayed.

21.2.5   Visualization of Address Space

The shift in the address space from IPv4 to IPv6 requires new visualization tools for viewing network activity [BvO09]. The authors of [BvO09] observe that interesting information can be drawn from patterns in the number of IPv6 packets that go between same source and destination addresses. Since the size of each address is 128 bits, it is not trivial to visualize this information. However, IPv6 addresses are allocated in a standardized hierarchical manner in order to keep global routing tables efficient. The work of [BvO09] exploits this assignment pattern to visualize packet information from 4.5 hours of network traffic using treemaps.

In Figure 21.9, the destination addresses of IPv6 packets are displayed. Each rectangle is split into levels that represent the hierarchical nature of the addresses. The size of every rectangle is determined by the number of packets routed to this address. The protocols of the packets are distinguished with colors, e.g., dark gray represents TCP. This visualization of the address space can be used to find frequent destinations, sources that have similar destination behavior, as well as patterns in the type of traffic routed.

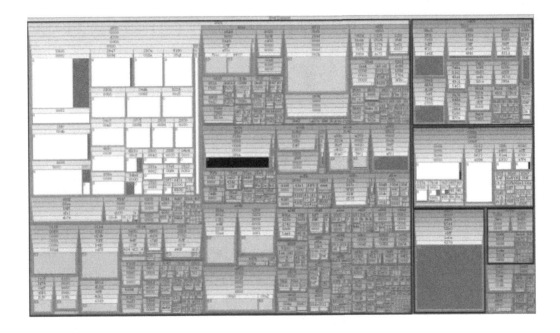

Figure 21.9 A treemap visualization of IPv6 source addresses, destination ports, and packet count of network traces. The colors are used to display the protocol of each packet (thumbnail of image from [BvO09]).

21.2.6   Visualization of Name Server Migration

Recall that a name server finds the IP address of a domain name queried by a computer. To find this IP address, a name server queries other name servers, including root servers. A root server is responsible for the root zone of the domain name space. Specifically, root servers keep a database of the authoritative name servers for the top-level domains (e.g., .com, .edu, .net, .org, etc.). Since the number of root servers is small and their role is very important in resolving domain names, each of them is implemented via a number of computers, called instances, that provide efficiency and resilience for that root server. Hence, when a name server is trying to answer a query, its request is sent to one of the instances depending on the status of the routing. Migration happens when the same name server is served by different instances of the same root server across time.

The authors of [DSN12] describe an animated visualization of migration of name servers between instances of a single root server, the K-root server. We refer to name servers that query the K-root server as clients. The migration process is visualized between instances by measuring the number of clients served and the total number of queries received by each instance. This information allows to monitor changes in migration patterns and in the workload of each instance over time. Moreover, it also helps to observe any anomalies in changes. For example, one instance is suddenly flooded by requests or clients of a specific Internet Service Provider change root servers in a suspicious pattern.

Two interesting animated visualization techniques are proposed to observe migration between the instances. Both visualizations are based on a *migration graph* G where each node is an instance of the root server and two nodes are connected if migration between the two is considered to be usual. The first visualization technique, *country map*, uses a geography-based layout to show the migration of clients (see Figure 21.10(a)). Each instance is represented as a bounded region of a distinct color, and its size is proportional to the number of clients that it currently serves. If two instances are exchanging clients, then they are adjacent on the map. Unusual migration of clients is represented via a flow traversed by bubbles with size proportional to the amount of flow. The second visualization, *octopus map*, represents instances as circles connected by "tentacles" to show the flow of usual migration (see Figure 21.10 (b)). The width of a tentacle is proportional to the amount of flow between the two instances it connects, while the color is related to the colors of the corresponding instances. Unusual changes are represented by arrows connecting non-adjacent instances.

Drawing of the first visualization consists of constructing a planar graph from the migration graph G (if G is not planar already) as a backbone for the final graph. A straight-line drawing of the backbone preserving its planar topology is drawn in such a way that each vertex has enough area around it to fit the average number of the clients that it serves in a given time period. This is achieved by using a spring embedder algorithm [DETT99] where the charge and the lengths depend on the number of clients an instance and its adjacent instances serve. For smoother visual transition between time intervals in the animation, the drawing of the backbone is modified to maximize the angles between adjacent edges. This step is done by adding new edges to the drawing and using constrained Delaunay triangulation. The skeleton obtained from this step is then adjusted for each specific time interval during animation.

Octopus map drawing involves computing a topology for the migration graph G such that the number of crossings between its edges is minimized. Then a straight-line drawing for G respecting the computed topology is built. During the animation steps, the vertices and edges of G are substituted by circles and tentacles, respectively, and any intersections between the shapes are removed. The challenge is then to scale the drawing to fit into an area where the animation is projected. For this purpose, a constrained spring embedder [DLR11] is used to preserve the original planar topology and ensure that no intersections between the shapes appear.

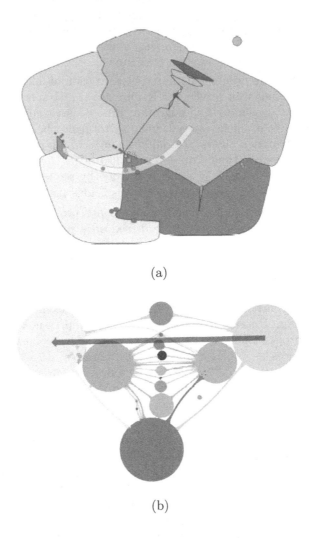

(a)

(b)

Figure 21.10 Snapshots of animations of client migration between instances of the K-root server. (a) A country map drawing where instances are represented as shapes of distinct colors and borders signify usual exchanges of clients between the instances. Unusual migration is displayed via a flow traversed by bubbles of size proportional to the number of clients migrated. (b) An octopus map drawing of the same data as in the country map drawing above. Each circle is an instance of the K-root server, a tentacle represents an expected migration and its width is proportional to the number of clients exchanged, while a gray arrow shows an unusual migration (thumbnails of images from [DSN12]).

21.3   Border Gateway Protocol

The *Border Gateway Protocol* (*BGP*) manages the routing of IP packets across different Autonomous Systems (AS), which can be informally viewed as collections of hosts under the same administrative control. In this section, we survey selected visualization methods for the Border Gateway Protocol that can be used to discover attacks, anomalies, and faults in the routing network.

21.3.1   Topology of Autonomous Systems

VAST (Visualizing Autonomous System Topology) [OKB06] is a tool that uses 3D straight-line drawings to display the BGP interconnection topology of Autonomous Systems (see Figure 21.11). The goal of the tool is to allow security researchers to quickly extract relevant information from raw routing datasets.

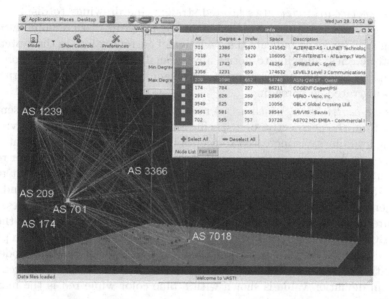

Figure 21.11 Some large autonomous systems in the Internet visualized with VAST (thumbnail of image from [OKB06]).

VAST employs a quad-tree to show information about an Autonomous System and an octo-tree to represent relationships between multiple Autonomous Systems. The visualization allows users to efficiently detect routing malfunctions and sensitive points, including the following ones:

- address-space hijacking attacks, where an Autonomous System maliciously sends routing announcements crafted to attract to itself traffic destined to IP addresses that belong to a different Autonomous System, e.g., to create alternative versions of popular websites;

- anomalous routing announcements, which accidentally cause portions of the Internet to become temporarily unreachable; and

- critical portions of the Internet topology, which are essential for its reliable operation.

The authors have also developed another tool, called *Flamingo*, that uses the same graphical engine as VAST but is used for real-time visualization of network traffic.

21.3.2 BGP Monitoring

BGP Eye [TRNC06] visualizes in real time the status of BGP activity with easy-to-read layouts (see Figure 21.12) and supports root-cause analysis of BGP anomalies. Its main objective is to track the healthiness of BGP activity, raise an alert when an anomaly is detected, and indicate its most likely cause. In particular, the authors show how BGP Eye can be used to analyze two Internet anomalies. First, they use the tool to study a worm outbreak, detecting the ASes that contributed the most to the spread of the infection. In the second use case, BGP Eye visualizes how prefix hijacking affects various ASes and routing tables over time.

BGP Eye provides two different types of visualization of BGP dynamics:

- The *Internet-centric view* uses layered straight-line drawings to display the interaction between Autonomous Systems (AS) in terms of BGP announcements exchanged. The colors indicate the deviation of current values in the system from the historic ones, allowing to notice any anomalies. Displaying the global view also helps to track the propagation of a problem through the entire Internet, e.g., its growing rate and spreading.

- The *Home-centric view*, which uses a radial drawing, is designed to present BGP activity from the perspective of a specific Autonomous System. In this visualization, the granularity is increased to the router level. The inner ring contains the routers of an Autonomous System and the outer ring contains their peer routers, belonging to other Autonomous Systems. In the outer ring, the layout method groups together routers belonging to the same Autonomous System and uses a placement algorithm that reduces the distance between connected nodes.

  In Figure 21.12 (b) the size of each AS represents the moving average of the number of BGP events originated by the AS. The thickness of AS-AS links represents the number of BGP events traversing this link. The color of lines and nodes gives information on the deviation of the current sample from its historical trend. The minimum deviation value is shown with a blue color while red is the maximum deviation.

21.3.3 BGP Evolution

BGPlay [BMPP04] and *iBGPlay* provide animated graphs of the BGP routing announcements for a certain IP prefix within a specified time interval (see Figure 21.13). Both visualization tools are targeted to Internet Service Providers. Each node represents an Autonomous System, and paths are used to indicate the sequence of Autonomous Systems needed to be traversed to reach a given destination according to a given announcement. The resulting graphs can be used to discover faults in the links traversed by the traffic flows and to check the consistency of router configurations.

BGPlay shows the paths to the chosen destination (prefix) that appear in announcements collected by observation points spread over the Internet. iBGPlay shows data privately collected by one ISP. The ISP can obtain from iBGPlay visualizations of outgoing paths from itself to any destination. The drawing algorithm is a modification of the force-directed approach that aims at optimizing the layout of the paths.

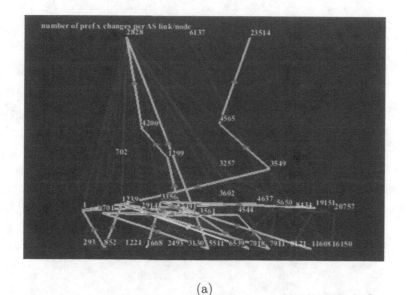

(a)

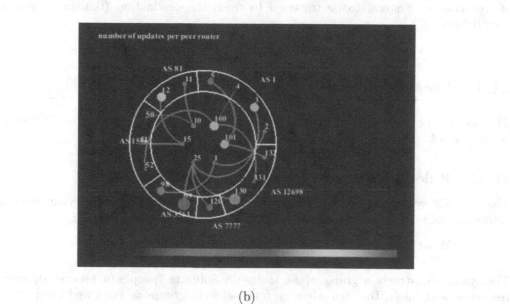

(b)

Figure 21.12 Visualizations in BGP Eye: (a) Internet-centric view; and (b) Home-centric view (thumbnails of images from [TRNC06]).

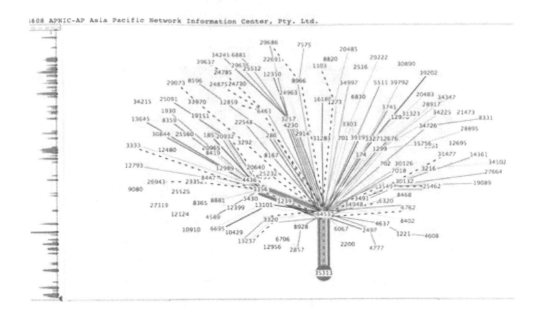

Figure 21.13 In BGPlay, nodes represent Autonomous Systems, and paths are sequences of Autonomous Systems to be traversed to reach the destination (thumbnail of image from [BMPP04]).

21.4   Access Control

This section considers selected graph-based visualization techniques for several aspects of access control.

21.4.1   Rule-Based Access Control

The RubaViz system [MFG+06] is a graphical tool for managing and querying rule-based access control systems (see Figure 21.14). RubaViz makes it easy to answer questions like

"What group has access to which files during a given time span?"

The system constructs a graph whose nodes are subjects (people or processes), groups, resources, and rules. Directed edges go from subjects/groups to rules and from rules to resources to display allowed accesses. The layout is straight-line and upward.

21.4.2   File System Access-Control

TrACE [HPPT08] is a tool for visualizing file permissions in the NTFS file system (Figure 21.15). TrACE allows a user or administrator to gain a global view of the permissions in a file system, thus simplifying the detection and repair of incorrect configurations leading to unauthorized accesses.

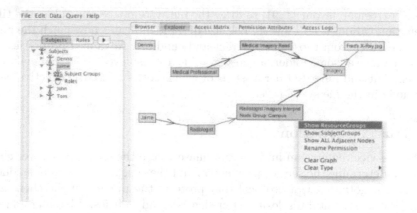

Figure 21.14 The RubaViz system for rule-based access control (thumbnail of image from [MFG+06]).

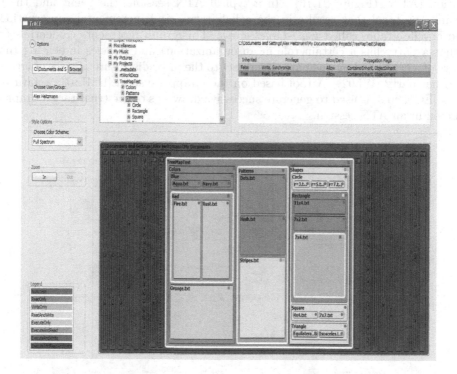

Figure 21.15 Visualization of permissions in the NTFS file system with TrACE (thumbnail of image from [HPPT08]).

In the NTFS file system, there are three types of permissions:

- *explicit* permissions are set by each user or members of a group;
- *inherited* permissions are dynamically inherited from the explicit permissions of the ancestor folders; and
- *effective* permissions are obtained by combining the explicit and inherited permissions.

TrACE uses a treemap layout [JS91] to draw the file system tree and colors the tiles with a palette denoting various access levels. The size of a tile indicates how much the permissions of a folder/file differ from those of its parent and children. Advanced properties, such as a break of inheritance at some folder, are also graphically displayed. The tool makes it easy to figure out explicit and inherited permissions of which nodes affect the effective permissions of a given node in the file system tree.

21.4.3 Trust Negotiation

Using a Web service requires an initial setup phase where the client and server enter into a negotiation to determine the service parameters and the cost by exchanging credentials and policies. Trust negotiation is a protocol that protects the privacy of the client and server by enabling the incremental disclosure of credentials and policies. Planning and executing an effective trust negotiation strategy can be greatly aided by tools that explore alternative scenarios and show the consequences of possible moves.

In [YSTW05], the authors use a layered upward drawing to visualize automated trust negotiation (ATN) (Figure 21.16). In a typical ATN session, the client and the server engage in a protocol that results in the collaborative and incremental construction of a directed acyclic graph, called *trust-target graph*. This graph represents credentials (e.g., a proof that a party has a certain role in an organization) and policies indicating that the disclosure of a credential by one party is subject to the prior disclosure of a set of credentials by the other party [WL02]. A tool based on the Grappa system [BML97], a Java port of Graphviz [EGK+04], is used to generate successive drawings of the trust-target graph being constructed in an ATN session.

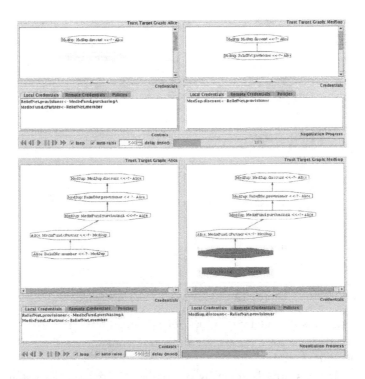

Figure 21.16 Drawing of the trust-target graph generated by a trust negotiation session (thumbnail of image from [YSTW05]).

21.4.4 Privacy Settings in Social Networks

User privacy in social networks is of concern to the users and companies providing this service. To this end, social networking companies are creating more and more tools to help users manage their privacy settings. The authors of [MLA12] describe a visual tool for users of social networks to assess the visibility of their data among their friends in a more accessible way. The tool parses the data of the user and creates a graph of groups and subgroups from a list of friends of the user, where friends are split into groups using modularity optimization. Hence, a node in the final graph represents a group of friends. The user can then query this graph via zooming or direct queries to see the visibility of his data among his friends. Authors chose a force-directed approach to display nodes in the graph, and the color of the nodes represents the privacy level (see Figure 21.17).

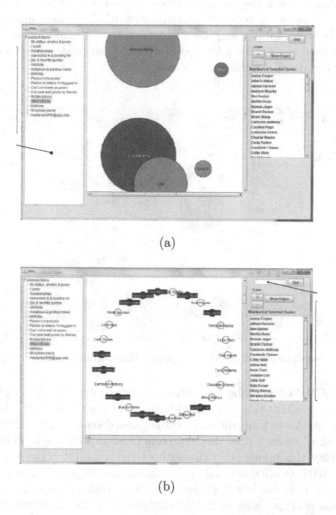

(a)

(b)

Figure 21.17 Visualization of privacy settings in a social network. (a) Circles represent groups of friends that have the same visibility of user's data and the color shows the privacy level. (b) The granularity is increased to the level of specific information of a user, e.g., a phone number, and its visibility among the user's friends (thumbnails of images from [MLA12]).

21.5   Attack Graphs

21.5.1   Model

Given a network and a database of known vulnerabilities that apply to certain machines of the network, one can construct a directed graph where each node is a machine (or group of machines) and an edge denotes how a successful attack on the source machine allows to exploit a vulnerability on the destination machine. This graph, called attack graph, can be rather large and complex. Thus, it is essential to use automated tools to analyze attack graphs.

21.5.2   Tools

A tool for visualizing attack graphs is described in [NJKJ05] (Figure 21.18). The system clusters machines in order to reduce the complexity of the attack graph (e.g., machines that belong to the same subnet may be susceptible to the same attack). The Graphviz tool [EGK⁺04] is used to produce a layered drawing of the clustered attack graph. Similar layered drawings for attack graphs are proposed in [NJ04].

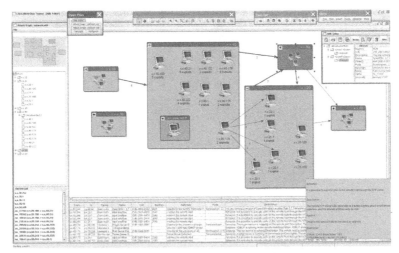

Figure 21.18   Visualization of an attack graph (thumbnail of image from [NJKJ05]).

The authors of [CIL⁺10] describe Navigator, another tool for visualizing attack graphs for displaying server-side, client-side, credential-based, and trust-based attacks in the network. Navigator groups machines from the same subnet based on similar vulnerabilities but also gives an "asset value" to each host that represents the importance level of this host in the network. To display this information the authors use a modification of the strip treemap algorithm by [BSW02] (see Figure 21.19). Navigator can display different types of attacks that can be brought from one entity of the network to another. In Figure 21.19, rectangles represent host groups of the same subnet and arrows show the steps that the attacker could take to progress through the network. The background colors of the entities represent the compromise levels achievable in the displayed attack scenarios. For example, red means that the root is compromised while green stands for no compromise. The color of the arrows shows the depth of the attack. The visualization tool can also show how the attack from one subnet affects the rest of the network.

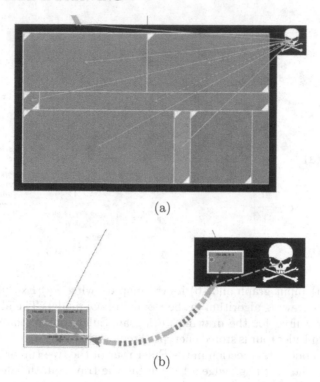

(a)

(b)

Figure 21.19 Visualization of (a) an attack graph with importance levels of the hosts, (b) multiple attacks between subnets, displayed as a hybrid edge of multiple colors (thumbnails of images from [CIL+10]).

Given that there are multiple attacks and that the type and the depth of these attacks can vary, the authors propose a way to aggregate this information into hybrid edges to avoid clutter (see Figure 21.19(b)). For example, the solid part of an edge shows server-side attacks, while client-side attacks are displayed as dashed segments.

The authors of [CIL+10] have modified the strip treemap algorithm to achieve visualizations that are aesthetically pleasing and easier to navigate. Their algorithm sets minimum width and height for the hosts within a subnet to avoid very long and thin or very short and wide host representations. Any extra space that is accumulated due to the minimum rectangle requirements is then propagated and scaled to top layers. The modification also preserves the order of the hosts when the user zooms in a host group.

21.6 Private Graph Drawing

In previous sections, graph drawing was used to visualize the data in a way that can help to observe any anomalies in the network, privacy settings, or access control to a filesystem. The authors of [GOT12] raise instead a privacy concern related to the process of drawing a graph. With the recent shift of data storage to a cloud-based storage, one can no longer assume that the input data for a graph drawing algorithm is stored locally. Hence, to draw a graph, one accesses the data remotely. This raises concerns about the privacy of the outsourced data. In [GOT12], a new model for graph drawing algorithms is proposed that fits the cloud computing paradigm and preserves data privacy and its access pattern.

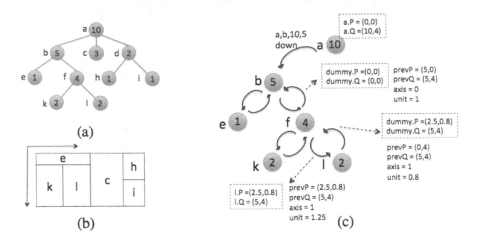

(a)

(b)

(c)

Figure 21.20 (a) Input graph and (b) its treemap drawing. (c) Execution of a privacy-preserving treemap drawing algorithm in the cloud storage model. The algorithm performs the computation required for the drawing during an Euler tour of the graph. The edge representation of an Euler tour is stored encrypted remotely. Each edge contains information about its adjacent nodes, its direction in the Euler tour of the tree (up or down); and a link to the next edge in the tour (e.g., edge a-b). During the traversal, the algorithm maintains several variables: unit, the unit length of current rectangle; prevP and prevQ, location of a previously drawn rectangle; and axis, the direction of the drawing which takes a value of x or y. When the traversal is going down the tree, each node on the way is assigned x and y coordinates of its corresponding rectangle, variables P and Q. After reaching a leaf node the algorithm follows the Euler tour up the tree. The algorithm cannot reveal when it sees the nodes it has already assigned since it would expose the height and the width of the tree. Hence, it writes dummy coordinates for all nodes it encounters when going up the tree, dummy.P and dummy.Q in the figure. (thumbnails of images from [GOT12]).

21.6.1   Compressed Scanning

In the *compressed-scanning* model, data is stored encrypted on the cloud and is permuted using a pseudo-random permutation. A graph drawing algorithm is then split into rounds. During each round, the algorithm scans remotely-stored data related to the graph. For example, a data item can be a node or an edge of the graph. Once the data item is retrieved, a small computation, which could potentially modify the item, is performed in local private memory. The item is then re-encrypted and written back. A small private memory is used to store information during each round. After each round, the data is re-permuted according to a new permutation.

The number of rounds is specific to the graph drawing algorithm and does not depend on the data. The privacy of the model comes from using data encryption and accessing data in a nonrevealing manner. Note that during each round every data item is read only once, and a new permutation is created so that nobody observing the access pattern of the algorithm can deduce information about the graph, including its layout and depth. Hence, the only information that is leaked from running a graph drawing algorithm is the size of the graph, but not its topology. An algorithm satisfying this privacy property is said to be *data-oblivious*.

21.6.2   Data-Oblivious Drawing Algorithms

The authors show how to modify four classical graph drawing algorithms to fit the compressed-scanning model, including symmetric straight-line drawings and treemaps [JS91], drawings of trees, dominance drawings of planar acyclic digraphs [DTT92], and Δ-drawings of series-parallel graphs [BCD$^+$94]. These algorithms work over trees or construct a tree representation, e.g., spanning trees for acyclic digraphs.

The main technique is based on representing the graph via an Euler tour of the tree, hence every edge is stored twice and traversal of the tree involves accessing each edge only once. If the number of nodes in the graph is n, then three algorithms in [GOT12] use only a constant amount of private storage during each round. While drawing a tree with bounding rectangles uses $\log n$ private memory. For an illustration of the technique applied to treemap drawings, see the examples in Figure 21.20.

The authors show that graph drawing methods that fit the compressed-scanning model require $T(A)\mathsf{sort}(n)$ accesses to remote data storage. $T(A)$ is the number of rounds specific to the graph drawing algorithm A and does not depend on n. The factor $\mathsf{sort}(n)$ is the number of rounds that is required to securely order the data according to a new permutation after each round.

The work of [GS12] expands this line of work and shows how to simulate parallel algorithms data-obliviously using the compresses-scanning model. Given a CRCW PRAM algorithm B that runs in $T(B)$ steps using a memory of size M and $P \leq M$ processors, the authors show how to simulate B sequentially in a data-oblivious fashion in $O(T(B)M \log M)$ accesses to remote data storage. This result can be used to obliviously compute st-numberings, visibility representations, upward grid drawings, and orthogonal grid drawings of planar graphs that are stored remotely.

Acknowledgments

This chapter is an extended version of a conference paper on graph drawing methods for security visualization [TPP09]. Charalampos Papamanthou contributed to this chapter while he was at Brown University. This work was supported in part by the U.S. National Science Foundation under grant OCI–0724806. We are indebted to Massimo Rimondini for detailed comments on an earlier version of this chapter. We also thank Giuseppe Di Battista, Michael Goodrich, and Ioannis Tollis for useful suggestions.

References

[BCD⁺94]   P. Bertolazzi, R. F. Cohen, G. Di Battista, R. Tamassia, and I. G. Tollis.
           How to draw a series-parallel digraph. *International Journal of Computational Geometry and Applications*, 4:385–402, 1994.

[BFN04]   R. Ball, G. A. Fink, and C. North.   Home-centric visualization of network traffic for security administration.   In *Proceedings of the 2004 ACM Workshop on Visualization and Data Mining for Computer Security*, VizSEC/DMSEC '04, pages 55–64, New York, NY, USA, 2004. ACM.

[BML97]   N. Barghouti, J. Mocenigo, and W. Lee.   Grappa: A GRAPh Package in JAVA.   In Giuseppe Di Battista, editor, *Graph Drawing*, volume 1353 of *Lecture Notes in Computer Science*, pages 336–343. Springer Berlin Heidelberg, 1997.

[BMPP04]   G. Battista, F. Mariani, M. Patrignani, and M. Pizzonia. iBGPlay: A system for visualizing the interdomain routing evolution. In Giuseppe Liotta, editor, *Graph Drawing*, volume 2912 of *Lecture Notes in Computer Science*, pages 295–306. Springer Berlin Heidelberg, 2004.

[BSW02]   B. B. Bederson, B. Shneiderman, and M. Wattenberg. Ordered and quantum treemaps: Making effective use of 2D space to display hierarchies. *ACM Transactions on Graphics*, 21(4):833–854, 2002.

[BvO09]   D. Barrera and P. C. van Oorschot. Security visualization tools and IPv6 addresses. In *Proceedings of the 6th International Workshop on Visualization for Computer Security*, VizSec '09, pages 21–26, 2009.

[Cha96]   M. Chalmers. A linear iteration time layout algorithm for visualising high-dimensional data. In *Proceedings of the 7th conference on Visualization*, VIS '96, pages 127–ff., Los Alamitos, CA, USA, 1996. IEEE Computer Society Press.

[CIL⁺10]   M. Chu, K. Ingols, R. Lippmann, S. Webster, and S. Boyer. Visualizing attack graphs, reachability, and trust relationships with NAVIGATOR. In *Proceedings of the 7th International Symposium on Visualization for Cyber Security*, VizSec '10, pages 22–33, New York, NY, USA, 2010. ACM.

[Con07]   G. Conti. *Security Data Visualization*. No Starch Press, San Francisco, CA, USA, 2007.

[DETT99]   G. Di Battista, P. Eades, R. Tamassia, and I. G. Tollis. *Graph Drawing*. Prentice Hall, Upper Saddle River, NJ, 1999.

[DLR11]   W. Didimo, G. Liotta, and S. A. Romeo. Topology-driven force-directed algorithms. In Ulrik Brandes and Sabine Cornelsen, editors, *Graph Drawing*, volume 6502 of *Lecture Notes in Computer Science*, pages 165–176. Springer Berlin Heidelberg, 2011.

[DSN12]   G. Di Battista, C. Squarcella, and W. Nagele. How to visualize the K-Root name server. *Journal of Graph Algorithms and Applications*, 2012. In print.

[DTT92]   G. Di Battista, R. Tamassia, and I. G. Tollis. Area requirement and symmetry display of planar upward drawings. *Discrete & Computational Geometry*, 7:381–401, 1992.

[Ead84]   P. Eades. A heuristic for graph drawing. *Congressus Numerantium*, 42:149–160, 1984.

[EGK+04]   J. Ellson, E.R. Gansner, E. Koutsofios, S.C. North, and G. Woodhull.
Graphviz and dynagraph – static and dynamic graph drawing tools. In
Michael Jünger and Petra Mutzel, editors, *Graph Drawing Software*, Mathematics and Visualization, pages 127–148. Springer Berlin Heidelberg,
2004.

[FMK+08]   F. Fischer, F. Mansmann, D. A. Keim, S. Pietzko, and M. Waldvogel.
Large-scale network monitoring for visual analysis of attacks. In *Proceedings of the 5th International Workshop on Visualization for Computer Security*, VizSec, pages 111–118, Berlin, Heidelberg, 2008. Springer-Verlag.

[FR91]   T. Fruchterman and E. Reingold. Graph drawing by force-directed placement. *Software: Practice and Experience*, 21(11):1129–1164, November
1991.

[GB98]   L. Girardin and D. Brodbeck. A visual approach for monitoring logs. In
Proceedings of the 12th Conference on Systems Administration, LISA '98,
pages 299–308, Berkeley, CA, USA, 1998. USENIX Association.

[GOT12]   M. Goodrich, O. Ohrimenko, and R. Tamassia. Graph drawing in the cloud:
Privately visualizing relational data using small working storage. In *Graph
Drawing*, 2012. To appear.

[GS12]   M. Goodrich and J. Simons. More graph drawing in the cloud: Data-
oblivious st-numbering, visibility representations, and orthogonal drawing
of biconnected planar graphs. In *Graph Drawing*, 2012. Poster.

[GT11]   M. Goodrich and R. Tamassia. *Introduction to Computer Security*.
Addison-Wesley, 2011.

[HPPT08]   A. Heitzmann, B. Palazzi, C. Papamanthou, and R. Tamassia. Effective
visualization of file system access-control. In *Proceedings of the 5th International Workshop on Visualization for Computer Security*, VizSec '08,
pages 18–25, Berlin, Heidelberg, 2008. Springer-Verlag.

[JS91]   B. Johnson and B. Shneiderman. Tree-maps: A space-filling approach to
the visualization of hierarchical information structures. In *IEEE Visualization*, pages 284–291, 1991.

[Mar08]   R. Marty. *Applied Security Visualization*. Addison-Wesley, 2008.

[MFG+06]   J. Montemayor, A. Freeman, J. Gersh, T. Llanso, and D. Patrone. Information visualization for rule-based resource access control. In *Proceedings
of the 2nd Symposium on Usable Privacy and Security*, SOUPS '06, 2006.

[MFK+09]   Florian Mansmann, Fabian Fischer, Daniel A. Keim, Stephan Pietzko, and
Marcel Waldvogel. Interactive analysis of netflows for misuse detection
in large IP networks. In *DFN-Forum Kommunikationstechnologien*, pages
115–124, 2009.

[MLA12]   A. Mazzia, K. LeFevre, and E. Adar. The PViz comprehension tool for
social network privacy settings. In *Proceedings of the 8th Symposium on
Usable Privacy and Security*, SOUPS '12, pages 13:1–13:12, New York, NY,
USA, 2012. ACM.

[MMB05]   C. Muelder, K. Ma, and T. Bartoletti. A visualization methodology for
characterization of network scans. In *Proceedings of the IEEE Workshops
on Visualization for Computer Security*, VIZSEC '05, pages 4–, Washington, DC, USA, 2005. IEEE Computer Society.

[MMK07]   F. Mansmann, L. Meier, and D. Keim. Graph-based monitoring of host
          behavior for network security. In *Proceedings of the 4th International Work-
          shop on Visualization for Computer Security*, VizSec '07, 2007.

[NJ04]   S. Noel and S. Jajodia. Managing attack graph complexity through visual
          hierarchical aggregation. In *Proceedings of the 2004 ACM workshop on
          Visualization and data mining for computer security*, VizSEC/DMSEC '04,
          pages 109–118, New York, NY, USA, 2004. ACM.

[NJKJ05] S. Noel, M. Jacobs, P. Kalapa, and S. Jajodia. Multiple coordinated views
          for network attack graphs. In *Proceedings of the IEEE Workshops on Vi-
          sualization for Computer Security*, VIZSEC '05, pages 12–, Washington,
          DC, USA, 2005. IEEE Computer Society.

[Noa04]   A. Noack. An energy model for visual graph clustering. In Giuseppe Liotta,
          editor, *Graph Drawing*, volume 2912 of *Lecture Notes in Computer Science*,
          pages 425–436. Springer Berlin Heidelberg, 2004.

[OKB06]   J. Oberheide, M. Karir, and D. Blazakis. VAST: visualizing autonomous
          system topology. In *Proceedings of the 3rd International Workshop on
          Visualization for Computer Security*, VizSec '06, pages 71–80, New York,
          NY, USA, 2006. ACM.

[TN00]   J. Tölle and O. Niggemann. Supporting intrusion detection by graph clus-
          tering and graph drawing. In *Proceedings of 3rd International Workshop
          on Recent Advances in Intrusion Detection*, RAID '00, Toulouse, France,
          2000.

[Tol]   J. Toledo.   EtherApe a live graphical network monitor tool.
          http://etherape.sourceforge.net/.

[TPP09]   R. Tamassia, B. Palazzi, and C. Papamanthou. Graph drawing for security
          visualization. In Ioannis G. Tollis and Maurizio Patrignani, editors, *Graph
          Drawing*, volume 5417 of *Lecture Notes in Computer Science*, pages 2–13.
          Springer Berlin Heidelberg, 2009.

[TRNC06] S. T. Teoh, S. Ranjan, A. Nucci, and C. Chuah. BGP eye: a new visu-
          alization tool for real-time detection and analysis of BGP anomalies. In
          *Proceedings of the 3rd International Workshop on Visualization for Com-
          puter Security*, VizSec '06, pages 81–90, New York, NY, USA, 2006. ACM.

[WL02]   W. H. Winsborough and N. Li. Towards practical automated trust ne-
          gotiation. In *Proceedings of the 3rd International Workshop on Policies
          for Distributed Systems and Networks*, POLICY '02, pages 92–103. IEEE
          Computer Society Press, June 2002.

[XMB+06] I. Xydas, G. Miaoulis, P.-F. Bonnefoi, D. Plemenos, and D. Ghazanfar-
          pour. 3D graph visualization prototype system for intrusion detection: A
          surveillance aid to security analysts. In *Proceedings of the 9th International
          Conference on Computer Graphics and Artificial Intelligence*, 3IA, pages
          153–165, 2006.

[Yee06]   G. Yee. Visualization for privacy compliance. In *Proceedings of the 3rd
          International Workshop on Visualization for Computer Security*, VizSec
          '06, pages 117–122, New York, NY, USA, 2006. ACM.

[YSTW05] D. Yao, M. Shin, R. Tamassia, and W. H. Winsborough. Visualization of
          automated trust negotiation. In *Proceedings of the IEEE Workshops on
          Visualization for Computer Security*, VIZSEC '05, pages 8–, Washington,
          DC, USA, 2005. IEEE Computer Society.

[YYT+04] X. Yin, W. Yurcik, M. Treaster, Y. Li, and K. Lakkaraju. VisFlowConnect: netflow visualizations of link relationships for security situational
awareness. In *Proceedings of the 2004 ACM Workshop on Visualization
and Data Mining for Computer Security*, VizSEC/DMSEC '04, pages 26–
34, New York, NY, USA, 2004. ACM.

22

Graph Drawing for Data Analytics

Stephen G. Eick
VisTracks and U. Illinois at Chicago

22.1 Introduction

Over the last decade graph drawing and network visualization has emerged as an exciting research area that is addressing a significant problem: how to make sense of the ever increasing amounts of relational information that has become widely available. With the growth of networking and decreasing cost of storage it has become technically feasible and cost effective to store and access vast sets of information. The academic, business, and government challenge is how to make sense of this information and translate the insights into value-producing activities.

As a new emerging field, there will certainly be opportunities for network visualization and graph drawing technology. There have already been some early successes and also many prototypes that have been research successes but have not led to successful deployments. Unfortunately, not all network visualizations create enough value so that users will switch over from conventional user interfaces to adopt new visual interfaces. The goal for this chapter is to present a simple framework that predicts problem areas where network visualization will achieve utilization and result in successful business applications—that is, be useful enough so that users will adapt new visual interfaces. For our academic colleagues our framework is not intended to identify what network problems are interesting for research nor is it intended to identify high-quality results. Rather it attempts to predict which application areas might lead to commercially successful applications.

Network visualizations are exciting and the demos inevitably generate interest among potential users. Unfortunately, however, visualization, as exciting as it is, only involves the user interface or presentation layer in a technology stack. Useful network applications solve problems that involve collecting relational data, manipulating it, organizing it, performing calculations, and finally presenting the results to users. The value of the application is captured by the complete system. It is often the case that each system component individually is not particularly useful. For example, tires are not useful without a car, but better tires improve a car's performance. The presentation layer, like beauty, is only "skin deep" and the usefulness of the application comes from the whole solution and not just the "lipstick."

Thus, by itself, network visualization is naturally a feature of system and rarely is a complete application by itself. This, unfortunately, makes commercial utilization of a new technique or novel method difficult. With a few exceptions, the technology must be part of an application to capture sustainable value. Network visualization "makes it better" but, except in rare situations, does not make it. The network visualization value stack challenge is to find applications where network visualization creates enough value, either by itself or as part of an applications, to support utilization where it is a key part of the value proposition.

Throughout this chapter, the term *network visualization* refers to methods for visualizing graphs and networks that make use of graph drawing techniques.

22.2   Where Network Visualization Creates High Value

At its most basic level, network visualization is a technique for helping analysts understand structure and relationships. This section describes seven broad classes of information problems, illustrated by examples, where network visualizations create significant value.

22.2.1   User Interface

In certain cases, the *user interface* is essentially a complete application. The canonical example of this is *computer games* which are innovative and sophisticated user interfaces that involve, relatively speaking, little computation and no data integration. Successful games must have a great user interface that challenges and engages prospective players within the first few seconds.

Perhaps the closest network visualization application where the interface is the application involves graph drawing and layout. Arguably, the most successful application in this space is Microsoft's Visio™. Visio is perhaps the most widely used graph drawing package and is distributed as part of Microsoft Office.

22.2.2   Visual Presentation and Branding

Visual presentation and *branding* involves creating custom 3D displays of networks for presentations that are visually exciting. It frequently incorporates aspects of branding and has a high glitz and wow factor. Typical presentation and branding techniques include animation, colorful 3D networks, and visualization that have a high "wow" factors.

Figures 22.1 and 22.2 show two visually exciting examples of network visualizations for presentation and branding. The visualization on the left shows worldwide Internet traffic and the image on the right shows Internet traffic betwen countries. These images have been used on the covers of multiple books, magazines, and as raw material for art work. Their use is really for branding. See, for example, Praba Pilar's network visualization art gallery [Pil05].

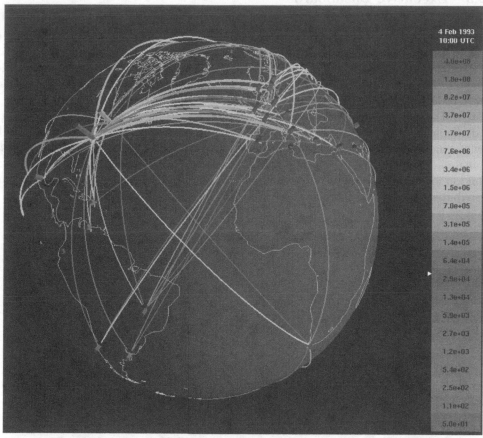

Figure 22.1 3D Internet Network visualizations for presentation and branding.

Figure 22.2 Internet traffic flows between countries used for presentation and branding.

22.2.3   Executive Dashboards

Executive dashboards provide decision-makers with instant access to key metrics that are relevant for particular tasks.   Much of the intellectual content in dashboards is in the choices of metrics, organization of information on the screen, and access to supporting, more detailed information. Network visualization techniques can improve this presentation, as shown in Figure 22.3. Executive dashboards may include the ability to export result-sets to other tools for deeper analysis.

State-of-the-art implementations of active executive dashboards are web-based, interactive, dynamic, involve no client-side software to install, and often include *action alerts* that fire when pre-defined events occur.   End user customizations include *sorting*, *subsetting*, *rearranging layouts* on the screen, and the ability to *include* or *exclude* various metrics. It is common for visual reports to be distributed via email, published on a corporate intranet, or distributed through the internet.

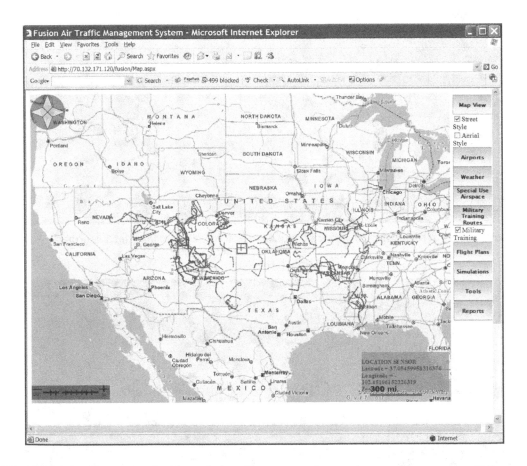

Figure 22.3 Executive dashboard showing a network superimposed on a geospatial map.

22.2.4 Real-Time Visual Reports

Real-time visual reports are related to executive dashboards but provide an active presentation of an information set consumable at a glance. Although the distinction is subtle, visual reports usually involve fat client-side software and thus can provide richer presentations of the information. Visual reports exploit the idea that a picture is worth a thousand words and, in particular, for many tasks a picture is more useful than a large table of numbers.

Visual reporting systems are:

1. Easy to use for both sophisticated and non-sophisticated user communities;
2. Suitable for broad deployments; and
3. Provide capabilities for flexible customization;

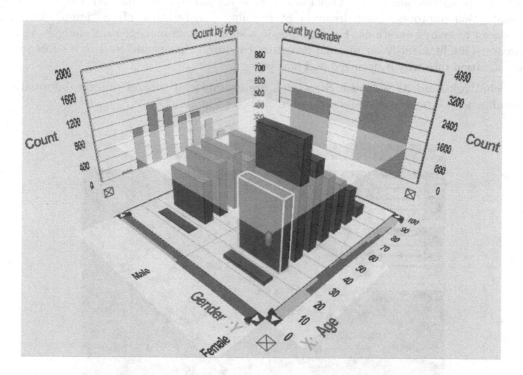

Figure 22.4 Real-time 3D visual report.

Visual reports, as with all reports, are a tool for *assumptive-based* analysis. Reports answer "point questions": How much of a particular item is in stock? Where is it? How long will it take to get more? Reports are ideal for operational tasks, but do not provide full analytics, or enable an analyst to automatically discover new information that a user has not thought to ask about.

This is a well-known characteristic of all report-based analytical solutions. The reports pre-assume relationships that are reported upon. The difficulty with this approach is that most environments are too complex for a pre-defined report or query to be exactly right. The important issues will undoubtedly be slightly, but significantly different. This is particularly true for complex, turbulent, environments where the future is uncertain. There are two common solutions to this problem. The first is to create literally hundreds of reports that are distributed out to an organization, either using a push distribution mechanism such as email or a pull mechanism involving a web-based interface. The second involves adding a rich customization capability to the reporting interface that increase UI complexity. Unfortunately, neither works particularly well. Although a report containing novel information might exist, finding it is like finding a needle in a haystack. Adding UI features makes the reporting system difficult to use for non specialists.

22.2.5   Visual Discovery for Deep Analysis

Visual discovery-based analysis addresses the shortcomings of assumptive-based analytics by providing a rich environment to support novel discovery. Systems supporting visual discovery are used by analysts and frequently combine data mining, aspects of statistics, and also predictive analytics. Visual discovery is domain specific and iterative. Network visualization improves visual discovery by enabling discoveries to often "jump" out and may lead to "why" questions. For example, in a supply chain management analysis, visual discovery might identify an unusual inventory condition that would lead to a subsequent investigation into why it occurred and how to fix it.

NicheWorks and its successor StarGraph are examples of general purpose information visualization system for visual discovery [Eic00] (see Figure 22.5).

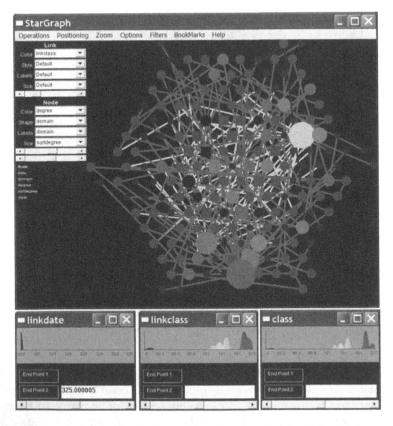

Figure 22.5 Network visual discovery and analysis tool. The linked histograms function as interactive filters to control display complexity.

It consisted of a workspace with standard data acquisition capabilities, and a set of visual metaphors, e.g., views, each of which showed data in a particular way. Some of the views were conventional, (e.g., geographical networks, abstract networks, barcharts, linecharts, piecharts) and some were novel (Data Constellations, Multiscape, Data Sheet). For visual analysis, the views could be combined into fixed arrangements called perspectives. Within any perspective the views could be linked in four ways: by *color*, *focus*, *selection* and *exclusion*. Components linked by color used common color scales and those linked by *focus*, *selection* and *exclusion* were tied by data table row state using a *case-based* model [EW95].

There are three important ideas in this general class of visual discovery and analysis tools. First, *perspectives* extend general linked view analysis systems by reducing complexity for non-expert users. Perspectives are "authored" by "power users" who are experts. Analysts who are domain experts, but not power users, use the perspectives as a starting point for analysis and as a guiding framework. The output from their analysis, visual reports, may be published and distributed for use by casual users, executives, and decision-makers. The user model is similar to that employed by spreadsheets where there are spreadsheet authors, users, and consumers.

Second, *visual design patterns* are recurring patterns within perspectives that are broadly useful and apply to many similar problems. Following the object-oriented programming community [GHJV95], recognizing, cataloging, and reusing design patterns have the potential for significantly improving network visualizations.

Examples of design patterns are Shneiderman's *information-seeking mantra*:

overview first, zoom and filter, then details on demand [CMS99]

The *overview* shows the entire dataset, e.g., all movies in the dataset, and supports the ability to *zoom* in on interesting movies and query the display with the mouse to extract additional details. This design pattern incorporates interactive filters, frequently bar and pie charts, that enable you to *filter* out uninteresting folders so that you display only the data that is interesting. Filtering might be by category, numeric range, or even selected value.

Another design pattern, called *linked bar charts*, is particularly strong for data tables containing categorical data. Categorical data, sometimes called contingency tables, involves counts of the number of data items organized in various bins or subcategories. This design pattern employs one bar plot for each categorical column with the height of the bar tied to the number of rows having that particular value. In statistical terms each of the bar charts shows a marginal distribution. As the user selects an individual bar, the display recalculates to show one-way interactions. Using exclusion and selection shows two-way interactions.

Third, *details on demand* is a feature set where the system provides tooltips and other details when the user mouses over any particular item on the screen. The idea is to provide immediate access to fine-grain information when it is needed without unnecessarily cluttering the interface.

22.2.6 Searching and Exploration

Network visualizations focused on *visual searching* involves undirected *knowledge discovery* against massive quantities of uncategorized, heterogeneous relationship data with varying complexity. This scenario is typical of web searching where users recognize information when they find it. Searches are iterative, intuitive, and involve successive refinements.

The key measures for the performance visual searching systems revolve around the amount of information per unit of search effort expended. The search effort may be measured in user time, number searches, personal energy, etc. The results, or information found, may be measured in articles, references, relevance, novelty, ease of understanding, etc. Different systems exploit various design points trading off these factors.

22.2.7   Domain Task-Specific Visualizations

Task-specific visualizations help users solve critical, high-value tasks. Examples include visualizations to:

1. Design and layout complex circuits;
2. Identify relationships in product purchases;
3. Trace calling patterns among subscribers (Figures 22.6 and 22.7);
4. Manage huge communications networks (Figure 22.8); and
5. Study relationships in a complex social network.

These visualizations are tuned to particular problems often delivered as part of a complex system. They are highly valuable, frequently involve fusing of a large number of information streams, and serve both as an output presentation for information display and also control panel and input interface for user operations.

22.3   Network Visualization Sweet Spot

One very simply way to characterize network visualization problems uses three dimensions:

1. *Dataset size* is a measure of the total amount of data to be analyzed. Although some might disagree, sophisticated network visualization techniques are not needed for small datasets containing tens of observations. In these cases reports, spreadsheet graphics, and standard techniques work fine. More powerful techniques are unneeded.

   Conversely, network visualization techniques do not scale to analyze massive datasets containing gigabytes of information. The basic problem is that network visualization is a technique that makes human analysts more efficient and human scalability is quite limited. The exact scalability limits of network visualization are subject to debate and are an active research area [EK02, Eic04]. Most researchers would agree, however, that massive datasets containing hundreds of thousands to millions of observations are too big and need to be subdivided, aggregated, or in some way reduced before the information can be presented visually. Network visualization, it would seem, cannot be applied to analyze massive image databases containing millions of images, but might be applied to meta data associated with the images.

2. *Dataset complexity* can be measured by the number of dimensions, structure, or richness of the data. Network visualizations are not needed for (even large) simple datasets with low-dimensional complexity. Statistical reduction tools such as regression work fine and are sufficient in this situation.

   Conversely, datasets of massive complexity containing thousands of dimensions are too complex for humans and thus for network visualizations. Some have argued that network visualizations can cope with as many as fifty dimensions, although a more practical upper limit is say half to a dozen dimensions.

3. *Dataset change rate* is a measure of how frequently the underlying problem changes. Static problems, even for very complex problems, can eventually be solved by developing an algorithmic solution. The algorithmic solution has a

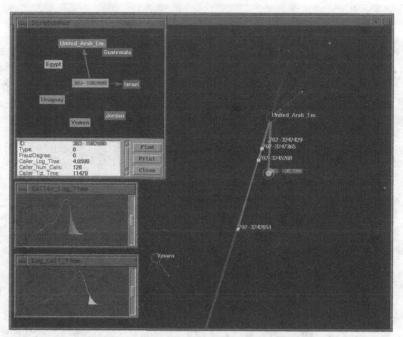

Figure 22.6 Calling patterns among subscribers in a massive international network.

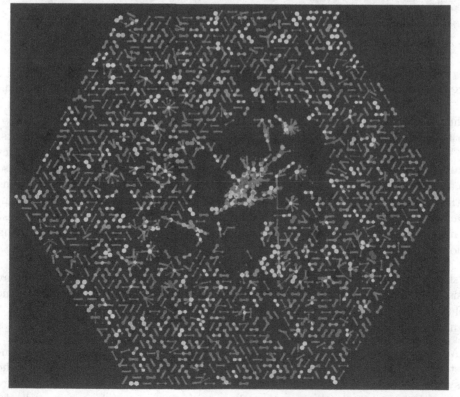

Figure 22.7 Relationship and calling patterns among subscribers in a network.

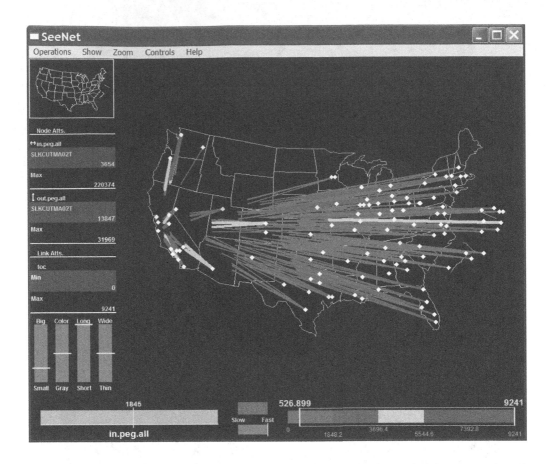

Figure 22.8   Network visualization showing traffic patterns after California earthquake.

huge advantage over an information visualization-based solution since the algorithm can be applied repeatedly without the need for expensive human analysts. Conversely, analysis problems involving change or other dynamic characteristics are extremely difficult to automate because the problem keeps moving. In these cases, human insight is essential. Humans, however, cannot cope with problems that change too quickly. We are incapable of instantaneous responses. Human analytical problem solving occurs on a time scale of minutes to months. We must automate problems needing faster response and partition problem those involving longer time scales.

As shown in Table 22.1 the application sweet spot for network visualization involves analysis problems of moderate data sizes, rich, but not overwhelming, dimensional structure, that change, are not easily automated, or for some reason need human involvement.

Examples of prototypical applications include:

- *Network management* for complex networks where the system is dynamic, constantly changing with new protocols, new devices, and new applications. The systems are instrumented and collect alarms with complex dimensional structure. It is frequently the case that the number of events (alarms) exceeds the capacity of network visualizations and must be algorithmically reduced.

Attribute	Low Value	High Value
Dataset size	10^1 to 10^2	10^4 to 10^6
Dataset complexity	2 or 3 dimensions	50 dimensions
Dataset change rate	minutes	months

Table 22.1 Dimensions and bounding ranges for network visualization sweet spot.

- *Customer behavior* involving human buying patterns and transaction analysis is an ideal candidate for network visualizations. Human behavior is complex, unpredictable, and dynamic. Furthermore, although aggregate numbers of transactions are large, for any individual or set of individuals the numbers of transactions are not overwhelming and easily suitable for analysis.
- *Intelligence analysis* is an ideal candidate for network visualization. It is difficult to automate, involves complex dimensional data, is dynamic, and necessarily involves human analysts.

22.4 Customers for Network Visualization Software

There are three broad classes of potential network visualization users: *scientists, analysts* (including both intelligence and commercial analysts), and *business users.*

- *Scientists* have deep needs for network visualization, are extremely technical, and work on the most significant problems. They want powerful tools for cutting-edge analyses.
- *Analysts*, particularly in commercial companies, also have a strong need for network visualization, but tend to have specialized needs. They are not as sophisticated as scientists and will not tolerate raw software packages.
- *Business users* need simple network visualizations and are easily frustrated by complex software. Business users are numerous, have budget, but need solutions to problems and are not inherently interested in the complexly that excites scientists and analysts.

These three classes of users have different needs and varying tolerances for complex software. As a scientists and analysts want complex rich software that is full featured. Scientists are often willing to use flaky software that is cutting edge and incorporates the latest features. However, there are not many scientists and analysts, and they tend not to have large budgets. Thus the addressable market is not particularly large. Business users, however, have budget, have problems, but do not have patience for leading-edge software that is not robust. The business challenge is to create software that is sophisticated enough to solve scientific problems and yet easy to use for business users. These dynamics shape the market for network visualziation software.

22.5 Business Models for Network Visualization

Successfully deploying network visualizations involves solving a technical problem and creating a business model that supports widespread utilization. Broadly speaking, there are several classes of business models for software companies, as discussed below.

22.5.1   Custom Software

Custom software is written to solve a specific problem, usually for a single customer. The problem being addressed must be significant, valuable, important, and yet specialized enough so that general solutions do not exist. The projects often involve next generation technology and new approaches to problems.

Typical price points for custom software projects usually start at $250K. Custom software is sold directly by the vendor with six months to two year sales cycle. The sales team is highly specialized and the sales process frequently involves company executives.

Organizations involved with customer software include universities, government labs, large commercial organizations, and boutique specialty shops. Although it might seem surprising to some, research universities and government labs act as custom software developers where the funding agencies effectively hire university principal investigators using BAAs and solicitations to solve important custom problems. In this setting, the principal investigators function as both sales professionals and also lead fulfillment efforts with "graduate student" development teams.

In the large organizations that sponsor customer software development there are commonly multiple roles. It is often the case, particularly with government-sponsored projects, that the funding organization is not the organization that will eventually use the software and the users of the software may not receive the value from its use. These separate organizational roles complicate the software sales process. For example, the National Science Foundation funds research to build software for scientists to use. The scientists use the software to solve important national problems. Thus, citizens are the ultimate beneficiary. In the commercial environment, the CFO (Chief Financial Officer) funds a project that is implemented by the CIO (Chief Information Officer) for a business unit. Thus, three organizations are involved.

22.5.2   Enterprise Software

Enterprise software, sold by commercial companies, is essentially a flexible template that is "implemented" on site, either by the vendor or a "business partner." In the implementation phase, the template is customized for a particular customer by means of tasks that include connecting up data sources, defining the specific reports a company needs, and populating tables (e.g., inserting employee names into a payroll file). For an enterprise application, data integration is essential. Since enterprise software is reusable, it can be sold more economically than custom software. Generally price points for enterprise software range from $25K to $250K. The sales model for enterprise software may be direct at the higher price points, e.g., SAP, or through local business partners who are "certified" by the vendor.

22.5.3   Shrink-Wrapped Software

Shrink-wrapped software is highly functional software that solves a specific problem very well. The software usually is customer installed and provides for little or no customization. Customer support, if provided, is usually self-serve via a web site or perhaps with limited help desk support.

Shrink-wrap software is almost always sold by distributors or OEMed to the hardware vendors and sold as part of a bundle. For example, Microsoft, the largest producer of shrink-wrap software, sells essentially all of its software through distributors. As a mass-market item, the price point for shrink-wrap software is less than $25K and more frequently less than $1K.

22.5.4 Open Source Software

Open source network visuazliation software supported by services is one of the newer emerging software business models. In the open source model software is developed by volunteers working on donated time and made available at no charge through the internet. However, support and other customizations are offered as service by companies using the open source model. For operating systems and major applications this model appears viable. It is too soon, however, to predict how well the open source model will do for targeted applications and specialized technologies such as network visualization software. In general, the open source experience for visualization software has been mixed. There have been a few successes but many other projects have not gotten widespread traction.

22.5.5 Cloud Computing

Cloud computing solutions via Web portals and network visualization services are another possible business and distribution model for network visualization software. There is a strong push in corporations away from client software because of its high cost of ownership. As a result an increasing number of applications are moving toward a cloud computing model.

22.5.6 Network Visualization Deployments

Relating the business models back to the visualization deployments, most of the demand for network visualization has been met with custom research software built by universities, government labs, and large communications companies. The customers are the military, intelligence community, biomedical researchers, and other highly specialized users. Demand for network visualization within the research community is healthy.

Within the enterprise category we might expect network visualization-enabled applications to emerge. In this category the value is provided by the whole application and a network visualization presentation layer could be described as a software feature or add-on product.

In a related field, Business Intelligence, there have been some early successes for visualization-enabled applications. Perhaps the most notable success has been Cognos Visualizer. Cognos sold 300K[1] units of Cognos Visualizer, an add-on for Cognos PowerPlay, at $695 per unit and some of the other "Business Intelligence" software vendors have had similar experiences.

The "Gorilla" analytic application within shrink-wrap category for network visualization is Microsoft Visio. It is generally considered to be good enough for 90% of problems and essentially everybody has it.

22.6 Thin-client Network Visualization

One of the challenges in creating a successful business model for network visualization software involves deployment. The problem is that rich network visualization clients run on desktop machines which means that they are deployed and managed through IT organizations for many institutions. The cost of maintaining and deploying desktop software

---

[1]For comparison, a software application that sold 2,000 to 5,000 units would generally be considered successful.

restricts their use and potential application to all but the most important problems. One way around this involves web-based deployment.

The advantage of web-based interfaces is that they are simple to deploy, have proliferated rapidly, and are quickly becoming the de facto standard for accessing information. The disadvantage of web-based interfaces is that desktop applications have a richness and responsiveness that has not been possible on the web. Recently, several new web applications have appeared that provide a rich user experience on the web that previously was only available in desktop applications. Examples include Google Maps, Microsoft Virtual Earth, and Google Suggest. The applications are examples of a new approach to web development that combines Asynchronous JavaScript, XML, and DHTML and represents a fundamental shift in what is possible on the web. On Microsoft's Virtual Earth, for example, you use your cursor to grab the map and scroll it around. On Google Suggest the system automatically attempts to complete your search query. This all occurs almost instantly, without waiting for pages to reload. This programming paradigm is often called Web 2.0 or *AJAX* in the popular press.

There are several technologies, each flourishing in its own right, that can be combined in powerful new ways to create the next generation web-based visualization capability. These are:

- Standards-based presentation using XHTML and CSS;
- Web-based 2D graphics using Scalar Vector Graphics (SVG);
- Dynamic display and interaction using JavaScript to manipulate the Document Object Model (DOM);
- Data interchange and manipulation using XML and XSLT;
- Asynchronous data retrieval using JavaScript's XMLHttpRequest;
- DHTML (JavaScript) binding everything together.

Traditional web applications work on a client-server model. The client, a web browser, issues an http request to a server for a new page when the user clicks on a link. The web server, usually Apache or IIS, does some processing, retrieves information from legacy systems, does some crunching, and sends a formatted page of hypertext back to the client for display. This approach is the simplest technically, but does not make much sense from the user perspective. The problem is that the user waits while the server does its thing for the next page to reload.

The new model enabled by these new technologies eliminates start-stop-start-stop nature of web applications. Instead, information is asynchronously downloaded to the client in using XML. JavaScript code in the browser caches this information when it is received from the server and displays it upon user request. Since the information is cache, the system can provide instantaneous responses. JavaScript code in the browser also handles other interactions with the user such as panning, zooming, scaling, and data validation. The advantage of the asynchronous requests for XML data is that users can continue working with the application while data is downloading.

The application shown in Figure 22.3 is an example of a thin-client interactive geospatial network visualization. It is written using SVG and interaction is done via manipulating each page's DOM. Although the programming is quite difficult, the result is stunning. It is able to provide the richness of a desktop application without the hassles of desktop software with the flexibility and rich hyperlinking that is only possible in browsers.

22.7 Discussion and Summary

This chapter attempts to define opportunities where network visualizations create significant business value. Network visualization involves the presentation layer with is naturally a feature of many products. By itself, it usually has insufficient value to support widespread usage and deployment. It is generally a feature of an application and a critical component of a solution.

The chapter identifies various types of network applications and develops a simple model that characterizes an opportunity space for an application. The target for the model is to help identify commercial opportunities network visualization applications.

Although this chapter is not expected to be particularly exciting for researchers or scientists who are interested in pushing the state-of-the-art, it should help practitioners identify opportunities for successful applications. Unfortunately, business and other pragmatic issues often dominate the technical issues when it comes to determining which network visualization applications will achieve commercial success.

References

[CMS99] Stuart K. Card, Jock D. Mackinlay, and Ben Shneiderman. *Readings in Information Visualiation: Using Vision to Think.* Morgan Kaufman, San Francisco, California, 1999.

[Eic00] Stephen G. Eick. Visual discovery and analysis. *IEEE Transactions on Computer Graphics and Visualization,* 6(1):44–59, January–March 2000.

[Eic04] Stephen G. Eick. Scalable network visualization. In Christopher R. Johnson and Charles D. Hansen, editors, *Visualization Handbook,* pages 819–831. Academic Press, 2004.

[EK02] Stephen G. Eick and Alan F. Karr. Visual scalability. *Journal of Computational Graphics and Statistics,* 11(1):22–43, March 2002.

[EW95] Stephen G. Eick and Graham J. Wills. High interaction graphics. *European Journal of Operational Research,* 81:445–459, 1995.

[GHJV95] Erich Gamma, Richard Helm, Ralph Johnson, and John Vlissides. *Design Patterns.* Addison-Wesley, 1995.

[Pil05] Monica Praba Pilar. *Cyber.Labia: Gendered Thoughts and Conversations on Cyber Space.* Tela Press, 2005. Available at http://www.prabapilar.com/pages/projects/cyberlabia.htm.

Graph Drawing and Cartography

Alexander Wolff
University of Würzburg

23.1 Introduction

Graph drawing and cartography come together when networks whose elements have geographic locations, that is, geometric networks, have to be visualized. Examples of such networks are street, subway, river, or cable networks. Often it helps to visualize the underlying network for analyzing certain network parameters. For example, traffic on a road network can be visualized by drawing each road as a rectangle whose width is proportional to the amount of traffic going through that road.

One of the main problems in map production is a process called *generalization*. Given cartographic data that has been collected at large scale, this data must be simplified in order to produce maps at small scale. In order to obtain readable maps, detail must be reduced and spacing must be enlarged. Traditionally this has been done manually by cartographers, but increasingly semi-automated and even automated methods are in use, particularly in conjunction with geographic information systems (GIS) [Ass96]. Cartographers have identified a number of generalization *operators* such as displacement, size exaggeration, size reduction, and deletion in order to cope with the many constraints that govern the generalization process. The main difficulty in automating generalization is the interdependency of these operators.

Saalfeld [Saa95], both geodetic and computer scientist, pointed out (in one of the first editions of the graph drawing conference) that map generalization can be seen as a graph drawing problem—if one accepts that a cartographic map is but a straight-line drawing of a graph in the plane. Then the process of redrawing a map at smaller scale can be interpreted as a sequence of modifications of both the graph and its drawing. Graph elements must be contracted or removed, and the drawing must be modified to reflect the graph reductions. Moreover, the drawing must be modified in that "old" graph elements must be moved, for example, due to distance constraints. Saalfeld is an early advocate of continuous generalization: his ultimate goal is a map with a slider bar for scale. (For a rather restricted continuous generalization problem, see Section 23.2.2.) Saalfeld points to the key issue: address the "big picture"—take feature interaction into account. He challenges the graph drawing community to "design and implement an efficient and effective automated map generalization system for the line network of a digital map."

Note that general graph-drawing algorithms cannot be used ad hoc for drawing geometric networks since they do not respect the geometry that comes with the vertices and edges. A good drawing of a geometric network must reflect geometry since a user typically has an intuitive notion of the underlying geometry, in other words, a *mental map* [ELMS91]. For example, the user of a metro system expects stations in the north to appear on the top of maps that depict the metro system. Thus the "art" of drawing geometric networks is to find a good compromise between distorting geometry and maximizing aesthetics. This will be the leitmotif of this chapter, which also explains why we will not touch *point-set embeddability* problems. Recall that, in a point-set embeddability problem, one is given not just a graph but also a set of points in the plane (or on a line) and the aim is to find a mapping between vertices and points such that the edges can be drawn under some drawing convention. For example, Gritzmann et al. [GMPP91] showed that any n-vertex outerplanar graph can be embedded on any set of n points in the plane (in general position) such that edges are represented by straight-line segments connecting the respective points and no two edge representations cross. In the type of problem we are interested here, in contrast, the mapping between vertices and points is part of the input and, in many cases, we may move the points to some extent.

We focus on node-link representations of geometric networks, that is, we insist on representing vertices by points or small icons such as disks or squares and edges by some linear features (Jordan curves, in general). This excludes contact or intersection representations (such as rectangular cartograms) where edges are represented implicitly; by the contact or intersection behavior of the "large" geometric objects that represent the vertices. For such representations, see Chapter 10 on rectangular drawings and Chapter 11 on simultaneous drawings.

Note, however, that additional requirements come into play in a geographic context. For example, the relative position or the relative sizes of the geometric objects representing the vertices are often prescribed by the user. As an example for this additional difficulty, take Koebe's beautiful theorem [Koe36] that says that every planar graph can be represented as a coin graph, that is, as a set of interior-disjoint disks, two of which touch if and only if the corresponding vertices are adjacent in the given graph. If one now introduces geometric constraints by prescribing a set of "anchor" points and a bijection between points and vertices (and, hence, disks), and by insisting that each disk contains its point, then realizability of a given planar graph as a *cover contact graph* becomes NP-hard [AdCC⁺12].

In this chapter, we give an overview of the main types of geometric networks that are being visualized in an automated fashion, using node-link diagrams. For each network type, we consider the application-dependent aesthetic constraints. We group the network types according to the graph class to which they belong: paths (simplified, schematized and generalized in Section 23.2), matchings (used in boundary labeling in Section 23.3), trees (as in flow maps; see Section 23.4), (near-) plane graphs (such as street or metro maps; see Section 23.5), and other graphs (such as timetable graphs, the Internet multicast backbone, or social networks; see Section 23.6). Note that we use the term *plane* to stress that the graphs are given with a planar embedding. For example, a self-intersecting polygonal line can be considered a path and hence a planar graph, but it is not a plane graph.

23.2 Paths

When drawing paths nicely, the main problem is data reduction: which points can be dropped while maintaining the important features of a polygonal line? Due to its many applications, polygonal line simplification has been identified as an important problem both in cartography and in computational geometry. Since Douglas and Peucker [DP73] presented a simple and frequently used algorithm, cartographers have devised solutions of higher cartographic quality [VW93, Saa99, LL99], while geometers have given a more efficient implementation of the Douglas-Peucker algorithm [HS94] and have designed new algorithms for specialized error criteria [AV00, BCC+06] or for a restricted number of orientations [Ney99, MG07, DHM08]. Still, finding a near-linear time solution for polygonal line simplification is listed as problem 24 in the *Open Problems Project* [MO01].

23.2.1 Simplifying and Schematizing Polygonal Paths

The path-drawing problem that Agrawala and Stolte [AS01] considered has more of a graph-drawing flavour. Their *route maps* help car drivers to get from A to B. While most route planners draw routes using a fixed-scale map as background, Agrawala and Stolte suggested to draw edges of the path (that is, roads between turns) as straight-line segments which are usually *not* to scale. Instead, their system LineDrive exaggerates the length of short road segments in order to label them properly with street name and real length, see Figure 23.1.

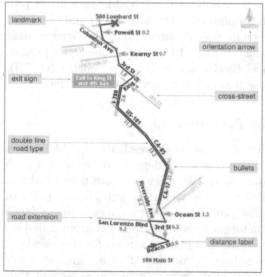
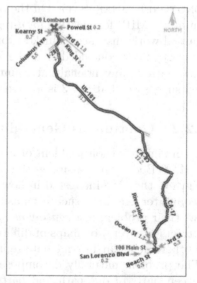

| (a) map generated with LineDrives | (b) constant-scale map (for comparison) |

Figure 23.1   LineDrives generates driving directions. Sketches taken from [AS01].

In the resulting drawings, angles at turns are mostly kept, except at very sharp turns. Roads that are close to being vertical or horizontal are usually made vertical or horizonatal, respectively. The LineDrive system, which is based on simulated annealing, was publicly available for some period of time and received very positive response from most users.

Later on, the path-drawing problem of Agrawala and Stolte [AS01] inspired research in the graph-drawing community. Brandes and Pampel [BP13] showed, by reduction from MONOTONE3SAT, that the rectilinear (orthogonal) case is NP-hard; more precisely, it is NP-hard to decide whether a given polygonal path has a simplification that consists exclusively of horizontal and vertical segments and preserves the *orthogonal order* (that is, horizontal and vertical order) of the vertices along the path. The ordering constraint is meant to help the user maintain his mental map.

On the positive side, Delling et al. [DGNP10] showed that, given an polygonal path and a set $\mathcal{C}$ of directions, they can efficiently compute a simplification such that (a) all edge directions are in $\mathcal{C}$ and (b) the orthogonal order of the vertices is preserved—if the input path is x-monotone. Their algorithm finds a simplification of minimum cost, which they define to be the sum over the costs of all edges. The cost of an edge, in turn, is defined to be the angle between the edge in the output and the direction in $\mathcal{C}$ that is closest to the direction of the edge in the input. The algorithm is based on a clever characterization of optimum solutions and on dynamic programming. When the number of directions, $|\mathcal{C}|$, is considered a constant, their algorithm runs in $O(n^2)$ time and uses $O(n)$ space, where n is the number of vertices of the input path. Using a linear-programming formulation (of linear size), Delling et al. can even find, among all simplifications with a fixed direction for each edge, one of minimum total length. In addition, they present a heuristic for dealing with the non-monotone case.

A natural generalization of the rectilinear case considered by Brandes and Pampel [BP13] is the *d-regular* case, where the set of directions consists of multiples of $90°/d$. Delling et al. [DGNP10] established their positive result for x-monotone paths for *any* set of directions (actually, any set containing the multiples of $90°$); in particular, their result holds for the d-regular case for any $d \geq 1$. Gemsa et al. [GNPR11] generalized the negative result of Brandes and Pampel from $d = 1$ to any $d \geq 1$, using a different reduction (from MONOTONEPLANAR3SAT). On the other hand, they presented a mixed-integer linear programming (MIP) formulation for d-regular path simplification (for any $d \geq 1$) and evaluated it on real-world instances (quickest routes between random destinations in the German road network). They concluded that the MIP runs fast enough if the road geometry is preprocessed with a conventional path simplification method (such as Douglas-Peucker [DP73]). They suggested that $d = 3$ is a good compromise between accuracy and abstraction.

23.2.2   Continuous Generalization for Polygonal Lines

A path simplification problem of a rather different flavor was investigated by Merrick et al. [MNWB08]. They assumed that both a detailed and a less detailed drawing of a path are given; they are interested in how to get from one to the other in a continuous fashion. In computer graphics, such a transition is called a *morph*. From a cartographic point of view, their problem is a *continuous generalization* problem: given two linear objects (such as streets or rivers) on maps of different scale, deform one representation continuously into the other such that intermediate representations are valid generalizations for their scale.

The problem naturally decomposes into two subproblems: first, find a correspondence between parts of one path and parts of the other path; second, define a movement that moves the parts of one path onto the corresponding parts of the other path. Merrick et al. focused on the first subproblem and solve the second subproblem by simply moving the vertices of one path on *linear* trajectories to their counterparts. The first subproblem can again be subdivided into two tasks: first, find *characteristic* points on both paths; second, find a good correspondence between the subpaths defined by consecutive corresponding points. The idea behind the characteristic points is not only data reduction, but detecting such points and treating them with special care makes it more probable that the viewers of the resulting morph keep their mental map during the animation.

For the first task, Merrick et al. incrementally fitted cubic Bézier curves to a growing part of the given polygonal path. When the distance between the current subpath and the curve surpasses a pre-specified error bound $\varepsilon > 0$, Merrick et al. viewed the point added last as a characteristic point, and repeat the fitting process with the subpath starting at that point. The distance between subpath and curve is approximated by sampling both with a relatively large number of points, measuring the distances only between corresponding points and taking the maximum over these point-to-point distances.

Figure 23.2 shows a mountain road in the French Alps and the characteristic points that were detected using the Bézier-fitting method of Merrick et al. for two different values of the error bound ε; 1 and 25. Subfigure (c) shows the same road with manually selected characteristic points. The automatically detected set of characteristic points for $\varepsilon = 25$ and the manually detected set are quite similar.

For the second task, Merrick et al. presented a dynamic program that computes a correspondence between the two paths, in $O(nm)$ time, where n and m are the numbers of subpaths of the first and second path, respectively. The correspondence is optimal with respect to the distance function defined by the user; the authors make a number of suggestions for such functions.

Figure 23.3 shows snapshots morphs between two representations of the road in Figure 23.2. The more detailed representation is from a BD(R) Carto map at scale 1:50,000; the less detailed, generalized representation of the same road at scale 1:100,000 is from an IGN Carto2001 TOP100 map. The example road was chosen because it is represented by three serpentines on the detailed scale but only by two serpentines in the less detailed scale. Each morph is based on a different choice of characteristic points; *linear interpolation* (variant (c)) is a simple ad-hoc method that matches each point on one polyline to the point at the same relative distance from the start on the other polyline. The middle snapshot produced by this method shows its weakness, especially in the part of the polyline labeled "Region A." While the two other morphs (in subfigures (a) and (b)) keep the "amplitude" of the serpentines while merging the first two, linear interpolation first reduces the amplitude and then increases it again.

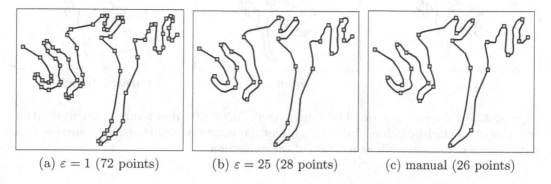

(a) $\varepsilon = 1$ (72 points)   (b) $\varepsilon = 25$ (28 points)   (c) manual (26 points)

Figure 23.2   Selection of characteristic points according to Merrick et al. [MNWB08]. The polyline is a mountain road from the French Alps; it consists of 155 vertices.

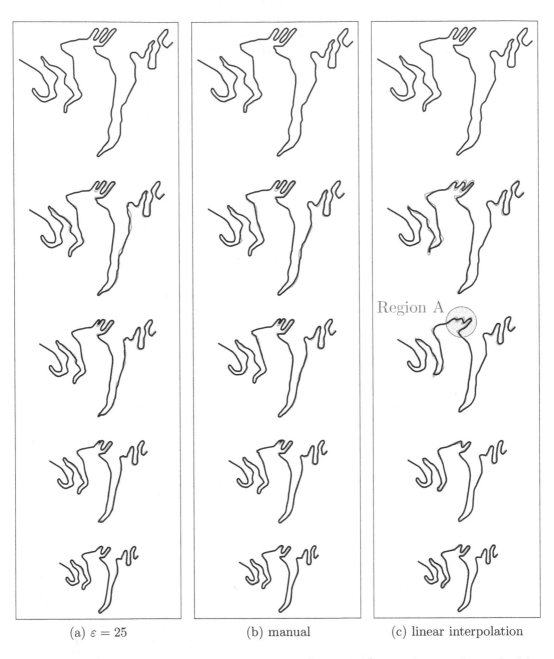

(a) $\varepsilon = 25$ (b) manual   (c) linear interpolation

Figure 23.3 Morphs generated by Merrick et al. [MNWB08] depending on the method for selecting characteristic points. In each snapshot, previous frames are shown in increasingly light shades of gray to assist perception of the animation.

For this road, which has 190 and 155 vertices on the 1:50K and the 1:100K maps, respectively, it took less than 0.01 seconds to compute the characteristic points, 1.39 seconds to compute the optimal correspondence for $\varepsilon = 1$, and 0.59 seconds for $\varepsilon = 25$. The road is part of a map sheet with 382 roads consisting of 13345 and 10869 vertices on the two maps, which were reduced (for $\varepsilon = 25$) to 2742 and 2387 characteristic points, respectively. For the whole 1:50K map sheet, this reduction took 0.69 seconds; computing the correspondence then took 13.17 seconds. The experiments were performed on an AMD Athlon XP 2600+ PC with 1.5 GB main memory running under SuSE Linux 10.1. These running times are acceptable since tasks can be considered pre-processing. Only the resulting simple linear morph needs to be executed in real time. In order to solve the continuous generalization problem for complete street or river networks across large scale intervals, the line-simplification algorithm sketched here must be combined with a topology-simplification algorithm, which yet has to be devised.

23.3 Matchings

Matchings do not appear to be an exciting graph class for graph drawing, but they have an interesting application that brings cartography and graph drawing together: so-called *boundary labeling*. In boundary labeling, one is given a set of point sites on a rectangular map and, for each site, a rectangular label that contains, for example, textual information about the site. Other than in normal point labeling, labels are not placed next to the site they label, either because the point set is too dense with respect to the label sizes or because the map background must not be covered by the labels. Instead, labels are placed outside the map such that they touch the map boundary with one side. In order to visualize the mapping between sites and labels, each site is connected to its label with a polygonal line, the so-called *leader*. For three real-world examples with different leader types, see Figure 23.4.

The boundary labeling problem was introduced by Bekos et al. [BKSW07]. For a given rectangle R (for example, a cartographic map), a set P of point sites in R and, for each site s in S, a rectangular label L_s, Bekos et al. define a *feasible leader-label placement* to be a placement of the labels and a drawing of the leaders that fulfills the following requirements:

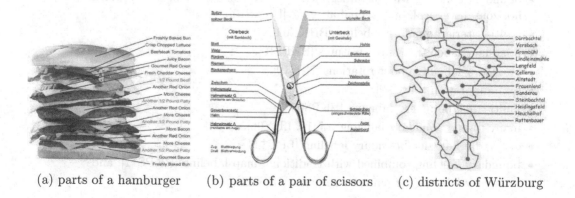

(a) parts of a hamburger   (b) parts of a pair of scissors   (c) districts of Würzburg

Figure 23.4   Examples of boundary labeling.

(B1) Labels are disjoint.

(B2) Labels lie outside (the interior of) R such that, for each label, one of its edges is contained in one of the edges of R.

(B3) Each point is connected to its unique label by a leader.

(B4) Leaders may not intersect other leaders, points or labels.

(B5) The point where a leader touches a label is called *port*; ports may be *fixed* (for example, to the centers of the label edges) or *sliding* (that is, arbitrary).

(B6) Labels either have fixed positions or can slide along an edge of R.

(B7) Labels can be attached to one, two or all four edges of R. The resulting problems are called one-side, two-side and four-side leader-label placements.

In addition to feasible leader-label placements, mainly the following objective functions have been considered:

(O1) small ink consumption (minimize total leader length),

(O2) straightness (minimize number of bends).

These are typical graph drawing objectives; they help to keep the visual complexity of the resulting drawing low.

Several types of leaders have been considered; until now all of them are polygonal with up to two bends. Generally, a leader type is denoted by a word from the set $\{s, \{p, o, d\}^*\}$; the letters refer to the direction of the line segments that form the leader, starting at the point to be labeled and ending at the port that lies on some edge e of R. The leader type s refers to straight-line leaders; their direction is arbitrary. Leader segments labeled p are *parallel* with e, segments labeled o are *orthogonal* to e, and segments labeled d are *diagonal*, that is, they form an angle of $45°$ or $-45°$ degrees with e.

Two-sided boundary labeling with labels of *non-uniform* height is NP-hard; the reduction from PARTITION is obvious. Therefore, most references focus on uniform labels, that is, all labels are unit-height rectangles. In Table 23.1, we summarize the running times of the best known algorithms (in big-Oh-Notation) for various versions of the boundary labeling problem.

The following variants and extensions of the boundary labeling problem have been considered:

- boundary labeling with *octilinear* leaders, that is, leaders whose segments are horizontal, vertical, or diagonal at $\pm 45°$ [BKNS10],
- multi-criteria boundary labeling [BHKN09],
- boundary labeling for area features [BKPS10],
- boundary labeling under rotations [NPS10],
- text annotation [LWY09],
- multi-stack boundary labeling [BKPS06],
- many-to-one boundary labeling [Lin10, LKY08],
- one-and-a-half-side boundary labeling [LPT⁺11],
- boundary labeling combined with traditional map labeling [BKPS11], and
- boundary labeling for panorama images [GHN11].

In order to give the reader at least a flavor of this variety of results, we review some of the early algorithms for type-s and type-po leaders. In the case of one- and two-side problems, we attach labels to the right edge and both vertical edges of R, respectively.

leader type	# map edges with labels	feasible solution	bend-minimal solution	length-minimal solution		
				fixed ports	sliding ports	reference
s	1	$n \log n$	N/A	$n^{2+\varepsilon}$	n^3	[BKSW07]
s	4	$n \log n$	N/A	$n^{2+\varepsilon}$	n^3	[BKSW07]
po	1		n^3	$n \log n$	$n \log n$	[BHKN09]
po	2			n^2	n^2	[BKSW07]
po	2		n^8			[BHKN09]
opo	1	$[n \log n]$	$[n^2]$	$n \log n$	$[n^2]$	[BKSW07]
opo	2		open	$n^2 \quad [nH^2]^\star$	n^2	[BKSW07]
opo	4	$n \log n$	open	$n^2 \log^3 n$	n^3	[BKSW07]
do	1		n^5	n^2	n^2	[BHKN09]
do	2		n^{14}			[BHKN09]
$\{do, pd\}$	1		open	$n^3 \; [\text{—}]^\star$	n^3	[BKNS10]
$\{od, pd\}$	1	$n \log n$	open	n^3	n^3	[BKNS10]
$\{do, pd\}$	2		open	n^3	n^3	[BKNS10]
$\{od, pd\}$	2	$n \log n$	open	n^3	n^3	[BKNS10]
$\{od, pd\}$	4	$n \log n$	open	n^3	n^3	[BKNS10]

Table 23.1 Running times of the best known algorithms (in big-Oh-Notation) for various versions of boundary labeling, where ε is an arbitrarily small positive constant and n is the number of sites. The time bounds in square parentheses refer to the case of non-uniform labels. The problems marked by $\star$ are NP-hard. The pseudo-polynomial algorithm for 2-sided opo-type leader-label placement assumes that label heights and the height H of the bounding rectangle are integers. N/A stands for non-applicable. Entries in column "Feasible solution" are filled only if there is a feasible solution that is asymptotically faster than a bend- or length-optimal solution.

23.3.1 Boundary Labeling with Type-s Leaders

In the case of fixed ports and fixed labels, a type-s label-leader placement or total length L corresponds to a Euclidean perfect bipartite matching of cost L. For the case of sliding ports (and fixed labels), the problem can also be reduced to a matching problem, albeit at a somewhat higher computational cost.

Theorem 23.1 *[[BKSW07]] Given a set S of n point sites, a one-side type-s leader-label placement of minimum total leader length for fixed labels can be computed in $O(n^{2+\varepsilon})$ time for any $\varepsilon > 0$ in the case of fixed ports and in $O(n^3)$ time in the case of sliding ports.*

Proof: In the case of fixed ports, we have a set P of n ports. Then a Euclidean minimum-cost perfect bipartite matching in the set $S \cup P$ yields a feasible leader-label placement of minimum total leader length. Feasibility follows from two properties of the Euclidean plane; the triangle inequality and the fact that the distances from the endpoints of a line segment to a point on the segment add up to the length of the segment. Indeed, suppose that two leaders would intersect then swapping the matching locally would decrease its cost; see Figure 23.5. (For the same reason, any solution to the Euclidean traveling salesperson

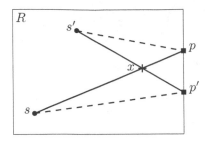

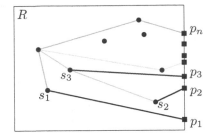

Figure 23.5   A minimum-length Euclidean matching is plane.

Figure 23.6   Feasible type-s leader layout via dynamic convex-hull.

problem forms a simple polygon unless all points lie on a line.) A Euclidean minimum-cost perfect bipartite matching can be computed $O(n^{2+\varepsilon})$ time for any $\varepsilon > 0$ [AES99].

For the case of sliding ports, the time complexity increases since we now need a *general* minimum-cost perfect bipartite matching in the complete bipartite graph on the set S of n points and the set L of the n label positions; the weight of an edge $(s, \ell) \in S \times L$ is the Euclidean distance of s to its closest point on ℓ. Since we assume that labels are attached to the right edge e of R, the point on ℓ closest to s is either the top or bottom point of ℓ or the orthogonal projection of s on e. A general minimum-cost perfect bipartite matching can be computed in $O(n^3)$ time [Law76]. □

If we content ourselves with a *feasible* leader-label placement, we can, in the case of fixed labels with fixed ports, speed up the computation.

Theorem 23.2   *[[BKSW07]] Given a set S of n point sites, a feasible one-side type-s leader-label placement for fixed labels with fixed ports can be computed in $O(n \log n)$ time.*

Proof: We assume that the set of ports, $P = \{p_1, \ldots, p_n\}$, is sorted according to increasing y-coordinate. Let H be the convex hull of the set $S \cup P$. Consider the edge of H that connects the bottommost point p_1 in P to a site. Call this site s_1 and make the line segment $s_1 p_1$ a leader; see Figure 23.6. Remove s_1 from S and p_1 from P. Repeat until each site is matched to a port. Since no two ports have the same y-coordinate, in each step, the convex hull of the diminished set $S \cup P$ is disjoint from the line segment connecting the site and the port that were removed last. Hence, the resulting leader-label placement is feasible.

To make our algorithm run in $O(n \log n)$ time, we just need a semi-dynamic convex-hull data structure that preprocesses a set of n points in $O(n \log n)$ time to allow for neighbor queries and point deletions in $O(\log n)$ time. Hershberger and Suri [HS92] provided such a data structure. □

23.3.2   Boundary Labeling with Type-*po* Leaders

We start with the simplest possible variant of the problem; the algorithm for this variant is illustrated in Figure 23.7. The idea behind the algorithm will turn out to be useful for two generalizations.

Theorem 23.3   *[[BKSW07]] Given a set S of n point sites, a feasible one-side type-po leader-label placement for fixed labels with fixed ports can be computed in $O(n^2)$ time.*

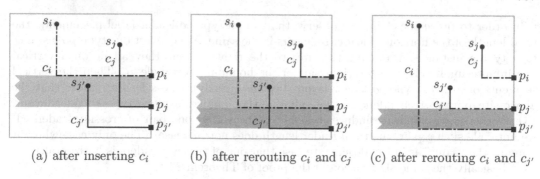

(a) after inserting c_i (b) after rerouting c_i and c_j (c) after rerouting c_i and $c_{j'}$

Figure 23.7   Rerouting type-po leaders.

Proof: We first sort sites and ports such that $s_1, \ldots, s_n$ and $p_1, \ldots, p_n$ are indexed in order of non-decreasing y-coordinates. For $i = 1, \ldots, n$, we connect s_i to p_i by a po-leader c_i that consists of a (possibly zero-length) vertical line segment incident to s_i and a horizontal line segment incident to p_i. We assume that the previously placed leaders $c_1, \ldots, c_{i-1}$ are pairwise disjoint, and we show that we can add c_i such that this assumption continues to hold.

In the following, we treat the case that s_i lies above p_i; the other case can be analyzed analogously. If c_i does not intersect any of the (pairwise disjoint) leaders $c_1, \ldots, c_{i-1}$, we are done. Otherwise, let s_j be the rightmost site with $j < i$ whose leader intersects c_i; see Figure 23.7(a). We reroute the leaders such that s_j is connected to p_i and s_i to p_j; see Figure 23.7(b).

After the rerouting, the new leader c_j does not intersect any other leader since (i) its vertical segment is shorter than before and (ii) its horizontal segment used to belong to c_i, which—due to the choice of s_j—did not intersect any other leader to the right of s_j. Hence, in this process, we remove the intersections of other leaders with the horizontal segment of c_i one by one, even if new intersections occur, as in the step from Figure 23.7(a) to Figure 23.7(b).

It remains to observe that the growing vertical segment of c_i never intersects other leaders. This is true since, initially, c_i goes to the top-most port p_i and, after each rerouting operation, c_i is prolonged by a vertical sub-segment that used to "belong" to a leader to the right of c_i; the sub-segments move within the gray horizontal strips in Figure 23.7. Thus, if a leader was to intersect the new vertical sub-segment of c_i, it would have earlier intersected one of the other leaders, contradicting our above assumption.   □

As it turns out, the feasible leader layout that the algorithm in the proof of Theorem 23.6 computes is already length-minimal.

Theorem 23.4   *[[BKSW07]] Given a set S of n point sites, a minimum-length one-side type-po leader-label placement for fixed labels with fixed ports can be computed in $O(n^2)$ time.*

Proof: Consider the site–port correspondence that we used in proof of Theorem 23.6: going through sites and ports from bottom to top, we connected the i-th site s_i to the i-th port p_i. We claim that the type-po leader layout induced by this correspondence has minimum total length among all type-po leader layouts (including the layouts with crossings). Combining this claim with the simple observation that rerouting does not change the total length of the leaders (see Figure 23.7), yields the theorem.

In order to prove the claim, we observe that, in all type-*po* leader-label placements, the total length of the horizontal leader segments is the same. We convert our type-*po* instance to a type-*s* instance by moving the sites to the right so that they all lie on a vertical line infinitesimally close to the right side of the boundary rectangle R. Then the vertical segments of a given type-*po* leader layout become (nearly) type-*s* leaders. Note that the above site–port correspondence is the only one that induces a plane type-*s* leader layout. Every other correspondence induces a layout with at least one pair of crossing leaders. If we untangle such a pair, the total leader length does not increase. (The only case where it remains the "same" is in the degenerate case that one of the two leaders is horizontal.) We used basically the same observation in the proof of Theorem 23.1.   □

The same result holds if labels are attached to two (opposite) sides of the bounding rectangle R.

Theorem 23.5 *[[BKSW07]] Given a set S of n point sites, a minimum-length two-side type-po leader-label placement for fixed labels with fixed ports can be computed in $O(n^2)$ time.*

Proof: As in the proofs of Theorems 23.3 and 23.4, we first compute a minimum-length layout without caring about crossings. For the one-side case, this was trivial; for the two-side case, we employ a simple dynamic program. Specifically, we use a two-dimensional table; table entry (l, r) contains the minimum total leader length for the $l + r$ lowest sites under the condition that l are connected to labels at the left side of R and the remaining r to labels at the right side. Since each entry in the table can be filled in constant time, the dynamic program runs in $O(n^2)$ total time.

Again, as in the proofs of the two preceding theorems, we then apply our rerouting scheme in order to remove all crossings. Recall that this does not change the total leader length. It remains to observe that leaders going to different sides of R never cross in this process; if they did cross, rerouting them would decrease the total leader length. This, in turn, would contradict the minimality of the total leader length of the original layout.   □

For the one-sided case, Benkert et al. [BHKN09] have observed that a length-minimal leader layout has a structure that can be exploited in order to speed-up its computation. The rectangular map R can be partitioned in horizontal strips such that all sites within a strip have horizontal leaders, have upward-going leaders (as in Figure 23.7(c)), or have downward-going leaders. Strips of upward- or downward-going leaders are always separated by strips with horizontal leaders, which can be detected easily. (In the case of fixed ports, "horizontal" means here that, as in the case of sliding ports, the site lies in the vertical range of the label.) Benkert et al. determine these strips in a first pass through the instance. Then, in a second pass, they determine the leader layout for the sites within a strip using a sweep-line algorithm. In total, their algorithm takes $O(n \log n)$ time. They show that this running time is worst-case optimal; sorting reduces to length-minimal leader layout. All in all, Benkert et al. have the following result.

Theorem 23.6 *[[BHKN09]] Given a set S of n point sites, a minimum-length one-side type-po leader-label placement for fixed labels with fixed or sliding ports can be computed in $\Theta(n \log n)$ time.*

23.4 Trees

In economy and social sciences, a common problem is to visualize the flow of goods or people from or into a specific destination. It makes sense to require that the flow between two nodes of the underlying network is depicted by curves whose width is proportional to the amount of flow. Usually these curves are drawn on the background of a regular map. For visualization purposes, the network is drawn as a tree—although, in general, the actual flow network is a rooted directed acyclic graph. The drawing of such a network is called a *flow map*.

Henry Drury Harness [Har38] is being cited [Rob55, FD01] for having created the first flow maps; in an atlas accompanying a report of the Railway Commissioners concerning population and movement of goods in Ireland in 1837. A few years later, Charles Joseph Minard, a French civil engineer, made flow maps mostly on economic topics, depicting, for example, the amount of wine export from France, but also, in 1869, the location and size of Napoleon's army during its 1812/13 Russian campaign; see Figure 23.8. Tufte [Tuf01, p. 40] says that this map "may well be the best statistical graphic ever drawn."

Drawing flow maps automatically was first studied by Tobler [Tob87] who used straight-line arrows of appropriate width. The restriction to straight-line edges causes a lot of visual clutter; see Figure 23.9(a).

Nearly twenty years later, Phan et al. [PXY+05] set out to improve on Tobler's result by taking advantage of clustering and curved edges. Given the positions of the network nodes, they first compute an agglomerative hierarchical clustering—independent of the position of the root. The binary tree that corresponds to the clustering captures the spacial distribution of the input. Then, they transform this unrooted binary tree into a tree rooted at the given root node. In this process, the root can get several children. The layout of the flow tree follows this tree recursively. A tree edge connecting a node to its child is routed from the position of the node to the closest corner of the bounding box of the cluster that corresponds to the child. The routing detours boxes containing sibling clusters. To make the final layout of the flow map more aesthetically pleasing, the polygonal paths that represent the edges are drawn as *Catmull–Rom splines*, that is, as special cubic curves that go through the given points. For the resulting layout, see Figure 23.9(b). Under the (rather strong)

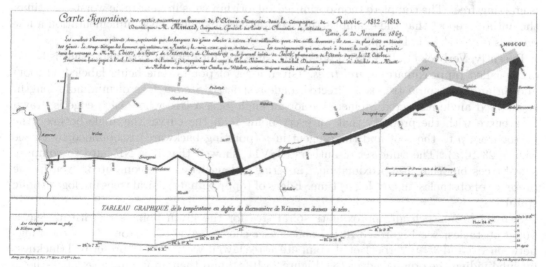

Figure 23.8 Minard's map of Napoleon's Russian campaign of 1812/13 [Min69].

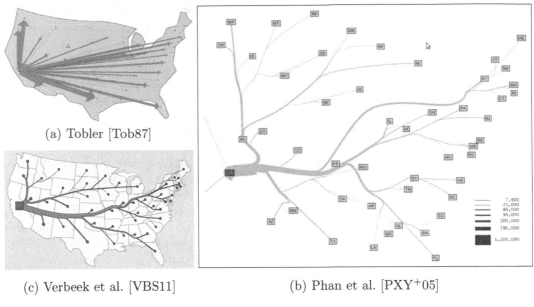

(a) Tobler [Tob87]

(c) Verbeek et al. [VBS11]

(b) Phan et al. [PXY$^+$05]

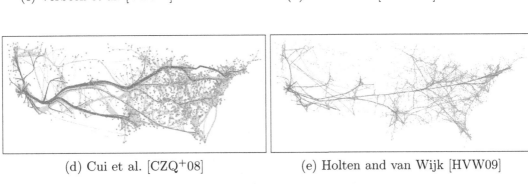

(d) Cui et al. [CZQ$^+$08]

(e) Holten and van Wijk [HVW09]

Figure 23.9   Flow maps showing migration leaving California in the years 1995–2000.

assumption that the boxes of child clusters are pairwise disjoint, the (polyline) tree layout is crossing-free. The complete (non-optimized) algorithm runs in quadratic worst-case time; the authors report that the examples they computed took their Java implementation a few seconds on a 1.4-GHz laptop.

Recently, Verbeek et al. [VBS11] presented a method for drawing flow maps that is based on so-called (approximate) *spiral trees*. Given a set of points (one being labeled as root) and an angle, a spiral tree is a directed angle-restricted Steiner tree of minimum length. A directed angle-restricted Steiner tree for an angle α is a tree where each edge is drawn as a curve with the property that, in every point p on the curve, the angle between the vector from p to the root and the tangent in p (pointing backward) is bounded by α; see Figure 23.10(a). The same set of authors [BSV11] showed that it is NP-hard to compute spiral trees but that 2-approximations (in terms of length) can be computed, even in the presence of obstacles, in $O(n \log n)$ time. Edges of (approximate) spiral trees are logarithmic spirals.

Starting from such an approximate spiral tree for the given point set (with all leaves being obstacles; see Figure 23.10(b)) and a user-chosen value of α (roughly in the range between 15° and 35°), Verbeek et al. compute a tree layout with edges of prescribed thickness by subdividing the original edges (see Figure 23.10(c)) and then improving a set of aesthetic

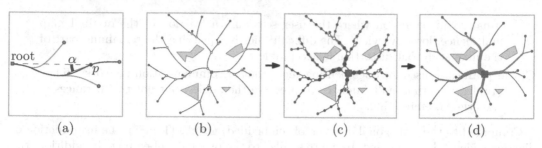

Figure 23.10 From spiral tree to tree map [VBS11]: (a) defining a directed angle-restricted Steiner tree. Workflow: (b) (approximate) spiral tree, (c) thickening and subdividing edges, (d) optimizing aesthetic criteria using the method of deepest descent.

parameters in order to smooth and straighten the tree, to avoid obstacles, to balance and to maintain its original angles (see Figure 23.10(d)). The authors model these parameters by defining cost functions; they apply the method of deepest descent in order to minimize the global cost function, which is the weighted sum of the individual cost functions. In order to ensure that no crossings are introduced in the optimization process, the algorithm checks for intersections before each move. In case an intersection would occur, the movement vector is repeatedly divided by 2 until the movement is safe. The edges are drawn as a new type of cubic Hermite splines that approximates logarithmic spirals well.

For an example output of the method of Verbeek et al., see Figure 23.9(c). For comparison, the results of two other, more general methods (by Cui et al. [CZQ+08] and Holten and van Wijk [HVW09]) are also depicted; see Figures 23.9(d) and (e). The input to these methods is a graph (with vertex positions) rather than a tree; in the output, the curved edges are *bundled* in order to better reflect the structure of the graph. Concerning running time, Verbeek et al. report that their algorithm drew most flow maps in less than a minute on a (dual-core) Pentium-D 3-GHz processor with 1 GB of RAM, whereas world maps required a few minutes.

23.5 Plane and Near-Plane Graphs

There are a number of applications where plane or near-plane graphs have to be drawn. We differentiate between four different types of applications. In all four types, the original embedding can be made planar by introducing few extra nodes where roads or tracks cross, for example, at bridges. The topology of the original embedding must be preserved and edges are drawn as polygonal lines. In most cases it is desirable to keep vertices roughly in the same place as in the original embedding or to at least preserve the relative position of vertices (for example, left/right, above/below). This helps the user to keep his mental map.

The application types that we consider in this section are as follows.

Schematic road maps are used for road or transportation networks. They try to keep vertices (that is, cities or junctions) at or close to their original location. Edges (that is, roads or tracks) can have diagonal segments.

Metro maps also use diagonals, but other than schematic maps they use very different scales for downtown versus suburban areas. Relative position is important. Another special feature of metro maps is that they usually have many degree-2 nodes.

Street maps with focus regions do not restrict edge directions but allow the user to *select* a region that is displayed at larger scale. This is different from the usual zoom operation where the user sees only a fraction of the original map and, hence, loses overview. The difficulty lies in squeezing the remaining part of the map such that distortion is acceptable.

Cable plans are used for documenting the layout of communication networks. They are drawn orthogonally and try to preserve the angles, but not the distances of the original embedding.

Compared to the orthogonal drawing of (embedded) graphs [Tam87], the introduction of diagonals yields drawings that are more similar to the original embedding. In addition, the maximum node degree increases from 4 to 8. In a sense, however, the problem becomes more difficult as Bodlaender and Tel [BT04] point out. They define a planar graph to be *d-linear* if it can be embedded such that all angles are multiples of $2\pi/d$. The *angular resolution* of a plane straight-line drawing is the minimum angle between edges incident to a common vertex, over all vertices. Bodlaender and Tel show that, for $d = 4$, an angular resolution of $2\pi/d$ implies d-linearity and that this is not true for any $d > 4$.

In what follows, we refer to the set of directions that are given by the two coordinate axes and their two bisectors as the *octilinear directions*.

23.5.1   Schematic Road Maps

Schematic maps usually try to preserve the position of vertices as much as possible while simplifying the polygonal lines that represent edges without changing the topology of the original drawing. Edges are drawn as x- and y-monotone paths that consist of usually no more than three horizontal (H), diagonal (D), and vertical (V) line segments. Cabello et al. [CdBvD+01] have given an algorithm that decides in $O(n \log n)$ time whether a node-embedded graph can be drawn such that each edge follows one of a given set of allowed segment sequences (such as {HVH, VDV}, for example). If an edge embedding of the required type exists, the algorithm finds it.

While Avelar and Müller [AM00] also try to make edges octilinear, they use a very different method that moves vertices based on local decisions. They guarantee that the topology of the original network is kept, but they do not guarantee that every edge in the final layout is actually octilinear. They first use a polygonal line simplification method to simplify all edges (that is, polygonal lines) of the original embedding. In order to preserve topology, a more involved method like Saalfeld's [Saa99] must be used instead of the classical method of Douglas and Peucker [DP73] mentioned above.

After the simplification, each street junction and each bend of a street is considered a vertex. Hence, edges are straight-line segments. Avelar and Müller iteratively go through all vertices and compute new destinations based on the current (imperfect) directions of the incident edges. They do this as follows. For each vertex v and each vertex w incident to v, they compute an offset for v that would make the edge vw confirm to one of the allowed directions. The arithmetic mean of these offsets yields a tentative new position for v. Before actually moving a vertex, Avelar and Müller check the topology of the resulting embedding. If topology would change, they restrict the vertex movement accordingly. They continue to change vertex positions until all edges follow one of the desired four directions or until the number of iterations has reached a fixed threshold.

23.5.2   Metro Maps

The problem of drawing maps of subways and other means of public transportation is an interesting compromise between schematic maps where vertex positions are (mostly) fixed and "conventional" graph drawing where vertices can go anywhere. The first approach maximizes (user) orientation, the second aesthetics.

We now define the problem in graph-drawing terms. Let $G = (V, E)$ be the input graph. We assume that G is *plane*, that is, G comes with a planar embedding. We actually assume that we know the geographic location $\Pi(v)$ of each vertex $v \in V$ in the plane and that the straight-line embedding induced by the vertex locations is plane. In case some edges cross others, we simply introduce dummy vertices that represent the crossings. Let $\mathcal{L}$ be a *line cover* of G, that is, a set of paths of G such that each edge of G belongs to at least one element of $\mathcal{L}$. An element $L \in \mathcal{L}$ is called a *line* and corresponds to a metro line of the underlying transport network. We refer to the pair $(G, \mathcal{L})$ as the *metro graph*. The task is now to find a drawing Γ of $(G, \mathcal{L})$ according to a set of rules (which we will discuss later).

In the last few years, a number of methods for automating the drawing of metro maps have been suggested. The author [Wol07] surveyed the area earlier, with an emphasis on experimental comparison. Our treatment here is more compact, but adds some recent development. Before we go into the methods, let us quickly turn to the origins of the problem.

History. While metro networks were small in size, it made perfect sense to draw them geographically. This was easy for the graphic designers and gave map users a sense of distance, for example, between stations that are close to each other in the above-ground street network but far in the underground metro network: sometimes it is indeed faster to walk a little more than to reach the metro stop closest to one's destination. Electrical draftsman Henry Beck was the first to draw a metro network in a schematic way. His rationale was that connection information and the number of stops on a line are more important information for the network user than geographic distances. His design was so revolutionary that the London Transport Authority, in 1931, rejected his first proposal and only in 1933 dared to print and sell Beck's map. Therefore, Berlin got the honor of having the first printed schematic metro map (in 1931). While the Nazis in Berlin soon moved back to a geographic layout [Pol06], Beck's tube map was an instant success and became the basis of all subsequent official maps of the London Underground. In 2006, his original map was elected, right after the supersonic airplane Concorde, the second-most popular British design icon of the twentieth century [Wik12]; it has an interesting history in its own right [Gar94]. In the meantime, graphic designers have invented different layout styles all over the world (see the book of Ovenden [Ove03]), but the use of the octilinear set of directions for drawing is still prevailing.

Complexity. Using eight edge directions seems to be a good compromise between an unrestricted drawing and the restriction to the four orthogonal (or rectilinear) edge directions predominant in circuit diagrams, VLSI layout, and—traditionally—in graph drawing. As it turns out, the additional freedom that an octilinear layout gives the designer compared to a rectilinear layout comes at a price. Nöllenburg [Nöl05b] proved, by means of a visually very appealing reduction from PLANAR3SAT, that it is NP-hard to decide whether a plane graph has an octilinear drawing. This is in sharp contrast to the rectilinear case, for which Tamassia [Tam87] showed that the same question can in fact be answered efficiently. In his seminal paper, the theoretical foundation of orthogonal graph drawing, Tamassia reduced the problem to a network flow problem, which yields an orthogonal drawing with the minimum number of bends and small area.

Curve evolution. The first attempt to automate the drawing of metro maps was made by Barkowsky et al. [BLR00]. They use an algorithm for polygonal line simplification, which they call *discrete curve evolution* [LL99], to treat the lines of the Hamburg subway system. Their algorithm, however, neither restricts edge directions nor does it increase station distances in the crowded downtown area. Stations are labeled but no effort is made to avoid label overlap.

Force-directed layout. Hong et al. [HMdN06] give five methods for the metro-map layout problem. The most refined of these methods modifies PrEd [Ber99], a topology-preserving spring embedder, such that edge weights are taken into account and such that additional magnetic forces draw edges toward the closest octilinear direction. Edges are drawn as straight-line segments connecting the corresponding vertices. Relative position is only taken into account implicitly by using the original embedding as initial layout.

In a preprocessing step, Hong et al. simplify the metro graph by contracting each edge that is incident to a degree-2 vertices. After performing all contractions, the weight of each remaining edge is set to the number of original edges it replaces. After the final layout has been computed, all degree-2 vertices are re-inserted into the corresponding edges in an equidistant manner. Due to this preprocessing the numbers of vertices and edges decrease by a factor of 3 to 8, and all networks (with 22 to 92 vertices and 32 to 317 edges after contraction) were solved within 0.2 to 22 seconds. Station labels are placed in one out of eight directions using the interactive LabelHints system [dNE03]. While label–label overlaps are avoided, diagonally placed labels sometimes intersect network edges.

The results of Hong et al. [HMdN06] are clearly superior to those of Barkowsky et al. [BLR00]. However, they are still not very similar to commercial maps drawn by graphic designers. The main deficiency is that most edges in the final layouts are close to, but not quite octilinear. This seems to be due to the fact that the magnetic forces that determine the layout are the sum of many conflicting terms.

Local optimization. Stott et al. [SRMW11] draw metro maps using multicriteria optimization based on hill climbing. For a given layout they define metrics for evaluating the octilinearity and the length of edges, the angular resolution at vertices and the straightness of metro lines. The quality of a layout is the sum over these four metrics. Their optimization process is iterative. They start with a layout on the integer grid that is obtained from the original embedding. In each iteration they go through all vertices. For each vertex they consider alternative grid positions within a certain radius that shrinks with each iteration. For each of these grid positions they compute the quality of the modified layout. If any of the positions improves the quality of the layout, they move the current vertex to the position with the largest improvement among those positions where the topology of the layout does not change. After implementing their algorithm they observed a typical problem of local optimization: overlong edges are often not shortened since this would need moving several vertices at the same time. For a *bridge*, that is, an edge whose removal disconnects the graph, this can easily be fixed by moving all nodes of the smaller component closer to the larger component. They run this fix after each iteration for all bridges.

Stott et al. have experimented with enforcing relative position, but report that the results were disappointing as there were many situations where a better layout could only be found by violating the relative position of some vertices. They can label stations, but do not check for overlaps other than with the edges incident to the current station. They use the same contraction method as Hong et al. [HMdN06] to preprocess the input graph. Even with this preprocessing their algorithm is much slower. For example, an earlier version of their algorithm [SR05] drew the simplified Sydney CityRail network in about 4 minutes and the unsimplified network in 28 minutes; the new algorithm (in Java 1.6 on a 1.4-GHz Celeron M machine with 1.5 GB RAM under Windows XP) needs about two hours for the labeled network. This compares with the 7.6 seconds that Hong et al. need for the simplified, but labeled network. The drastic increase in running time, however, is worth it—in the resulting maps nearly all edges are octilinear, which makes the maps more legible.

Global optimization. Nöllenburg and Wolff [NW11] draw metro maps using the toolbox of mathematical programming. They approach the problem by setting up the following list of design rules which are based on the design of real-world metro maps.

(R1) Restrict the drawing of edges to the octilinear directions.

(R2) Do not change the geographical network topology. This is crucial to support the mental map of the passengers.

(R3) Avoid bends along individual metro lines, especially in interchange stations, to keep them easy to follow for map readers. If bends cannot be avoided, obtuse angles are preferred over acute angles.

(R4) Preserve the relative position between stations to avoid confusion with the mental map. For example, a station being north of some other station in reality should not be placed south of it in the metro map.

(R5) Keep edge lengths between adjacent stations as uniform as possible with a strict minimum length. This usually implies enlarging the city center at the expense of the periphery.

(R6) Stations must be labeled and station names should not obscure other labels or parts of the network. Horizontal labels are preferred and labels along the track between two interchanges should use the same side of the corresponding path if possible.

(R7) Use distinctive colors to denote the different metro lines. This means that edges used by multiple lines are drawn thicker and use colored copies for each line.

A subset of these rules has also been listed by Hong et al. [HMdN06].

Nöllenburg and Wolff divide their rules into strict requirements, also called *hard constraints*, and into aesthetic optimization criteria, also called *soft constraints*. Their hard constraints are:

(H1) *Octilinearity:* For each edge e, the line segment $\Gamma(e)$ in the output drawing must be octilinear.

(H2) *Topology preservation:* For each vertex v, the circular order of its neighbors must agree in Γ and the input embedding.

(H3) *Minimum length:* For each edge e, the line segment $\Gamma(e)$ must have length at least ℓ_e.

(H4) *Minimum distance:* Each edge e must have distance at least $d_{\min} > 0$ from each non-incident edge in Γ.

Constraint (H1) models the octilinearity requirement (R1). It is this constraint that makes the problem NP-hard [Nöl05a], see the discussion in the paragraph on complexity above. Constraint (H2) models the topology requirement (R2), (H3) models the minimum edge length in (R5), and (H4) avoids introducing additional edge crossings and thus also models a part of (R2). This is because two intersecting edges would have distance $0 < d_{\min}$.

The soft constraints should hold as tightly as possible. They determine the quality of Γ and are as follows:

(S1) *Straightness:* The lines in $\mathcal{L}$ should have few bends in Γ, and the bend angles ($< 180°$) should be as large as possible.

(S2) *Geographic accuracy:* For each pair of adjacent vertices (u, v), their relative position should be preserved, that is, the angle $\angle(\Gamma(u), \Gamma(v))$ should be similar to the angle $\angle(\Pi(u), \Pi(v))$, where $\angle(a, b)$ is the angle between the x-axis and the line through a and b.

(S3) *Size:* The total edge length of Γ should be small.

Clearly, constraint (S1) models minimizing the number and "strength" of the bends (R3) and (S2) models preserving the relative position (R4). The uniform edge length rule (R5) is realized by the combination of a strict lower bound of unit length (H3) and a soft upper bound (S3) for the edge lengths. Rule (R4) for the relative position can be interpreted as both a soft and a hard constraint, for example, by restricting the angular deviation to at most 90° as a hard constraint and charging costs for smaller deviations as a soft constraint.

Nöllenburg and Wolff then show that the existence of a drawing that fulfills the hard constraints (H2)–(H4) and optimizes a weighted sum of the soft constraints can be formulated as a mixed-integer linear program (MIP). The basic idea behind their formulation is as follows. Each edge has a number of binary variables that correspond to its feasible octilinear directions. Exactly one of these variables must be 1. All other constraints regarding an edge, such as its minimum length and minimum distance from other edges, are expressed for each feasible direction. The constraints are designed such that they are trivially fulfilled if the edge has a different direction. Angles are "measured" in multiples of 45°, for example, in soft constraint (S 1), an angle is punished proportionally to its degree of acuteness: the bend of the edges uv and vw incident to a vertex v can be of size 180°, 135°, 90°, or 45°. The bend cost of this bend is 0, 1, 2, or 3, respectively. Expressing this with linear constraints is somewhat tricky, but it can be done using the directions of the edges uv and vw and two new binary variables per bend.

In general it is NP-hard to solve a MIP, but highly optimized commercial solvers such as CPLEX or Gurobi can solve relatively large MIPs relatively quickly. Consider a medium-sized metro system such as the CityRail network of Sydney with 10 lines and 174 stations. For this network, the MIP of Nöllenburg and Wolff as sketched above consists of roughly 38,000 variables and 150,000 constraints—assuming that one applies the obvious data reduction trick of replacing each path of k degree-2 vertices by a single edge of length at least $k \cdot d_{\min}$. Actually, Nöllenburg and Wolff proposed to keep up to two vertices between each pair of neighboring interchange stations so as to have some flexibility for making bends; this helps to be more accurate in terms of relative position (geographic accuracy). Solving such a MIP to optimality can take days.

Therefore, Nöllenburg and Wolff described a number of ways in order to further reduce the size of the MIP. Their fastest approach is based on the so-called *callback function* of the CPLEX solver. It allows them to set up the MIP without any planarity constraints according to hard constraint (H4), check any intermediate feasible solution for crossings and then add constraints needed to forbid the specific crossings at hand. For the reduced Sydney example, this yields a MIP with roughly 4800 variables and 3500 constraints; constraints for just three edge pairs were added during optimization. Still, computing the layout in Figure 23.11(c) from the geographic input depicted in Figure 23.11(a) took about 23 minutes. For a comparison with the work of a professional graphic designer, see Figure 23.11(b).

Things get worse when drawing maps with station labels that can change sides with respect to metro lines. Even when aggregating all labels between two interchanges into one big label (that is then modeled as a dummy metro line) and taking advantage of the callback functionality, the MIP ends up having nearly 93,000 variables and 22,000 constraints. Computing the layout in Figure 23.11(d) took 10.5 hours; in both cases an optimality gap of about 16% remained, that is, the solver knows that the unknown objective value of an optimal solution is at most 16% less than that of the layouts in Figures 23.11(c) and 23.11(d).

Least squares. Wang and Chi [WC11] presented a system for octilinear on-demand focus-and-context metro maps that highlight routes returned by a route planning

(a) geographic layout (by John Shadbolt)

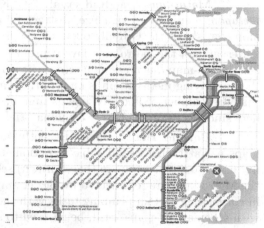

(b) corresponding clipping of the official map [Syd08]

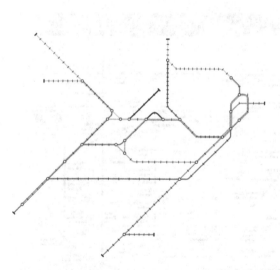

(c) unlabeled layout of Nöllenburg and Wolff

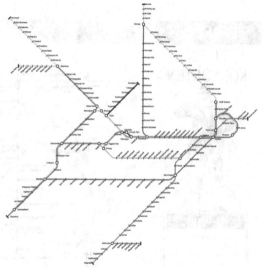

(d) labeled layout of Nöllenburg and Wolff

Figure 23.11 The Sydney CityRail network.

system while showing the rest of the network as less important context information. It can also be used to draw non-focused metro maps. They deform the given geographic map by the conjugate gradient method [HS52] in a least-squares sense, minimizing a set of energy terms that model the aesthetic constraints. Labeling is performed independently. Their method is both fast and creates good layouts, e.g., for mobile devices.

Metaphor.   Sandvad et al. [SGSK01] and Nesbitt [Nes04] use the *metro-map metaphor* as a way to visualize abstract information. A particularily nice example is the map that shows the O'Reilly open source product lines [O'R03], see Figure 23.12.

Research of the metro-map layout problem triggered the investigation of a new subproblem, metro-map *line crossing minimization*. In that problem, one assumes that the layout of the underlying metro graph is known; the aim is to order the metro lines on each edge such that the number of line crossings is minimized [BNUW07]. We do not treat the problem here since its nature is purely combinatorial, not geometric.

Beyond Henry Beck   Recently, a completely different style for drawing metro maps has attracted considerable attention: the *curvilinear* style. Roberts et al. [RNL+11] did user studies to compare (hand-drawn) schematized maps to (hand-drawn) maps where the Metro lines are represented by Beziér curves. Surprisingly, users were up to 50 % faster in completing certain planning tasks with the new and unfamiliar Beziér maps rather than with schematized maps. Still, being used to schematized maps, they liked them better.

These findings prompted Fink et al. [FHN+13] to investigate ways to automate the process of drawing metro maps with Beziér curves; see Figure 23.13. They use a force-directed approach. Starting with a straight-line or octilinear input drawing (see Figure 23.13(a)),

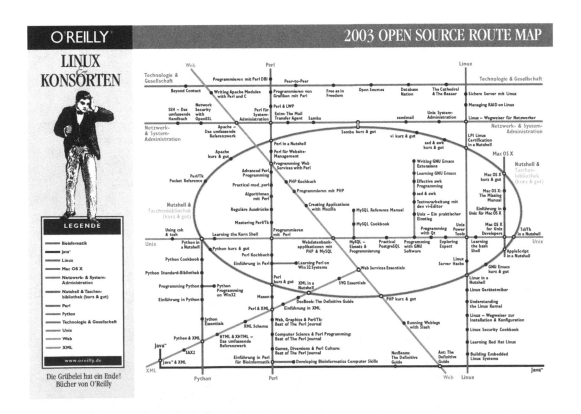

Figure 23.12   O'Reilly's open source product lines 2003.

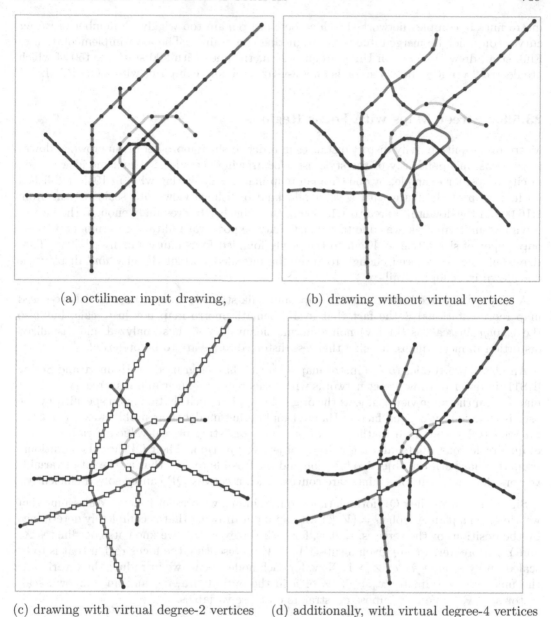

(a) octilinear input drawing, (b) drawing without virtual vertices

(c) drawing with virtual degree-2 vertices (d) additionally, with virtual degree-4 vertices

Figure 23.13 Metro network of Vienna drawn using Bézier curves [FHN+13].

the authors go through each metro line and replace each line segment by a nearly-straight cubic Bézier curve that shares tangents with its predecessor and successor. Then they apply attracting and repulsive forces to vertices, but also to tangents. The aim is to merge as many consecutive Bézier curves on each metro line as possible in order to reduce the visual complexity; compare Figures 23.13(b), (c), and (d). Vertices that are incident to merged edges only are called *virtual*; forces can no longer be applied to them. In all but the last iteration, merges happen only at degree-2 vertices. In the final iteration, degree-4 vertices are handled, too.

Whereas the results are quite nice for small networks (such as Montreal or Vienna), long metro lines in complex networks (such as London) remain too wiggly. A number of Bézier curves could not be merged due to contradicting constraints. The Java implementation of Fink et al. drew the London Underground (20 metro lines with 200 stations, 150 of which are degree-2 vertices) in 224 seconds on a 3-GHz dual-core computer with 4 GB RAM.

23.5.3   Street Maps with Focus Regions

Metro maps quite heavily distort distances in order to show more details in crowded downtown areas, independently of the style used for drawing the edges. The same idea is used in city maps, for example, by the German map maker Falk-Verlag who, in 1945, published its first map of Hamburg with a very mild kind of fisheye view with scale varying from 1:16.000 in the downtown area to 1:18.500 in the suburbs. Interestingly enough, the idea to use a non-uniform scale was due to the fact that the post-war military government allotted only paper of size 60 cm × 40 cm to the newly founded four-man company [Hol95]. That size would not have been enough to cover the intended part of the city *and* display the downtown in enough detail.

A major difference between metro maps and Falk-style city maps (that is, fisheye-based map representations) is the fact that in a schematic metro map not just scale, but also the *change* in scale is (highly) non-uniform. Jenny [Jen06] has analyzed and visualized distortion in metro maps, arguing that less distorted maps are to be preferred.

An idea more similar to the metro-map approach has been used by Haunert and Sering [HS11] in order to draw street networks with focus regions. Their aim is to redraw a street map within the same view frame as the original map, but such that a region specified by the user is enlarged by a given factor. Haunert and Sering model their problem as a *quadratic program* (QP), that is, a mathematical program consisting of real-valued variables, a set of quadratic constraints, and a quadratic objective function. Their QP has the additional property that both the objective function and the feasible region, that is, the set of variable vectors that fulfill all constraints, are convex. Such a convex QP can be solved efficiently.

Since the core of their QP formulation is quite simple, we present it here. We assume that we are given a plane graph $G = (V, E)$ with an input drawing that is completely determined by the positions of the vertices, that is, for each vertex $v \in V$, we know its coordinates X_v and Y_v. Moreover, we are given a subset $V' \subseteq V$ representing the focus region that is to be scaled up by a *zoom factor* $Z > 1$. Now, for each node $u \in V$, we introduce three variables: the unknown coordinates $x_u, y_u \in \mathbb{R}$ of u in the output drawing and an unknown scale factor $s_u \in \mathbb{R}^+$. We now impose constraints on these variables.

First, we define a constraint to ensure that the output drawing remains within the bounding box of the input drawing.

$$\min_{v \in V}\{X_v\} \leq x_u \leq \max_{v \in V}\{X_v\}$$
$$\min_{v \in V}\{Y_v\} \leq y_u \leq \max_{v \in V}\{Y_v\} \qquad \text{for each } u \in V \qquad (23.1)$$

Second, we fix the scale factor for each node in the focus region:

$$s_u = Z \qquad\qquad \text{for each } u \in F \qquad\qquad (23.2)$$

For a node $u \notin F$, we don't know its scale factor s_u; we will determine it through the optimization, together with the coordinates of u in the output map. It remains to ensure that the scale factor s_u is valid for the neighborhood of u.

Suppose that we would express the idea of a locally valid scale factor with the constraint

$$
\begin{aligned}
s_u(X_v - X_u) &= (x_v - x_u) \\
s_u(Y_v - Y_u) &= (y_v - y_u)
\end{aligned}
\qquad \text{for each } u \in V, \, v \in \mathrm{Adj}(u), \qquad (23.3)
$$

where $\mathrm{Adj}(u)$ is the set of neighbors of u in G. With constraint (23.3), the star-shaped subgraph of G that contains u and its neighbors is scaled by s_u. For two adjacent nodes i and j, however, we can only satisfy this constraint if we set $s_i = s_j$. Therefore, if G is connected, we would have to select the same scale factor for all nodes in V. With constraint (23.3), it is thus impossible to design a *variable*-scale map.

In order to allow for different scale factors in different parts of the map, we introduce a relaxed version of constraint (23.3). We do not require that the neighborhood of node u is *exactly* mapped to scale. Instead, we allow for small distortions, which we measure based on residuals δx_{uv} and δy_{uv}. For this purpose, we introduce, for each edge $uv \in E$, auxiliary variables δx_{uv} and δy_{uv} into our model. Relaxing constraint (23.3) simply yields

$$
\begin{aligned}
\delta x_{uv} &= s_u(X_v - X_u) - (x_v - x_u) \\
\delta y_{uv} &= s_u(Y_v - Y_u) - (y_v - y_u)
\end{aligned}
\qquad \text{for each } u \in V, \, v \in \mathrm{Adj}(u). \qquad (23.4)
$$

If both u and v lie in the focus region F, we require

$$
\delta x_{uv} = \delta y_{uv} = 0 \quad \text{for each } u, v \in F, v \in \mathrm{Adj}(u). \qquad (23.5)
$$

This makes sure that edges in the focus region indeed become enlarged by the zoom factor Z.

Our objective is to minimize the *weighted* square sum of the residuals:

$$
\text{Minimize} \sum_{u \in V} \sum_{v \in \mathrm{Adj}(u)} \left(\big(w(u,v) \cdot \delta x_{uv}\big)^2 + \big(w(u,v) \cdot \delta y_{uv}\big)^2 \right) \qquad (23.6)
$$

with $w(u,v) = 1/\sqrt{(X_v - X_u)^2 + (Y_v - Y_u)^2}$. With this weight setting, we express that the validity of the scale factor s_u decreases with increasing distance from node u. This finishes the description of the core of the QP. All its constraints are linear; its objective function is convex since it doesn't contain mixed terms and all the weights are positive. Therefore, the core QP can be solved efficiently.

Unfortunately, the core QP does not prevent edge crossings. Crossings are unlikely to occur in triangulations but they do occur in less strongly connected networks. Hence, an obvious idea is to triangulate the given plane graph G. Experiments, however, show that this ad-hoc solution produces drawings with rather high distortion all over the network. The reason is that the additional edges make the network inflexible. Sparse regions that otherwise can help to balance the expansion of the focus regions are artificially made dense.

A more promising approach is to define, for each pair of edges, a line that separates the two edges and to add new variables (the parameters of the line) and new constraints to the QP. As it turns out, the necessary constraints are such that the set of feasible solutions is not convex any more. In order to stay in the realm of convex quadratic programming, the authors came up with a clever trick. They simply removed one degree of freedom from the separating line; by fixing its slope. Clearly, adding the corresponding constraints to the QP yields a new QP that is more constrained than actually necessary. By choosing a "good" slope, however, the negative impact of the additional restriction can be kept small. Haunert and Sering suggested to use the slope of the maximum-width strip that separates the two edges in the *input* drawing. The second trick they applied is to *not* add all new planarity constraints before solving the QP, but only in case the solution of the QP actually contains crossings. For each such crossing, exactly the constraints that forbid it are added to the QP, and the modified instance is given back to the QP solver. A similar trick was used by Nöllenburg and Wolff [NW11] in order to deal with planarity constraints in their MIP for drawing metro maps, see Section 23.5.2.

Concerning an example, consider the input instance depicted in Figure 23.14. This street network consists of 5864 vertices and 6675 edges. Applying the QP-based method to that input with the focus region represented by the black circle and a zoom factor of 2 took 51.8 seconds on a Windows PC with 3 GB main memory and a 3 GHz Intel dual-core CPU. The output is shown in Figure 23.15(a). For comparison, Figure 23.15(b) depicts the result of applying a fisheye transformation [YOT09] to the same input. Applying such a transformation takes only fractions of a second. Figure 23.15 also shows, in the small inlets on the right-hand side, the residuals of the street network, which can be seen as a measure for the deformation of the network. (The lower inlet has a legend that explains the color-coding.) Clearly, the method of Haunert and Sering yields very good solutions for drawing maps with focus regions. More work is needed, however, to come up with a method that is similarly good but much faster. This would be very interesting for all kinds of mobile applications.

Böttger et al. [BBDZ08] provide an interesting link between the schematized world of metro maps and the non-schematized world of city (street) maps. They show how to gradually morph a map showing both types of networks between a representation that is geographic and a representation where the map is distorted such that the metro network is schematized.

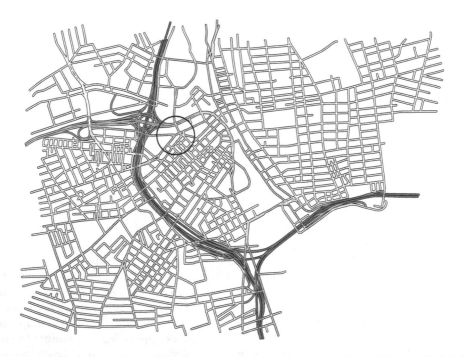

Figure 23.14   A street map (showing a detail of Providence, Rhode Island, U.S.A.) with a circular focus region that contains the conference site of InfoVis 2011.

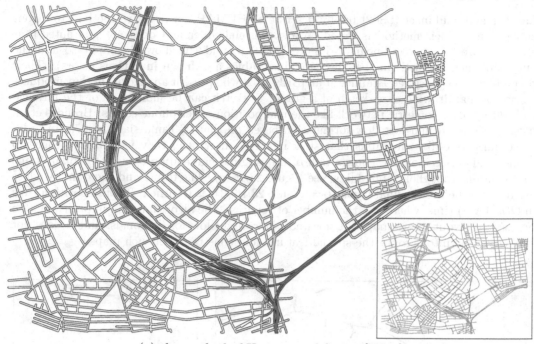

(a) the method of Haunert and Sering [HS11]

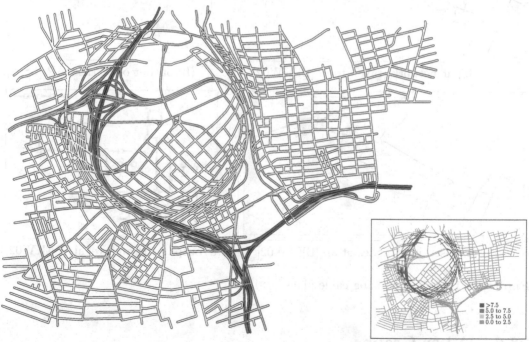

(b) the fish-eye transformation of Yamamoto et al. [YOT09]

Figure 23.15 The results of applying two deformation methods to the map in Figure 23.14. The inlets show edges with residuals in red.

23.5.4 Cable Plans

Lauther and Stübinger [LS02] briefly describe SCHEMAP, an iterative method to layout cable plans. Their method is based on a spring embedder and does not guarantee that all edges are drawn rectilinearly. Figures 23.16 (a) and (b) show the input to and the output of their method (in (b), the individual cables are drawn in various colors). Their preliminary work inspired Brandes at al. [BEKW02] who present an algorithm that produces an orthogonal drawing of a sketch of a graph. A sketch can be handmade or the physical embedding of a geometric network like the real position of telephone cables. Brandes et al. use a path-based min-cost flow formulation based on that of Tamassia [Tam87]. In order to stabilize tree-like subgraphs that stick into the outer face, they use dummy edges to connect all vertices on the convex hull of the original embedding to a rectangular frame that contains the whole embedding; see Figure 23.16 (c). The frame and the dummy edges are removed before the final layout (see Figure 23.16 (d)) is returned. Their algorithm runs in $O(n^2 \log n)$ time, where n is the number of vertices. The algorithm can, in principle, also be used to layout metro maps. It does not, however, allow for diagonals, and it does not explicitly take into account the special features and constraints of such maps.

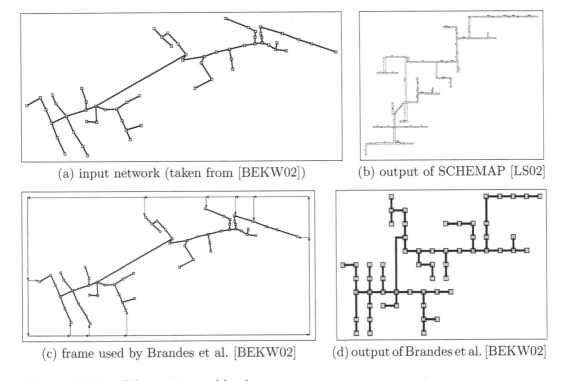

(a) input network (taken from [BEKW02])   (b) output of SCHEMAP [LS02]

(c) frame used by Brandes et al. [BEKW02] (d) output of Brandes et al. [BEKW02]

Figure 23.16 Schematizing cable plans.

23.6 Other Graphs

In this section we consider graphs that do not fit into the classes we have treated in the previous sections. We focus on two scenarios, one scenario that has a geographic background—graphs that describe train connections (see Section 23.6.1)—and one scenario that uses the cartographic-map metaphor to convey cluster information in (non-geographic) social networks such as collaboration graphs (see Section 23.6.3).

23.6.1   Timetable Graphs

A *timetable graph* has a vertex for each train station and an edge for each pair of stations connected by a train that does not stop in between. The graph is of interest to railway companies to check completeness and consistency of their schedules and to analyze changes between consecutive schedules. An obvious way to layout such graphs is to embed vertices at their geographic locations and edges as straight-line segments between them. However, this causes many edge crossings and small angles between edges along the same train line. Instead, Brandes and Wagner [BW00] introduce the concept of *minimal* and *transitive* edges. An edge $\{u, v\}$ is minimal if it corresponds to a piece of track that does not contain a station served by some other train. The remaining transitive edges correspond to through trains.

Whereas Brandes and Wagner use straight-line edges for minimal edges and long transitive edges, they suggest to use cubic Bézier curves [Béz72] to draw all other edges. Vertices are kept at their geographic location to allow for easy orientation. Then the layout problem consists of placing two control points for each Bézier curve. The authors define attractive and repulsive forces between control points and train stations within a local neighborhood. Using the random field layout framework [Bra99] and a customized version of the force-directed Fruchterman-Reingold method [FR91], they managed to draw even large timetable graphs nicely within minutes.

In subsequent work, Brandes et al. [BST00] explored ways to speed up their method and, at the same time, achieve perfect (or any prescribed) angular resolution in drawings of timetable graphs (and the Internet multicast backbone). They show that the flexibility of cubic Bézier curves allows them to optimize a number of criteria (with respect to the given straight-line drawing) in linear time by considering each vertex separately. This reduces the running time of their method on the same graphs as above from minutes to fractions of seconds. They refer to their new method as the *rotation* method. For a sample output, see Figure 23.17(b) and compare to the straight-line layout in Figure 23.17(a).

Unfortunately, due to the locality of the rotation method, the number of edge crossings and S-shaped edges increases. With a slightly different set of authors, Brandes et al.

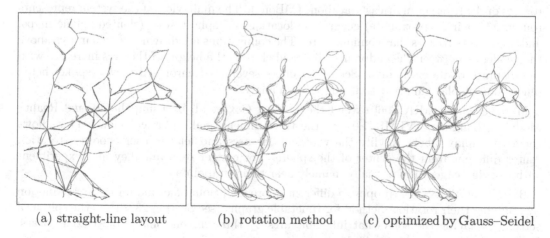

| (a) straight-line layout | (b) rotation method | (c) optimized by Gauss–Seidel |

Figure 23.17   Timetable graphs of the sourroundings of Venice.

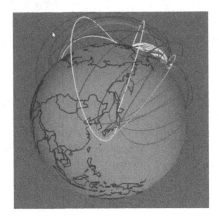

Figure 23.18   A comparison of elevated great circles (left) and Bézier curves output by the rotation method of Brandes et al. [BST00] (right).

[BSTW01] get a grip on these issues by using two new ingredients. First, the preprocess timetable graphs in order to make them more susceptible to the rotation method. Second, they introduce a new objective function that combines three criteria concerning the position and shape of edges, namely angular resolution, straightness, and roundness. They show that this objective function is a generalization of the layout function of Tutte's barycentric method [Tut63]. Therefore, the function has a unique minimum, which is the solution of a system of linear equations. Due to the size of their system, they resort in using the iterative Gauss–Seidel method, which in their case converges very fast. Their new algorithm is just about four times slower than the rotation method, and hence 50–100 times faster than the force-directed approach. In terms of aesthetics, the new method comes much closer to, but doesn't quite reach the force-directed approach; see Figure 23.17(c).

23.6.2   Internet Traffic

Clearly, computer scientists are interested in analyzing the structure of the largest man-made network, the Internet. Visualization plays an important role in this endeavor. Cox et al. [CEH96] created SeeNet3D, a tool that can be used to view and analyze traffic between routers of the Internet multicast backbone (MBone). The main view of the system represents routers at their (approximate) geographic locations on spherical or (slanted) plane maps, and it connects routers that communicate. The connections are drawn as circular arcs above the geodesics between the endpoints. To avoid clutter, the height of the arcs increases with the distance of its endpoints. SeeNet3D offers several synchronized views (spoke, helix, pincushion display) in order to facilitate data analysis.

Munzner et al. [MHCF96] extend the work of Cox et al. by using the Virtual Reality Modeling Language (VLMR 1.0) for the three-dimensional, spherical view. This allows them to display labels, modify the width of the arcs and let the user choose a rotation center different from the center of the sphere. For clutter removal, they also experiment with drawing edges only partially; namely near their endpoints.

Brandes et al. [BST00] propose a different, more traditional method for clutter reduction by applying their rotation method for timetable graphs (see Section 23.6.1) to the spherical setting, replacing the somewhat inflexible arcs by three-dimensional cubic Bézier curves. For a comparison, see Figure 23.18.

23.6.3 Social Networks

In order to get a grip on the problem of visualizing large graphs with vertex clusters, Gansner et al. [HGK10] came up with the idea of using the metaphor of a political map. In such maps, each country is colored such that no two neighboring countries use the same color. Gansner et al. take advantage of this well-known map style. Their tool *GMap* combines existing general-purpose graph drawing methods for visualizing the given graph (as a traditional node-link diagram) with new methods to create artificial maps whose countries correspond to the clusters in the graph. GMap also colors the countries, striving to make the color difference between adjacent countries large. Note that the GMap approach, while exploiting (the map-users exposure to) cartography, is about visualizing an *abstract* binary relation. Still, we found the idea of combining the drawing of graphs and maps so striking that we decided to discuss it in this chapter.

In their paper, Gansner et al. give specific solutions to two steps of the above approach, namely the steps of map making and of map coloring. The map-making step assumes that the given graph has been drawn and clustered; the authors suggest to use pairs of algorithms that have similar notions of distance, for example, multi-dimensional scaling [KW78] for drawing the graph and the k-means algorithm [Llo82] for clustering. The GMap implementation uses the GraphViz [GN00] spring embedder and modularity clustering [New06].

Making the map. Assuming a drawing of the given graph $G = (V, E)$, Gansner et al. first place vertex labels (with font size as some function of vertex weight). They use standard overlap removal techniques [GH10]. In order to subdivide the given rectangular map area such that each vertex v of G receives a cell that is large enough for its label ℓ_v, Gansner et al. use a Voronoi-based approach. Rather than directly computing the Voronoi diagram of the labels (which would give rise to rather artificial-looking regions), they select a set P_v of equidistant points on the boundary of ℓ_v and perturb them slightly; see the black dots in Figure 23.19. In order to avoid large regions with awkward shapes at the boundary of the given graph drawing, they insert random points in the "sea," that is, in the map region that is sufficiently far from the drawn graph; see the small circles in Figure 23.19. Then they compute the Voronoi diagram of the point set that they have constructed; see the gray tessellation in Figure 23.19. The region R_v that corresponds to a vertex v of G is the union of the Voronoi cells of the points in P_v. Let $V = \bigcup \mathcal{C}$ be the given clustering, that is, a partition of V. Then, for each cluster $C \in \mathcal{C}$, Gansner et al. simply define the

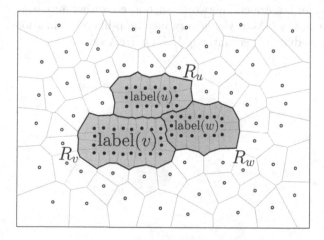

Figure 23.19 The Voronoi-based map-making step of GMap. Note that vertex v is more important than vertices u and w. Hence, v receives a label typeset in larger font size.

corresponding "country" to be $\bigcup_{v \in C} R_v$. This finishes the description of the map-making step.

For examples of maps that were generated with the method of Gansner et al., see Figures 23.20 and 23.21. The two graphs represent co-authorship of articles published at the International Symposium on Graph Drawing during the years 1994–2004 and 1994–2007. It is interesting to compare the traditional node-link diagram in Figure 23.20(a) with the corresponding map in Figure 23.20(b), which, technically, contains the same information—but in a much more accessible way. It is also interesting to observe the changes that occurred during the three additional years that were taken into account in Figure 23.21 as compared to Figure 23.20(b). Note that Figure 23.21 is a clipping of a slightly larger map that, apart from the "main land" has seven small islands (each with at most eight vertices)

Coloring the map. In the last step of their approach, Gansner et al. color the countries of the map that they have computed. While the famous Four-Color Theorem ensures that four colors always suffice for maps whose country adjacency graph $G_c = (V_c, E_c)$ is planar, this does not hold if countries have exclaves (such as the Kaliningrad district, which is not connected to Russia proper, or Steve North, who is part of the AT&T cluster in Figure 23.20 but lies in a region disconnected from the main body of the cluster).

Gansner et al. model the coloring problem as follows. In order to handle exclaves properly, they insist that *each* of the $k := |\mathcal{C}| = |V_c|$ countries actually receives a different color. They assume that a set of colors in a linear color space has been predetermined so that the difference between the colors is roughly equidistant. Hence, they simplify the problem by asking for a (bijective) assignment $c \colon V_c \to \{1, \ldots, k\}$ of the k vertices of G_c to the numbers $1, \ldots, k$ such that

$$\sum_{uv \in E_c} (c(u) - c(v))^2 \tag{23.7}$$

is maximized over all such assignments. The problem is NP-hard [HKV11]. Therefore, they solve the continuous version of the problem, where $c' \colon V_c \to \mathbb{R}$ must fulfill the additional requirement that $\sum_{v \in V} (c'(u))^2 = 1$. This problem is solved when c' is the eigenvector corresponding to the largest eigenvalue of the Laplacian of G_c. As a heuristic for the discrete version of the problem, they let $c(u)$ be the rank of $c'(u)$ in the sorted sequence of the c'-values. They suggest to apply, in a post-processing step, a 2-opt type greedy algorithm that swaps the c-values of pairs of vertices whenever this increases the term (23.7). The combination of the two methods seems to yield good results in practice.

Concluding, even for large social networks such as co-authorship graphs or Amazon book co-purchase networks, the GMap yields very nice map-like visualizations. Recently, GMap has been extended to dynamic scenarios [MKH11, HKV12].

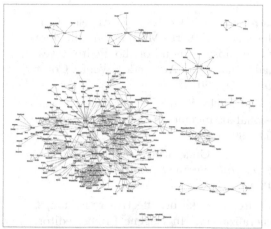 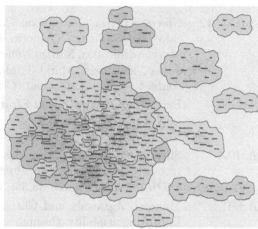

(a) traditional node-link diagram, label colors indicate cluster membership

(b) same graph drawing plus map background with countries corresponding to clusters

Figure 23.20 A portion of the co-authorship graph of articles published in the proceedings of the International Symposium on Graph Drawing in the years 1994–2004.

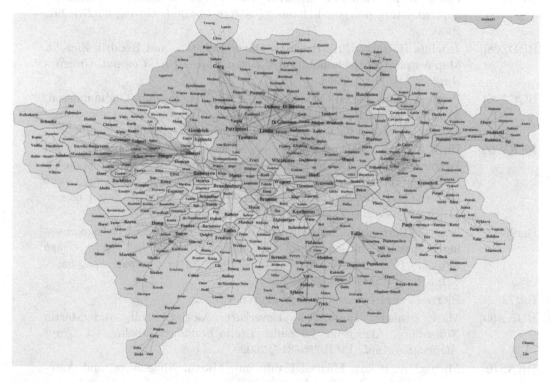

Figure 23.21 A portion of the co-authorship graph of articles published in the proceedings of the International Symposium on Graph Drawing in the years 1994–2007. The original map was clipped to increase the font size.

References

[AdCC⁺12] Nieves Atienza, Natalia de Castro, Carmen Cortés, M. Ángeles Garrido, Clara I. Grima, Gregorio Hernández, Alberto Márquez, Auxiliadora Moreno-González, Martin Nöllenburg, José Ramon Portillo, Pedro Reyes, Jesús Valenzuela, Maria Trinidad Villar, and Alexander Wolff. Cover contact graphs. *J. Comput. Geom.*, 3(1), 2012.

[AES99]   Pankaj K. Agarwal, Alon Efrat, and Micha Sharir. Vertical decomposition of shallow levels in 3-dimensional arrangements and its applications. *SIAM J. Comput.*, 29(3):912–953, 1999.

[AM00]   Silvania Avelar and Matthias Müller. Generating topologically correct schematic maps. In *Proc. 9th Int. Symp. Spatial Data Handling (SDH'00)*, pages 4a.28–4a.35, 2000.

[AS01]   Maneesh Agrawala and Chris Stolte. Rendering effective route maps: Improving usability through generalization. In Eugene Fiume, editor, *Proc. 28th Annu. Conf. Comput. Graphics Interactive Techniques (SIGGRAPH'01)*, pages 241–249. ACM Press, 2001.

[Ass96]   Association for Geographic Information, London. GIS dictionary. http://www.agi.org.uk/resources/dicitionary/content.htm, 1996.

[AV00]   Pankaj K. Agarwal and Kasturi R. Varadarajan. Efficient algorithms for approximating polygonal chains. *Discrete Comput. Geom.*, 23:273–291, 2000.

[BBDZ08]   Joachim Böttger, Ulrik Brandes, Oliver Deussen, and Hendrik Ziezold. Map warping for the annotation of metro maps. *IEEE Comput. Graphics Appl.*, 28(5):56–65, 2008.

[BCC⁺06]   Prosenjit Bose, Sergio Cabello, Otfried Cheong, Joachim Gudmundsson, Marc van Kreveld, and Bettina Speckmann. Area-preserving approximations of polygonal paths. *J. Discrete Algorithms*, 4(4):554–566, 2006.

[BEKW02]   Ulrik Brandes, Markus Eiglsperger, Michael Kaufmann, and Dorothea Wagner. Sketch-driven orthogonal graph drawing. In Stephen G. Kobourov and Michael T. Goodrich, editors, *Proc. 10th Int. Symp. Graph Drawing (GD'02)*, volume 2528 of *Lecture Notes Comput. Sci.*, pages 1–11. Springer-Verlag, 2002.

[Ber99]   François Bertault. A force-directed algorithm that preserves edge crossing properties. In Jan Kratochvíl, editor, *Proc. 7th Int. Symp. Graph Drawing (GD'99)*, volume 1731 of *Lecture Notes Comput. Sci.*, pages 351–358. Springer-Verlag, 1999.

[Béz72]   Pierre Bézier. *Numerical Control*. Wiley, 1972.

[BHKN09]   Marc Benkert, Herman J. Haverkort, Moritz Kroll, and Martin Nöllenburg. Algorithms for multi-criteria boundary labeling. *J. Graph Algorithms Appl.*, 13(3):289–317, 2009.

[BKNS10]   Michael A. Bekos, Michael Kaufmann, Martin Nöllenburg, and Antonios Symvonis. Boundary labeling with octilinear leaders. *Algorithmica*, 57(3):436–461, 2010.

[BKPS06]   Michael A. Bekos, Michael Kaufmann, Katerina Potika, and Antonios Symvonis. Multi-stack boundary labeling problems. In S. Arun-Kumar and Naveen Garg, editors, *Proc. Int. Conf. Foundat. Software Tech.*

Theor. Comput. Sci. (FSTTCS'06), volume 4337 of *Lecture Notes Comput. Sci.*, pages 81–92. Springer-Verlag, 2006.

[BKPS10]   Michael A. Bekos, Michael Kaufmann, Katerina Potika, and Antonios Symvonis. Area-feature boundary labeling. *Comput. J.*, 53(6):827–841, 2010.

[BKPS11]   Michael A. Bekos, Michael Kaufmann, Dimitrios Papadopoulos, and Antonios Symvonis. Combining traditional map labeling with boundary labeling. In Ivana Cerná, Tibor Gyimóthy, Juraj Hromkovic, Keith G. Jeffery, Rastislav Královic, Marko Vukolic, and Stefan Wolf, editors, *Proc. 37th Conf. Current Trends Theory Practice Comput. Sci. (SOF-SEM'11)*, volume 6543 of *Lecture Notes Comput. Sci.*, pages 111–122. Springer-Verlag, 2011.

[BKSW07]   Michael A. Bekos, Michael Kaufmann, Antonios Symvonis, and Alexander Wolff. Boundary labeling: Models and efficient algorithms for rectangular maps. *Comput. Geom. Theory Appl.*, 36(3):215–236, 2007.

[BLR00]   Thomas Barkowsky, Longin Jan Latecki, and Kai-Florian Richter. Schematizing maps: Simplification of geographic shape by discrete curve evolution. In C. Freksa, W. Brauer, C. Habel, and K. F. Wender, editors, *Proc. Spatial Cognition II—Integrating abstract theories, empirical studies, formal models, and practical applications*, volume 1849 of *Lecture Notes in Artificial Intelligence*, pages 41–53, 2000.

[BNUW07]   Marc Benkert, Martin Nöllenburg, Takeaki Uno, and Alexander Wolff. Minimizing intra-edge crossings in wiring diagrams and public transport maps. In Michael Kaufmann and Dorothea Wagner, editors, *Proc. 14th Int. Symp. Graph Drawing (GD'06)*, volume 4372 of *Lecture Notes Comput. Sci.*, pages 270–281. Springer-Verlag, 2007.

[BP13]   Ulrik Brandes and Barbara Pampel. Orthogonal-ordering constraints are tough. *J. Graph Algorithms Appl.*, 17(1):1–10, 2013.

[Bra99]   Ulrik Brandes. *Layout of Graph Visualizations*. PhD thesis, University of Konstanz, 1999. See http://nbn-resolving.de/urn:nbn:de:bsz:352-opus-2552.

[BST00]   Ulrik Brandes, Galina Shubina, and Roberto Tamassia. Improving angular resolution in visualizations of geographic networks. In *Proc. Joint Eurographics–IEEE TCVG Symp. Visual. (VisSym'00)*, pages 23–32, 2000.

[BSTW01]   Ulrik Brandes, Galina Shubina, Roberto Tamassia, and Dorothea Wagner. Fast layout methods for timetable graphs. In Joe Marks, editor, *Proc. 8th Int. Symp. Graph Drawing (GD'00)*, volume 1984 of *Lecture Notes Comput. Sci.*, pages 127–138. Springer-Verlag, 2001.

[BSV11]   Kevin Buchin, Bettina Speckmann, and Kevin Verbeek. Angle-restricted Steiner arborescences for flow map layout. In Takao Asano, Shin-Ichi Nakano, Yoshio Okamoto, and Osamu Watanabe, editors, *Proc. 22nd Int. Symp. Algorithms Comput. (ISAAC'11)*, volume 7074 of *Lecture Notes Comput. Sci.*, pages 250–259. Springer-Verlag, 2011.

[BT04]   Hans L. Bodlaender and Gerard Tel. A note on rectilinearity and angular resolution. *J. Graph Algorithms Appl.*, 8(1):89–94, 2004.

[BW00]   Ulrik Brandes and Dorothea Wagner. Using graph layout to visualize train connection data. *J. Graph Algorithms Appl.*, 4(3):135–155, 2000.

[CdBvD+01]   Sergio Cabello, Mark de Berg, Steven van Dijk, Marc van Kreveld, and Tycho Strijk. Schematization of road networks. In *Proc. 17th Annu. ACM Symp. Comput. Geom. (SoCG'01)*, pages 33–39, 2001.

[CEH96]   Kenneth C. Cox, Stephen G. Eick, and Taosong He. 3d geographic network displays. *ACM SIGMOD Record*, 25(4):50–54, 1996.

[CZQ+08]   W. Cui, H. Zhou, H. Qu, P. Wong, and X. Li. Geometry-based edge clustering for graph visualization. *IEEE Trans. Visual. Comput. Graph. (InfoVis'08)*, 14(6):1277–1284, 2008.

[DGNP10]   Daniel Delling, Andreas Gemsa, Martin Nöllenburg, and Thomas Pajor. Path schematization for route sketches. In Haim Kaplan, editor, *Proc. 12th Scand. Workshop Algorithm Theory (SWAT'10)*, volume 6139 of *Lecture Notes Comput. Sci.*, pages 285–296. Springer-Verlag, 2010.

[DHM08]   Tim Dwyer, Nathan Hurst, and Damian Merrick. A fast and simple heuristic for metro map path simplification. In George Bebis et al., editor, *Proc. 4th Int. Symp. Advances Visual Comput. (ISVC'08)*, volume 5359 of *Lecture Notes Comput. Sci.*, pages 22–30. Springer-Verlag, 2008.

[dNE03]   Hugo A. D. do Nascimento and Peter Eades. User hints for map labelling. In *Proc. 26th Australasian Comput. Sci. Conf.*, ACM Int. Conf. Proc. Series, pages 339–347, 2003.

[DP73]   David H. Douglas and Thomas K. Peucker. Algorithms for the reduction of the number of points required to represent a digitized line or its caricature. *Can. Cartogr.*, 10(2):112–122, 1973.

[ELMS91]   Peter Eades, Wei Lai, Kazuo Misue, and Kozo Sugiyama. Preserving the mental map of a diagram. In *Proc. Compugraphics*, pages 34–43, Sesimbra, Portugal, September 1991.

[FD01]   Michael Friendly and Daniel J. Denis. Milestones in the history of thematic cartography, statistical graphics, and data visualization. Web document, http://www.datavis.ca/milestones/. Entry on the first flow map: www.datavis.ca/milestones/index.php?group=1800%2B&mid=ms113, 2001. Accessed: Sept. 2012.

[FHN+13]   Martin Fink, Herman Haverkort, Martin Nöllenburg, Maxwell Roberts, Julian Schuhmann, and Alexander Wolff. Drawing metro maps using Bézier curves. In Walter Didimo and Maurizio Patrignani, editors, *Proc. 20th Int. Symp. Graph Drawing (GD'12)*, volume 7704 of *Lecture Notes Comput. Sci.*, pages 463–474. Springer-Verlag, 2013.

[FR91]   Thomas M. J. Fruchterman and Edward M. Reingold. Graph drawing by force-directed placement. *Software Pract. Exper.*, 21(11):1129–1164, 1991.

[Gar94]   Ken Garland. *Mr Beck's Underground Map*. Capital Transport Publishing, Harrow Weald, Middlesex, 1994.

[GH10]   Emden R. Gansner and Yifan Hu. Efficient, proximity-preserving node overlap removal. *J. Graph Algorithms Appl.*, 14(1):53–74, 2010.

[GHN11]   Andreas Gemsa, Jan-Henrik Haunert, and Martin Nöllenburg. Boundary-labeling algorithms for panorama images. In *Proc. 19th ACM SIGSPATIAL Int. Conf. Advances Geogr. Inform. Syst. (ACM-GIS'11)*, pages 289–298, 2011.

[GMPP91] Peter Gritzmann, Bojan Mohar, János Pach, and Richard Pollack. Embedding a planar triangulation with vertices at specified positions. *Amer. Math. Mon.*, 98:165–166, 1991.

[GN00] Emden R. Gansner and Stephen C. North. An open graph visualization system and its applications to software engineering. *Software Pract. Exper.*, 30:1203–1233, 2000.

[GNPR11] Andreas Gemsa, Martin Nöllenburg, Thomas Pajor, and Ignaz Rutter. On d-regular schematization of embedded paths. In Ivana Cerná, Tibor Gyimóthy, Juraj Hromkovic, Keith Jefferey, Rastislav Královic, Marko Vukolic, and Stefan Wolf, editors, *Proc. 37th Int. Conf. Current Trends Theory Practice Comput. Sci. (SOFSEM'11)*, volume 6543 of *Lecture Notes Comput. Sci.*, pages 260–271. Springer-Verlag, 2011.

[Har38] Henry Drury Harness. Atlas to accompany the second report of the railway commissioners, Ireland. Her Majesty's Stationery Office, 1838.

[HGK10] Yifan Hu, Emden R. Gansner, and Stephen Kobourov. Visualizing graphs and clusters as maps. *IEEE Comput. Graphics Appl.*, 30:54–66, 2010.

[HKV11] Yifan Hu, Stephen G. Kobourov, and Sankar Veeramoni. On maximum differential graph coloring. In Ulrik Brandes and Sabine Cornelsen, editors, *Proc. 18th Int. Symp. Graph Drawing (GD'10)*, volume 6502 of *Lecture Notes Comput. Sci.*, pages 274–286. Springer-Verlag, 2011.

[HKV12] Yifan Hu, Stephen G. Kobourov, and Sankar Veeramoni. Embedding, clustering and coloring for dynamic maps. In *Proc. 5th IEEE Pacific Visual. Symp. (PacificVis'12)*, pages 33–40, 2012.

[HMdN06] Seok-Hee Hong, Damian Merrick, and Hugo A. D. do Nascimento. Automatic visualisation of metro maps. *J. Visual Lang. Comput.*, 17(3):203–224, 2006.

[Hol95] Christine Holch. Der vielfältige Falk. Die Zeit (German newspaper), 15 December 1995.

[HS52] Magnus R. Hestenes and Eduard Stiefel. Methods of conjugate gradients for solving linear systems. *J. Research Nat. Bur. Standards*, 49(6):409–436, 1952.

[HS92] John Hershberger and Subhash Suri. Applications of a semi-dynamic convex hull algorithm. *BIT*, 32(2):249–267, 1992.

[HS94] John Hershberger and Jack Snoeyink. An $O(n \log n)$ implementation of the Douglas-Peucker algorithm for line simplification. In *Proc. 10th Annu. ACM Symp. Comput. Geom. (SoCG'94)*, pages 383–384, 1994.

[HS11] Jan-Henrik Haunert and Leon Sering. Drawing road networks with focus regions. *IEEE Trans. Vis. Comput. Graphics*, 17(12):2555–2562, 2011.

[HVW09] Danny Holten and Jarke J. Van Wijk. Force-directed edge bundling for graph visualization. *Comput. Graphics Forum*, 28(3):983–990, 2009.

[Jen06] Bernhard Jenny. Geometric distortion of schematic network maps. *Bulletin Soc. Cartogr.*, 40:15–18, 2006.

[Koe36] Paul Koebe. Kontaktprobleme der konformen Abbildung. *Ber. Sächs. Akad. Wiss. Leipzig, Math.-Phys. Klasse*, 88:141–164, 1936.

[KW78] Joseph B. Kruskal and Myron Wish. *Multidimensional Scaling*. Number 07-011 in Sage University Paper series on Quantitative Application in the Social Sciences. Beverly Hills and London: Sage Publications, 1978.

[Law76]　　　Eugene L. Lawler. *Combinatorial Optimization: Networks and Matroids.* Holt, Rinehart & Winston, New York, 1976.

[Lin10]　　　Chun-Cheng Lin. Crossing-free many-to-one boundary labeling with hyperleaders. In *Proc. IEEE Pacific Visual. Symp. (PacificVis'10)*, pages 185–192, 2010.

[LKY08]　　Chun-Cheng Lin, Hao-Jen Kao, and Hsu-Chun Yen. Many-to-one boundary labeling. *J. Graph Algorithms Appl.*, 12(3):319–356, 2008.

[LL99]　　　Longin Jan Latecki and Rolf Lakämper. Convexity rule for shape decomposition based on discrete contour evolution. *Comput. Vision Image Underst.*, 73(3):441–454, 1999.

[Llo82]　　　Stuart P. Lloyd. Least squares quantization in PCM. *IEEE Trans. Inform. Theory*, 28(2):129–137, 1982.

[LPT$^+$11]　　Chun-Cheng Lin, Sheung-Hung Poon, Shigeo Takahashi, Hsiang-Yun Wu, and Hsu-Chun Yen. One-and-a-half-side boundary labeling. In Weifan Wang, Xuding Zhu, and Ding-Zhu Du, editors, *Proc. 5th Int. Conf. Combin. Optim. Appl. (COCOA'11)*, volume 6831 of *Lecture Notes Comput. Sci.*, pages 387–398. Springer-Verlag, 2011.

[LS02]　　　Ulrich Lauther and Andreas Stübinger. Generating schematic cable plans using springembedder methods. In Petra Mutzel, Michael Jünger, and Sebastian Leipert, editors, *Proc. 9th Int. Symp. Graph Drawing (GD'01)*, volume 2265 of *Lecture Notes Comput. Sci.*, pages 465–466. Springer-Verlag, 2002.

[LWY09]　　Chun-Cheng Lin, Hsiang-Yun Wu, and Hsu-Chun Yen. Boundary labeling in text annotation. In *Proc. 13th Int. IEEE Conf. Inform. Vis. (IV'09)*, pages 110–115, 2009.

[MG07]　　　Damian Merrick and Joachim Gudmundsson. Path simplification for metro map layout. In Michael Kaufmann and Dorothea Wagner, editors, *Proc. 14th Int. Symp. Graph Drawing (GD'06)*, volume 4372 of *Lecture Notes Comput. Sci.*, pages 258–269. Springer-Verlag, 2007.

[MHCF96]　Tamara Munzner, Eric Hoffman, K. Claffy, and Bill Fenner. Visualizing the global topology of the MBone. In *Proc. IEEE Symp. Inform. Visual. (InfoVis'96)*, pages 85–92, 1996.

[Min69]　　　Charles Joseph Minard. Carte Figurative des pertes successives en hommes de l'Armée Française dans la campagne de Russie 1812–1813. http://en.wikipedia.org/wiki/File:Minard.png, 1869.

[MKH11]　　Daisuke Mashima, Stephen G. Kobourov, and Yifan Hu. Visualizing dynamic data with maps. *IEEE Trans. Visual. Comput. Graphics*, 18(9):1424–1437, 2011.

[MNWB08]　Damian Merrick, Martin Nöllenburg, Alexander Wolff, and Marc Benkert. Morphing polylines: A step towards continuous generalization. *Comput. Environ. Urban Syst.*, 32(4):248–260, 2008.

[MO01]　　　Joseph S. B. Mitchell and Joseph O'Rourke. Computational geometry column 42. *SIGACT News*, 32(3):63–72, 2001.

[Nes04]　　　Keith V. Nesbitt. Getting to more abstract places using the metro map metaphor. In *Proc. 8th Int. Conf. Inform. Visual. (IV'04)*, pages 488–493. IEEE Computer Society, 2004.

[New06]   Mark E. J. Newman. Modularity and community structure in networks. *Proc. Natl. Acad. Sci. USA*, 103:8577–8582, 2006.

[Ney99]   Gabriele Neyer. Line simplification with restricted orientations. In Frank K. Dehne, Arvind Gupta, Jörg-Rüdiger Sack, and Roberto Tamassia, editors, *Proc. 6th Int. Workshop Algorithms Data Struct. (WADS'99)*, volume 1663 of *Lecture Notes Comput. Sci.*, pages 13–24. Springer-Verlag, 1999.

[Nöl05a]   Martin Nöllenburg. Automated drawing of metro maps. Master's thesis, Fakultät für Informatik, Universität Karlsruhe, 2005. Available at `http://www.ubka.uni-karlsruhe.de/indexer-vvv/ira/2005/25`.

[Nöl05b]   Martin Nöllenburg. Automated drawings of metro maps. Technical Report 2005-25, Fakultät für Informatik, Universität Karlsruhe, 2005. Available at `http://www.ubka.uni-karlsruhe.de/indexer-vvv/ira/2005/25`.

[NPS10]   Martin Nöllenburg, Valentin Polishchuk, and Mikko Sysikaski. Dynamic one-sided boundary labeling. In *Proc. 18th Int. ACM Symp. Advances Geogr. Inform. Syst. (ACM-GIS'10)*, pages 310–319, 2010.

[NW11]   Martin Nöllenburg and Alexander Wolff. Drawing and labeling high-quality metro maps by mixed-integer programming. *IEEE Trans. Visual. Comput. Graphics*, 17(5):626–641, 2011.

[O'R03]   O'Reilly. Open source route map. `http://www.oreilly.de/artikel/routemap.pdf`, 2003.

[Ove03]   Mark Ovenden. *Metro maps of the world*. Harrow Weald: Capital Transport Publishing, 2nd edition, 2003.

[Pol06]   Klemens Polatschek. Die Schönheit des Untergrundes. Frankfurter Allgemeine Sonntagszeitung 28, 16 July 2006. Available via `http://fazarchiv.faz.net`.

[PXY⁺05]   Doantam Phan, Ling Xiao, Ron B. Yeh, Pat Hanrahan, and Terry Winograd. Flow map layout. In *Proc. IEEE Symp. Inform. Visual. (InfoVis'05)*, pages 219–224, 2005.

[RNL⁺11]   Maxwell J. Roberts, Elizabeth J. Newton, Fabio D. Lagattolla, Simon Hughes, and Megan C. Hasler. Objective versus subjective measures of metro map usability: Investigating the benefits of breaking design rules. Available at `http://privatewww.essex.ac.uk/~mjr/underground/Roberts_Metro.pdf`, August 2011.

[Rob55]   Arthur H. Robinson. The 1837 maps of henry drury harness. *Geogr. J.*, 121(4):440–450, 1955.

[Saa95]   Alan Saalfeld. Map generalization as a graph drawing problem. In Roberto Tamassia and Ioannis Tollis, editors, *Proc. 3rd Int. Symp. Graph Drawing (GD'94)*, volume 894 of *Lecture Notes Comput. Sci.*, pages 444–451. Springer-Verlag, 1995.

[Saa99]   Alan Saalfeld. Topologically consistent line simplification with the Douglas-Peucker algorithm. *Cartogr. Geogr. Inform. Sci.*, 26(1), 1999.

[SGSK01]   Elmer S. Sandvad, Kaj Grønbæk, Lennert Sloth, and Jørgen Lindskov Knudsen. A metro map metaphor for guided tours on the Web: the Webvise Guided Tour System. In *Proc. 10th Int. World Wide Web Conf. (WWW'01)*, pages 326–333. ACM Press, 2001.

[SR05]   Jonathan M. Stott and Peter Rodgers. Automatic metro map design techniques. In *Proc. 22nd Int. Cartogr. Conf. (ICC'05)*, La Coruña, Spain, 2005.

[SRMW11]   Jonathan Stott, Peter Rodgers, Juan Carlos Martínez-Ovando, and Stephen G. Walker. Automatic metro map layout using multicriteria optimization. *IEEE Trans. Visual. Comput. Graphics*, 17(1):101–114, 2011.

[Syd08]   http://www.cityrail.nsw.gov.au/networkmaps/network_map.pdf, 2008.

[Tam87]   Roberto Tamassia. On embedding a graph in the grid with the minimum number of bends. *SIAM J. Comput.*, 16(3):421–444, 1987.

[Tob87]   Waldo Tobler. Experiments in migration mapping by computer. *Amer. Cartogr.*, 14(2):155–163, 1987.

[Tuf01]   Edward R. Tufte. *The Visual Display of Quantitative Information*. Graphics Press, Cheshire, CT, 2nd edition, 2001.

[Tut63]   William T. Tutte. How to draw a graph. *Proc. London Math. Soc.*, 13(52):743–768, 1963.

[VBS11]   Kevin Verbeek, Kevin Buchin, and Bettina Speckmann. Flow map layout via spiral trees. *IEEE Trans. Visual. Comput. Graphics*, 17(12):2536–2544, 2011.

[VW93]   Mahes Visvalingam and J. D. Whyatt. Line generalisation by repeated elimination of points. *Cartogr. J.*, 30(1):46–51, 1993.

[WC11]   Yu-Shuen Wang and Ming-Te Chi. Focus+context metro maps. *IEEE Trans. Visual. Comput. Graphics*, 17(12):2528–2535, 2011.

[Wik12]   Wikipedia. Harry Beck, 2012. Accessed March 5, 2012.

[Wol07]   Alexander Wolff. Drawing subway maps: A survey. *Informatik – Forschung & Entwicklung*, 22(1):23–44, 2007.

[YOT09]   Daisuke Yamamoto, Shotaro Ozeki, and Naohisa Takahashi. Focus+Glue+Context: an improved fisheye approach for Web map services. In *Proc. 17th Annu. ACM Symp. Advances Geogr. Inferm. Syst. (ACM-GIS'09)*, pages 101–110, 2009.

24

Graph Drawing in Education

Stina Bridgeman
Hobart and William Smith Colleges

24.1 Introduction

Illustrations are a powerful explanatory tool, so one might expect a long history of the use of graph drawing in education. This history can be traced back to at least the Middle Ages, where squares of opposition (Figure 24.1) were used as pedagogical tools in logic and other fields [KMBW02]. Murdoch [Mur84] provides examples of both basic squares and more complex structures.

In mathematics, drawings of abstract graphs began to appear as illustrations in the late 18th century, 150 years after Euler's famous paper on the Königsberg bridges launched the field of graph theory [KMBW02]. Now commonplace, hand-drawn pictures of small graphs are often used as illustrations in math and computer science textbooks to describe a graph-related concept or to explain a graph algorithm—any graph theory, discrete math, or data structures text will contain many such pictures. Drawings of graphs are also used to illustrate graph-structured information, such as the topology of a computer network or a flow chart showing a program's execution.

The introduction of computers into the classroom has led to new applications of graph drawing, including algorithm animation, algorithm simulation, exercise systems, exploration systems, program visualization, and software visualization. Section 24.2 surveys these applications, with emphasis on tools specifically developed for or used in the classroom. Many

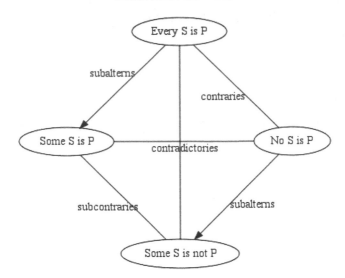

Figure 24.1   An Aristotelian square of opposition showing the relationships between the four logical forms (drawn using the `circo` algorithm from the Graphviz package [BCE⁺]).

of these applications place special requirements on the graph drawing algorithms used. Sections 24.3–24.6 address relevant graph drawing techniques.

24.2   Applications

24.2.1   Algorithm Animation

Algorithm animation deals with graphically illustrating the conceptual behavior of an algorithm or data structure.

Algorithm animation has been used in educational settings for many years. An early and well-known example is Baecker's 1981 video "Sorting Out Sorting" [Bae81], which animates and explains nine sorting techniques. The video illustrates how each of the sorting algorithms works by showing how bars of varying heights are gradually rearranged into increasing order, then makes an effective point about running time by showing a "race" between all of the algorithms.

In the classroom, instructor-prepared animations can be used as demonstrations during class to help explain a new concept—an animated version of the explanatory illustration. Animations used in class can be made available for students to pause, step, and replay so they can absorb the material at their own pace. Algorithm animation can also be used to engage students in the learning process—creating their own animations can deepen students' understanding of concepts, and incorporating the creation of animations into assignments can add interest to what might otherwise be a dry algorithm implementation task.

Animations of graph algorithms naturally make use of a drawing of a graph, using annotations, changing colors, or other visual effects to show the progression of the algorithm. Animations of data structure manipulations, such as inserting or removing elements from a binary tree, may also utilize a drawing of a graph or a tree. Support for automatic graph drawing frees the animation designer from having to specify the details of how the graph is drawn in each step, allowing her to focus on expressing the concept being illustrated.

Example Systems

Balsa BALSA [BS84, BS85] is one of the classic algorithm animation systems. It is a general-purpose system, designed for animating any kind of algorithm. BALSA introduced the idea of "interesting events," key points in the program where the visualization must be updated. Animations are created by implementing one or more graphical views and then augmenting the program code with calls to update those views when interesting events occur. Views are often created from scratch, though it is possible to create a reusable library of standard views. Graph layout algorithms are not provided, but can be implemented as part of a view. BALSA has been used to illustrate concepts in both mathematics and computer science courses, and for research in algorithm design and analysis.

Tango Tango [Sta90a, Sta90b] is another classic general-purpose algorithm animation system. Tango also utilizes the idea of interesting events, but provides a framework to aid in defining views. Four kinds of elements are provided as building blocks for animation scenes: basic graphical objects (shapes and text), locations of objects, transitions (movement, size, and color changes, etc.), and paths specifying how the transitions occur. Creating an animation involves three steps: defining a series of "animation scenes" (which may be a static view or an animated step), annotating the program with interesting events, and specifying the mapping of interesting events to animation scenes. Of note is Tango's support for (and emphasis on) smooth transitions between view states—many algorithm animation systems simply present a series of snapshots.

Samba Samba [Sta97] was designed to make it as easy as possible for students to create their own algorithm animations. Samba is a front-end for Polka [SK93], the successor to Tango; it reads in a command script and generates the animation from that script. Samba commands are deliberately kept simple; basic commands allow the creation of graphical objects (such as circles and lines) and the modification of existing objects (such as by moving them or changing their color). Animations are created by augmenting the program to be animated with instructions to output the Samba script. An advantage of Samba is that it does not require the animator to implement separate graphical views and link them to the code.

JAWAA JAWAA [PR98] is a web-based system for animating data structures. Animations are specified by writing a script in JAWAA's command language—unlike many algorithm animation systems, the algorithm being animated does not need to be implemented. Graphs and trees can be drawn using user-specified node positions, or can be drawn automatically using one of three built-in layout algorithms (circular layout, Tunkelang's force-directed layout [Tun94], and tree layout).

Swan Swan [SHY96] was designed specifically for visualizing graph algorithms and their related data structures. Animations are created by augmenting a C/C++ program with commands to build a graph representing the data structure to be visualized, specify visual parameters such as the shape and color of the node, and draw the graph. This means that the visualizations created are not tied to the physical representation of data structures in the program and can instead represent a conceptual view. Swan includes special "layout components" which perform automatic layout for specific types of data structures such as linked lists, arrays, trees, and general graphs. Layout components for general graphs include circular layout, Kamada and Kawai's force-directed layout [KK89], and a Sugiyama-style hierarchical layout.

24.2.2 Algorithm Simulation

Animations can only be viewed; in an algorithm simulation, a student experimenting with data structures or an instructor providing on-the-fly demonstrations in class can modify the data structure being animated or even carry out the algorithm's steps by hand.

Example Systems

Matrix Matrix [KM02] provides both algorithm animation and algorithm simulation. It can also be used to create visualizations of students' own implementations of data structures, and to perform visual testing. Multiple levels of abstraction are supported; for example, when carrying out an algorithm involving inserting an element into a balanced binary search tree, the student can perform the entire operation manually, allow the system to insert the element into the underlying binary tree but then perform rebalancing steps herself, or allow the system to do the entire insertion. Matrix provides supports several types of data structures and includes automatic tree layout. MatrixPro [KKMS04a, KKMS04b] utilizes the Matrix platform and provides a GUI tailored for instructor use in the classroom.

24.2.3 Exercise Systems

Exercise systems present students with exercises to solve and provide feedback on the students' answers. Such systems can be used for learning and practice—students can test their knowledge of an algorithm by trying exercises, and gain further understanding as the system provides feedback about their mistakes—or for assessment and grading. Key features of exercise systems include automatic generation of problem instances and automatic feedback and assessment, allowing students to practice on as many or as few problems as they wish without burdening a faculty member or teaching assistant with excessive problem-set creation or grading duties. The problem-generation feature can also be used to create individualized problem sets to help thwart cheating.

Intelligent tutoring systems also build up a model of the student's knowledge and understanding, and tailor the problems generated to address each student's individual weaknesses.

Incorporating automatic graph drawing into an exercise system is important if the system is to support graph- or tree-based problems, because each problem is randomly generated as needed.

Example Systems

PILOT PILOT [BGKT00] is a Web-based exercise system supporting trace-the-algorithm exercises. It supports automatic generation of exercise instances, feedback at each step of the tracing process, solution grading, and algorithm animation. PILOT's feedback and assessment mechanism is based on whether each step is consistent with correct execution of the algorithm at that point rather than simply checking if some final answer matches the correct solution. This allows PILOT to easily accommodate cases with multiple correct solutions and to provide meaningful feedback and reasonable partial credit when a single mistake is made early in the process. Several graph-based problems including minimum spanning tree, breadth-first and depth-first search, and shortest path algorithms have been implemented. Force-directed and hierarchical drawing algorithms provided by the Graph Drawing Server [BGT99] are used for graph layout.

TRAKLA2 TRAKLA2 [MKK+04] is a Web-based exercise system built on the Matrix [KM02] algorithm animation and simulation framework. Like PILOT, TRAKLA2 supports trace-the-algorithm exercises and includes automatic generation of exercise instances, solution grading, and animation of model solutions. Evaluation of a student's answer is limited to comparing the student's solution to a model solution, and the student only receives notice of how many steps were correct. However, TRAKLA2 contains a number of features making it useful for coursework including storage of students' grades and submitted answers, deadlines for exercises, and the ability to control whether the same instance of an exercise may be repeatedly submitted for feedback (a practice exercise) or if it must be reset with new input data each time (a graded assignment). Exercises involving a variety of data structures and algorithms, including graph algorithms, have been implemented.

AnimalSense AnimalSense [RMS11] takes a different approach. Instead of providing an environment where students manually trace the execution of an algorithm, AnimalSense supports questions that provide evidence of successful algorithm-tracing such as "Give the sorted order" or "Give your third chosen edge." This approach allows greater latitude in the types of exercises that can be supported — it can also accommodate questions like "Provide an array which uses 4 pivots to be sorted," which go beyond simply tracing and which require deeper thinking about the functioning of an algorithm. Algorithm animation is provided to aid in solving the problem and to help reveal the cause of a mistake. Exercises involving graph algorithms, searching algorithms, and sorting algorithms have been implemented.

24.2.4   Exploration Systems

Exploration systems support experimentation with graph structures and graph theory concepts. In the classroom, exploration systems can be used in a professor-led discussion to illustrate or animate examples or algorithms, or for student exploration or experimentation.

Support for automatic graph drawing frees the experimenter from having to find a reasonable layout, and can be important in revealing the structure of the graph being studied.

Example Systems

LINK LINK [BDG+00] is designed for education and research in discrete mathematics. It consists of a library of templated C++ classes for graphs and other data structures coupled with an interactive front-end for animation and visualization. The library also contains a collection of graph algorithms commonly used as building blocks and which are often covered in their own right in graph theory and computer science courses. To aid in visualization, LINK includes several simple graph layout algorithms (place vertices randomly, on a circle, or on a grid), a spring embedder, and several algorithms suited for particular applications (e.g., illustrating the results of a depth-first search, drawing the graph as a bipartite graph, and laying out each biconnected component of the graph separately to emphasize the components) [Ber].

GraphPack GraphPack [KOD+96] is a tool designed for experimenting with graphs and graph algorithms. It supports several 3D and 2D graph layout algorithms, contains a graph viewer, and can integrate functionality from other packages such as Mathematica, Maple, and Matlab. A novel feature is its ability to extract the graph structure from a black-and-white bitmap image of a drawing.

24.2.5   Program Visualization

Program visualization deals with visualizing a program's actual execution rather than a high-level conceptual view of an algorithm. Aspects of the program being visualized can include source code, data structures, and runtime behavior. Program visualization can be used to illustrate the functioning of an algorithm or data structure (as in algorithm animation), to gain an understanding of how the program works, to aid in debugging, and to evaluate and improve program performance.

In the classroom, program visualization can help students learn to program and debug by revealing what their programs are actually doing. This is more effective than systems which attempt to explain a bug (because explanations require understanding the underlying concept in the first place), try to guide the student to a particular way of solving the problem (ignoring other valid solutions), or are limited to a small set of toy problems [EPD92]. For more advanced students, visualizations can help explain the underlying semantics of the programming language, design patterns, and the workings of multithreaded programs [GJ05]. As with algorithm animation, instructors can also use program visualization to spice up an implementation assignment.

Automatic graph drawing is an essential component for program visualization systems which display graph-structured information because the particular graph depends on the runtime state of the program.

Example Systems

A simple form of program visualization—and one that is also suitable for algorithm animation—is to display graphical snapshots of the key data structures whenever the state of the structure changes.

GraphTree/GraphHeap Owen's GraphTree and GraphHeap subroutines [Owe86] were designed as a low-overhead animation system for illustrating binary tree and heap operations. The subroutines take the data structure to be visualized as a parameter, and are called when the animator wants to produce a graphical snapshot of the current state of the tree or heap. A simple layout algorithm is used: parent nodes are centered above their two children, with empty spaces for missing child nodes.

VisualGraph VisualGraph [LNR03] is a Java graph class which provides typical graph querying and manipulation operations, as well as visualization operations (highlighting and changing the color of edges and vertices) and related utility routines (random graph generation, graph layout using the force-directed method of Kamada [Kam89], and file I/O). Simple visualizations are created by augmenting the program code with calls to "print graph" whenever a picture of the current state of the graph is desired. VisualGraph is implemented as a front-end to an algorithm animation system—animation operations produce output in the ANIMALSCRIPT language [RF01], which can then be read and displayed by a system such as JHAVÉ [NEN00].

Visualiser Naps' Visualiser class [Nap98] supports multiple data structures, including trees and graphs. It parses a string representation of the data structure to be visualized rather than working directly with particular Java objects, so it can be extended to new implementations of data structures by providing a new "to string" routine. New data structures or visualization styles can be supported by adding new Visualiser subclasses.

JDSL Visualizer The JDSL Visualizer [BBG+99] does not require users to modify their code to generate visualizations of data structures, as snapshots are automatically generated before and after data structure operations. However, data structures must be implemented to a particular API. Several linear and binary-tree-based structures are supported.

LJV The Lightweight Java Visualizer (LJV) [Ham04] uses Java's reflection mechanism to determine the structure of a Java object and is thus suitable for use with any Java program. Visualizing an object requires only adding calls to a "display object" routine when an object is to be visualized. The resulting graph structure is drawn using GraphViz [BCE+]. More advanced users or instructors setting up the tool for a course can customize the appearance of particular classes, such as to hide the internal representation of the `String` class. Of note is that because the structure is derived directly from the object itself, both correct data structures and students' incorrect ones can be visualized. The tool is also effective for demonstrating aspects of the Java language which often cause confusion, such as the pervasive but hidden use of references and the meaning of static fields.

Other systems provide visualization of data structures without requiring the program to be modified.

UWPI The University of Washington illustrating compiler (UWPI) [HWF90] analyzes program source code (written in a subset of Pascal) and automatically constructs a visualization of the data structures used in the program. UWPI attempts to infer the abstract data type of each variable from its concrete data type and usage patterns in order to determine an appropriate visualization. Supported ADTs are numbers, arrays, and digraphs; graphs are converted to directed acyclic graphs and drawn using the methods of Sugiyama, Tagawa, and Toda [STT81] and Rowe et al. [RDM+87].

jGRASP jGRASP [HCIB04] is a Java development environment combining a debugger and visualization tools. Data structures to be visualized are extracted automatically from the program; "external viewers" specify how to render a visual representation of an object of that type. This architecture allows multiple views of a single data structure to be displayed simultaneously. New external viewers can be added, so the system can be used for creating animations as well for debugging. More recent versions of jGRASP include a "Data Structure Identifier" which automatically identifies the data structure being visualized and suggests appropriate viewers [CIHJB07]. jGRASP uses FLGL, a graph drawing library based on VCJ [MB98], to produce layouts of data structures. VCJ includes Walker's algorithm [Wal90] for drawing rooted trees, Kamada and Kawai's spring embedder [KK89] for undirected graphs, and clan-based graph drawing [MCS98] for directed graphs.

Program visualization can include visualization of more than just data structures.

Jeliot 3 Jeliot 3 [MMSBA04] is the fourth system in a series of program visualization systems designed for beginning programmers. Jeliot displays both object structures and control flow, providing a fine-grained animations of every step of the program's execution, including the evaluation of expressions. One drawback is that Jeliot does not support the full Java language.

JIVE JIVE [GJ05] is designed for the visualization of object-oriented programs (specifically, Java) and shows objects not just as data structures but also as execution environments. JIVE's views show an object's fields and its methods, structural links between objects, and the history of the method calls made as the program runs. This approach reveals much more of how Java actually works than data structure visualization approaches, and helps the viewer more thoroughly understand what is really going on when the program is run. Streib and Soma [SS10] discuss experiences using JIVE and the contour diagrams used by JIVE in introductory programming courses.

24.2.6 Software Visualization

The field of software visualization encompasses the visualization of all aspects of a software system, including its structure, execution, and evolution over time. Graph drawing plays an important role in software visualization as many aspects of a software system can be represented using graphs, including control-flow graphs, program call graphs, class diagrams, and dependency graphs.

While there has been a great deal of work in the field of software visualization, many of the software visualization tools designed for use in the classroom focus on algorithm animation or program visualization. BlueJ, described below, is one exception.

Example Systems

BlueJ BlueJ [KQPR03] is an integrated development environment (IDE) developed for the teaching of Java programming. BlueJ emphasizes class structure and design through UML class diagrams—a class diagram is displayed in the main window when a project is opened, and it is through the diagram that students can edit, compile, and create instances of classes.

24.3   Graph Drawing for Algorithm Animation

In algorithm animation (and in many program visualization applications), an abstract view of the data structure is both sufficient and desired. In many cases, standard graph drawing algorithms are suitable for this task. The most important criteria for drawings are following familiar conventions (such as placing the root of a tree at the top or directing edges downward) and readability, properties which are easily achieved by many standard algorithms. Examples of suitable algorithms include Walker's algorithm [Wal90] for rooted trees, Sugiyama-style layout [STT81, GKNV93] for directed graphs, and force-directed methods (e.g., Kamada-Kawai [KK89] and Tunkelang [Tun94]) for general graphs.

24.3.1   A Unified Approach to Drawing Data Structures

One drawback to using standard algorithms is that different algorithms must be chosen for linked lists, trees, directed graphs, and general graphs. For applications such as visual debuggers, which need to be able to visualize any data structure (including buggy or ill-formed ones) and where the type of data structure is not known in advance, a unified approach is needed.

Since data structure graphs are directed graphs and convention often places the root of the structure at the top, a hierarchical layout is a natural layout style for drawings of data structures. The classic Sugiyama algorithm [STT81] for producing a hierarchical layout of a directed graph consists of five phases:

- Cycle removal: If the graph to be drawn is not acyclic, one or more edges must be reversed in order to remove all directed cycles.
- Layer assignment: Nodes are assigned to layers, where all nodes on the same layer will have the same y-coordinate in the final drawing. Dummy nodes are inserted as needed so that edges only connect nodes on adjacent layers.
- Crossing reduction: The nodes in each layer are rearranged so as to reduce edge crossings between layers, typically through repeated passes in which the ordering of one layer is held fixed while the nodes in an adjacent layer are rearranged. One strategy for rearranging nodes is to sort them according to the average position of the adjacent nodes in the other layer (barycenter method).
- Coordinate assignment: The nodes in each layer are assigned x-coordinates, preserving the left-to-right ordering of each layer.

- Edge routing: Edges are commonly drawn as polylines, with bends introduced by the placement of dummy nodes. However, other routing strategies (such as splines [GKNV93] and edge bundling [PNK11]) have been introduced.

Constraints can then be added to respect specialized conventions for drawing particular kinds of data structures. Waddle [Wad01] identifies three types of constraints as the most important for data structures: "same-level" constraints defining nodes which must appear on the same level, left-to-right ordering constraints between nodes or paths, and edge-orientation constraints which preference edges for reversal during cycle removal. Adapting the Sugiyama algorithm to accommodate these constraints will be discussed below.

Same-Level Constraints

Same-level constraints may result in edges connecting nodes in the same level. Traditional layer assignment prevents same-level edges, and furthermore same-level edges cannot be handled by the traditional compute-barycenters-and-sort crossing reduction method. (Sorting requires a fixed barycenter for the duration of the sort, but the barycenter of a node with same-level neighbors will change as the neighbors are rearranged during the sorting process.)

Waddle's solution is a two-tier layer assignment and crossing reduction strategy. First, same-level constraints are used to define equivalence classes of nodes that must appear on the same level and layer assignment is performed using a single proxy node in place of each equivalence class. "Virtual layers" are then created within each layer and layer assignment is repeated for each equivalence class using the virtual layers. This results in nodes involved in same-level constraints being assigned to different virtual layers. During crossing reduction, a layer containing virtual layers is sorted by applying the usual crossing reduction procedure to the virtual layers.

Finally, all nodes within a layer (regardless of virtual layer) are assigned the same y-coordinate and same-level edges are routed around intervening nodes as needed. Böhrigner and Paulisch [BN90] add an additional constraint that same-level edges must connect consecutive nodes in order to avoid the need for edge routing.

Node Ordering Constraints

Node ordering constraints specify the left-to-right ordering of pairs of nodes in the same level. (After layer assignment, path ordering constraints can be converted to node ordering constraints involving pairs of nodes and dummy nodes along the extent of the paths.) Node ordering constraints are implemented in the crossing reduction phase.

A simple strategy for respecting node ordering constraints is to proceed with sorting nodes by their barycenters, but to disallow any swaps which would violate the ordering constraints.

Waddle [Wad01] uses a different strategy: the ordering constraints are checked after the barycenters have been computed and, if a constraint is violated, a new barycenter is assigned which places the node just to the right of the rightmost node which must precede it according to the constraints. The nodes are then sorted according to their revised barycenters.

Both of these approaches are fast and result in an ordering which satisfies the constraints, but may result in a large number of avoidable crossings.

A third strategy is the "penalty graph" approach [Fin01], which produces fewer crossings at the expense of a more complex algorithm and a higher running time. In this approach, the penalty graph contains the nodes of the layer to be reordered. A directed edge (u,v) indicates that placing u to the left of v results in fewer crossings than placing v to left of u. The weight of the edge (u,v) indicates by how much the number of crossings is improved.

An ordering constraint requiring u to be to the left of v can be imposed by assigning the edge (u,v) an infinite weight. The ordering of the layer is determined by applying a heuristic to find the minimum-weight set of arcs whose removal makes the penalty graph acyclic (the minimum weighted feedback arc set problem), and then performing a topological sort of the resulting acyclic penalty graph.

Forster [For04] gives a heuristic which combines the efficiency and simplicity of the barycenter approach with the quality of the penalty graph method. First, barycenters are computed for each node. Then, for each violated constraint, the nodes involved are replaced by a single proxy node and a new barycenter is computed for the proxy node based on the combined neighbors of the original nodes. Once all of the constraints have been accommodated, the nodes and proxy nodes are sorted by their barycenters. The final sorted layer is obtained by replacing each proxy node with the ordered collection of individual nodes that were grouped together.

Constraints must be considered in the correct order when creating proxy nodes or else it can become impossible to satisfy all of the constraints. The constraints to be satisfied can be represented by a constraint graph, which contains a directed edge (u,v) for each constraint of the form "u must be placed to the left of v." The next constraint to consider can be found by performing a topological sort of the constraint graph; as each node is visited, its incoming constraints are considered in reverse traversal order. The first violated constraint encountered is the next one to collapse into a proxy node. The constraint graph must be updated and the traversal restarted after each proxy node is created.

Forster's heuristic is based on the assumption that if the barycenter ordering causes vertex v to be placed to the left of u in violation of an ordering constraint, no vertices would be placed between u and v in the optimal solution with the correct ordering (u left of v). Though counterexamples can be easily found, the heuristic gives results that are nearly as good as the penalty graph approach in much less time.

Edge-Orientation Constraints

Since layer assignment requires an acyclic graph, the cycle removal phase reverses the direction of one or more edges in order to remove directed cycles. For some data structures, such as doubly-linked lists or trees where each node has both "child" and "parent" pointers, arbitrarily selecting edges for reversal may result in drawings that violate standard drawing conventions or have inconsistent edge orientations.

Waddle [Wad01] addresses the problem by tagging edges which may be reversed during cycle-breaking in the layer assignment phase. These edges will be reversed first, before untagged edges.

24.3.2 Special-Purpose Layouts

Space is a powerful visual variable, and an animation designer may choose to devise a custom layout algorithm which makes more effective use of space than a general-purpose algorithm. For example, Brown and Sedgewick [BS85] discuss the design of an animation involving binary search trees: noting that the simple recursive strategy of devoting half of the width of the current region to each of the left and right subtrees quickly leads to crowding even in trees of the size typically used in examples, they instead base the x coordinate of a node on the node's position in an in-order traversal of the tree. This ensures that each subtree has a width proportional to the number of nodes in that subtree, and also helps reinforce the organizational structure of the tree.

24.4 Graph Drawing for Program Visualization

Many program visualization applications focus on visualizing the objects in memory. These objects, along with their references to other objects, naturally form directed graphs.

Standard drawing algorithms for directed graphs can be used to produce layouts for object graphs. However, program visualization applications may have requirements that are not well-served by standard drawing algorithms. The rest of this section addresses specialized drawing techniques relevant for program visualization.

24.4.1 Complex Node Structures

Objects in programs are complex structures with multiple fields. Seeing this internal structure can be important for understanding the program's behavior, particularly in debugging applications.

The convention when drawing object structures is to show pointers or references as edges which end at distinct points inside the node. This can pose problems for standard drawing algorithms. For example, traditional crossing-reduction strategies used by Sugiyama-style layout algorithms assume that edges connect node centers and thus crossings can only occur between edges connecting different pairs of nodes. With complex nodes, edges may originate and terminate at any point within a node, and crossings can occur even when two edges are incident on the same node.

Waddle [Wad01] uses a Sugiyama-style approach for drawing object graphs, and accommodates complex nodes by using the coordinate of the edge's actual endpoint within the node instead of the node's center when computing barycenters for crossing reduction. Problems can still arise if a node contains several edges whose endpoints are vertically aligned because the adjacent nodes may end up with the same barycenter—and improper ordering of those nodes can result in edge crossings. This is addressed by assigning a secondary sort key (or "secondary barycenter") based on the vertical ordering of the endpoints. Figure 24.2 shows two ways to assign secondary barycenters.

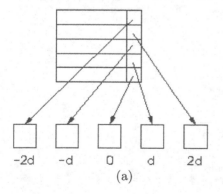 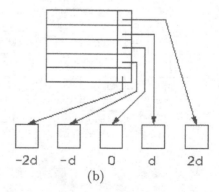

Figure 24.2 Two strategies for assigning secondary barycenters. The value of the secondary barycenters are shown below the nodes ($d > 0$). (a) Drawing with edge-node overlaps. (b) Drawing that avoids edge-node overlaps but involves additional edge routing.

24.4.2   Taking Structure into Account

Not all of the nodes in the object graph serve the same purpose — some are part of a data structure, such as a binary tree, while others are data fields. With this in mind, Gestwicki et al. [GJG04] identify two important aesthetic criteria for drawing object graphs:

- Leaf objects, which have exactly one incoming reference and no outgoing references, should be grouped with the objects (called aggregators) that reference them.
- Recursive structures should be clustered.

Figure 24.3 illustrates the benefits of this approach.

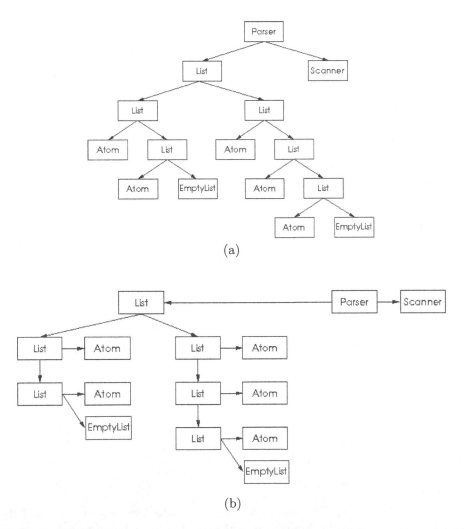

(a)

(b)

Figure 24.3 (a) Object graph for a simple expression parser drawn using a traditional Sugiyama-style layout algorithm. (b) Drawing taking the class structure into account. Example from [GJG04].

Gestwicki et al. [GJG04] use the program's class diagram to identify the important structures. A leaf class is a class with no outgoing associations—all of its fields, including inherited fields, are either primitive types or immutable wrappers around primitive types. A recursive type is defined by a directed cycle along generalization and aggregation relationships in the class diagram—all of the classes along the cycle are part of the recursive type. The simplest case is a single class containing a field of its own type.

The leaf classes and recursive types identified in the class diagram can then be used to identify interesting structures in the object graph. A leaf cluster consists of an aggregator node and its leaf-class children. (Note that an aggregator node may have other children in the object graph that are not part of the leaf cluster.) A recursive cluster is a connected subgraph containing objects belonging to a single recursive type and their leaf-class children, and with at most one node with incoming edges from outside the cluster.

Once the leaf and recursive clusters have been identified, the graph is drawn in three steps:

- Draw the leaf clusters.
- Replace the leaf clusters by single nodes, and draw the recursive clusters.
- Replace the recursive clusters by single nodes, and draw the remaining structure.

In order to avoid needlessly complicating the drawing with unnecessary detail, only nodes whose type is included in the class diagram are drawn. A variety of algorithms can be used in each stage, though the drawing algorithms chosen for the last two steps must be able to take into account the area needed to draw the collapsed cluster nodes. Using different layout techniques for each cluster, such as a radial layout for leaf clusters and a hierarchical layout for recursive clusters, emphasizes the distinct nature of each type of cluster.

The advantage of deriving leaf and recursive clusters from structures in the class diagram instead of basing them solely on the object graph is that the final drawing will reflect the correct semantics of the program—it will not be dependent on the current state in the program's execution. Consider, for example, the definition of a leaf cluster—it distinguishes between nodes in the object graph which currently have no outgoing edges and those which will never have any outgoing edges.

24.4.3   Drawing Execution Environments

Many program visualization systems show objects only as containers for data, but JIVE [GJ05] aims to give a more comprehensive view of the execution of object-oriented programs by showing objects both as containers for data and as environments for execution. In the most detailed view, objects are shown with both fields and methods; each active method is shown with its parameters and local variables. This structure may be multiple levels deep as contained objects may themselves contain fields and active methods. Inheritance relationships are also shown so each object's scope is clear. JIVE's object graphs present a challenge for graph drawing, as the graphs have large nodes containing complex internal structures, nested structures, multiple types of nodes and edges, and edges which connect to internal points within nodes.

The nested structure of objects-within-objects can be represented as a tree, and the object can be drawn by creating an HV-inclusion drawing of the nesting tree. In this drawing style, child nodes are drawn as rectangles within the rectangle devoted to the parent and are either arranged in a row or stacked vertically. Garg et al. [GGJ06] give a dynamic programming algorithm for computing minimum-area HV-inclusion drawings.

The rest of the graph structure in the object graph can be drawn using an algorithm for layered drawings of weighted multigraphs [GJ05].

24.4.4 Drawing Sequence Diagrams

In addition to displaying object graphs, JIVE [GJ05] uses a sequence diagram to show the program's execution history. In a sequence diagram, each method activation is represented by a vertical bar and all of the method activation bars belonging to a single object are drawn along the same vertical line. Method calls and returns are represented by arrows drawn from one activation bar to another.

Drawing sequence diagrams can be formulated as a graph drawing problem. A sequence graph contains a node for each method activation bar and a directed edge for each method call and return; the task is to find a left-to-right ordering for the object lines which minimizes edge length, the number of edges crossing activation bars, and the number of method-call edges directed to the left. Clustering constraints may also be applied to ensure that object lines for related objects are close together.

Garg et al. [GGJ06] give a simulated annealing algorithm for finding a left-to-right ordering of object lines which respects the desired clustering and optimizes an objective function incorporating the aesthetic criteria. Each object line is assigned a unique integer value; lines belonging to the same cluster receive consecutive labels. Two object lines are selected randomly and, with a probability related to the potential improvement in the objective function and the temperature of the system, the integer labels of either the lines (if the object lines belong to the same cluster) or the clusters (if the object lines belong to different clusters) are swapped.

24.5 Graph Drawing for Software Visualization

24.5.1 Drawing UML Class Diagrams

One challenge in drawing UML class diagrams is handling the multiple types of edges—generalizations and associations—because generalizations are hierarchical and associations are not. In addition, Purchase et al. [PAC01] have identified several aesthetic criteria that are important for UML class diagrams, including orthogonality, a consistent orientation for the edges, and joined inheritance arcs instead of separate edges. The traditional aesthetic criteria of few crossings and bends are also important.

Two-Pass Approach

Seemann [See97] prioritizes showing the different types of relationships over the other aesthetic criteria. A two-pass strategy is used: first the inheritance hierarchies are drawn using a variation of the Sugiyama algorithm, and then the association edges are drawn with an orthogonal style.

In the first phase of the algorithm, a modified Sugiyama layout is applied to just the generalization edges and their incident vertices. The initial layer assignment is adjusted to reduce the span of association edges: if a node has an association with a node in a lower layer, and moving the node to the lower layer does not violate the desired direction of any generalization edges, the node is moved. In addition, the crossing reduction stage is modified to attempt to place nodes with association edges between them next to each other in the layer.

Next, the remaining nodes are placed into levels. New nodes are added incrementally; in each pass, nodes which have not yet been placed but which are adjacent to nodes which have been placed are added. Let v be an already-placed node and S be the set of to-be-placed nodes adjacent to v. If $|S| \leq 2$ and the already-placed nodes to either side of v are not adjacent to v, the nodes of S can be placed to the right and left of v in v's layer. If there is not enough room to place the nodes of S next to v—either because S is too large, or v is already connected to the nodes next to it—the nodes of S are placed on a sublayer above or below v's layer. Once all of the nodes have been placed, the sublayers are used to further reduce crossings and bends due to assocation edges connecting non-consecutive nodes on a layer.

Finally, node sizes are computed, edges are routed, and x coordinates are calculated. Node sizes are based on the information that must be displayed inside the node. Generalization edges are drawn as straight lines, with connection ports evenly spaced along the bottom or top of a node. Association edges are drawn with an orthogonal style; to route edges connecting nodes in different layers, dummy nodes representing bends in the edge are added on one side of the nodes being connected. These dummy nodes are constrained to stay vertically aligned when x coordinates are assigned. Connection ports for association edges are evenly spaced along the left or right side of a node.

Integrated Approach

Gutwenger et al. [GJK+03] give a more complex drawing algorithm which respects all of the aesthetic criteria identified by Purchase et al. [PAC01] and additionally ensures that all generalization edges within the same class hierarchy are oriented in the same direction, generalization edges in different hierarchies do not cross, and hierarchies do not contain each other. The algorithm follows the topology-shape-metrics approach [DETT99]: first the graph is planarized, then the bends and angles are fixed, and finally edge lengths are computed.

Because the convention that inheritance arcs are drawn joined can result in additional crossings (Figure 24.4), the planarization phase begins with a preprocessing step which adds a new vertex for each join point. Consistency of direction of edges within a hierarchy is achieved by computing an upward planar representation for each class hierarchy, and separation of different hierarchies is achieved by treating each hierarchy as a cluster and applying a cluster planarization algorithm.

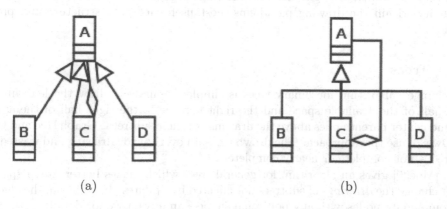

(a)   (b)

Figure 24.4 (a) A planar embedding. (b) With the same embedding, joining the generalization edges results in a crossing.

In the shape phase, vertices with degree greater than four are replaced by a "cage" containing a cycle of degree-3 vertices prior to computing an orthogonal representation.

Finally, two compaction steps are used in the metrics phase. After the first compaction step, the cages are replaced by the original high-degree vertices. Because the cages may be larger than the vertices they contain, additional bends may be needed in order to route edges within the cage. The second compaction step addresses this problem and removes unnecesssary bends.

24.6   Sequences of Drawings

Both algorithm animation and program visualization often involve graphs whose structure changes over the course of the visualization. In these cases, it is important to preserve the user's *mental map* [ELMS91]—that is, to maintain a degree of layout stability so the viewer can focus on what is really going on in the algorithm or program without being distracted by the side-effects of the layout algorithm. However, many standard layout algorithms assume complete freedom over the placement of nodes.

There are many models for the user's mental map. The most rigid is the "no change" model, where existing portions of the drawing are preserved exactly (e.g. [MHT93, PT98]). Böhringer and Paulisch [BN90] limit change to nodes within a certain graph distance of those directly affected by an update. Other strategies seek to preserve absolute vertex position, but allow some movement (e.g. [LMR98]). Misue et al. [MELS95] seek more generally to preserve the shape of the drawing and give several models for the mental map based on orthogonal ordering (the relative up/down/left/right relationships between nodes), proximity (nodes near each other should stay near each other), and topology (specifically, the dual graph). Specific metrics for measuring mental map preservation are given by Lyons et al. [LMR98], Bridgeman and Tamassia [BT98], and Brandes and Wagner [BW98]. Time can also be a factor, with the idea that it is more costly to the user's mental map when long-stable portions of the drawing are changed instead of relatively new sections [BW97].

Preserving the mental map typically leads to a tradeoff with drawing quality. Algorithms which more rigidly preserve the original layout result in drawings which are less good according to traditional aesthetic criteria such as drawing area, crossing minimization, and bend minimization. Some dynamic graph drawing algorithms allow user control over the relative weight given to each goal.

An overview of dynamic graph drawing and its application in several drawing paradigms is given by Branke [Bra01]. This section will address some strategies for maintaining layout stability within the drawing paradigms most useful for data structure and program visualization.

24.6.1   Trees

A "no change" algorithm for binary trees is simple: recursively draw the left subtree in the left half of the available space and the right subtree in the right half of the available space, and center parent nodes above the drawings of their subtrees. GraphTree and GraphHeap [Owe86] use this approach. The drawback is an overly-wide drawing and wasted space if the tree is not complete or nearly complete.

Moen [Moe90] gives an algorithm for general trees which makes better use of space and does not change the drawings of subtrees not affected by updates. In addition, the algorithm can accommodate nodes with any polygonal shape—an advantage for data structures with complex nodes. The algorithm is based on computing a contour around each subtree, which is then used to pack subtrees together as closely as possible. Contours are computed recursively. Making changes to the tree structure requires recomputing contours (only) for the subtrees containing the affected nodes.

A similar approach is used by Workman et al. [WBP04], with the drawing convention that trees are laid out horizontally (children next to parents instead of below) and parents are placed on the same level as the first child.

24.6.2 Force-Directed Layout

In the force-directed model, layout stability is most commonly achieved by incorporating additional forces into the model. Varying the strength of the stability forces provides a convenient way to balance layout stability and drawing quality.

Absolute vertex positions can be maintained by adding forces that attract nodes to their former positions [LMR98, BW97].

Relative distances between nodes can be maintained by adding springs whose natural length is the desired distance [BW97]. Stiffening the springs makes the distances more rigid. Stiffening entire subgraphs can help maintain the shape of the drawing.

Clustering can be maintained by adding attractive forces toward the center of the cluster and repulsive forces between clusters [Tam98].

It is also possible to incorporate some hard constraints. For example, Tamassia [Tam98] mentions truncating a node's movement each time forces are applied in order to keep it within the desired region. In addition, the shape of a subgraph can be preserved exactly (up to translation and rotation) by treating it as a single rigid body when computing forces.

24.6.3 Sugiyama-Style Hierarchical Layout

Within the Sugiyama framework, several basic approaches can be used: incremental techniques, in which the existing drawing is modified to accommodate the changes; constraint-based techniques, in which a new layout is computed subject to constraints meant to preserve the user's mental map; and cost-based techniques, in which stability is encouraged by assigning a cost to changes that affect the user's mental map.

Incremental Techniques

North [Nor96] describes an incremental heuristic for maintaining both geometric (position) and topological (ordering) stability in Sugiyama-style layouts. It is assumed that changes are made to the graph one at a time, so the algorithm only needs to accommodate the addition or removal of a single node or edge.

A new node is assigned to the highest possible level consistent with maintaining a downward orientation for edges, and existing nodes are shifted to lower levels as needed. New level assignments are determined by depth-first search. Nodes are moved downward one level at a time. At each step, the node is shifted into its correct horizontal position in the level according to the median of its neighbors' positions. Finally, a linear program is used to assign horizontal coordinates to the nodes. An additional cost is introduced to penalize moving nodes to new positions.

Constraint-Based Techniques

Böhringer and Paulisch [BN90] maintain layout stability by adding constraints to maintain the level assignment of nodes and the ordering of nodes within a level. When a node or edge is added or removed, constraints are weakened (more likely to be deactivated in the case of contradictory constraints) or removed in the vicinity of the changes. They define "vicinity" in terms of graph distance, but other notions (such as Euclidean distance in the drawing) could be used.

Waddle's algorithm for drawing data structures [Wad01] also handles layout stability by adding constraints. He focuses on maintaining the relative ordering of subgraphs rather than fixing the layer assignment, adding

- node-ordering constraints between root nodes in the top level,
- edge-ordering constraints between edges incident on root nodes, and
- edge-ordering constraints between downward edges.

New elements added to the graph are initially unconstrained; constraints which are rendered invalid by the removal of elements are updated or deleted. Böhringer and Paulisch's [BN90] scheme of weakening constraints in the vicinity of changes could also be applied.

Section 24.3.1 outlines how the basic Sugiyama algorithm can be modified to accommodate these constraints.

Cost-Based Techniques

North and Woodhull [NW02] reduce the layer assignment and coordinate assignment phases to integer linear programs. Layout stability can be incorporated by adding terms to the objective functions to penalize movement to a different layer (layer assignment) or a different position (coordinate assignment). An advantage of this approach is that the tradeoff between drawing quality and layout stability can be managed by adjusting the cost of movement.

In the crossing reduction phase, median sort and transposition sort are used to reduce crossings. Only new or modified nodes and edges and edges incident on new or modified nodes are considered to be movable during sorting.

24.6.4   Offline Dynamic Graph Drawing

In "canned" animations, both the graph and the sequence of changes being made to the graph are known in advance. In this situation, an offline drawing algorithm—which takes into account future graph states when producing a layout—can be used to increase layout stability.

Force-Directed Layout

Erten et al. [EHK+04] combine the individual snapshot graphs into a single aggregrate graph, adding edges between corresponding vertices in different snapshots. At a minimum, between-snapshot edges should be added between corresponding vertices in consecutive snapshots. Global layout stability can be increased by adding edges between more distant snapshots.

The aggregate graph is drawn using the Kamada-Kawai algorithm [KK89], modified so that there are no repulsive forces between vertices in different snapshots. The balance between layout stability and readability can be controlled by adding weights to the between-snapshot edges. To accommodate weights, the Kamada-Kawai forces are modified to use the ideal distance between vertices (based on the weights of edges between them) instead of the graph distance between them.

Foresighted Layout

Diehl, Görg, and Kerren [DGK01] give a more general strategy which they call "foresighted layout." In the simplest case, the individual snapshot graphs are combined to create a supergraph containing every node and every edge present in at least one individual graph. A layout is then computed for the supergraph, and each individual graph is drawn using the subset of the supergraph layout information.

A drawback to this approach is that the supergraph can be quite large if the graph structure changes significantly over the course of the animation, leading to wasted space in the individual layouts. To save space, a reduced version of the supergraph in which nodes and edges with disjoint "live times" are grouped together is used instead. (Since elements with disjoint live times do not occur in the same snapshot graph, they can occupy the same position in different snapshots.)

As a final step, layout adjustment strategies can be used to improve the quality of each snapshot layout at the expense of layout stability [DG02].

24.6.5 Smooth Animation

When viewing a series of drawings, animation can be used to help the viewer see and understand what has changed from one drawing to the next.

A simple scheme is to move each vertex along a straight line between its starting and ending positions. However, this can lead to very poor animations which confuse rather than reveal the structure of the changes, particularly in cases where part or all of the drawing has been rotated or flipped.

With this problem in mind, Friedrich and Eades [FE02] identify several properties of a good animation:

- Uniform motion—groups of vertices with similar relative positions at the beginning and the end should move together.
- Separation—vertices with different motion paths should not be too close together.
- Rigid motion—movements should be consistent with 2D projections of the motion of 3D rigid objects, to exploit human perceptual strengths.
- No misleading layouts—unfortunate overlaps, such as a vertex lying on an edge, can lead to incorrect conclusions about the graph's structure.
- Short motion paths—vertices should travel as short a route as possible, to make the motion easier to follow.

They give a four-step algorithm for animation designed to satisfy these properties:

- fade out vertices and edges not present in the end drawing,
- apply a rigid transformation (composed of translation, rotation, scaling, flipping, and/or shearing) to the entire graph to move the elements of the graph as close as possible to their positions in the end drawing,
- complete the movement of vertices to their final positions, and
- fade in vertices and edges not present in the beginning drawing.

The transformation is chosen to minimize the sum of the squared distances between the transformed nodes and their positions in the end drawing. To reduce the effect of outliers, a (weighted) centroid can be included in the node set, or several random subsets of nodes can be chosen and the best transformation used. Once the transformation has been found, smooth animation paths for the nodes can be computed by extracting the rotational part of the transformation using polar matrix decomposition, then simultaneously interpolating the angle of rotation and the entries of the non-rotational part of the transformation matrix. Rotation around the center of the drawing can be achieved by incorporating a translation to the origin and back into the transformation matrix before the decomposition is done.

Simple linear interpolation can be used to move the transformed nodes to their final positions, but a more pleasing result can be obtained with a force-directed approach where nodes repel each other but are attracted to their final positions instead of their neighbors.

If the drawings contain subgraphs which move in different ways, the animation can be improved by applying the middle steps separately to each distinct subgraph. Friedrich and Houle [FH02] suggest two strategies for clustering nodes into groups with common transformations—k-means and eliminating edges in the Delaunay triangulation to merge triangles with sufficiently similar transformations—and note that both strategies can produce good results though they also have limitations.

References

[Bae81]   R. M. Baecker. Sorting out sorting, 1981. 30 minute color sound film, Dynamic Graphics Project, University of Toronto, excerpted and reprinted in SIGGRAPH Video Review 7, 1983.

[BBG⁺99]   Ryan S. Baker, Michael Boilen, Michael T. Goodrich, Roberto Tamassia, and B. Aaron Stibel. Testers and visualizers for teaching data structures. In *Proceedings of the 30th SIGCSE Technical Symposium on Computer Science Education*, SIGCSE '99, pages 261–265, New York, NY, USA, 1999. ACM.

[BCE⁺]   A. Bilgin, D. Caldwell, J. Ellson, E. Gansner, Y. Hu, and S. North. GraphViz. http://www.graphviz.org/.

[BDG⁺00]   J. Berry, N. Dean, M. K. Goldberg, G. E. Shannon, and S. Skiena. LINK: a system for graph computation. *Software: Practice and Experience*, 30(11):1285–1302, 2000.

[Ber]   J. Berry. LINK online manual. http://dimacs.rutgers.edu/ berryj/manual/.

[BGKT00]   S. Bridgeman, M. T. Goodrich, S. G. Kobourov, and R. Tamassia. PILOT: an interactive tool for learning and grading. In *SIGCSE 2000*, pages 139–143, March 2000.

[BGT99]   Stina Bridgeman, Ashim Garg, and Roberto Tamassia. A graph drawing and translation service on the World Wide Web. *Internat. J. Comput. Geom. Appl.*, 9(4–5):419–446, 1999.

[BN90]   K. Bohringer and F. Newbery Paulisch. Using constraints to achieve stability in automatic graph layout algorithms. In *Proc. ACM Conf. on Human Factors in Computing Systems*, pages 43–51, 1990.

[Bra01]   Jürgen Branke. Dynamic graph drawing. In Michael Kaufmann and Dorothea Wagner, editors, *Drawing Graphs*, volume 2025 of *Lecture Notes in Computer Science*, pages 228–246. Springer Berlin / Heidelberg, 2001.

[BS84]   Marc H. Brown and Robert Sedgewick. A system for algorithm animation. *SIGGRAPH Comput. Graph.*, 18:177–186, January 1984.

[BS85]   M. H. Brown and R. Sedgewick. Techniques for algorithm animation. *IEEE Softw.*, 2(1):28–39, January 1985.

[BT98]   Stina Bridgeman and Roberto Tamassia. Difference metrics for interactive orthogonal graph drawing algorithms. In *Journal of Graph Algorithms and Applications*, pages 57–71. Springer-Verlag, 1998.

[BW97]   Ulrik Brandes and Dorothea Wagner. A bayesian paradigm for dynamic graph layout. In G. Di Battista, editor, *Graph Drawing (Proc. GD '97)*, volume 1353 of *Lecture Notes Comput. Sci.*, pages 236–247. Springer-Verlag, 1997.

[BW98]   Ulrik Brandes and Dorothea Wagner. Dynamic grid embedding with few bends and changes. In *Proceedings of the 9th International Symposium on Algorithms and Computation*, ISAAC '98, pages 89–98. Springer-Verlag, 1998.

[CIHJB07]   James H. Cross II, T. Dean Hendrix, Jhilmil Jain, and Larry A. Barowski. Dynamic object viewers for data structures. In *Proceedings of the 38th SIGCSE Technical Symposium on Computer Science Education*, SIGCSE '07, pages 4–8, New York, NY, USA, 2007. ACM.

[DETT99] G. Di Battista, P. Eades, R. Tamassia, and I. G. Tollis. *Graph Drawing*. Prentice Hall, Upper Saddle River, NJ, 1999.

[DG02] Stephan Diehl and Carsten Görg. Graphs, they are changing. In *Revised Papers from the 10th International Symposium on Graph Drawing*, GD '02, pages 23–30. Springer-Verlag, 2002.

[DGK01] Stephan Diehl, Carsten Görg, and Andreas Kerren. Preserving the mental map using foresighted layout. In *In Proceedings of Joint Eurographics IEEE TCVG Symposium on Visualization VisSym'01*, pages 175–184. Springer Verlag, 2001.

[EHK+04] Cesim Erten, Philip Harding, Stephen Kobourov, Kevin Wampler, and Gary Yee. GraphAEL: Graph animations with evolving layouts. In Giuseppe Liotta, editor, *Graph Drawing*, volume 2912 of *Lecture Notes in Computer Science*, pages 98–110. Springer Berlin / Heidelberg, 2004.

[ELMS91] P. Eades, W. Lai, K. Misue, and K. Sugiyama. Preserving the mental map of a diagram. In *Proceedings of Compugraphics 91*, pages 24–33, 1991.

[EPD92] Marc Eisenstadt, Blaine A. Price, and John Domingue. Software visualization as a pedagogical tool. *Instructional Science*, 21:335–364, 1992.

[FE02] Carsten Friedrich and Peter Eades. Graph drawing in motion. *Journal of Graph Algorithms and Applications*, 6:2002, 2002.

[FH02] Carsten Friedrich and Michael Houle. Graph drawing in motion II. In Petra Mutzel, Michael Jünger, and Sebastian Leipert, editors, *Graph Drawing*, volume 2265 of *Lecture Notes in Computer Science*, pages 122–125. Springer Berlin / Heidelberg, 2002.

[Fin01] I. Finocchi. Layered drawings of graphs with crossing constraints. In *COCOON '01*, pages 357–367, 2001.

[For04] M. Forster. A fast and simple heuristic for constrained two-level crossing reduction. In *GD '04*, pages 206–216, 2004.

[GGJ06] Ashim Garg, Paul V. Gestwicki, and Bharat Jayaraman. Interactive program visualization and graph drawing. In M. Sethumadhavan, editor, *Discrete Mathematics and Its Applications*, pages 36–52. Narosa Publishing House Pvt. Ltd., 2006.

[GJ05] Paul Gestwicki and Bharat Jayaraman. Methodology and architecture of JIVE. In *Proceedings of the 2005 ACM Symposium on Software Visualization*, SoftVis '05, pages 95–104, New York, NY, USA, 2005. ACM.

[GJG04] P. V. Gestwicki, B. Jayaraman, and A. Garg. From class diagrams to object diagrams: A systematic approach. Technical Report 2004-21, University at Buffalo, State University of New York, December 2004.

[GJK+03] Carsten Gutwenger, Michael Jünger, Karsten Klein, Joachim Kupke, Sebastian Leipert, and Petra Mutzel. A new approach for visualizing UML class diagrams. In *Proceedings of the 2003 ACM Symposium on Software Visualization*, SoftVis '03, pages 179–188, New York, NY, USA, 2003. ACM.

[GKNV93] E. R. Gansner, E. Koutsofios, S. C. North, and K. P. Vo. A technique for drawing directed graphs. *IEEE Trans. Softw. Eng.*, 19:214–230, 1993.

[Ham04]   J. Hamer. Visualising Java data structures as graphs. In *CRPIT '30: Proceedings of the 6th Conference on Australian Computing Education*, pages 125–129. Australian Computer Society, Inc., 2004.

[HCIB04]   T. Dean Hendrix, James H. Cross II, and Larry A. Barowski. An extensible framework for providing dynamic data structure visualizations in a lightweight IDE. In *Proceedings of the 35th SIGCSE Technical Symposium on Computer Science Education*, pages 387–391, New York, NY, USA, 2004. ACM.

[HWF90]   Robert R. Henry, Kenneth M. Whaley, and Bruce Forstall. The University of Washington illustrating compiler. In *Proceedings of the ACM SIGPLAN Conference on Programming Language Design and Implementation*, pages 223–233, New York, NY, USA, 1990. ACM.

[Kam89]   T. Kamada. *Visualizing Abstract Objects and Relations*. World Scientific Series in Computer Science, 1989.

[KK89]   T. Kamada and S. Kawai. An algorithm for drawing general undirected graphs. *Inform. Process. Lett.*, 31:7–15, 1989.

[KKMS04a]   V. Karavirta, A. Korhonen, L. Malmi, and K. Stalnacke. MatrixPro—a tool for demonstrating data structures and algorithms ex tempore. In *Proc. IEEE Int. Conf. on Advanced Learning Technologies*, pages 892–893, 2004.

[KKMS04b]   Ville Karavirta, Ari Korhonen, Lauri Malmi, and Kimmo Stlnacke. MatrixPro – a tool for on-the-fly demonstration of data structures and algorithms. In *Proceedings of the Third Program Visualization Workshop*, pages 26–33. Department of Computer Science, University of Warwick, UK, July 2004.

[KM02]   Ari Korhonen and Lauri Malmi. Matrix: concept animation and algorithm simulation system. In *Proceedings of the Working Conference on Advanced Visual Interfaces*, AVI, pages 109–114, New York, NY, USA, 2002. ACM.

[KMBW02]   E. Kruja, J. Marks, A. Blair, and R. C. Waters. A short note on the history of graph drawing. In *GD '01*, pages 272–286. Springer-Verlag, 2002.

[KOD⁺96]   M. S. Krishnamoorthy, F. Oxaal, U. Dogrusoz, D. Pape, A. Robayo, R. Koyanagi, Y. Hsu, D. Hollinger, and A. Hashimi. GraphPack: Design and features. In P. Eades and K. Zhang, editors, *Software visualization*, pages 83–99. World Scientific, 1996.

[KQPR03]   Michael Kolling, Bruce Quig, Andrew Patterson, and John Rosenberg. The BlueJ system and its pedagogy. *Journal of Computer Science Education, Special issue on Learning and Teaching Object Technology*, 13(4):249–268, December 2003.

[LMR98]   Kelly A. Lyons, Henk Meijer, and David Rappaport. Algorithms for cluster busting in anchored graph drawing. *J. Graph Algorithms Appl.*, 2(1):1–24, 1998.

[LNR03]   J. Lucas, T. L. Naps, and G. Rößling. VisualGraph—a graph class designed for both undergraduate students and educators. In *SIGCSE 2003*, pages 167–171, February 2003.

[MB98]   Carolyn McCreary and Larry Barowski. VGJ: Visualizing graphs through java. In Sue Whitesides, editor, *Graph Drawing*, volume 1547 of *Lecture Notes in Computer Science*, pages 454–455. Springer, 1998.

[MCS98]   Carolyn McCreary, Richard Chapman, and Fwu-Shan Shieh. Using graph parsing for automatic graph drawing. *IEEE Transactions on Systems, Man, and Cybernetics, Part A*, pages 545–561, 1998.

[MELS95]   K. Misue, Peter Eades, W. Lai, and K. Sugiyama. Layout adjustment and the mental map. *J. Visual Lang. Comput.*, 6(2):183–210, 1995.

[MHT93]   K. Miriyala, S. W. Hornick, and R. Tamassia. An incremental approach to aesthetic graph layout. In *Proc. Internat. Workshop on Computer-Aided Software Engineering*, 1993.

[MKK$^+$04]   Lauri Malmi, Ville Karavirta, Ari Korhonen, Jussi Nikander, Otto Seppl, and Panu Silvasti. Visual algorithm simulation exercise system with automatic assessment: TRAKLA2. In *Informatics in Education*, page 048, 2004.

[MMSBA04] Andrés Moreno, Niko Myller, Erkki Sutinen, and Mordechai Ben-Ari. Visualizing programs with Jeliot 3. In *Proceedings of the Working Conference on Advanced Visual Interfaces*, AVI '04, pages 373–376, New York, NY, USA, 2004. ACM.

[Moe90]   S. Moen. Drawing dynamic trees. *IEEE Softw.*, 7:21–28, 1990.

[Mur84]   J. E. Murdoch. *Album of Science: Antiquity and the Middle Ages*. Charles Scribner's Sons, New York, 1984.

[Nap98]   Thomas L. Naps. A Java visualiser class: incorporating algorithm visualisations into students' programs. In *Proceedings of the 6th Annual Conference on the Teaching of Computing and the 3rd Annual Conference on Integrating Technology into Computer Science Education: Changing the Delivery of Computer Science Education*, ITiCSE '98, pages 181–184, New York, NY, USA, 1998. ACM.

[NEN00]   T. L. Naps, J. R. Eagan, and L. L. Norton. JHAVÉ: An environment to actively engage students in web-based algorithm visualizations. In *SIGCSE '00: Proceedings of the 31st SIGCSE Technical Symposium on Computer Science Education*, page 109–113, Austin, Texas, 2000. ACM Press.

[Nor96]   S. North. Incremental layout in DynaDAG. In F. J. Brandenburg, editor, *Graph Drawing (Proc. GD '95)*, volume 1027 of *Lecture Notes Comput. Sci.*, pages 409–418. Springer-Verlag, 1996.

[NW02]   Stephen C. North and Gordon Woodhull. Online hierarchical graph drawing. In *Revised Papers from the 9th International Symposium on Graph Drawing*, GD '01, pages 232–246. Springer-Verlag, 2002.

[Owe86]   G. S. Owen. Teaching of tree data structures using microcomputer graphics. In *SIGCSE '86: Proceedings of the 17th SIGCSE Technical Symposium on Computer Science Education*, pages 67–72. ACM Press, 1986.

[PAC01]   Helen C. Purchase, Jo-Anne Allder, and David A. Carrington. User preference of graph layout aesthetics: A UML study. In *Proceedings of the 8th International Symposium on Graph Drawing*, GD '00, pages 5–18. Springer-Verlag, 2001.

[PNK11]   Sergey Pupyrev, Lev Nachmanson, and Michael Kaufmann. Improving layered graph layouts with edge bundling. In *Proceedings of the 18th International Conference on Graph Drawing*, GD'10, pages 329–340. Springer-Verlag, 2011.

[PR98]   W. C. Pierson and S. H. Rodger. Web-based animation of data structures using JAWAA. In *SIGCSE '98*, pages 267–271, 1998.

[PT98]   A. Papakostas and I. G. Tollis. Interactive orthogonal graph drawing. *IEEE Trans. Comput.*, C-47(11):1297–1309, 1998.

[RDM+87]   L. A. Rowe, M. Davis, E. Messinger, C. Meyer, C. Spirakis, and A. Tuan. A browser for directed graphs. *Softw. – Pract. Exp.*, 17(1):61–76, 1987.

[RF01]   G. Rößling and B. Freisleben. Program visualization using AnimalScript. In *Proceedings of the First Program Visualization Workshop, PVW'00*, page 41–52, Porvoo, Finland, 2001. University of Joensuu Press, University of Joensuu Press.

[RMS11]   Guido Rößling, Mihail Mihaylov, and Jerome Saltmarsh. AnimalSense: combining automated exercise evaluations with algorithm animations. In *Proceedings of the 16th Annual Joint Conference on Innovation and Technology in Computer Science Education*, ITiCSE '11, page 298–302, New York, NY, USA, 2011. ACM.

[See97]   Jochen Seemann. Extending the sugiyama algorithm for drawing UML class diagrams: Towards automatic layout of object-oriented software diagrams. In G. Di Battista, editor, *Graph Drawing (Proc. GD '97)*, volume 1353 of *Lecture Notes Comput. Sci.*, pages 415–424. Springer-Verlag, 1997.

[SHY96]   C. A. Shaffer, L. S. Heath, and J. Yang. Using the Swan data structure visualization system for computer science eduction. In *SIGCSE '96*, pages 140–144, February 1996.

[SK93]   John T. Stasko and Eileen Kraemer. A methodology for building application-specific visualizations of parallel programs. *J. Parallel Distrib. Comput.*, 18:258–264, June 1993.

[SS10]   James T. Streib and Takako Soma. Using contour diagrams and JIVE to illustrate object-oriented semantics in the Java programming language. In *Proceedings of the 41st ACM Technical Symposium on Computer Science Education*, SIGCSE '10, pages 510–514, New York, NY, USA, 2010. ACM.

[Sta90a]   J. T. Stasko. Simplifying algorithm animation with tango. In *Proc. IEEE Workshop on Visual Languages*, pages 1–6, 1990.

[Sta90b]   J. T. Stasko. Tango: a framework and system for algorithm animation. *IEEE Computer*, 23(9):27–39, 1990.

[Sta97]   John T. Stasko. Using student-built algorithm animations as learning aids. In *Proceedings of the 28th SIGCSE Technical Symposium on Computer Science Education*, SIGCSE '97, pages 25–29, New York, NY, USA, 1997. ACM.

[STT81]   K. Sugiyama, S. Tagawa, and M. Toda. Methods for visual understanding of hierarchical systems. *IEEE Trans. Syst. Man Cybern.*, SMC-11(2):109–125, 1981.

[Tam98]   R. Tamassia. Constraints in graph drawing algorithms. *Constraints*, 3(1):89–122, 1998.

[Tun94]   D. Tunkelang. A practical approach to drawing undirected graphs. Technical Report CMU-CS-94-161, School Comput. Sci., Carnegie Mellon University, June 1994.

[Wad01]   V. E. Waddle. Graph layout for displaying data structures. In *GD '00: Proceedings of the 8th International Symposium on Graph Drawing*, pages 241–252. Springer-Verlag, 2001.

[Wal90]   J. Q. Walker II. A node-positioning algorithm for general trees. *Softw. – Pract. Exp.*, 20(7):685–705, 1990.

[WBP04]   David Workman, Margaret Bernard, and Steven Pothoven. An incremental editor for dynamic hierarchical drawing of trees. In Marian Bubak, Geert Dick van Albada, Peter M. A. Sloot, and Jack J. Dongarra, editors, *Computational Science—ICCS 2004*, volume 3038 of *Lecture Notes in Computer Science*, pages 986–995. Springer-Verlag, 2004.

25

Computer Networks

Giuseppe Di Battista
Roma Tre University

Massimo Rimondini
Roma Tre University

25.1 Introduction

Communication systems are nowadays fundamental to support various applications, and this is especially true for computer networks as their utmost expression. Some examples include information interchange for critical operations, such as bank transfers or military data, as well as commonly used services such as the web, email, or streaming of multimedia contents. It is therefore essential to be able to ensure an uninterrupted and efficient operation of a computer network.

However, the task of maintaining a computer network may get considerably harder as the complexity of the network increases, either in terms of the topology or in terms of the enabled services. Therefore, as it often happens in other contexts, a significant aid in the maintenance comes from the ability to obtain a visual representation of the network.

25.1.1 Benefits of Visualizing Computer Networks

The availability of a methodology and a tool to visualize computer networks brings benefits both to network administrators and to researchers that work on studying network related phenomena.

Network administrators can exploit visualization tools to accurately design a network before deploying it. This includes, for example, defining the topology as well as tuning link bandwidths. At a higher level of abstraction, a visual representation of the interconnections between Internet Service Providers (ISPs) can help to better plan future commercial relationships among them. On the other hand, once the network is operational, a visualization system can significantly help in maintaining it, by providing graphical monitoring facilities that also simplify troubleshooting potential problems.

The research community can also benefit from the existence of a methodology to visualize networks. For example, it is possible to validate a theoretical model by spotting anomalies in the generated layout at a glance. This is particularly useful for the case of techniques to infer network topologies based on a limited amount of information and for the case of random network generators. Also, a graphical representation of a network can support reconstructing the root cause and the impact of a particular routing event.

25.2 The Very Basics of Computer Networking

This section briefly recalls some basic concepts about the operation of computer networks.

25.2.1 A Network Model

A computer network essentially consists of an interconnection of devices (computers, printers, routers, etc.) that exchange information with each other. In order to do this, a device encodes data in a format that other devices can understand, which is called *protocol*. To support a flexible configuration and ensure a good scalability, several encodings are usually stacked upon each other, so that data are first encoded (*encapsulated*) using a certain protocol, then the encoded data are encapsulated using another protocol, the resulting information is again encoded using a different protocol, and so on. After the last encapsulation step, information is actually sent to the destination, which performs the steps in reverse order: interprets the protocol used in received information and decapsulates the data, then again the decoded information is analyzed to interpret a different protocol and its payload is decapsulated, etc. After the last decapsulation step, the information that was originally forwarded by the sender is available for processing by the receiver.

This mechanism allows to consider and configure separately the different features of a network. For example, a high level protocol such as HTTP (HyperText Transfer Protocol), that is typically used to transfer web pages, can be configured and used independently of the actual protocol spoken on the transmission medium (copper cable, fiber, wireless link, etc.).

This kind of operation is defined in the ISO/IEC Open Systems Interconnection (OSI) model [fS, Com, ISO]. That is, computer networks usually operate according to a layered model, where each layer corresponds to an encapsulation/decapsulation step and is associated with a specific protocol used to communicate with the corresponding layer at the destination. According to the ISO OSI model, network protocols are organized as a *stack* consisting of 7 layers. Therefore, the setup of a network link consists at least of the specification of the stack of protocols to be used.

The following sections briefly describe some of the most commonly adopted network protocols.

25.2.2 Interconnection Technologies

Network links can be implemented using different physical media, usually wire, fiber, or air. The usage of a physical medium rather than another implies a choice for the implementation of the *physical layer*, the lowest layer of the ISO OSI protocol stack. At this layer, the implementation contains the specification of parameters such as electrical signals or frequencies, that are used to encode the information transmitted on the physical medium. Examples of protocols at the physical layer are SDH and DWDM [SS96, IMN84].

Often, the choice of a physical medium is associated with the choice of a *data link layer* protocol that exploits the medium in such a way as to provide a reasonably fast and reliable communication channel. For example, wired communication can happen by using Ethernet on a local network or SDLC on a wide area network [Sta07, Sys09b]. On the other hand, wireless communication can take place using either the IEEE802.11 protocol or the IEEE802.16/WiMAX protocol, the choice depending on parameters such as the distance between antennas, the transmitting power, or the desired communication speed [IEE09, wim09]. Communication over optical fiber often takes place on Gigabit Ethernet links [Nor02].

25.2.3 Routing and Routing Protocols

Once the protocols for the physical and data link layers have been chosen, there must be a mechanism to allow the delivery of a piece of information from any source in the network to any destination. This is accomplished by *network layer* protocols such as the well-known Internet Protocol [Pos81b] (IP), which are usually run by devices known as *routers*.

Actually, network layer protocols support the reachability of remote destinations based on the knowledge of a previously built *routing table*, a data structure stored on the routers that specifies the physical port to be used to forward information to a given destination. Since network topologies frequently change, there must also be a way to update the routing tables without human intervention. This is achieved by running *routing protocols* that are designed for this purpose, such as OSPF, IS-IS, and BGP [Moy94, Ora90, YTS06]. While these protocols often exploit layers of the ISO OSI stack that are higher than the network layer, their function is still to support the operation of the network layer.

25.2.4 The Internet Structure

The structure of the Internet, both from the point of view of the topology and from the point of view of the configuration, is rather complex. The Internet embraces lots of different, yet interacting, physical media and data link protocols. The only commonly adopted standard is on the network layer protocol which, for almost all network nodes, can be assumed to be IP [Pos81b]. Actually, it is very common to hear about the TCP/IP stack: this is just a shortcut to indicate that the routing layer is implemented by IP and the transport layer (which is on top of the routing layer) is implemented by TCP.

Further complexity is brought about by the fact that different ISPs may be interested in adopting their own routing mechanisms, protocols, and policies, and they may establish different economical agreements with neighboring ISPs. This poses a big challenge because different parties, using different protocols, and adopting different configuration policies, must be enabled to communicate with each other.

For this reason, Internet devices are usually grouped into *Autonomous Systems* (ASes), such that within each AS a single routing protocol and consistent routing configurations are adopted. Routing between different ASes (*interdomain routing*) is made possible by the Border Gateway Protocol (BGP) [YTS06]. BGP has been conceived to support routing optimization as well as the specification of political or economical constraints on the routing. Because of its features, BGP is the ideal solution to implement, for example, commercial relationships.

Two BGP-speaking routers that are configured to exchange BGP routing information are said to have a *peering*. Routing information exchanged over a peering includes at least data about the reachability of blocks of contiguous network addresses called *network prefixes*. Because of the relevance to the operation of Internet routing, a common requirement for both operators and researchers is to get a visual representation of the Internet topology at the level of peerings between Autonomous Systems.

25.2.5 The User's Point of View

The configuration of a network, even of a local area network, consists of a plenty of settings. However, depending on the type of usage and on the skill level of a network user, only a few settings may be relevant and, therefore, interesting candidates for visualization.

An ordinary user is typically only interested in getting plug-and-play connectivity, if possible without having to explicitly configure any parameter. In this case, getting a graphical representation of the network would be helpful for the user, because he could get an automatically generated representation of the topology without having to fiddle with any devices or advanced settings. Some modern operating systems provide this feature out of the box, usually as a support to automatic network troubleshooting procedures [Mic09].

Network administrators and operators have full knowledge of network settings and may also be responsible for its design. In this case, the purpose of a graphical representation would be to make maintenance easier, and for this reason the visualization should be enriched with additional information such as link usage and capacities, commercial relationships, routing changes, etc. Sometimes the visualization system may allow the administrator to interact with the network, so that he can quickly reconfigure some settings without using uncomfortable router interfaces.

Researchers have strong interests in studying the behavior of a network in order to build models that capture well network events and design algorithms and protocols that support a more efficient and robust operation. For this purpose, researchers may resort to examining different kinds of data, depending on the kind of analysis to be performed. Therefore, a visual representation that is augmented with additional information from those data can aid in pointing out interesting patterns and validating existing models against real world data.

25.3 A Taxonomy of Visualization Methods and Tools

It has been shown in previous sections that several aspects in the operation of a network can be better investigated by taking advantage of a visualization system. For this reason, a variety of methodologies and tools have been made available, each designed to address a certain kind of analysis.

The literature about the visualization of computer networks includes some interesting survey contributions. A paper by Withall et al. [WPP07] first considers some basic guidelines that network visualization systems should obey, then it distinguishes between contributions where the network topology is laid out using real geographic coordinates for network nodes and contributions where this constraint is relaxed. The same paper also surveys techniques focused on the visualization of data from a single point in the network, typically in the form of a plot. In 2002, Dodge et al. published a comprehensive book [KD01] that classifies visualization systems according to the type of information to be displayed. In particular, the book introduces visualization systems putting them in a historical perspective. Next, it spans over techniques for visualizing network topologies augmented with traffic volumes, the relationships among web pages, the structure of social networks, and some cues for futurist visualizations of the Internet. A report by Vandenberghe et al. [VT06] briefly reviews some layout algorithms, then shows their effectiveness on a reference network topology. The same paper also provides a quick comparative evaluation of the described algorithms.

In this section we review methodologies and tools for the visualization of computer networks by proposing a classification according to some fundamental coordinates: the set of data being visualized, the drawing paradigm that is used to display information, and the features of the tools that implement visualization methodologies. Consider that some visualization methodologies may be conceived to work on nearly arbitrary data sets and to support different drawing paradigms. Such methodologies do not fit this classification and in the following will be gathered in groups called "Other." The following conventions are adopted in the classification:

- If a contribution adopts multiple approaches or, in general, falls into multiple classes, its citation is repeated for all the applicable values of the affected classification coordinate. For example, [BEW95] supports visualization at different scales and therefore may appear multiple times in the same table.

- If the methodology adopted in a paper or in a tool applies to arbitrary values of some classification coordinate, the contribution is considered in the Arbitrary or Customizable class.

- Contributions which do not fit any of the proposed values for a classification coordinate appear in the Other class. There are also contributions which we could not classify along some coordinate, for example because the approach is scarcely documented, and they appear in the Unknown class. Moreover, contributions that cannot be perfectly fit into a value of a classification coordinate because of some specificities (e.g., visualization at the level of granularity of countries instead of Autonomous Systems, or visualization of the nodes of a circuit-switched network instead of the routers of a packet-switched one) are accommodated in the best reasonably fitting class.

- All the contributions that are relevant within a certain section appear in every table in that section. For example, Tables 25.1 through 25.7 provide the reader with a general classification of the literature and therefore each of them considers all the contributions about visualization. Instead, each of the Tables from 25.10 to 25.13 considers all the contributions about Internet-scale visualization.

Some of the classification coordinates that we introduce in the following could be further refined. For example, the scale could consider the span of time considered in visualizations of historical data, or the number of packets that are transmitted during an observation period. In order to make the classification simpler and better understandable, we have picked values for the classification coordinates that offer a compromise between precision of the classification and ease of lookup.

25.3.1 Visualized Data

Depending on the specific goal, a network visualization system may visualize different kinds of data. We classify these data according to the following coordinates.

Scale : The visualization may span a local network, the entire network of an Internet Service Provider, or even the whole Internet. Table 25.1 classifies the contributions in the literature with respect to the scale of the visualization. As shown in the table, a significant number of contributions consider the whole Internet, and there are some that support visualization at different scales.

Scale	**Internet**	[BBGW04], [BBP07], [BEW95], [HPF07], [PH99], [PN99], [Jac99], [CAI09], [Dit09], [CBB00], [CDD+00], [CDM+06], [CDM+05], [CE95], [CEH96], [CRC+08], [GGW07], [GKN04], [GT00], [kc97], [GMO+03], [LMZ04], [LMZ06], [MB95], [OCP+07], [OCLZ08], [Piz07], [YSS05], [Sii01], [RIS09], [Oli09], [OC07], [oANTtE08], [AS06], [Gro00], [RTU09], [Che07], [Pro02], [LUM], [ea05], [Des09], [Map08], [DLV97], [Cor09], [Bou02], [Tel09], [Lim09], [Vis09], [Aug03], [WS04], [YGM05], [YMMW09]
	ISP	[AGL+08], [AGN99], [BCD+04], [BEW95], [Mei00], [kcH97], [EHH+00], [FNMT94], [GMN03], [KMG88], [KNK99], [KNTK99], [Kvi03], [MFKN07], [MHkcF96], [MKN+07], [Piz07], [SMW04], [Sal00], [3Co09], [oANTtE08], [Ent09], [RTU09], [IBM09], [Dar09], [Net09c], [WAN08], [EHH+05], [Com09b], [CP], [Hew09], [Tec09], [Sof09b], [UNI09], [Jon09]
	Local	[AGL+08], [EHH+00], [EW93], [KGS07], [PIP05], [Mic09], [WCH+03], [3Co09], [Ent09], [TvAG+06], [Net09b], [Hir07], [Net09c], [Vol09], [EHH+05], [NoCSNCTUT09], [Wyv09], [net09a], [Tec09], [Sof09a], [Tec05], [Jon09], [Ips09], [WH09], [ZW92]
	Arbitrary	[AHDBV05b], [AHDBV05a], [BMB00], [HNkc97], [HJWkc98], [GH02], [McR99], [Mun97], [Hyu05], [Bro01], [AHDBV], [Cor]

Table 25.1   A classification of the state of the art in terms of visualization scale.

Granularity : Determines the level of detail at which information is made available. A network visualization system may display single routers and hosts, the Points of Presence (POPs, introduced in Section 25.6), or the OSPF areas in the network of an Internet Service Provider, or the Autonomous Systems that, all together, constitute the Internet. It is interesting to correlate the granularity of the visualization with the scale. From Table 25.2 it is possible to notice that, even if the most natural granularities are perhaps Autonomous System for the Internet, POP for ISPs, and Router/Host for Local networks, there are several works that explore different choices. For example the router-level granularity, typically adopted for local networks, is often used in the visualization of POPs or even the Internet. Of course, some of the contributions that allow the visualization with an arbitrary scale and granularity trade quality of the representation for flexibility.

Additional displayed information : Displayed network topologies may be augmented with auxiliary information that better describe the features of nodes and links. Such information may include bandwidth, delays, traffic volumes, TCP ports, geographical locations, etc., and are typically encoded by using different sizes, colors, and labels for both vertices and edges. Table 25.3 classifies contributions in the literature according to their ability to enrich visualized network topologies with additional information. It can be easily noticed that only a small number of visualization approaches is limited to the representation of the sole topology. The row Unknown accounts for those contributions for which it could not be determined whether additional information is also displayed.

Data source : Typical sources of information about the operation of a network are collections of routing data. Since there are lots of different ways to obtain such data, instead of classifying the literature along this coordinate we separately pro-

		Scale			
		Internet	ISP	Local	Arbitrary
Granularity	**Autonomous System**	[BBGW04], [BBP07], [CAI09], [CDD+00], [CDM+06], [CDM+05], [CRC+08], [GGW07], [kc97], [GMO+03], [LMZ04], [LMZ06], [OCP+07], [OCLZ08], [RIS09], [Oli09], [OC07], [Gro00], [Pro02], [Des09], [Cor09], [Bou02], [WS04], [YGM05], [YMMW09]			
	POP	[CE95], [CEH96]	[AGN99], [kcH97], [KNK99], [KNTK99], [SMW04]		
	Router/Host	[BEW95], [PN99], [Jac99], [Dit09], [CBB00], [GT00], [Piz07], [YSS05], [Sii01], [oANTtE08], [AS06], [RTU09], [Che07], [LUM], [ea05], [Vis09], [Aug03]	[AGL+08], [BCD+04], [BEW95], [Mei00], [EHH+00], [FNMT94], [KMG88], [Kvi03], [MHkcF96], [Piz07], [SMW04], [Sal00], [3Co09], [oANTtE08], [Ent09], [RTU09], [IBM09], [Dar09], [Net09c], [WAN08], [EHH+05], [Com09b], [CP], [Hew09], [Tec09], [UNI09], [Jon09]	[AGL+08], [EHH+00], [PIP05], [Mic09], [WCH+03], [3Co09], [Ent09], [TvAG+06], [Net09b], [Hir07], [Net09c], [Vol09], [EHH+05], [NoCSNCTUT09], [Wyv09], [net09a], [Tec09], [Sof09a], [Tec05], [Jon09], [Ips09], [WH09], [ZW92]	[McR99]
	Arbitrary	[HPF07], [PH99], [Map08], [Tel09], [Lim09]	[MFKN07], [MKN+07], [Sof09b]	[EW93]	[AHDBV05b], [AHDBV05a], [BMB00], [HNkc97], [HJWkc98], [GH02], [Mun97], [Hyu05], [Bro01], [AHDBV], [Cor]
	Other	[GKN04], [MB95], [DLV97]	[GMN03]	[KGS07]	

Table 25.2 A classification of the state of the art that compares visualization granularity against scale.

	Yes	[AGN99], [BBP07], [BEW95], [BMB00], [Mei00], [HPF07], [PH99], [PN99], [Jac99], [CAI09], [kcH97], [HNkc97], [HJWkc98], [McR99], [Mun97], [Hyu05], [CBB00], [CDD$^+$00], [CDM$^+$06], [CE95], [CEH96], [EHH$^+$00], [GMN03], [GMO$^+$03], [KGS07], [KNK99], [KNTK99], [Kvi03], [LMZ04], [LMZ06], [MB95], [MFKN07], [MKN$^+$07], [OCP$^+$07], [OCLZ08], [Piz07], [Sal00], [Sii01], [Bro01], [Oli09], [OC07], [TvAG$^+$06], [Gro00], [Hir07], [RTU09], [IBM09], [Dar09], [Che07], [Vol09], [Cor], [LUM], [WAN08], [EHH$^+$05], [Wyv09], [Tec09], [ea05], [DLV97], [Cor09], [Sof09a], [Tel09], [Sof09b], [UNI09], [Lim09], [Vis09], [Jon09], [Ips09], [Aug03], [WS04]
Additional Displayed Information	No	[AHDBV05b], [AHDBV05a], [BBGW04], [BCD$^+$04], [GH02], [Dit09], [CDM$^+$05], [CRC$^+$08], [FNMT94], [GGW07], [GKN04], [GT00], [kc97], [KMG88], [MHkcF96], [PIP05], [SMW04], [YSS05], [Mic09], [WCH$^+$03], [AS06], [Pro02], [AHDBV], [NoCSNCTUT09], [Com09b], [CP], [Bou02], [WH09], [YGM05]
	Unknown	[AGL$^+$08], [EW93], [3Co09], [RIS09], [oANTtE08], [Ent09], [Net09b], [Net09c], [net09a], [Hew09], [Des09], [Map08], [Tec05], [YMMW09], [ZW92]

Table 25.3 Capability of displaying additional information besides the network topology.

vide in Section 25.4 a detailed description of the data sources used in visualization systems.

Collection rate : The displayed data can be gathered on demand, at the moment in which the visualization is requested by the user, or on a periodical basis. This coordinate of the classification determines how often the data is reloaded. Consider that low (e.g., periodical) collection rates may imply a skew between the depicted network and its actual status.

Collection strategies : Visualized information can be retrieved by using different methodologies including, for example, active probing of a network and passive monitoring. Table 25.4 classifies the state of the art according to the rate and strategy adopted to collect the information to be visualized. Most of the contributions fall in the class Other because either the collection process is undocumented, or they provide already drawn network topologies without details about the way they have been laid out.

25.3.2   Graph Drawing Conventions and Methodologies

We focus on visualization methodologies and systems that display network information using a graph. Whereas this is a commonly adopted metaphor, different data sets are better visualized using different graph drawing conventions and methodologies. We classify the graph drawing paradigm according to the following coordinates:

Graph drawing convention : A drawing convention is a basic rule that the drawing must satisfy to be admissible. In this chapter we mainly focus on different conventions for edge representation. Hence, we distinguish among straight-line drawings, where edges are drawn as segments, curved-line drawings, where edges are drawn as curves (e.g., parametric curves), and orthogonal drawings, where edges are represented with polygonal lines composed by horizontal and

		Collection Strategy			
		Passive Monitoring	Active Probing	Customizable	Other
Collection Rate	Periodic	[CDD+00], [CRC+08], [OCP+07], [OCLZ08], [Piz07], [RIS09], [Oli09], [OC07], [Gro00], [RTU09]	[CBB00], [Kvi03], [Sii01], [Che07], [LUM], [ea05], [UNI09]		
	On Demand	[TvAG+06]	[Mic09], [3Co09], [Ent09], [Net09b], [Hir07], [IBM09], [Net09c], [Vol09], [Cor], [Wyv09], [Vis09], [Aug03]		[LMZ04], [LMZ06], [Des09]
	Customizable	[CDM+06], [CDM+05], [Com09b], [Tec05], [ZW92]	[Ips09]	[AHDBV05b], [AHDBV05a], [BMB00], [HPF07], [PH99], [HNkc97], [HJWkc98], [GH02], [Mun97], [Hyu05], [EW93], [KMG88], [Bro01], [AHDBV], [Lim09], [Jon09]	[CE95], [CEH96], [Dar09], [Hew09]
	Other	[BCD+04], [Sof09a], [YMMW09]	[Mei00], [PN99], [Jac99], [CAI09], [Dit09], [GT00], [GMO+03], [MHkcF96], [SMW04], [Sal00], [YSS05], [AS06], [Tec09]		[AGL+08], [AGN99], [BBGW04], [BBP07], [BEW95], [kcH97], [McR99], [EHH+00], [FNMT94], [GGW07], [GKN04], [GMN03], [kc97], [KGS07], [KNK99], [KNTK99], [MB95], [MFKN07], [MKN+07], [PIP05], [WCH+03], [oANTtE08], [Pro02], [WAN08], [EHH+05], [NoCSNCTUT09], [net09a], [CP], [Map08], [DLV97], [Cor09], [Bou02], [Tel09], [Sof09b], [WH09], [WS04], [YGM05]

Table 25.4 A classification of the state of the art according to the rate and strategy by which visualized information is collected.

Graph Drawing Convention	Straight-Line	[AHDBV05b], [AHDBV05a], [AGL+08], [BBGW04], [BBP07], [BEW95], [BMB00], [Mei00], [PH99], [PN99], [Jac99], [CAI09], [kcH97], [HNkc97], [HJWkc98], [GH02], [Mun97], [Hyu05], [CBB00], [CDM+06], [CDM+05], [CE95], [CEH96], [CRC+08], [EHH+00], [EW93], [FNMT94], [GGW07], [GKN04], [GMN03], [GT00], [kc97], [GMO+03], [KGS07], [KMG88], [Kvi03], [LMZ04], [LMZ06], [MB95], [MFKN07], [MKN+07], [OCP+07], [OCLZ08], [PIP05], [Piz07], [SMW04], [Sal00], [YSS05], [Sii01], [WCH+03], [3Co09], [RIS09], [Bro01], [Oli09], [OC07], [oANTtE08], [AS06], [TvAG+06], [Net09b], [RTU09], [Dar09], [Che07], [Pro02], [Net09c], [AHDBV], [Cor], [LUM], [WAN08], [EHH+05], [NoCSNCTUT09], [CP], [Hew09], [Tec09], [ea05], [Des09], [DLV97], [Cor09], [Bou02], [Sof09a], [Tec05], [UNI09], [Vis09], [Jon09], [Ips09], [WH09], [WS04], [YMMW09], [ZW92]
	Curved-Line	[CE95], [CEH96], [MHkcF96], [oANTtE08], [Hir07], [Com09b], [Jon09], [Aug03], [YGM05]
	Orthogonal-Line	[BCD+04], [CDD+00], [Mic09], [Gro00], [Dar09], [Vol09], [Wyv09], [Tec09]
	Other/Unknown	[AGN99], [HPF07], [Dit09], [McR99], [KNK99], [KNTK99], [Ent09], [IBM09], [net09a], [Map08], [Tel09], [Sof09b], [Lim09]

Table 25.5   Graph drawing conventions adopted in different approaches for the visualization of computer networks.

vertical segments. Table 25.5 shows that the vast majority of the literature adopts a straight-line convention, probably because of its simplicity. The row Other/Unknown includes those contributions for which either the graph drawing convention is arbitrary (e.g., configurable), or it is undocumented. For example, in [Sof09b] the convention can be selected by the user, while in commercial visualization systems such as [Ent09] there is no publicly available specification of the graph layout engine. It is very interesting to point out that most of the visualization methodologies do not consider obtaining a planar drawing as a priority objective.

Readers interested in more details about specific graph drawing conventions may refer to the applicable chapter in this handbook.

Spatial dimension : While most methodologies build the graphical representation on a plane, some systems provide the user with the ability to explore a three-dimensional view of the network. Table 25.6 shows how contributions in the literature are distributed by adopted spatial dimension. For example, [MHkcF96] proposes a visualization of the MBone, namely the Multicast experimental backbone that was deployed around the early 1990s. Figure 25.1 shows an example of this visualization, obtained using tools described in [HNkc99].

Graph drawing methodologies : The graph drawing methodologies adopted in network visualization may include, for example, hierarchical and upward planar drawings, circular layouts, and force-directed methods (we refer in this way to a variety of methods, including spring embedders, magnetic fields, barycenter methods, etc.). See [DETT99] for a survey of existing methodologies. The usage of these methodologies will be discussed in the sections devoted to the different scales of the visualization.

	2D	[BBP07], [BCD⁺04], [BEW95], [Mei00], [HPF07], [PH99], [PN99], [Jac99], [CAI09], [kcH97], [HNkc97], [HJWkc98], [GH02], [Dit09], [McR99], [CBB00], [CDD⁺00], [CDM⁺06], [CDM⁺05], [CRC⁺08], [EHH⁺00], [EW93], [FNMT94], [GGW07], [GKN04], [GMN03], [GT00], [KGS07], [KMG88], [Kvi03], [LMZ04], [LMZ06], [MFKN07], [MKN⁺07], [OCP⁺07], [OCLZ08], [PIP05], [Piz07], [SMW04], [Sal00], [Sii01], [Mic09], [WCH⁺03], [3Co09], [RIS09], [Oli09], [OC07], [oANTtE08], [Ent09], [TvAG⁺06], [Net09b], [Gro00], [Hir07], [RTU09], [IBM09], [Dar09], [Che07], [Pro02], [Net09c], [Vol09], [Cor], [LUM], [WAN08], [EHH⁺05], [NoCSNCTUT09], [Com09b], [Wyv09], [net09a], [Hew09], [Tec09], [Des09], [Map08], [DLV97], [Cor09], [Bou02], [Sof09a], [Tec05], [Tel09], [UNI09], [Vis09], [Jon09], [Ips09], [WH09], [WS04], [YMMW09], [ZW92]
Dimensions of the Visualization	3D	[AHDBV05b], [AHDBV05a], [AGL⁺08], [AGN99], [BBGW04], [BMB00], [Mun97], [Hyu05], [CE95], [CEH96], [kc97], [GMO⁺03], [KNK99], [KNTK99], [MB95], [MHkcF96], [YSS05], [Bro01], [AS06], [AHDBV], [CP], [ea05], [Aug03], [YGM05]
	Customizable	[Sof09b], [Lim09]

Table 25.6 Spatial dimension adopted in different visualization approaches.

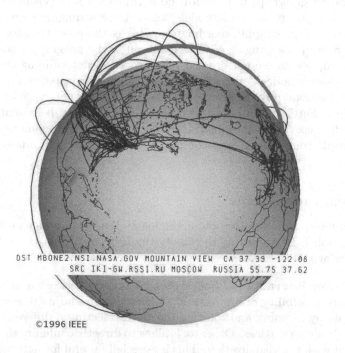

DST MBONE2.NSI.NASA.GOV MOUNTAIN VIEW CA 37.39 -122.08
SRC IKI-GW.RSSI.RU MOSCOW RUSSIA 55.75 37.62

©1996 IEEE

Figure 25.1 A three-dimensional visualization of the global topology of the MBone, the Internet's multicast backbone, as it appeared in 1996. The picture is taken from [MHkcF96].

Uses absolute geographic coordinates	**Yes**	[AGN99], [BEW95], [Mei00], [HPF07], [PH99], [PN99], [Jac99], [kcH97], [HNkc97], [HJWkc98], [Dit09], [CE95], [CEH96], [KNK99], [KNTK99], [Kvi03], [MHkcF96], [SMW04], [oANTtE08], [Dar09], [CP], [Bou02], [Tel09], [UNI09], [Lim09], [Vis09], [Aug03], [YGM05], [YMMW09]
	No	[AHDBV05b], [AHDBV05a], [AGL$^+$08], [BBGW04], [BBP07], [BCD$^+$04], [BMB00], [CAI09], [GH02], [McR99], [Mun97], [Hyu05], [CBB00], [CDD$^+$00], [CDM$^+$06], [CDM$^+$05], [CRC$^+$08], [EHH$^+$00], [EW93], [FNMT94], [GGW07], [GKN04], [GMN03], [GT00], [kc97], [GMO$^+$03], [KGS07], [KMG88], [LMZ04], [LMZ06], [MB95], [MFKN07], [MKN$^+$07], [OCP$^+$07], [OCLZ08], [PIP05], [Piz07], [Sal00], [YSS05], [Sii01], [Mic09], [WCH$^+$03], [RIS09], [Bro01], [Oli09], [OC07], [AS06], [TvAG$^+$06], [Net09b], [Gro00], [Hir07], [RTU09], [IBM09], [Che07], [Pro02], [Net09c], [AHDBV], [Vol09], [LUM], [WAN08], [EHH$^+$05], [NoCSNCTUT09], [Com09b], [Wyv09], [net09a], [Hew09], [Tec09], [ea05], [Des09], [DLV97], [Cor09], [Sof09a], [Tec05], [Ips09], [WH09], [WS04], [ZW92]
	Unknown	[3Co09], [Ent09], [Cor], [Map08], [Sof09b], [Jon09]

Table 25.7 A classification of visualization methodologies according to the usage of absolute geographic coordinates for the placement of network nodes.

Detailed information about specific graph drawing methodologies can be found in the applicable chapters of this handbook.

Usage of absolute geographic location coordinates : If the visualization includes the network topology, some methodologies envisage arranging network nodes on the basis of real geographic coordinates. This is the case, e.g., for popular visual traceroute tools [Aug03, Vis09]. Quite often the geographic coordinates of network components are not deemed relevant. Also, determining the actual position of network nodes is not an easy task. Therefore, as shown in Table 25.7, most contributions rely on layout approaches that do not consider geographic coordinates. Further, it may happen that the entities to be visualized do not have specific geographic coordinates. For example, an Autonomous System can span multiple countries or even continents, having routers in many geographic locations.

25.3.3 Visualization Tools

For the cases in which the visualization methodology has been implemented in the form of a visualization tool, it is interesting to classify the tool according to the functions offered by the user interface.

Possibility of user interaction : Some tools display the network as a static image, without any possibility of interaction. Others allow to adjust the visualization, for example by zooming and rotating the view, selecting a different layout, and manually dragging vertices. Other tools allow to directly configure the network by interacting with the visualization. This is especially useful for network operators, and is often possible only with commercial tools.

Static/Dynamic(Animated) : If the tool is aimed at visualizing dynamically changing information (e.g., routing data), the changes may be displayed by animation.

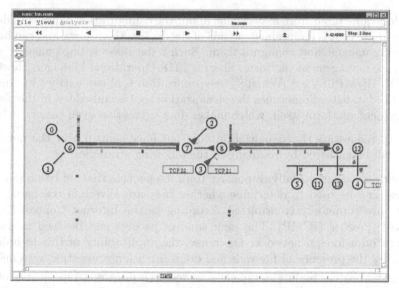

Figure 25.2 A screenshot of the NAM Network Animator, taken from [EHH+05].

Consider that this poses very interesting challenges in the optimization of the layout. The classification of the state of the art according to the ability to animate the visualization is further discussed in the following sections. Figure 25.2 shows a screenshot of a well-known tool for network simulation, that is able to display the flow of packets along the pipes.

Visualization tools are made available as software pieces in different flavors:

Type of tool : The tool may consist of a library of drawing functions, a standalone application, or an applet that can be embedded in a web browser. If a certain approach only consists of a methodological contribution and we were not able to identify a publicly available implementation, we classify that contribution separately.

License : Depending on the target users, the tool may be distributed commercially or freely, possibly with an open source license.

25.4 Data Sources

The information to be visualized can be gathered from different data sources. The available sources can vary depending on whether the network to be visualized is managed by the same entity requesting the visualization or not. In the first case, one can use two main strategies:

- Routers maintain a rich database containing information about their current status. This database is called Management Information Base (MIB) and can be accessed by a widely available protocol called Simple Network Management Protocol (SNMP) [MR91, CFSD90]. A visualization tool can query the MIB of routers via SNMP in order to collect information about the network. This is the approach adopted by the Polyphemus tool [BCD+04].

- Providers quite often store information about the configuration of network devices in a database that is maintained outside of the devices themselves and can be used to quickly manage and configure them. Such a database is implemented within a network management platform, like, e.g., HP Operations Manager [Hew09] or WANDL IP/MPLSView [WAN08]. A visualization tool can extract information from this database. Sometimes the visualization tool is embedded in the network management platform itself, which makes data extraction even easier.

If the entity requesting the visualization does not have control over the network, information can still be gathered by resorting to the following strategies:

- Network devices are usually responsive to probe packets that, in troubleshooting sessions, can be used to determine whether they are alive and reachable. Such packets are commonly transmitted according to the Internet Control Message Protocol [Pos81a] (ICMP). The same kind of packets can be used to discover the links on a foreign network. Of course, the applicability of this technique is limited by the presence of firewalls and other intrusion prevention systems.

- Some organizations maintain, typically for research purposes, repositories of publicly available information collected from routers. This is the case, for example, of the RIPE Routing Information Service [RIP] (RIS) or the Oregon Route Views project [oO]. These services collect information about the reachability of Autonomous Systems on a periodic basis. Information about router configurations is also maintained in a worldwide Internet Routing Registry [NCC] (IRR), which is also used to extract network topologies at the Autonomous System level. Unfortunately, the IRR is not always up-to-date with actually deployed configurations.

From another point of view, it is possible to consider the strategy adopted to collect the visualized information. Roughly speaking, there are two possible ways of achieving this: actively probing or passively monitoring the network. An example of active probing strategy is the usage of the traceroute tool to discover the routers along paths to certain destinations. This is the approach adopted in the CAIDA Archipelago measurement infrastructure [CAI07]. On the other hand, passive monitoring may consist in observing the status of the network by accessing routing information stored at the routers.

Table 25.8 shows the relationship between the visualization scale and the adopted collection strategy. It is clear from the table that the approaches for Internet visualization exploit many different collection strategies, while data for visualizing local and ISP-scale networks is often collected using active probing.

The collection strategy also influences the timing of data accesses. In particular, data may be collected periodically or just when it is needed to generate a visualization (see Table 25.9). Interestingly, collecting data on a periodic basis for the sole visualization purpose is rather unusual on local networks. We recall that, depending on the adopted collection rate, the displayed information may not reflect the current status of the network.

Table 25.4 puts in evidence the relationships between the collection strategy and the collection rate. As shown in the table, those tools whose collection strategy can be customized usually allow to customize the collection rate as well. Works for which the collection rate and/or the strategy are unspecified typically focus on a single snapshot of a network (e.g., a graph of an ISP's infrastructure at a given time instant) or simply provide a visualization methodology without considering the data sets to be displayed.

		Collection Strategy			
		Passive Monitoring	Active Probing	Customizable	Other
Scale	Internet	[CDD$^+$00], [CDM$^+$06], [CDM$^+$05], [CRC$^+$08], [OCP$^+$07], [OCLZ08], [Piz07], [RIS09], [Oli09], [OC07], [Gro00], [RTU09], [YMMW09]	[PN99], [Jac99], [CAI09], [Dit09], [CBB00], [GT00], [GMO$^+$03], [YSS05], [Sii01], [AS06], [Che07], [LUM], [ea05], [Vis09], [Aug03]	[HPF07], [PH99], [Lim09]	[BBGW04], [BBP07], [BEW95], [CE95], [CEH96], [GGW07], [GKN04], [kc97], [LMZ04], [LMZ06], [MB95], [oANTtE08], [Pro02], [Des09], [Map08], [DLV97], [Cor09], [Bou02], [Tel09], [WS04], [YGM05]
	ISP	[BCD$^+$04], [Piz07], [RTU09], [Com09b]	[Mei00], [Kvi03], [MHkcF96], [SMW04], [Sal00], [3Co09], [Ent09], [IBM09], [Net09c], [Tec09], [UNI09]	[KMG88], [Jon09]	[AGL$^+$08], [AGN99], [BEW95], [kcH97], [EHH$^+$00], [FNMT94], [GMN03], [KNK99], [KNTK99], [MFKN07], [MKN$^+$07], [oANTtE08], [Dar09], [WAN08], [EHH$^+$05], [CP], [Hew09], [Sof09b]
	Local	[TvAG$^+$06], [Sof09a], [Tec05], [ZW92]	[Mic09], [3Co09], [Ent09], [Net09b], [Hir07], [Net09c], [Vol09], [Wyv09], [Tec09], [Ips09]	[EW93], [Jon09]	[AGL$^+$08], [EHH$^+$00], [KGS07], [PIP05], [WCH$^+$03], [EHH$^+$05], [NoCSNCTUT09], [net09a], [WH09]
	Arbitrary		[Cor]	[AHDBV05b], [AHDBV05a], [BMB00], [HNkc97], [HJWkc98], [GH02], [Mun97], [Hyu05], [Bro01], [AHDBV]	[McR99]

Table 25.8 Data collection strategies adopted for different visualization scales.

		Collection Rate			
		Periodic	On Demand	Customizable	Other
Scale	Internet	[CBB00], [CDD$^+$00], [CRC$^+$08], [OCP$^+$07], [OCLZ08], [Piz07], [Sii01], [RIS09], [Oli09], [OC07], [Gro00], [RTU09], [Che07], [LUM], [ea05]	[LMZ04], [LMZ06], [Des09], [Vis09], [Aug03]	[HPF07], [PH99], [CDM$^+$06], [CDM$^+$05], [CE95], [CEH96], [Lim09]	[BBGW04], [BBP07], [BEW95], [PN99], [Jac99], [CAI09], [Dit09], [GGW07], [GKN04], [GT00], [kc97], [GMO$^+$03], [MB95], [YSS05], [oANTtE08], [AS06], [Pro02], [Map08], [DLV97], [Cor09], [Bou02], [Tel09], [WS04], [YGM05], [YMMW09]
	ISP	[Kvi03], [Piz07], [RTU09], [UNI09]	[3Co09], [Ent09], [IBM09], [Net09c]	[KMG88], [Dar09], [Com09b], [Hew09], [Jon09]	[AGL$^+$08], [AGN99], [BCD$^+$04], [BEW95], [Mei00], [kcH97], [EHH$^+$00], [FNMT94], [GMN03], [KNK99], [KNTK99], [MFKN07], [MHkcF96], [MKN$^+$07], [SMW04], [Sal00], [oANTtE08], [WAN08], [EHH$^+$05], [CP], [Tec09], [Sof09b]
	Local		[Mic09], [3Co09], [Ent09], [TvAG$^+$06], [Net09b], [Hir07], [Net09c], [Vol09], [Wyv09]	[EW93], [Tec05], [Jon09], [Ips09], [ZW92]	[AGL$^+$08], [EHH$^+$00], [KGS07], [PIP05], [WCH$^+$03], [EHH$^+$05], [NoCSNCTUT09], [net09a], [Tec09], [Sof09a], [WH09]
	Arbitrary		[Cor]	[AHDBV05b], [AHDBV05a], [BMB00], [HNkc97], [HJWkc98], [GH02], [Mun97], [Hyu05], [Bro01], [AHDBV]	[McR99]

Table 25.9   Collection rates adopted for different visualization scales.

25.5 Visualization of the Internet

The visualization of the whole Internet poses important challenges, because the amount of information to be displayed is generally very large and still the drawing presented to the user must preserve readability. To get a feeling of how complex the Internet can be, consider that its size can be roughly estimated in more than 20,000 Autonomous Systems and more than 60,000 links between them, and these sizes keep growing with time [MKF+06, RTM08]. Given that an ISP can span over several Autonomous Systems, managing tens of thousands of routers on its own [SMW04], and that there exist tens of thousands of ISPs, it can be easily imagined how complex it is to provide a complete yet useful visualization that spans the whole Internet.

In order to provide an overview of the most commonly adopted approaches for Internet visualization, Table 25.10 proposes a comparison of the graph drawing conventions against the graph drawing methodologies used in the literature about this field. There is a clear predominance in the adoption of the straight-line convention, and many approaches rely on force-directed methods for vertex placement. Figures 25.3(a) and 25.3(b) contain screenshots of two well-known systems for visualizing Internet routing at the Autonomous System level: BGPlay [CDM+05] and LinkRank [LMZ06]. Both tools use a spring embedder as node placement algorithm. Also, these tools make the choice of presenting only a small portion of the Internet. BGPlay focuses on how a network prefix is seen from a collection of vantage points, while LinkRank tries to put in evidence links that may have undergone faults. CAIDA's Walrus [Mun97, Hyu05] is a tool that exploits a three-dimensional hyperbolic geometry to display graphs under a fisheye-like distortion. Walrus is able to visualize network topologies consisting of more than 500,000 nodes and more than 600,000 links.

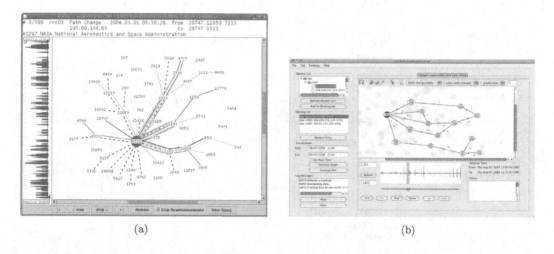

(a) (b)

Figure 25.3 The BGPlay [CDM+05] ((a)) and LinkRank [LMZ06] ((b)) tools provide an animated visualization of routing changes at the Autonomous System level. The screenshots are taken from [Rom09] and [Lab09], respectively.

Table 25.11 matches graph drawing methodologies against the ability to provide an animated visualization. Most of the tools that provide an animation rely on a force-directed drawing methodology. In order to achieve a smooth transition between different frames of the animation, sometimes the placement of nodes is influenced by constraints.

		Graph Drawing Convention			
		Straight-Line	**Curved-Line**	**Orthogonal-Line**	**Other/Unknown**
Graph Drawing Methodology	**Force Directed**	[BBGW04], [BEW95], [CBB00], [CDM⁺06], [CDM⁺05], [GGW07], [LMZ04], [LMZ06], [OCP⁺07], [OCLZ08], [Piz07], [Sii01], [RIS09], [Oli09], [OC07], [RTU09], [Che07], [Pro02], [LUM]			
	Circular	[AHDBV05b], [AHDBV05a], [CAI09], [HNkc97], [HJWkc98], [Mun97], [Hyu05], [CE95], [CEH96], [AHDBV], [Cor09]	[CE95], [CEH96]		
	Clustering	[GMO⁺03], [WS04]		[CDD⁺00], [Gro00]	
	Layering	[BBGW04], [CRC⁺08], [DLV97]			
	Topology-Shape-Metrics			[CDD⁺00], [Gro00]	
	Customizable/ Various	[GH02], [GT00], [kc97], [ea05]			[Lim09]
	Other/Unknown	[BMB00], [PH99], [PN99], [Jac99], [GKN04], [YSS05], [Bro01], [oANTtE08], [AS06], [Cor], [Des09], [Bou02], [Vis09], [YMMW09]	[oANTtE08], [Aug03], [YGM05]		[HPF07], [Dit09], [McR99], [Map08], [Tel09]

Table 25.10   Graph drawing conventions and methodologies adopted for the visualization of the Internet.

		Static	Animated
Graph Drawing Methodology	Force Directed	[BBGW04], [CBB00], [GGW07], [Sii01], [Che07], [Pro02], [LUM]	[BEW95], [CDM+06], [CDM+05], [LMZ04], [LMZ06], [OCP+07], [OCLZ08], [Piz07], [RIS09], [Oli09], [OC07], [RTU09]
	Circular	[AHDBV05b], [AHDBV05a], [CAI09], [HNkc97], [Mun97], [Hyu05], [AHDBV], [Cor09]	[HJWkc98], [CE95], [CEH96]
	Clustering	[CDD+00], [GMO+03], [Gro00], [WS04]	
	Layering	[BBGW04], [CRC+08], [DLV97]	
	Topology-Shape-Metrics	[CDD+00], [Gro00]	
	Customizable/Various	[GH02], [GT00], [kc97], [ea05]	[Lim09]
	Other/Unknown	[PH99], [PN99], [Jac99], [Dit09], [McR99], [GKN04], [YSS05], [oANTtE08], [AS06], [Cor], [Map08], [Bou02], [Tel09], [Vis09], [Aug03], [YGM05], [YMMW09]	[BMB00], [HPF07], [Bro01]

Table 25.11 Visualization of the Internet: ability to animate the visualizations with respect to the adopted graph drawing methodology.

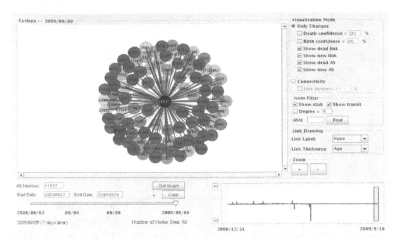

Figure 25.4 The Cyclops tool provides an animated visualization of the changes in the relationships between Autonomous Systems. This is a snapshot of the running tool available at [OC07].

Animations are typically used to convey information about historical data sets (see, e.g., [BBP07]), and often facilitate the exploration of topological changes. An example of a tool providing an animation of Internet routing changes is Cyclops [OCP+07], whose screenshot is provided in Figure 25.4. The tool visualizes disappeared and newly detected peerings between Autonomous Systems over a user-selected time period. The time instant the visualization refers to can be picked by using a slider. Although Table 25.11 shows that many contributions take advantage of animation, in our opinion the obtained results are not always satisfactory. Hence, the animation of network topologies, that combines methodological challenges and practical relevance, is one of the promising fields of research for future developments.

Despite its sheer size, there are efforts to map and visualize the Internet not just at the Autonomous System level, but also at finer levels of granularity. Table 25.12 provides an overview of the graph drawing methodologies exploited in the visualization with different granularities. It can be easily seen that the Internet is sometimes visualized at the router level, like it happens in [CBB00]. More often, network nodes are aggregated to provide a visualization at the Autonomous System level. The Internet topology maps provided by CAIDA [CAI09] are a famous example of this kind of visualization.

It is interesting to observe that some proposals display the Internet topology combined with other topological aspects. As an example, the approach in [CDM+06] allows to visualize the Internet topology in the context of a customer-provider hierarchy. Actually, Autonomous Systems establish commercial agreements in order to gain connectivity. In these agreements each ISP plays the role of provider of certain ISPs and is in turn a customer of others. In [CDM+06] the Internet is displayed within a topographic metaphor that visually renders the customer-provider hierarchy in a pretty intuitive way.

For most of the visualization methodologies, an implementation in the form of a software tool is also available. As shown in Table 25.13, most of the implementations can be accessed freely. Among them, there are several tools that are made available in the form of Java applets, so that the user can benefit from the visualization without having to install any application or to obtain source data.

		Granularity				
		Autonomous System	POP	Router/Host	Arbitrary	Other
Graph Drawing Methodology	**Force Directed**	[BBGW04], [CDM⁺06], [CDM⁺05], [GGW07], [LMZ04], [LMZ06], [OCP⁺07], [OCLZ08], [RIS09], [Oli09], [OC07], [Pro02]		[BEW95], [CBB00], [Piz07], [Sii01], [RTU09], [Che07], [LUM]		
	Circular	[CAI09], [Cor09]	[CE95], [CEH96]		[AHDBV05b], [AHDBV05a], [HNkc97], [HJWkc98], [Mun97], [Hyu05], [AHDBV]	
	Clustering	[CDD⁺00], [GMO⁺03], [Gro00], [WS04]				
	Layering	[BBGW04], [CRC⁺08]				[DLV97]
	Topology-Shape-Metrics	[CDD⁺00], [Gro00]				
	Customizable/ Various	[kc97]		[GT00], [ea05]	[GH02], [Lim09]	
	Other/Unknown	[Des09], [Bou02], [YGM05], [YMMW09]		[PN99], [Jac99], [Dit09], [McR99], [YSS05], [oANTtE08], [AS06], [Vis09], [Aug03]	[BMB00], [HPF07], [PH99], [Bro01], [Cor], [Map08], [Tel09]	[GKN04]

Table 25.12 Graph drawing methodologies adopted for the visualization of the Internet at different levels of granularity.

		License		
		Free	Commercial	Other/Unknown
Type of Tool	Application	[AHDBV05b], [AHDBV05a], [BEW95], [BMB00], [HPF07], [PN99], [Jac99], [CAI09], [HNkc97], [HJWkc98], [GH02], [McR99], [Mun97], [Hyu05], [CDM$^+$06], [CDM$^+$05], [CE95], [CEH96], [LMZ04], [LMZ06], [YSS05], [Bro01], [AS06], [AHDBV], [ea05], [DLV97], [Aug03]	[Cor], [Vis09]	[GMO$^+$03], [YMMW09]
	Java Applet	[PH99], [Dit09], [CDD$^+$00], [CDM$^+$06], [CDM$^+$05], [CRC$^+$08], [OCP$^+$07], [OCLZ08], [Piz07], [RIS09], [Oli09], [OC07], [Gro00], [RTU09], [Pro02], [DLV97]	[Vis09]	
	None Publicly Available		[Map08], [Tel09]	[BBGW04], [BBP07], [CBB00], [GGW07], [GKN04], [GT00], [MB95], [Sii01], [Che07], [LUM], [WS04]
	Other/Unknown		[Des09], [Cor09]	[kc97], [oANTtE08], [Bou02], [Lim09], [YGM05]

Table 25.13 License policies used for Internet visualization tools.

Granularity	**Autonomous System**	
	POP	[AGN99], [kcH97], [KNK99], [KNTK99], [SMW04]
	Router/Host	[AGL$^+$08], [BCD$^+$04], [BEW95], [Mei00], [McR99], [EHH$^+$00], [FNMT94], [KMG88], [Kvi03], [MHkcF96], [Piz07], [SMW04], [Sal00], [3Co09], [oANTtE08], [Ent09], [RTU09], [IBM09], [Dar09], [Net09c], [WAN08], [EHH$^+$05], [Com09b], [CP], [Hew09], [Tec09], [UNI09], [Jon09]
	Arbitrary	[AHDBV05b], [AHDBV05a], [BMB00], [HNkc97], [HJWkc98], [GH02], [Mun97], [Hyu05], [MFKN07], [MKN$^+$07], [Bro01], [AHDBV], [Cor]
	Other	[GMN03]

Table 25.14   A classification of the state of the art on the visualization of ISP networks according to the granularity of the visualization.

25.6   Visualization of an Internet Service Provider Network

The difficulty of visualizing the network of a single ISP is comparable to the one of visualizing the Internet at the Autonomous System level. In fact an ISP has roughly as many network devices as the number of Autonomous Systems in the Internet. In some cases ISP networks may be displayed at a coarser granularity, taking into account the Points of Presence (POP). A POP is a set of network devices housed at a certain location that are used to provide access to the Internet. CAIDA's Mapnet [kcH97] provides a visualization of the interconnections between the POPs, for different ISPs. More often, ISP-scale visualization considers every router or host on the network. This is the case, for example, of the commercial system InterMapper [Dar09], which visualizes routers and their interconnections superimposed on a geographical map. Table 25.14 provides a classification of the literature according to the granularity adopted in the visualization.

Table 25.15 classifies the graph drawing conventions and methodologies adopted within approaches for the visualization of ISP-scale networks. Most notably, almost every contribution adopts a straight-line convention. A very popular visualization tool is Otter [HNkc97], which is at the basis of the famous Internet maps provided by CAIDA. Node placement in Otter happens on the basis of geographic coordinates. Those nodes for which the coordinates are not available are laid out in semi-circles around their parent node. Otter has been exploited to generate several visualizations, including, e.g., the multicast backbone of an ISP.

Some of the contributions exploit clustered drawings. This is the case, for example, of [Sal00], where different virtual communication channels are grouped together to highlight the interconnections established by the ATM protocol [Sta07]. In [CP] the authors visualize the NSFnet, a wide-area network developed by the National Science Foundation (NSF). In a three-dimensional space, they arrange backbone nodes on a higher layer, and client nodes that utilize the backbone on a lower layer. Everything is displayed in the context of a geographical map. The Systrax community proposes a prototype of a tool for visualizing traffic flows collected by the NetFlow tool [Com09b, Sys09a]. The visualization exploits curved lines to represent device interconnections. The Polyphemus tool [BCD$^+$04] visualizes the routing of an ISP by collecting information about the OSPF protocol [Moy94], and exploits an orthogonal drawing with a topology-shape-metrics methodology.

		Graph Drawing Convention			
		Straight-Line	**Curved-Line**	**Orthogonal-Line**	**Other/Unknown**
Graph Drawing Methodology	**Force Directed**	[BEW95], [GMN03], [Piz07], [RTU09], [Hew09]			
	Circular	[AHDBV05b], [AHDBV05a], [HNkc97], [HJWkc98], [Mun97], [Hyu05], [KMG88], [AHDBV]	[Com09b]		
	Clustering	[AGL$^+$08], [SMW04], [Sal00]			
	Layering	[CP]			
	Topology-Shape-Metrics			[BCD$^+$04]	
	Customizable/ Various	[GH02], [EHH$^+$00], [EHH$^+$05], [Tec09]		[Tec09]	[IBM09]
	Other/Unknown	[BMB00], [Mei00], [kcH97], [FNMT94], [Kvi03], [MFKN07], [MKN$^+$07], [SMW04], [3Co09], [Bro01], [oANTtE08], [Dar09], [Net09c], [Cor], [WAN08], [UNI09], [Jon09]	[MHkcF96], [oANTtE08], [Jon09]	[Dar09]	[AGN99], [McR99], [KNK99], [KNTK99], [Ent09]

Table 25.15   Graph drawing conventions and methodologies adopted in the visualization of the network of an ISP.

		Uses absolute geographic coordinates		
		Yes	No	Unknown
Graph Drawing Convention	Straight-Line	[BEW95], [Mei00], [kcH97], [HNkc97], [HJWkc98], [Kvi03], [SMW04], [oANTtE08], [Dar09], [CP], [UNI09]	[AHDBV05b], [AHDBV05a], [AGL+08], [BMB00], [GH02], [Mun97], [Hyu05], [EHH+00], [FNMT94], [GMN03], [KMG88], [MFKN07], [MKN+07], [Piz07], [Sal00], [Bro01], [RTU09], [Net09c], [AHDBV], [WAN08], [EHH+05], [Hew09], [Tec09]	[3Co09], [Cor], [Jon09]
	Curved-Line	[MHkcF96], [oANTtE08]	[Com09b]	[Jon09]
	Orthogonal-Line	[Dar09]	[BCD+04], [Tec09]	
	Other/Unknown	[AGN99], [KNK99], [KNTK99]	[McR99], [IBM09]	[Ent09]

Table 25.16   Graph drawing conventions used in the visualization of ISP-scale networks when geographic location coordinates are or are not considered.

Most of the approaches in which network nodes are placed according to their geographic location make use of a straight-line drawing convention. This is highlighted in Table 25.16.

There are few contributions that support an animated visualization of the ISP under consideration. Table 25.17 relates these contributions with the graph drawing methodology they adopt. Interestingly, some of the tools that arrange nodes according to geographic coordinates also provide support for animated visualizations. For example, the system described in [KNK99, AGN99] displays traffic flows over time, where traffic sources are placed over a geographic map.

Finally, Table 25.18 shows the license policies under which ISP-scale visualization tools are distributed. These tools are sometimes released as standalone applications within network management suites (see, e.g., [Hew09, IBM09, WAN08]).

		Static	Animated
Graph Drawing Methodology	**Force Directed**		[BEW95], [GMN03], [Piz07], [RTU09]
	Circular	[AHDBV05b], [AHDBV05a], [HNkc97], [Mun97], [Hyu05], [KMG88], [AHDBV], [Com09b]	[HJWkc98]
	Clustering	[AGL$^+$08], [SMW04], [Sal00]	
	Layering	[CP]	
	Topology-Shape-Metrics	[BCD$^+$04]	
	Customizable/Various	[GH02], [IBM09], [Tec09]	[EHH$^+$00], [EHH$^+$05]
	Other/Unknown	[Mei00], [kcH97], [McR99], [FNMT94], [Kvi03], [MHkcF96], [SMW04], [3Co09], [oANTtE08], [Ent09], [Dar09], [Net09c], [Cor], [WAN08], [UNI09]	[AGN99], [BMB00], [KNK99], [KNTK99], [MFKN07], [MKN$^+$07], [Bro01], [Jon09]

Table 25.17   Graph drawing methodologies adopted for static/animated visualizations of an ISP's network.

		License		
		Free	**Commercial**	**Other/Unknown**
Type of Tool	**Application**	[AHDBV05b], [AHDBV05a], [BCD$^+$04], [BEW95], [BMB00], [HNkc97], [HJWkc98], [GH02], [McR99], [Mun97], [Hyu05], [Bro01], [AHDBV], [Jon09]	[3Co09], [Ent09], [IBM09], [Dar09], [Net09c], [Cor], [WAN08], [Hew09], [Tec09]	[AGL$^+$08], [EHH$^+$00], [GMN03], [SMW04], [Sal00], [EHH$^+$05]
	Java Applet	[kcH97], [Piz07], [RTU09], [Com09b]		[Sal00]
	None Publicly Available	[CP]		[AGN99], [FNMT94], [KMG88], [KNK99], [KNTK99], [Kvi03], [MFKN07], [MHkcF96], [MKN$^+$07], [UNI09]
	Other/Unknown			[Mei00], [oANTtE08]

Table 25.18   Licensing policies for different kinds of ISP-scale visualization tools.

		Graph Drawing Convention			
		Straight-Line	Curved-Line	Orthogonal Line	Other/Unknown
Graph Drawing Methodology	Force Directed				
	Circular	[AHDBV05b], [AHDBV05a], [HNkc97], [HJWkc98], [Mun97], [Hyu05], [TvAG⁺06], [AHDBV]			
	Clustering	[AGL⁺08]			
	Layering				
	Topology-Shape-Metrics				
	Customizable/ Various	[GH02], [EHH⁺00], [Net09b], [EHH⁺05], [Tec09]		[Tec09]	
	Other/Unknown	[BMB00], [EW93], [PIP05], [WCH⁺03], [3Co09], [Bro01], [Net09c], [Cor], [NoCSNCTUT09], [Sof09a], [Tec05], [Jon09], [Ips09], [WH09], [ZW92]	[Hir07], [Jon09]	[Mic09], [Vol09], [Wyv09]	[McR99], [Ent09], [net09a]

Table 25.19 Graph drawing conventions and methodologies adopted for the visualization of local networks.

25.7   Visualization of Local Networks

A local network typically consists of a few hundreds of devices, hence it is meaningful to visualize it as a whole.

Table 25.19 shows that most of the contributions to the visualization of local networks adopt a straight-line drawing convention and different, sometimes customizable methodologies. There are some notable exceptions to this rule: for example the "Full Map View" embedded in the Microsoft Windows Vista™ operating system (see Figure 25.5) and the LanTopolog [Vol09] discovery tool adopt an orthogonal convention, while the Weathermap application [Jon09] exploits curved lines for the visualization.

Methodologies targeted at the visualization of local networks may also support animated displays. This is useful, for example, to monitor traffic exchanges among network devices or the distribution of bandwitdh usage over time. Tables 25.20, 25.21, and 25.22 classify the literature by correlating the ability of animating the visualization with the graph drawing methodology, the rate, and the strategy by which visualized data are collected, respectively. Some interesting considerations can be derived from these tables.

First of all, providing an animated view is not a fundamental requirement, as there are fewer contributions that have this capability.

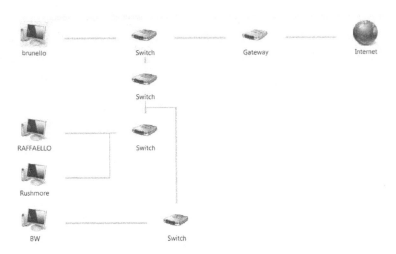

Figure 25.5 A snapshot of the Windows Vista™ "Full Map View" feature.

		Static	Animated
Graph Drawing Methodology	**Force Directed**		
	Circular	[AHDBV05b], [AHDBV05a], [HNkc97], [Mun97], [Hyu05], [AHDBV]	[HJWkc98], [TvAG⁺06]
	Clustering	[AGL⁺08]	
	Layering		
	Topology-Shape-Metrics		
	Customizable/ Various	[GH02], [Net09b], [Tec09]	[EHH⁺00], [EHH⁺05]
	Other/Unknown	[McR99], [EW93], [PIP05], [Mic09], [WCH⁺03], [3Co09], [Ent09], [Hir07], [Net09c], [Vol09], [Cor], [NoCSNCTUT09], [Wyv09], [net09a], [Sof09a], [Ips09], [WH09]	[BMB00], [Bro01], [Tec05], [Jon09], [ZW92]

Table 25.20 Graph drawing methodology adopted for static and animated visualizations of local networks.

		Static	Animated
Collection Rate	**Periodic**		
	On Demand	[Mic09], [3Co09], [Ent09], [Net09b], [Hir07], [Net09c], [Vol09], [Cor], [Wyv09]	[TvAG$^+$06]
	Customizable	[AHDBV05b], [AHDBV05a], [HNkc97], [GH02], [Mun97], [Hyu05], [EW93], [AHDBV], [Ips09]	[BMB00], [HJWkc98], [Bro01], [Tec05], [Jon09], [ZW92]
	Other	[AGL$^+$08], [McR99], [PIP05], [WCH$^+$03], [NoCSNCTUT09], [net09a], [Tec09], [Sof09a], [WH09]	[EHH$^+$00], [EHH$^+$05]

Table 25.21 Data collection rate for static and animated visualizations of local networks.

		Static	Animated
Collection Strategy	**Passive Monitoring**	[Sof09a]	[TvAG$^+$06], [Tec05], [ZW92]
	Active Probing	[Mic09], [3Co09], [Ent09], [Net09b], [Hir07], [Net09c], [Vol09], [Cor], [Wyv09], [Tec09], [Ips09]	
	Customizable	[AHDBV05b], [AHDBV05a], [HNkc97], [GH02], [Mun97], [Hyu05], [EW93], [AHDBV]	[BMB00], [HJWkc98], [Bro01], [Jon09]
	Other	[AGL$^+$08], [McR99], [PIP05], [WCH$^+$03], [NoCSNCTUT09], [net09a], [WH09]	[EHH$^+$00], [EHH$^+$05]

Table 25.22 Strategies by which data are collected for static and animated visualizations of local networks.

		License		
		Free	**Commercial**	**Other/Unknown**
Type of Tool	**Application**	[AHDBV05b], [AHDBV05a], [BMB00], [HNkc97], [HJWkc98], [GH02], [McR99], [Mun97], [Hyu05], [WCH$^+$03], [Bro01], [TvAG$^+$06], [AHDBV], [Vol09], [NoCSNCTUT09], [Jon09], [WH09]	[Mic09], [3Co09], [Ent09], [Net09b], [Hir07], [Net09c], [Cor], [Wyv09], [net09a], [Tec09], [Sof09a], [Ips09]	[AGL$^+$08], [EHH$^+$00], [EHH$^+$05], [Tec05], [ZW92]
	Java Applet			
	None Publicly Available			[EW93], [PIP05]
	Other/Unknown			

Table 25.23   License policies applied for visualization tools targeted at local networks.

Second, the graph drawing methodology adopted for animated visualizations is often undocumented. An exception to this rule is, for example, the well-known network animator Nam [EHH$^+$00] (Figure 25.2), which provides a dynamic visualization of the packets that traverse the links of a network. The tool keeps network nodes in a fixed position while the animation is displayed.

Another interesting observation based on the tables is that tools for animated visualizations usually do not collect data on their own ("Periodic" or "On Demand" collection rate), but rather are fed with data sets that have been gathered separately and contain enough information to support the animation. This is the case, for example, of the Cichlid visualization tool [BMB00], which provides three-dimensional views of resource utilization in a network over time. The gathering process is typically based on a passive observation of the network. This is the approach pursued in the Etherape tool [TvAG$^+$06], which displays network nodes and links with different sizes depending on their network activity.

Table 25.23 shows that most of the tools aimed at the visualization of local networks are distributed, either freely or under a commercial license, in the form of standalone applications. A possible motivation is that Java applets are more suited for the case in which information about the network to be visualized is only available remotely.

25.8   Visualization of Basic Internet Services and Specific Network Contexts

As highlighted in the previous sections, the topology of computer networks can be considered at different levels of granularity. Besides the topology itself, computer networks very often consist of overlapping logical infrastructures called *overlays*, that are set up in order to provide additional services and optimize network usage. These logical infrastructures only exist in the form of configuration statements on board the network devices, and support commonly used services such as peer-to-peer content sharing.

There is a class of visualization systems that, instead of visualizing the topology of a network, aim at displaying the exchanges of information that are happening on an overlay network.

For example, [Mai02], [BB07], and [FHN+07] propose a visualization of email exchanges between users. The first one is an orthogonal drawing showing the relationships among Internet domains that generate spam. The second one proposes to visualize email exchanges as an example of application of a layout algorithm of graphs with different levels of detail. The last one is a paper describing the visualization of mail exchange patterns with different drawing methodologies. Other contributions focus on monitoring the usage of distributed services like the Domain Name System (DNS). For example, the approach in [DSN12] allows to observe how the workload shifts between different name servers, thus showing how clients migrate from one server to another and simplifying the recognition of unusual operational patterns.

An emerging hot topic is the visualization of relationships in social networks. Users registered on Facebook [web09] can visualize their friendship relationships using a tool called Nexus [Net09d]. In [Tri06] and [BdM06] the authors describe methodologies and tools to display communication relationships between entities (for example, a student and a teacher). In particular, the tool SoNIA [MBd07] also offers animated views of these relationships.

Peer-to-peer networks are more and more widespread, mainly due to the simplicity of setup and to the effectiveness for content sharing. Thanks to the fact that participating devices are self-organizing, the network attempts to preserve connectivity and performance levels even in the presence of frequent topology changes. The problem of displaying the topology of a peer-to-peer network has been considered in [JD08], where a method to generate animated visualization of simulated networks is proposed.

The new version of the Internet Protocol, IPv6 [DH98], has received much attention in the last years due to the technical challenges associated with its deployment. Besides studying transition mechanisms [6ne05], researchers have also focused on visualizing the growth of some test networks. In [NSM99] the authors illustrate the design and implementation of a tool for three-dimensional visualization of the topography of an IPv6 network and of the hierarchy of its address space. The paper proposes sample drawings of an experimental IPv6 network in Japan.

Wireless and sensor networks, due to their continuously changing topology, have also caught the interest of the community interested in visualization. For example, in [GK05] the authors describe a graph drawing algorithm that is based on inter-sensor communication and exploits a force-directed layout. A visualization of the connectivity graph of simulated sensor networks is proposed in [MBS08]. There are also collaborative projects aimed at collecting information about the presence of wireless access points around the world: they usually exploit contour-like maps to visualize network coverage rather than a graph of the topology [oKIC02, com09a].

Another attractive field of research is the analysis of relationships between pages in the World Wide Web. Lots of efforts have been put in analyzing the logical topology implied by hyperlinks between web pages, also known as *web graph*. Even though most of the literature aims at building a compact and efficient representation of the web graph, there are contributions also on the visualization side. In [MB95] the authors used a custom web spider to construct graphical representations of sections of the web graph in 3D hyperbolic space. The structure of the web graph has also been graphically analyzed in a historical perspective in [TK05]. The authors of [YDZ04] apply data mining and visualization techniques to analyze large web data sets. The TouchGraph Google Browser available at [Tou09] offers the user a visual representation of the results of web searches in the form of a clustered graph, where pages are grouped by similarity.

A number of papers are devoted to the visualization of anomalies that are a symptom of intrusion attempts into a networked system. A survey on techniques to visualize security-related information as a graph is provided in [Tam08]. The three-dimensional visualization tool Flamingo has been exploited as an engine to visualize network topologies [OKB06] and to perform network monitoring [OGK06], and is also at the basis of anomaly detection approaches [TJKMW04]. In [WJA05] the authors propose a visualization of BGP routing that aids in detecting abnormal changes. The visualizations exploit the Graphviz [Res09] library and can also be animated. VisFlowConnect [YYT04] is a tool that visualizes traffic exchanges between hosts in order to detect interesting and potentially anomalous patterns. Chapter 20 overviews related work on the visualization of network security aspects.

There also exist tools that provide drawing functionalities. Instead of automatically laying out the topology of a network, these tools provide an environment with ready-to-use symbols that users can exploit to draw networks on their own. An example of a commercial tool that supports creating network diagrams is SmartDraw [Sma09]. Such contributions are outside the scope of this chapter.

References

[3Co09]   3Com. H3C intelligent management center, 2009.

[6ne05]   6net. Large-scale international IPv6 pilot network. `http://www.6net.org/`, 2005.

[AGL⁺08]   P. Abel, P. Gros, D. Loisel, C. R. Dos Santos, and J.P. Paris. CyberNet: A framework for managing networks using 3D metaphoric worlds. *Annals of Telecommunications*, 55(3–4), 2008.

[AGN99]   J. Abello, E. R. Gansner, and S. C. North. Large-scale network visualization. *Computer Graphics*, 33:13–15, 1999.

[AHDBV]   I. Alvarez-Hamelin, L. Dall'Asta, A. Barrat, and A. Vespignani. Large network visualization tool (LaNet-vi). `http://xavier.informatics.indiana.edu/lanet-vi/`.

[AHDBV05a]   I. Alvarez-Hamelin, L. Dall'Asta, A. Barrat, and A. Vespignani. k-core decomposition: a tool for the visualization of large scale networks. ArXiv Computer Science e-prints, 2005.

[AHDBV05b]   I. Alvarez-Hamelin, L. Dall'Asta, A. Barrat, and A. Vespignani. Large scale networks fingerprinting and visualization using the k-core decomposition. In *Proc. NIPS 2005*, 2005.

[AS06]   S. Asafi and M. Sandler. DIMES4DVisualizer, 2006.

[Aug03]   B. Augustsson. Xtraceroute. `http://www.dtek.chalmers.se/~d3august/xt/`, 2003.

[BB07]   M. Baur and U. Brandes. Multi-circular layout of micro/macro graphs. In *Proc. GD 2007*, 2007.

[BBGW04]   M. Baur, U. Brandes, M. Gaertler, and D. Wagner. Drawing the AS Graph in 2.5 dimensions. In *Proc. Graph Drawing 2004*, pages 43–48, 2004.

[BBP07]   K. Boitmanis, U. Brandes, and C. Pich. Visualizing Internet evolution on the Autonomous System level. In *Proc. Graph Drawing 2007*, pages 364–376, 2007.

[BCD⁺04]   G. Barbagallo, A. Carmignani, G. Di Battista, W. Didimo, and M. Pizzonia. Polyphemus and Hermes – Exploration and visualization of computer networks. In *Graph Drawing Software*, Mathematics and Visualization Series, pages 341–364. 2004.

[BdM06]   S. Bender-deMoll and D. McFarland. The art and science of dynamic network visualization. *Journal of Social Structure*, 7(2), 2006.

[BEW95]   R. A. Becker, S. G. Eick, and A. R. Wilks. Visualizing network data. *IEEE Trans. on Visualization and Computer Graphics*, 1(1):16–28, 1995.

[BMB00]   J. A. Brown, A. J. McGregor, and H. W. Braun. Network performance visualization: Insight through animation. In *Proc. PAM '00*, pages 33–41, 2000.

[Bou02]   P. Bourcier. Rootzmap - Mapping the Internet. `http://sysctl.org/rootzmap/`, 2002.

[Bro01]   J. Brown. Cichlid data visualization software. `http://moat.nlanr.net/Software/Cichlid/`, 2001.

[CAI07]   CAIDA. Archipelago measurement infrastructure. Network measurement tool, 2007.

[CAI09]   CAIDA. Visualizing IPv4 and IPv6 Internet topology at a macroscopic scale. `http://www.caida.org/research/topology/as_core_network/`, 2009.

[CBB00]   B. Cheswick, H. Burch, and S. Branigan. Mapping and visualizing the Internet. In *Proc. USENIX '00*, pages 1–12, 2000.

[CDD+00]   A. Carmignani, G. Di Battista, W. Didimo, F. Matera, and M. Pizzonia. Visualization of the Automous Systems interconnections with HERMES. In *Proc. Graph Drawing 2000*, pages 150–163, 2000.

[CDM+05]   L. Colitti, G. Di Battista, F. Mariani, M. Patrignani, and M. Pizzonia. Visualizing interdomain routing with BGPlay. *Journal of Graph Algorithms and Applications, Special Issue on the 2003 Symposium on Graph Drawing, GD '03*, 9(1):117–148, 2005.

[CDM+06]   P. F. Cortese, G. Di Battista, A. Moneta, M. Patrignani, and M. Pizzonia. Topographic visualization of prefix propagation in the Internet. *IEEE Transactions on Visualization and Computer Graphics*, 12(5):725–732, 2006.

[CE95]   K. C. Cox and S. G. Eick. 3D displays of Internet traffic. In *Proc. INFOVIS '95*, page 129, 1995.

[CEH96]   K. C. Cox, S. G. Eick, and T. He. 3D geographic network displays. *SIGMOD Record*, 25(4):50–54, 1996.

[CFSD90]   J. D. Case, M. Fedor, M. L. Schoffstall, and J. Davin. Simple Network Management Protocol (SNMP). RFC 1157 (Historic), May 1990.

[Che07]   B. Cheswick. Internet mapping project. `http://www.cheswick.com/ches/map/`, 2007.

[Com]   International Electrotechnical Commission. Web site. `http://www.iec.ch/`.

[com09a]   Mimezine community. Wigle – wireless geographic logging engine. `http://www.wigle.net/gps/gps/main`, 2009.

[Com09b]   Plixer International Systrax Community. Visualization of NetFlow data, 2009.

[Cor]   Lumeta Corporation. IPsonar MapViewer.

[Cor09]   Renesys Corporation. Routing intelligence tools. `http://www.renesys.com/`, 2009.

[CP]   D. Cox and R. Patterson. NSFNET growth until 1995. `http://www.caida.org/projects/internetatlas/gallery/nsfnet/data.xml`.

[CRC+08]   L. Cittadini, T. Refice, A. Campisano, G. Di Battista, and C. Sasso. Measuring and visualizing interdomain routing dynamics with BGPath. In *Proc. ISCC 2008*, 2008.

[Dar09]   LLC Dartware. Intermapper. `http://www.intermapper.com/products/intermapper/`, 2009.

[Des09]   Packet Design. Route explorer. `http://www.packetdesign.com/`, 2009.

[DETT99]   G. Di Battista, P. Eades, R. Tamassia, and I. G. Tollis. *Graph Drawing*. Prentice Hall, 1999.

[DH98]   S. Deering and R. Hinden. Internet Protocol, Version 6 (IPv6) Specification. RFC 2460 (Draft Standard), December 1998.

[Dit09]   W. Dittmer. World-wide mapping of reverse traceroute and looking glass servers, 2009.

[DLV97]   G. Di Battista, R. Lillo, and F. Vernacotola. Ptolomaeus. http://www.dia.uniroma3.it/~ptolemy/, 1997.

[DSN12]   G. Di Battista, C. Squarcella, and W. Nagele. How to visualize the K-Root name server. *Journal of Graph Algorithms and Applications, Special Issue of GD 2011*, 2012.

[ea05]   B. Lyon et al. The Opte project. http://www.opte.org/, 2005.

[EHH+00]   D. Estrin, M. Handley, J. Heidemann, S. McCanne, Y. Xu, and H. Yu. Network visualization with the VINT network animator Nam. *IEEE Computer*, 33(11):63–68, 2000.

[EHH+05]   D. Estrin, M. Handley, J. Heidemann, S. McCanne, Y. Xu, and H. Yu. Nam: Network Animator. http://isi.edu/nsnam/nam/, 2005.

[Ent09]   Enterasys. NetSight Atlas Console, 2009.

[EW93]   S. G. Eick and G. J. Wills. Navigating large networks with hierarchies. In *Proc. Visualization '93*, pages 204–209, 1993.

[FHN+07]   X. Fu, S. H. Hong, N. S. Nikolov, X. Shen, Y. Wu, and K. Xu. Visualization and analysis of email networks. In *Proc. IEEE Asia Pacific Symposium on Visualization (APVIS '07)*, pages 1–8, 2007.

[FNMT94]   H. Fuji, S. Nakai, H. Matoba, and H. Takano. Real-time bifocal network-visualization. In *Proc. NOMS 1994*, 1994.

[fS]   International Organization for Standardization. Web site. http://www.iso.org/.

[GGW07]   R. Görke, M. Gaertler, and D. Wagner. LunarVis – Analytic visualizations of large graphs. In *Proc. Graph Drawing 2007*, pages 352–364, 2007.

[GH02]   J. Gallagher and B. Huffaker. PlotPaths. http://www.caida.org/tools/visualization/plotpaths/, 2002.

[GK05]   C. Gotsman and Y. Koren. Distributed graph layout for sensor networks. *Journal of Graph Algorithms and Applications*, pages 1–15, 2005.

[GKN04]   E. Gansner, Y. Koren, and S. North. Topological fisheye views for visualizing large graphs. In *Proc. INFOVIS '04*, pages 175–182, 2004.

[GMN03]   E. R. Gansner, J. M. Mocenigo, and S. C. North. Visualizing software for telecommunication services. In *Proc. SOFTVIS 2003*, pages 151–157, 2003.

[GMO+03]   P. A. Aranda Gutiérrez, P. Malone, M. ÓFoghlú, S. Michaelis, and J. Seger. Acquisition, modelling and visualisation of inter-domain routing data. In *Proc. IPS 2004*, 2003.

[Gro00]   Roma Tre University Computer Networks Research Group. Hermes.
             `http://tocai.dia.uniroma3.it/~hermes/`, 2000.

[GT00]   R. Govindan and H. Tangmunarunkit. Heuristics for Internet map
             discovery. In *Proc. INFOCOM 2000*, 2000.

[Hew09]   Hewlett-Packard. HP Operations Manager (formerly OpenView),
             2009.

[Hir07]   Hirschmann. Industrial HiVision, 2007.

[HJWkc98]   B. Huffaker, J. Jung, D. Wessels, and kc claffy. Visualiza-
             tion of the growth and topology of the NLANR caching hier-
             archy. `http://www.caida.org/tools/visualization/plankton/`
             `Paper/plankton.xml`, Mar 1998.

[HNkc97]   B. Huffaker, E. Nemeth, and kc claffy. Otter: A general-
             purpose network visualization tool. `http://www.caida.org/`
             `tools/visualization/otter/`, 1997.

[HNkc99]   B. Huffaker, E. Nemeth, and kc claffy. Tools to visualize the internet
             multicast backbone. In *Proc. INET 99*, 1999.

[HPF07]   B. Huffaker, J. Polterock, and M. Fomenkova. Cuttlefish
             geographic visualization tool. `http://www.caida.org/tools/`
             `visualization/cuttlefish/`, Jun 2007.

[Hyu05]   Y. Hyun. Walrus – Graph visualization tool. `http://www.caida.`
             `org/tools/visualization/walrus/`, 2005.

[IBM09]   IBM. Tivoli network manager topology visualization, 2009.

[IEE09]   IEEE. Working group for 802.11 WLAN standards, 2009.

[IMN84]   H. Ishio, J. Minowa, and K. Nosu. Review and status of wavelength-
             division-multiplexing technology and its application. *Journal of
             Lightwave Technology*, 2(4), 1984.

[Ips09]   Ipswitch, Inc. WhatsUpGold network mapping. `http://www.`
             `whatsupgold.com/`, 209.

[ISO]   ISO/IEC. Open Systems Interconnection – basic reference model.
             ISO/IEC standard 7498.

[Jac99]   V. Jacobson.   Gtrace.   `http://www.caida.org/tools/`
             `visualization/gtrace/`, 1999.

[JD08]   K. Juenemann and J. Dinger. Ovlvis: Visualization of peer-to-peer
             networks in simulation and testbed environments. In *Proc. Com-
             munications and Networking Simulation Symposium (CNS 2008)*,
             2008.

[Jon09]   H.   Jones.   Network   weathermap.   `http://www.`
             `network-weathermap.com/`, 2009.

[kc97]   kc claffy. IETF presentation BGP MPEG animations. `http://www.`
             `caida.org/publications/presentations/Ietf199712/Movie/`,
             1997.

[kcH97]   kc claffy and B. Huffaker. Mapnet, macroscopic Internet vi-
             sualization and measurement. `http://www.caida.org/tools/`
             `visualization/mapnet/`, 1997.

[KD01]   R. Kitchin and M. Dodge. *Atlas of Cyberspace*. Addison-Wesley,
             2001.

[KGS07]   H. Kang, L. Getoor, and L. Singh. Visual analysis of dynamic group membership in temporal social networks. *SIGKDD Explorations Newsletter*, 9(2), 2007.

[KMG88]   G. Kar, B. Madden, and R. Gilbert. Heuristic layout algorithms for network management presentation services. In *Proc. IEEE Network 1988*, 1988.

[KNK99]   E. E. Koutsofios, S. C. North, and D. A. Keim. Visualizing large telecommunication data sets. *IEEE Computer Graphics and Applications*, 19:16–19, 1999.

[KNTK99]   E. E. Koutsofios, S. C. North, R. Truscott, and D. A. Keim. Visualizing large-scale telecommunication networks and services. In *Proc. Visualization '99*, pages 457–461, 1999.

[Kvi03]   O. Kvittem. Scaling network management tools. In *Proc. NANOG 2003*, 2003.

[Lab09]   UCLA Internet Research Lab. LinkRank visualization. `http://linkrank.cs.ucla.edu/`, 2009.

[Lim09]   M. Lima. VisualComplexity.com Internet visualizations. `http://www.visualcomplexity.com/vc/index.cfm?domain=Internet`, 2009.

[LMZ04]   M. Lad, D. Massey, and L. Zhang. LinkRank: A graphical tool for capturing BGP routing dynamics. In *Proc. NOMS 2004*, 2004.

[LMZ06]   M. Lad, D. Massey, and L. Zhang. Visualizing Internet routing changes. *IEEE Transactions on Visualization and Computer Graphics, Special Issue on Visual Analytics*, 12(6), 2006.

[LUM]   LUMETA. Internet mapping project. `http://www.lumeta.com/internetmapping/`.

[Mai02]   Clueless Mailers. Spamdemic map poster. `http://www.cluelessmailers.org/spamdemic/index.html`, 2002.

[Map08]   Peacock Maps. Internet posters. `http://www.peacockmaps.com/`, 2008.

[MB95]   T. Munzner and P. Burchard. Visualizing the structure of the World Wide Web in 3d hyperbolic space. In *Proc. VRML 1995*, pages 33–38, 1995.

[MBd07]   D. McFarland and S. Bender-deMoll. Sonia – social network image animator, 2007.

[MBS08]   C. Morrell, T. Babbitt, and B. K. Szymanski. Visualization in sensor network simulator, sense and its use in protocol verification. Technical Report cs-08-13, Department of Computer Science, Rensselaer Polytechnic Institute, 2008.

[McR99]   D. W. McRobb. Skping. `http://www.caida.org/tools/measurement/skitter/skping/`, 1999.

[Mei00]   M. Meiss. Abilene weather map. `http://www.caida.org/projects/internetatlas/gallery/mping/`, 2000.

[MFKN07]   F. Mansmann, F. Fischer, D. A. Keim, and S. C. North. Visualizing large-scale IP traffic flows. In *Proc. International Workshop on Vision, Modeling, and Visualization, 2007*, 2007.

[MHkcF96]   T. Munzner, E. Hoffman, kc claffy, and B. Fenner. Visualizing the
              global topology of the MBone. In *Proc. INFOVIS '96*, pages 85–92,
              1996.

[Mic09]   Microsoft. Windows Vista™ Full Map View, 2009.

[MKF⁺06]   P. Mahadevan, D. Krioukov, M. Fomenkov, B. Huffaker, X. Dim-
              itropoulos, kc claffy, and A. Vahdat. The internet as-level topology:
              Three data sources and one definitive metric. *ACM SIGCOMM
              Computer Communication Review*, Jan 2006.

[MKN⁺07]   F. Mansmann, D. A. Keim, S. C. North, B. Rexroad, and D. Shele-
              heda. Visual analysis of network traffic for resource planning, in-
              teractive monitoring, and interpretation of security threats. *IEEE
              Trans. on Visualization and Computer Graphics*, 13(6):1105–1112,
              2007.

[Moy94]   J. Moy. OSPF Version 2. RFC 1583 (Draft Standard), March 1994.

[MR91]   K. McCloghrie and M. Rose. Management Information Base for Net-
              work Management of TCP/IP-based internets:MIB-II. RFC 1213
              (Standard), March 1991.

[Mun97]   T. Munzner. H3: Laying out large directed graphs in 3D hyperbolic
              space. In *Proc. INFOVIS 1997*, pages 2–10, Oct 1997.

[NCC]   RIPE NCC. The Internet Routing Registry: History and Purpose.
              `http://www.ripe.net/db/irr.html`.

[net09a]   neteXpose. DNA network topology visualization, 2009.

[Net09b]   Brocade (Foundry Networks). IronView network manager, 2009.

[Net09c]   Juniper Networks. Network and security manager, 2009.

[Net09d]   Ludios Networks. Nexus facebook graph. `http://nexus.ludios.
              net/`, 2009.

[NoCSNCTUT09] Network and System Laboratory Dep.t of Computer Science Na-
              tional Chiao Tung University Taiwan. NCTUns. `http://nsl10.
              csie.nctu.edu.tw/`, 2009.

[Nor02]   M. Norris. *Gigabit Ethernet Technology and Applications*. Artech
              House Publishers, 2002.

[NSM99]   S. Nakamae, Y. Sekiya, and J. Murai. A study into a visualization
              of an ipv6 network. In *Proc. INET 99*, 99.

[oANTtE08]   DANTE (Delivery of Advanced Network Technology to Europe).
              Topology maps. `http://www.dante.net/`, 2008.

[OC07]   R. Oliveira and Y. J. Chi. Visualizing Internet topology dynamics
              with Cyclops. `http://irl.cs.ucla.edu/cyclops/`, 2007.

[OCLZ08]   R. Oliveira, Y. J. Chi, M. Lad, and L. Zhang. Cyclops: the AS-Level
              connectivity observatory. Presentation at NANOG 43, 2008.

[OCP⁺07]   R. Oliveira, Y. J. Chi, I. Pefkianakis, M. Lad, and L. Zhang. Vi-
              sualizing Internet topology dynamics with Cyclops. Poster at SIG-
              COMM 2007, 2007.

[OGK06]   J. Oberheide, M. Goff, and M. Karir. Flamingo: Visualizing internet
              traffic. In *Proc. NOMS 2006*, pages 150–161, 2006.

[OKB06]   J. Oberheide, M. Karir, and D. Blazakis. Vast: Visualizing autonomous system topology. In *Proc. Workshop on Visualization for Computer Security (VizSEC '06)*, pages 71–80, 2006.

[oKIC02]   University of Kansas Information and Telecommunication Technology Center. Wireless network visualization project. `http://www.ittc.ku.edu/wlan/`, 2002.

[Oli09]   R. Oliveira. Cyclops: the AS-Level connectivity observatory. `http://cyclops.cs.ucla.edu/`, 2009.

[oO]   University of Oregon. Route views project. `http://www.routeviews.org/`.

[Ora90]   D. Oran. OSI IS-IS Intra-domain Routing Protocol. RFC 1142 (Informational), February 1990.

[PH99]   R. Periakaruppan and B. Huffaker. Geoplot. `http://www.caida.org/tools/visualization/geoplot/`, 1999.

[PIP05]   N. Patwari, A. O. Hero III, and A. Pacholski. Manifold learning visualization of network traffic data. In *Proc. SIGCOMM 2005*, 2005.

[Piz07]   M. Pizzonia. From BGPlay to iBGPlay: Graphical inspection of your routing data. In *55th Re'seaux IP Europe'ens Meeting (RIPE 55)*, 2007.

[PN99]   R. Periakaruppan and E. Nemeth. GTrace – A graphical traceroute tool. In *Proc. LISA '99*, Nov 1999.

[Pos81a]   J. Postel. Internet Control Message Protocol. RFC 792 (Standard), September 1981.

[Pos81b]   J. Postel. Internet Protocol. RFC 791 (Standard), September 1981.

[Pro02]   M. Prodanovic. Java Autonomous System Path Visualization Interface (JASPVI). `http://lab.verat.net/Jaspvi/`, 2002.

[Res09]   AT&T Research. Graphviz graph visualization software. `http://www.graphviz.org/`, 2009.

[RIP]   Réseaux IP Européens – RIPE. Routing information service. `http://www.ripe.net/ris/`.

[RIS09]   RIPE RIS. BGPviz. `http://www.ris.ripe.net/bgpviz/`, 2009.

[Rom09]   Roma Tre University, Computer Networks Research Group. BGPlay @ Route Views. `http://bgplay.routeviews.org/bgplay/`, 2009.

[RTM08]   M. Roughan, S. J. Tuke, and O. Maennel. Bigfoot, sasquatch, the yeti and other missing links: What we don't know about the as graph. In *Proc. ICM 2008*, 2008.

[RTU09]   Computer Networks Research Group Roma Tre University. iBGPlay. `http://www.ibgplay.org/`, 2009.

[Sal00]   T. Salo. Real-time visualization of IP over connection-oriented WANs. In *Proc. NANOG 2000*, 2000.

[Sii01]   K. Siil. Using topological mapping to manage and secure large networks. In *Proc. NANOG 2001*, 2001.

[Sma09]   SmartDraw. Network topology drawing software. `http://www.smartdraw.com/`, 2009.

[SMW04]   N. Spring, R. Mahajan, and D. Wetherall. Measuring ISP topologies
             with RocketFuel. *IEEE Trans. on Networking*, 12(1):2–16, 2004.

[Sof09a]   Inc. SoftConcept. SeeNet. `http://www.softconcept-inc.com/`
             `seenetweb/`, 2009.

[Sof09b]   Tulip Software. Tulip. `http://www.tulip-software.org/`, 2009.

[SS96]   C. A. Siller and M. Shafi. *SONET/SDH: A Sourcebook of Syn-*
             chronous Networking. Wiley – IEEE Press, 1996.

[Sta07]   W. Stallings. *Data and Computer Communications*. Prentice Hall,
             2007.

[Sys09a]   Cisco Systems. NetFlow. `http://www.cisco.com/go/netflow`,
             2009.

[Sys09b]   Cisco Systems. Synchronous Data Link Control (SDLC) and deriva-
             tives, 2009.

[Tam08]   R. Tamassia. Graph drawing for security visualization. GD 2008,
             2008.

[Tec05]   BBN Technologies. BBN SpyGlass animation of application traf-
             fic flow. `http://www.dist-systems.bbn.com/tech/spyglass/`,
             2005.

[Tec09]   OPNET Technologies. NetMapper. `http://www.opnet.com/`, 2009.

[Tel09]   TeleGeography. Cartographic maps. `http://www.telegeography.`
             `com/maps/index.php`, 2009.

[TJKMW04]   S. T. Teoh, T. J. Jankun-Kelly, K.-L. Ma, and S. F. Wu. Visual
             data analysis for detecting flaws and intruders in computer network
             systems. *Computer Graphics and Applications, Special Issue on*
             Visual Analytics, 24(5):27–35, 2004.

[TK05]   M. Toyoda and M. Kitsuregawa. A system for visualizing and an-
             alyzing the evolution of the web with a time series of graphs. In
             Proc. ACM Conference on Hypertext and Hypermedia, pages 151–
             160, 2005.

[Tou09]   LLC TouchGraph. Touchgraph google browser. `http://www.`
             `touchgraph.com/TGGoogleBrowser.html`, 2009.

[Tri06]   M. Trier. Towards a social network intelligence tool for visual analy-
             sis of virtual communication networks. In *Virtuelle Organisationen*
             und Neue Medien, 2006.

[TvAG+06]   J. Toledo, V. van Adrighem, R. Ghetta, E. Mann, and F. Peters.
             EtherApe. `http://etherape.sourceforge.net/`, 2006.

[UNI09]   UNINETT. Operations center network maps. `http://drift.`
             `uninett.no/index.en.html`, 2009.

[Vis09]   Visualware. Visualroute. `http://visualroute.visualware.com/`,
             2009.

[Vol09]   Y. Volokitin. Lantopolog physical network discovery software.
             `http://www.lantopolog.com/`, 2009.

[VT06]   G. Vandenberghe and J. Treurniet. Automating the presentation
             of computer networks. Technical Report ADA477079, Defence Re-
             search and Development, Canada, Ottawa, 2006.

[WAN08]   Wide Area Network Design Laboratory – WANDL. IP/MPLSView. `http://www.wandl.com/html/mplsview/`, 2008.

[WCH⁺03]   S. Y. Wang, C. L. Chou, C. H. Huang, C. C. Hwang, Z. M. Yang, C. C. Chiou, and C. C. Lin. The design and implementation of the NCTUns 1.0 network simulator. *Computer Networks*, 42(2):175–197, 2003.

[web09]   Facebook. `http://www.facebook.com/`, 2009.

[WH09]   S.Y. Wang and Y.M. Huang. *NCTUns Tool for Innovative Network Emulations*, chapter 13. Nova Science Publishers, 2009.

[wim09]   WiMAX forum, 2009.

[WJA05]   T. Wong, V. Jacobson, and C. Alaettinoglu. Internet routing anomaly detection and visualization. In *Proc. DSN 2005*, 2005.

[WPP07]   M. S. Withall, I. W. Phillips, and D. J. Parish. Network visualization: A review. *Loughborough University IET Communications*, 1(3), 2007.

[WS04]   A. Wool and G. Sagie. A clustering approach for exploring the Internet structure. In *Proc. IEEE Convention of Electrical and Electronics Engineers in Israel (IEEEI '04)*, pages 149–152, 2004.

[Wyv09]   Wyvernsoft. Netpalpus. `http://www.wyvernsoft.com/`, 2009.

[YDZ04]   A. H. Youssefi, D. J. Duke, and M. J. Zaki. Visual web mining. In *Proc. WWW 2004*, 2004.

[YGM05]   B. Yip, S. Goyette, and C. Madden. Visualising Internet traffic data with three-dimensional spherical display. In *Proc. APVIS 2005*, 2005.

[YMMW09]   H. Yan, D. Massey, E. McCracken, and L. Wang. BGPMon and NetViews: Real-time BGP monitoring system. Demo at INFOCOM 2009, 2009.

[YSS05]   Y. Shavitt and E. Shir. DIMES: Let the Internet measure itself. *Computer Communication Review*, 35(5), 2005.

[YTS06]   Y. Rekhter, T. Li, and S. Hares. A Border Gateway Protocol 4 (BGP-4). RFC 4271, Jan 2006.

[YYT04]   X. Yin, W. Yurcik, and M. Treaster. Visflowconnect: Netflow visualizations of link relationships for security situational awareness. In *Proc. Workshop on Visualization and Data Mining for Computer Security (VizSEC/DMSEC '04)*, pages 26–34, 2004.

[ZW92]   J. A. Zinky and F. M. White. Visualizing packet traces. In *Proc. SIGCOMM 1992*, 1992.

26

Social Networks

Ulrik Brandes
University of Konstanz

Linton C. Freeman
University of California, Irvine

Dorothea Wagner
Karlsruhe Institute of Technology

Social networks provide a rich source of graph drawing problems, because they appear in an incredibly wide variety of forms and contexts. After sketching the scope of social network analysis, we establish some general principles for social network visualization before finally reviewing applications of, and challenges for, graph drawing methods in this area. Other accounts more generally relating to the status of visualization in social network analysis are given, e.g., in [Klo81, BKR+99, Fre00, Fre05, BKR06]. Surveys that are more comprehensive on information visualization approaches, interaction, and network applications from social media are given in [CM11, RF10, CY10].

26.1   Social Network Analysis

The fundamental assumption underlying social network theory is the idea that seemingly autonomous individuals and organizations are in fact embedded in social relations and interactions [BMBL09]. The term *social network* was coined to delineate the relational perspective from other research traditions on social groups and social categories [Bar54].

In general, a social network consists of *actors* (e.g., persons, organizations) and some form of (often, but not necessarily: social) *relation* among them. The network structure is usually modeled as a graph, in which vertices represent actors, and edges represent *ties*, i.e., the existence of a relation between two actors. Since traits of actors and ties may be important, both vertices and edges can have a multitude of attributes. We will use graph terminology for everything relating to the data model, and social network terminology when referring to substantive aspects.

While attributed graph models are indeed at the heart of formal treatments, it is worth noting that theoretically justified data models are not as obvious as it may seem [But09]. In fact, social network analysis is maturing into a paradigm of distinct structural theories and associated relational methods. General introductions and methodological overviews can be found in [WB88, WF94, Sco00, CSW05, BE05], a historic account in [Fre04a], and a comprehensive collection of influential articles in [Fre08].

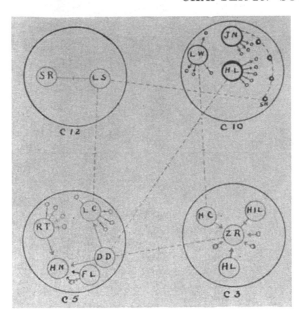

Figure 26.1   A sociogram from [Mor53, p. 422] showing a graph with fourteen highlighted vertices and four clusters.

In social network reseach it is important to clarify whether the networks are considered dependent or explanatory variables. In the former case the interest is in why and how networks form the way they do, and in the latter case the interest is in why and how networks influence other outcomes. For convenience, we will refer to the former as network theory (studying *network formation*) and to the latter as network analysis (studying *network effects*). A major distinction from non-network approaches is that the unit of analysis is the *dyad*, i.e., a pair of actors (may they be linked or not) rather than a monad (a singleton actor).

The methodological toolbox can be organized into the following main compartments.

Indexing The assignment of values to predetermined substructures of any size. Most common are vertex, edge, and graph indices such as vertex centrality and graph centralization [Fre79], but sometimes the interest is also in evaluating larger substructures (e.g., group centrality) or the distribution of scores (e.g., degree distribution).

Grouping The identification of substructures and membership in them. Most common are decomposition into relatively dense subgraphs, partitions into equivalent positions [Ler05], and, more generally, blockmodeling [DBF05]. Other examples include subgraph counts (e.g., triad census) and various forms of domination and brokerage.

Modeling The use of statistical models for assessment and inference. Most common are modeling attempts to reproduce networks statistics, parameter estimation, and regression-type analyses.

Concrete examples of such methods are considered later in this chapter. Other important types of variation arise from special types of data such as longitudinal (temporal), multimode (multiple actor types), multiplex (multiple relation types), or multi-level (hierarchies of actors) data.

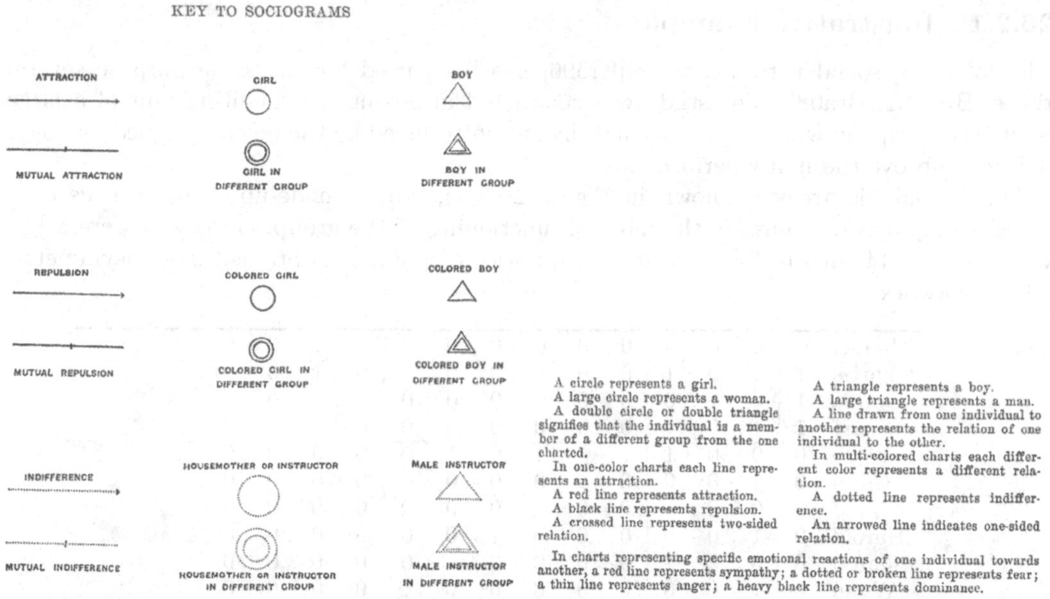
KEY TO SOCIOGRAMS
ATTRACTION
MUTUAL ATTRACTION
REPULSION
MUTUAL REPULSION
INDIFFERENCE
MUTUAL INDIFFERENCE
GIRL
GIRL IN DIFFERENT GROUP
COLORED GIRL
COLORED GIRL IN DIFFERENT GROUP
HOUSEMOTHER OR INSTRUCTOR
HOUSEMOTHER OR INSTRUCTOR IN DIFFERENT GROUP
BOY
BOY IN DIFFERENT GROUP
COLORED BOY
COLORED BOY IN DIFFERENT GROUP
MALE INSTRUCTOR
MALE INSTRUCTOR IN DIFFERENT GROUP
A circle represents a girl.
A large circle represents a woman.
A double circle or double triangle signifies that the individual is a member of a different group from the one charted.
In one-color charts each line represents an attraction.
A red line represents attraction.
A black line represents repulsion.
A crossed line represents two-sided relation.
A triangle represents a boy.
A large triangle represents a man.
A line drawn from one individual to another represents the relation of one individual to the other.
In multi-colored charts each different color represents a different relation.
A dotted line represents indifference.
An arrowed line indicates one-sided relation.
In charts representing specific emotional reactions of one individual towards another, a red line represents sympathy; a dotted or broken line represents fear; a thin line represents anger; a heavy black line represents dominance.

Figure 26.2 A notational system for sociograms [Mor53, p. 136].

Visualization has been instrumental in the study of social networks from the very beginning, and some historical examples are based on surprisingly sophisticated designs. The example in Figure 26.1 is from Moreno's book [Mor53], which is a rich source in this regard. In fact, he even specified a visual notation standard reproduced in Figure 26.2 (although neither graphical notation nor labeling are applied fully consistently), and introduced the terms *sociogram* (for a graphical representation of a social network) and *sociomatrix* (for a matrix representation of a social network).

26.2 Visualization Principles

Let us first establish a frame of reference for social network visualization based on a few organizing principles. We will then elaborate on various visualization approaches and the graph drawing problems they pose in Section 26.3.

The utility of a diagram is dependent on purpose and context. The two main purposes of network visualizations are *exploration* of data and *communication* of findings. The potential of using diagrammatic representations in the research process itself was stressed already by Moreno.

> "A process of charting has been devised by the sociometrists, the sociogram, which is more than merely a method of presentation. It is first of all a method of exploration." [Mor53, p. 95f]

A network diagram should therefore be designed to display the information relevant for an analytic perspective. As a consequence, there cannot be a single best way of representing social networks graphically, which in turn creates lots of opportunities for visualization and algorithm design. For concreteness, we give one striking example.

26.2.1 Illustrative Example

The following social network study [Kra96] has been used for the same purpose several times [BRW01, Bra08]. The study was conducted in an internal auditing unit of a large industrial company when organizational changes introduced by the newly assigned manager did not improve the unit's performance.

The formal hierarchy is shown in Figure 26.3(a), where made-up names are used to identify employees. To assess the internal functioning of the group, employees were asked who they would turn to for work related questions. The data obtained is an asymmetric advice network,

Manuel	.	0	0	0	0	0	0	0	0	0	0	1	0	0	*manager*
Charles	1	.	0	1	0	0	0	0	0	0	0	1	0	0	
Donna	1	0	.	0	0	0	0	0	0	0	0	1	0	0	*supervisors*
Stuart	1	1	0	.	0	0	0	0	0	0	0	1	0	0	
Bob	0	0	0	1	.	0	1	0	1	0	0	0	0	0	
Carol	0	1	0	0	0	.	0	0	0	0	0	0	0	0	
Fred	0	0	0	1	0	0	.	0	0	0	0	0	0	0	
Harold	0	1	0	0	0	0	0	.	0	0	0	0	0	0	*auditors*
Sharon	0	0	0	1	0	0	0	0	.	0	0	0	0	0	
Wynn	0	1	0	0	0	0	0	0	0	.	0	0	0	0	
Kathy	0	0	1	0	0	0	0	0	0	0	.	1	0	1	
Nancy	0	0	1	0	0	0	0	0	0	0	0	.	0	0	
Susan	0	0	1	0	0	0	0	0	0	0	1	0	.	1	*secretaries*
Tanya	0	0	1	0	0	0	0	0	0	0	1	1	0	.	

and the resulting directed network is shown in Figure 26.3(b). While the data is represented with clarity, there are no obvious implications. The situation becomes much more comprehensible when we understand that the rationale for looking at the advice network is the identification of an informal work hierarchy. Rearranging the vertices such that a maximum number of edges is directed upwards (and thus aligns with what can be assumed to be an informal work hierarchy) as in Figure 26.3(c) yields a strikingly clear picture with a single relation not in accordance with the informal hierarchy (but the formal): everyone, directly or indirectly, and including the manager himself, is seeking advice from Nancy, a secretary that was dismissive of the changes introduced. (Of course, after seeing a picture similar to this, the manager sat down with her, discussed her reservations and made sure that she understood his good intentions and the long-term benefits of his plans, thus turning the situation around.)

26.2.2 Substance, Design, Algorithm

The example above illustrates the importance of considering three key aspects in social network visualization [BKR⁺99].

Substance In general, the information to be conveyed in a network visualization is more than just the underlying graph. The substantive interest of those who collected network data typically necessitates the inclusion of attributes. Moreover, additional data may have been generated as the result of an analysis. In the above example, the substance of interest is the informal hierarchy within a business unit, and only by considering it in the design of the visualization, the diagram becomes informative. Through the appreciation of relevant substance, i.e., the application-specific contexts and interests, data visualizations are turned into information visualizations.

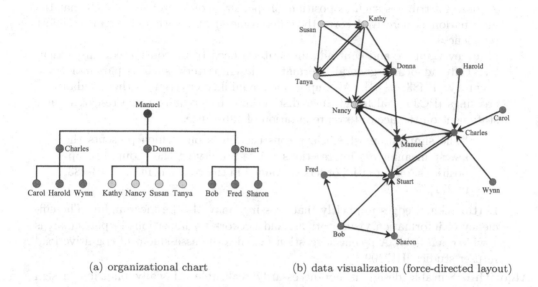

(a) organizational chart   (b) data visualization (force-directed layout)

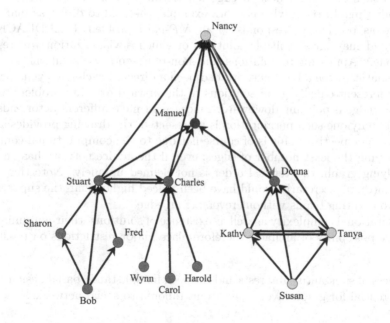

(c) information visualization (upward layout)

Figure 26.3 Organizational chart and advice network in a business unit (adapted from [Kra96]).

Design Visualization design is the specification of a mapping from an information space to its graphical representation. The core choices are in assigning graphical elements such as points, lines, and areas to data objects, and in defining their graphical attributes such as position, shape, size, color, and so on such that the information is perceived correctly (effectiveness) and with low cognitive effort (efficiency).

Many overlapping and contradicting criteria need to be considered. In particular, Tufte advocates general information design criteria such as parsimony and accuracy [Tuf83, p. 51]. An important readability criterion is the avoidance of crossings [PCJ97], although finer distinctions may require more research: in a statement on the accurate representation of substance,

> "The simplest, most efficient construction is one which presents the fewest meaningless intersections, while preserving the groupings, oppositions, or potential orders contained in the component ..." [Ber83, p. 271],

Bertin acknowledges implicitly that crossings may also be meaningful. The efficiency of information visualizations, and network visualizations in particular, is a wide open field. A recent suggestion includes the assessment of cognitive load in user studies [HEH09].

Algorithm Suitable design is not necessarily realizable. Locally plausible design choices such as certain desired edge lengths may be interdependent and even contradicting. In the advice network example, the goal to direct as many edges upward as possible corresponds to an $\mathcal{NP}$-hard problem (FEEDBACK ARC SET) and may have multiple solutions of which a visualization will represent only one. Approximate solutions and non-representative solutions are highly problematic. If the advice network had been a directed cycle, all cyclic permutations represent equally good solutions to the upward drawing problem, because only one edge is pointing downward. Since each time a different actor ends up on top and only one such permutation is represented, the drawing provides a rather selective perspective. This is of course not due to the computational complexity of reversing the least number of edges; even if the advice network has an acyclic underlying graph, the vertical order is not defined uniquely. Note that in Figure 26.3(c), all secretaries could have been placed higher than the supervisors of the two auditing teams without reversing an edge.

Computational complexity as well as existence of solutions, their non-uniqueness, and the possibility of artifacts therefore place major restrictions on possible designs.

The richness of substantive interests and the need for substance-based designs thus creates immense potential for graph layout algorithms tailored to social networks.

26.3   Substance-Based Designs

Depending on data, substantive interest, and presentation context, very different designs are required for effective and efficient exploration and communication of social network information. Depending on the point of view, this is either a major burden or a horn of plenty for algorithmic and design challenges.

In this section, we review examples of substance-based designs and corresponding graph drawing approaches to demonstrate the richness of both, existing approaches and open

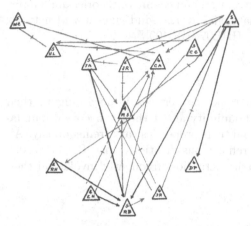

(a) Sympathy and antipathy among players in an American football team. Layout according to lineup, though slightly distorted to allow for straight edges [Mor53]

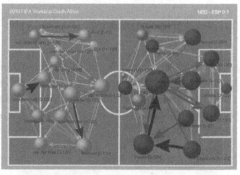

(b) Passes among players during the FIFA World Cup 2010 final. Layout according to (assumed) tactical lineup [PNK10]

(c) Friendship choices among villagers. Layout in 3D using indegree, outdegree, and a status attribute as coordinates [Cha50]

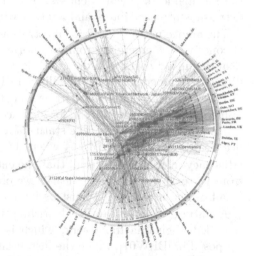

(d) Peering among autonomous systems of the Internet. Radial layout according to degree and latitude [CAI]

Figure 26.4 Network visualizations in which coordinates are not defined by a graph drawing algorithm.

problems. Our selection of examples is, of course, heavily biased by their algorithmic interestingness. Figure 26.4 shows examples for designs we have excluded, because they do not require graph layout algorithms (although labeling and edge routing may be an issue).

We concentrate on two important analytical concepts (prominence and cohesion), and two data categories (two-mode and dynamic data). The designs we deal with are largely based on intuition and plausibility rather than perceptual and cognitive theories and empirical evidence. Among the few attempts to evaluate effectiveness and efficiency of network visualization design are [Win94, PCJ97, MBK97, HHE07, HEH09].

26.3.1   Prominence

Prominence indices $p : V \to \mathbb{R}_{\geq 0}$ are used to rank the vertices V according to their structural importance [KB83]. Since there is no unanimity about their conceptual foundations [Fre79, Fri91], numerous such indices exist, and their properties vary tremendously. Although terminology is not well defined and more refined classifications exist [Bor05, BE06], we distinguish only two groups based on geometric metaphors frequently invoked in their interpretation.

Status

It is commonplace to differentiate status into "high" and "low," so that it seems almost mandatory to exploit this geometric interpretation for network visualizations. Not surprisingly, the apparent correspondence between substantive and geometric intuition has been used in the design of network diagrams already in times when no layout algorithms were available to social scientists. Figure 26.5 shows two historic examples, one with an extrinsic status attribute, and another with an intrinsic, structural one.

The advice network used for illustration above is an example in which a status hierarchy is conceived as emerging from an informal advice-seeking relation. Indeed, status is often analyzed in networks with directed edges, and because of how status indices are defined, the direction of an edge is generally aligned with the difference in status between the endpoints.

The simplest example of a structural status index is indegree, which was generalized in numerous ways. Katz [Kat53], for instance, defines status by taking into account all directed walks ending at a vertex. The status of a vertex $v \in V$ in a graph $G = (V, E)$ is defined by $p(v) = \sum_{u \in V} \left(\sum_{k=1}^{\infty} (\alpha A)^k \right)_{uv}$, where $1/\alpha$, $0 < \alpha < 1$, is an attenuation factor, and A the adjacency matrix of G. Recall that the entries $(A^k)_{uv}$ give the number of walks of length k from vertex u to vertex v. Clearly, we obtain the same ranking as with indegree for very small α, and attentuation must be large enough to make sure that the sum converges.

A natural class of layout algorithms that can be adapted for status drawings is the so-called Sugiyama framework, which is described in detail in Chapter 13. Its use was proposed in [BRW01], where the instantiation employs one-dimensional clustering of status scores for layer assignment and standard approaches for crossing reduction and horizontal coordinate assignment. Clustering is necessary since vertical coordinates are fixed and differences can be quite small, resulting in very close layers between which edges run almost horizontally. However, clustering worsens another, more general, open problem, namely how to accomodate intra-layer edges in layered layouts.

Two different approaches are less sensitive to this kind of problem and in addition more scalable. The first and simpler one is to fix y-coordinates and to determine the x-coordinates using a one-dimensional layout algorithm [BC03b], possibly taking the fixed dimension into account [KH05]. The second, more flexible, and likely to be more effective one is based on constrained optimization of stress using a gradient projection method [DKM09].

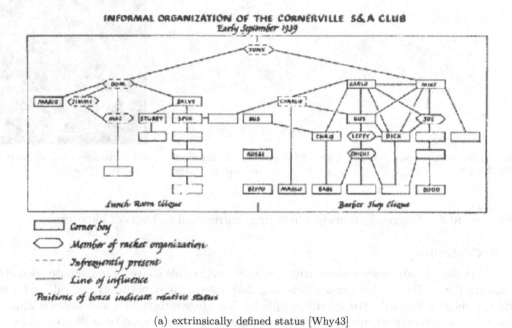

(a) extrinsically defined status [Why43]

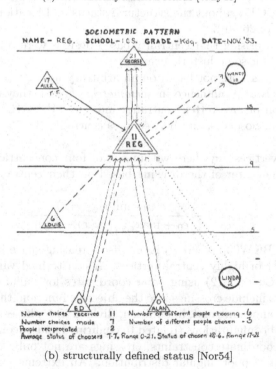

(b) structurally defined status [Nor54]

Figure 26.5 Two status diagrams using the high-low metaphor.

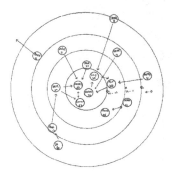

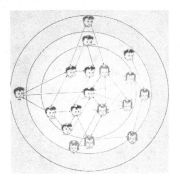

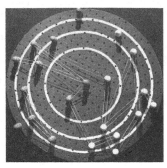

(a) Sociometric choice quartiles [Nor40]

(b) Grant's background gradient emphasizing the center [Nor52]

(c) McKenzie's board for manual layout [Nor52]

Figure 26.6 Target diagrams in which rings correspond to levels of importance.

Centrality

Similar to indices assessing status, centrality indices also have an immediate geometric connotation. A large value usually indicates that a vertex is structurally central, and a low value indicates that it is structurally peripheral. Just how strong, and sometimes confusing, the relationship between spatial metaphors and formal concepts can be is illustrated by the controversy of [Cha50, CJ51] about the structural status and location of prominent actors in the network of Figure 26.4(c).

It is therefore not surprising that centrality-based designs were proposed already in the 1940s. See Figure 26.6 for some historic examples.

While the early designs were not based on an arbitrary index, but on indegree quartiles (representing four levels of prominence in *sociometric choice*), they are easily generalized to exact representation of any vertex centrality index $c : V \to \mathbb{R}$. The most frequently used indices are degree, closeness, and betweenness centrality [Fre79] as well as eigenvector centrality [Bon72].

Instead of placing vertices anywhere within one of four concentric rings, we can define their distance from the center of the drawing based on their centrality score, for instance using radii

$$r(v) = 1 - \frac{c(v) - \min_{x \in V} c(x)}{c_0 + \max_{x,y \in V} [c(x) - c(y)]} \, ,$$

as layout constraints [BKW03], where c_0 is an offset creating space in the center and may depend on the number of highly central vertices. A constrained variant of the Kamada-Kawai approach (see Chapter 12) using polar coordinates for radial drawings is described in [Kam89]. In order to include crossings into the objective function, the method of [BKW03] is based on simulated annealing and, in addition, divided into phases in which more weight is placed on certain subconfigurations (related to confirmed and unconfirmed relationships).

Due to the radial coordinate constraints, crossings are not only a readability problem, but also an indication of poor angular distribution. An extreme example of this kind is presented in Figure 26.7(a), where a cutvertex and a separation pair are clearly visible by virtue of a circular ordering with few crossings. A second class of radial drawing algorithms is therefore based on combinatorial approaches that focus on crossing reduction in circular layouts to determine the angular coordinate (see Chapter 9 and [BB04]). More generally,

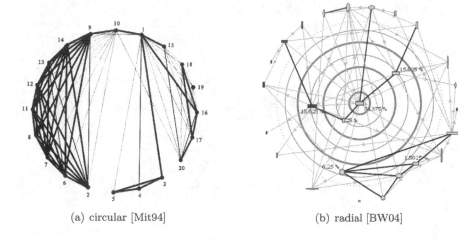

(a) circular [Mit94] (b) radial [BW04]

Figure 26.7 Circular and radial drawing.

approaches for layered layout can be adapted to radial levels (Chapter 13 and [Bac07]).

The most recent approach uses stress minimization with a penalty term [BP11] measuring deviation from the assigned radii. By increasing the relative weight of the penalty term during iterative stress reduction, vertices are gradually forced to lie on their respective circle. While it appears that the less severe restrictions to which intermediate layouts are subjected may provide an advantage over gradient-projection methods, thorough experimentation is needed to determine which methods are most practical for centrality layouts.

To explore the differences between various centrality indices, methods extending the comparison based on scatterplot matrices [KS04] have been proposed in [DHK+06]. Among them is a force-directed method for the joint layout of stacked radial drawings of the same network with varying radial constraints. A visual variational approach to centrality within a network is introduced in [CCM12].

26.3.2 Cohesion

Cohesion is broadly defined as strong interconnectedness of a group of actors, although the formal structural definition of interconnectedness may vary according to type of relations and substantive interest. In the extreme, cohesion is defined in terms of cliques [LP49], but weaker definitions such as the degree-based cores [Sei83] or connectivity-based λ-sets [BES90] exist.

In Gestalt Theory [Wer44], the law of proximity suggests that cohesive, and in fact all types of groups can be represented effectively by placing their members closer to each other than to other actors. The friendship network of pupils in the 4th grade shown in Figure 26.8, for instance, is divided according to gender, and this striking correspondence between a vertex attribute and structural cohesion is made evident by spatial separation. However, there is little empirical evidence whether and which kind of cohesion is represented effectively by spatial proximity or separation in graphical representations of networks [MBK97, HHE07].

While the clumping of densely connected subgraphs is an implicit objective of force-directed and spectral layout algorithms (see Chapter 12), a layout algorithm should be generic and suitable for a range of cohesion measures.

In the previous section we assumed that the result of an analysis, a vertex index $c : V \to \mathbb{R}$ representing prominence, is part of the input for a graph drawing algorithm. Similarly, let

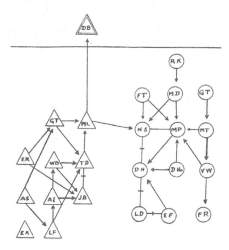

Figure 26.8   Friendship network of a 4th grade school class [Mor53, p. 163]. For graphical notation see Figure 26.2; note, though, the horizontal line delineating the class and the missing tick on the edge between NS and MP. The strong homophily effect is conveyed effectively through spatial separation.

us now assume that a cohesion analysis resulted in a decomposition that can be described in the following way.

A *(hierarchically) clustered graph* (G, T) is a graph $G = (V, E)$ together with a rooted tree T, the *cluster tree*, such that the leaves of T correspond to a partition of V and each inner node is the union of the vertex sets of its children. Consequently, the root corresponds to the entire vertex set V. A clustered graph (G, T) is called *flat*, if T has height one, i.e., it is equivalent to a graph $G = (V, E)$ together with a partition of V.

Note that cohesion analysis may result in other types of data. But one example is set covers of the vertices, which can be viewed as flat clustered graphs with overlapping clusters. They are equivalent to hypergraphs, which in turn are treated in Section 26.3.3.

Clustered drawings

In an *inclusion drawing* of a (cluster) tree, vertices are represented as areas, and the parent-child relation is represented by area inclusion. A straightforward representation of clustered graphs consists of an inclusion drawing of the cluster tree overlayed on a drawing of the underlying graph such that vertices are inside their cluster boundaries and edges cross cluster boundaries at most once. Such a representation is called a *clustered drawing*, and at least topologically implements the idea that vertices of the same group belong together. Figures 26.1 and 26.9 provide examples.

Often, especially when the notion of cohesion and the implicit criteria of general layout algorithms coincide sufficiently well, clustered drawings are obtained by adding boundary curves to a layout obtained without consideration of the cluster tree. A typical example is the application of multidimensional scaling to a distance matrix with the addition of hierarchical clusters based on connectedness as shown in Figure 26.9(b).

Multidimensional scaling based on stress minimization and, in fact, all force-directed approaches can be customized to clustered graphs by adding cluster vertices that are connected to cluster members via short edges, and to other cluster vertices via long edges or even repulsion (e.g., [WM96, PNR08]). Alternatively, cohesion-based proximity can be ensured by a combination of space-filling and force-directed techniques that explicitly consider a cluster

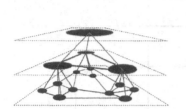
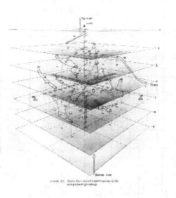
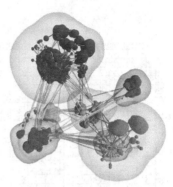

(a) Multilevel representation of a clustered graph [EF97]

(b) Clusters outlined in 2D perspective projection of 3D drawing [LG66]

(c) 3D drawing of a clustered graph with implicit surfaces [BD07]

Figure 26.9   Clustered graphs.

tree [IMMS09]. Edge bundling along the cluster tree has been proposed as a method to reduce visual clutter [Hol06].

To avoid meaningless crossings, every edge should cross only boundaries of clusters on the unique path in the cluster tree that connects the leaves containing its endvertices. These crossings are called necessary. A clustered graph is called *c-planar* (*cluster planar*) if it can be drawn such that, simultaneously, there are no edge-edge crossings (i.e., the graph is planar) and there are no edge-region crossings (except those necessary) [EFN99]. The graph in Figure 26.1 is c-planar, although the drawing is not. Whether c-planarity can be tested efficiently is an interesting open problem [CDB05].

A conceptually different visualization approach is based on clustering via *semantic substrates* [SA06], where regions are prescribed for vertices belonging to an extrinsically defined cluster (most often by sharing selected attribute values), and layout is carried out using any method respecting region boundaries.

Sociomatrices

In addition to the commonly used graph representation, there is also a tradition of depicting social relations in matrices. To distinguish them from socigrams, Moreno uses the term *sociomatrix* [Mor53]. Using the example in Figure 26.10 sociomatrices were advocated, e.g., in [FK46] (see also the interesting discussion that followed [Mor46, Kat47]), because they appear to be more effective at visualizing cohesion [GFC05]. Moreover, matrix cells are well defined and compactly organized locations for information associated with the edges [vHSD09].

The main degree of freedom is the ordering of rows and columns, and its effect on visualization is illustrated in Figure 26.11. While Bertin [Ber83] appears to have coined the term *reorderable matrix* and reordering is already discussed in [FK46], the idea has been introduced much earlier [Pet99, Cze09]. Most relevant ordering problems are $\mathcal{NP}$-hard, though. They have been researched extensively under various names including *seriation* and *linear layout* [DPS02]. Often, the underlying ordering objectives aim at reducing the span of edges so that well-clustered graphs lead to visible blocks along the diagonal. For a clustered graph, an optimal ordering can be determined efficiently if the maximum degree of the cluster tree is bounded by a constant (see, e.g., [BDW99, Bra07]).

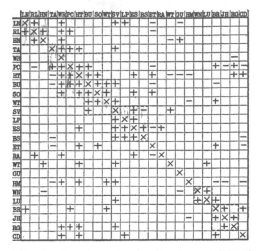

(a) ordered sociomatrix of a signed graph [FK46]

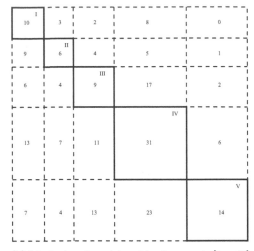

(b) blocked sociomatrix with edge counts [Lon48]

Figure 26.10   Sociomatrix and block partition.

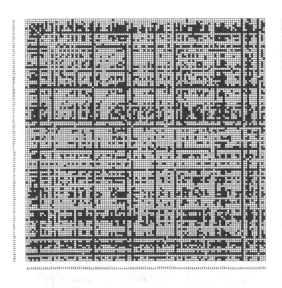

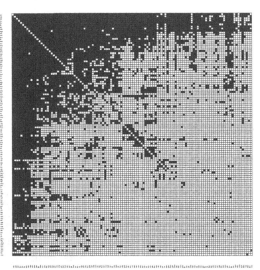

Figure 26.11   Trade between countries reordered according to a hierachical clustering (reproduced from [BM04]).

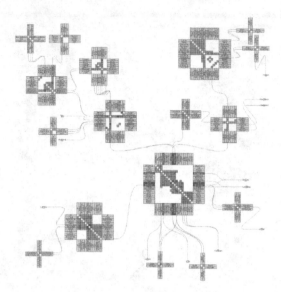

Figure 26.12 Integration of sociogram with sociomatrices for cohesive subgroups (reproduced from [HFM07]).

A system offering coordinated sociogram and sociomatrix views is MatrixExplorer [HF06]. The integrated view exemplified in Figure 26.12 uses both representations simultaneously and the decision about which representation is used for which subgraph is based on the observation that matrix representations are especially suitable for dense (sub)graphs [HFM07]. Other augmentations of matrix representations to ease the recognition of paths include [HF07, SM07], and a matrix representation for layered graphs that has been applied to genealogies [BDF+10].

An integrated representation that is not based on a matrix of adjacencies, but a grid layout of vertex attribute levels, are PivotGraphs [Wat06]. They generalize attribute-defined layouts (cf. Figure 26.4(d)) and are particularly suited for the interactive exploration of associations between vertex attributes and edges.

26.3.3 Two-Mode Networks

The networks considered so far are actually *one-mode* networks, because their vertices represent elements of the same mode or category such as persons. Quite frequently, however, the relation of interest is between elements of different categories such as persons and groups [Bre74]. Such networks can be represented in rectangular matrices with rows and columns indexed by the respective categories, and they are referred to as *two-mode* networks.

Two-mode networks are often visualized like one-mode networks, with different appearances for vertices from the two categories. However, their distinctive characteristic of being bipartite with a prescribed bipartition of vertices can also guide a layout algorithm.

As a variant of spectral layout for one-mode networks, left and right singular vectors of the rectangular adjacency matrix (or other matrices derived from it) can be used for coordinates. An entire family of related techniques is reviewed in [dLM00]. See also [Bre09] for a closely related analytic technique and [WG98] for a comparison with graphical representations described in the following subsection.

More combinatorial approaches are those developed for bipartite graphs. These include, in particular, drawings in which the vertices of the two modes are placed on different parallel

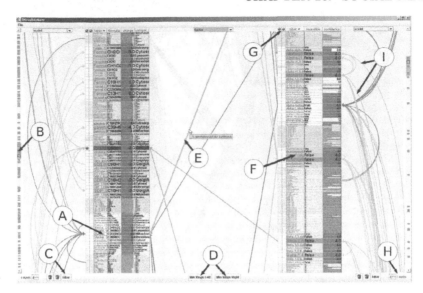

Figure 26.13 A two-mode network with straight-line edges drawn between attribute tables of the two node sets, and its one-mode projections drawn with curved edges on the sides (reproduced from [SJUS08]).

lines (2-level drawings) or, more generally, in separable regions. As usual, graph drawing research in this area has focused on conditions under which the resulting drawings can be made planar [Bie98, BKM98, CSW04, DGGL08] and on the difficulty of crossing minimization [ZSE05]. It would be interesting to identify criteria for informative visualization of bipartite graphs.

An example of several methods combining ideas of spatial separation and relative placement are the anchored maps of [Mis07]. While one set of the bipartition is arranged on a circle, vertices in the other are placed relative to their neighbors. An interesting mixture of 2-layer drawing and tabular representation is exemplified in Figure 26.13.

Hypergraphs

If it makes sense to consider the elements of one of the two modes as subsets of the other (as with groups and persons such as company boards and directors), a two-mode network can be treated as a hypergraph. In addition to those associated with the above bipartite graph model, several other graphical representations are available.

A straightforward variant is the *edge standard*, which is based on an ordinary layout of the bipartite graph representation of the hypergraph, but with a different rendering of the induced star subgraphs that represent hyperedges. This star may be substituted for a tree to shorten the total length of the hyperedge. For directed hypergraphs, layout constraints can be used to enable directed edges to be rendered confluently [Mäk90].

Subdivision drawings [KvKS09] are subdivisions of the plane such that each vertex corresponds to a region and the set of regions corresponding to a hyperedge is connected. This requires that the hypergraph has a planar *support*, i.e., the existence of a planar graph on the same vertices such that each of the hyperedges of the original hypergraph induces a connected subgraph. Deciding whether a hypergraph has a planar support is $\mathcal{NP}$-complete [JP87]. Tree supports, on the other hand, are characterized by the existence of an elimination ordering in which vertices contained in only one hyperedge, or in a subset of the hyperedges containing some other vertex, are removed iteratively. The main open

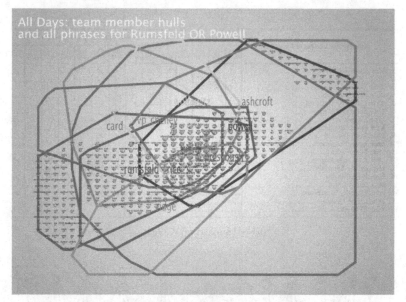

Figure 26.14 Post-hoc delineation of clusters with polygons (reproduced from [JK04]).

problem is whether the existence of an outerplanar support can be decided in polynomial time [BvKM+10, BCPS11a]. Supports with more restrictive constraints on the subgraphs induced by hyperedges are introduced in [BCPS11b].

The more general *subset standard* yields drawings also known as Euler diagrams [RZF08]. Each hyperedge is represented as a simple closed curve containing exactly the vertices of that edge. Note that this is also the usual convention for cluster boundaries in flat overlapping clustered graphs [DGL08]. For the example in Figure 26.14, cluster boundaries were drawn as convex polygons simply after the underlying graph had been laid out [JK04]. A more comprehensive postprocessing approach is proposed in [CPC09], and a restricted variant in which hyperedges are drawn as paths through already placed vertices is studied in [ARRC11]. The resulting visualization looks similar to the familiar metro map designs, and indeed layout algorithms for metro maps can be used to draw hypergraphs by first ordering the vertices in each hyperedge [Wol07].

As is common for problems that are difficult on general instances, many variant force-directed approaches have been devised [BE00, OS07, ST10, SAA09, KZ09]. While most approaches are based on dummy vertices and/or additional forces for the hyperedges, the approach of [SAA09] is based on the *intersection graph*, which is a line graph of the hypergraph. It is constructed by creating a vertex for each hyperedge and an edge between any two of them, if the corresponding hyperedges overlap.

Lattices

Inspired by their use in formal concept analysis [GW98], *Galois lattices* have been proposed as an alternative representation for two-mode networks [FW93]. An overview of the potential of lattices in data analysis and a standard tool, GLAD, are provided by Duquenne [Duq99].

Figure 26.15 shows an example of a two-mode network represented in a matrix, a bipartite graph, and a Galois lattice. In the Galois lattice representation, a node simultaneously represents a subset of women and a subset of events. The women are exactly those attending all of the corresponding events, and the events are exactly those attended by all of the

Names of Participants of Group I	Code Numbers and Dates of Social Events Reported in *Old City Herald*													
	(1) 6/27	(2) 3/2	(3) 4/12	(4) 9/26	(5) 2/25	(6) 5/19	(7) 3/15	(8) 9/16	(9) 4/8	(10) 6/10	(11) 2/23	(12) 4/7	(13) 11/21	(14) 8/3
1. Mrs. Evelyn Jefferson.....................	X	X	X	X	X	X		X	X					
2. Miss Laura Mandeville...................	X	X	X		X	X	X	X						
3. Miss Theresa Anderson.................		X	X	X	X	X	X	X	X					
4. Miss Brenda Rogers......................	X		X	X	X	X	X	X						
5. Miss Charlotte McDowd.................			X	X	X		X							
6. Miss Frances Anderson..................			X		X	X		X						
7. Miss Eleanor Nye.........................					X	X	X	X						
8. Miss Pearl Oglethorpe..................						X		X	X					
9. Miss Ruth DeSand........................					X		X	X	X					
10. Miss Verne Sanderson.................							X	X	X			X		
11. Miss Myra Liddell.......................								X	X	X		X		
12. Miss Katherine Rogers................								X	X	X		X	X	X
13. Mrs. Sylvia Avondale..................							X	X	X	X		X	X	X
14. Mrs. Nora Fayette......................						X	X		X	X	X	X	X	X
15. Mrs. Helen Lloyd........................							X	X		X	X	X		
16. Mrs. Dorothy Murchison.............								X	X					
17. Mrs. Olivia Carleton...................									X		X			
18. Mrs. Flora Price.........................									X		X			

(a) data matrix [DGG41]

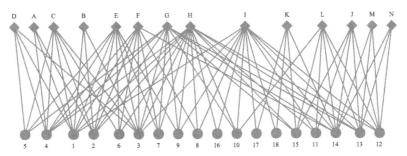

(b) two-layer drawing of bipartite network

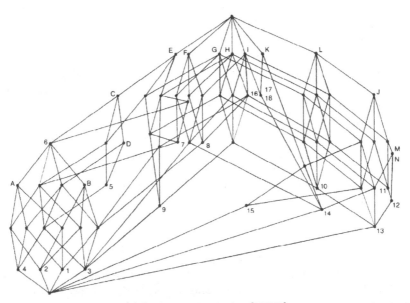

(c) lattice representation [FW93]

Figure 26.15 The Southern Women data of Davis, Gardner and Gardner [DGG41] is a two-mode network of 18 women's (labeled 1–18) attendance of 14 events (labeled A–N).

corresponding women. Nodes are ordered by set-inclusion, i.e., for each node the set of women consists of those encountered on downward paths to the bottom, and the set of events consists of those encountered on upward paths to the top. A study essentially concluded that users do understand such diagrams [EDB04].

Approaches for drawing lattices range from layer-constrained force-directed layout in three dimensions [Fre04b] to enumeration and decomposition approaches [RCE06, BPS11]. See the review in [MH01] for relations with other combinatorial structures.

26.3.4 Dynamics

Temporal aspects are considered explicitly in *longitudinal* social network analysis. Mostly, this is concerned with panel data, i.e., cross-sectional states of networks observed at discrete time points. This is immediate when data is collected in waves. Even when the situation is more accurately described in terms of relational events (such as phone calls), however, these are often aggregated over time intervals into cross-sectional graph representations to ensure applicability of the wide range of methods developed for (static) graphs.

In-between observations or aggregations, the sets of vertices and edges, as well as attribute values may be subject to change. Typical research questions are: how do actor characteristics affect structural change (*social selection*), and how do structural conditions affect actor behavior (*social influence*)? In addition, the subject of interest may actually be a process such as the diffusion of information taking place on a (possibly changing) network.

By combinatorial explosion, this leads to numerous problem variants. The variant most extensively researched in graph drawing, however, focuses on *dynamic graphs* which consist of a sequence of interrelated graphs $G^{(1)}, \ldots, G^{(T)}$, called *states*. In social network analysis these arise from panel data on social structure (network evolution). Research on a *streaming* scenario in which a single graph becomes available one edge at a time was initiated only recently [BBDB$^+$10], but may soon become relevant for dyadic event data.

There are two main scenarios for visualizing dynamic graphs, *online* and *offline* dynamic graph drawing. Layout approaches for these are considered in more detail below. In both cases, a solution consists of a sequence of layouts, one for each $G^{(t)}$ with $t = 1, \ldots, T$, and two conflicting criteria are used to evaluate the quality of a solution.

On the one hand, each layout in the sequence should be acceptable with respect to the criteria of a static graph drawing problem. We refer to this requirement as *layout quality*, and assume that the related static layout problem is fixed. On the other hand, the degree of change between consecutive layouts should be indicative of the degree of change between the corresponding graphs. This criterion is referred to as *layout stability* and generally motivated by preservation of a user's mental map [MELS95].

Note that the stability requirement applies to the difference between consecutive layouts and, depending on the visualization media, also to the transition from one layout to the next. These two aspects are referred to as the *logical* and the *physical* update, respectively [Nor96]. Difference metrics for pairs of layouts are treated in [BT00], and animation between layouts is the subject of [FE02]. The interpolation approach of [BFP07] implements the physical update as a refinement of the logical update.

Online Scenario

In an *online* scenario, a dynamic graph is presented one state at a time, and the layout of a state is to be determined before the next state is known. Stability can therefore only be introduced with respect to layouts of previous states.

Since iterative layout algorithms are very common in applied graph drawing in general, and social network visualization in particular, a simple approach to online drawing is the initialization of the layout algorithm for a state with the layout of the previous state [MMBd05, HEW98]. This is a convenient, though indirect, approach to address the stability requirement. It does work fairly well when the iterations of the layout algorithm are used for the physical updates of an animation.

Explicit consideration of stability by defining a layout objective that trades off quality and stability using a static objective and a difference metric is proposed in the Bayesian framework of [BW97]. A recent application of this is [FT08]. Since the employed difference metric is penalizing vertex movement via (weighted) distances from previous positions, this is a case of what is called anchoring [LMR98].

An unusual variant of online drawing is introduced in [DDBF+99], where drawings of a state must keep the drawing of the previous state intact, and expectations (rather than knowledge) about the additional subgraph to accomodate are available. However, this situation has so far been considered only for descending traversals of trees.

Offline Scenario

In an *offline* scenario, the input given is a dynamic graph and the output sought is a layout for each of its states. Except for streaming event data and some applications involving social networking sites, the offline scenario is the typical scenario in empirical social network analysis. Since the entire sequence of states is known before any layout needs to be determined, the layout of a state may be determined with subsequent states in mind. In other words, we may use knowledge about the future.

Note that methods for online scenarios can be applied in offline scenarios, although at the possible expense of both quality and stability, but the reverse is generally not possible.

In a recent review [BIM12], three primary approaches to offline dynamic graph drawing are distinguished:

Aggregation. All graphs in the sequence are aggregated into a single graph that has one vertex for each actor. The position of each occurrence of vertex in a state is fixed by the layout of the aggregated graph. Variants of this approach are considered, e.g., in [BC03a, DG04, MMBd05], and it is referred to as the *flip-book approach* in the last reference.

Anchoring. Using auxiliary edges, vertices are connected to immobile copies fixed to a desired location which may be, for instance, the previous position as in an online scenario, or a reference position determined from an aggregate layout in an offline scenario. This approach is used, e.g., in [LMR98, BW97, FT08].

Linking. All graphs in the sequence are combined into a single graph that has one vertex for each occurrence of an actor, and an edge is created between vertices representing the same actor in consecutive graphs. A layout of this graph directly yields positions for all vertex instances in the sequence. This approach is used, e.g., in [DG04, EKLN04, DHK+06].

Algorithmic experimental evidence [BM12] suggests that, at least for methods based on stress minimization [GKN04] and general conditions, linking dominates anchoring in terms of stability and quality. On the other hand, anchoring is computionally cheaper and especially suited when a dynamic graph has rather persistent global structure. Evidence from user experiments, on the other hand, is inconclusive about the actual value of stability in dynamic graph animation [PS08], and even animation itself [APP11].

week	stable	no stability	compromise

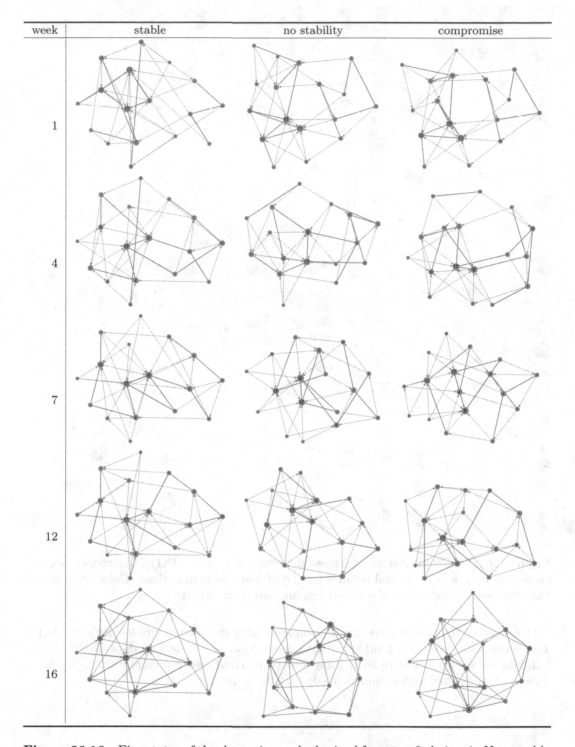

Figure 26.16 Five states of the dynamic graph obtained from top-3 choices in Newcomb's fraternity data. Layout obtained by stress minimization on an aggregate graph (stable), initialized by previous layout (no stability), and with linking of consecutive layouts (compromise).

Figure 26.17 Gestaltmatrix of Newcomb's fraternity data [BN11]. Matrix cells show evolution of rank (length) and balance (angle) of pairwise nominations. Color on diagonal indicates average deviation of received nominations from expected value.

Online and offline approaches can be compared using the small multiples in Figure 26.16. The networks shown form a subset of the famous Newcomb fraternity data [New61]. The full data are shown in Figure 26.17 using a static matrix-based representation for dynamic directed graphs that have a numerical edge attribute [BN11].

26.4   Trends and Challenges

While graphs arising from social networks across application domains exhibit some general tendencies such as sparseness and local clustering, there is no formal characterization in terms of structural requirements that delineates this class of graphs from others. Likewise, the information to be conveyed in a social network visualization differs by available data, interest, and domain.

Layout of social networks is therefore contingent on many factors, and comparison of approaches is possible only if scope and purpose are defined precisely. Identification of practically relevant and pragmatic tasks remains a challenge, though.

Despite the range of approaches presented in this chapter, force-directed methods – much like in other areas of applied graph drawing – are most commonly used for social network layout. This is likely because of their generality, simplicity, adaptability, and above all their availability. While force-directed methods generally perform well in separating clusters in graphs with varying local density, these methods are particularly troubled by small distances and skewed degree distributions [BP09]. A fundamental challenge is therefore to identify representations and layout criteria that allow to deal with such structures [ACL07].

Another very general challenge involves the interplay between methods for hierachical clustering of graphs and clustered graph layout. Especially for large graphs, hierarchical clustering is frequently used as a tool in multilevel layout algorithms, but the artifacts resulting from the choice of clustering or filter methods are not yet understood well (see, e.g., [JHGH08, vHW08, HN07]).

Two research directions that are more closely related to the type of data (rather than its properties) encountered in social network analysis are the genuine treatment of two-mode networks (Section 26.3.3) and visual means to support stochastic network modeling [BIM12].

Finally, there is also at least one example where artistic drawings of social networks inspired a new graph drawing convention, namely Lombardi drawings [Hob03, DEG+12].

Many software packages are available already for the analysis and visualization of social networks, and many more are in introduced for specific application domains. Among social scientists, UCINET [BEF99] is the most widely known. At the time of this writing, Pajek [BM04] is likely to be the most widely used across all disciplines, and visone [BW04] the social network analysis tool with the most sophisticated graph drawing features. Other comprehensive and popular tools include Tulip [Aub04], NodeXL [HSS10], Gephi [BHJ09], and ORA [CRSC11]. A recent software review can be found in [HvD11], and a comparative evaluation of some tools is attempted in [XTT+10]. A comprehensive list of software for social networks is maintained in Wikipedia.[1]

Very likely the most dominant force driving visualization research on social networks over the next decade will be online social and other networks derived from social media [Fur10]. This is in part because they combine virtually all the current challenges of size and dynamics with the more specific challenges that arise from multivariate complexity. Moreover, such research can be of economic relevance, draw large audiences, and make use of easily accessible data [LPA+09]. It will be exciting to witness whether graph drawing can make significant contributions to this area and thus challenge currently reigning adaptations of its oldest methods (see [HB05] for one solid example). An example in this direction is a layout algorithm for digital social networks tailored to smartphone displays [DLDBI12].

---

[1]http://en.wikipedia.org/wiki/Social_network_analysis_software

References

[ACL07]   Reid Andersen, Fan Chung, and Linyuan Lu. Drawing power law graphs
            using a local/global decomposition. *Algorithmica*, 47(4):379–397, 2007.

[APP11]   Daniel Archambault, Helen C. Purchase, and Bruno Pinaud. Animation,
            small multiples, and the effect of mental map preservation in dynamic
            graphs. *IEEE Transactions on Visualization and Computer Graphics*,
            17(4):539–552, 2011.

[ARRC11]   Basak Alper, Nathalie Riche, Gonzalo Ramos, and Mary Czerwinski. De-
            sign study of linesets, a novel set visualization technique. *IEEE Transac-
            tions on Visualization and Computer Graphics*, 17(12):2259–2267, 2011.

[Aub04]   David Auber. Tulip – a huge graph visualisation framework. In Jünger
            and Mutzel [JM04], pages 105–126.

[Bac07]   Christian Bachmaier. A radial adaptation of the Sugiyama framework for
            visualizing hierarchical information. *IEEE Transactions on Visualization
            and Computer Graphics*, 13(3):583–594, 2007.

[Bar54]   John A. Barnes. Class and committees in a Norwegian island parish.
            Human Relations, 7:39–58, 1954.

[BB04]   Michael Baur and Ulrik Brandes. Crossing reduction in circular layouts. In
            Juraj Hromkovič, Manfred Nagl, and Bernhard Westfechtel, editors, *Pro-
            ceedings of the 30th International Workshop on Graph-Theoretical Con-
            cepts in Computer Science (WG'04)*, volume 3353 of *Lecture Notes in
            Computer Science*, pages 332–343. Springer-Verlag, 2004.

[BBDB+10]   Carla Binucci, Ulrik Brandes, Giuseppe Di Battista, Walter Didimo,
            Marco Gaertler, Pietro Palladino, Maurizio Patrignani, Antonios Symvo-
            nis, and Katharina A. Zweig. Drawing trees in a streaming model.
            In *Proceedings of the 17th International Symposium on Graph Drawing
            (GD'09)*, volume 5849 of *Lecture Notes in Computer Science*, pages 292–
            303. Springer-Verlag, 2010.

[BC03a]   Ulrik Brandes and Steven R. Corman. Visual unrolling of network evo-
            lution and the analysis of dynamic discourse. *Information Visualization*,
            2(1):40–50, 2003.

[BC03b]   Ulrik Brandes and Sabine Cornelsen. Visual ranking of link structures.
            Journal of Graph Algorithms and Applications, 7(2):181–201, 2003.

[BCPS11a]   Ulrik Brandes, Sabine Cornelsen, Barbara Pampel, and Arnaud Sal-
            laberry. Blocks of hypergraphs (applied to hypergraphs and outerpla-
            narity). In *Proceedings of the 21st International Workshop on Combi-
            natorial Algorithms (IWOCA 2010)*, volume 6460 of *Lecture Notes in
            Computer Science*, pages 201–211. Springer-Verlag, 2011.

[BCPS11b]   Ulrik Brandes, Sabine Cornelsen, Barbara Pampel, and Arnaud Sal-
            laberry. Path-based supports for hypergraphs. In *Proceedings of the 21st
            International Workshop on Combinatorial Algorithms (IWOCA 2010)*,
            volume 6460 of *Lecture Notes in Computer Science*, pages 20–33. Springer-
            Verlag, 2011.

[BD07]   Michael Balzer and Oliver Deussen. Level-of-detail visualization of clus-
            tered graph layouts. In *Proceedings of the 6th International Asia-Pacific
            Symposium on Visualisation (APVis '07)*, pages 133–140. IEEE, 2007.

[BDF⁺10] Anastasia Bezerianos, Pierre Dragicevic, Jean-Daniel Fekete, Juhee Bae, and Ben Watson. GeneaQuilts: A system for exploring large genealogies. *IEEE Transactions on Visualization and Computer Graphics*, 16(6):741–748, 2010.

[BDW99] Rainer Burkhard, Vladimir G. Deĭneko, and Gerhard Woeginger. The travelling salesman and the PQ-tree. *Mathematics of Operations Research*, 24(1):262–272, 1999.

[BE00] François Bertault and Peter Eades. Drawing hypergraphs in the subset standard. In *Proceedings of the 8th International Symposium on Graph Drawing (GD'00)*, volume 1984 of *Lecture Notes in Computer Science*, pages 164–169. Springer-Verlag, 2000.

[BE05] Ulrik Brandes and Thomas Erlebach, editors. *Network Analysis: Methodological Foundations*, volume 3418 of *Lecture Notes in Computer Science*. Springer-Verlag, 2005.

[BE06] Stephen P. Borgatti and Martin G. Everett. A graph-theoretic perspective on centrality. *Social Networks*, 28:466–484, 2006.

[BEF99] Stephen P. Borgatti, Martin G. Everett, and Linton C. Freeman. *UCINET 6.0 Version 1.00*. Analytic Technologies, 1999.

[Ber83] Jacques Bertin. *Semiology of Graphics: Diagrams, Networks, Maps*. University of Wisconsin Press, 1983.

[BES90] Stephen P. Borgatti, Martin G. Everett, and Paul R. Shirey. LS sets, lambda sets and other cohesive subsets. *Social Networks*, 12(4):337–357, 1990.

[BFP07] Ulrik Brandes, Daniel Fleischer, and Thomas Puppe. Dynamic spectral layout with an application to small worlds. *Journal of Graph Algorithms and Applications*, 11(2):325–343, 2007.

[BHJ09] Mathieu Bastian, Sebastien Heymann, and Mathieu Jacomy. Gephi: an open source software for exploring and manipulating networks. In *Proceedings of the 3rd International AAAI Conference on Weblogs and Social Media (ICWSM '09)*, pages 361–362, 2009.

[Bie98] Therese C. Biedl. Drawing planar partitions I: LL-drawings and LH-drawings. In *Proceedings of the 14th Annual ACM Symposium on Computational Geometry (SoCG'98)*, pages 287–296, 1998.

[BIM12] Ulrik Brandes, Natalie Indlekofer, and Martin Mader. Visualization methods for longitudinal social networks and stochastic actor-oriented modeling. *Social Networks*, 34(3):291–308, 2012.

[BKM98] Therese C. Biedl, Michael Kaufmann, and Petra Mutzel. Drawing planar partitions II: HH-drawings. In *Proceedings of the 24th International Workshop on Graph-Theoretical Concepts in Computer Science (WG'98)*, volume 1517 of *Lecture Notes in Computer Science*, pages 101–114. Springer-Verlag, 1998.

[BKR⁺99] Ulrik Brandes, Patrick Kenis, Jörg Raab, Volker Schneider, and Dorothea Wagner. Explorations into the visualization of policy networks. *Journal of Theoretical Politics*, 11(1):75–106, 1999. Reprinted in Linton Freeman, ed., *Social Network Analysis*, vol. I (Data, Mathematical Models, and Graphics), Sage, 2007.

[BKR06]   Ulrik Brandes, Patrick Kenis, and Jörg Raab. Explanation through net-
            work visualization. *Methodology*, 2(1):16–23, 2006. Spanish translation in
            REDES 9(6), 2005.

[BKW03]   Ulrik Brandes, Patrick Kenis, and Dorothea Wagner. Communicating cen-
            trality in policy network drawings. *IEEE Transactions on Visualization
            and Computer Graphics*, 9(2):241–253, 2003.

[BM04]   Vladimir Batagelj and Andrej Mrvar. Pajek – analysis and visualization
            of large networks. In Jünger and Mutzel [JM04], pages 77–103.

[BM12]   Ulrik Brandes and Martin Mader. A quantitative comparison of stress-
            minimization approaches for offline dynamic graph drawing. In *Proceed-
            ings of the 19th International Symposium on Graph Drawing (GD 2011)*,
            volume 7034 of *Lecture Notes in Computer Science*, pages 99–110.
            Springer-Verlag, 2012.

[BMBL09]   Stephen P. Borgatti, Ajay Mehra, Daniel J. Brass, and Giuseppe Labi-
            anca. Network analysis in the social sciences. *Science*, 323(5916):892–895,
            2009.

[BN11]   Ulrik Brandes and Bobo Nick. Asymmetric relations in longitudinal social
            networks. *IEEE Transactions on Visualization and Computer Graphics*,
            17(12):2283–2290, 2011.

[Bon72]   Phillip Bonacich. Factoring and weighting approaches to status scores
            and clique identification. *Journal of Mathematical Sociology*, 2:113–120,
            1972.

[Bor05]   Stephen P. Borgatti. Centrality and network flow. *Social Networks*, 27:55–
            71, 2005.

[BP09]   Ulrik Brandes and Christian Pich. An experimental study on distance-
            based graph drawing. In *Proceedings of the 16th International Symposium
            on Graph Drawing (GD'08)*, volume 5417 of *Lecture Notes in Computer
            Science*, pages 218–229. Springer-Verlag, 2009.

[BP11]   Ulrik Brandes and Christian Pich. More flexible radial layout. *Journal of
            Graph Algorithms and Applications*, 15(1):157–173, 2011.

[BPS11]   Anne Berry, Romain Pogorelcnik, and Alain Sigayret. Vertical decomposi-
            tion of a lattice using clique separators. In *Proceedings of the 8th Interna-
            tional Conference on Concept Lattices and their Applications (CLA'11)*,
            pages 15–29, 2011.

[Bra96]   Franz J. Brandenburg, editor. *Proceedings of the 3rd International Sym-
            posium on Graph Drawing (GD '95)*, volume 1027 of *Lecture Notes in
            Computer Science*. Springer, 1996.

[Bra07]   Ulrik Brandes. Optimal leaf ordering of complete binary trees. *Journal
            of Discrete Algorithms*, 5(3):546–552, 2007.

[Bra08]   Ulrik Brandes. Social network analysis and visualization. *IEEE Signal
            Processing Magazine*, 25(6):147–151, 2008.

[Bre74]   Ronald L. Breiger. The duality of persons and groups. *Social Forces*,
            53(2):181–190, 1974.

[Bre09]   Ronald L. Breiger. On the duality of cases and variables. In David Byrne
            and Charles C. Ragin, editors, *The SAGE Handbook of Case-Based Meth-
            ods*, pages 243–259. Sage, 2009.

[BRW01]   Ulrik Brandes, Jörg Raab, and Dorothea Wagner. Exploratory network visualization: Simultaneous display of actor status and connections. *Journal of Social Structure*, 2(4), 2001.

[BT00]   Stina S. Bridgeman and Roberto Tamassia. Difference metrics for interactive orthogonal graph drawing algorithms. *Journal of Graph Algorithms and Applications*, 4(3):47–74, 2000.

[But09]   Carter T. Butts. Revisiting the foundations of network analysis. *Science*, 325(5939):414–416, 2009.

[BvKM+10] Kevin Buchin, Marc van Kreveld, Henk Meijer, Bettina Speckmann, and Kevin Verbeek. On planar supports or hypergraphs. In *Proceedings of the 17th International Symposium on Graph Drawing (GD'09)*, volume 5849 of *Lecture Notes in Computer Science*, pages 345–356. Springer-Verlag, 2010.

[BW97]   Ulrik Brandes and Dorothea Wagner. A Bayesian paradigm for dynamic graph layout. In Giuseppe Di Battista, editor, *Proceedings of the 5th International Symposium on Graph Drawing (GD '97)*, volume 1353 of *Lecture Notes in Computer Science*, pages 236–247. Springer, 1997.

[BW04]   Ulrik Brandes and Dorothea Wagner. visone – analysis and visualization of social networks. In Michael Jünger and Petra Mutzel, editors, *Graph Drawing Software*, pages 321–340. Springer-Verlag, 2004.

[CAI]   CAIDA. Cooperative association for internet data analysis. http://www.caida.org/research/topology/as_core_network/.

[CCM12]   Carlos D. Correa, Tarik Crnovrsanin, and Kwan-Liu Ma. Visual reasoning about social networks using centrality sensitivity. *IEEE Transactions on Visualization and Computer Graphics*, 18(1):106–120, 2012.

[CDB05]   Pier Francesco Cortese and Giuseppe Di Battista. Clustered planarity. In *Proceedings of the 21st Annual ACM Symposium on Computational Geometry (SoCG'05)*, pages 32–34, 2005.

[Cha50]   F. Stuart Chapin. Sociometric stars as isolates. *American Journal of Sociology*, 56(3):263–267, 1950.

[CJ51]   Joan H. Criswell and Helen Hall Jennings. A critique of Chapin's "Sociometric stars as isolates". *American Journal of Sociology*, 57(3):260–264, 1951.

[CM11]   Carlos D. Correa and Kwan-Liu Ma. Visualizing social networks. In Charu C. Aggarwal, editor, *Social Network Data Analytics*, pages 307–326. Springer-Verlag, 2011.

[CPC09]   Christopher Collins, Gerald Penn, and Sheelagh Carpendale. Bubble sets: revealing set relations with isocontours over existing visualizations. *IEEE Transactions on Visualization and Computer Graphics*, 15(6):1009–1016, 2009.

[CRSC11]   Kathleen M. Carley, Jeff Reminga, Jon Storrick, and Dave Columbus. ORA user's guide 2011. Technical Report CMU-ISR-11-107, Carnegie Mellon University, Institute for Software Research, 2011.

[CSW04]   Sabine Cornelsen, Thomas Schank, and Dorothea Wagner. Drawing graphs on two and three lines. *Journal of Graph Algorithms and Applications*, 8(2):161–177, 2004.

[CSW05]   Peter J. Carrington, John Scott, and Stanley Wasserman, editors. *Models and Methods in Social Network Analysis*. Cambridge University Press, 2005.

[CY10]   Ing-Xiang Chen and Cheng-Zen Yang. Visualization of social networks. In *Handbook of Social Network Technologies and Applications* [Fur10], pages 585–610.

[Cze09]   Jan Czekanowski.   Zur Differentialdiagnose der Neandertalgruppe. *Korrespondenz-Blatt der Deutschen Gesellschaft für Anthropologie, Ethnologie and Urgeschichte*, 40(6/7):44–47, 1909.

[DBF05]   Patrick Doreian, Vladimir Batagelj, and Anuška Ferligoj. *Generalized Blockmodeling*. Cambridge University Press, 2005.

[DDBF+99]   Camil Demetrescu, Giuseppe Di Battista, Irene Finocchi, Giuseppe Liotta, Maurizio Patrignani, and Maurizio Pizzonia. Infinite trees and the future. In *Proceedings of the 7th International Symposium on Graph Drawing (GD'99)*, volume 1731 of *Lecture Notes in Computer Science*, pages 379–391. Springer-Verlag, 1999.

[DEG+12]   Christian A. Duncan, David Eppstein, Michael T. Goodrich, Stephen G. Kobourov, and Martin Nöllenburg. Lombardi drawings of graphs. *Journal of Graph Algorithms and Applications*, 16(1):85–108, 2012.

[DG04]   Tim Dwyer and David R. Gallagher. Visualising changes in fund manager holdings in two and a half-dimensions. *Information Visualization*, 3(4):227–244, 2004.

[DGG41]   Allison Davis, Burleigh B. Gardner, and Mary R. Gardner. *Deep South: A Social Anthropological Study of Caste and Class*. University of Chicago Press, 1941.

[DGGL08]   Emilio Di Giacomo, Luca Grilli, and Giuseppe Liotta. Drawing bipartite graphs on two parallel convex curves. *Journal of Graph Algorithms and Applications*, 12(1):97–112, 2008.

[DGL08]   Walter Didimo, Francesco Giordano, and Giuseppe Liotta. Overlapping cluster planarity. *Journal of Graph Algorithms and Applications*, 12(3):267–291, 2008.

[DHK+06]   Tim Dwyer, Seokhee Hong, Dirk Koschützki, Falk Schreiber, and Kai Xu. Visual analysis of network centralities. In *Proceedings of the 2006 Asia-Pacific Symposium on Information Visualisation (APVis '06)*, pages 189–197. Australian Computer Society, 2006.

[DKM09]   Tim Dwyer, Yehuda Koren, and Kim Marriott. Constrained graph layout by stress majorization and gradient projection. *Discrete Mathematics*, 309(7):1895–1908, 2009.

[DLDBI12]   Giordano Da Lozzo, Giuseppe Di Battista, and Francesco Ingrassia. Drawing graphs on a smartphone. *Journal of Graph Algorithms and Applications*, 16(1):109–126, 2012.

[dLM00]   Jan de Leeuw and George Michailides. Graph layout techniques and multivariate data analysis. In F. Thomas Bruss and Lucien Le Cam, editors, *Game theory, optimal stopping, probability and statistics: Papers in honor of Thomas S. Ferguson*, pages 219–248. Beachwood, 2000.

[DPS02]   Josep Diaz, Jordi Petit, and Maria Serna. A survey of graph layout problems. *ACM Computing Surveys*, 34:313–356, 2002.

[Duq99]   Vinvent Duquenne. Latticial structures in data analysis. *Theoretical Computer Science*, 217:407–436, 1999.

[EDB04]   Peter Eklund, Jon Ducrou, and Peter Brawn. Concept lattices for information visualization: Can novices read line-diagrams? In *Proceedings of the 2nd International Conference on Formal Concept Analysis (ICFCA'04)*, volume 2961 of *Lecture Notes in Computer Science*, pages 235–236. Springer-Verlag, 2004.

[EF97]   Peter Eades and Qing-Wen Feng. Multilevel visualization of clustered graphs. In *Proceedings of the 5th International Symposium on Graph Drawing (GD'97)*, volume 1190 of *Lecture Notes in Computer Science*, pages 101–112. Springer-Verlag, 1997.

[EFN99]   Peter Eades, Qing-Wen Feng, and Hiroshi Nagamochi. Drawing clustered graphs on an orthogonal grid. *Journal of Graph Algorithms and Applications*, 3(4):3–29, 1999.

[EKLN04]   Cesim Erten, Stephen G. Kobourov, Vu Le, and Armand Navabi. Simultaneous graph drawing: Layout algorithms and visualization schemes. In *Proceedings of the 11th International Symposium on Graph Drawing (GD'03)*, volume 2912 of *Lecture Notes in Computer Science*, pages 437–449. Springer-Verlag, 2004.

[FE02]   Carsten Friedrich and Peter Eades. Graph drawing in motion. *Journal of Graph Algorithms and Applications*, 6(3):353–370, 2002.

[FK46]   Elaine Forsyth and Leo Katz. A matrix approach to the analysis of sociometric data: Preliminary report. *Sociometry*, 9:340–347, 1946.

[Fre79]   Linton C. Freeman. Centrality in social networks: Conceptual clarification I. *Social Networks*, 1:215–239, 1979.

[Fre00]   Linton C. Freeman. Visualizing social networks. *Journal of Social Structure*, 1(1), 2000.

[Fre04a]   Linton C. Freeman. *The Development of Social Network Analysis: A Study in the Sociology of Science*. Empirical Press, 2004.

[Fre04b]   Ralph Freese. Automated lattice drawing. In *Proceedings of the 2nd International Conference on Formal Concept Analysis (ICFCA'04)*, volume 2961 of *Lecture Notes in Computer Science*, pages 112–127. Springer-Verlag, 2004.

[Fre05]   Linton C. Freeman. Graphic techniques for exploring social network data. In Carrington et al. [CSW05], pages 248–269.

[Fre08]   Linton C. Freeman, editor. *Social Network Analysis*. Sage, 2008. Volumes I–IV.

[Fri91]   Noah E. Friedkin. Theoretical foundations for centrality measures. *American Journal of Sociology*, 96(6):1478–1504, May 1991.

[FT08]   Yaniv Frishman and Ayellet Tal. Online dynamic graph drawing. *IEEE Transactions on Visualization and Computer Graphics*, 14(4):727–740, 2008.

[Fur10]   Borko Furht. *Handbook of Social Network Technologies and Applications*. Springer-Verlag, 2010.

[FW93]   Linton C. Freeman and Douglas R. White. Using Galois lattices to represent network data. *Sociological Methodology*, 23:127–146, 1993.

[GFC05]   Mohammad Ghoniem, Jean-Daniel Fekete, and Philippe Castagliola. On the readability of graphs using node-link and matrix-based representations: a controlled experiment and statistical analysis. *Information Visualization*, 4(2):114–135, 2005.

[GKN04]   Emden R. Gansner, Yehuda Koren, and Stephen C. North. Graph drawing by stress majorization. In *Proceedings of the 12th International Symposium on Graph Drawing (GD'04)*, volume 3383 of *Lecture Notes in Computer Science*, pages 239–250. Springer-Verlag, 2004.

[GW98]   Bernhard Ganter and Rudolf Wille. *Formal Concept Analyis: Mathematical Foundations*. Springer-Verlag, 1998.

[HB05]   Jeffrey Heer and Danah Boyd. Vizster: Visualizing online social networks. In *Proceedings of IEEE Symposium of Information Visualization (InfoVis'05)*, pages 32–39, 2005.

[HEH09]   Weidong Huang, Peter Eades, and Seokhee Hong. Measuring effectiveness of graph visualizations: A cognitive load approach. *Information Visualization*, 8(3):139–152, 2009.

[HEW98]   Mao Lin Huang, Peter Eades, and Junhu Wang. On-line animated visualization of huge graphs using a modified spring algorithm. *Journal of Visual Languages and Computing*, 9(6):623–645, 1998.

[HF06]   Nathalie Henry and Jean-Daniel Fekete. MatrixExplorer: a dual-representation system to explore social networks. *IEEE Transactions on Visualization and Computer Graphics*, 12(5):677–684, 2006.

[HF07]   Nathalie Henry and Jean-Daniel Fekete. MatLink: Enhanced matrix visualization for analyzing social networks. In *Proceedings of the 11th IFIP Conference on Human-Computer Interaction (INTERACT 2007)*, volume 4663 of *Lecture Notes in Computer Science*, pages 288–302. Springer-Verlag, 2007.

[HFM07]   Nathalie Henry, Jean-Daniel Fekete, and Michael J. McGuffin. NodeTrix: a hybrid visualization of social networks. *IEEE Transactions on Visualization and Computer Graphics*, 13(6):1302–1309, 2007.

[HHE07]   Weidong Huang, Seokhee Hong, and Peter Eades. Effects of sociogram drawing conventions and edge crossings in social network visualization. *Journal of Graph Algorithms and Applications*, 11(2):397–429, 2007.

[HN07]   Mao Lin Huang and Quang Vinh Nguyen. A space efficient clustered visualization of large graphs. In *Proceedings of the 4th International Conference on Image and Graphics (ICIG '07)*, pages 920–927, 2007.

[Hob03]   Robert Hobbs. *Mark Lombard: Global Networks*. Independent Curators International, New York, 2003.

[Hol06]   Danny Holten. Hierarchical edge bundles: Visualization of adjacency relations in hierarchical data. *IEEE Transactions on Visualization and Computer Graphics*, 12(5):741–748, 2006.

[HSS10]   Derek Hansen, Ben Shneiderman, and Marc A. Smith. *Analyzing Social Media Networks with NodeXL: Insights from a Connected World*. Morgan Kaufmann, 2010.

[HvD11]   Mark Huisman and Marijtje A. J. van Duijn. A reader's guide to SNA software. In John Scott and Peter J. Carrington, editors, *The SAGE Handbook of Social Network Analysis*, pages 578–600. Sage, 2011.

[IMMS09]   Takayuki Itoh, Chris Muelder, Kwan-Liu Ma, and Jun Sese. A hybrid space-filling and force-directed layout method for visualizing multiple-category graphs. In *Proceedings of the IEEE Pacific Visualization Symposium (PacificVis'09)*, pages 121–128, 2009.

[JHGH08]   Yuntao Jia, Jared Hoberock, Michael Garland, and John C. Hart. On the visualization of social and other scale-free networks. *IEEE Transactions on Visualization and Computer Graphics*, 14(6):1285–1292, 2008.

[JK04]   Jeffrey C. Johnson and Lothar Krempel. Network visualization: The "Bush Team" in Reuters news ticker 9/11–11/15/01. *Journal of Social Structure*, 5(1), 2004.

[JM04]   Michael Jünger and Petra Mutzel, editors. *Graph Drawing Software*. Springer-Verlag, 2004.

[JP87]   David S. Johnson and Henry O. Pollak. Hypergraph planarity and the complexity of drawing Venn diagrams. *Journal of Graph Theory*, 11(3):309–325, 1987.

[Kam89]   Tomihisa Kamada. *Visualizing Abstract Objects and Relations*. World Scientific, 1989.

[Kat47]   Leo Katz. On the matric analysis of sociometric data. *Sociometry*, 10:233–241, 1947.

[Kat53]   Leo Katz. A new status index derived from sociometric analysis. *Psychometrika*, 18(1):39–43, 1953.

[KB83]   David Knoke and Ronald S. Burt. Prominence. In Ronald S. Burt and Michael J. Minor, editors, *Applied Network Analysis*, pages 195–222. Sage Publications, 1983.

[KH05]   Yehuda Koren and David Harel. One-dimensional layout optimization, with applications to graph drawing by axis separation. *Computational Geometry*, 32(2):115–138, 2005.

[Klo81]   Alden S. Klovdahl. A note on images of networks. *Social Networks*, 3:197–214, 1981.

[Kra96]   David Krackhardt. Social networks and the liability of newness for managers. In C. L. Cooper and D. M. Rousseau, editors, *Trends in Organizational Behavior*, volume 3, pages 159–173. John Wiley & Sons, 1996.

[KS04]   Dirk Koschützki and Falk Schreiber. Comparison of centralities for biological networks. In *Proceedings of the German Conference on Bioinformatics 2004*, volume P-53 of *Lecture Notes in Informatics*, pages 199–206. Springer-Verlag, 2004.

[KvKS09]   Michael Kaufmann, Marc van Kreveld, and Bettina Speckmann. Subdivision drawings of hypergraphs. In *Proceedings of the 16th International Symposium on Graph Drawing (GD'08)*, volume 5417 of *Lecture Notes in Computer Science*, pages 396–407. Springer-Verlag, 2009.

[KZ09]   Pushpa Kumar and Kang Zhang. Node overlap removal in clustered directed acyclic graphs. *Journal of Visual Languages and Computing*, 20(6):403–419, 2009.

[Ler05]   Jürgen Lerner. Role assignments. In Brandes and Erlebach [BE05], pages 216–252.

[LG66]   Edward O. Laumann and Louis Guttman. The relative associational con-
            tiguity of occupations in an urban setting. *American Sociological Review*,
            31:169–178, 1966.

[LMR98]   Kelly A. Lyons, Henk Meijer, and David Rappaport. Algorithms for clus-
            ter busting in anchored graph drawing. *Journal of Graph Algorithms and
            Applications*, 2(1):1–24, 1998.

[Lon48]   T.Wilson Longmore. A matrix approach to the analysis of rank and status
            in a community in peru. *Sociometry*, 11(3):192–206, 1948.

[LP49]   R. Duncan Luce and Albert Perry. A method of matrix analysis of group
            structure. *Psychometrika*, 14:95–116, 1949.

[LPA+09]   David Lazer, Alex Pentland, Lada Adamic, Sinan Aral, Albert-László
            Barabási, Devon Brewer, Nicholas Christakisand Noshir Contractor,
            James Fowler, Myron Gutmann, Tony Jebara, Gary King, Michael Macy,
            Deb Roy, and Marshall Van Alstyne. Computational social science. *Sci-
            ence*, 323(5915):721–723, 2009.

[Mäk90]   Erkki Mäkinen. How to draw a hypergraph. *International Journal of
            Computer Mathematics*, 34(3-4):177–185, 1990.

[MBK97]   Cathleen McGrath, Jim Blythe, and David Krackhardt. The effect of
            spatial arrangement on judgments and errors in interpreting graphs. *Social
            Networks*, 19(3):223–242, 1997.

[MELS95]   Kazuo Misue, Peter Eades, Wei Lai, and Kozo Sugiyama. Layout adjust-
            ment and the mental map. *Journal on Visual Languages and Computing*,
            6(2):183–210, 1995.

[MH01]   Matthias Müller-Hannemann. Drawing trees, series-parallel digraphs, and
            lattices. In Michael Kaufmann and Dorothea Wagner, editors, *Drawing
            Graphs: Methods and Models*, volume 2025 of *Lecture Notes in Computer
            Science*, pages 46–70. Springer, 2001.

[Mis07]   Kazuo Misue. Anchored maps: Visualization techniques for drawing bi-
            partite graphs. In *The Human-Computer Interaction International Con-
            ference Proceedings (HCII 2007)*, volume 4551 of *Lecture Notes in Com-
            puter Science*, pages 106–114. Springer-Verlag, 2007.

[Mit94]   J. Clyde Mitchell. Situational analysis and network analysis. *Connections*,
            17(1):16–22, 1994.

[MMBd05]   James Moody, Daniel A. McFarland, and Skye Bender-deMoll. Dynamic
            network visualization. *American Journal of Sociology*, 110(4):1206–1241,
            2005.

[Mor46]   Jakob L. Moreno. Sociogram and sociomatrix: a note to the paper by
            Forsyth and Katz. *Sociometry*, 9:348–349, 1946.

[Mor53]   Jakob L. Moreno. *Who Shall Survive? Foundations of Sociometry, Group
            Psychotherapy, and Sociodrama*. Beacon House, 1953. Originally pub-
            lished in 1934.

[New61]   Theodore M. Newcomb. *The Acquaintance Process*. Holt, Rinehart &
            Winston, 1961.

[Nor40]   Mary L. Northway. A method for depicting social relationships obtained
            by sociometric testing. *Sociometry*, 3(2):144–150, 1940.

[Nor52]   Mary L. Northway. *A Primer of Sociometry*. University of Toronto Press,
            1952.

[Nor54]   Mary L. Northway. A plan for sociometric studies in a longitudinal pro-
            gramme of research in child development. *Sociometry*, 17(3):272–281,
            1954.

[Nor96]   Stephen C. North. Incremental layout with DynaDag. In Brandenburg
            [Bra96], pages 409–418.

[OS07]   Hiroki Omote and Kozo Sugiyama. Method for visualizing complicated
            structures based on unified simplification strategies. *IEICE Transactions
            on Information and Systems*, E90-D(10):1649–1656, 2007.

[PCJ97]   Helen C. Purchase, Robert F. Cohen, and Murray James. An experimen-
            tal study of the basis for graph drawing algorithms. *ACM Journal of
            Experimental Algorithmics*, 2(4), 1997.

[Pet99]   W. M. Flinders Petrie. Sequences in prehistoric remains. *Journal of the
            Anthropological Institute of Great Britain and Ireland*, 29(3/4):295–301,
            1899.

[PNK10]   Ruth Pfosser, Helmut Neundlinger, and Harald Katzmair. Die un-
            verbrüchliche Solidarität in einer Extremsituation und die Fähigkeit, sich
            zu befreien. Das ist Spanien. *Der Standard*, 13. Juli, 2010.

[PNR08]   Christian Pich, Lev Nachmanson, and George G. Robertson. Visual anal-
            ysis of importance and grouping in software dependency graphs. In *Pro-
            ceedings of the ACM 2008 Symposium on Software Visualization (SOFT-
            VIS 2008)*, pages 29–32, 2008.

[PS08]   Helen C. Purchase and Amanjit Samra. Extremes are better: Investigat-
            ing mental map preservation in dynamic graphs. In *Proceedings of the 5th
            International Conference on Diagrammatic Representation and Inference
            (Diagrams 2008)*, volume 5223 of *Lecture Notes in Computer Science*,
            pages 60–73. Springer-Verlag, 2008.

[RCE06]   Jon Ducrou Richard Cole and Peter Eklund. Automated layout of small
            lattices using layer diagrams. In *Proceedings of the 4th International Con-
            ference on Formal Concept Analysis (ICFCA'06)*, volume 3874 of *Lecture
            Notes in Computer Science*, pages 291–305. Springer-Verlag, 2006.

[RF10]   Nathalie Riche and Jean-Daniel Fekete. Novel visualizations and inter-
            actions for social networks exploration. In *Handbook of Social Network
            Technologies and Applications* [Fur10], pages 611–636.

[RZF08]   Peter Rodgers, Leishi Zhang, and Andrew Fish. General Euler diagram
            generation. In *Proceedings of the 5th International Conference on the The-
            ory and Application of Diagrams (Diagrams'08)*, volume 5223 of *Lecture
            Notes in Artificial Intelligence*, pages 13–27. Springer-Verlag, 2008.

[SA06]   Ben Shneiderman and Aleks Aris. Network visualization by semantic
            subtrates. *IEEE Transactions on Visualization and Computer Graphics*,
            12(5):733–740, 2006.

[SAA09]   Paolo Simonetto, David Auber, and Daniel Archambault. Fully automatic
            visualisation of overlapping sets. *Computer Graphics Forum*, 28(3):967–
            974, 2009.

[Sco00]   John Scott. *Social Network Analysis: A Handbook*. Sage, 2nd edition,
            2000.

[Sei83]   Stephen B. Seidman. Network structure and minimum degree. *Social
            Networks*, 5:269–287, 1983.

[SJUS08]   Hans-Jörg Schulz, Mathias John, Andrea Unger, and Heidrun Schu-
            mann. Visual analysis of bipartite biological networks. In *Proceed-
            ings of the Eurographics Workshop on Visual Computing for Biomedicine
            (VCBM 2008)*, pages 135–142. Eurographics, 2008.

[SM07]   Zeqian Shen and Kwan-Liu Ma. Path visualization for adjacency matri-
            ces. In *Proceedings of the 9th Eurographics/IEEE-VGTC Symposium on
            Visualization (EuroVis '07)*, pages 83–90. Eurographics, 2007.

[ST10]   Rodrigo Santamaría and Roberto Therón. Visualization of intersecting
            groups based on hypergraphs. *IEICE Transactions on Information and
            Systems*, E93-D(7):1957–1964, 2010.

[Tuf83]   Edward R. Tufte. *The Visual Display of Quantitative Information*. Graph-
            ics Press, 1983.

[vHSD09]   Frank van Ham, Hans-Jörg Schulz, and Joan M. Dimicco. Honeycomb: Vi-
            sual analysis of large scale social networks. In *Proceedings of the 13th IFIP
            Conference on Human-Computer Interaction (INTERACT 2009)*, volume
            5727 of *Lecture Notes in Computer Science*, pages 429–442. Springer-
            Verlag, 2009.

[vHW08]   Frank van Ham and Martin Wattenberg. Centrality based visualization
            of small world graphs. *Computer Graphics Forum*, 27(3):975–982, 2008.

[Wat06]   Martin Wattenberg. Visual exploration of multivariate graphs. In *Proceed-
            ings of the SIGCHI Conference on Human Factors in Computing Systems
            (CHI '06)*, pages 811–819, 2006.

[WB88]   Barry Wellman and Stephen D. Berkowitz, editors. *Social Structures: A
            Network Approach*. Cambridge University Press, 1988.

[Wer44]   Max Wertheimer. *Gestalt Theory*. Hayer Barton Press, 1944.

[WF94]   Stanley Wasserman and Katherine Faust. *Social Network Analysis: Meth-
            ods and Applications*. Cambridge University Press, 1994.

[WG98]   Karl Erich Wolff and Siefgried Gabler. Comparison of visualizations in for-
            mal concept analysis and correspondence analysis. In Michael Greenacre
            and Jörg Blasius, editors, *Visualization of Categorical Data*, pages 85–97.
            Academic Press, 1998.

[Why43]   William F. Whyte. *Street Corner Society*. University of Chicago Press,
            1943.

[Win94]   William D. Winn. Contributions of perceptual and cognitive processes to
            the comprehension of graphics. In W. Schnotz and R.W. Kulhavy, editors,
            Comprehension of Graphics, number 108 in Advances in Psychology, pages
            3–27. Elsevier, 1994.

[WM96]   Xiaobo Wang and Isao Miyamoto. Generating customized layouts. In
            Brandenburg [Bra96], pages 504–515.

[Wol07]   Alexander Wolff. Drawing subway maps: Asurvey. *Informatik - Forschung
            und Entwicklung*, 22(1):23–44, 2007.

[XTT⁺10]   Kaikuo Xu, Changjie Tang, Rong Tang, Ghulam Ali, and Jun Zhu. A
            comparative study of six software packages for complex network research.
            In *Proceedings of the 2nd International Conference on Communication
            Software and Networks (ICCSN '10)*, pages 350–354, 2010.

[ZSE05]   Lanbo Zheng, Le Song, and Peter Eades. Crossing minimization problems
            of drawing bipartite graphs in two clusters. In *Proceedings of the 2005*

Asia-Pacific Symposium on Information Visualisation (APVis '05), pages 33–37. Australian Computer Society, 2005.

Index

Printed in the United States
by Baker & Taylor Publisher Services

Printed in the United States
by Baker & Taylor Publisher Services